# *Celebrating* a CENTURY of *Concours d'Elegance*

DANIEL CABART
DENIS COHIGNAC
PATRICK ROLLET
GAUTAM SEN

# Celebrating a Century of Concours d'Elegance

**DANIEL CABART**
**DENIS COHIGNAC**
**PATRICK ROLLET**
**GAUTAM SEN**

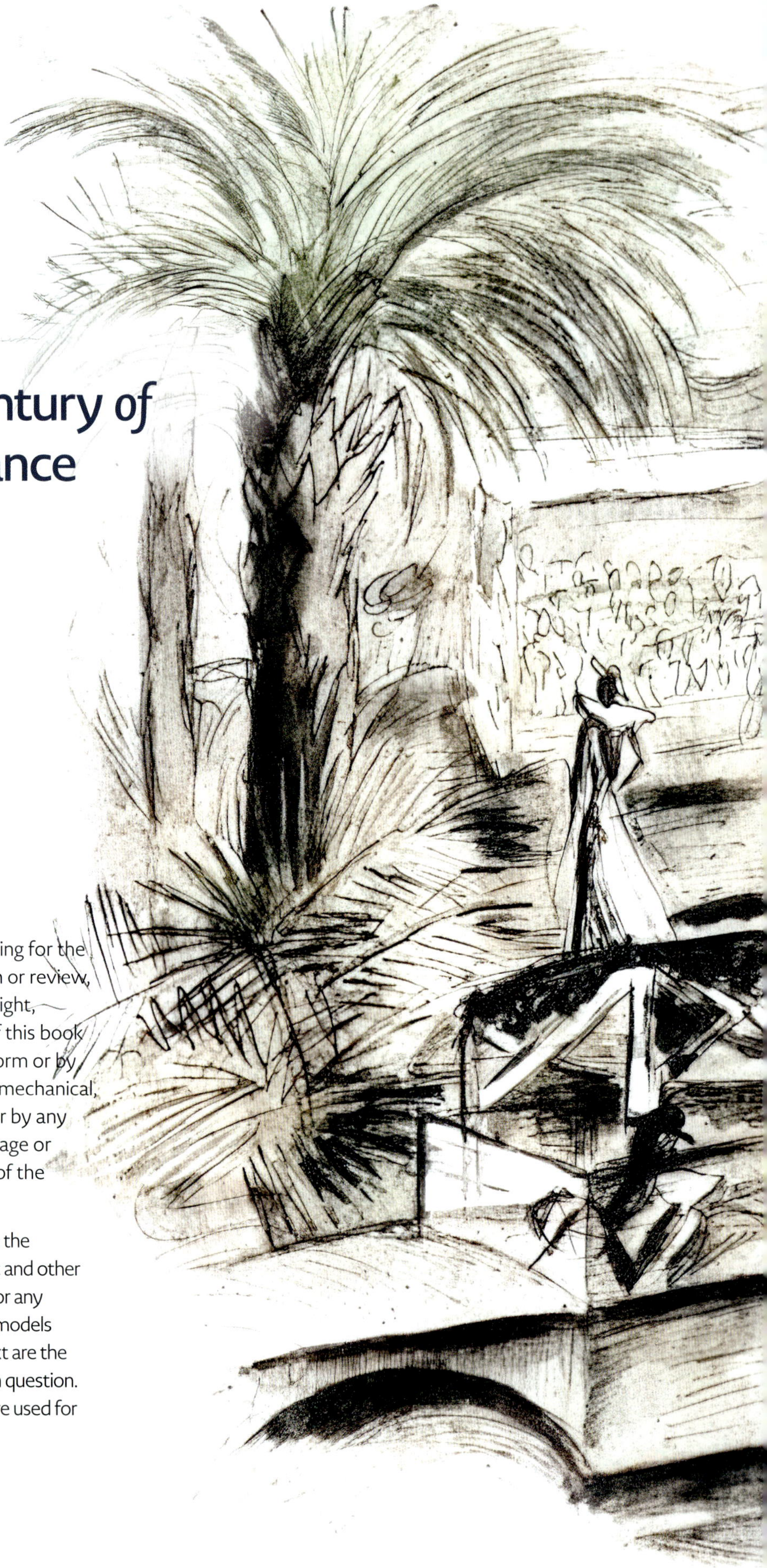

Published 2026

ISBN: 978-1-956309-22-5



Printed by Interpress Ltd., Hungary

for the publisher

Dalton Watson Fine Books
*Glyn and Jean Morris*
Deerfield, IL 60015 USA
**www.daltonwatson.com**

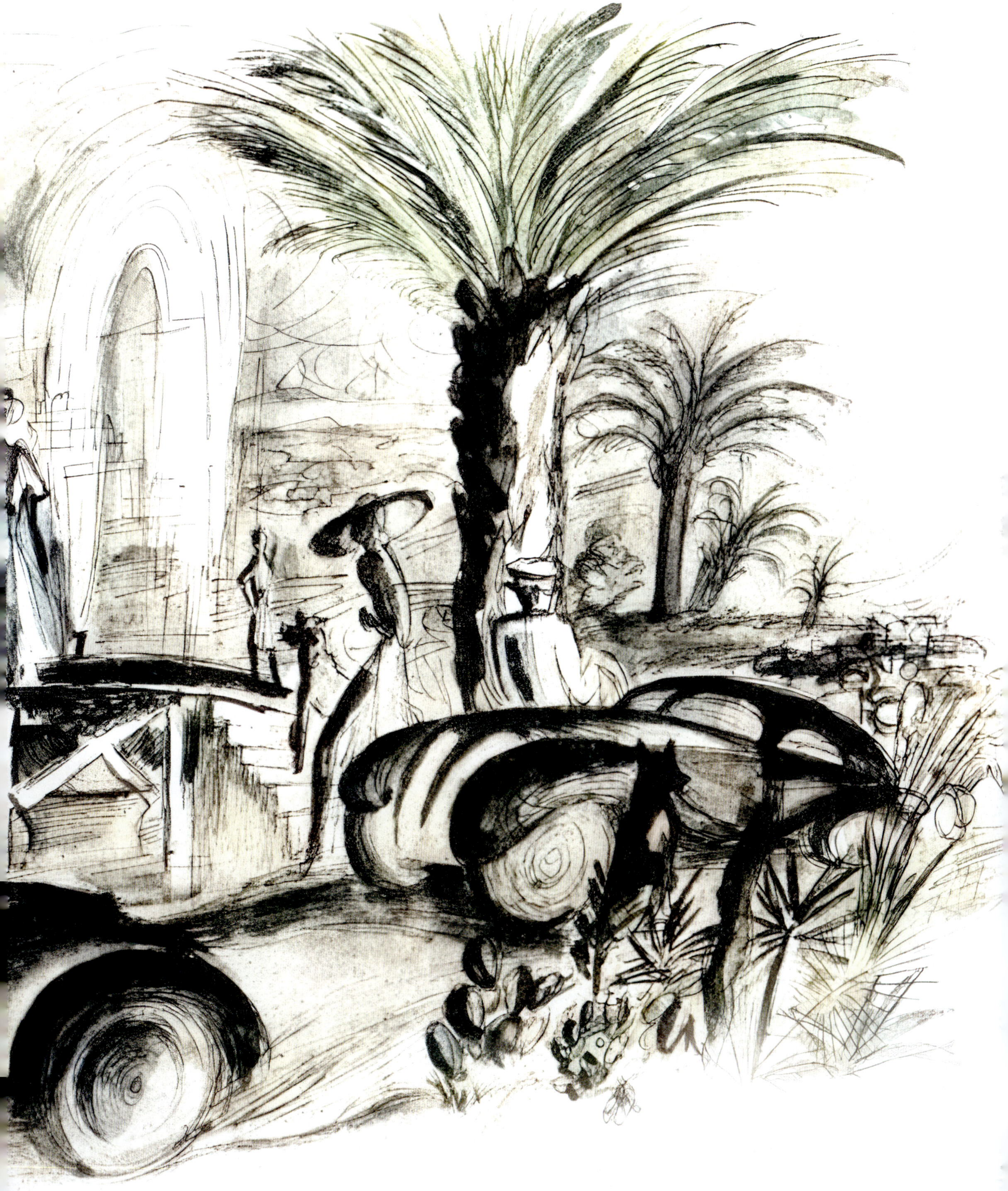

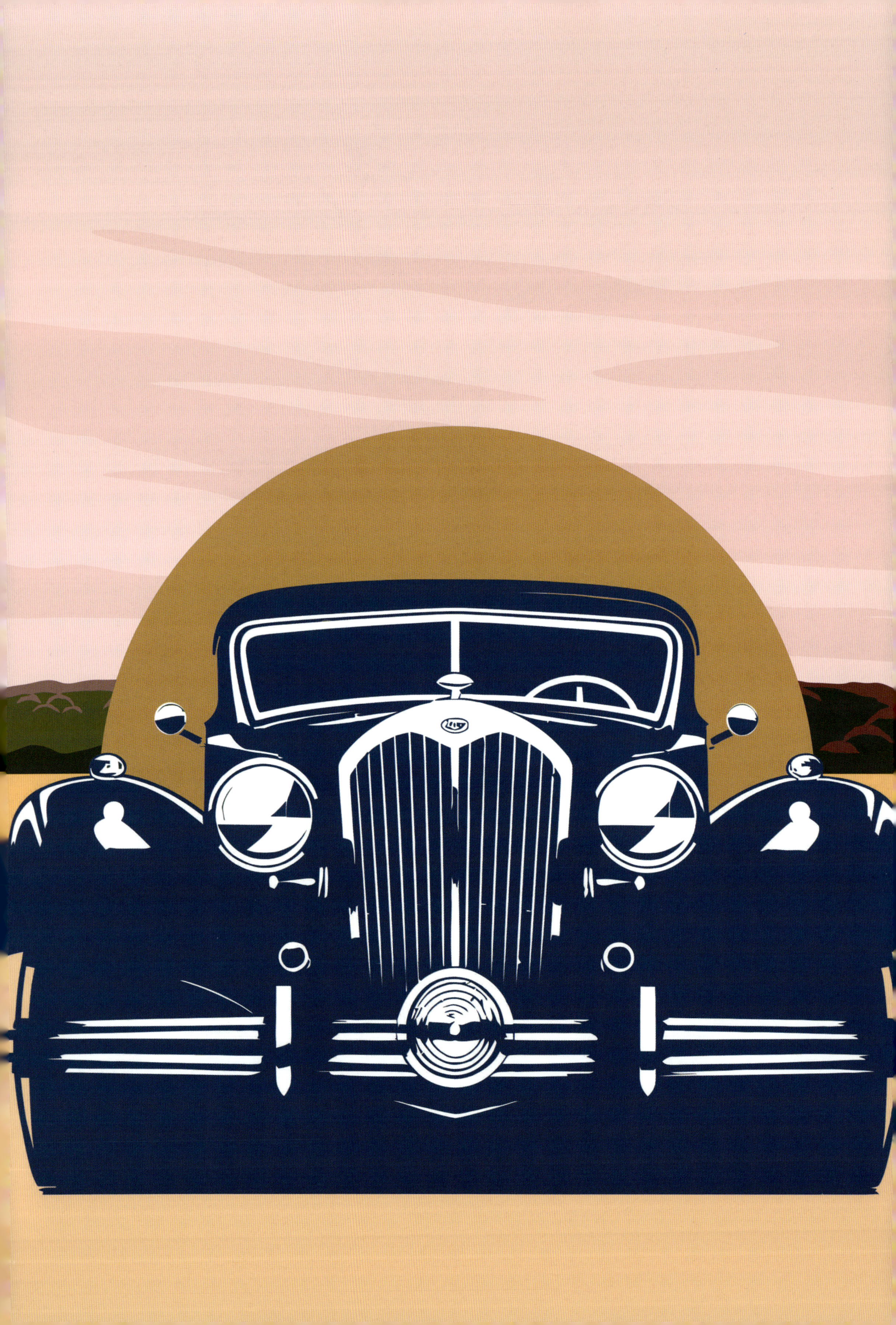

# TABLE *of* CONTENTS

# PREFACE

In recent decades, concours d'elegance events have regained popularity within the historic vehicle community. While modern interpretations may differ, the fundamental spirit of these elegant competitions persists. However, both the purpose and perception of these events have evolved considerably over the past century.

The term concours d'elegance may date back to 18th century France, when aristocrats paraded through Parisian parks in ornate horse-drawn carriages to showcase wealth and sophistication. With the advent of the automobile, these parades transformed into competitions between owners of extravagant cars – most famously, the Pebble Beach Concours d'Elegance in the United States, and Concorso d'Eleganza Villa d'Este in Italy.

While several events around the world claim to be the oldest running concours, the concept itself originated in France in the early 1920s. Renowned for blending classical traditions with emerging technologies, France hosted the first true concours d'elegance, as defined by André Becq de Fouquières, a prominent figure in Parisian and seaside society, around 1920–21. Among the earliest was the gathering in the seaside resort town of Dinard on September 4, 1921, where over 60 vehicles took part. The event was so well received that it became an annual tradition until the outbreak of World War II.

Soon, other French resort towns – La Baule, Deauville, Biarritz, and Monte Carlo, among others – adopted the concept. These gatherings celebrated not just automotive beauty, but fashion, culture, and leisure. They brought together elegant cars, haute couture, purebred dogs, sculptures, and art in a vibrant social atmosphere.

The 1920s also marked a golden era for coachbuilding. Leading French marques such as Bugatti, Delage, Delahaye, Hispano-Suiza, and Voisin offered rolling chassis to be "dressed" by coachbuilders, according to the buyer's personal taste. This era gave rise to legendary names in carrosserie – Figoni et Falaschi, Chapron, Gangloff, Guilloré, Kellner, Kelsch, Labourdette, Pourtout, Letourneur et Marchand, Saoutchik, and others – who showcased their creations at concours events.

It was likely André de Fouquières who codified the rules of the concours d'elegance à la française. His judging criteria included the car's design (typically brand-new), comfort, equipment, and the harmony between the automobile, the presenter's attire, and accessories. The result was a glamorous platform where automakers, coachbuilders, and fashion houses collaborated to present not just vehicles, but coordinated artistic statements – bringing together the "three Cs": Constructeur, Carrossier, and Couturier.

The onset of World War II brought an abrupt end to these events, and although they resurfaced in the post-war years, it wasn't until the 1960s that they saw a modest revival. In the past two decades, however, concours d'elegance have returned to the limelight with renewed vigor, thanks to prestigious events like Pebble Beach, The Amelia, Chantilly, and others across the globe – from Adelaide to Zoute.

Yet, the modern version of the concours differs fundamentally. Today, they focus not on new cars, but on historic vehicles – many of which won concours nearly a century ago. What many now refer to as concours d'elegance is more accurately what the French call concours d'état – a "competition of condition."

Originally, the concours d'elegance was a celebration of style, design, and presentation. Vehicles (and occasionally motorcycles) were judged not just on looks, but also on how well the presenter's outfit, hat, gloves, and even accompanying dog matched the elegance of the vehicle. The jury typically included automotive experts and fashion designers, blending technical and aesthetic judgment.

In contrast, today's concours tend to be closed, static displays judged by motoring experts, with an emphasis on authenticity and preservation over flair. While such scrutiny helps ensure historical accuracy and discourages inappropriate modifications, it often comes at the cost of emotion and storytelling. The International Chief Judge Advisory Group (ICJAG), for instance, uses a highly technical scoring system that some critics feel lacks soul – it overlooks historical context and emotional impact.

To address this, newer systems like the M100 have introduced scoring for "emotion," albeit just 5 out of 100 points. Chantilly Arts & Elegance goes further, assigning equal weight to design, history, authenticity, condition, and functionality – treating elegance as more than just mechanical purity.

Still, the use of "elegance" to describe technical perfection – such as flawless engine mounts or unblemished nuts and bolts – may feel out of place. In this book, we aim to explore the deeper roots of the original concours d'elegance, tracing its transformation and its essential place in the world of historic motoring – recognizing, too, that the term itself may no longer fully capture its original essence.

*– Daniel Cabart, Denis Cohignac, Patrick Rollet, Gautam Sen*

**OPPOSITE:** *2025 City of Dinard Award, 1970 Lamborghini Miura P 400 S. (Manoir de L'automobile de Lohéac)*

In order to achieve consistency of expression in this English language book, all references to events termed Concours d'Elégance, Concours d'Elegance, and similar have been standardized to Concours d'Elegance, and all references to the generic form of the concours, rather than official event names, such as concours d'élégance, have been standardized to concours d'elegance.

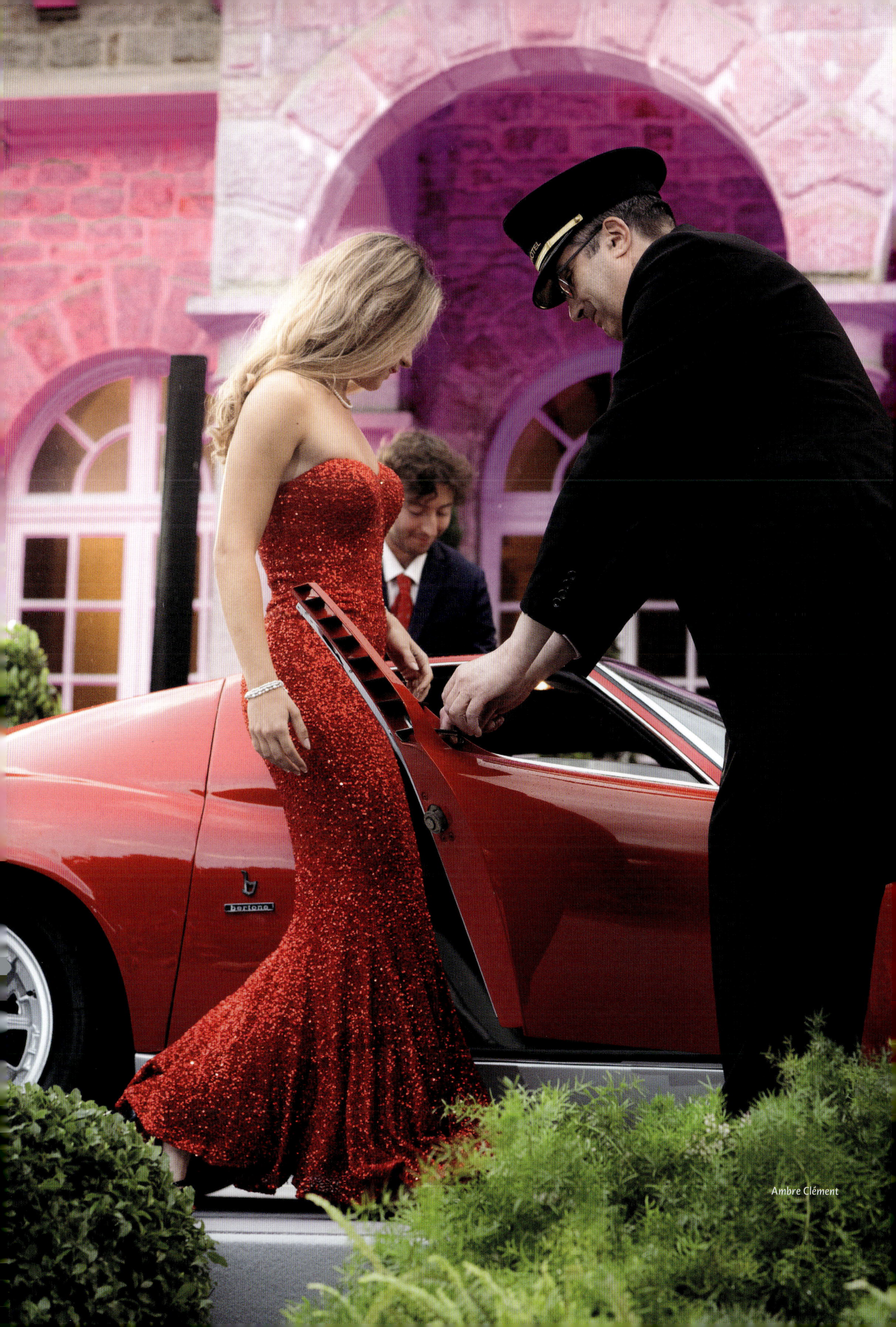
Ambre Clément

# FOREWORD

The horse-drawn carriage is the foundation of nearly every story about the automobile. Not everyone knows, however, that the same is true for automotive events, including the origin story of concours d'elegance, those shimmering showcases of automotive artistry and passion that are known and loved today. Concours events are rooted in 18th century France when aristocrats would parade elegant horse-drawn carriages through the parks of Paris during summer weekends, for all to see and admire.

In this wonderful book, Daniel Cabart (Delage historian), Denis Cohignac (concours organizer), Patrick Rollet (former FIVA President) and Gautam Sen (veteran automotive journalist and writer), do a masterful job tracing the rich history of concours from those genteel pre-motor days and the first motorized concours in 1921 in Dinard, France (won by a Gillotte-bodied Avions Voisin) to the rise, and later revival, of Villa d'Este and on into the present-day galaxy of concours events around the world. It is a deeply detailed and satisfying read for anyone who enjoys automotive history or has ever spent a perfect Sunday afternoon on a sunny show field marveling at vintage Rolls-Royces, Duesenbergs, Bugattis, Ferraris and other automotive treasures.

For many, this impressive editorial team's depiction of the early days of the pastime will be particularly fascinating, as they paint a picture of an optimistic new moment in history. Traveling to beautiful resorts and exotic locations was central to this new trend. *"In the early 1920s,"* they wrote, *"a fresh wave of change swept across the world, with Paris at its heart, radiating an atmosphere of euphoria and liberation that extended to fashionable seaside resorts. Everything was in flux, as architects, artists, fashion designers, and industrialists competed to express their creativity and reinvent themselves. Society, too, was eager for new experiences, embracing the latest leisure activities and sports, many of which came from England or the United States. Amid this whirlwind of innovation, the automobile quickly took center stage. Not only was it seen as a symbol of freedom and adventure, but it also became an integral part of elegant living."* It wasn't just about what new activity or trend was coming to a city near you; it was often about transporting your imagination to another place or traveling to a dreamy location to "become" someone new.

For me, as a lifelong collector and long-time judge, the chapter about the evolution of the Pebble Beach Concours d'Elegance and the visionary changes ushered in by Lorin Tryon and Jules Heumann (and later Sandra Button) stood out. The pair refocused the concours on classic cars, adding new marques and models to their lineup and establishing an annual tradition of spotlighting special classes. "Their foresight led them to introduce new classes, including one for Ferrari in the mid-1970s, and later for hot rods and microcars by the late 1990s. Additionally, Tryon and Heumann conceived and formalized the Tour d'Elegance, integrating it into the concours judging process, a lasting contribution to the event's legacy." With its grounding firmly in automotive history but also its annual gathering of modern concept cars and new vehicle launches, Pebble Beach ticks all the boxes for a concours event, but it, too, evolved through the years. We are reminded of an idealized past but are stimulated by what is to come.

There is top-quality reporting and commentary like that throughout the book, but what ultimately sets it apart from other histories, in my mind, is the editorial team's lively discussion about the present and future of concours. "The passion for historic vehicles remains strong," they wrote, "and revenues from this market are projected to rise by almost ten percent through 2026 (Statista 2024). The number of competitions continues to expand, with the Hothouse Media-published *Concours Year* featuring some 50 major events worldwide in 2024."

But challenges remain. While popular, most concours find themselves wrestling with big questions about how to maintain relevance moving forward. The editorial team's solutions are both innovative and common sense: upgrade displays with modern projection and lighting, do more to tell the stories behind these amazing cars (every car has one), experiment with new formats and new classes, feature newer cars, and much, much more.

Traditionalists might shudder but preserving the essence of concours d'elegance events does not mean limiting their reach. "In fact, expanding their audience is crucial not only for growth but also for long-term sustainability. The collector car community is diverse, often separated into distinct cultural groups, and the concours d'elegance stands at the intersection of this diversity, presenting an opportunity to bridge these worlds."

As the CEO of a company (Hagerty) that operates two popular US concours – The Amelia and the Greenwich Concours d'Elegance – and is a patron, sponsor and attendee of countless others, I couldn't agree more. We are clearly at a pivotal moment in the history of concours. If we do not meet the challenge of broadening our base, it is entirely conceivable that events could dwindle or even disappear, meaning many of the world's most special cars would seldom, if ever, be seen again in public. That would be an utter shame.

But if the challenge is met – and I have no doubt that we will – many more will come to understand why we love these cars and events so very much.

Please enjoy this marvelous book. Like me, I suspect you will come away from it pining for the start of another concours season and the sometimes-distant shores where the automotive masterpieces of history gather to strut their stuff!

McKeel Hagerty
*CEO and Chairman, Hagerty Inc.*

**OPPOSITE:** McKeel Hagerty.

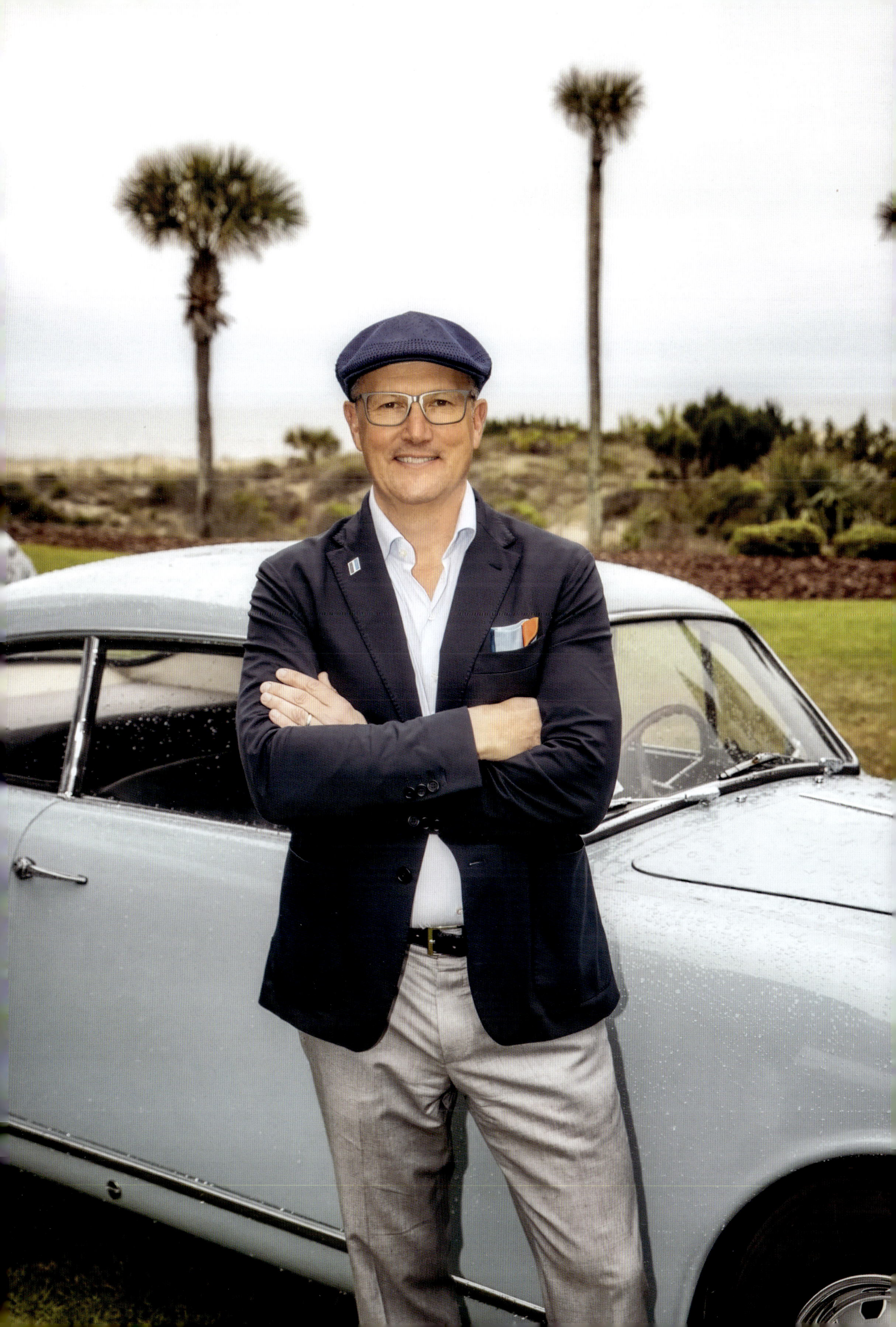

# *Preamble*

## BEFORE 1900

## When Did It All Begin?

Who can pinpoint exactly when all of this began? Who first devised the idea, and where was the inaugural Concours d'Elegance Féminine en Automobile held? As is often the case with historical research, questions of precedence arise, and events unfold gradually, making it difficult to define a clear starting point.

The terms elegance, automobile, concours, and féminine are key here, and it wasn't until much later in the 20th century that these four concepts would come together in a single event. Before diving into that, it's essential to understand the significance of these words. The phrase 'concours d'elegance,' in particular, has evolved, leading to countless interpretations.

**Elegance:** In the 19th century, this word was on everyone's lips. It was a marker of a certain 'aristocracy' of taste, flaunted at every opportunity. Whether strolling along the boulevards of Paris' upscale neighborhoods, in the Bois de Boulogne, or attending 'Drag Day' at the racecourses of Longchamp and Auteuil, elegance was always on display. And, of course, it reigned supreme on the Riviera during the season, where British and other European aristocrats set the tone long before this trend spread to the rest of France.

**Automobile:** Initially, elegance and the automobile seemed worlds apart. In the early years (1890s-1900s), people were slow to embrace the change, often reacting harshly to these 'horseless carriages.' What beauty or style could be found in these vehicles, stripped of the elements that once defined their elegance?

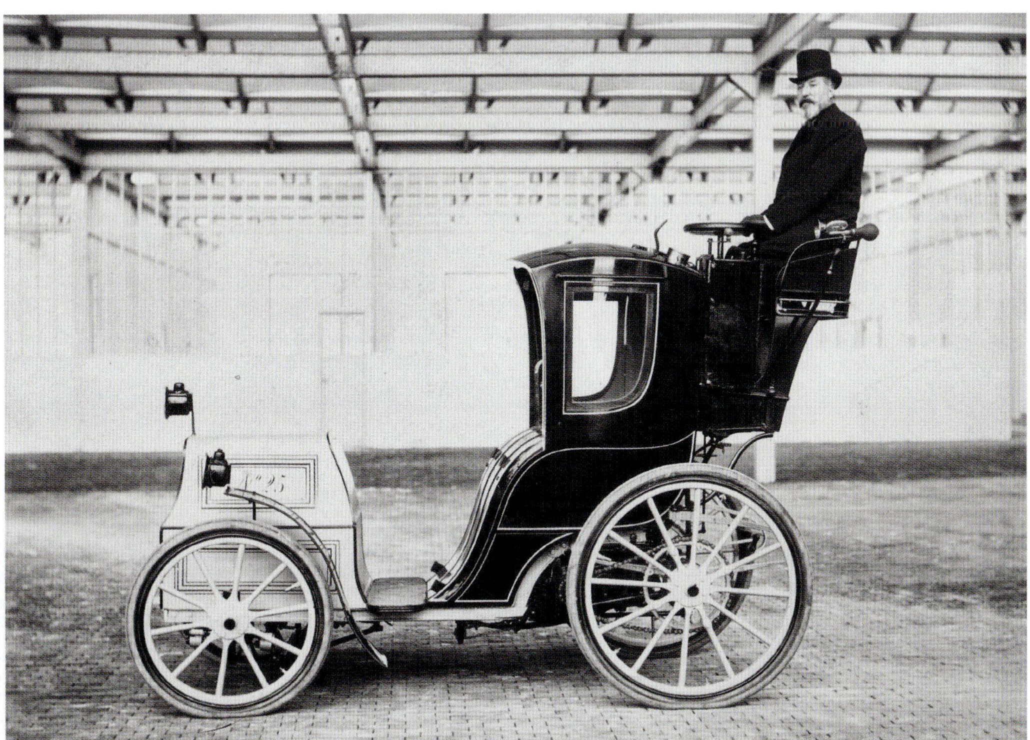

*"Where did the horse go?" One can easily picture the passerby's astonishment upon seeing this carriage. Charles Jeantaud (1840-1906), a pioneer of the electric car, designed and built vehicles in the late 19th century. Here, he is seen at the helm of a two-seater cab that competed in the automobile hackney carriage competition held in Paris in June 1898. (BnF Gallica)*

**Concours:** Whether centered on elegance or other qualities, competitions have always added flair to the seasons at seaside resorts and Parisian racecourses. There has always been a need to showcase something or someone, often through organized competitions between participants.

Nearly every activity gave rise to some form of competition. When the automobile emerged as a competitive object, the organizers' creativity knew no bounds. Beyond decoration contests, they introduced challenges related to driving skills, cleanliness, accessibility, braking, hill starts, and more. This included the certification of new motorized hackney carriages (precursors to modern taxis) that were replacing horse-drawn vehicles. When it came to performance, mechanical quality, or other technical aspects, competitions became synonymous with races, or what was known as the 'Competition of Speed.'

# Le Petit Journal

TOUS LES JOURS
Le Petit Journal
5 Centimes

SUPPLÉMENT ILLUSTRÉ
Huit pages : CINQ centimes

TOUS LES DIMANCHES
Le Supplément illustré
5 Centimes

Cinquième année | LUNDI 6 AOUT 1894 | Numéro 194

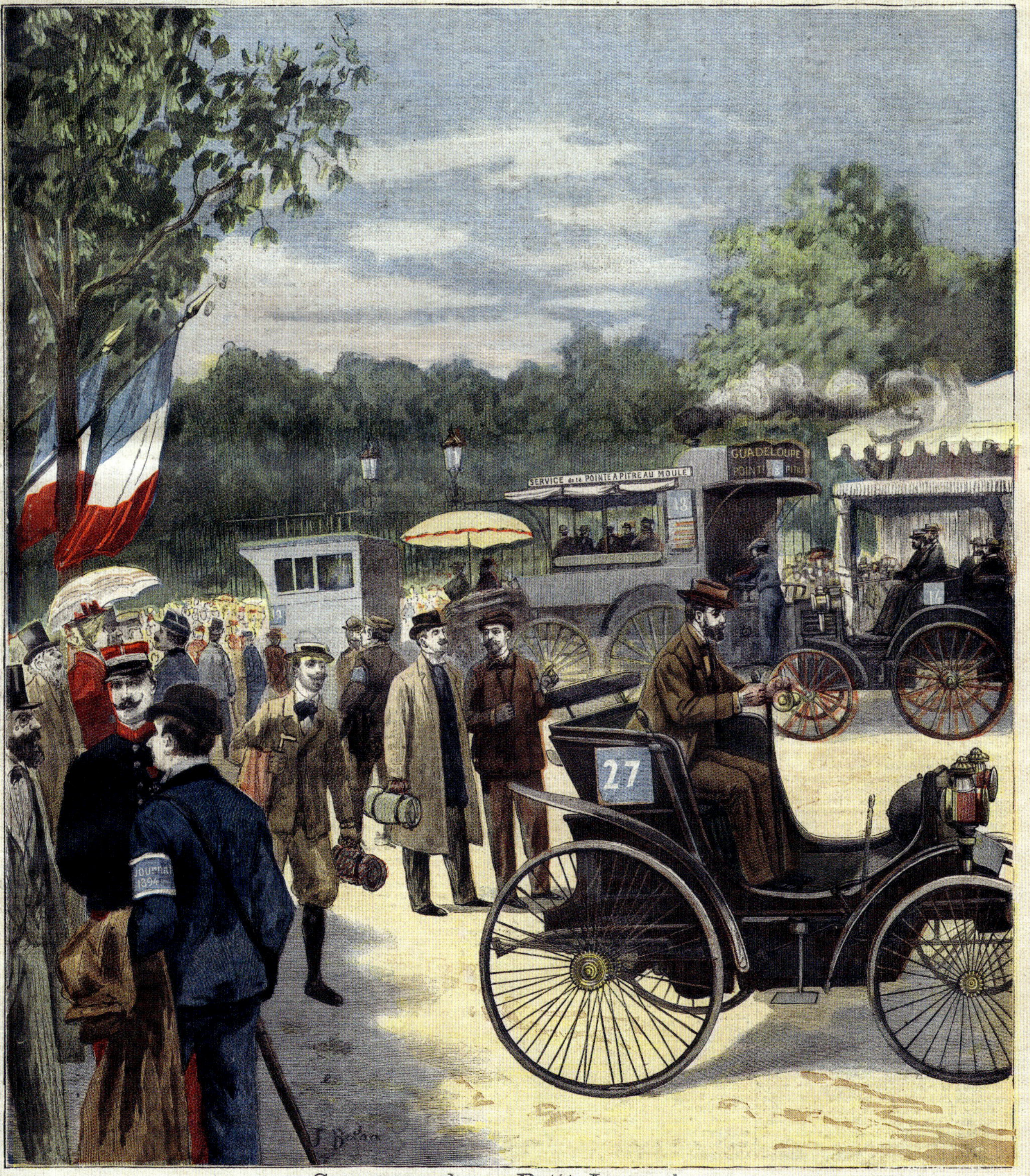


Concours du « Petit Journal »
LES VOITURES SANS CHEVAUX

*By 1894, competitions were already becoming more frequent. The first automobile competition in the world was organized at the initiative of Pierre Giffard, editor of Le Petit Journal. Held in July 1894 between Paris and Rouen, the race was won by Peugeot and Panhard et Levassor, who tied for first place. In the foreground, the Peugeot is easily recognizable by its spoked wheels.* (Daniel Cabart Archives)

**Féminine:** Finally, the presence of women emerged. Although they were frequently seen as passengers, and less often as drivers, in all the events mentioned, the journey toward women's full participation in competitions – where their elegance would stand alongside the automobile, receiving as much attention and recognition – was a lengthy one.

## *La Journée des Drags* (The Day of the Drags)

### La Vie au Grand Air, June 15, 1898

On Saturday, June 4, the weather was particularly pleasant, attracting a large number of our most elegant sportsmen and sportswomen to the invitation of the owners of the mail coaches. By one o'clock, twenty of these stylish cars were lined up at Place de la Concorde, much to the astonishment of onlookers and the delight of photographers eager to capture the moment. The starting signal was given by Mr. de La Haye-Jousselin, and the parade proceeded in perfect order, with the crews maintaining a few meters' distance from each other. Upon arrival at Auteuil, the cars lined up in front of the reserved stands, greeted by the excellent music of the 6th Line, which played a triumphant march. Among the participants were notable figures such as the Count de Lariboisière, Viscount and Viscountess A. de Contades, Count and Countess de Bourg, Mr. and Madame Jacques de Waru, Count de Narbonne-Lara, and Countess de Lariboisière. On the Duke of Noailles' mail, attendees included Count and Countess J. Murat, and Duke de Luynes.

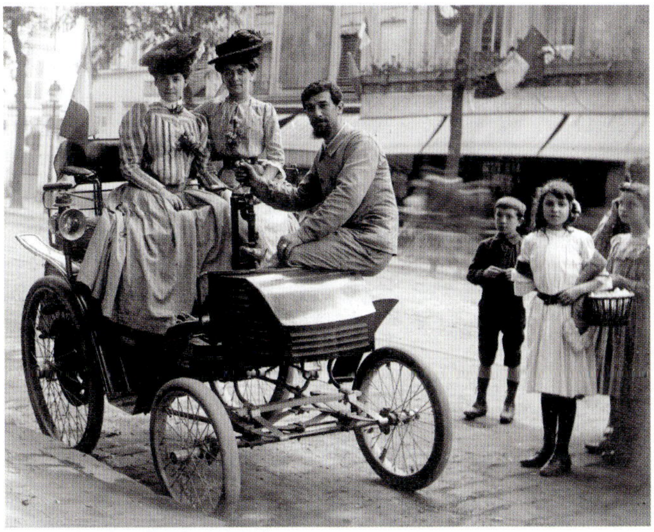

**LEFT:** *Two elegant women take their seats aboard the vehicle, though the privilege of driving has yet to be granted to them. (Daniel Cabart Archives)*

**BOTTOM:** *It would be years before the term "elegance" applied equally to both the passenger (or driver) and the vehicle's bodywork. In the case of the latter, as we will see, the word takes on an entirely different meaning. This Renault Type JV, a top-of-the-line model, features a 6-cylinder, 40 HP engine. Produced in only 42 units between 1921 and 1922, this particular example was bodied by Kellner as a landaulet coupé. The car is captured in front of the Château de Longchamp in the Bois de Boulogne. (Daniel Cabart Archives)*

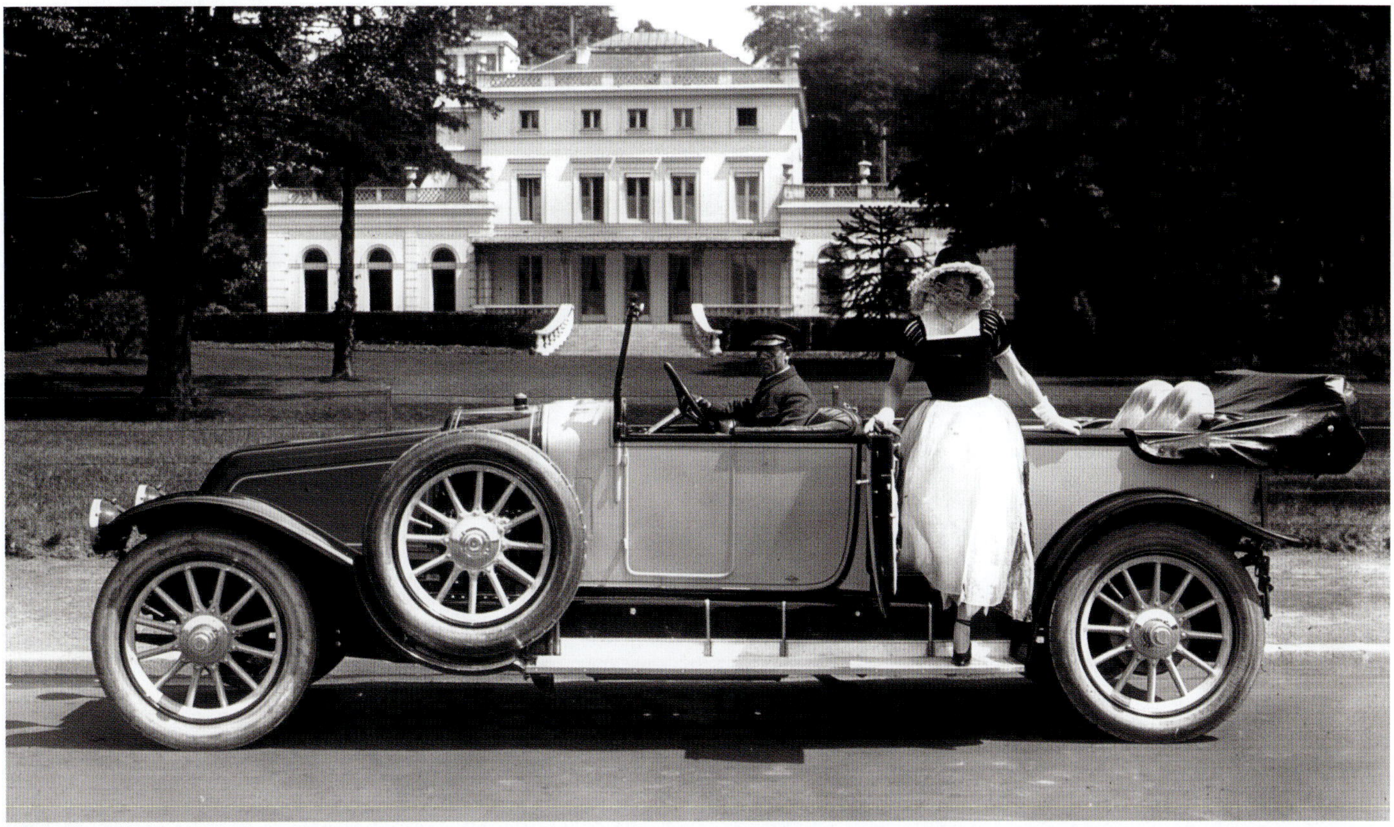

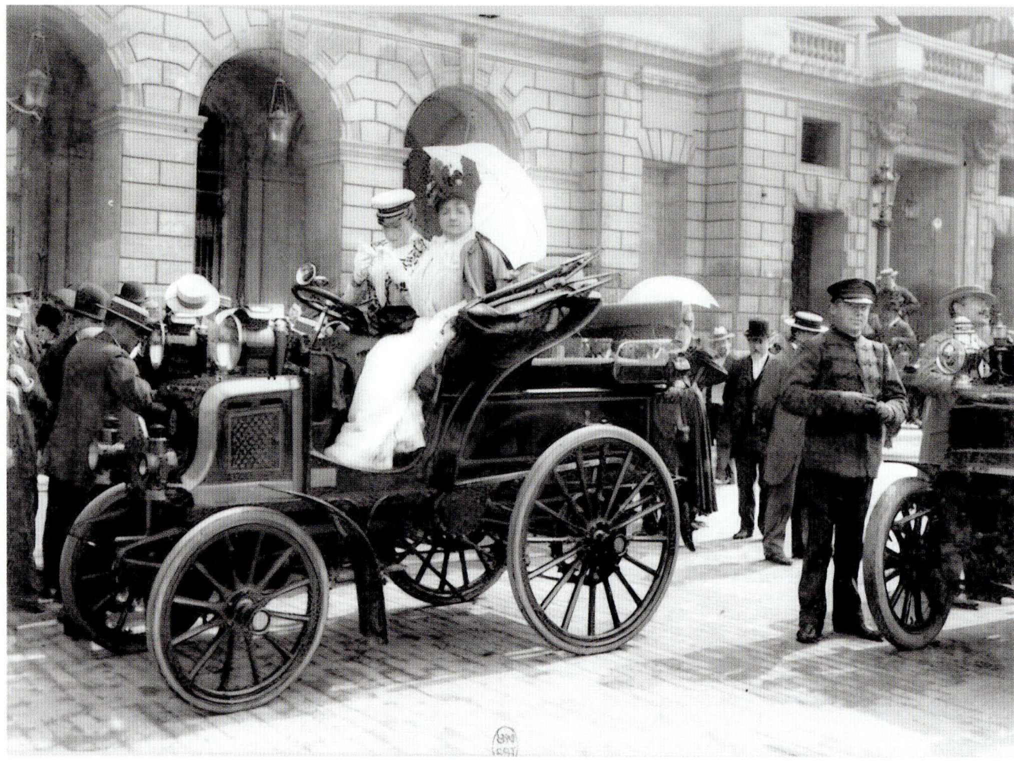

*During the "Drag Day" – a sort of precursor to the concours d'elegance – it was customary to arrive at the Auteuil racecourse dressed in elegant attire, dazzling onlookers with refined style. Pictured here is a Panhard et Levassor 6 CV Type M2F from 1898, featuring a char à banc body, parked in front of the Automobile Club de France (ACF) at Place de la Concorde in Paris. (Daniel Cabart Archives)*

## *Le Concours de fiacres Automobiles* (The Hackney Carriage Competition)

### *La Vie au Grand Air*, June 15, 1898

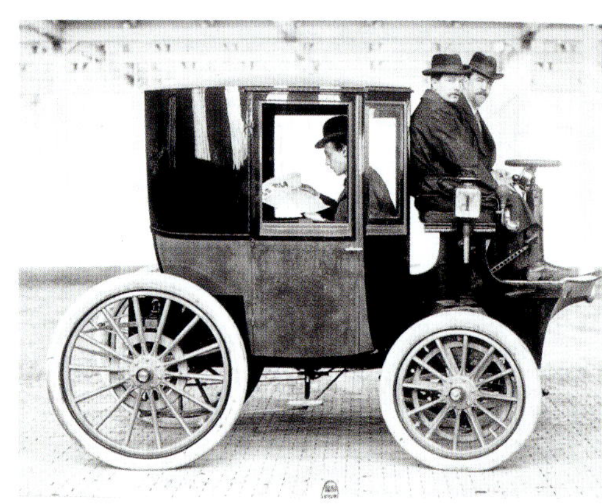

*A wide variety of "creations" were featured at the hackney carriage competition, organized by the Automobile Club de France. The trials took place over a week, covering Paris and its suburbs. Kriéger, a leading manufacturer of electric cars, built this three-seater coupe in 1898. Notably, its front axle housed the motors, making it one of the earliest front-wheel-drive vehicles. (BnF Gallica)*

The car carriage competition in Paris began on June 1, organized by the Automobile Club de France. This event was particularly significant, as it aimed to determine whether motorized carriages could provide reliable service. The well-organized test, overseen by Mr. Forestier, an esteemed inspector of bridges and roads, focused on the following criteria:

- The cost of operating a hackney carriage over a normal service day, completing a varied route of at least 60 kilometers within sixteen hours.
- The comfort and maneuverability of the car.
- The frequency of refueling, along with the importance and ease of repairs.

All vehicles equipped with engines, regardless of the system (editor's note: electric or petroleum), were admitted to the competition and classified into various categories:

a) Two-seater closed cars.

b) Two-seater open cars with a hood.

c) Mixed two-seater cars that can be instantly closed or uncovered.

d) Four-seater closed cars with luggage racks (30 kilos per passenger).

After twelve days of navigating the streets of the capital and the surrounding roads, the automobile hackney carriages demonstrated their capability to provide regular service through satisfactory tests. This result is promising, and we look forward to seeing a significant number of these new vehicles operating on the streets of Paris soon.

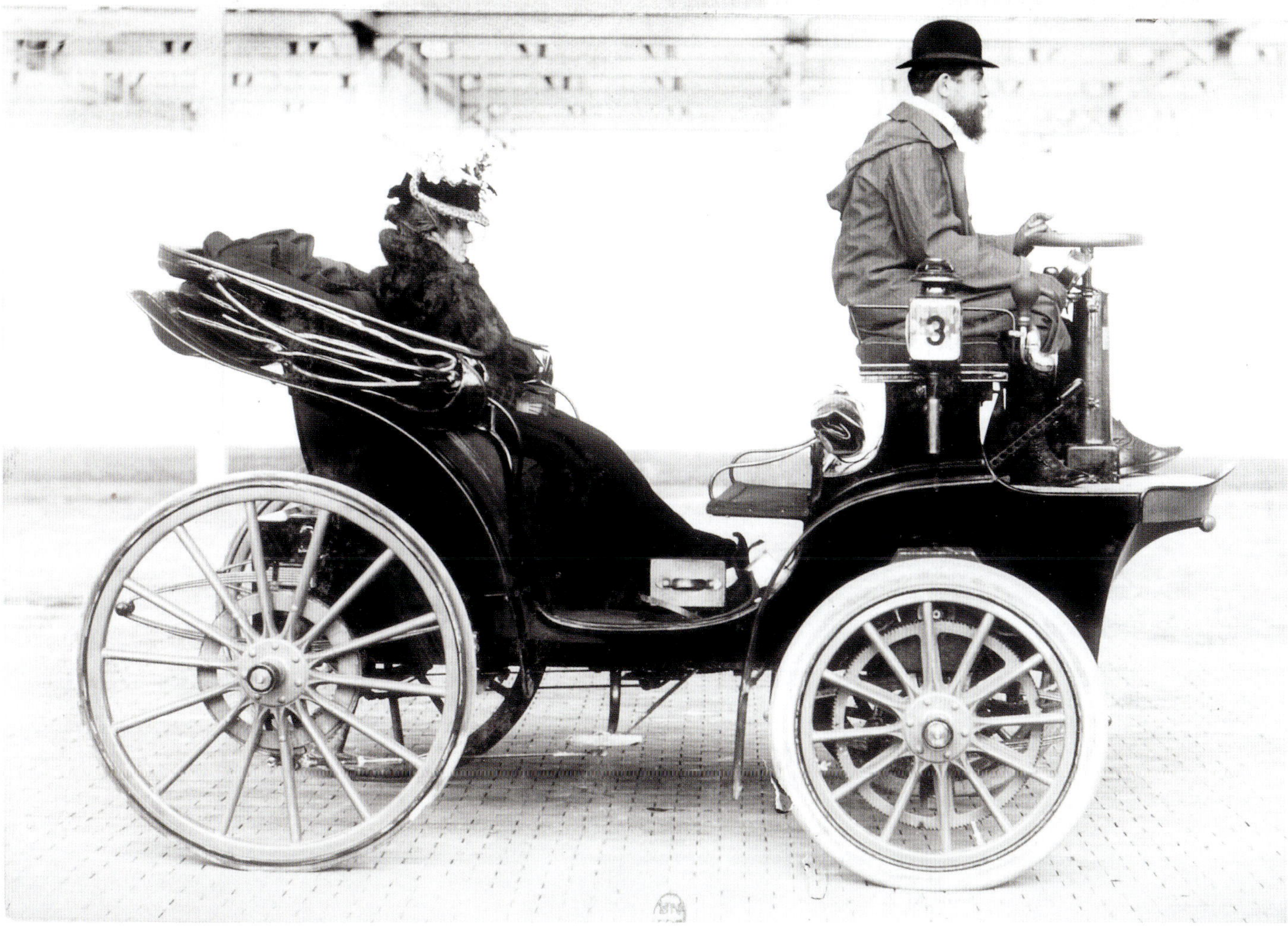

## Carriage Competition in the Bois de Boulogne

*"A committee led by Baron de Zuylen organized an event in the Bois de Boulogne last week, featuring a flower parade and a gymkhana, a kind of 'flowery petticoat' display."* (La Vie au Grand Air, October 29, 1899)

### The Emergence of Automobile Elegance Competitions 1890/1900

This development did not happen without resistance or criticism, far from it! What could be admirable about parading and marveling at these horseless carriages? Where was the elegance?

In 1890, *Le Petit Journal* introduced the idea of a competition for the best horseless carriages. There was little doubt that this new form of transportation would soon be widely adopted. However, one obstacle seemed certain to deter the wealthy, who valued elegance, from using these vehicles – their unsightly design. As one critic remarked, *"All that has been done is to remove the horses, and the eye instinctively searches for them in front of the coachman's seat."* (Le Figaro, November 17, 1894)

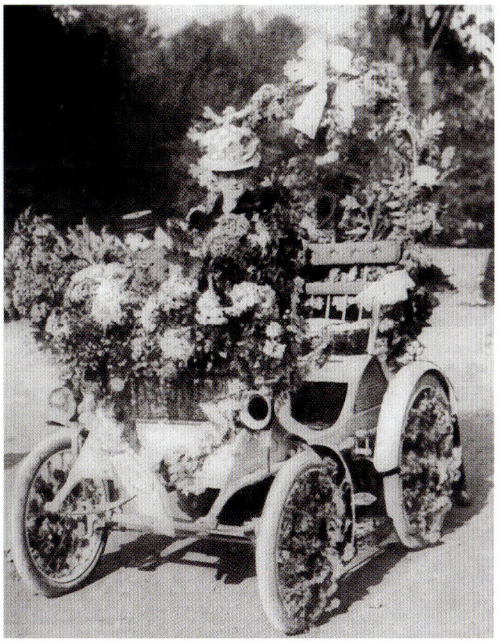

**ABOVE:** *These machines were considered suitable for regular service in Paris, though initial skepticism was inevitable. Another electric vehicle by Kriéger, a four-seater vis-à-vis, also participated in the 1898 automobile hackney carriage competition. At the wheel was Louis Kriéger himself.* (BnF Gallica)

**LEFT:** *Here, elegance defined the entire scene, with the driver almost completely concealed behind the flowers. The buggy competition in the Bois de Boulogne, held in October 1899, concluded with a grand flower parade, where Claire Price was awarded First Prize.* (Daniel Cabart Archives)

A few years later, in 1898, Maurice Leblanc – author of the character, Arsène Lupin – captured the prevailing sentiments toward the automobile in an article for *La Vie au Grand Air* titled "The Aesthetics of Automobiles." At that time, these new machines, stripped of their animal component, were viewed as not only dangerous, dirty, and foul-smelling but also aesthetically unappealing.

"Not long ago," Leblanc[1] wrote, "*I asked: why is the automobile so ugly, and why does it evoke in us a feeling of unease? It's because it hides its mechanisms like something shameful. We don't understand its purpose; it seems unfinished...*" He admitted that in writing this critical view, he fell into the same trap as those who opposed bicycles on aesthetic grounds. "*I was guilty of judging something without considering its intended purpose, comparing it to things that bear no relation to it, and being unable to free myself from preconceived notions and familiar images.*" Whether the design of automobiles had actually improved or he had simply grown accustomed to them through frequent exposure, Leblanc concluded that his opinion had changed. "*No, it is not ugly; no, it should no longer make us uneasy.*"

Leblanc further remarked, "*When we talk about the elegance of a* Victoria *or a landau, aren't we referring to a luxurious carriage drawn by two horses?*" For a long time, elegance had been tied to horse-drawn transport. As horses gradually disappeared, however, the notion that automobiles could also be elegant began to take hold in public consciousness.

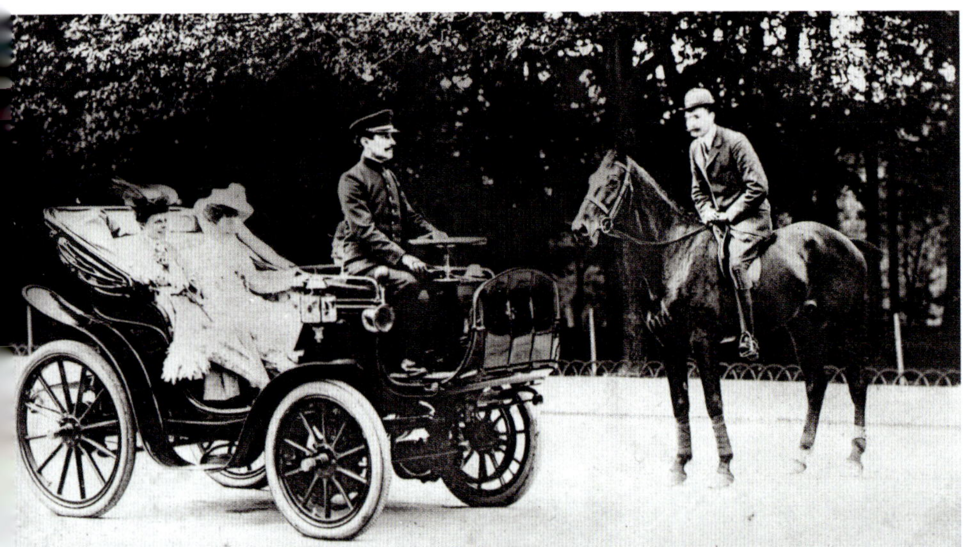

A journalist attending one of the first automobile competitions in the Bois de Boulogne in 1898 echoed this shifting perspective. He noted that "*no one seemed to miss the absence of horses.*" He observed that people's eyes were growing accustomed to carriages running without horses and that "*one can begin to see a certain elegance in motor cars, even when they're neither decorated with flowers nor drawn by a rose-covered swan...*" (*La Vie au Grand Air*, July 1, 1898)

And so, in places like Longchamp and Avenue du Bois, horseless carriages would soon compete in elegance, gradually replacing the traditional dorsay, landaulet, Victoria, cabriolet, and coachmen. Despite the disappearance of horse-drawn vehicles, their names continued to linger, often

**THIS PAGE:** *Actresses, social luminaries, and even queens and princesses – everyone wanted to be seen on Avenue du Bois aboard a "horseless carriage." Whether powered by electricity or petrol, like this city hackney carriage built on a 1904 Renault Type UA chassis (bottom), the appeal of these vehicles was undeniable. (Daniel Cabart Archives)*

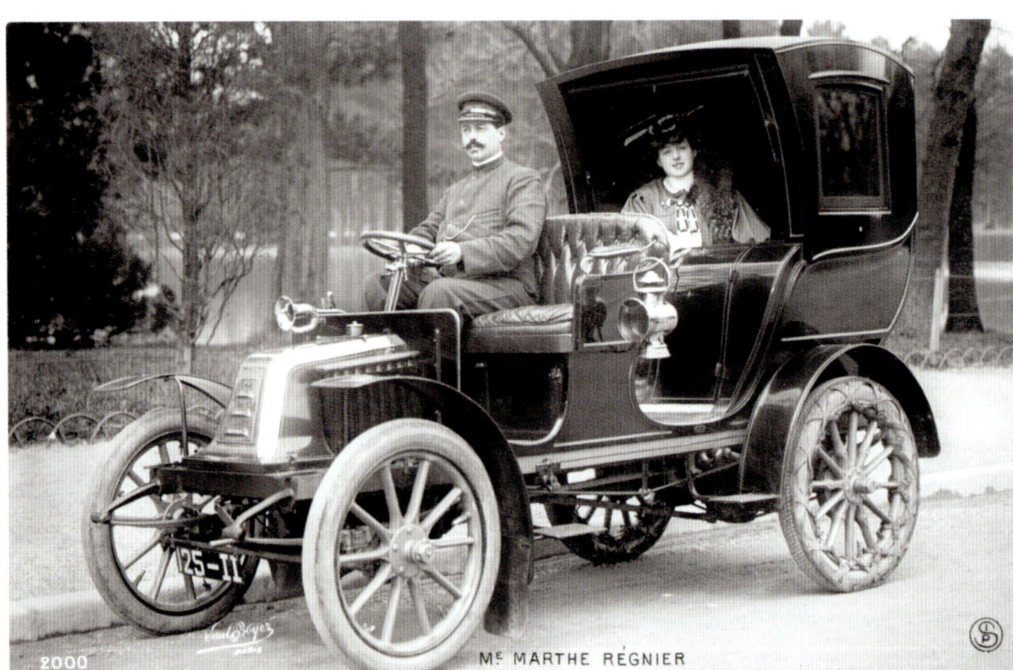


[1] *Maurice Leblanc in* Voici des ailes, 1898

L'usine de Hondouville

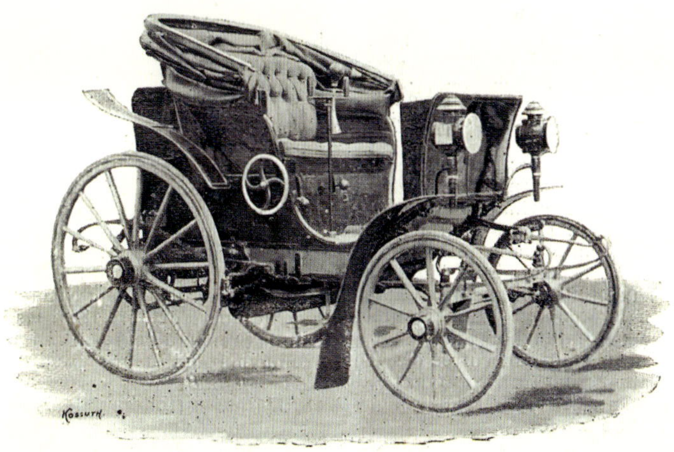


Duc de la Compagnie M. L. B.

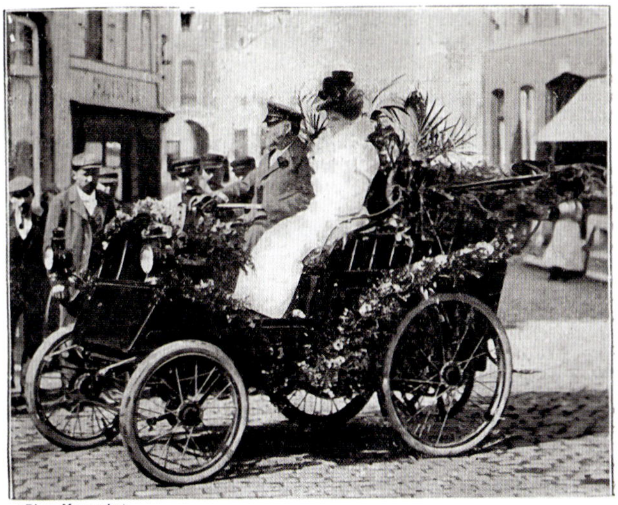

(Phot. Massardo.)
M. MORS, DEUXIÈME DE LA COURSE BRUXELLES-SPA.

adorning the new automobiles crafted by the same renowned coachbuilders (Kellner, Labourdette, Grümmer, Binder, and others).

In an article published in *La Locomotion Automobile* in March 1897, the idea was finally acknowledged: the automobile, once viewed with skepticism, was now seen as a tool of aesthetic progress, with elegance in its design. The embrace of the automobile by the "gentry" played a significant role in this shift, as demonstrated by the following excerpt:

> *"...The official acceptance of automobiles in the Bois marks the beginning of a glorious chapter in their history and will do more to enhance their design than any competitions created for that purpose. We are entering a time when, in the bustling paths of High Society, the outdated luxury of traditional carriages will be overshadowed by the splendor of the automobile convoys... the next flower festival will dazzle with this new backdrop. This is what drivers desire, and the lady chauffeurs see no objection."*

Naturally, these parades and demonstrations evolved into competitions – a way to inspire rivalry among owners, and later, among coachbuilders and eventually, fashionable women.

In 1898, for instance, Nice hosted an event called the *Concours d'Elegance et de Confort*. When looking back on this history, the Automobile Club of Nice proudly claimed in 1939 that it had organized "the first elegance competitions" in 1898. Neighboring towns along the Riviera soon followed.

As a key feature of the social season spanning December to April, the automobile elegance competition became one of many prestigious events alongside horse racing, sailing, motor yachting, and pigeon shooting. The competition grew fierce, and in March 1899, Nice disapproved of nearby Monaco's decision to host its own *Concours d'Elegance et de Confort*.

Littoral Mondain reported, *"On March 24, for a single day, our hearts and leisure activities swung between yachting and motoring, and as if such attractions weren't enough, the Société des Bains de Mer de Monaco organized a double competition for automobile comfort and elegance on the same day (what will the spirit of invention come up with next?). However, I find it puzzling that this competition for automobile comfort and elegance is taking place in Monte Carlo rather than Nice. It seems to me that Nice, with its splendid promenades, wide avenues, and spacious squares, would be far more suitable for this type of sporting event than the narrow plateau of Monte Carlo, accessible only by relatively tight roads. A parade of vehicles along a broad, straight path would present a much more impressive spectacle than navigating the steep, winding roads leading to a space that is too confined."*

The concours held on March 24, 1899, in Monaco was briefly recounted by *La Locomotion Automobile* (April 1899 issue), offering a glimpse into what these early "events" might have been like at the turn of the century, particularly in terms of what was presented and highlighted. The success of the event appeared limited, with only a "few remarkable vehicles" noted. Only the owner and body type of vehicles were mentioned, with no references to the coachbuilders. It's also noteworthy that

**ABOVE LEFT:** *"Duc" and "Petit-Duc" were the ideal cars for female customers to learn how to drive. Manufactured by Moteurs Landry et Beyroux (M.L.B.) from 1898 to 1902 in Hondouville, located in the Eure department of France, these cars featured a 4 HP single-cylinder engine at the rear. (Daniel Cabart Archives)*

**ABOVE RIGHT:** *For a final touch of elegance, Madame took her place in the parade at the end of the event. Émile Mors participated in the Brussels-Ardennes-Spa race from June 25 to 26, 1898, with one of his cars. He finished second in the Car category, completing the 210.5 km course in 10 hours, 17 minutes, and 20 seconds. (Daniel Cabart Archives)*

the body types listed were still reminiscent of those used in the not-so-distant past of horse-drawn transportation: phaeton, landaulet, station wagon, petit-duc, and Victoria. Furthermore, there was little mention of female participants!

### The Monte Carlo Elegance Competition

*On Saturday afternoon, the elegance and comfort competition took place in the square of the Monte Carlo Casino. The lack of sunshine was felt more than ever, and towards the end of the event, rain began to fall, prompting the drivers to retreat to Nice, where they encountered slick roads drenched by the afternoon downpour. There were few truly remarkable cars present. The awards were as follows:*

**2-seater cars**

1st *Mr. Roubert, small motor*
2nd *Madame Dhasty, Victoria*
3rd *Mr. Santos-Dumont, small red runabout*

**4- and 6-seater cars**

1st *Mr. C. Blanc, electric Phaeton*
2nd *Mr. Brunetta d'Alsany, petrol Phaeton*
3rd *Mr. A. de Rothschild, Phaeton*
4th *Mr. Reverdy, Omnibus station wagon*
5th *Mr. Meyan, Phaeton*
6th *Mr. Bouhey, Landaulet*
7th *Baron de Zuylen, petrol Dog-cart*
8th *Mr. P. Chauchard, Wagonette*
9th *Mr. Pinson, Wagonette*

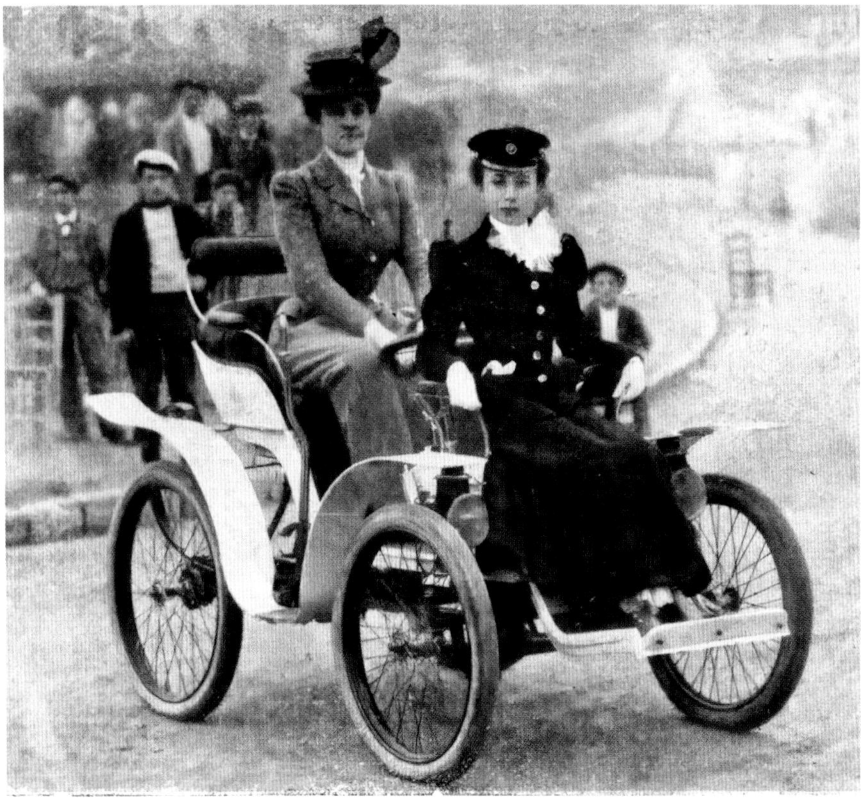

*"A well-stocked little car" represented the modest beginnings of feminine elegance in automobiles. This Phébus, manufactured by Noé Boyer in Suresnes, participated in the Bordeaux-Biarritz race in October 1899 and then the concours d'elegance that marked the conclusion of the event. (Daniel Cabart Archives)*

However, the seaside resorts of the Côte d'Azur were not the only places hosting these competitions. Events were also organized in Biarritz and Bordeaux, particularly during the Paris-Bordeaux race or at an "automobile week" in 1899.

> *"For four days, Bordeaux hosted a significant gathering centered around the Quinconces exhibition, where an elegance competition showcased truly stylish automobiles. The local coachbuilders in Bordeaux displayed remarkable craftsmanship, as evidenced by the fact that only they received awards, even though some exquisite cars had come from Paris. Not a single prize-winning vehicle conformed to standard catalog designs; all the competitors exhibited delightful creativity in their models."* (La Vie au Grand Air, June 4, 1899)

Concours also took place internationally, wherever automobiles made their presence known. For instance, in Richmond, England, "elegance and comfort competitions" were held, with prizes awarded for the best-designed motor cars, as well as for their finish and appearance. Similarly, Ostend in Belgium was one of the pioneers of this genre, and even in America, events emerged. *"Recently, in Newport, American drivers organized a car festival featuring a flower parade, a skill contest, and an elegance competition. The steering contest involved maneuvering vehicles through cutout silhouettes like those used for training Parisian electric coachmen. The only incident reported was when a car driven by Madame Hermann Oelrichs toppled after a sudden turn to avoid a silhouette. Fortunately, the speed was low, and the passengers escaped with minor bruises. The elegance prize was awarded to Mr. Oelrichs, among others."* (Littoral Mondain, February, 1899)

As the 19th century ended, the last decade marked a significant shift in the perception of this new mode of transportation. While it wasn't quite "goodbye horses" yet (that would take time), it was certainly "hello horseless carriage." Regarding automobile concours d'elegance, the transition into the next century would not bring about many changes.

DALTON WATSON FINE BOOKS

# NEWEST RELEASES

Dalton Watson Fine Books
*Glyn and Jean Morris*
**www.daltonwatson.com**
+1 847 274 5874
info@daltonwatson.com

## MASERATI 250F: A LEGENDARY FORMULA 1 CAR

*Walter Bäumer and Jean-François Blachette* 2 volumes • 848 pages • 667 illustrations **$395**

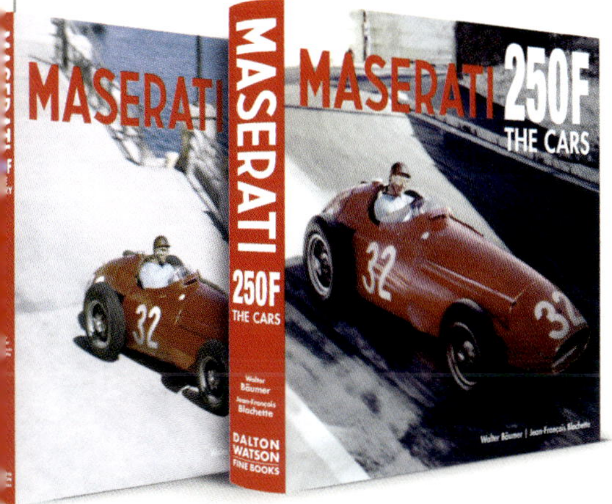


The Maserati 250F is widely regarded as the most iconic race car ever produced by the marque. Celebrated for its balanced handling and driver-friendly nature, it reached the pinnacle of success by winning the 1957 Formula 1 World Championship in the hands of Juan Manuel Fangio.

Driven throughout the 1950s by many of the era's greatest names—including Stirling Moss, Jean Behra, Phil Hill, and Fangio—the 250F also found favor with privateers worldwide. Elegant, effective, and enduring, it remains one of the most beautiful and influential Formula 1 cars ever built.

This definitive book on the Maserati 250F offers readers an in-depth journey through its development, racing triumphs, and enduring influence.

## ONE SUBJECT. TWO LANGUAGE EDITIONS.

ENGLISH AND FRENCH EDITIONS SOLD SEPARATELY.

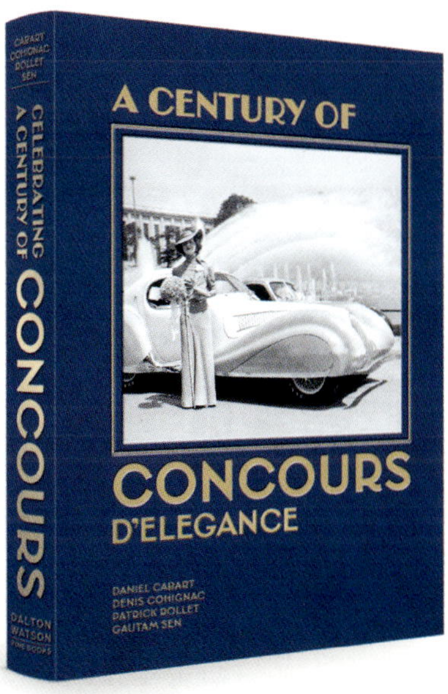


### A CENTURY OF CONCOURS D'ELEGANCE

*Daniel Cabart, Denis Cohignac, Patrick Rollet, Gautam Sen* 400 pages • 661 illustrations **$150**

Born in France in the 1920s, concours d'elegance evolved from refined showcases of the most innovative automobiles into true international celebrations of automotive heritage and craftsmanship. Initially focused on the latest creations of manufacturers, coachbuilders, and fashion houses, these events became cultural institutions that honor the classic automobile as an icon of industrial heritage — preserved, restored, and revered.

This book traces that remarkable journey, exploring the earliest concours and their close ties to women, fashion, and lifestyle, while also examining how the tradition was reimagined in the New World. There, it evolved into a vibrant and dynamic festival culture, embracing both the exceptional and the unexpected, and breathing new vitality and renewed meaning into the concours d'elegance tradition.

### UN SIÈCLE DE CONCOURS D'ELÉGANCE

*Daniel Cabart, Denis Cohignac, Patrick Rollet, Gautam Sen* 400 pages • 661 illustrations **$150**

Née en France dans les années 1920, les concours d'élégance ont évolué, passant de vitrines raffinées mettant en scène les automobiles les plus novatrices à de véritables célébrations internationales du patrimoine automobile et du savoir-faire artisanal. D'abord centrés sur les créations les plus récentes des constructeurs, des carrossiers et des maisons de couture, ces événements sont devenus des institutions culturelles honorant l'automobile classique comme une icône de l'héritage industriel — préservée, restaurée et vénérée.

Cet ouvrage retrace ce parcours remarquable, explorant les premiers concours et leurs liens étroits avec les femmes, la mode et l'art de vivre, tout en analysant la manière dont cette tradition fut réinventée dans le Nouveau Monde. Elle s'y est transformée en une culture festive et dynamique, accueillant aussi bien l'exceptionnel que l'inattendu, insufflant ainsi une nouvelle vitalité et un sens renouvelé à la tradition des concours d'élégance.

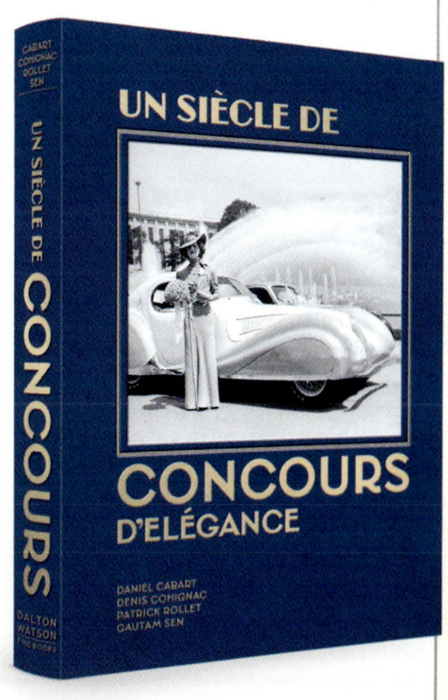

## Tracing the Origins

For centuries, horses were the primary means of transportation for people and goods. This dominance began to wane at the start of the 20th century as automobiles gradually replaced them. In some places, this transition was remarkably swift, as illustrated by two photographs of New York City's Fifth Avenue: one from 1900, showing a single car amidst countless horses, and another from 1913, where a single horse is surrounded by automobiles.

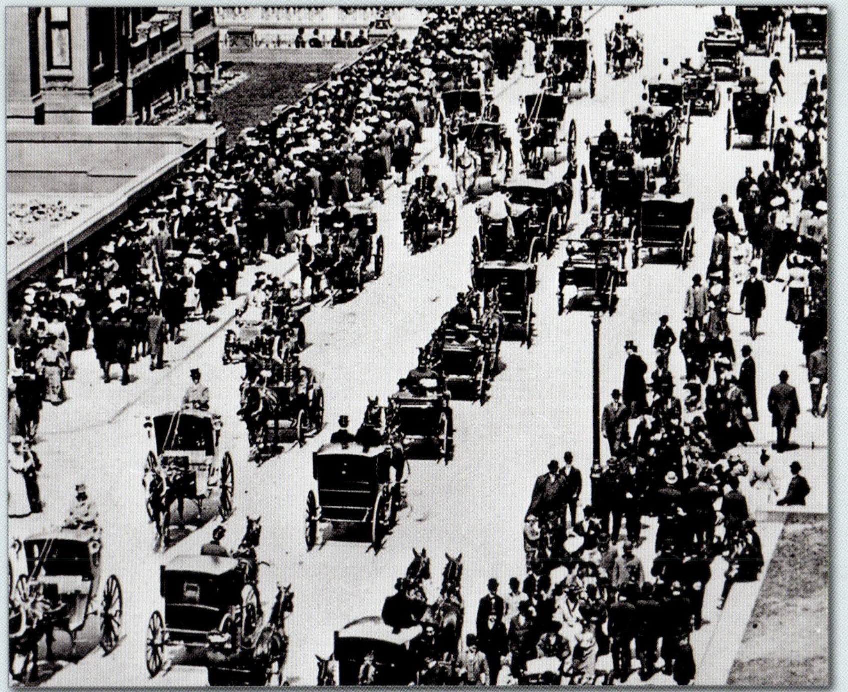

**THIS PAGE:** *Fifth Avenue, New York City. Taken on Easter morning, 1900, the top image shows a busy street bustling with horse-drawn traffic and very few cars. The bottom photograph is Easter 1913, capturing the bustle of the city with mostly cars and one lone horse and carriage. (Alamy)*

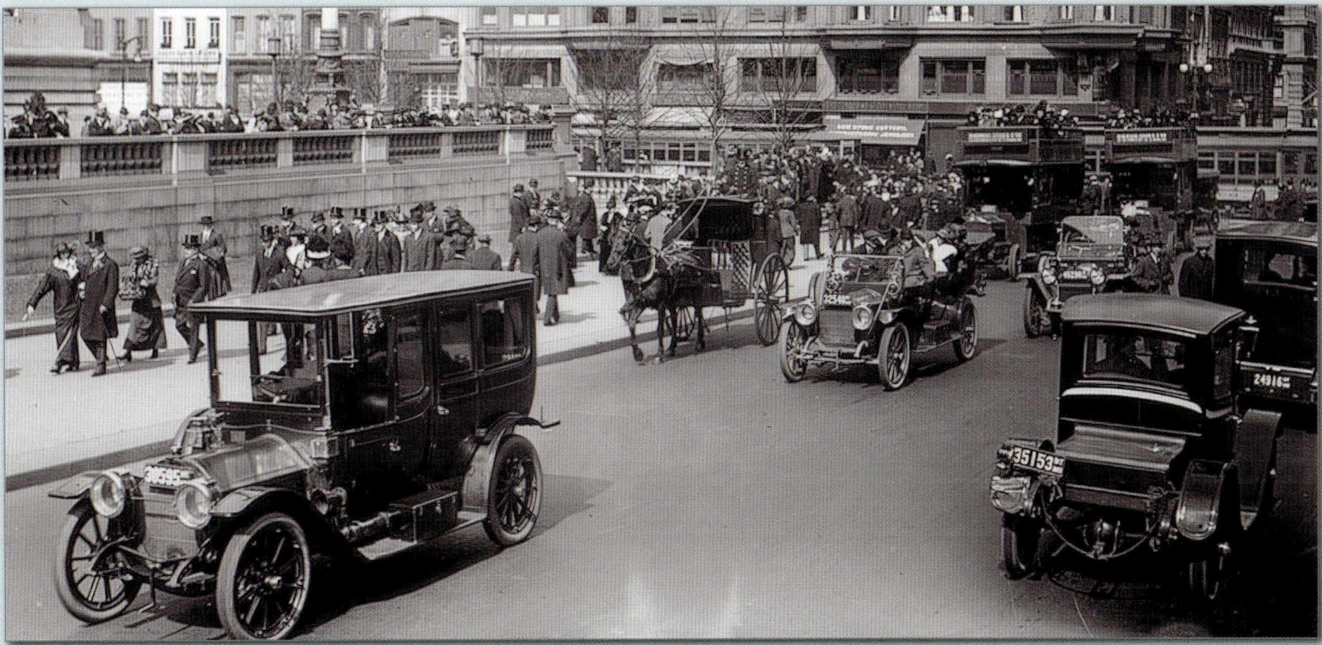

The equestrian world left a significant legacy for the burgeoning automobile era, starting with its terminology. Many words still used in the automotive industry – such as coupé, convertible, sedan, phaeton, and limousine – originate from horse-drawn carriages. Furthermore, numerous early automobile designers, like Binder and Labourdette, began their careers crafting equestrian carriages. Even the term "coachbuilder" reflects this shared history.

Two specific aspects of this rich heritage directly connect to the theme of automobile elegance competitions: model and gait competitions and horse drags.

## Model and Gait Competitions

Model and gait competitions evaluate horses based on their physical form (model) and movement (gait). These events traditionally served two purposes: bringing together breeders and buyers and promoting selective breeding by exhibiting the best examples of each breed. Over time, evaluation criteria evolved:

**Antiquity:** Chariot races and endurance events emphasized strength and speed.

**Middle Ages:** Tournaments and jousts celebrated robust warhorses capable of carrying armored riders.

**Renaissance:** Royal courts spurred interest in selective breeding, establishing studbooks to document horse pedigrees.

**18th-19th Centuries:** Agricultural fairs began hosting model and gait competitions, particularly for draft horses, which were essential to industrial transportation. In France, the National Stud Farms played a pivotal role in these events.

**Modern Day:** Standardized rules govern model and gait competitions under organizations like the International Equestrian Federation (FEI), chaired for years by the Duke of Edinburgh.

A parallel can be drawn between these equestrian competitions and historic vehicle shows, which focus on the authenticity of restoration or preservation. Both involve detailed scoring sheets and expert judges. For example, the French Institute of Horses and Riding (IFCE) developed criteria for horse evaluation that could easily be adapted to automobile competitions.

## Horse Drags

If model and gait competitions resemble automobile preservation contests, horse drags were the precursor to automobile elegance competitions. These events, popular among high society, merged equestrian tradition with social exclusivity.

The term "drag" originally referred to horse-drawn carriages used for transporting goods or passengers. As roads improved and industry advanced, drags became a leisure activity for the wealthy, blending elegance and performance.

In France, horse drag culture flourished during the mid-19th century, partly thanks to Baron Haussmann, who designed Paris's Bois de Boulogne and opened it to the public in 1852. Inspired by London's parks like Hyde Park, Haussmann created a space where the elite could showcase their wealth. Parisian society gathered on the Avenue des Acacias in ornate carriages, flaunting the craftsmanship of master coachbuilders, fine horses, and luxurious attire.

During the Belle Époque, drag outings became formalized with regular events and dedicated clubs such as the Paris Drag Club, which hosted parades and elegance competitions. These outings adhered to strict etiquette: women donned chic dresses and accessories, while men wore suits and top hats. The Bois de Boulogne symbolized Parisian refinement, attracting not just locals but also international visitors.

Although horse drag culture declined after the World Wars due to urbanization and the rise of automobiles, its legacy endures. Enthusiasts and equestrian clubs continue to organize drag gatherings, preserving this unique aspect of cultural heritage.

One memorable record of these events is *Sem au Bois* (1908) by Georges Goursat, known as Sem, a renowned illustrator and social commentator. His work humorously caricatures the vanity and extravagance of the era's elite.

## From Drags to Automobiles

Horse drags paved the way for the first demonstrations of automobile elegance, such as Sunday parades, flower-decorated cars, and early competitions. These events laid the foundation for the grand automobile elegance competitions that would become central to the social seasons of France's chic resorts.

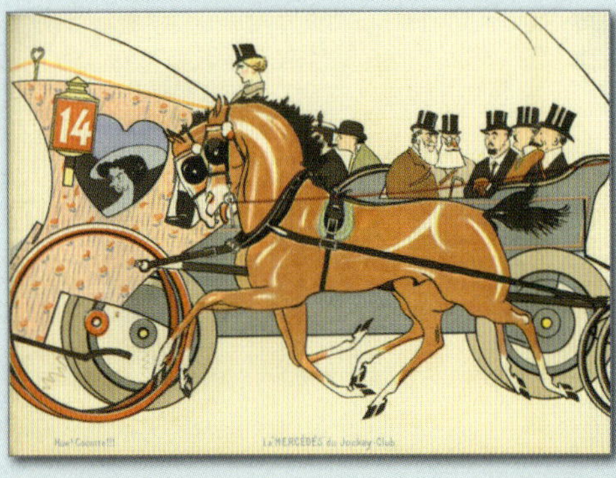

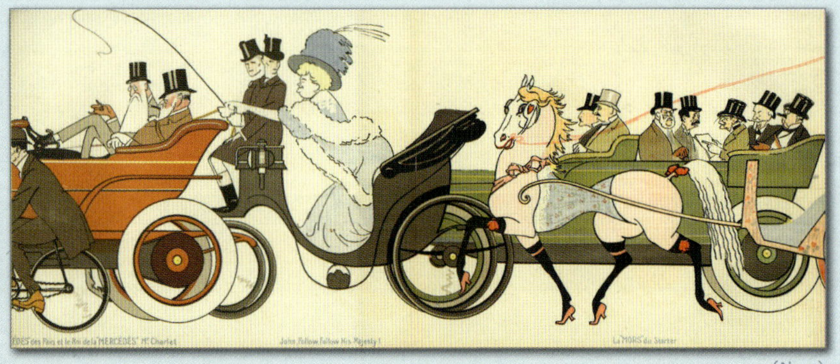

(Alamy)

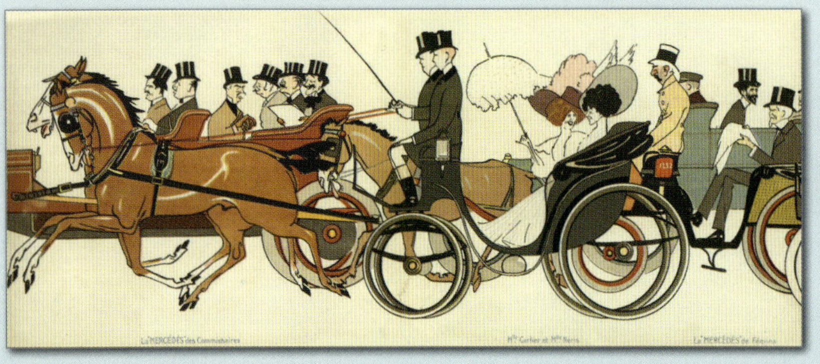

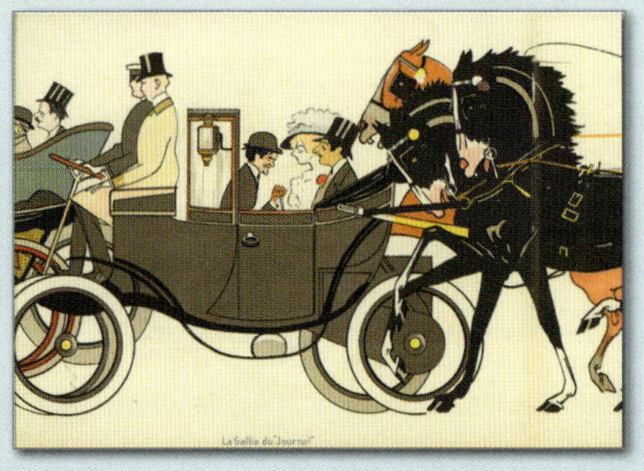

## Profile examination

Stand approximately 6 to 7 meters away from the horse, positioning yourself perpendicular to it on level ground. Ideally, choose an open space to ensure a clearer field of vision.

*Start by observing the horse as a whole, assessing its overall shape, look and proportions.*

| | To look for | To avoid |
|---|---|---|
| Hair | Shiny | Dull, pitted |
| Overall look | Good | Too skinny or fat |
| Proportions of the body (between forehand, body and hindquarters) | Harmonious, well balanced | Looking like three separate parts |
| Direction (please note that a foal normally has its rump higher than its withers) | Sloping slightly (withers higher than the rump) | Sloping downwards (rump higher than withers = tendency to lean is on its shoulders) |

*Detail the different parts of the body*

| | To look for | To avoid |
|---|---|---|
| Head | Proportionate, "intelligent" eye | Heavy sunken eye, hanging lips, split mouth |
| Neckline (top edge orientation) | Quite long, horizontal, convex | Short, slender, vertical, very concave |
| Withers | Clearly detached | Hollow or atrophied, sharp (high-withered) |
| Shoulder | Long, oblique | Short, straight |
| Legs at front | Long, oblique | Short, straight |
| Chest | Deep | Cylindrical |
| Back | Short and tight | Too long, saddled |
| Junction with rump | Short, well attached | Looking disconnected |
| Rump | Inclined, quite long | Horizontal, short |
| Hind limbs | Wide, long | Short, straight |
| Limbs | Ample, correct | Unsteady |
| Feet and hooves | Healthy | Damaged, tight, flat |

## Assessing from the front

Stand in front of the horse 3-4 meters away

| | To look for | To avoid |
|---|---|---|
| Chest | Wide open | Flat |
| Knee | | Narrow |
| Aplombs | Open, correct | Tight, knock-kneed |

## Assessing from the rear

Stand behind the horse 3-4 meters away

| | To look for | To avoid |
|---|---|---|
| Rump | Wide, rounded | Narrow, sharp |
| Thigh | Large, long | Slender, short, narrow |
| Hock | Large, correct | Narrow, arched, closed |
| Aplombs | Open, correct | Tight, too wide, knock-kneed |

This morphological assessment should be complemented by a movement test – conducted in hand, at liberty, on a lunge line, under saddle, or in harness – depending on the horse's training level and intended use.

# 1900~
# *La Belle Époque!*

## A NEW CENTURY, YET NOT MUCH PROGRESS

With the dawn of the twentieth century, the concept of elegance evolved differently between regions with the more "feminine" aspect of elegance yet to emerge.

In 1901, the Monaco Concours followed the tradition of previous years. Nothing changed as fifty cars paraded before a distinguished, aristocratic jury. *"These vehicles, notably free of floral decorations (in contrast to the frequent "flower parades" of the time), circled the square and halted in front of the judges. The event drew large crowds and was considered a great success."* (L'Auto, March 30, 1901)

In the evening, a *Corso d'Automobiles Illuminées* took place. A dazzling lantern-lit parade followed, equally successful. Numerous prizes were awarded. Yet once again, only the owners were acknowledged, with no mention of car marques or body styles.

By 1902, the rivalry between Nice and Monaco had subsided. The Monaco competition was now scheduled to follow the *Grande Quinzaine de Nice* after events like the hill climb at La Turbie, a mile race in Nice, and a Nice-Abbazia-Nice rally. This was the grand finale of the season for the Riviera's winter events. The competing cars, having departed from Paris on March 30, arrived in Nice by April 4.

> *"As usual, the Monte Carlo Concours d'Elegance was a resounding success. After a gray morning, the sun's warm rays shone on the Casino Square, where a large crowd gathered to watch the elegant cars parade. Each vehicle, it should be noted, exuded true elegance. Among those highlighted were vehicles owned by Messrs. Howard-Johnston, P. Chauchard, Captain Ferber, Baron Vuillet, A. de Rothschild, Plassard, Schiff, Baron de Bellet, and Lord Wolverton, as well as the phaeton of Prince Lubecki, the saloon car of Mr. Van Zuylen, and the milord of Mr. Levinstein.*
>
> *"In total, around sixty vehicles took part in the parade. The Prince of Monaco presided over the jury, alongside the Duke and Duchess of Urach of Württemberg, Prince Strozzi, and Baron A. de Rothschild. Various awards were presented, including art pieces, to notable participants such as Van Zuylen, de Bary, Georges Prades, Siegfried, and Madame Gobron.*
>
> *"The delightful gathering concluded around 5 p.m., and it is worth mentioning that even the heavy vehicles, which had crossed the La Turbie hill that morning in about 1 hour and 30 minutes, were present at the Concours d'Elegance."* (L'Auto, April 8, 1902)

**As the press reported, this was not a flower parade, where the car is overshadowed by bouquets and garlands. Instead, in Monte Carlo in 1904, it was a true parade focused on judging the design of the car bodies. (BnF Gallica)**

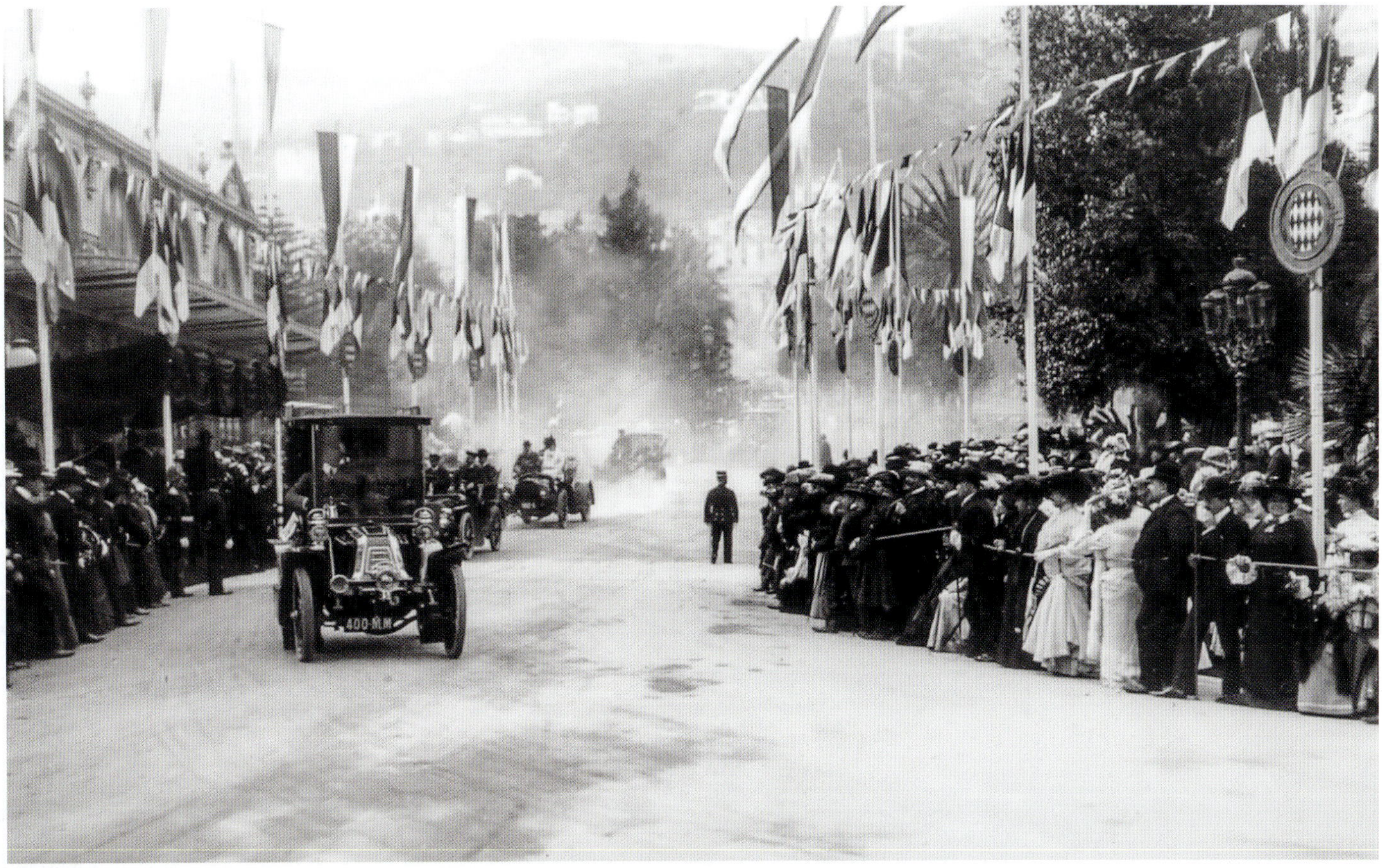

**RIGHT:** *An advertisement for the Delahaye, which made its mark in various competitions with notable performances.* (Daniel Cabart Archives)

**BELOW:** *At a time when women drivers were still a rarity, the wife of the manufacturer Gobron is pictured confidently at the wheel.* (Daniel Cabart Archives)

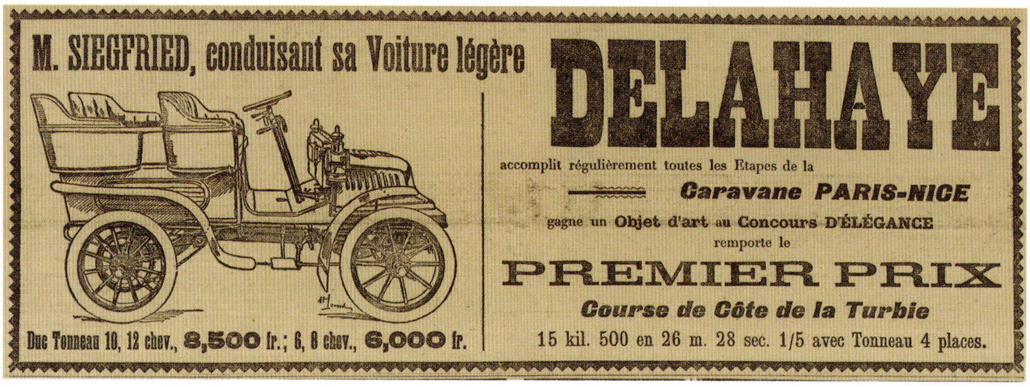


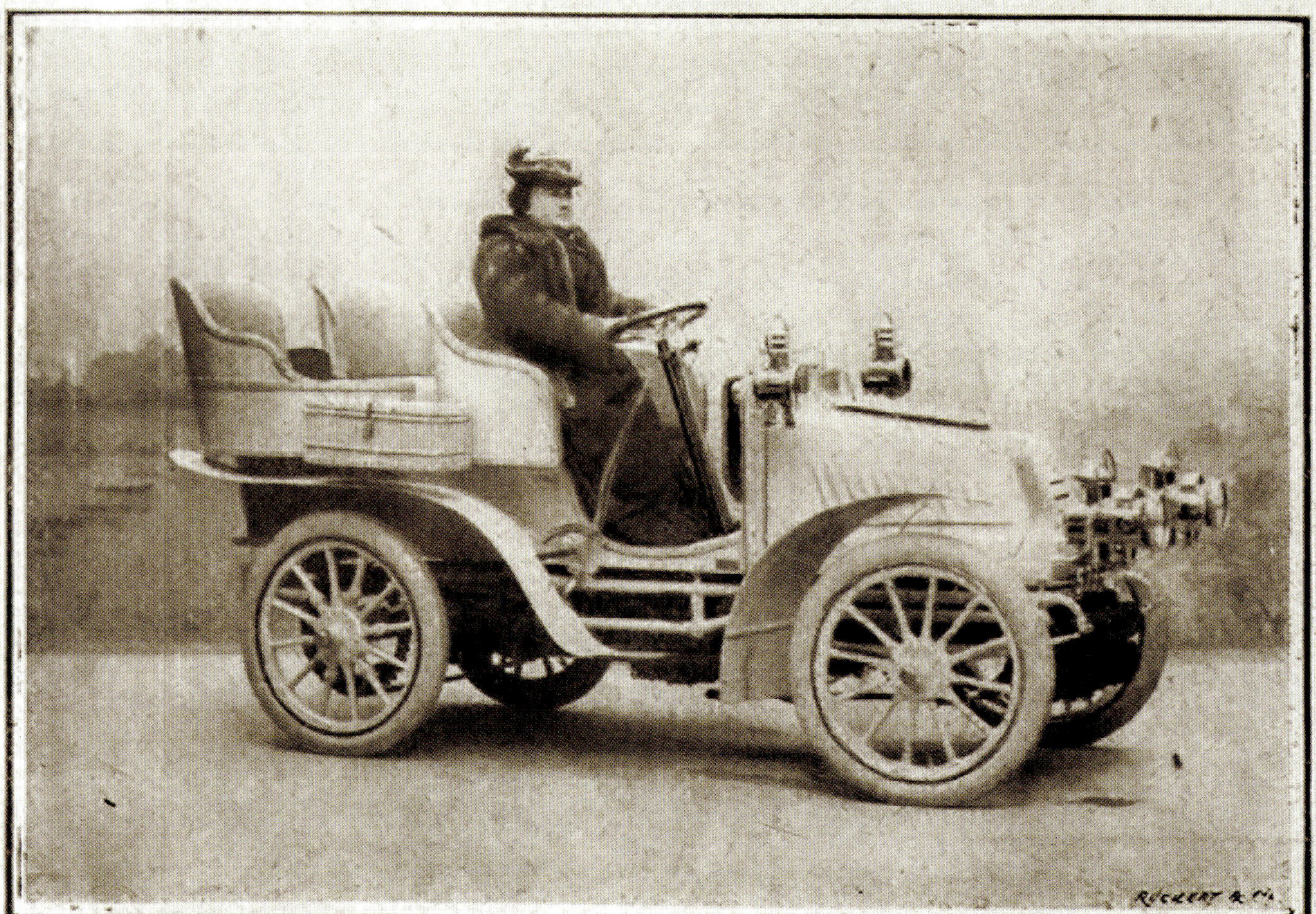

(Cliché J. Barenne).

Mme Gobrón sur sa voiture *Gobron-Brillié*, munie de pneumatiques *Continental*, qu'elle a conduite de Paris à Nice.

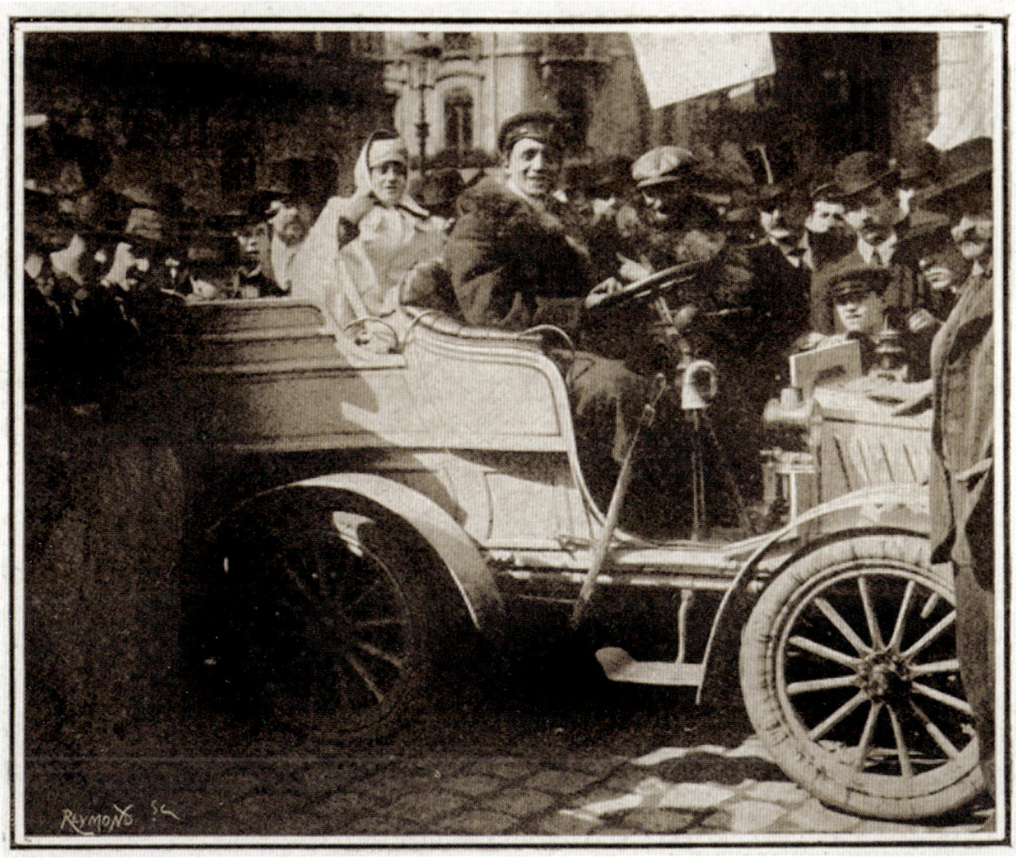

**LEFT:** *The wife of the journalist responsible for the report took her place at the finish line. She likely did not travel the distance from Nice to Paris in the car, however.* (Daniel Cabart Archives)

**BELOW:** *During the Paris-Nice rally in 1902, a stop in Lyon provided the mechanics with a chance to rest and the crew with much-needed refreshment after all their hard work.* (Daniel Cabart Archives)

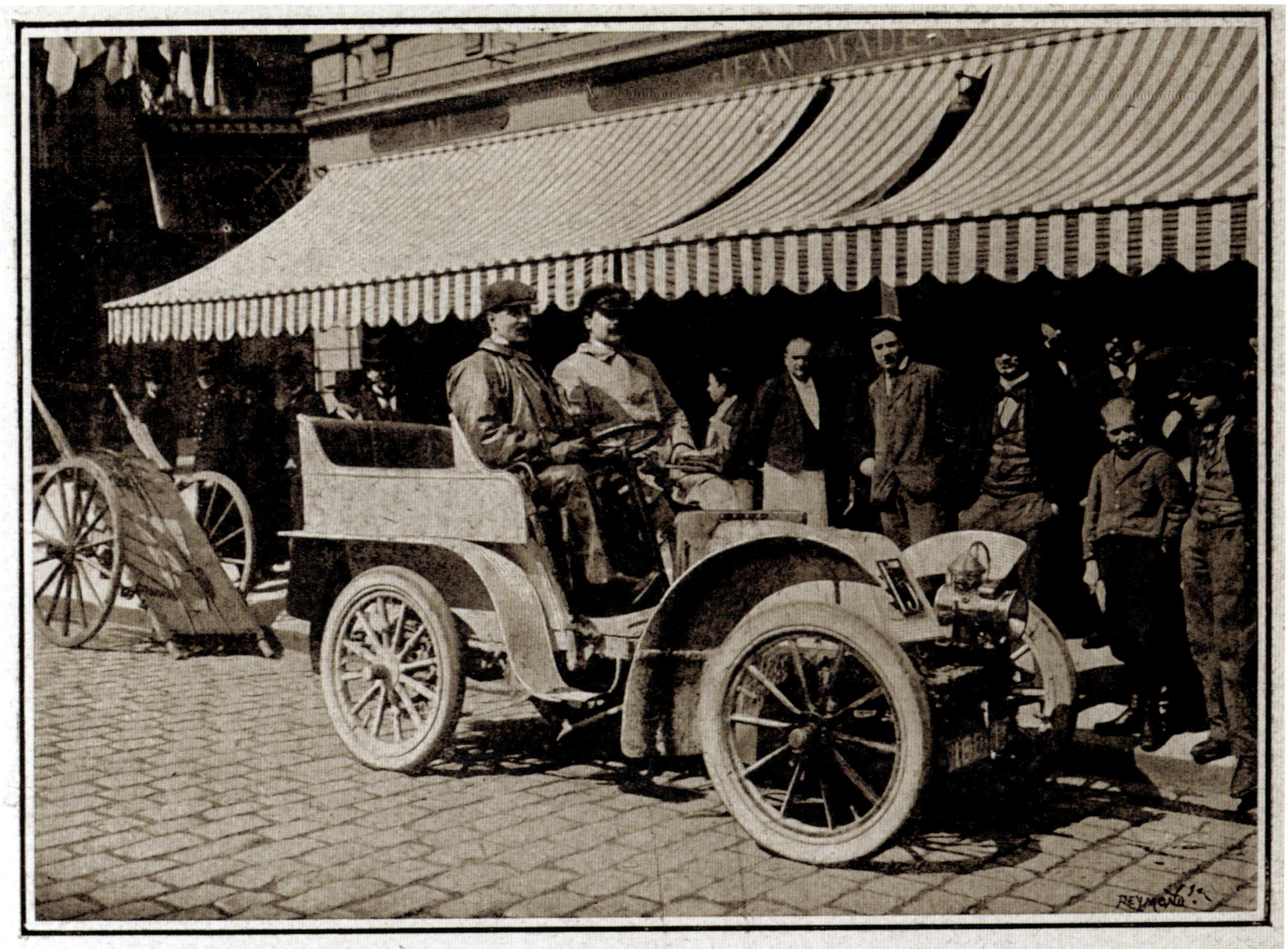

**M. Tranchant à Lyon dans sa voiture *Gladiator* à moteur *Aster*, munie de pr eumatiques *Michelin*.**

## December Motor Show: A Competition of Elegance

At the end of 1902, the organizers of the Paris Motor Show, not wanting to fall behind, hosted their own competition. This included a contest focused on the elegance of the vehicle bodies on display.

> *"The aim was to encourage manufacturers to strive for innovation in the design of automobiles and to showcase and reward advancements, new designs, and vehicle comfort. The competition would focus on the overall form of the car, including its engine and all necessary accessories, with a particular emphasis on the bodywork. Exhibitors willing to submit their vehicles for review would have their products evaluated by a jury appointed for the occasion."* (L'Auto, November 15, 1902)

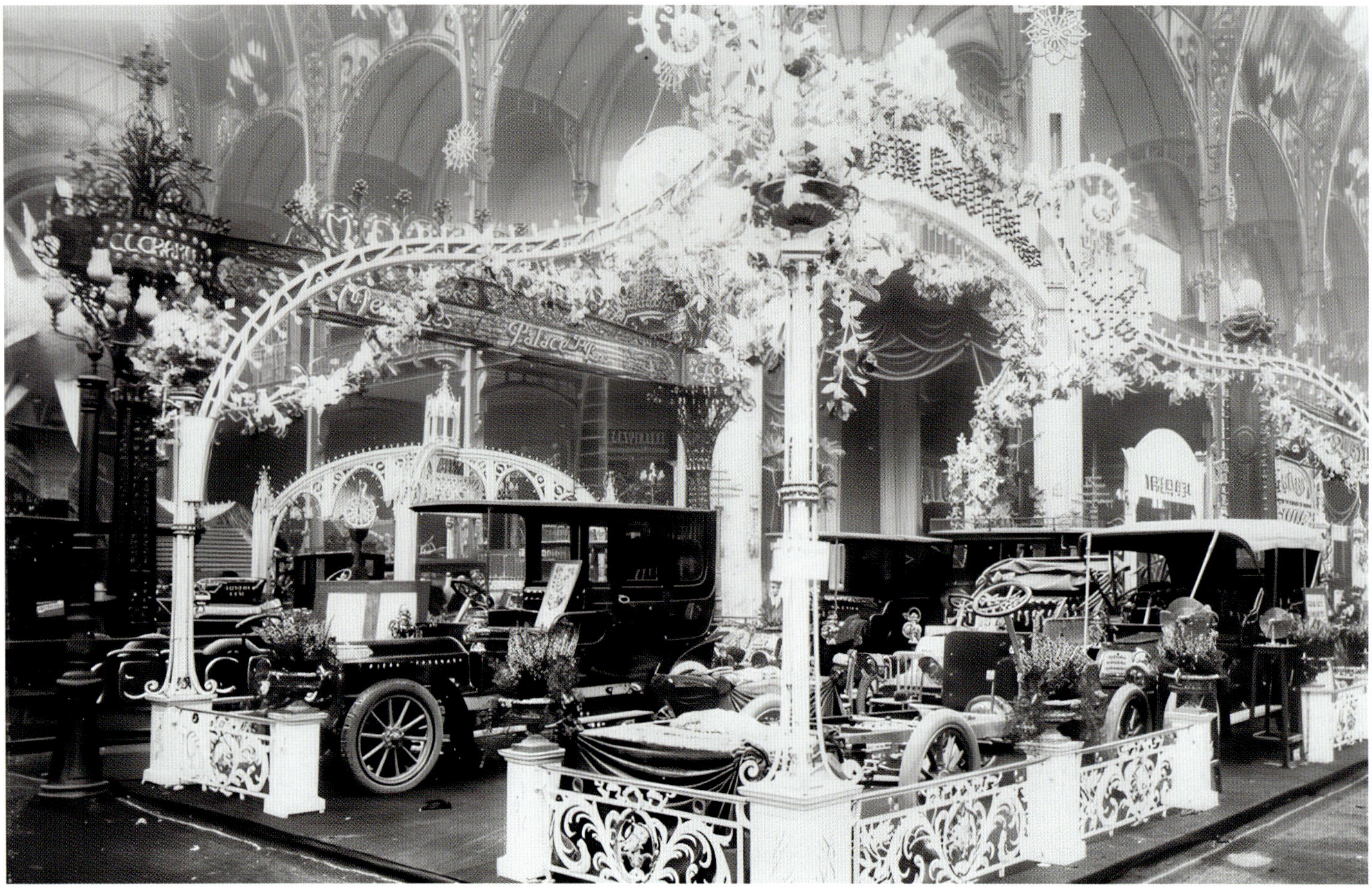

*This was essentially a concours d'elegance at a motor show. A jury judged the aesthetics and presentation of the cars, which defined the "elegance" of a stand, in this case, De Dion-Bouton. (BnF Gallica)*

## 1903 Monte Carlo Competition

While automotive competitions were becoming increasingly common throughout France, they often resembled flower parades more than true concours d'elegance.

Once again, one of the highlights of the Riviera's "season" this year was the Semaine de Nice, which featured a variety of competitions, including speed races, braking tests, and fuel consumption trials or the *Concours du bidon*. The event culminated in the traditional concours d'elegance held in Monaco.

It appears that, as in previous iterations, the concept of "elegance" remained elusive, with no clear guidelines or instructions provided to the jury – assuming there were any. However, according to the article reproduced on page 29, 1903 might have marked a shift in this practice.

Firstly, the event's popularity with the gentry, bourgeoisie, and aristocracy was evident, with a remarkable 65 cars registered. The competition report offers some insight into how the event unfolded, though it remains unclear what criteria led to the selection of 30 cars from the 65 entrants.

The question persists: how did the "selection committee" and the jury distinguish between the "comfortable elegance" of such different vehicle types – whether a petit-duc, a phaeton, or a landaulet?

What rationale could justify eliminating some while retaining others, given the significant differences between them?

Why this choice? *L'Auto* (April 1903) described the jury's decision as *"rigorous,"* yet *"a priori always debatable."*

After the parade of 30 cars – without any apparent opportunity for detailed inspection or close examination – the jury announced its verdict: four cars tied for first place, another four tied for second, and the remaining 22 vehicles were deemed "equal prizewinners."

Despite the somewhat opaque selection process, there was a consensus that the event was a success. Many felt it was a significant improvement over previous competitions, with results that were more serious and credible than in the past. However, no further details about the jury's criteria or decision-making process were made available.

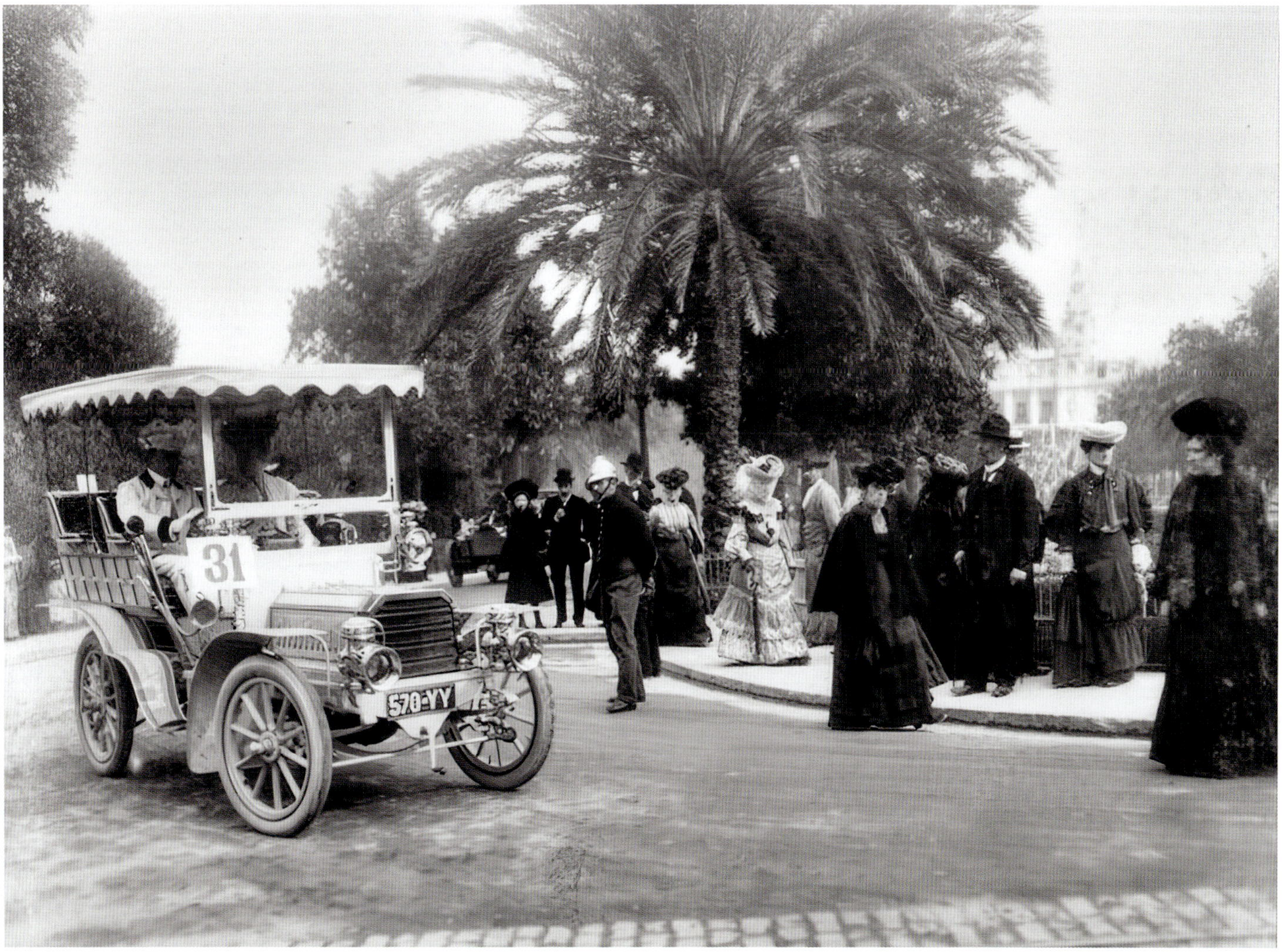


As explained, during the early years of the century, the competitions organized along the Riviera – primarily in Nice and Monte Carlo – appeared to be the most successful. While other seaside resorts also hosted events, the expansion of such festivities progressed slowly. A search through contemporary press reveals little of significance until the onset of World War I. It seems that most of these competitions were social events tied closely to the Riviera. Even Paris, where this movement had somewhat begun, did not appear to have organized any competitions in the years leading up to the war.

*Following the morning's thorough examination by the jury, the prize-winning vehicles paraded before the public.* (BnF Gallica)

Notably, there were still some prominent competitions held in places like Le Touquet (at the end of Boulogne Week), Ostend (Belgium), and Biarritz. However, these were often small-scale events linked to local festivals and accompanied by gymkhanas, lacking the grandeur of their Riviera counterparts.

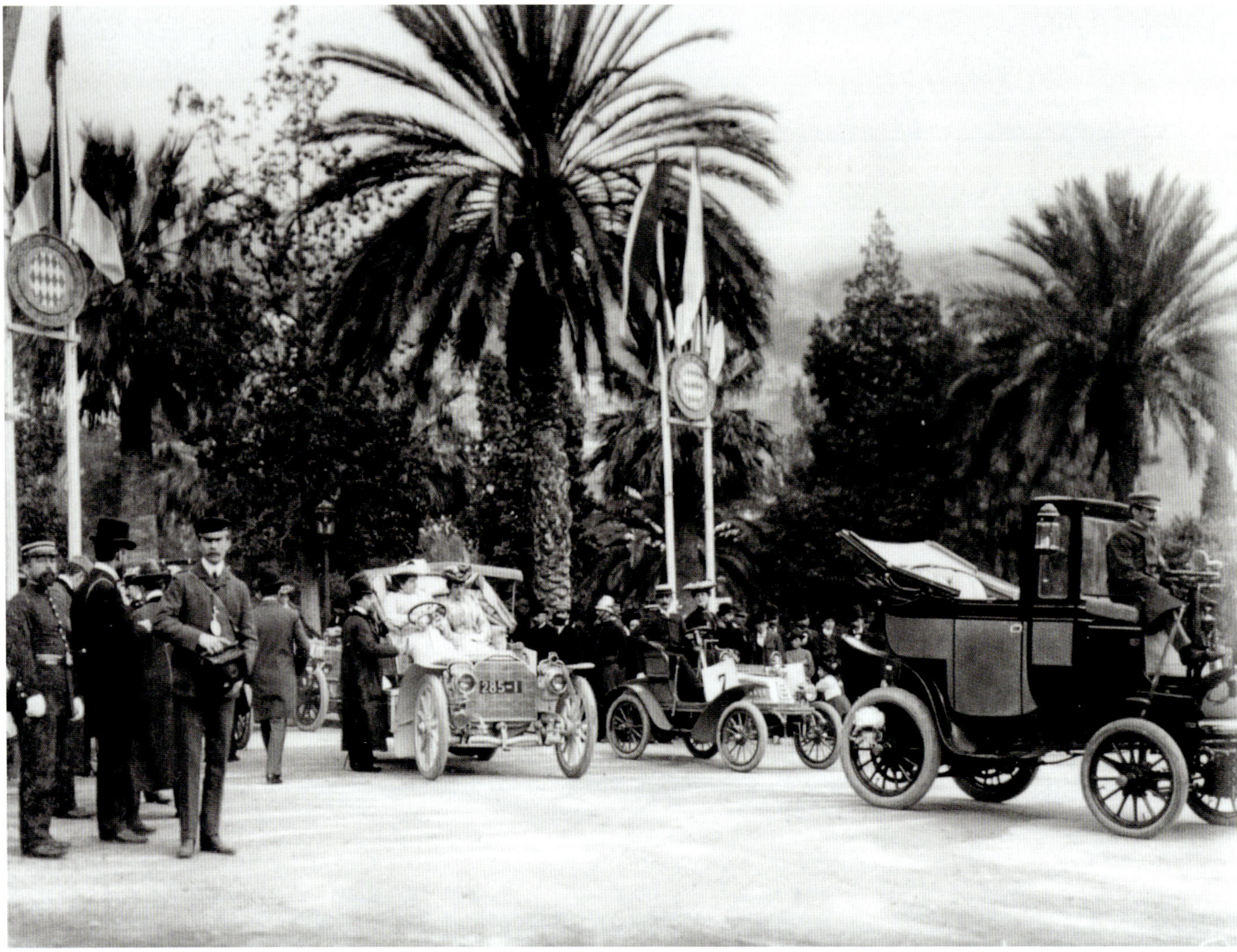

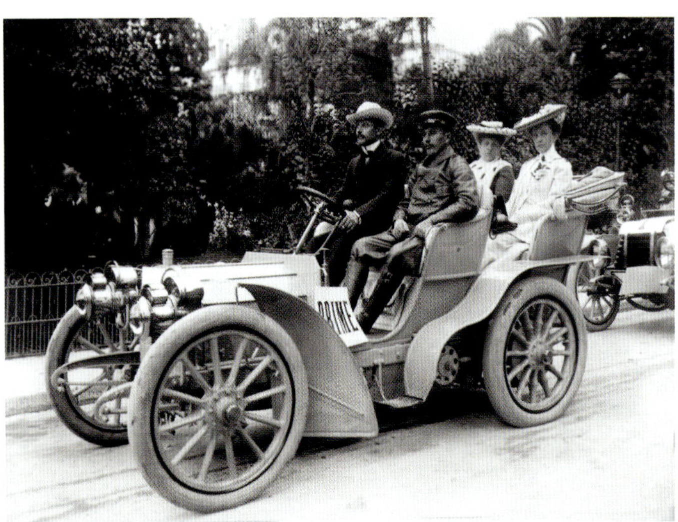

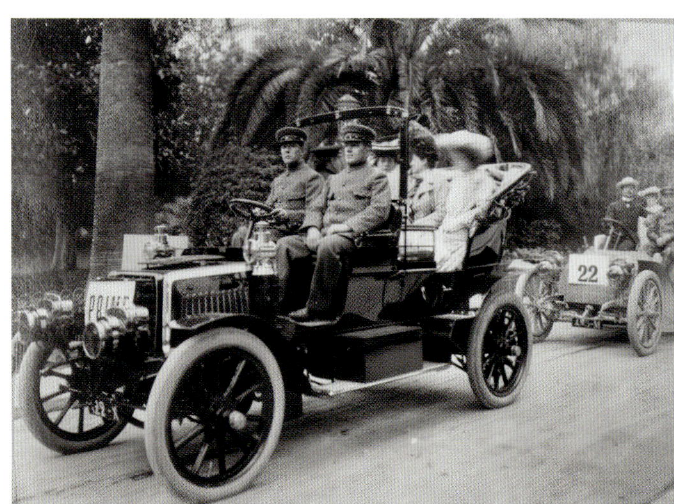

**TOP:** *The Mercedes spider with Rothschild bodywork, owned by Baroness de Zuylen. It was quite rare at the time for women to drive their own cars, but the Baroness drove it herself.* (BnF Gallica)

**ABOVE LEFT:** *Another Mercedes, this one driven by Count Zichy.* (BnF Gallica)

**ABOVE RIGHT:** *Mr. Levinstein's Panhard et Levassor, with Labourdette bodywork, is also one of the award-winning vehicles that the jury struggled to choose between.* (BnF Gallica)

## The Concours d'Elegance in Monte Carlo (*L'Auto*, April 3, 1903)

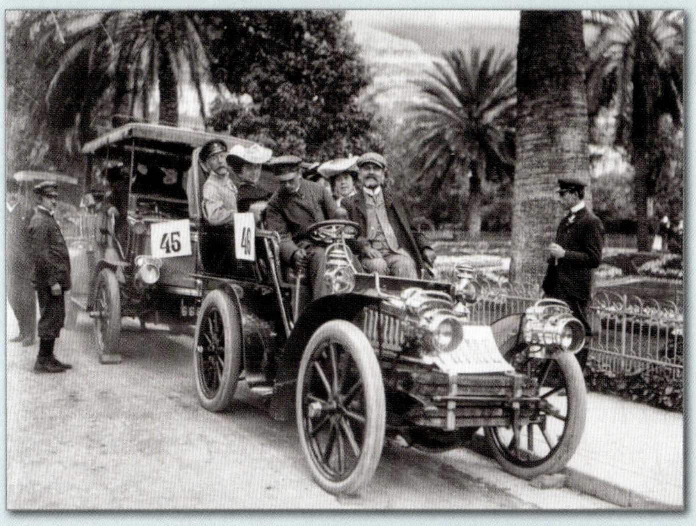


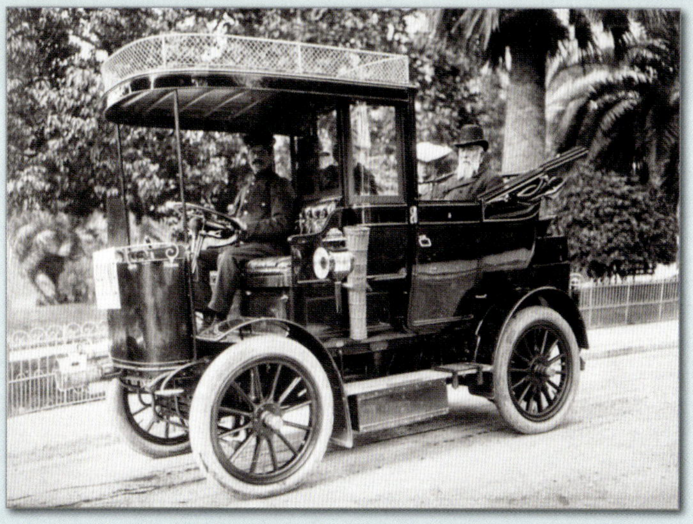

We all know the typical concours d'elegance: around fifty pristinely cleaned cars circle by, judged quickly, followed by a grand parade with banners. But this year was different.

The jury spent three hours rigorously evaluating the 65 vehicles submitted for the competition. Out of these, they selected 30 cars, which were the only ones to participate in the final parade. Unlike previous events, this time there was a thorough assessment of the vehicles' comfort and elegance. Although any such selection process can always be debated, this year's event clearly involved a more serious and practical examination of automotive design.

Four vehicles received top honors, ranked first in the 1903 Monte Carlo Concours d'Elegance. The winners, listed by their registration numbers, were as follows:

- No. 14 – Baroness de Zuylen (Mercedes Spider), Rothschild bodywork
- No. 53 – H. H. Prince Don Jaime de Bourbon (rear-entry tonneau), Rothschild bodywork
- No. 57 – Madame Camille Blanc (John Bullet landaulet), Belvallette bodywork
- No. 58 – Dr. Ladmiral (Delahaye saloon), Delahaye bodywork

Four vehicles tied for second place:

- No. 1 – Madame Berlier (Panhard tonneau), Rothschild coachwork
- No. 21 – Count Zichy (Mercedes tonneau and bodywork)
- No. 33 – Levinstein (Panhard vis-à-vis), Labourdette coachwork
- No. 46 – Mr. and Madame Schwab (Mors tonneau), Wiederkehr coachwork

Additionally, 22 vehicles were ranked equally, including:

- No. 2 – Colonel d'Arnoldi (Serpollet landau), Kellner coachwork

### The Parade

The procession took place around the massif in front of the Casino. The jury members were seated in the grandstand, with Prince Albert of Monaco and Prince Louis in the front row, accompanied by Baron de Zuylen. The event drew a distinguished audience, including notable figures such as Prince Orloff and Prince Galitzine. The chauffeurs were in large attendance, and the overall atmosphere was one of elegance. The day, marked by the seriousness of the results, was considered an improvement over previous shows.

**TOP LEFT:** *A Mors tourer with Wiederkehr bodywork, presented by Mr. Schwab. No fewer than six people took their places on board for the final parade. (BnF Gallica)*

**TOP RIGHT:** *Four vehicles tied for first place, and 23 others were classified together. It was a kind of "non-classification," though, as the press reported, this competition was considered a significant event and "more serious in terms of results than it had been until now." Colonel d'Arnoldi, with his Kellner landau on a Serpollet steam chassis, was one of the 22 vehicles classified equally. (BnF Gallica)*

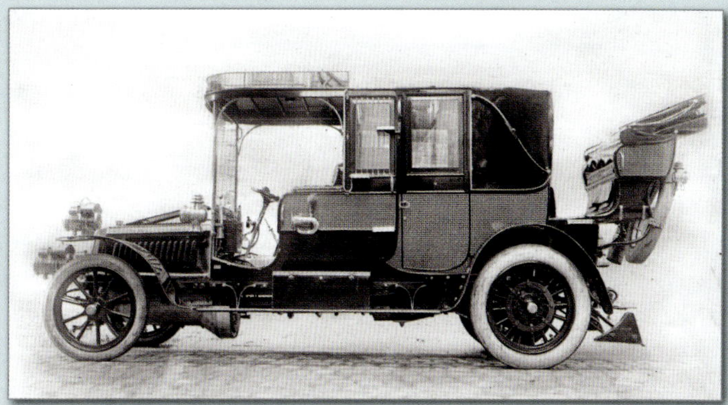

*Prince Orloff's very large Panhard et Levassor car did not participate in the concours d'elegance competition that day, but its owner certainly caught attention as one of the members of the jury. (Daniel Cabart Archives)*

## The Concours d'Elegance Competition (L'Auto, March 30, 1904)

In 1904, following the *Semaine de Nice*, the competition was once again held in Monte Carlo, this time under the name "Elegance and Cleanliness Competition." While it's unclear if the results were considered as "serious" as the previous year, the jury faced such difficulty in declaring a clear winner that they made a bold decision. Here's how the event unfolded:

> *"This annual festival, which consistently draws significant attention, was blessed with beautiful weather this year, contributing to its success. The competition, originally established to inspire coachbuilders to design a motor car that combined speed, comfort, and elegance, had yet to fully achieve this objective. Nonetheless, considerable progress has been made in recent years, as evidenced by the 38 stylish cars that gathered in Monte Carlo on Tuesday.*
>
> *"The jury, tasked with evaluating these vehicles, narrowed the field down to 31. However, faced with the challenge of ranking them – since each car offered its own distinct advantages – the jury awarded all 31 vehicles equally. Among those recognized were Dr. Ladmiral's saloon, Madame Jeanne Furstenberg's limousine, and Mr. Kriéger's double phaeton.*
>
> *It's clear why ranking them posed a dilemma: the 31 winners represented a diverse array of body styles, including twelve distinct types, ranging from bucket cars to phaetons, and from saloon-landaus to demountable limousines and the Roi des Belges type of bodywork.*
>
> *Among the honored coachbuilders were many who had earned renown during the heyday of horse-drawn carriages, such as Kellner, Million-Guiet, Rothschild, Labourdette, and Belvallette. Their presence signaled the enduring prestige of traditional craftsmanship within the evolving world of motor vehicles."*

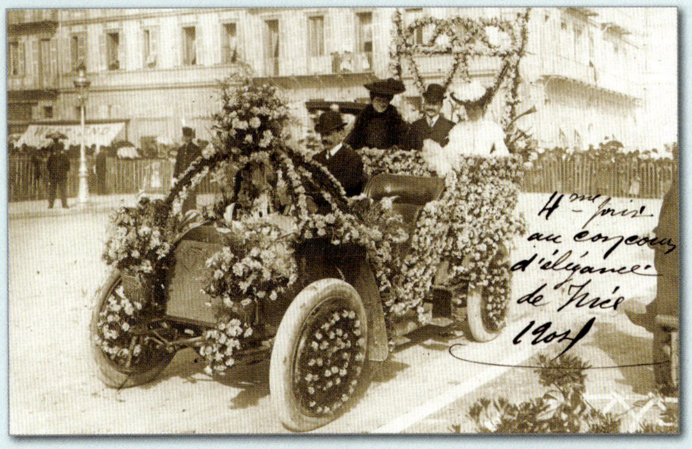


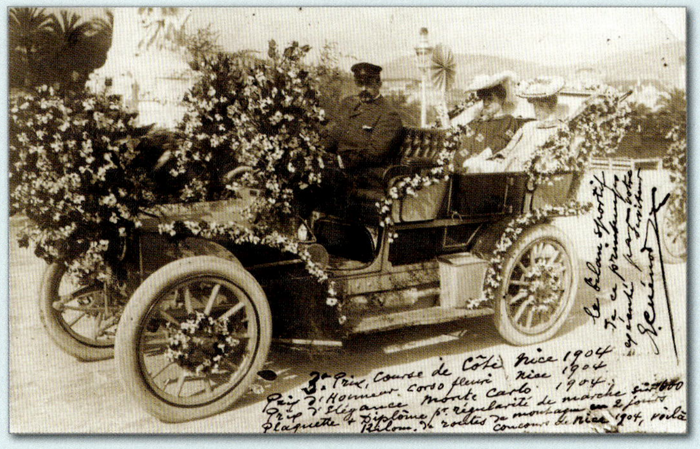


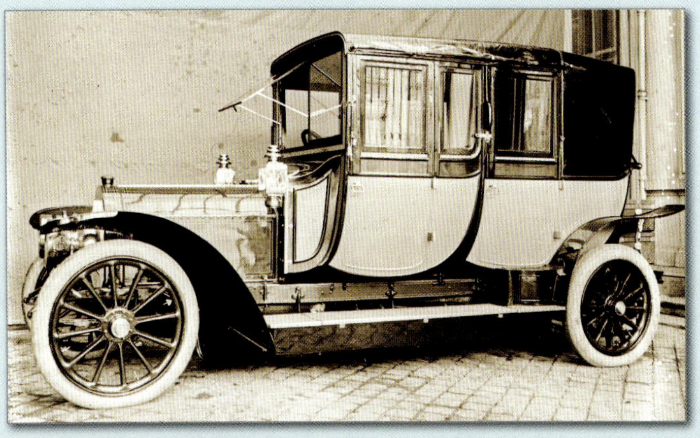

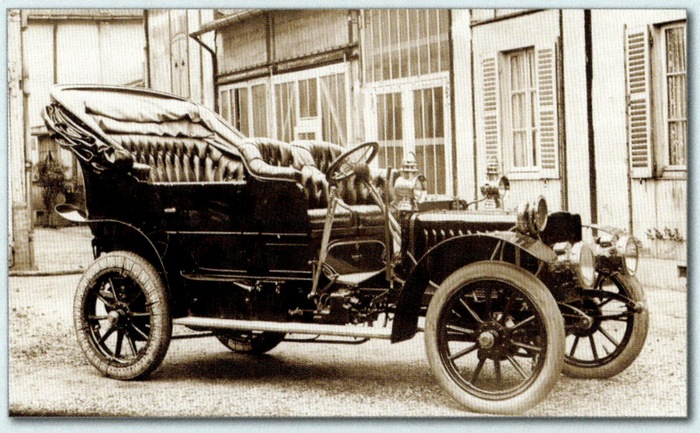

**TOP LEFT:** *Nice, 1904. One of the five Mercedes cars present at this concours d'elegance competition, which resembled more of a flower parade than a true display of elegance, whether automotive or feminine. "The first banner is awarded to Mr. and Madame Walter Munro and Mr. and Madame Schwab, for the superb decoration of their car with carnations, cornflowers, and white wallflowers." (Daniel Cabart Archives)*

**TOP RIGHT:** *At the same Corso Fleuri in Nice in 1904, Mr. Cuénod, transporting Madame Mauser and Madame Georges Prades, won a prize at the wheel of his Martini, which was decorated with "narcissi and anemones." But as we can see by the results written on the card, the same car could be steady, efficient and fast, as well as elegant. (Daniel Cabart Archives)*

**ABOVE LEFT:** *Kellner, a renowned name in French coachbuilding since the 19th century, crafted this Mercedes landaulet. It was considered an elegant car, both in terms of its design and the arrangement of its accessories. (Daniel Cabart Archives)*

**ABOVE RIGHT:** *A truly elegant body, "royal," as it represents what was known as the "Roi des Belges" phaeton. The driver's seat is narrower than that of the "royal" passenger, who insists on sitting at the front rather than the back seats, as was customary. This Weyher et Richemond, manufactured in Pantin, near Paris, was a particularly luxurious steam car. (Daniel Cabart Archives)*

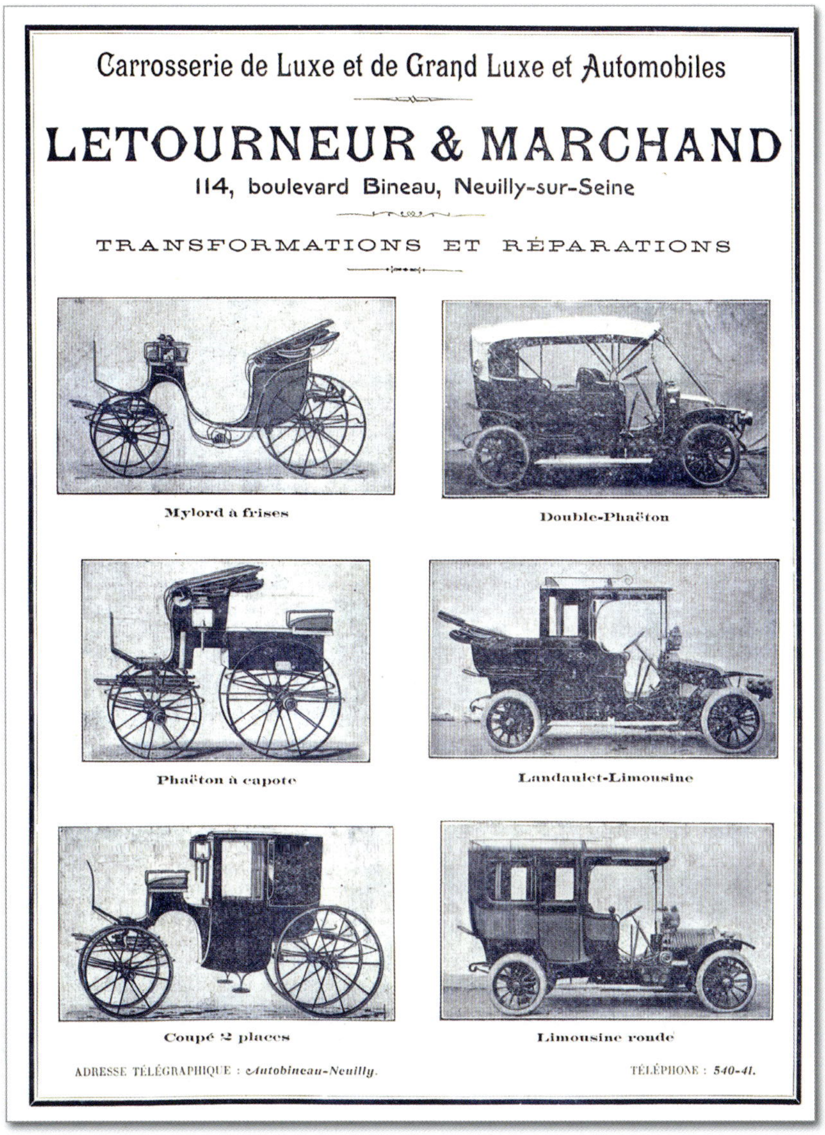


**LEFT:** *Letourneur et Marchand, which went on to become one of the most renowned French coachbuilders, like several others, preserved the connection between horse-drawn and automobile coachwork. In the top right, a Renault double-phaeton is displayed. (Daniel Cabart Archives)*

**BELOW:** *In 1906, a Tourand automobile was featured in the first advertisement – illustrated with a photograph – for the Letourneur et Marchand coachbuilder, established the previous year. Pictured here is a luxurious double-phaeton. (Daniel Cabart Archives)*

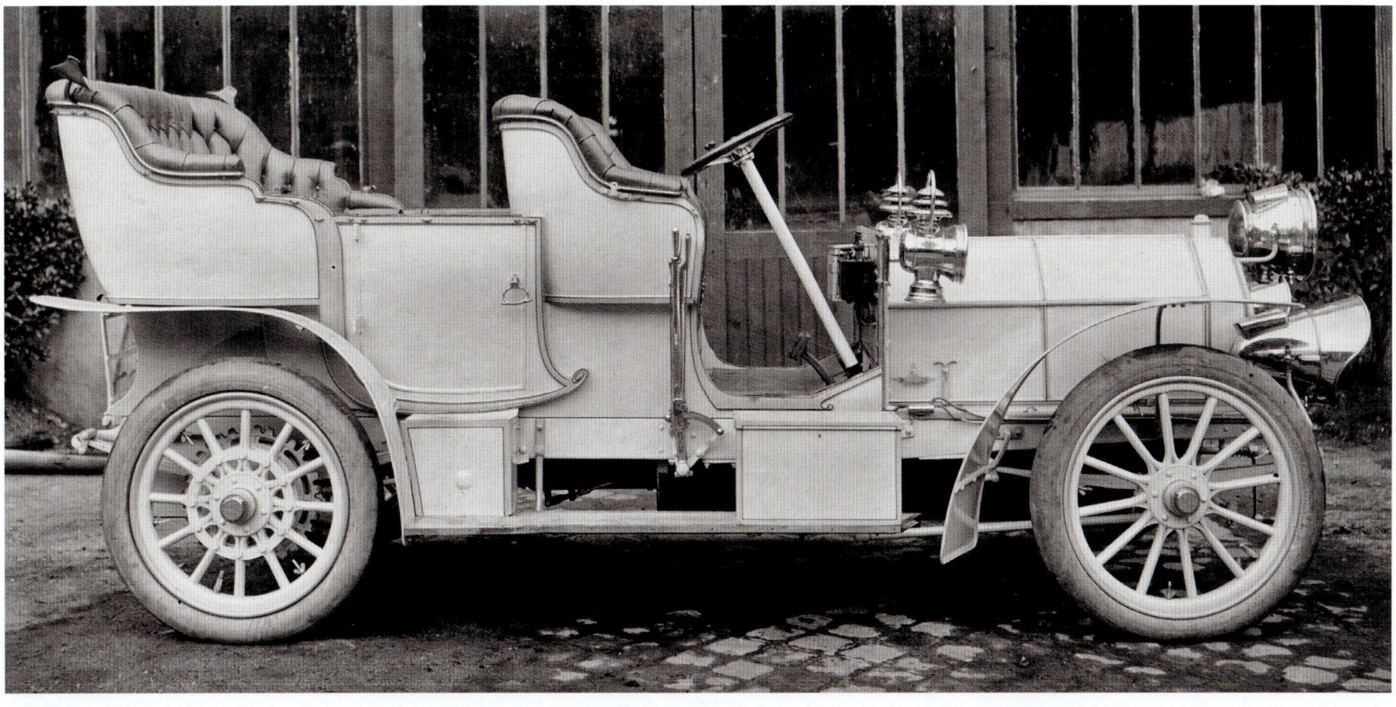

In 1906, a new variation of the Concours d'Elegance emerged. This iteration took place in Paris under the auspices of the sports daily *L'Auto* and was titled the *Concours d'Elegance et de bonne tenue*[1] (for mechanics and automobiles). This event can be likened to the previous year's *Concours d'Elegance et de propreté* (concours of elegance and cleanliness) held in Monaco. Here is an overview of the event's presentation from *L'Auto, December 26, 1906*:

### Concours d'Elegance et de bonne tenue

"We must promote the cleanliness of motor vehicles and their drivers. To this end, *L'Auto* is organizing a Concours d'Elegance et de bonne tenue for mechanics and cars on Sunday, January 20. The rules for this exciting event are outlined below.

### Statement of Reasons

The days when a good mechanic was defined by dirty hands are long gone. In those outdated times, a car owner was considered proficient only if they were covered in a thick layer of mud and grease. Accumulating a certain number of kilometers was something to boast about, with road dust being worn like a badge of honor.

Those times are over. We no longer refer to drivers as having "sooty or grimy faces." On the contrary, there is now genuine pride in maintaining a vehicle that is as meticulously cared for as the finest master coupé. This competition aims to reflect that shift, celebrating the elegance of the cars and the professionalism of those who operate them."

*Continued on the following page...*

**THIS PAGE:** *Hérald, a rare brand whose vehicles were "appreciated by true connoisseurs," consistently won the top honors in both racing and concours d'elegance competitions, including the city of Tours and on the Riviera. (Daniel Cabart Archives)*

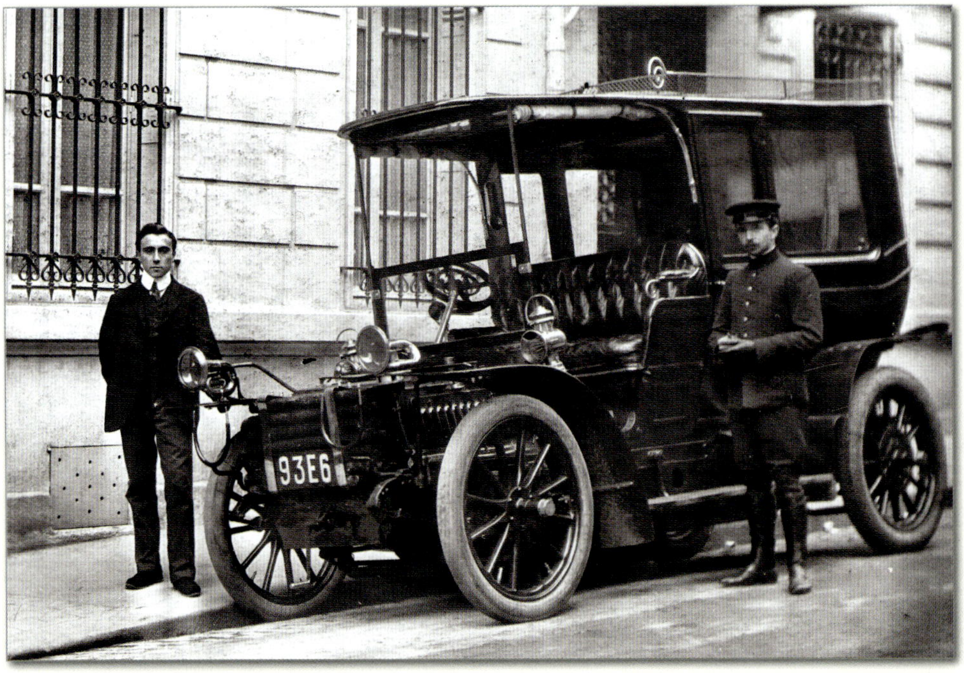


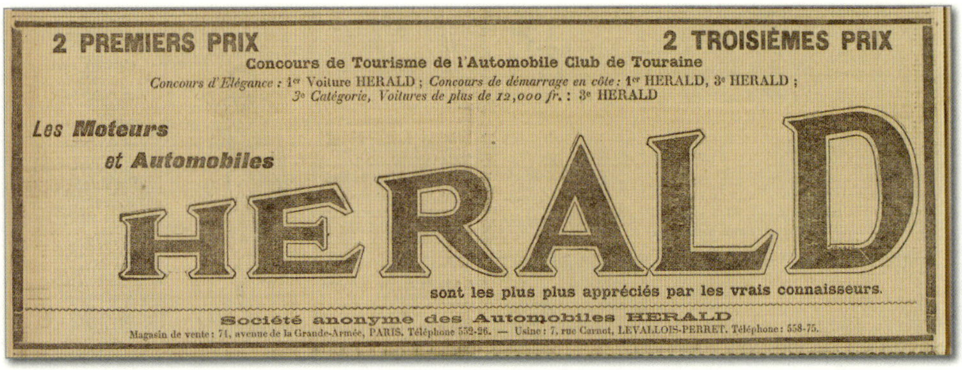


[1] The phrase "Concours d'Elegance et de bonne tenue" translates to: "Elegance and good conduct contest" or "Elegance and proper demeanor competition."

This type of event typically judges participants not only on appearance (e.g., style, grooming, poise) but also on behavior, posture, or how well they carry themselves.

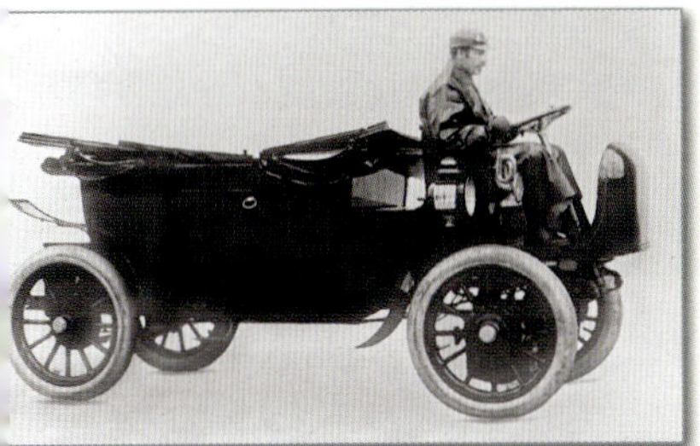

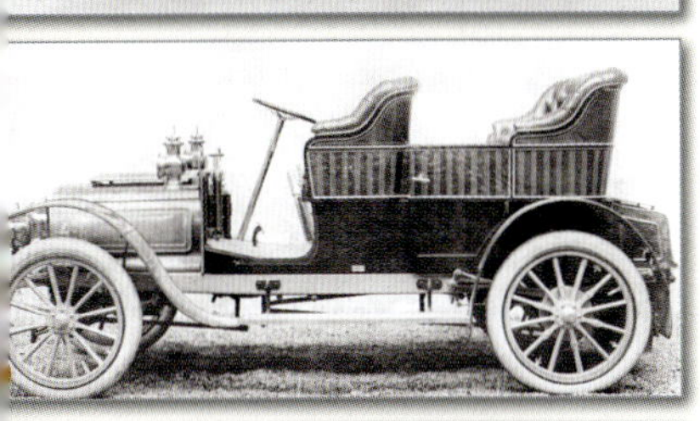

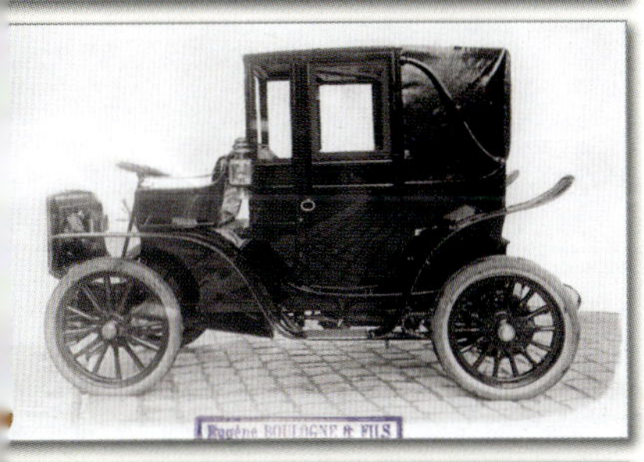

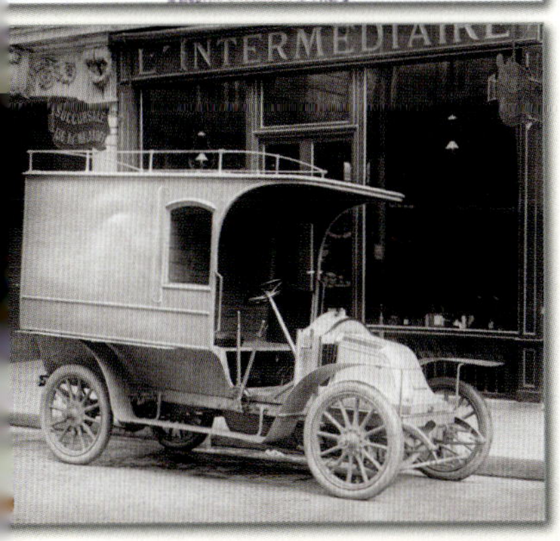


## Let's Encourage...

Since then, and gradually due to changing circumstances, cleanliness in motoring has become increasingly valued. Nowadays, it's not uncommon to encounter well-maintained cars driven by tidy mechanics, especially in Paris. We cannot emphasize these habits enough.

In this spirit, *L'Auto* has decided to promote driver elegance and vehicle cleanliness by hosting a competition. This event aims to reward the mechanics of both private and public motor cars who, alongside their vehicles, present the best appearance. A jury composed of the "arbiters of Parisian elegance" will oversee the assessments.

## Rules

**Art. 2:** The competition will be open to four categories of vehicles not driven by their owners:

- Electric cars
- Petrol or steam cars intended for tourism or city use
- Cabs
- Department store delivery cars

**Art. 5:** The jury will assign two scores to each automobile crew, ranging from 0 to 10: one for the driver-mechanic and the other for the vehicle itself.

**Art. 7:** Importantly, the jury will not be tasked with determining the prettiest car or the most elegant bodywork. Even the simplest vehicles will have a fair chance of success. The jury's evaluation will focus solely on the overall condition of the vehicles.

In essence, while this is a competition of elegance, it is specified that the jury will not prioritize the most aesthetically pleasing car or the most sophisticated design, but rather the vehicle that presents "the best appearance."

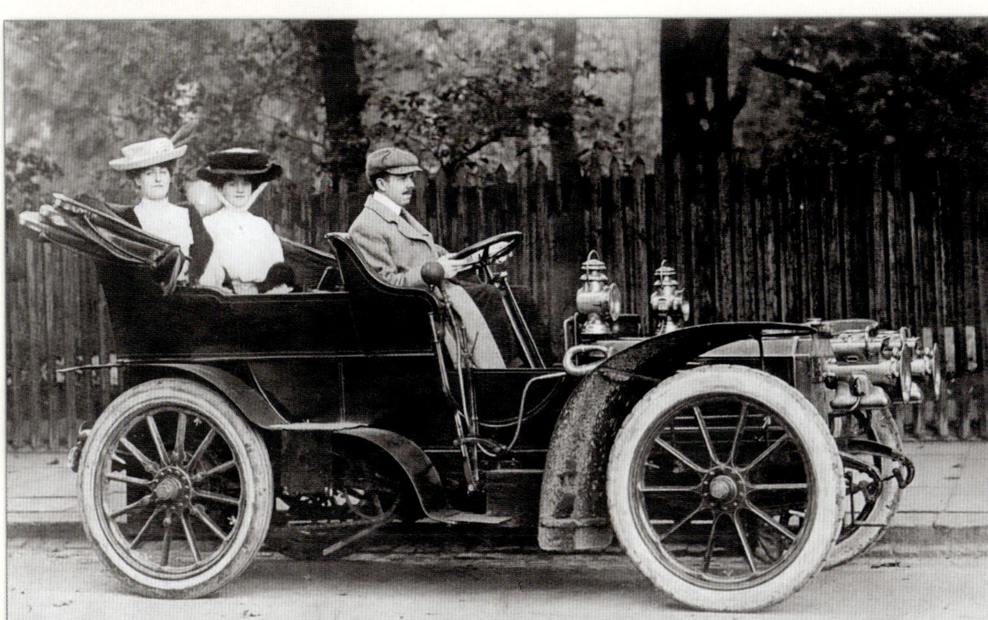

❶ *This landau from the Kriéger company was powered by an electric motor, with each front wheel having its own motor. The heritage of horse-drawn carriages is clearly evident in this type of vehicle, a "summer landau" typically reserved for official events. (Daniel Cabart Archives)*

❷ *Gardner-Serpollet manufactured steam cars until 1907. This double phaeton, featuring a side door, boasts a sober and elegant design. Notably, the boiler is clearly visible at the rear of the vehicle. (Daniel Cabart Archives)*

❸ *An elegant hackney carriage built on a Panhard et Levassor chassis. While not yet a taxi, it served similar functions. The Eugène Boulogne coachwork already had a long-standing tradition of crafting horse-drawn carriages since the beginning of the 20th century. (Daniel Cabart Archives)*

❹ *In this concours, where vehicles with the 'best appearance' would be rewarded, entries ranged from hackney carriages and landaus to luxurious phaetons and humble delivery cars. According to the regulations, every vehicle – regardless of whether it was a commercial workhorse or an aristocratic means of conveyance – had a chance to win a prize. (Daniel Cabart Archives)*

**LEFT:** *Whether it was a concours d'elegance or a pleasure outing, taking a seat aboard such a Mercedes Simplex 40 HP required the ladies to be impeccably dressed, and the driver (likely the owner in this case) had to present himself flawlessly attired too. (Daniel Cabart Archives)*

As we have just seen, during these early years of the century, only the concours held on the Riviera–primarily in Nice and Monte Carlo – seem to have achieved any real renown. Concours were organized in other seaside resorts, but the expansion of the concept progressed very slowly. A thorough search of the contemporary press reveals nothing of real significance until the First World War. It therefore appears that the vast majority of these concours were essentially social events associated with the Côte d'Azur. Even Paris, where the movement had more or less begun, does not seem to have staged any notable concours during the years immediately preceding the First World War.

There were, nevertheless, some fine concours at Le Touquet, held at the conclusion of Boulogne Week, as well as in Ostend (Belgium) and Biarritz. Most often, however, these were modest events incorporated into local festivities and accompanied by a gymkhana, and they could not compete with their Riviera counterparts.

**OPPOSITE:** *1912, a* Concours d'Elegance *graces the cover of a car magazine for the first time. This appears to be an unprecedented occurrence. The event was one of the highlights planned in San Sebastian, Spain as the Monte Carlo Rally passed through this fashionable seaside resort. The winning Delage, in an elegant aubergine color, was owned by an Amsterdam entrant. (Daniel Cabart Archives)*

**BELOW:** *Although two women posed on board for the photo, the true star was the Delage with a Beuzelin body, designed by Repusseau. Its elegance was primarily judged by its expansive glass surfaces and the luxury of its interior fittings. (Daniel Cabart Archives)*

## 1912 San Sebastian Competition

One competition, however, received considerable press coverage, possibly because it was held in a region favored by affluent international clientele, including the elite from Biarritz, Saint-Jean-de-Luz, and San Sebastian – though not on the Mediterranean Riviera. This noteworthy event was organized in 1912 in San Sebastian, Spain, during a stage of the Monte Carlo Rally. Dubbed "an event within the event," it garnered significant attention due to its connection to the prestigious rally and the glamorous setting.

As the years passed by, there appeared to be little evolution in the definition of elegance in automobiles. A few excerpts from Charles Faroux's article in *L'Auto* on August 29, 1912, covering the Concours d'Elegance et de voiturette in Boulogne-sur-Mer illustrate this point:

> *"The day began with the operations of the Concours d'Elegance, the technical results of which can be found below. The competitors were divided into three categories: open bodies, closed bodies, and a special class for those who had participated in various events during the Boulogne meeting. In the closed bodies category, the first prize was justly awarded to an admirable 20 HP Panhard owned by Madame Dior, luxuriously bodied by Driguet. Meanwhile, the jury unanimously awarded the first prize for open bodies to a Renault Torpédo of exceptional elegance, harmoniously designed by Letourneur et Marchand, who created a body worthy of the chassis. Among other competitors noted for the elegance of their bodywork and maintenance, I must mention the Hispano-Suiza belonging to Mr. Early."*

Harmoniously designed and meticulously finished – this was the essence of elegance!

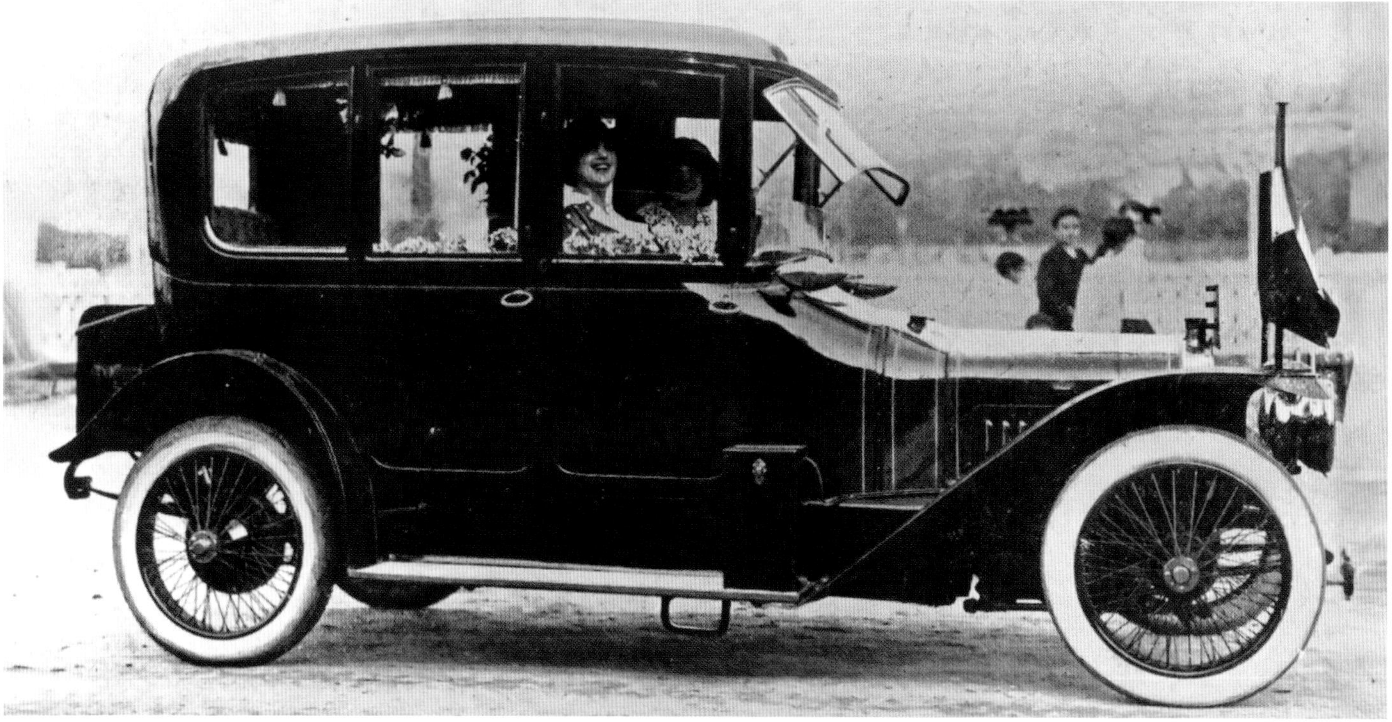

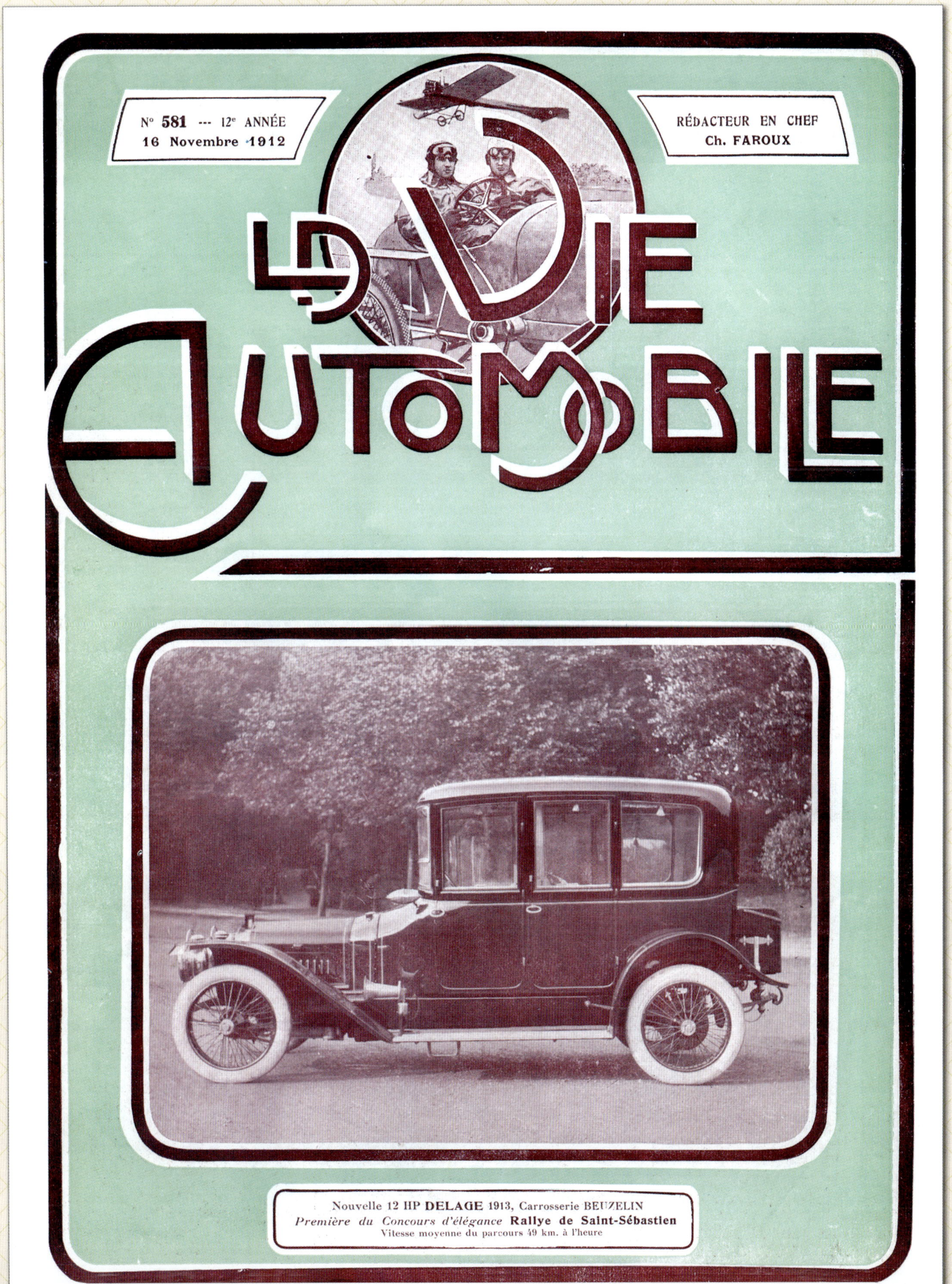

N° 581 --- 12e ANNÉE
16 Novembre 1912

RÉDACTEUR EN CHEF
Ch. FAROUX

# LA VIE AUTOMOBILE

Nouvelle 12 HP **DELAGE** 1913, Carrosserie BEUZELIN
*Première du Concours d'élégance* **Rallye de Saint-Sébastien**
Vitesse moyenne du parcours 49 km. à l'heure

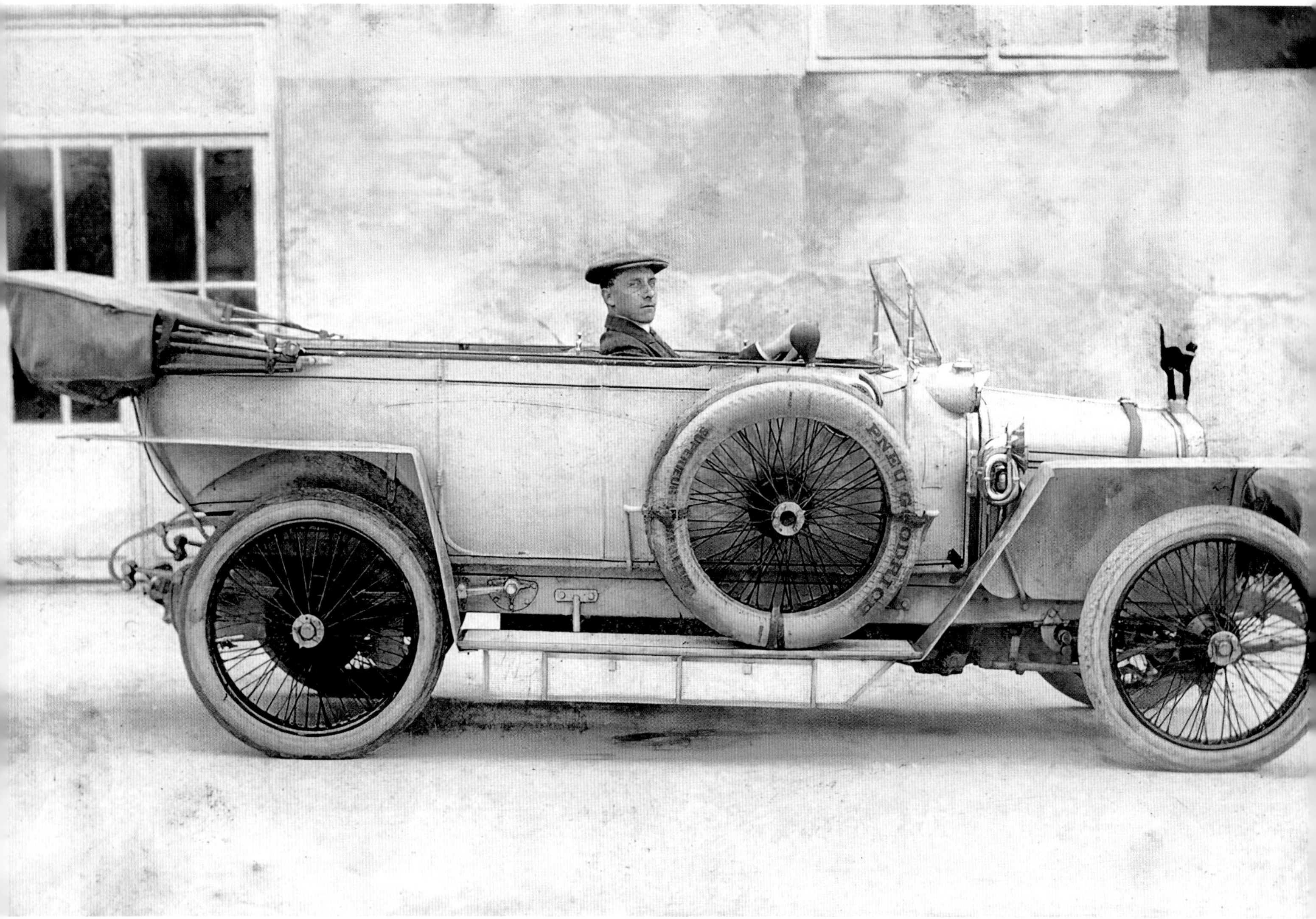

*1912, Vittel. On the way to the Concours d'Elegance, the driver of this Hispano-Suiza has chosen a black cat as a mascot, hoping to ward off bad luck! A few years later, the brand would adopt the stork – mascot of the French legendary World War I flying ace Georges Guynemer[2] – which would go on to adorn all Hispano-Suiza radiators. (Daniel Cabart Archives)*

## 1913-1914

In Nice, Monaco, and Cannes, nothing changed. The only variation was in the composition of the jury, which remained drawn from the international aristocracy or affluent foreign residents, including bankers and prominent Anglo-Saxons, Russians, or American magnates. The automobile events continued to present their gleaming bodywork, inspected in the morning by the jury and paraded in the afternoon before a crowd eager to admire the "best of breed" among residents and the finest creations from leading coachbuilders. In the end, nearly all the cars were ranked equally, as it seemed crucial not to offend anyone!

In 1914, during the speed races in Tours organized by the Automobile Club de l'Ouest (ACO), *L'Auto*, on May 27, 1914, gave further insight into what an automobile elegance competition entailed on the eve of the Great War. Notably, there was a clear evolution in the categorization of vehicles, reflecting the increasing number of participants. Gone were the days of only two categories – open cars and closed cars.

[2] *Georges Guynemer was a legendary French flying ace and national hero during World War I. He became one of the most celebrated fighter pilots of the war, known for his extraordinary skill, courage, and devotion to France.*

### The Elegance Competition

The vehicles participating in the elegance competition would now be divided into twelve distinct categories:

- Limousines
- Coupés
- Six-seater with closed body
- Four-seater with closed body
- Passenger cars with hoods
- Convertible open and closed cars

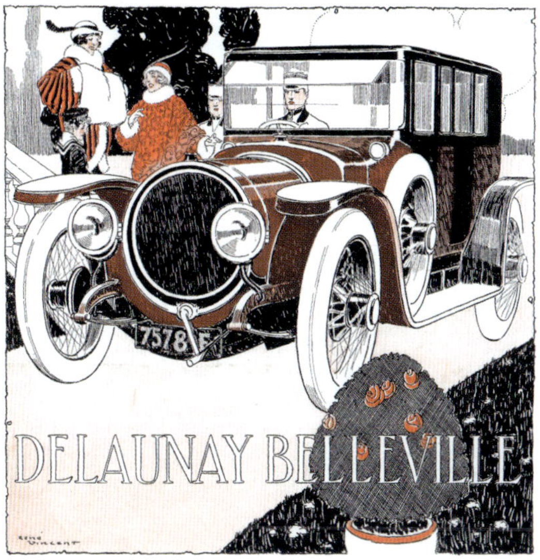


Points were to be awarded as follows:

- General appearance of the car (exterior): 6 points
- General appearance of the car (interior): 6 points
- Hood: 6 points
- Paint: 6 points
- Fenders, running boards, luggage racks, headlight racks, tire racks: 6 points

**Total: 30 points**

This new system undoubtedly aimed to provide a more nuanced judgment, accommodating the diverse array of vehicles presented. However, it may also have served to appease participants eager to return home from the competition "with a prize," as suggested in an ironic text from the 1920s (see page 87). Regardless, there was still little indication of any distinctly "feminine" influence on the proceedings.

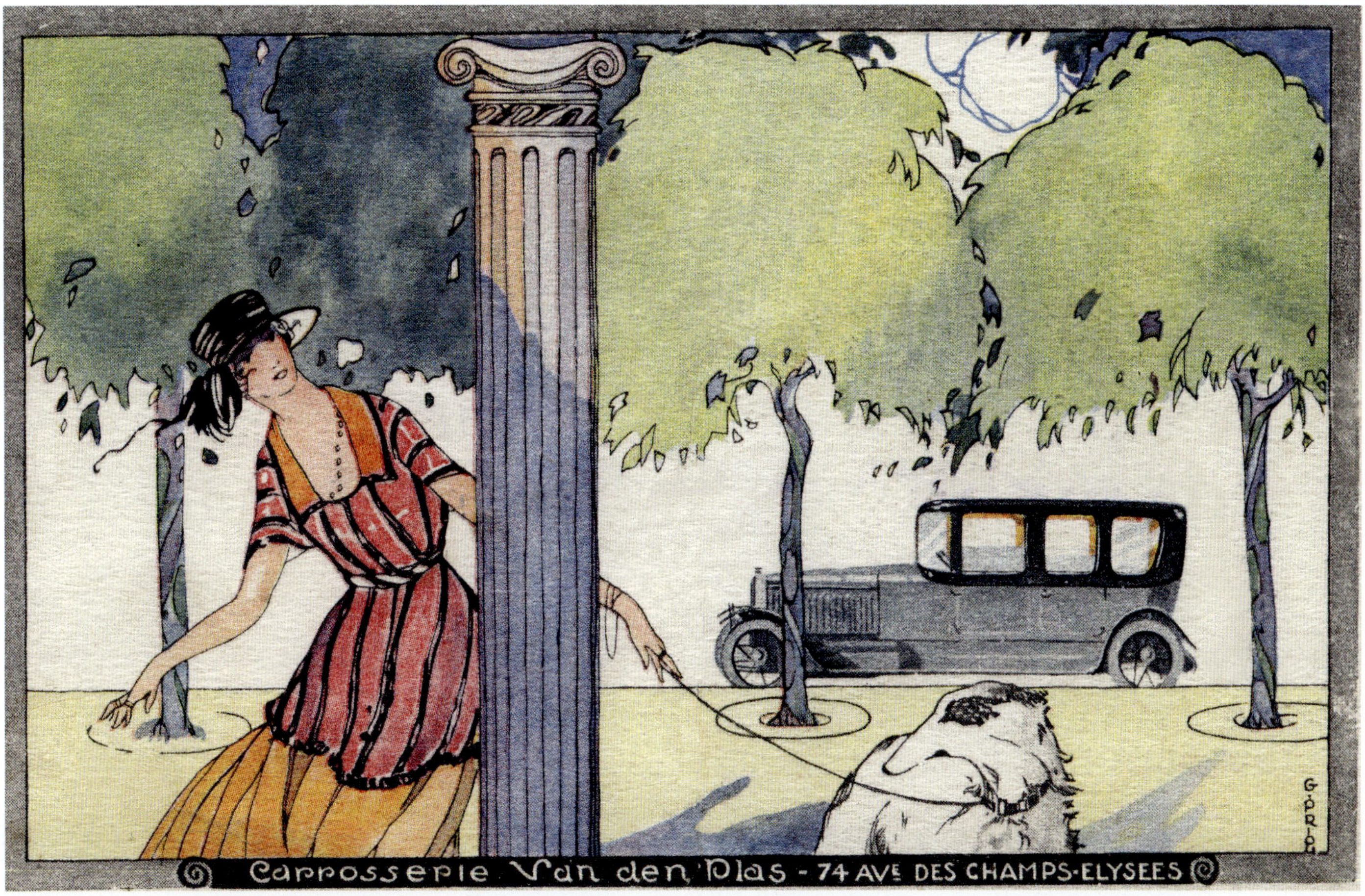

# *The first Concours d'Elegance at Dinard*

## THE AUTOMOBILE AS A SYMBOL OF MODERN ELEGANCE

In the early 1920s, a fresh wave of change swept across the world, with Paris at its heart, radiating an atmosphere of euphoria and liberation that extended to fashionable seaside resorts. Everything was in flux, as architects, artists, fashion designers, and industrialists competed to express their creativity and reinvent themselves.

Society, too, was eager for new experiences, embracing the latest leisure activities and sports, many of which came from England or the United States. Amid this whirlwind of innovation, the automobile quickly took center stage. Not only was it seen as a symbol of freedom and adventure, but it also became an integral part of elegant living.

But what did "elegance" mean in this context?

Once viewed as an innate aristocratic quality, elegance, as Honoré de Balzac described it, was *"a perfect harmony between the color of the clothing and all its accessories, between the nature of the attire and the occasion or setting, be it Sunday or weekday, on foot or in a car."*

For Balzac, living elegantly was about *"a high thought of order and harmony, designed to bring poetry to everyday things."* This concept of elegance deeply resonated with André Becq de Fouquières, who organized his first concours d'elegance in the early 1920s.

## The First Concours d'Elegance in Dinard, September 4, 1921

André de Fouquières, a long-time visitor of the chic resort of Dinard since 1890, created a grand automobile rally, in collaboration with the town's municipality, the Automobile Club de France (ACF), and the Automobile Club de l'Ouest (ACO). The highlight of this rally would be a concours d'elegance.

The rally began simultaneously from seven towns in western France, with participation from key figures in the worlds of art, fashion, and industry. Out of the 50 vehicles entered in the competition, 24 different French marques were represented, demonstrating the vitality of the French automobile industry at the time.

1924 Dinard Concours d'Elegance. (Dinard Elegance)

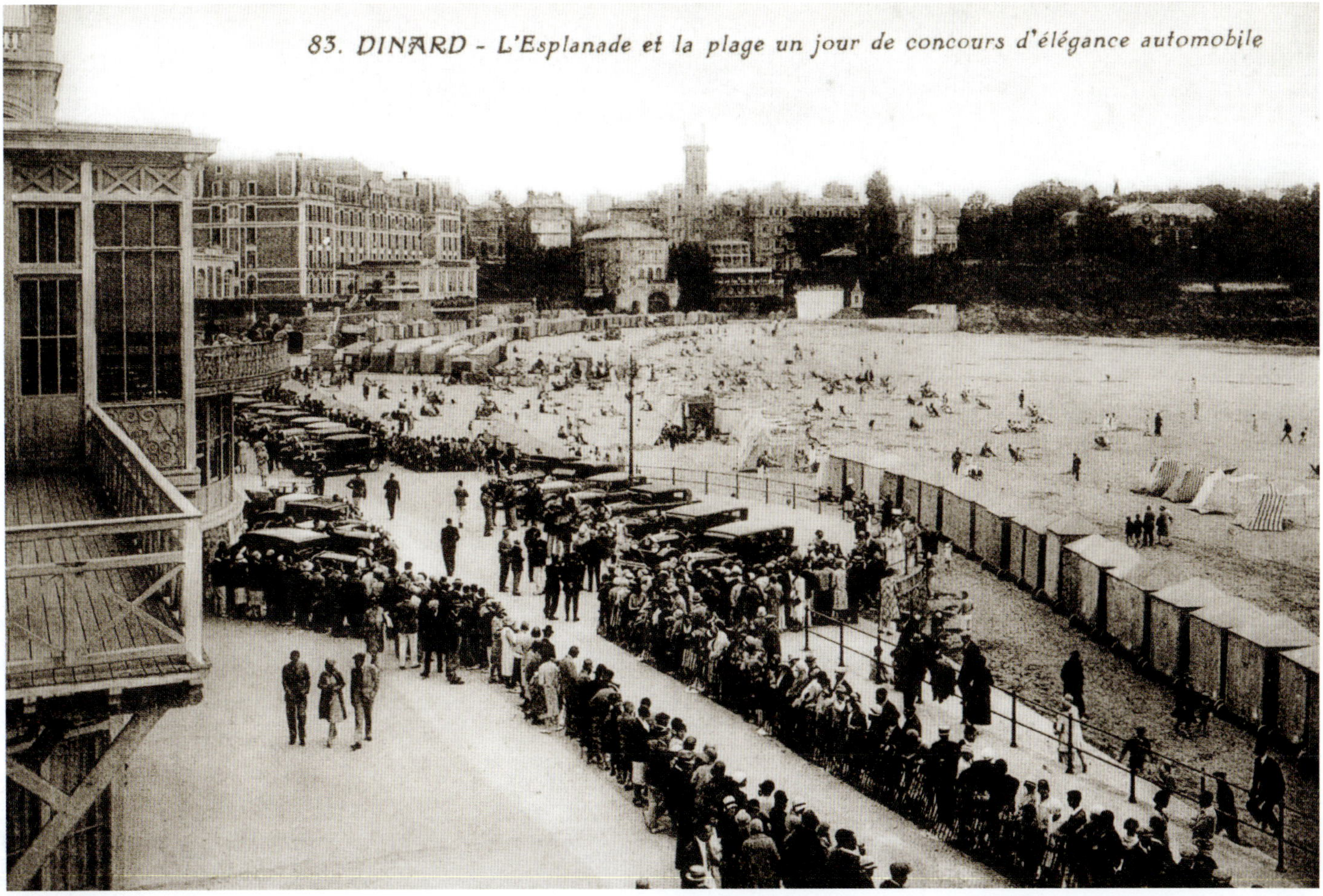

Several prominent manufacturers took part, including Léon Laisné from Nantes, presenting one of his new models, and Mr. Maurice Françon, who arrived from Paris in a Françon cyclecar. Among the participants was Madame Georgina Citroën, entering with a luxury Citroën Type A in the torpédo (tourer) category. Her husband, the visionary André Citroën, understood the promotional power of concours d'elegance events and seized the opportunity to enhance his brand's prestige.

Delage also made a significant impact. Louis Delâge, who owned a holiday villa in nearby Saint-Briac, ensured his brand was a regular feature at the Dinard Concours. In the first year, Henri Laraque, the son of a prominent promoter from Saint-Lunaire, won an elegance prize with his Delage convertible, featuring a custom body by Kelsch.

The 1921 Dinard Concours d'Elegance marked the beginning of a tradition that combined the allure of automobiles with the art of elegance, making it a defining cultural event of its era.

The first concours d'elegance in Dinard in 1921 attracted the *crème de la crème* of society, quickly establishing itself as the latest "place to be." Among the notable participants was Mr. D'Albert-Lake, a regular at the Thomelin wooden carriage competitions, who transitioned to the automotive world with his Peugeot landaulet. The event drew an impressive crowd, filling the beach and seafront, which had to be closely monitored by vigilant security.

**RIGHT:** *Madame Valton in her winning 1933 Delage D6-11 S, photographed among the assembled cars at the seaside event. (Daniel Cabart Archives)*

**BELOW:** *In 1937, it was a Delage D6-70 race car, bodied by Letourneur et Marchand and presented by Mlle. Jacqueline Watney (daughter of Walter Watney), that won the Dinard Concours d'Elegance. (Daniel Cabart Archives)*

**OPPOSITE TOP:** *A 1937 Lancia Belna with Pourtout bodywork. (Dinard Elegance)*

**OPPOSITE BOTTOM:** *A 1935 Renault Vivasport. (Dinard Elegance)*

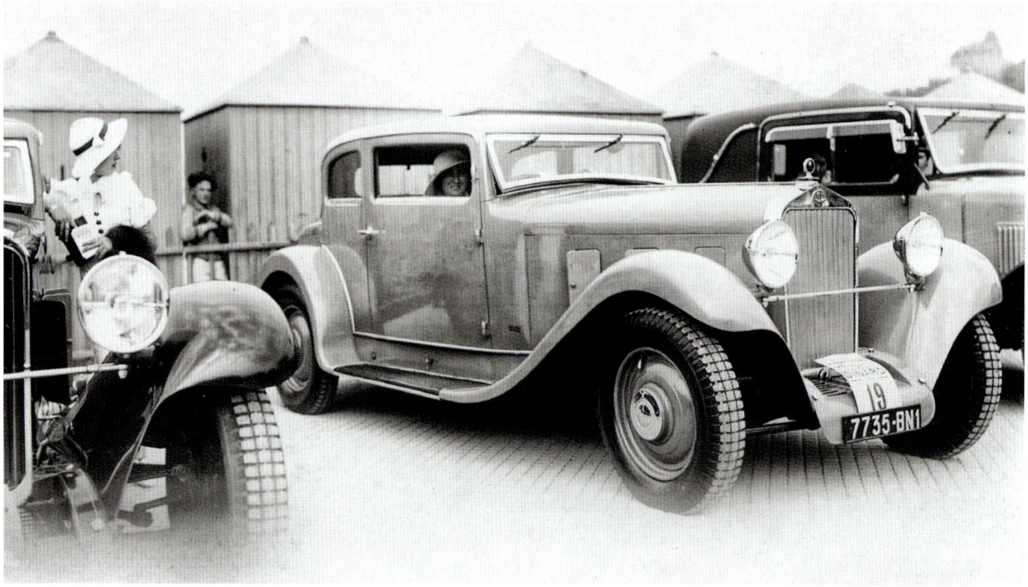


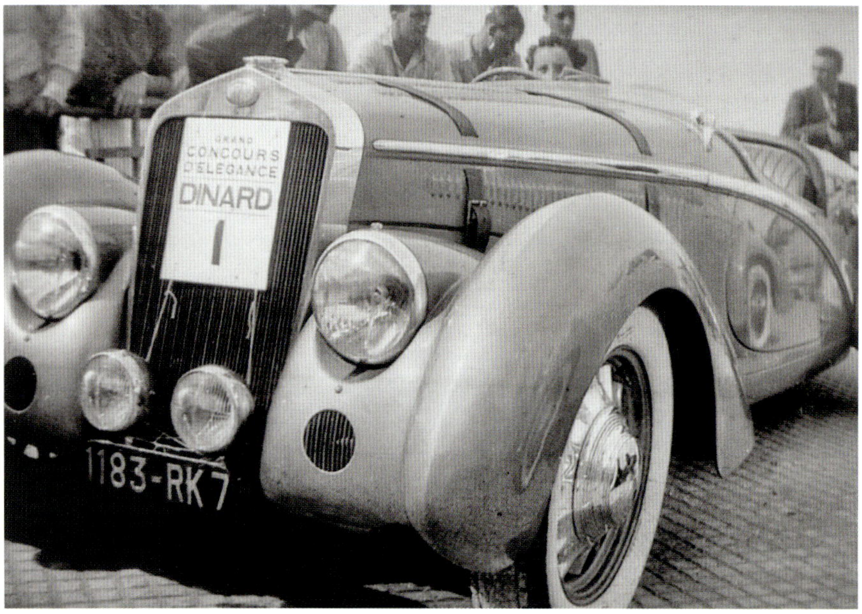


*"The astonished audience enthusiastically applauded, while connoisseurs remarked that the French automotive industry was shining brighter than ever in terms of both technology and elegance. The select jury, comprising influential figures from the aristocracy, politics, and business, faced the difficult task of judging the competitors based on regulations still heavily centered around the automobiles themselves.*

*The Grand Prix d'Honneur went to Mr. Pierre Durand-Ruel, the descendant of the famed art dealer who supported the Impressionist painters, for his performance with an Avions Voisin. The awards ceremony took place in the opulent lounges of the Casino High Life, in a festive and sophisticated atmosphere."*

*(L'Ouest-Éclair, September 7, 1921)*

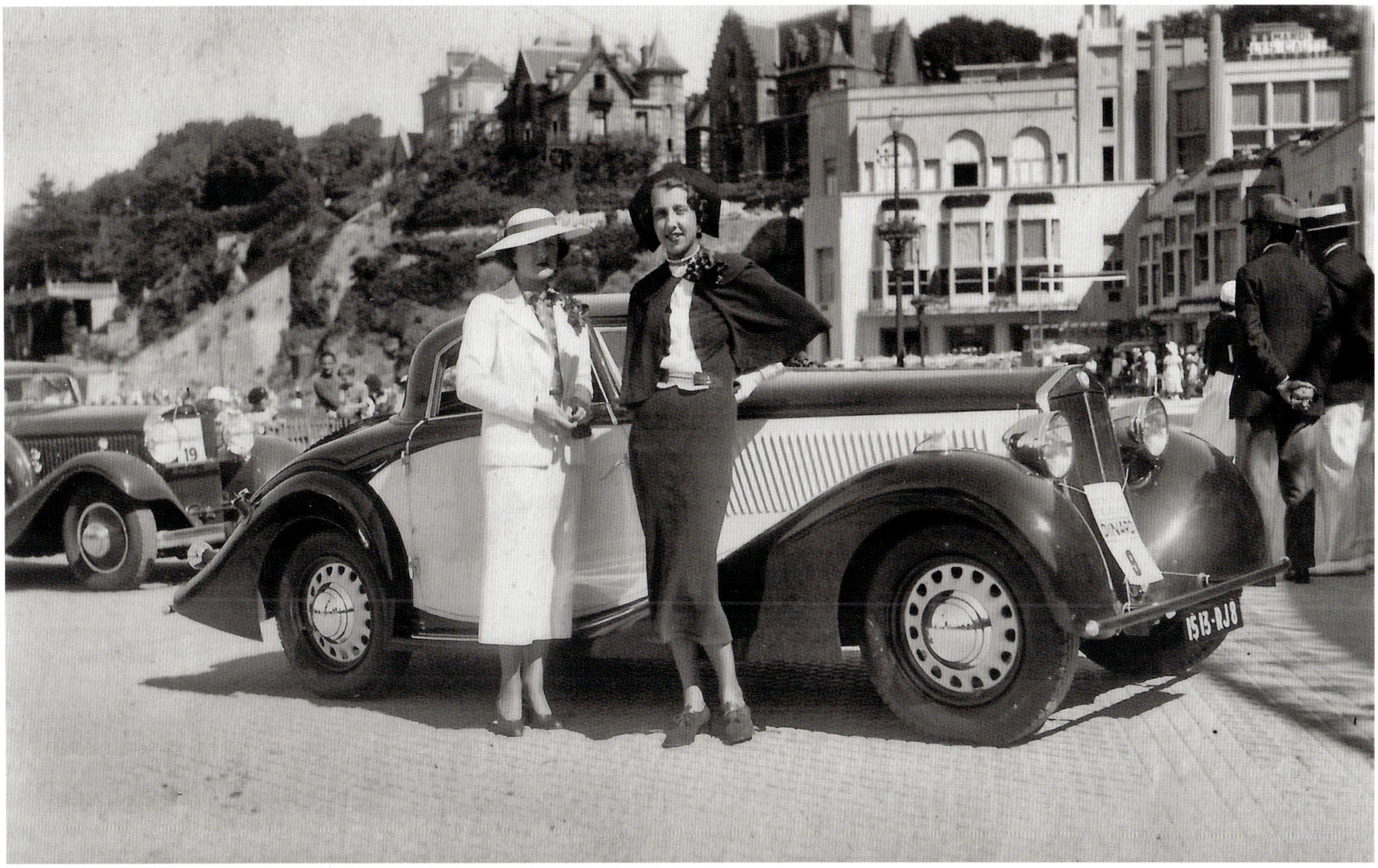
DINARD
9
1513-RJ8

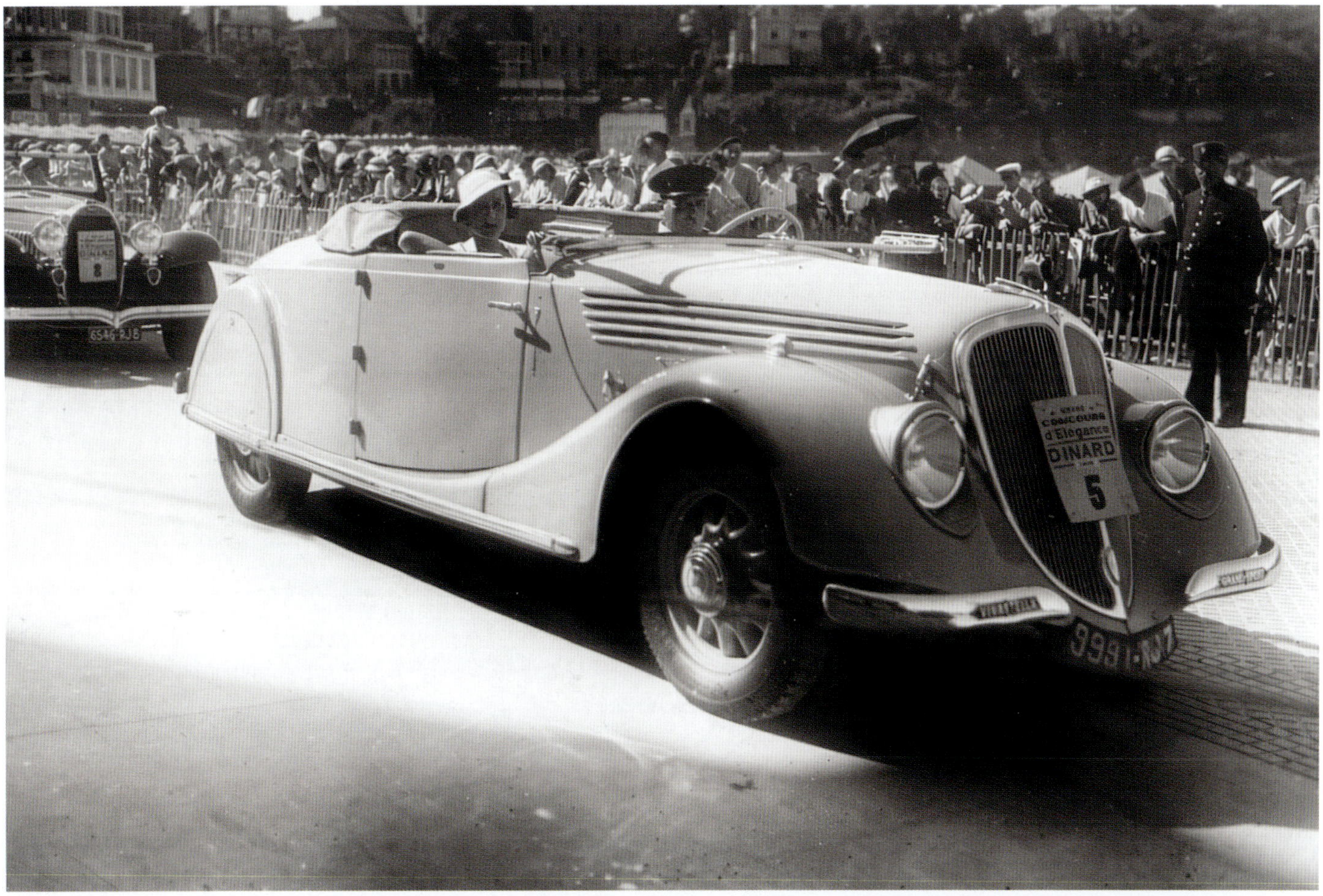
Concours
d'Elegance
DINARD
5

Throughout the 1920s, the concours d'elegance in Dinard followed this model, but with notable developments. Crowds grew significantly, with the press reporting that over 20,000 people attended the 1924 event, which featured not only the concours but also other automotive spectacles. As the event grew in scale, it became more organized and professionalized.

Prominent artists and personalities were invited to participate, with revue leaders like Madame Laure Diana from the Casino de Paris joining in. Artists such as painter Foujita were tasked with awarding the Trophée Féminin du concours (Women's Trophy). By 1929, fashion magazines like *L'Officiel de la Mode* were advising their readers on appropriate attire for attending the concours, while also reporting on the event's social buzz. The magazine noted that the Pullman train and its luxurious sleeping cars were insufficient to accommodate the influx of travelers from Paris eager to attend.

Over time, the worlds of art and fashion increasingly intertwined with the concours d'elegance, transforming what was initially seen as a purely automotive event into a broader celebration of style. In 1924, magazines like *Fémina* and *L'Intransigeant* even established their own women's automobile elegance competitions, further solidifying the event's cultural significance.

*The world of the arts is represented here by Mlle. Diana. (Daniel Cabart Archives)*

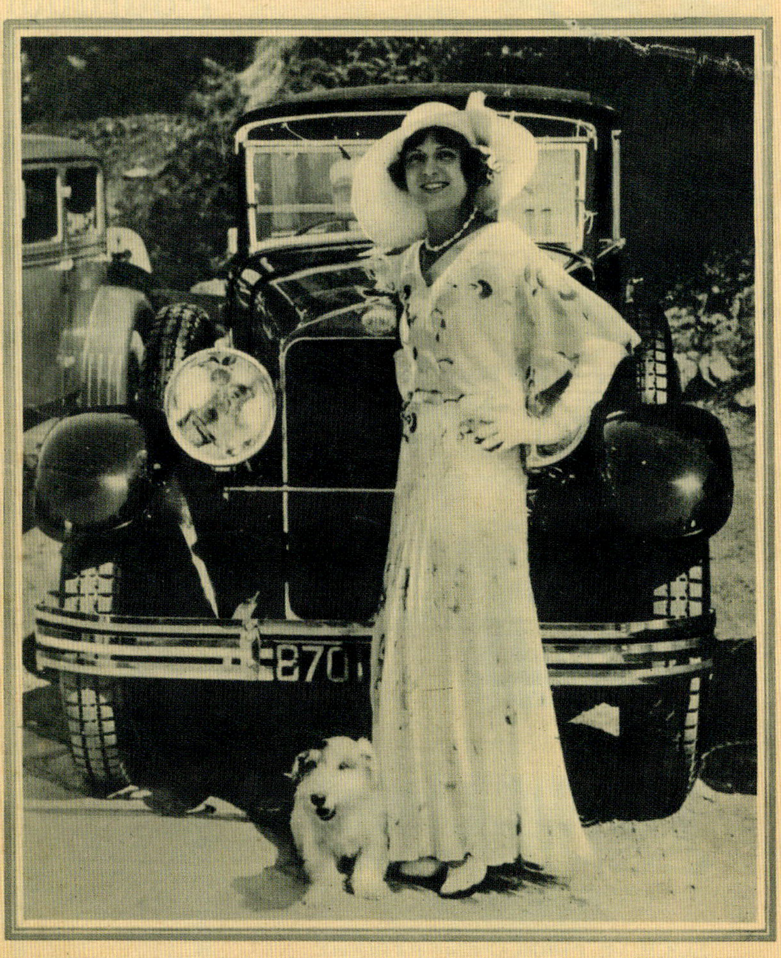

NUMÉRO 134 — 1 franc — 31 AOUT 1930

le miroir de la route

ÉDITÉ PAR Le Petit Parisien ~ RÉDACTION & ADMINISTRATION 13, rue d'Enghien, PARIS-X[e] ~ DIMANCHE-AUTO

L'AUTOMOBILE A LA PLAGE



(Photo Manuel Frères.)

M[lle] Diana à qui fut décerné le Grand Prix d'Honneur du concours d'élégance automobile de Dinard.

*Dans ce numéro :* DE MORLAIX A BREST PAR LA CEINTURE DORÉE ET LA COTE SAUVAGE

## The 1930s: A shift to the feminine for the *concours d'elegance en automobile*

By 1930, the town of Dinard was more committed than ever to supporting the concours d'elegance, the season's premier event in both sporting and social circles. Despite the challenges posed by the Great Depression, the allure of the concours endured. The creativity of haute couture designers, luxury car manufacturers, and master coachbuilders remained strong, continuing to captivate a select clientele that craved refinement and exclusivity.

A striking development that year was the dominant presence of women behind the wheel. Out of the 60 entrants, only a few male drivers participated. Women took center stage, driving the elegant automobiles with style and skill. As part of the presentations, these women would step out of their cars and circle around them with slow, measured steps, evoking the regal elegance of queens from antiquity.

Under the watchful eye of the jury, every detail was carefully scrutinized, with the ever-charismatic André de Fouquières offering commentary on the vehicles' design, the attire of the female drivers, and the overall harmony of the presentation. While the fashion of the 1930s had become more subdued, accessories remained crucial. Hats, gloves, brooches, and handbags played a pivotal role in helping contestants secure valuable points. The originality of the car mascots, elevated to the status of jewelry by de Fouquières, was particularly admired. Pedigree dogs, another accessory favored by the public, also earned competitors additional points. Meanwhile, if a driver was present, he was merely considered an accessory, with his attire expected not to detract from the overall presentation.

During this period, Bugatti cars dominated the competition. Anne-Cécile Rose-Itier, part of a new generation of independent women, won an award for excellence in her Bugatti Type 46 in 1930. A pioneer in her own right, Itier was not only an elegant and sophisticated figure but also a professional racing driver. She participated in the 24 Hours of Le Mans six times and the Monte Carlo Rally five times. Itier drove a range of high-performance vehicles, including a supercharged Bugatti Type 37A and a Bugatti Type 51 with dual overhead camshafts and a supercharger.

*André de Fouquières presents an award to Madame Lucienne Radisse, virtuoso cellist, pilot of airplanes and race cars. (Daniel Cabart Archives)*

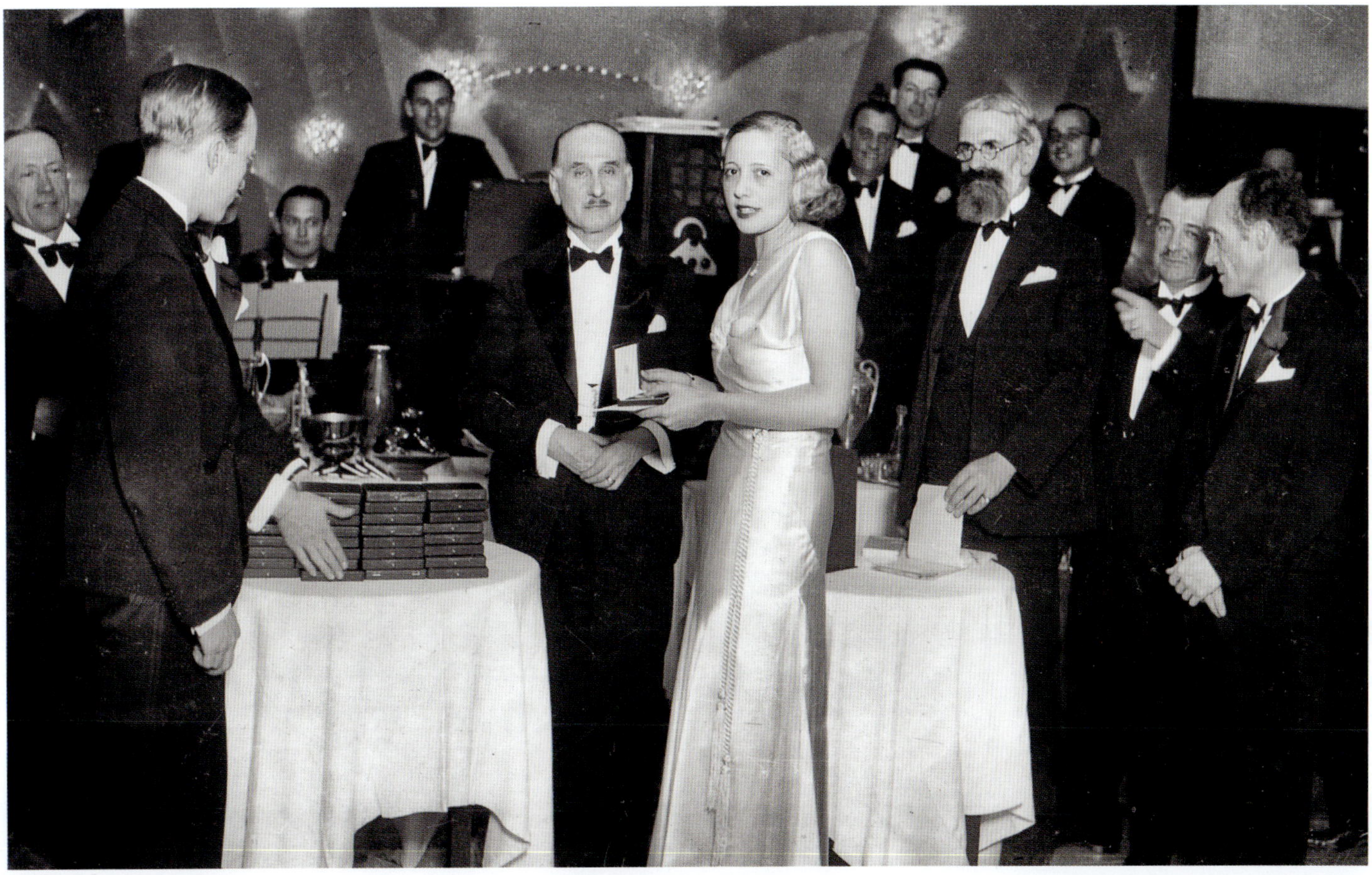

The *Grand Prix d'Honneur* that year (1930), however, went to Madame Gaillard, who impressed the jury with a unique Bugatti Type 46 roadster featuring factory bodywork. The car was owned by Maurice Goddet, co-director of the newspaper *L'Auto*, and was widely admired for its elegance and craftsmanship.

The 1930s thus marked a significant shift toward the increasing role of women in the concours d'elegance, both as drivers and as icons of automotive elegance, reflecting broader societal changes during the era. To the delight of fans of the elegant lifestyle, the magical era of the Dinard Concours d'Elegance extended for nearly another decade, with each year unveiling innovations.

In 1933, the celebrated actress and singer Betty Spell captured the spotlight, winning the competition in a white Delage D8 with bodywork by Letourneur et Marchand (see the same car presented by Mistinguett, page 153). As the muse of the marque, Spell was often dressed by the esteemed couturier Worth, further enhancing her elegance. Otto Lucas, the official milliner to the British crown and Greta Garbo, also received special recognition for his participation in a Swallow SS, earning *La grande mention du concours* (special mention of the competition).

The following year, in 1934, Betty Spell secured first prize once again, this time with another stunning Delage (see photograph on page 129, top). The grandstand audience was equally thrilled, as Queen Marie of Romania attended to support her niece, Princess Kira, who paraded in a Chenard et Walcker Aiglon.

In 1935, the concours revived the tradition of adorning cars with flowers, while also introducing the crowd to a groundbreaking innovation: the mechanical Éclipse sunroof, designed by Parisian coachbuilder Pourtout. This feature was demonstrated on the Peugeot 402 and the Lancia Belna, adding a modern touch to the event.

By 1937, Grand Duke Vladimir and Grand Duchess Kirill of Russia were awarded a First Prize for their performance behind the wheel of a Delahaye. However, the *Grand Prix d'Honneur* was claimed by a Delage once again, and in 1938, Madame Richer-Delavau, a regular at Parisian concours d'elegance events, won the top prize with her presentation of another Delage.

The concours d'elegance during the 1930s perfectly embodied elegance, luxury, and impeccable taste. Prestigious car marques like Bugatti, Delahaye, Delage, Isotta Fraschini, and Hispano-Suiza were elevated to new heights by the artistic visions of master coachbuilders like Labourdette, Figoni, Saoutchik, and Chapron. These artisans continuously pushed the boundaries of creativity, competing for recognition at concours d'elegance events across France.

**BELOW:** *The success was complete for this Bugatti Type 46 presented by Madame Gaillard. (Dinard Elegance photograph, restored and colorized in 2025 by H. Mercier.)*

**OPPOSITE BOTTOM:** *Fiat marks its territory at the 1934 Dinard Concours d'Elegance. (Les enfants terribles)*

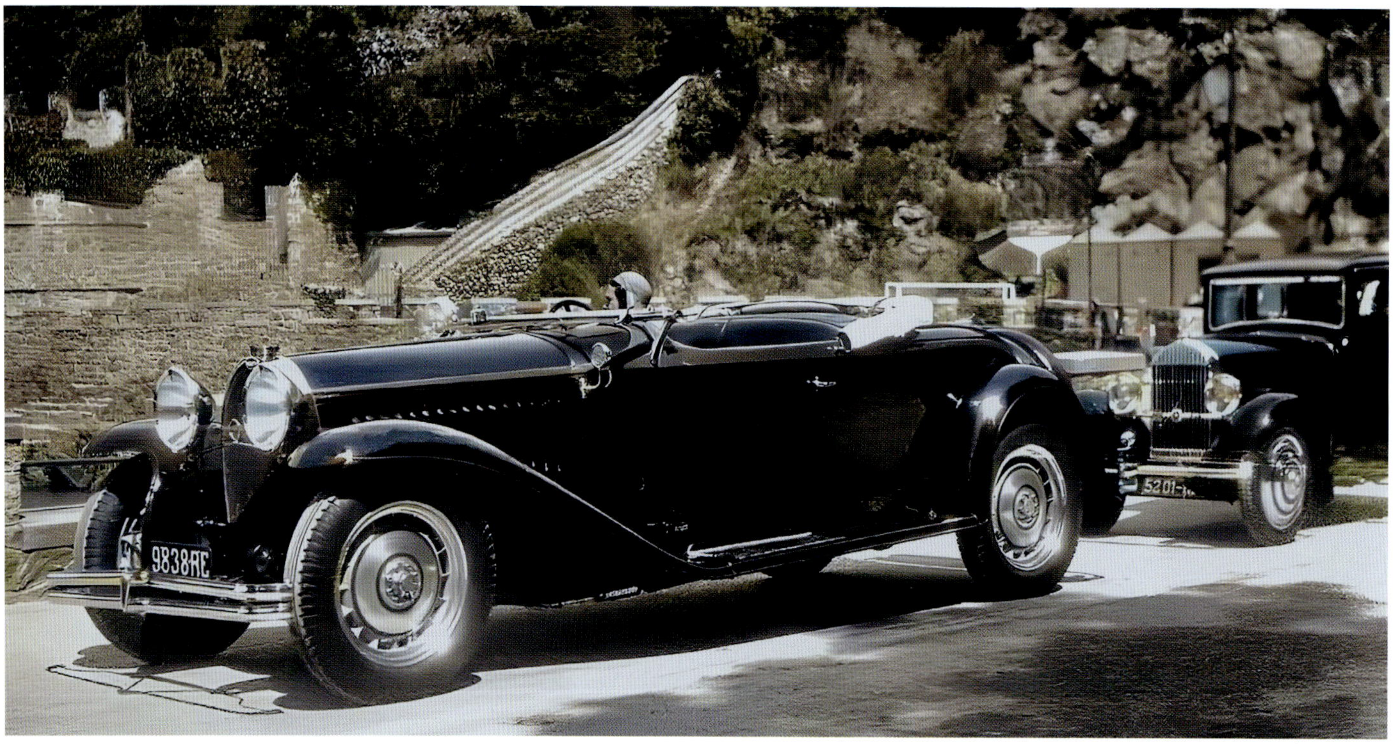

## André Becq de Fouquières

In 1928, André Becq de Fouquières expressed his thoughts on the concours d'elegance in the magazine *Adam*, highlighting the deep connection between fashion and automobiles:

> *"Our eyes were also drawn to the colors of the automobiles and the pretty, charming faces – also colored in various ways, but with an art of nuance that modern women practice so judiciously. Is not art found in the harmony of colors and lines, for the woman who enters such a competition must ensure that her appearance and her car form a cohesive whole. She must also take care to be accompanied by a friend whose attire complements hers or by a mechanic of impeccable propriety. In this context, soft collars, colorful caps, and bare hands are to be banished. Only a few might be tempted to forget that this is a concours d'elegance in automobiles, not one of engines or bodywork."*

De Fouquières' reflections capture the essence of the concours d'elegance, emphasizing that it was not merely about showcasing cars but about the harmony between the vehicle, the driver, and their entire presentation. The fusion of fashion, style, and subtle details defined the true spirit of the concours.

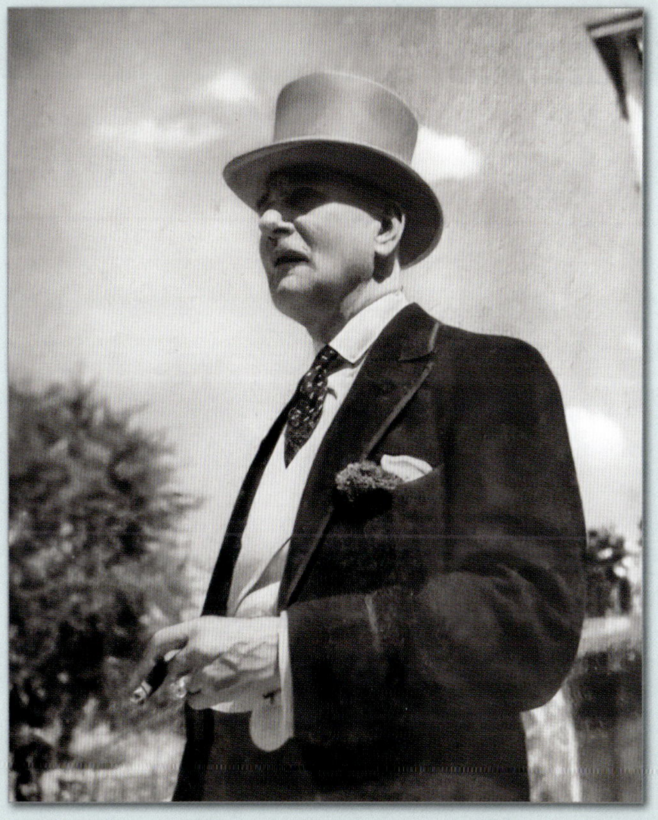

**RIGHT:** *André de Fouquières, a figure of high society.* (BnF Gallica)

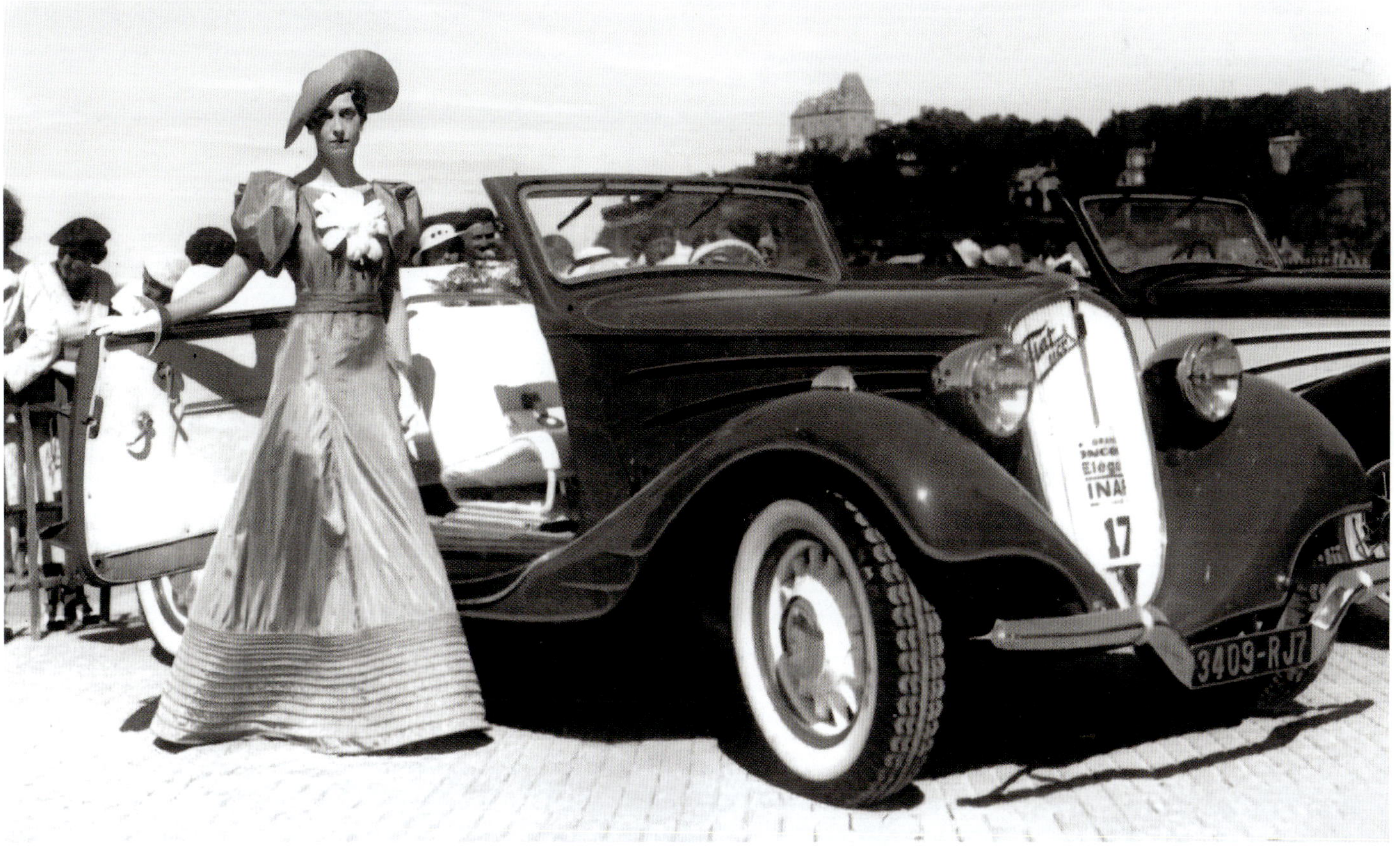

As André de Fouquières noted in his writings, the concours underwent a profound transformation during the 1930s. Women assumed central roles, representing the spirit of modernity, grace, and beauty. This fusion of the automobile world and high fashion created a new aesthetic, blending luxury and romance into a sublime living experience that echoed a different era and way of life.

André de Fouquières was born on May 17, 1875, in Paris's 8th arrondissement, into an old aristocratic family from Artois. He was the son of essayist and historian Louis de Fouquières and the great-nephew of Alfred de Dreux, a painter known for capturing 19th century elegance. Throughout his life, André de Fouquières sought to embody the ideals of French social etiquette.

Educated by Jesuits, he earned a diploma in oriental languages and displayed a talent for writing. Early in life, he ventured into politics, joining the Molé-Tocqueville Conference and founding the Comité de l'Œillet Blanc at just 20, a royalist group that supported the exiled Duke of Orléans in Brussels. His aristocratic connections flourished, and he became a leading social figure, sought after for his advice on guest lists, buffets, and musical events.

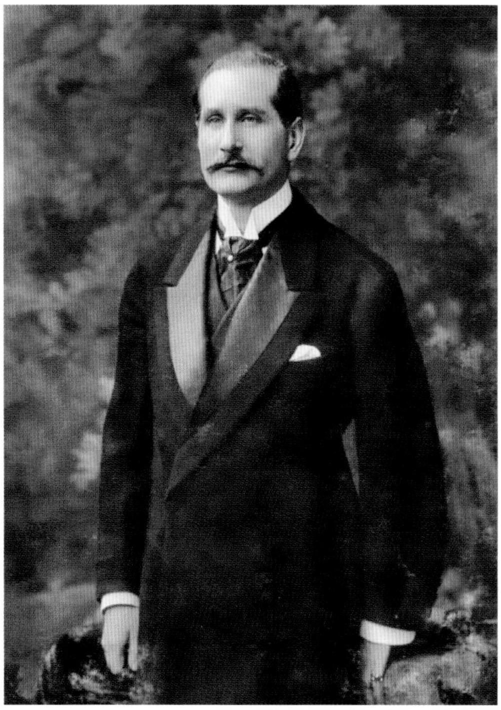

**André de Fouquières (BnF Gallica)**

After unsuccessful bids in the 1900 Paris municipal elections and a later legislative race in Bagnères-de-Bigorre, de Fouquières stepped back from politics. He became closely associated with the Countess of Rohan-Chabot, who ran an esteemed literary salon, where he honed his intellectual and writing skills.

In 1905 and 1906, he collaborated on several plays with Georges Casella and Jacques Monnier, and by 1907-1908, he embarked on a lecture series that culminated in his 1910 essay *De l'art, de l'élégance et de la charité* ("The Art and Elegance of Charity"), earning him a reputation as an arbiter of elegance, reminiscent of Georges Brummel. Passionate about preserving traditional customs, he upheld the standards of French etiquette at various social gatherings, both in Paris and seaside resorts.

In his book *Cinquante ans de panache* ("Fifty Years of Panache"), de Fouquières reflected on his fondness for France's Côte d'Émeraude (Emerald Coast), likely sparked by his friendship with American socialite Mrs. Hughes Hallet, whom he met in 1900. Known as the "Queen of Dinard," she hosted grand parties at her villa *Mon Plaisir*, with de Fouquières serving as her trusted advisor, earning the title of her "knight." His influence on the social scene in Dinard was unmatched, as he became a central figure in organizing soirées attended by French, Anglo-American, and Central European aristocracy. Their close relationship endured until her death in 1914, the same year the world was thrust into the turmoil of World War I.

During the Roaring Twenties, as automobiles became a symbol of social status, de Fouquières recognized the potential of blending automotive design with elegance. For him, the concept of harmony was paramount. On September 4, 1921, in Dinard, he organized the first meeting of automobiles and fashion – a precursor to the concours d'elegance competitions that would dominate French luxury culture for two decades. These events, held in Paris and seaside resorts like La Baule, Deauville, Biarritz, and spa towns such as Aix-les-Bains and Vichy, became a showcase for high society.

Though the landscape changed after World War II, de Fouquières continued to serve as the prominent host of concours d'elegance events for nearly a decade. Elected president of the Parisiens de Paris, he remained a cultural and social icon until his death at 84, passing away in his Boulevard Haussmann apartment in Paris's upscale 9th arrondissement.

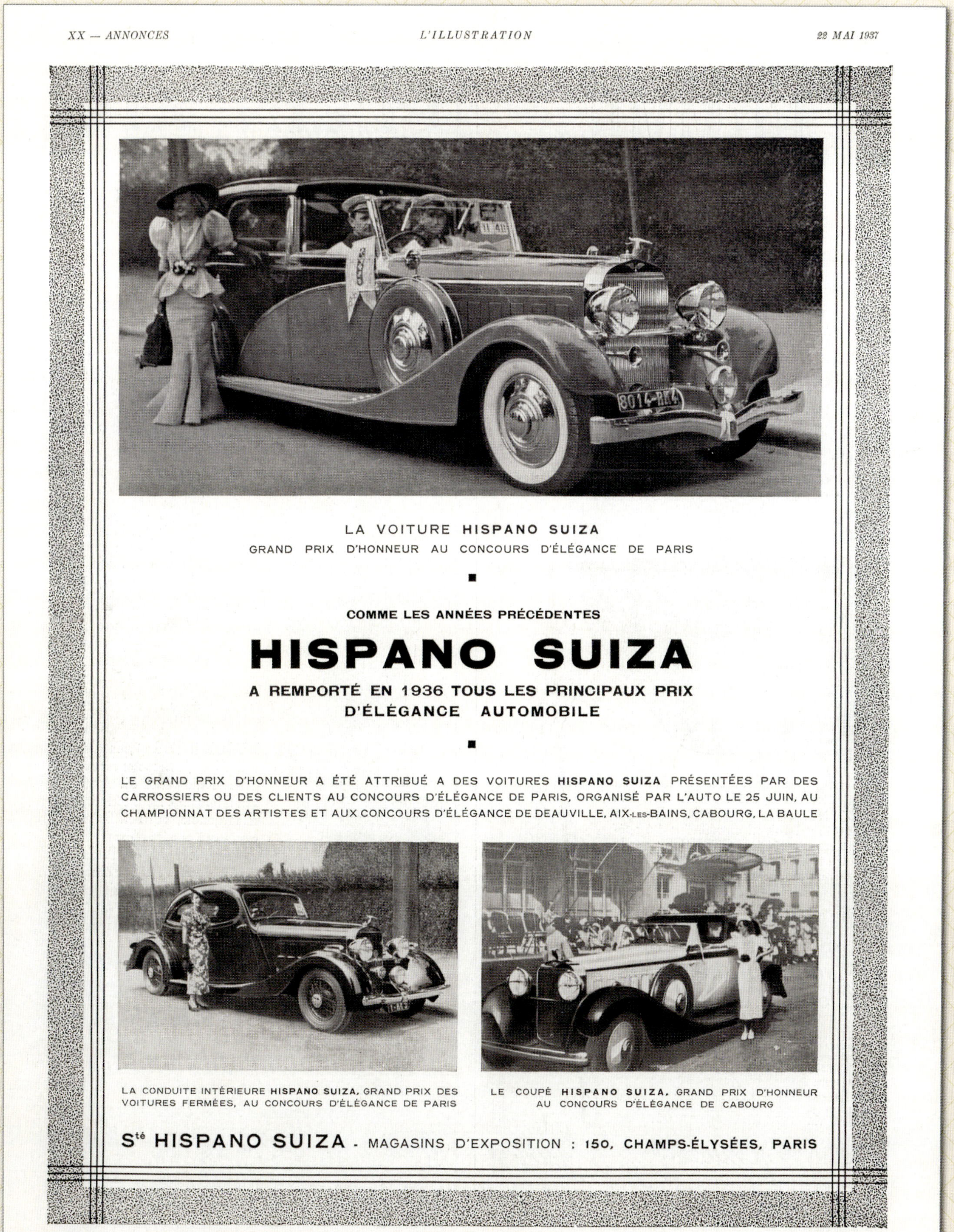

XX — ANNONCES L'ILLUSTRATION 22 MAI 1937

LA VOITURE **HISPANO SUIZA**
GRAND PRIX D'HONNEUR AU CONCOURS D'ÉLÉGANCE DE PARIS

**COMME LES ANNÉES PRÉCÉDENTES**

**HISPANO SUIZA**

**A REMPORTÉ EN 1936 TOUS LES PRINCIPAUX PRIX D'ÉLÉGANCE AUTOMOBILE**

LE GRAND PRIX D'HONNEUR A ÉTÉ ATTRIBUÉ A DES VOITURES **HISPANO SUIZA** PRÉSENTÉES PAR DES CARROSSIERS OU DES CLIENTS AU CONCOURS D'ÉLÉGANCE DE PARIS, ORGANISÉ PAR L'AUTO LE 25 JUIN, AU CHAMPIONNAT DES ARTISTES ET AUX CONCOURS D'ÉLÉGANCE DE DEAUVILLE, AIX-LES-BAINS, CABOURG, LA BAULE

LA CONDUITE INTÉRIEURE **HISPANO SUIZA**, GRAND PRIX DES VOITURES FERMÉES, AU CONCOURS D'ÉLÉGANCE DE PARIS

LE COUPÉ **HISPANO SUIZA**, GRAND PRIX D'HONNEUR AU CONCOURS D'ÉLÉGANCE DE CABOURG

**Ste HISPANO SUIZA** - MAGASINS D'EXPOSITION : **150, CHAMPS-ÉLYSÉES, PARIS**

*Hispano-Suiza in the spotlight.* (*L'illustration* 1937)

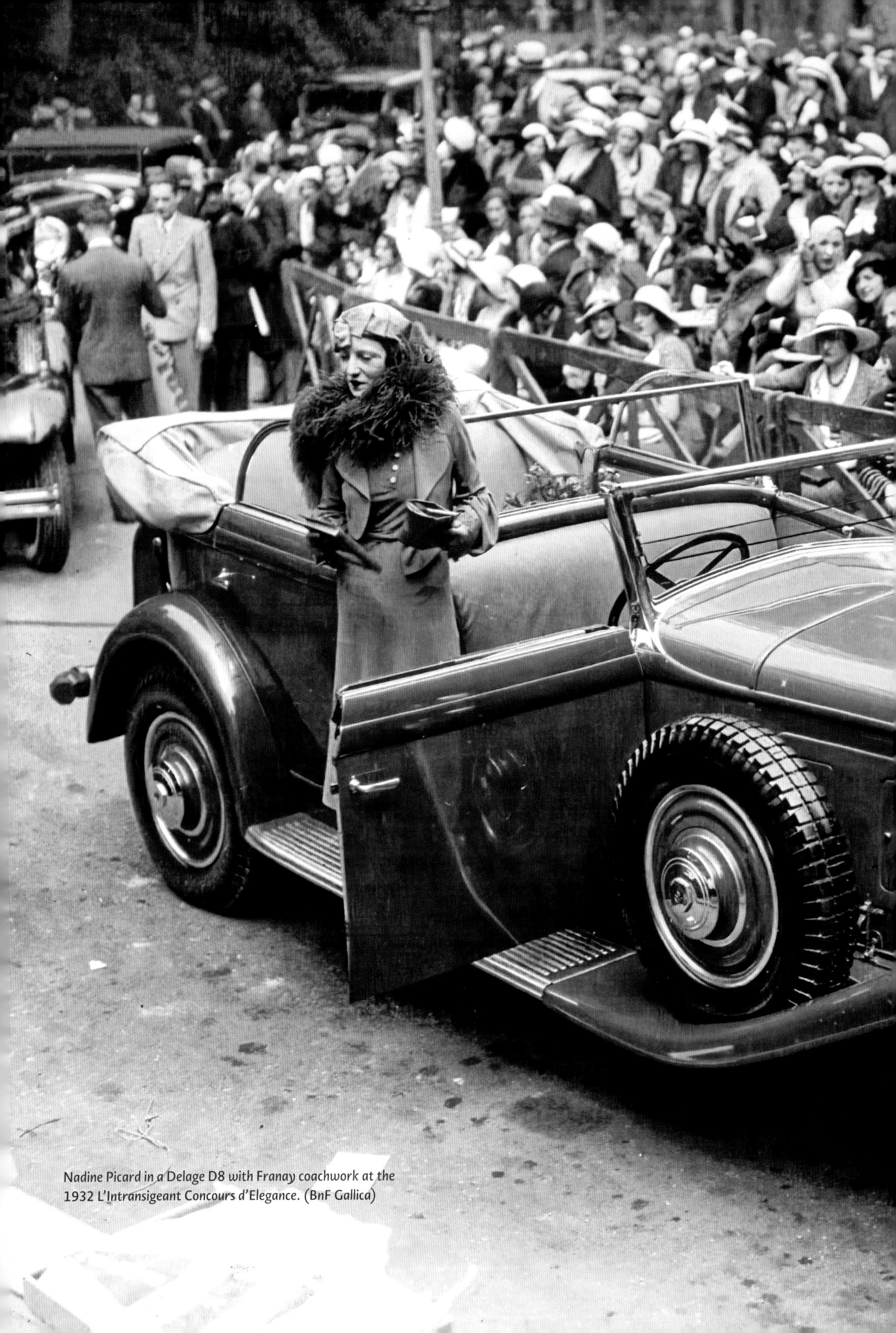

Nadine Picard in a Delage D8 with Franay coachwork at the 1932 L'Intransigeant Concours d'Elegance. (BnF Gallica)

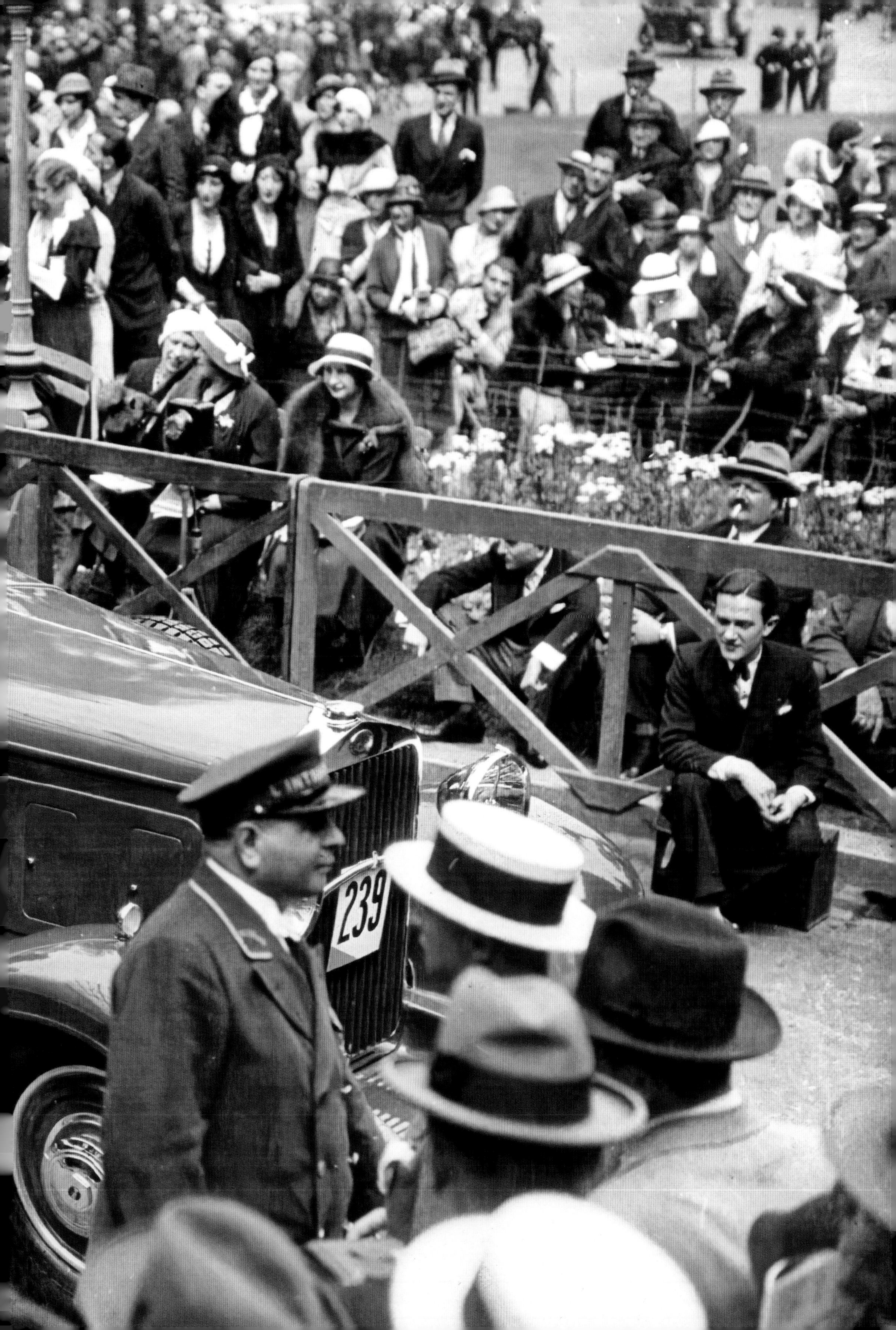
239

# *Carmaker, Coachwork, Couture*

# THE THREE CS OF CONCOURS D'ELEGANCE

## Carmaker and Fusion of Innovation and Elegance

Just before the Great War, senior leaders of the automobile industry gathered at the *Chambre Syndicale de L'Automobile et des Industries* under the chairmanship of Jules-Albert de Dion. The group aimed to chart a future path for the industry, facing a key question: should the automobile be regarded as a sophisticated object – a product of skilled trades excelling in multifaceted innovation, as de Dion believed? Or was it merely an industrial output – essentially a machine with an engine, chassis, and four wheels, as Robert Peugeot argued, according to Anne Françoise Garçon in *L'automobile, Son monde et ses réseaux*.[1]

[1] Anne Françoise Garçon, Gaëlle Beaussaron, *L'automobile, Son monde et ses réseaux*, Presses universitaires de Rennes, 1998, p.26-27, (Archives Comité des Constructeurs Français d'Automobiles CCFA, May 1998)

Across the Atlantic, Henry Ford adopted Taylorism in 1908, favoring standardization and efficiency. In Europe, however, political complexities hindered clear decisions, leading to divisive viewpoints. Though the automobile was seen as a symbol of progress, Europe's fragile economic and social fabric resisted the idea of mass consumption.

In the early 1920s, this paradox deepened. While Europe, particularly France, had pioneered the automobile, no brand had yet embraced mass production, putting the industry nearly a decade behind the US. Nonetheless, France's inventive spirit led to a unique trajectory: rather than competing directly with American mass producers, it fostered distinguished marques like Delage, Hispano-Suiza, Bugatti, and soon Citroën and Renault. According to historian Jean Loubet, this divergence enriched the French automobile industry during this period.

A recurrent debate also emerged among the *Chambre Syndicale* on whether to continue participating in costly car shows. This question, unresolved at the time, highlighted the industry's uncertainties just before the onset of war.

The effectiveness of automotive information and marketing methods became a critical issue as rapid advances in technology and design reshaped the industry. To capture public interest, manufacturers needed to adapt quickly, despite limited media options and the financial constraints faced by industry players, whose operations demanded tight budgeting and constant risk management. Over time, industry engagement was channeled through three key event types: motor shows, motor sport, and concours d'elegance.

From the 1920s, these major events gained prominence, drawing large audiences and extensive media coverage across newspapers, magazines, and even the occasional radio broadcast. By the early 1930s, larger manufacturers, despite holding three-quarters of the national market, were struggling due to

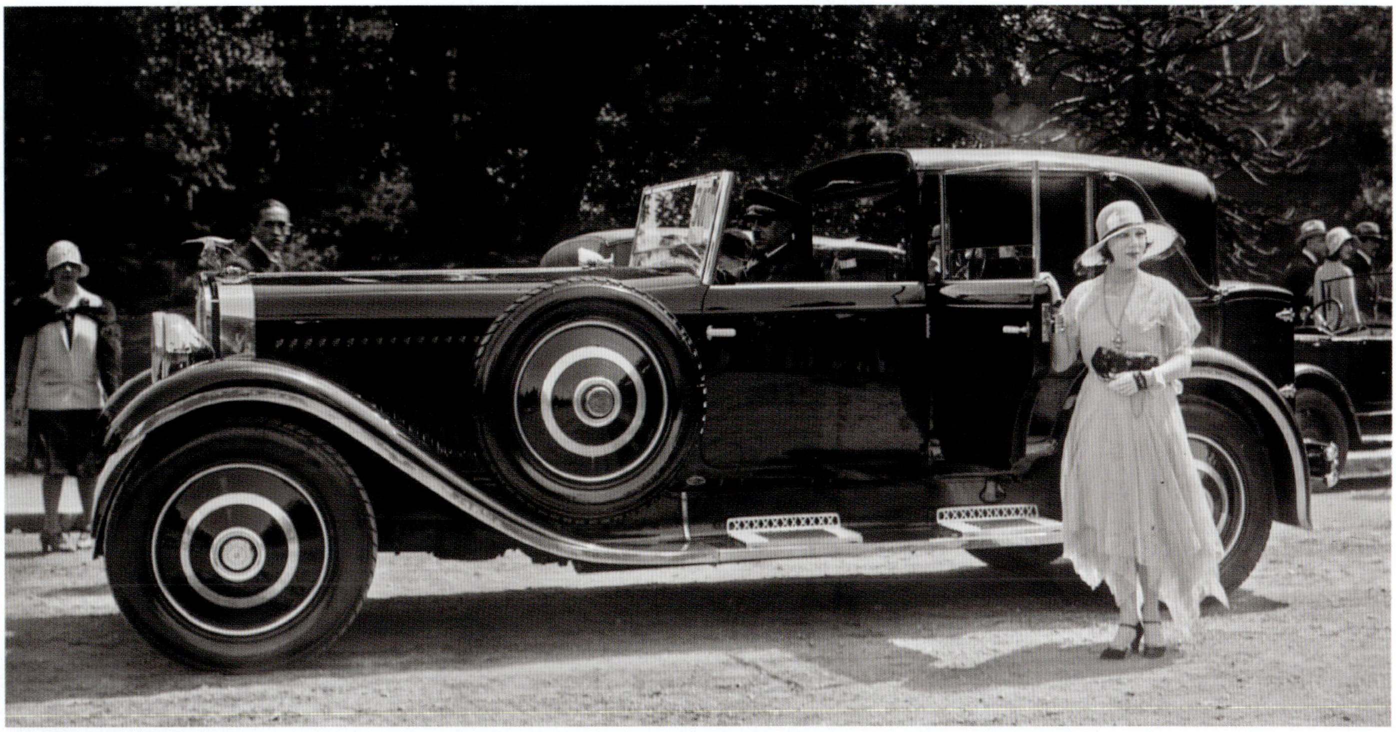

**Mlle. Raquel Meller with an Hispano-Suiza, Boulogne, 1928. (BnF Gallica)**

the Great Depression, with many on the brink of bankruptcy, prompting increased financial caution. Meanwhile, luxury marques and coachbuilders continued to stand out, thanks to their innovative designs and creativity.

Concours d'elegance events remained popular up to World War II, attracting a full spectrum of automotive industry players. Historic marques positioned themselves as torchbearers of French luxury, embodying the country's traditions of craftsmanship, textiles, and haute couture, and championing values of excellence and customization. Industrial manufacturers, drawing inspiration from American production models, sought to persuade the public that a more open market could drive mass production.

Simultaneously, foreign competitors leveraged concours d'elegance events to bolster their market presence. For marques like Fiat and Simca (a French marque with Italian roots), these events provided a platform to emphasize their commitment to the French market, supported by their local manufacturing. Smaller French marques, such as La Licorne and Rosengart, were highly

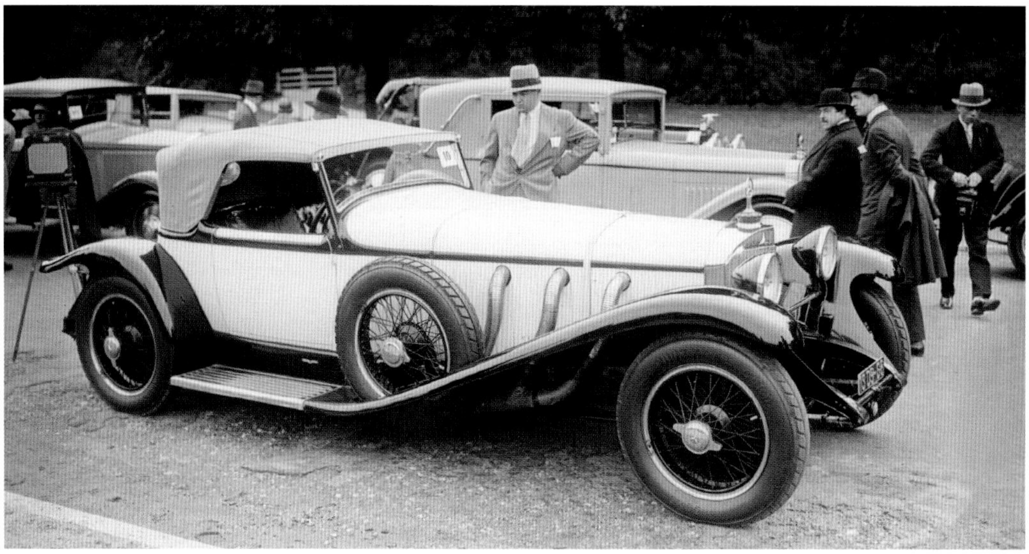

**RIGHT:** *Young Argentinian playboy, Martin de Alzaga Unzue, who raced Bugatti at the Indy 500 in 1923, presented this racy Mercedes-Benz Type S bodied by Hibbard & Darrin, the American coachbuilders established in Paris, at the Concours d'Elegance de L'Auto, on June 9, 1928. (BnF Gallica)*

**BELOW:** *Two Fiats at the Paris Concours d'Elegance, 1934. (Boris Lipnitzki / Roger-Viollet)*

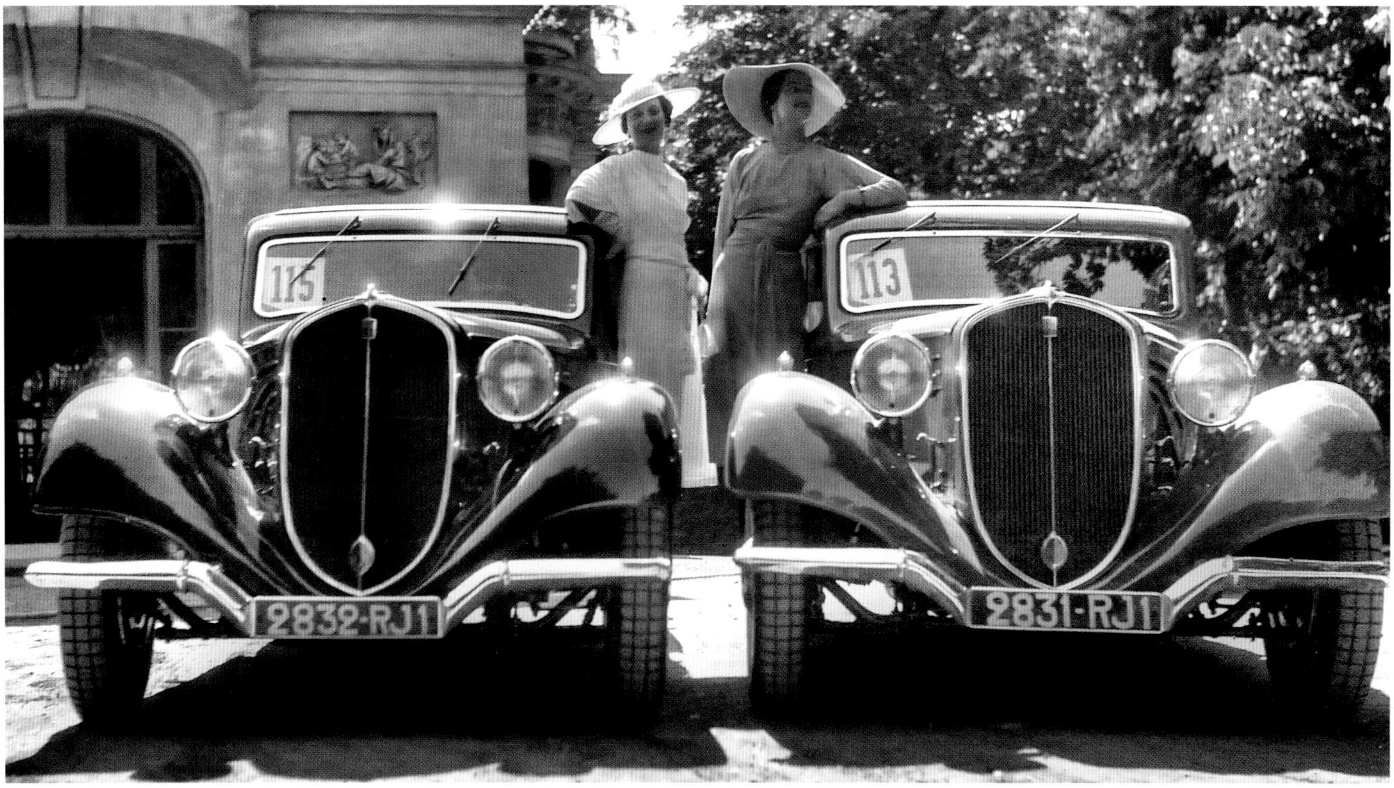

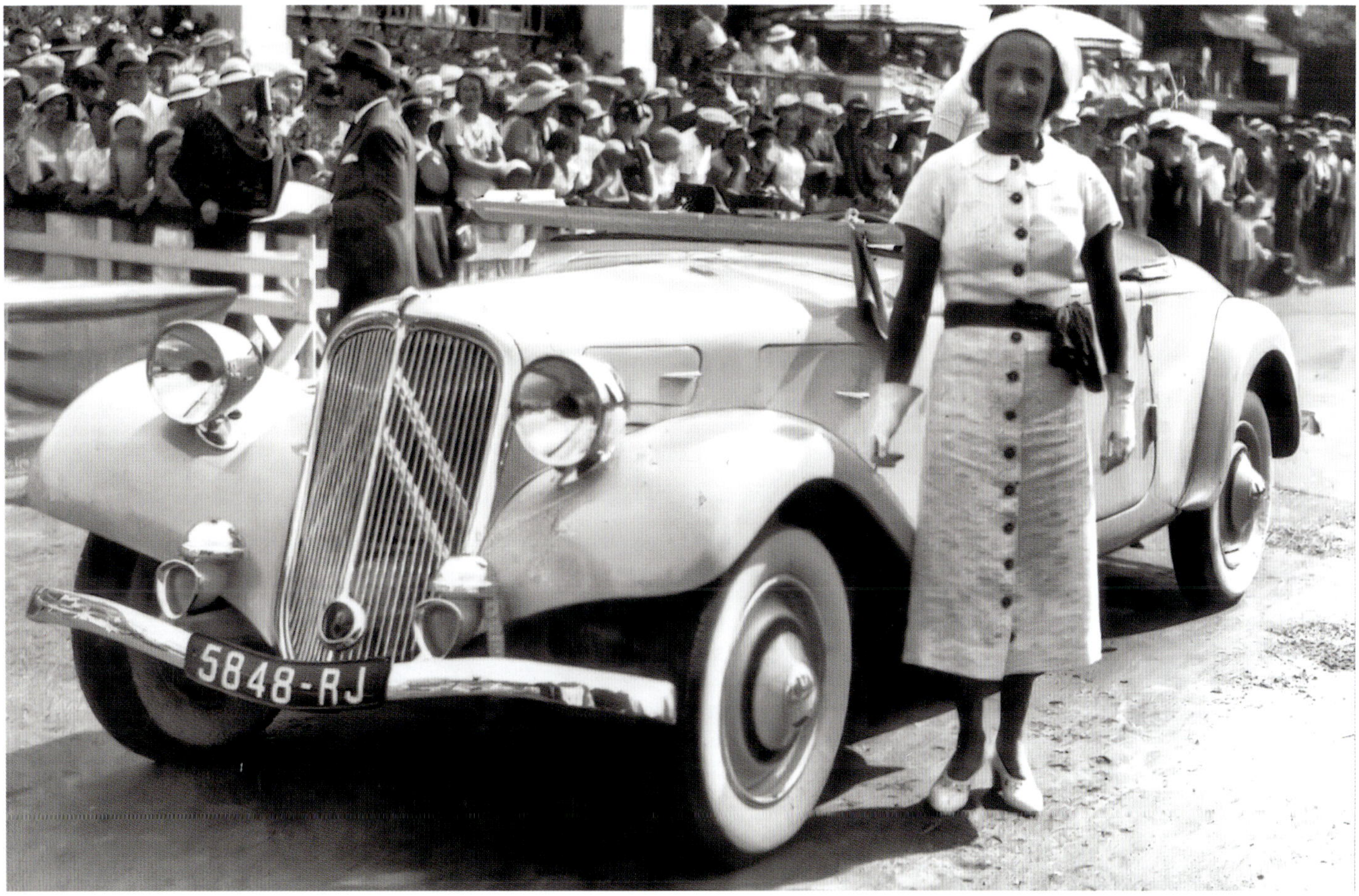


Jacqueline Citroën, daughter of the manufacturer, presents the Citroën 7 Traction Avant at the Deauville Concours d'Elegance. (Daniel Cabart Archives)

visible at concours events, enthusiastically promoting their successes and fully capitalizing on these opportunities for market communication.

Advertisements tied closely to the concours d'elegance events, highlighting the distinct strategies of each brand. Peugeot, prominently featured in the 1937 concours, used advertising to emphasize its achievements in these events with the tagline: *"They prove that mass production can also be a sign of perfection!"*

Meanwhile, Fiat leveraged its success at the 1933 Nice Concours d'Elegance with the message: *"Overall, a beautiful event worthy in every way of the reputation of elegance enjoyed throughout France by the French Fiat."*

André Citroën, the master of marketing innovation, understood the strategic advantage of placing his latest models alongside luxury marques at prestigious events. His wife, Georgina Citroën, had already pioneered this approach by presenting their newest model at Dinard in 1921.

Years later, on June 8, 1934, at the concours d'elegance in the Bois de Boulogne, Citroën introduced the striking Citroën 7 Traction Avant, a groundbreaking model that defied convention. Anticipation was high, and Citroën made a grand display. The factory selected seven young women, each driving a Citroën 7 Traction Avant "Special Concours d'Elegance Edition," to participate in the event. The women, dressed in couture by leading designers, included the Baroness de Rothschild, who embodied French elegance and refinement. Despite minor brake issues that briefly delayed the jury, four of the seven vehicles won prizes, marking a tremendous success for Citroën.

A few weeks later, at the sunny seaside setting of the Deauville Concours d'Elegance, Jacqueline Citroën, André's daughter, continued this strategy. The event offered a perfect backdrop to further the Traction Avant's appeal. Jacqueline captivated the audience as she drove across the podium in a beautiful white dress that complemented the car, blending elegance and modernity in an unforgettable moment.

Legendary French marques like Bugatti, Delage, and Delahaye attained iconic status through their celebrated appearances at concours d'elegance events, and they continue to captivate audiences even today. However, the achievements of esteemed international marques such as Isotta Fraschini, Duesenberg, Mercedes-Benz, and Rolls-Royce are equally noteworthy. These prestigious vehicles found an ideal stage at concours d'elegance events, where they displayed their elegance and innovation in full splendor.

Over time, these grand concours d'elegance events became true cultural institutions, shaping the collective imagination and the evolution of the automobile industry. Far beyond mere concours, they fostered a vibrant creative exchange among manufacturers, coachbuilders, and artisans, making them essential to the legacy and artistry of automotive design.

While luxury marques remained central to the golden age of concours d'elegance events, it's essential to also recognize the role of industrial manufacturers driven by a bold vision: that the future of the automobile lay in its accessibility to the masses. Visionaries like Henry Ford, Louis Renault, and André Citroën championed this ideal, using mass production to lower costs and bring the dream of car ownership within reach for a broader audience.

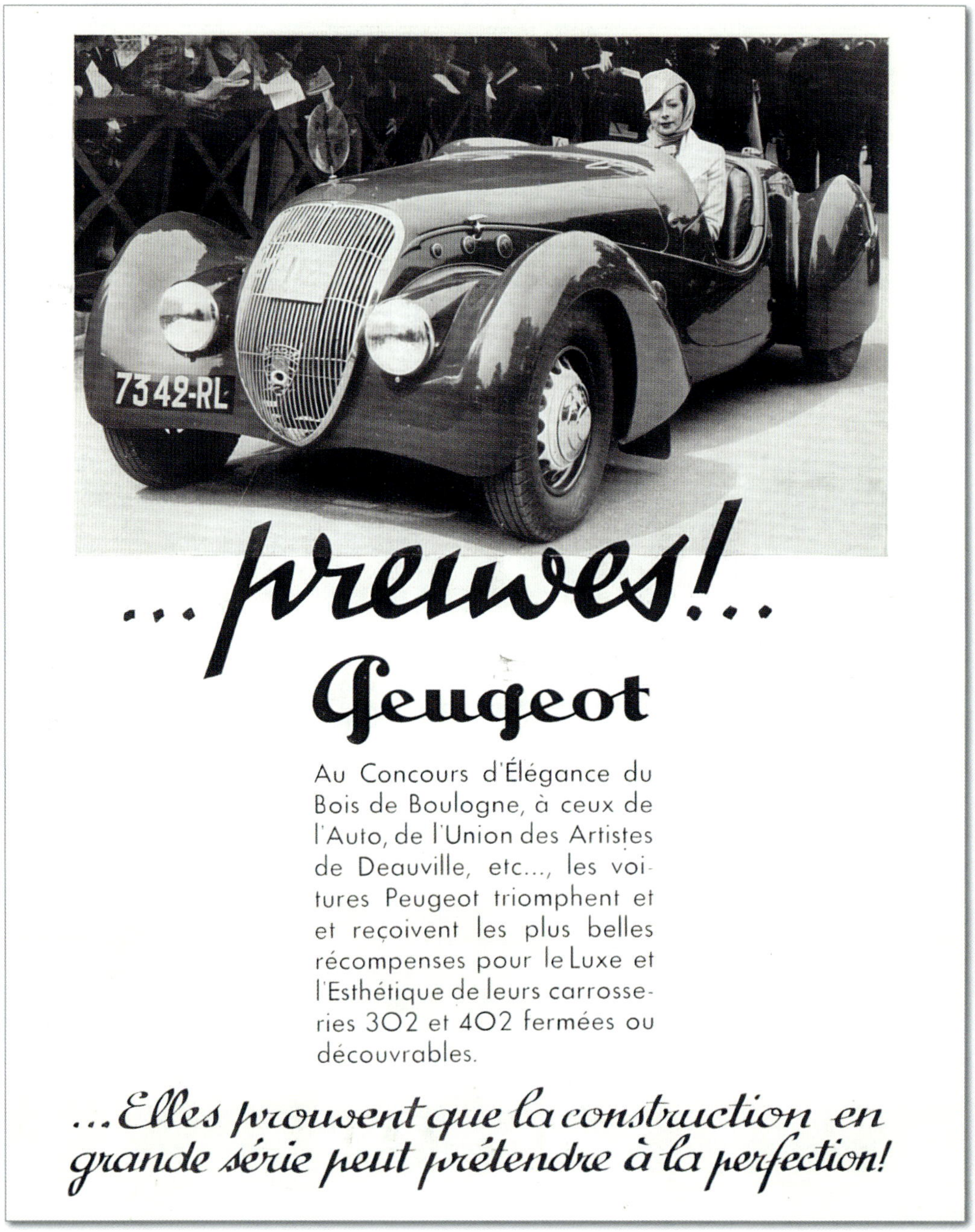


*This advertisement highlights Peugeot's success in prestigious automotive elegance contests, emphasizing that even mass-produced cars can achieve high levels of luxury and design. The translation follows below. (La Carrosserie)*

***...Proof!.. Peugeot***

*At the concours d'elegance of the Bois de Boulogne, at the Auto competition, the Union of Artists event in Deauville, etc., Peugeot cars triumph and receive the most prestigious awards for the Luxury and Aesthetics of their 302 and 402 models, both closed and convertible.*

***...They prove that mass production can aspire to perfection!***

During these years, concours d'elegance events welcomed manufacturers from all segments and drew audiences from diverse backgrounds, celebrating not only superb craftsmanship and rarity but also forward-thinking designs that would ultimately shape the future of the automobile industry.

## Coachwork: An Ode to Customization

In the 1920s, coachbuilders faced new challenges as the automobile, previously viewed primarily through a technological lens, began to embrace aesthetics and style. The role of the coachbuilder evolved, becoming more complex and demanding, requiring not just the design and construction of a vehicle but a holistic approach that considered a range of new factors. No longer was the task merely to fit a chassis with an engine and body; considerations of speed, visibility, safety, comfort, and the treatment of interiors and accessories became increasingly crucial.

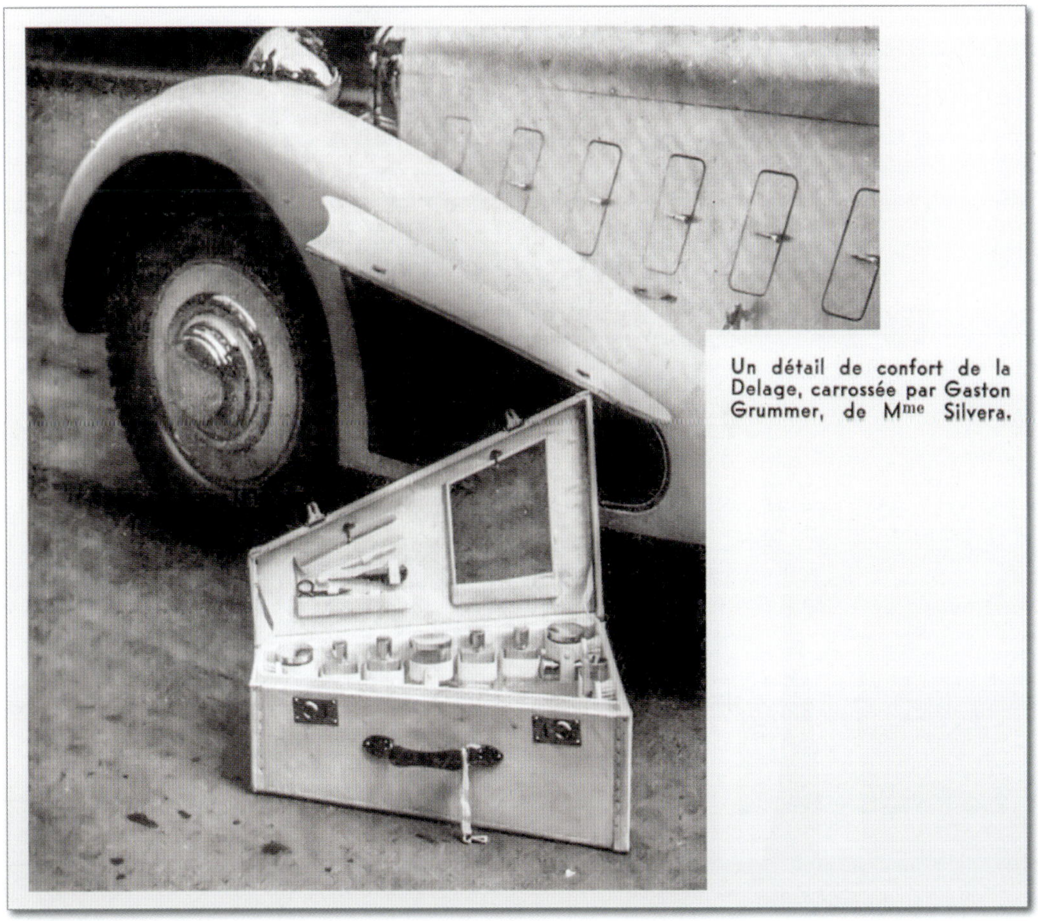


*Accessories are inseparable from high-end coachbuilding. (La Carrosserie)*

The Roaring Twenties were marked by a thirst for modernity, and the convergence of art and science was in full swing. The 1925 International Exhibition of Decorative Arts in Paris epitomized this movement, as the creative worlds began influencing new aesthetic codes. The concept of elegance itself was being redefined. Designers like Robert Mallet-Stevens, Émile-Jacques Ruhlmann, and Louis Süe moved beyond their traditional professions, adopting a more comprehensive approach that anticipated fashion trends and coordinated new disciplines to cater to a demanding clientele. The parallel with coachbuilders is clear – like these designers, coachbuilders were becoming interior decorators and trendsetters.

At the start of the decade, automobile designs were relatively simple and angular, with thin fenders and running boards as standard features. Interiors were often cozy, filled with inventive accessories. Concours d'elegance events provided the perfect platform for coachbuilders to showcase their latest innovations and win over clients who were sometimes resistant to change. Renowned coachbuilders like Kellner, Labourdette, Kelsch, Binder, and Gallé used these events to communicate their achievements and stand out in the field.

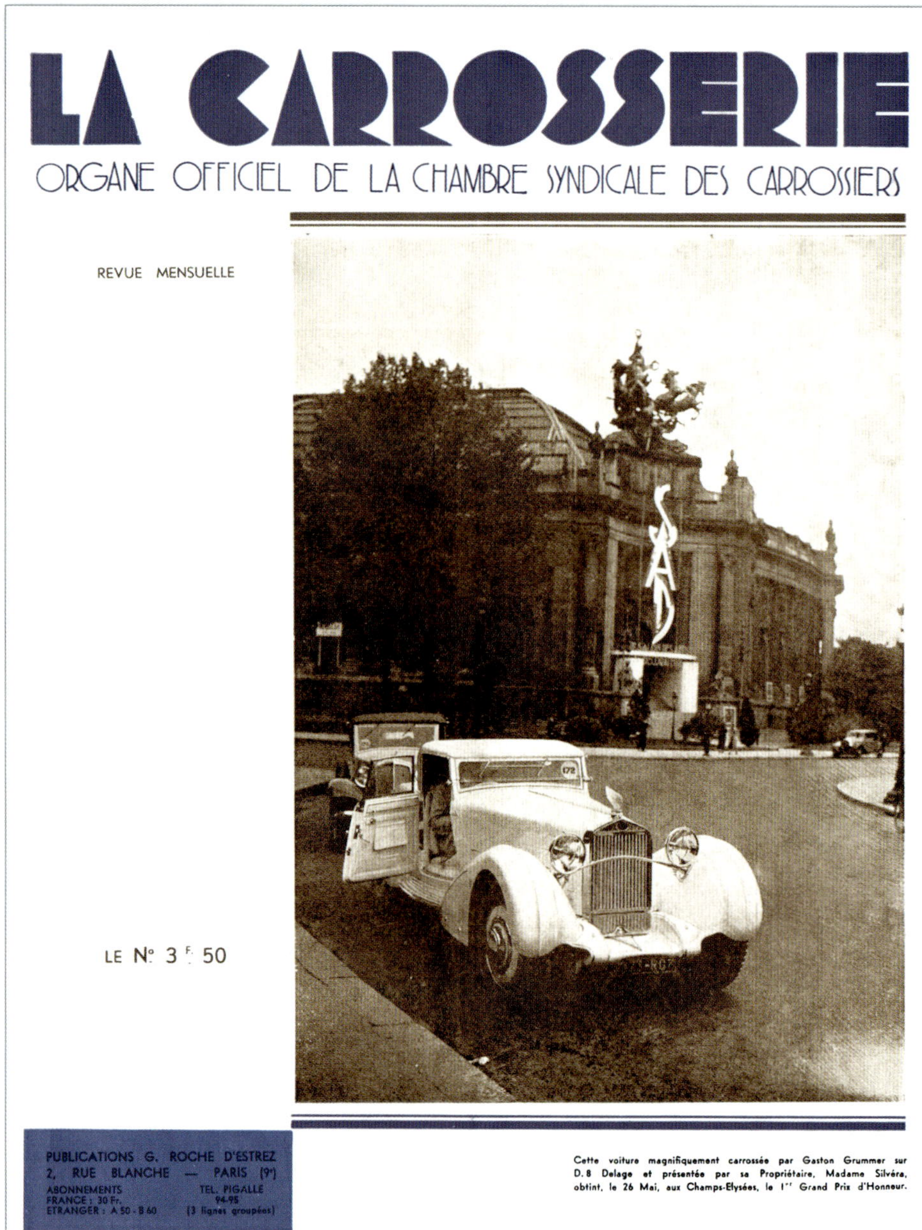

LA CARROSSERIE

ORGANE OFFICIEL DE LA CHAMBRE SYNDICALE DES CARROSSIERS

REVUE MENSUELLE

LE N° 3 F. 50

PUBLICATIONS G. ROCHE D'ESTREZ
2, RUE BLANCHE — PARIS (9e)
ABONNEMENTS FRANCE : 30 Fr. ETRANGER : A 50 - B 60
TEL. PIGALLE 94-95 (3 lignes groupées)

Cette voiture magnifiquement carrossée par Gaston Grummer sur D. 8 Delage et présentée par sa Propriétaire, Madame Silvéra, obtint, le 26 Mai, aux Champs-Elysées, le 1er Grand Prix d'Honneur.

**RIGHT:** *Special issue of the monthly magazine* La Carrosserie *dedicated to French coachbuilding. (La Carrosserie)*

**BELOW:** *With her ivory dress and black accessories, Baroness de Pitray, great-great-granddaughter of the Countess of Ségur, is in perfect harmony with her Delahaye, bodywork by Labourdette. Shown at the 1933 Champs-Élysées Concours d'Elegance. (La Carrosserie)*

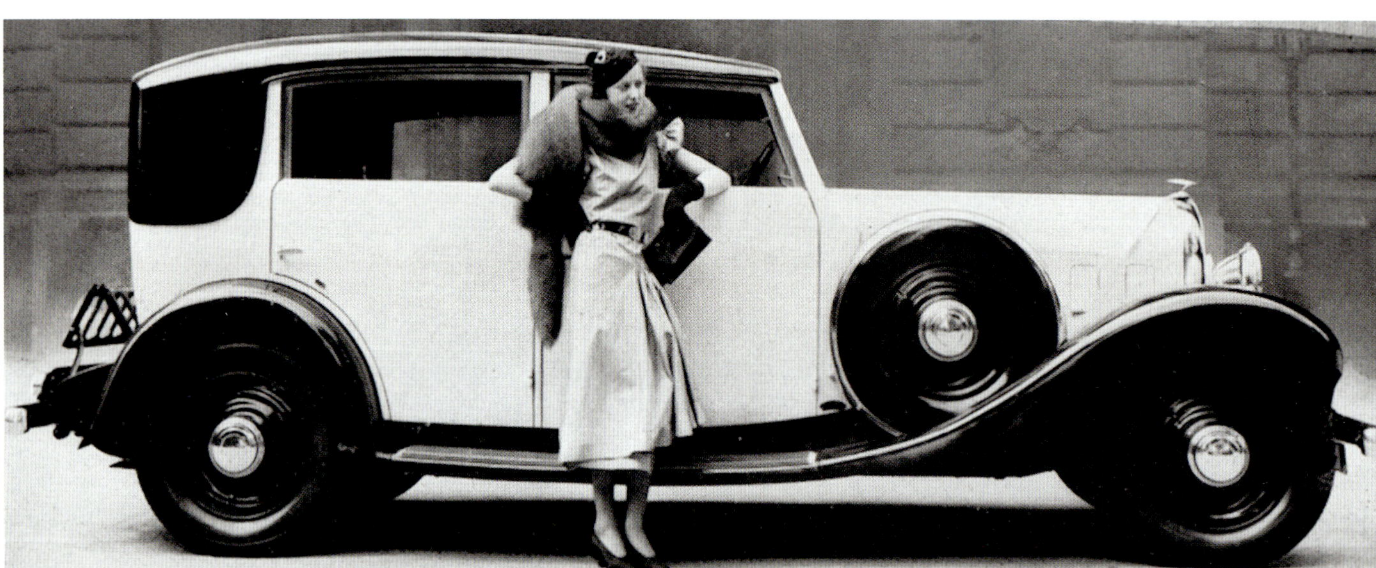

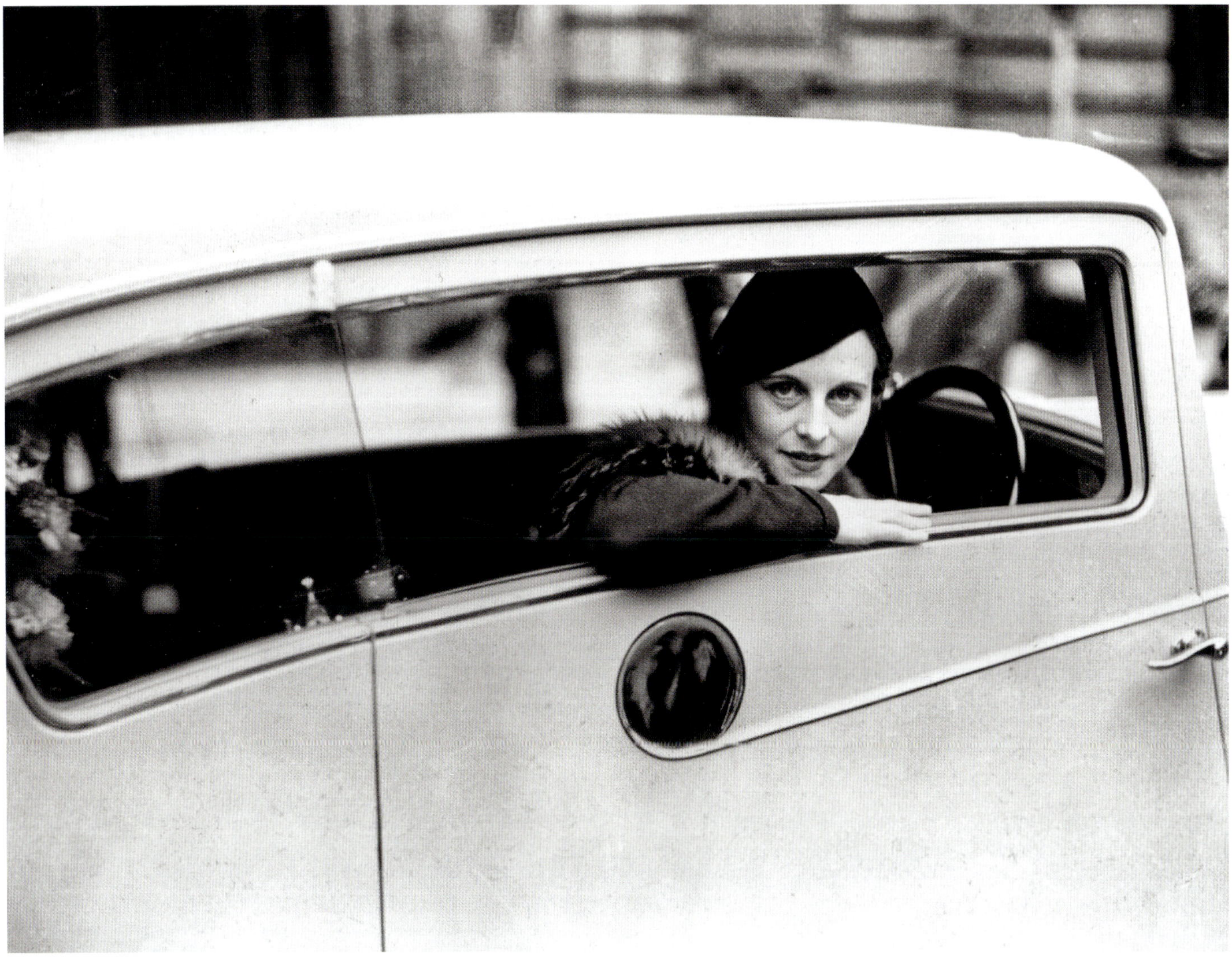

By the late 1920s, cars had become sleeker and lower to the ground. The beltline - a term that Jean Henri-Labourdette[2] likened to the waistline of a couture dress - was lowered and extended to the radiator, creating a more elongated appearance. Delicate moldings further emphasized this effect. Like Pullman train cars, the window contours became more rounded, and the slope of the rear aligned with the windshield's angle.

Chauffeur-driven coupés were becoming less common, and new models such as the "S" for *surbaissé* (lowered) emerged. Meanwhile, a more avant-garde clientele embraced the bold aesthetic codes of Avions Voisin, a brand that found great success in the early years of the concours d'elegance. While Le Corbusier called for decluttering home interiors, critics like Robert Mallet-Stevens[3] accused coachbuilders of clinging too closely to the traditions of horse-drawn carriages.

However, coachbuilders had a different vision for their future. Rather than abandoning their heritage, they sought to preserve their decades of craftsmanship, pushing luxury, sophistication, and customization even further. For a certain class of clientele, travel had to remain as opulent as first class, surrounded by neo-Baroque décor. Many of the artisans - wheelwrights, cabinetmakers, painters, and saddlers - continued to work on both automobiles and horse-drawn carriages, blending traditional skills with modern needs.

For coachbuilders, the standardization emerging from Detroit posed a significant threat. Citroën's adoption of all-steel bodies in 1924, abandoning wooden frames, signaled a shift. Yet, coachbuilders believed that by staying true to their craft and focusing on custom designs, they could secure their future. The tone in the industry's publications was clear: creativity and custom-made designs were the way forward.

***Madame Claude Revel, an icon of French elegance, at the wheel of a Delage D8-S Yoyo by Letourneur et Marchand. (Daniel Cabart Archives)***

2 *Jean Henri-Labourdette, Un siècle de carrosserie française, Édita Lazarus, 1972, p. 70*

3 *Revue de la Carrosserie, December 1932, p.10-12*

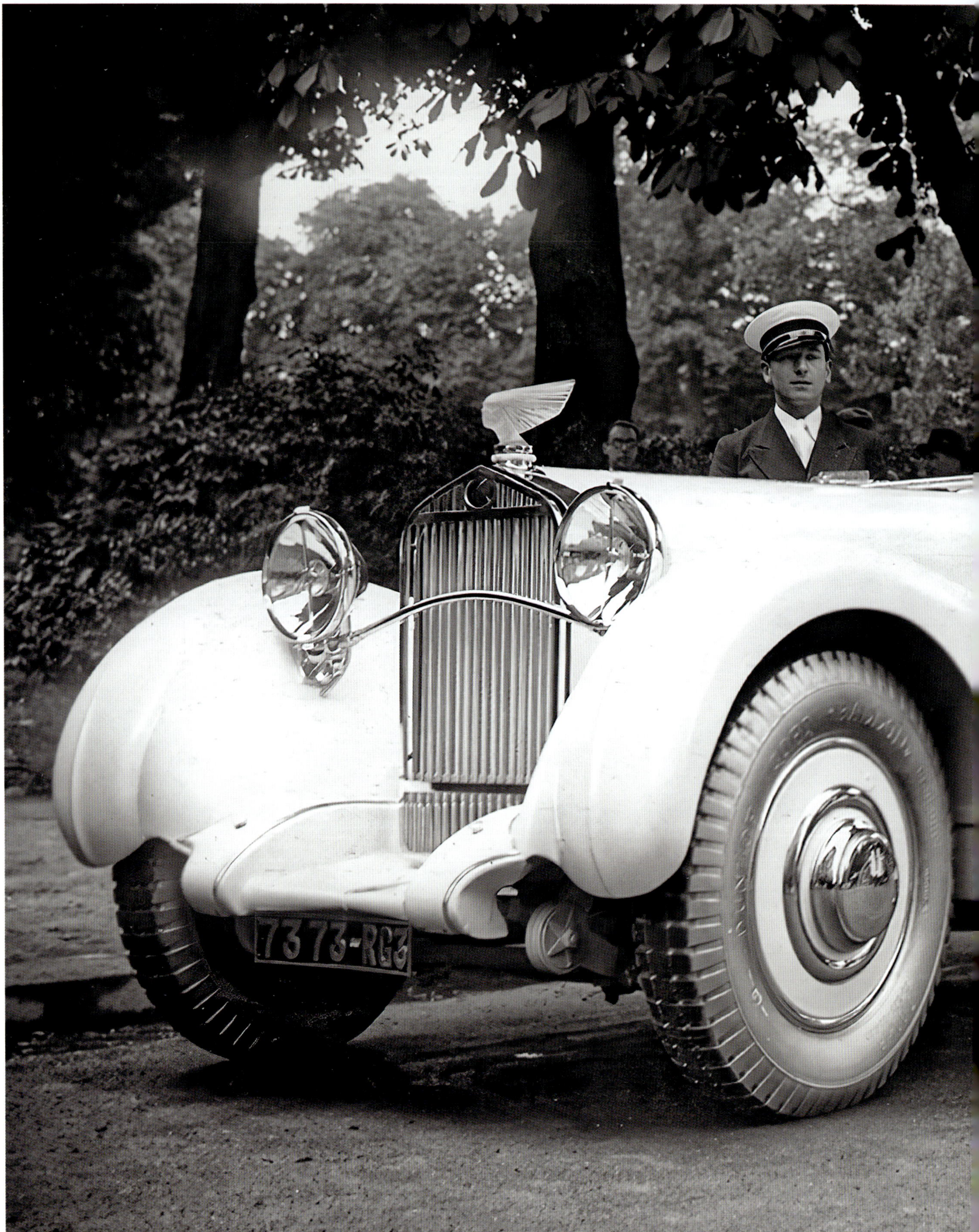
7373-RG3

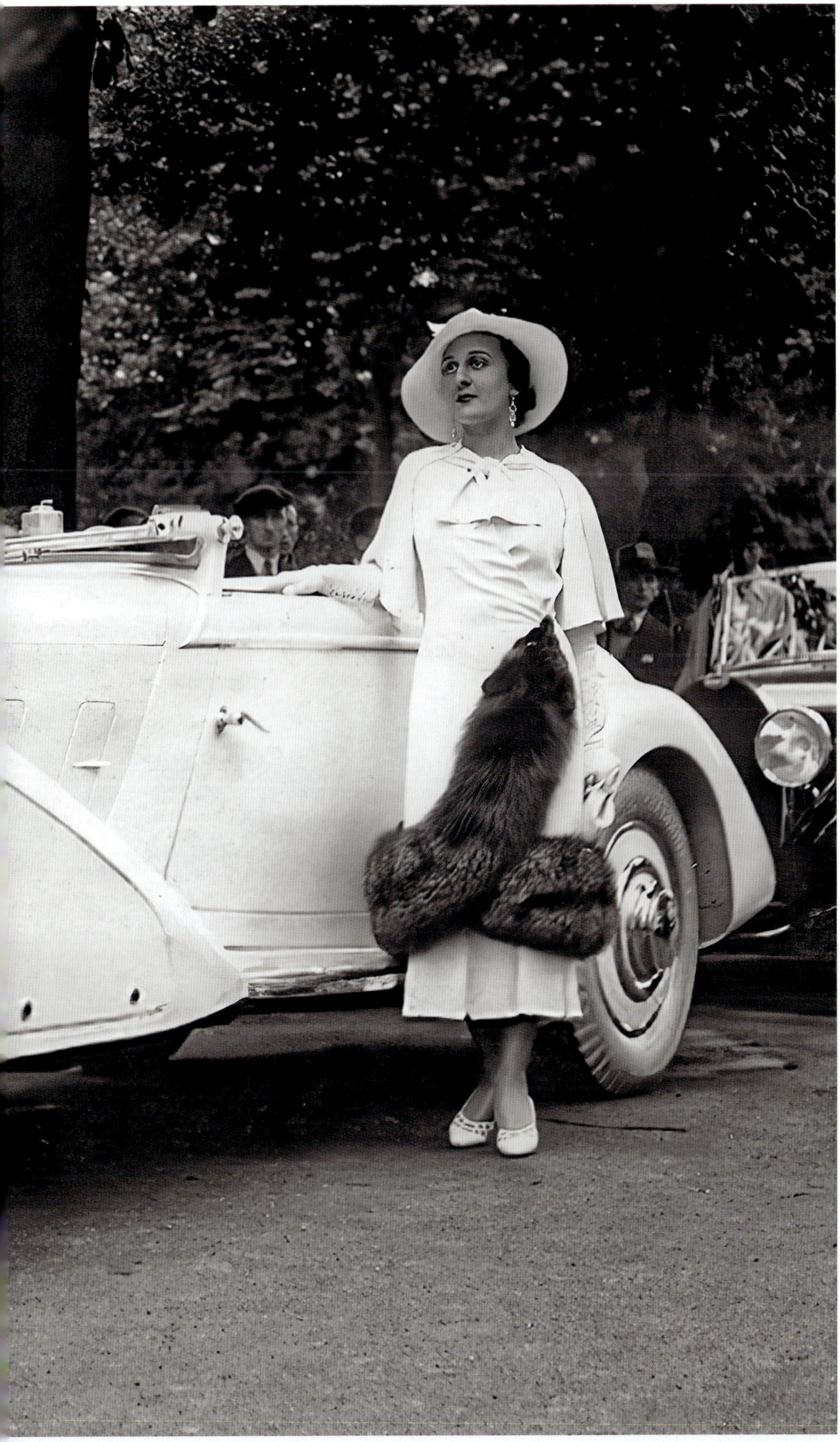

*Grümmer bodywork entirely in white, including the retractable hood, Connolly leather interior, dashboard and steering wheel. The dress, hat, gloves, shoes, and bag of Madame Silvera are also white. This ensemble will claim the* Grand Prix d'Honneur *at the 1933 Champs-Élysées Concours.*
(Daniel Cabart Archives)

## The 1933 Paris Motor Show: A Fusion of Aerodynamics and Aesthetics

The 1933 Paris Motor Show introduced significant innovations in aerodynamics and vehicle profiling, resulting in dramatic changes to car body designs. Vertical grilles became rounded, rear sections were elongated, windshields were angled, rooflines lowered, and fenders became more enveloping. Doors grew wider, with some models featuring reverse-opening designs, while chrome details such as window trims added a touch of sophistication. This marked the merging of aesthetics and technology, with form and function coming together in a striking new way.

As a result, concours d'elegance events grew in number and prominence, with major events attracting increasing attention. Also in 1933, the *Chambre Syndicale*, led by Jean Henri-Labourdette, organized its own grand-scale competition in the heart of Paris. The French bodywork virtuosos hosted five separate contests within a single event: the women's concours d'elegance automobile, contests for the most elegant, refined, and comfortable car, the open-top car concours, the city car concours, and the concours for the most beautiful paintwork. Their ambition to elevate and promote their profession through these concours was evident, as highlighted in Georges Callon's article in *La Carrosserie*.[4]

> *"Today, one can confidently say that no car owner is limited to a standard model. By simply consulting a coachbuilder, they can select from a wide range of French chassis, choosing the one that best suits their needs, and have a custom body tailored to their tastes and personality. For only a modest additional cost, they can own not just a mass-produced vehicle, but a truly personalized car – uniquely their own.*
>
> *The event of May 26, 1933, played a pivotal role in spreading this message, which remains unknown to many. It significantly advanced the cause of French coachbuilding."*

In the same issue, a letter of support for the event appeared from Georges Prades, Vice President of the General Council of the Seine, titled "*The Bodywork: Expression of Personality*." In it, Prades

[4] *Revue de la Carrosserie*, June 1933, p.11-23

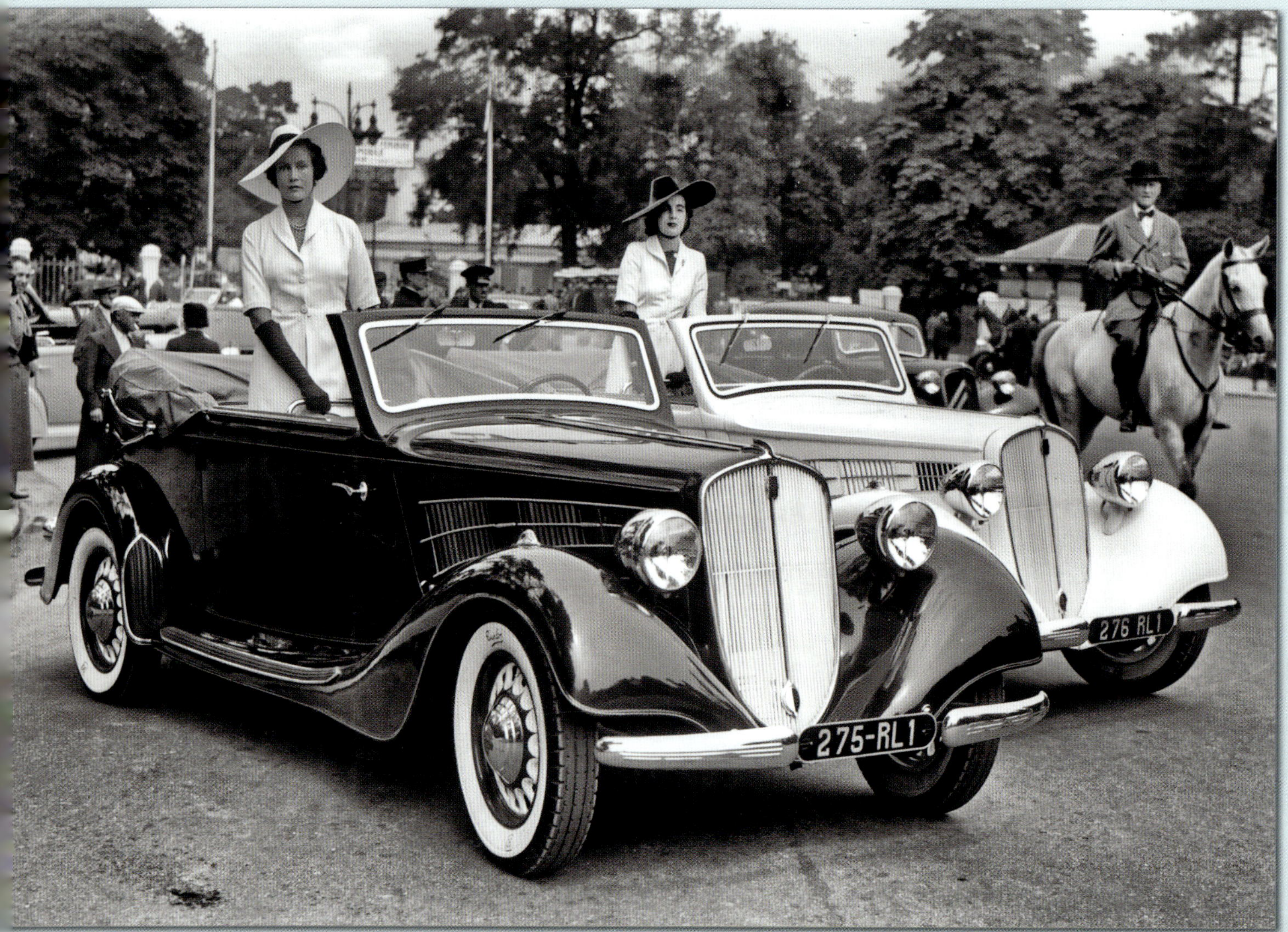


**Simca Fiats at the 1939 Paris Concours d'Elegance. (© AGIP)**

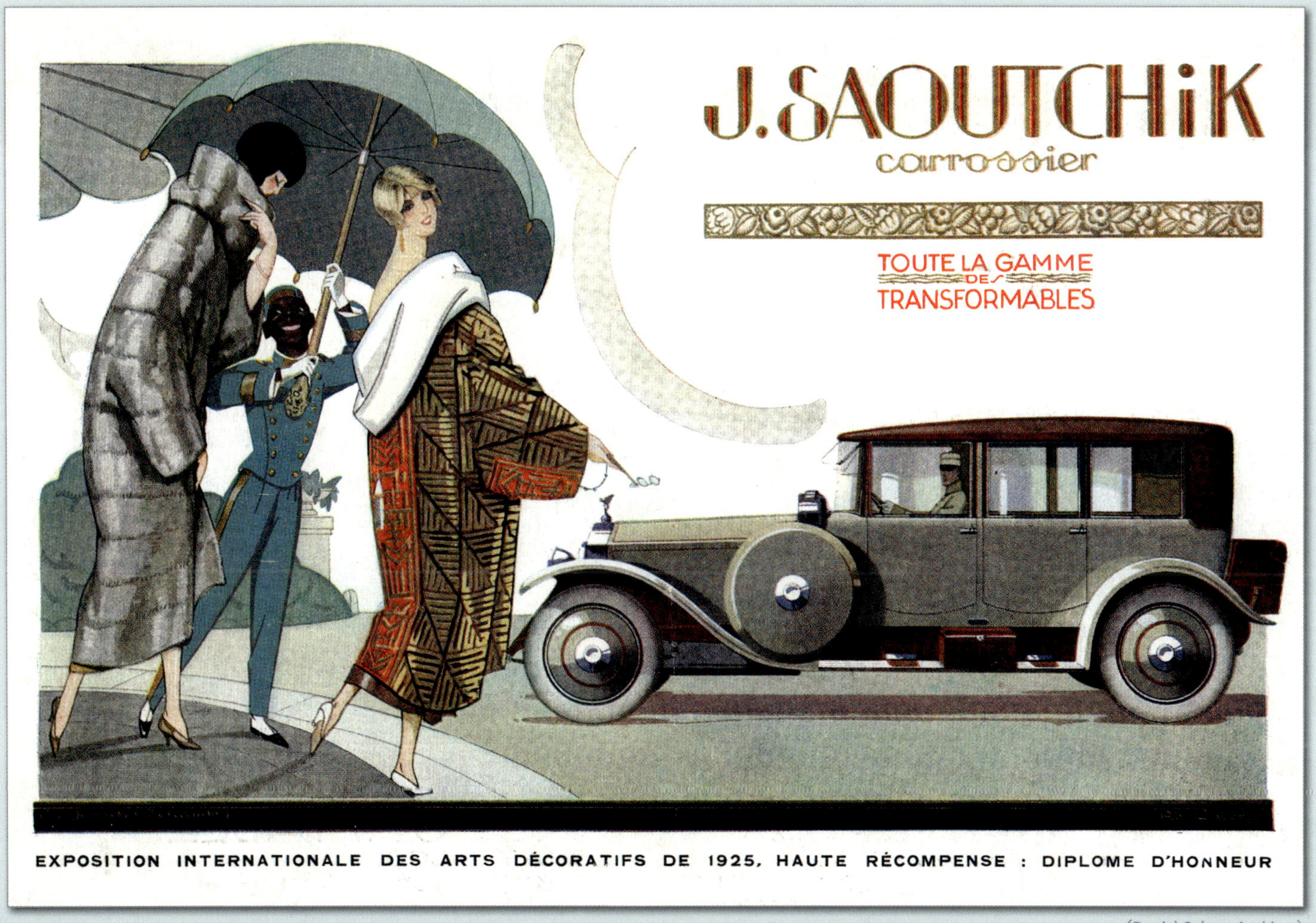


(Daniel Cabart Archives)

strongly criticized the industrialization efforts of the United States, condemning the standardization of car bodywork and passionately defended French craftsmanship. He wrote: "*What a magnificent protest from Paris, vibrant and confident in its own taste, against the mechanization of personality that has been forced upon it... The automobile has the unique ability to bring together the many facets of true elegance: naturally, couture, fashion, furs, perfumes, and jewelry all align with it. Car bodywork must remain individual, expressing the distinct preferences of each person. Let's not endure the indignity of recognizing our cars solely by their color – no, not even by their number!*"

This sentiment resonated with the distinguished, often conservative clientele of the concours d'elegance events, who saw the democratization of the automobile as a threat to their social status.

For these clients, coachbuilders were more than just craftsmen – they were custodians of memory, masters of the rare, the unique, and the beautiful. While technical expertise was crucial, it was the emotional impact of the lines and curves that truly mattered. Each new design represented a fresh journey into luxury, with concours d'elegance events offering judges the opportunity to appreciate the subtlety of the finest finishes and accessories. A simple press of a button might have revealed a concealed powder compact, a mirror, a lorgnette, or a retractable watch – no detail was overlooked for the fashionable woman of the 1930s.

Materials like mother-of-pearl, onyx, lacquer, pewter, and gold were transformed into unprecedented creations of absolute refinement. Designers like Ruhlmann, Dunan, and Lalique were often called upon to contribute their expertise. Each technical challenge offered a new opportunity for creative triumphs, and every new creation was a testament to the pursuit of uniqueness and excellence. As Victor Hugo famously said, "*The beautiful is as useful as the useful.*"

The craftsmanship of every coachbuilder evoked its own sense of emotion with each design. Legendary names like Labourdette, Binder, Grümmer, and Gallé continually pushed their limits, while the traditionalists like Chapron and Letourneur et Marchand provided steady, reliable elegance, contrasted with the bold creativity of Figoni or the extravagant flair of Saoutchik. Franay, for his part, aimed for the highest levels of luxury.

This period marked the golden age of French coachbuilding, an unparalleled chapter in automotive history. Coachbuilders of the era mastered every challenge, fulfilling the desires of an elite clientele with extraordinary precision and artistry.

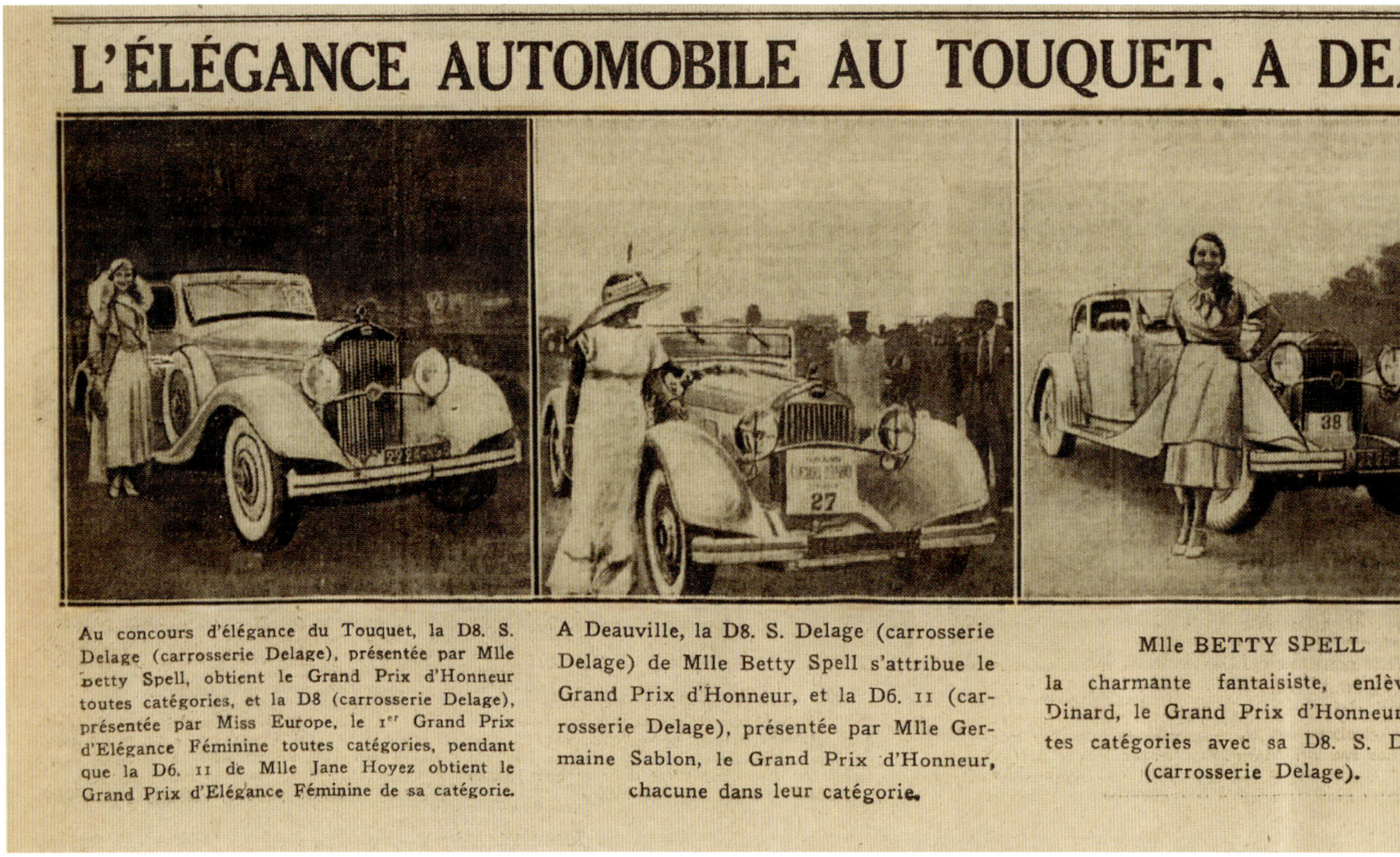

L'ÉLÉGANCE AUTOMOBILE AU TOUQUET, A DE

Au concours d'élégance du Touquet, la D8. S. Delage (carrosserie Delage), présentée par Mlle Betty Spell, obtient le Grand Prix d'Honneur toutes catégories, et la D8 (carrosserie Delage), présentée par Miss Europe, le 1er Grand Prix d'Elégance Féminine toutes catégories, pendant que la D6. 11 de Mlle Jane Hoyez obtient le Grand Prix d'Elégance Féminine de sa catégorie.

A Deauville, la D8. S. Delage (carrosserie Delage) de Mlle Betty Spell s'attribue le Grand Prix d'Honneur, et la D6. 11 (carrosserie Delage), présentée par Mlle Germaine Sablon, le Grand Prix d'Honneur, chacune dans leur catégorie.

Mlle BETTY SPELL

la charmante fantaisiste, enlève Dinard, le Grand Prix d'Honneur tes catégories avec sa D8. S. D (carrosserie Delage).

## Couture and Women Elevate Concours to New Heights

The history of haute couture in France predates even the rise of manufacturers and coachbuilders. In 1858, Charles Frederick Worth, considered the pioneer of the couture house, became the most sought after designer among Parisian high society. Despite the challenges of the Great War, haute couture would experience a golden age lasting nearly 50 years.

By 1925, at its height, fashion accounted for 15 percent of France's exports, ranking second overall. The haute couture industry employed more than 500,000 artisans and workers (according to Couture Federation Archives). Paul Poiret, an extravagant and inventive figure, led this era of fashion innovation. At the Paris Decorative Arts Exhibition, Poiret declared himself the champion of modern art and a leading figure in applied arts. His forward-thinking approach influenced every aspect of the luxury industry, including interior design, fragrances, packaging, illustrations, printing, tapestries, furniture, tableware, embroidery, trimmings, lace, as well as dresses and coats.

By 1921, his global success earned him the moniker "king of the world" in the press. A portrait of Poiret was even painted by the famed artist André Derain, known for his association with Bugatti cars (Grenoble Museum). During this era, Lanvin, Paquin, Vionnet, and later Chanel each made their indelible mark on haute couture.

One well-known encounter between Coco Chanel and Paul Poiret highlights the shifting tides in the fashion world. When Poiret mockingly asked Chanel, dressed in black, who she was mourning, she famously replied, "*For you, sir.*" The winds of change were blowing, and women were beginning to assert their influence in fashion.

At the same time, the automobile industry was recognizing the need to appeal to women – both as passengers and drivers. In January 1921, *Vogue* addressed this shift, stating, "*Few modern industries can afford to ignore the female perspective. The automobile industry has long understood this, designing cars with women's needs in mind.*" The distinction between sports and touring cars and city cars provided the opportunity to cater to "feminine delicacy."

As a result, advertisements from car manufacturers, coachbuilders, and equipment suppliers began to

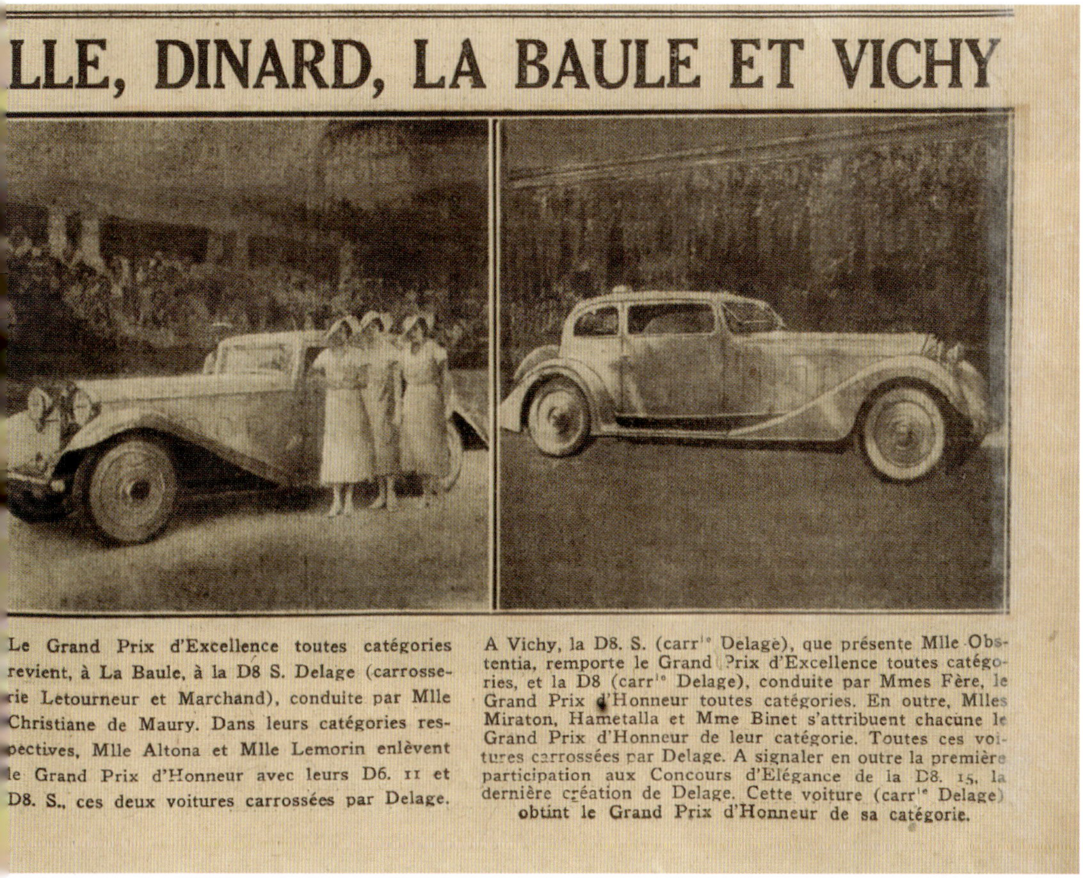

# LLE, DINARD, LA BAULE ET VICHY

Le Grand Prix d'Excellence toutes catégories revient, à La Baule, à la D8 S. Delage (carrosserie Letourneur et Marchand), conduite par Mlle Christiane de Maury. Dans leurs catégories respectives, Mlle Altona et Mlle Lemorin enlèvent le Grand Prix d'Honneur avec leurs D6. 11 et D8. S., ces deux voitures carrossées par Delage.

A Vichy, la D8. S. (carr[ie] Delage), que présente Mlle Obstentia, remporte le Grand Prix d'Excellence toutes catégories, et la D8 (carr[ie] Delage), conduite par Mmes Fère, le Grand Prix d'Honneur toutes catégories. En outre, Mlles Miraton, Hametalla et Mme Binet s'attribuent chacune le Grand Prix d'Honneur de leur catégorie. Toutes ces voitures carrossées par Delage. A signaler en outre la première participation aux Concours d'Elégance de la D8. 15, la dernière création de Delage. Cette voiture (carr[ie] Delage) obtint le Grand Prix d'Honneur de sa catégorie.

**LEFT:** *Harvest of victories in major competitions for Delage with its D8-S in 1933. (Excelsior)*

**BELOW:** *The curved lines created by the coachbuilder De Villars are echoed by the dress designer of Madame Paul Cartier. (Daniel Cabart Archives)*

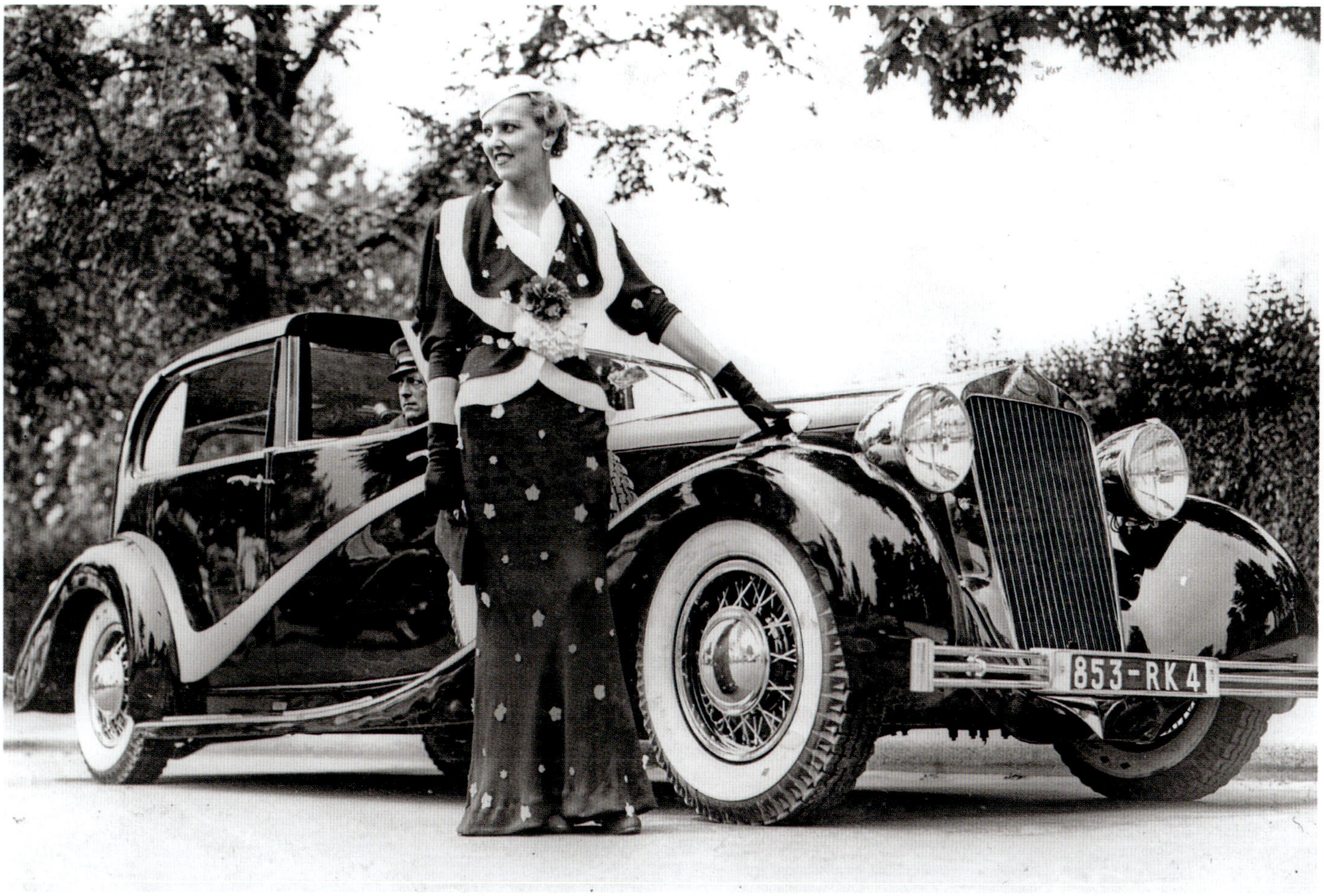

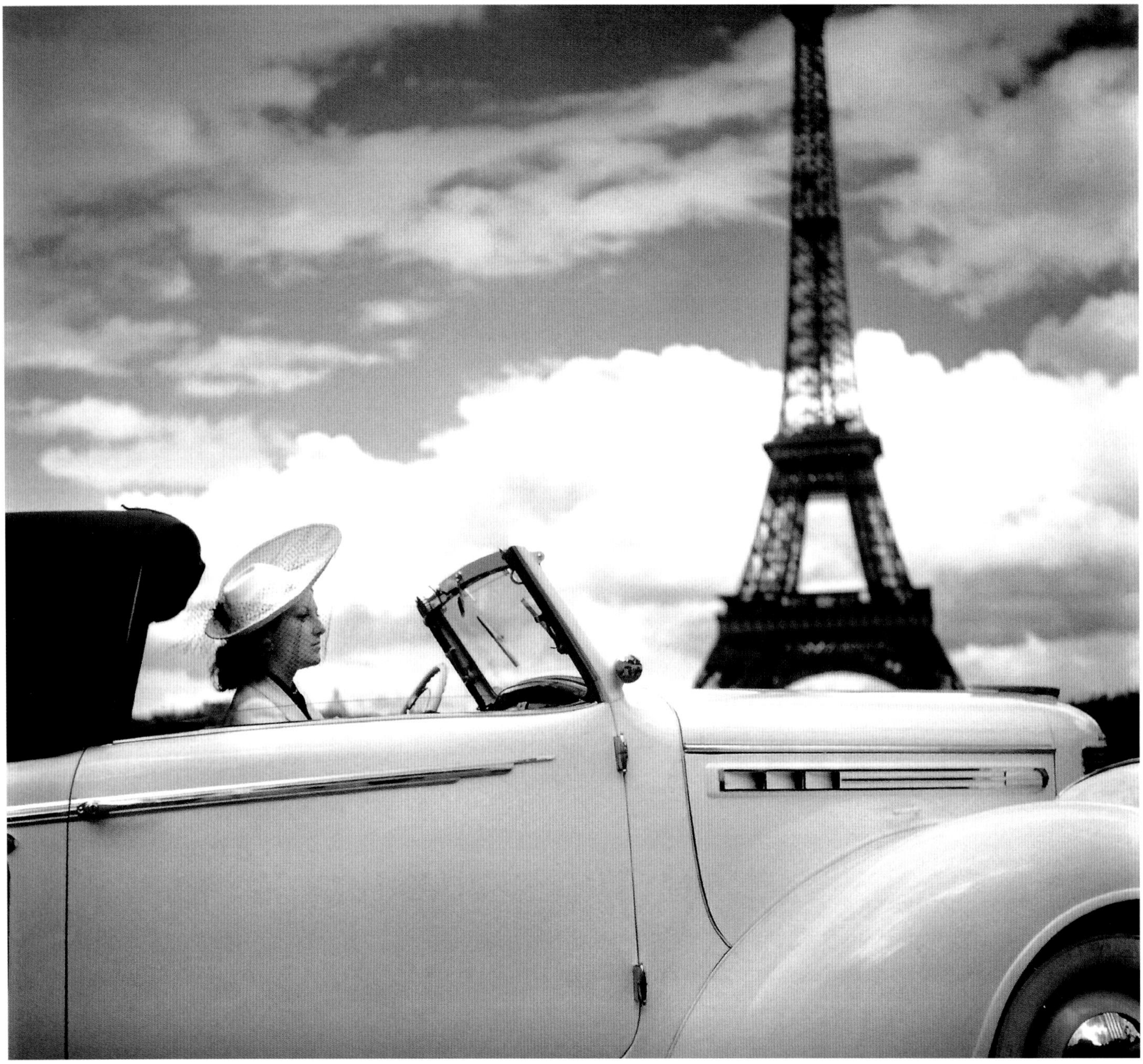

**Mlle. Claude May, dressed by the Maison Rosevienne, 1938 Fémina L'Intran Concours d'Elegance. (Boris Lipnitzki / Roger-Viollet)**

appear alongside those from fashion houses, perfumers, and milliners. Another *Vogue* article from the same year noted, "*The latest cars exude a refined elegance, much like the season's new fashions: knitted suits and travel hats are the must-haves of the spring.*"

Both industries were in sync, with similar trends and aspirations. In 1920, the automobile magazine *Omnia* devoted several articles to women's involvement in driving. One excerpt reads: "*The choice is yours: a wide-striped coat, an Egyptian-style garment in thick homespun or wool velvet, a straight-cut coat with a large cape, or a raglan made from charmingly patterned wool.*"

Automobile marques soon embraced female muses to promote their products, but as France Renucci[5] noted, "*the goal isn't to seduce* through *women, but to seduce women.*" Concours d'elegance events soon became a natural battleground for both fashion and automotive industries.

By the mid-1920s, under the influence of André de Fouquières and leading fashion magazines, concours d'elegance events began to be rebranded in some cases as "women's concours d'elegance automobile."

In June 1929, *Fémina* magazine, announcing the competition it was sponsoring, noted: "*The concours*

[5] *France Renucci, Autos-femmes: union imposée, rupture impensable, Les cahiers de médiologie 2001/2 N° 12, p. 227-237*

*d'elegance event, hosted by Intra and Fémina, which has seen growing success each year, has now become one of the traditional highlights of the grand Parisian week. No longer just 30 participants, but nearly 300, all equally charming and graceful, faced the judges..."*

For this event, prominent couturiers were in attendance, with Madeleine Vionnet designing the honorary banners. Instead of the traditional trophies or radiator caps, winners received jewelry created by the jeweler Curiel. For de Fouquières, the presence of women and fashion significantly enhanced the allure of concours d'elegance events.

The 1930s truly solidified the status of concours d'elegance competitions, which came to be regarded as the most spectacular event of the season, according to Guillaume Garnier[6]. Fashion designers eagerly embraced these concours as a platform to showcase their new collections and accessories, which played a significant role in the contests.

Maggy Rouff, who served as President of the *Chambre Syndicale de la Haute Couture* from 1929 to 1948, was deeply involved in these events. She dressed famous models, such as Hélène Ostrowska (wife of jeweler Louis Arpels), who won numerous accolades, including the prestigious *Grand Prix d'Honneur* at the 1937 automobile concours. Rouff was a staunch defender of the unique attributes of Parisian haute couture, remarking in an interview with Pierre Lazareff: "*At the core of our taste lies a sense of balance and harmony – a French quality that no one can deny.*"

*Madame Louis Arpels, wife of the famed jeweler, was awarded the Honor Prize for the elegance of her Delage D8 drophead at the 1937 L'Auto Concours d'Elegance at Bois de Boulogne near Paris. She is dressed by Maggy Rouff and wears a hat by Caroline Reboux. (Daniel Cabart Archives)*

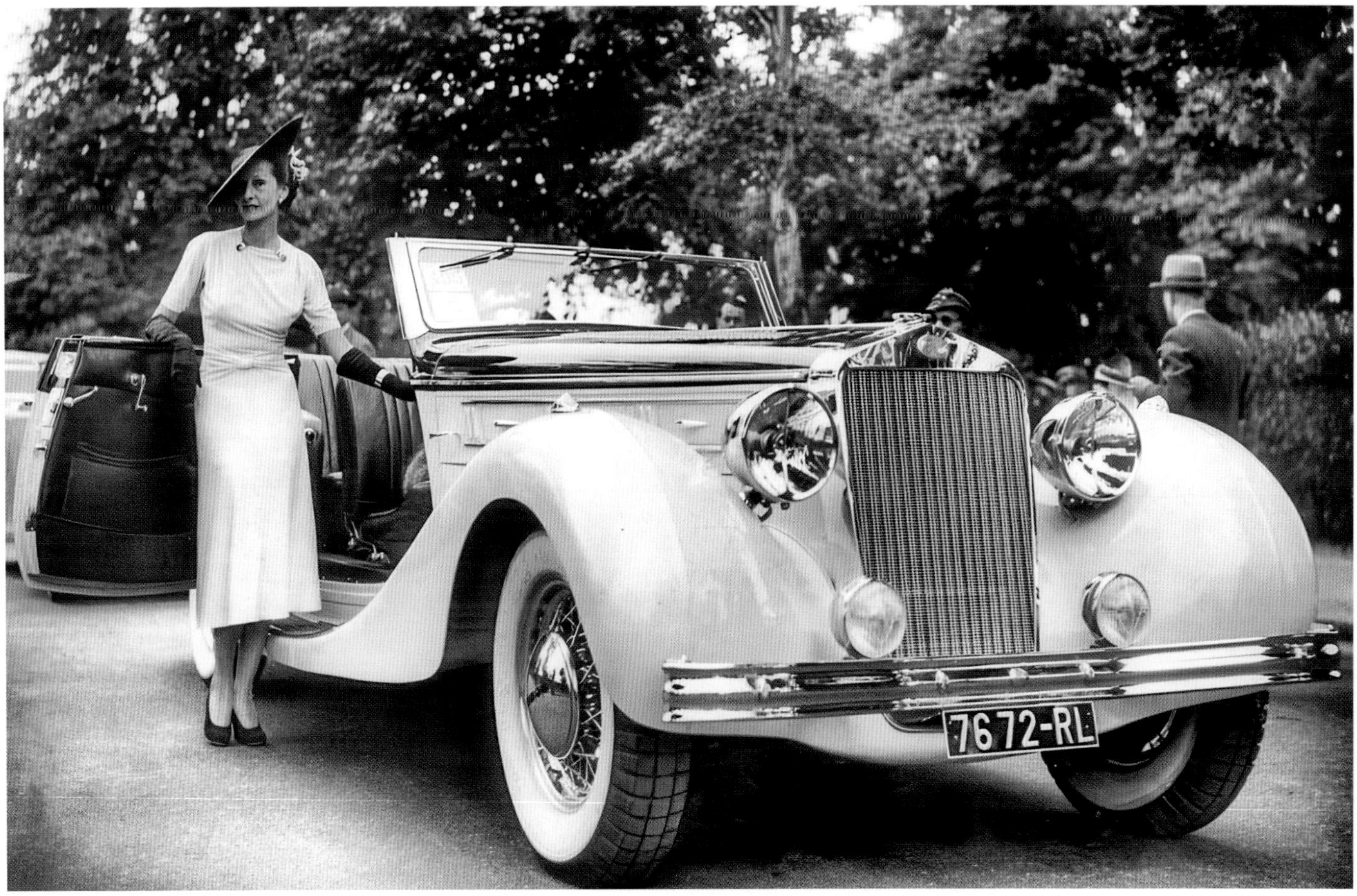


This philosophy aligned closely with that of Lucien Lelong, who similarly stated in *Art Goût Beauté*: "*In France, innovation can be bold without being eccentric.*"

In contrast, Elsa Schiaparelli took a different approach, with a focus on invention, audacity, and humor. For her, fashion was a spectacle, and concours d'elegance events were prime opportunities for her designs to shine. Schiaparelli, like Jean Patou, recognized that sports and leisure activities were booming among the wealthy elite. Her mastery of knitwear, particularly her famous sweaters, redefined them as chic garments, marking the early rise of what would later be known as "sportswear."

Coco Chanel, known for her sharp wit, distanced herself from the concours d'elegance events. In an

[6] *Guillaume Garnier, Paris Couture – Années Trente, Éditions Palais Galliera, 1987, p. 76*

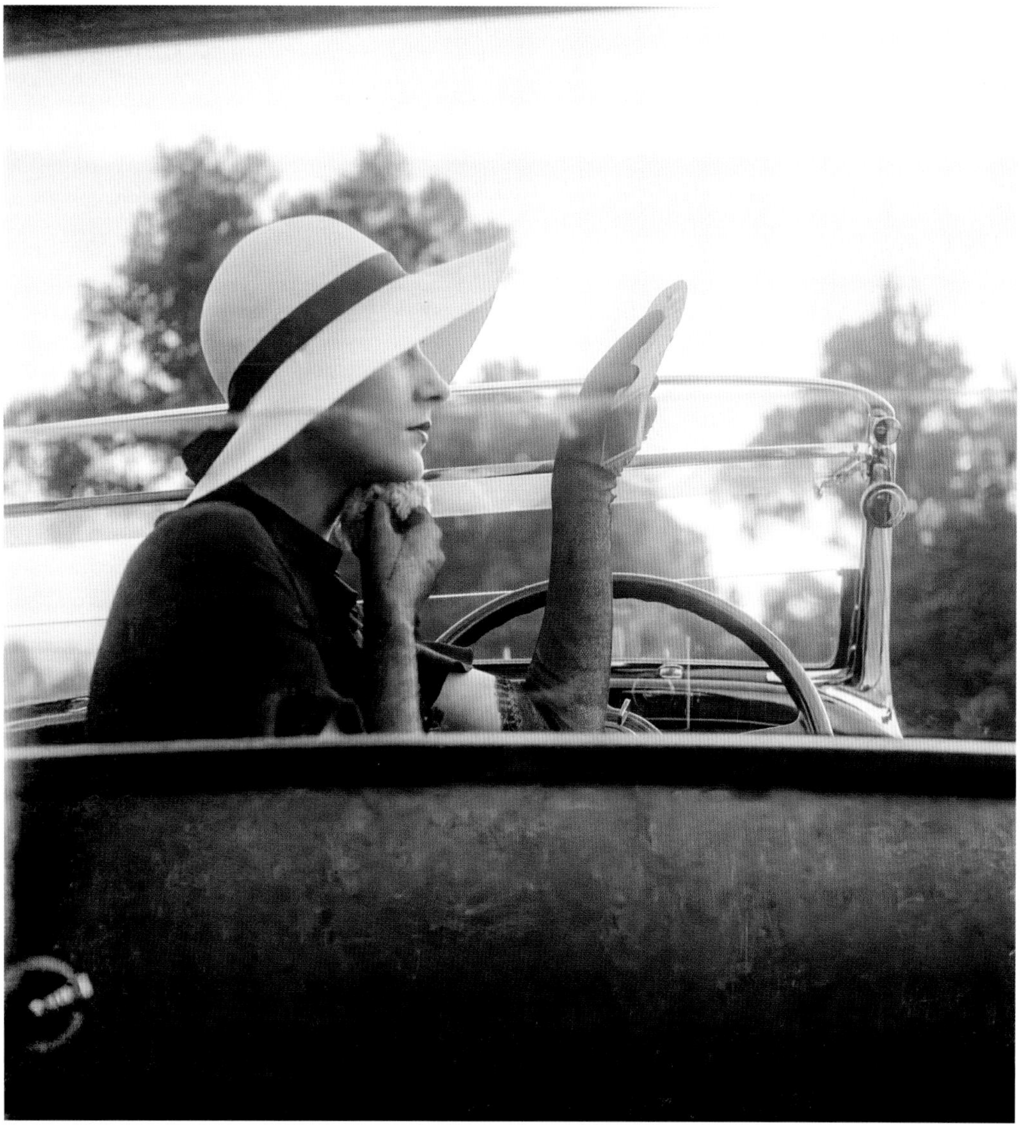

1934 Concours d'Elegance. *Every little detail plays a role in seducing the jury – especially now that the car's occupant holds as much importance as the car itself.* (Boris Lipnitzki / Roger-Viollet)

October 1934 article in *Ford Mathis*, she criticized the events, claiming that women were being used as mere props to highlight automobiles.

Geneviève Cortinovis, in her thesis on Parisian concours d'elegance events[7], noted the informal nature of the collaborations between designers and automobile manufacturers. There was little concrete archival evidence of official partnerships between fashion houses, coachbuilders, and car marques.

However, after 1935, some co-branded advertisements began to appear in leading fashion magazines. For instance, in 1938, René Gruau created two illustrations that featured a Lucien Lelong dress alongside a Chenard et Walcker car and a Schiaparelli outfit with a Rolls-Royce in the background.

Visionary figures like Robert Ricci were instrumental in creating professional communication strategies around concours d'elegance events. With a passion for graphic design and advertising, Ricci founded his own agency, Pro-Publicité, and managed Maison Nina Ricci, co-directing it with his mother from 1932. He was also one of the founders of the renowned Studio Harcourt in 1934.

Geneviève Cortinovis highlights the *Nina Ricci* advertisements that appeared in fashion magazines of the 1930s, such as *L'Officiel de la Mode* and *La Couture*. Many of these full-page advertisements prominently featured both "Talbot 4 liters" and "Nina Ricci," illustrating the strategic communication system that capitalized on the concours d'elegance events. Raymonde Ricci, Robert's wife, frequently participated in the major Parisian concours alongside the *Maison* muses, exclusively entering Talbot-

[7] *Geneviève Cortinovis, Thesis (2010): Presenting Parisian Chic at the Concours d'Elegance 1936-1939*

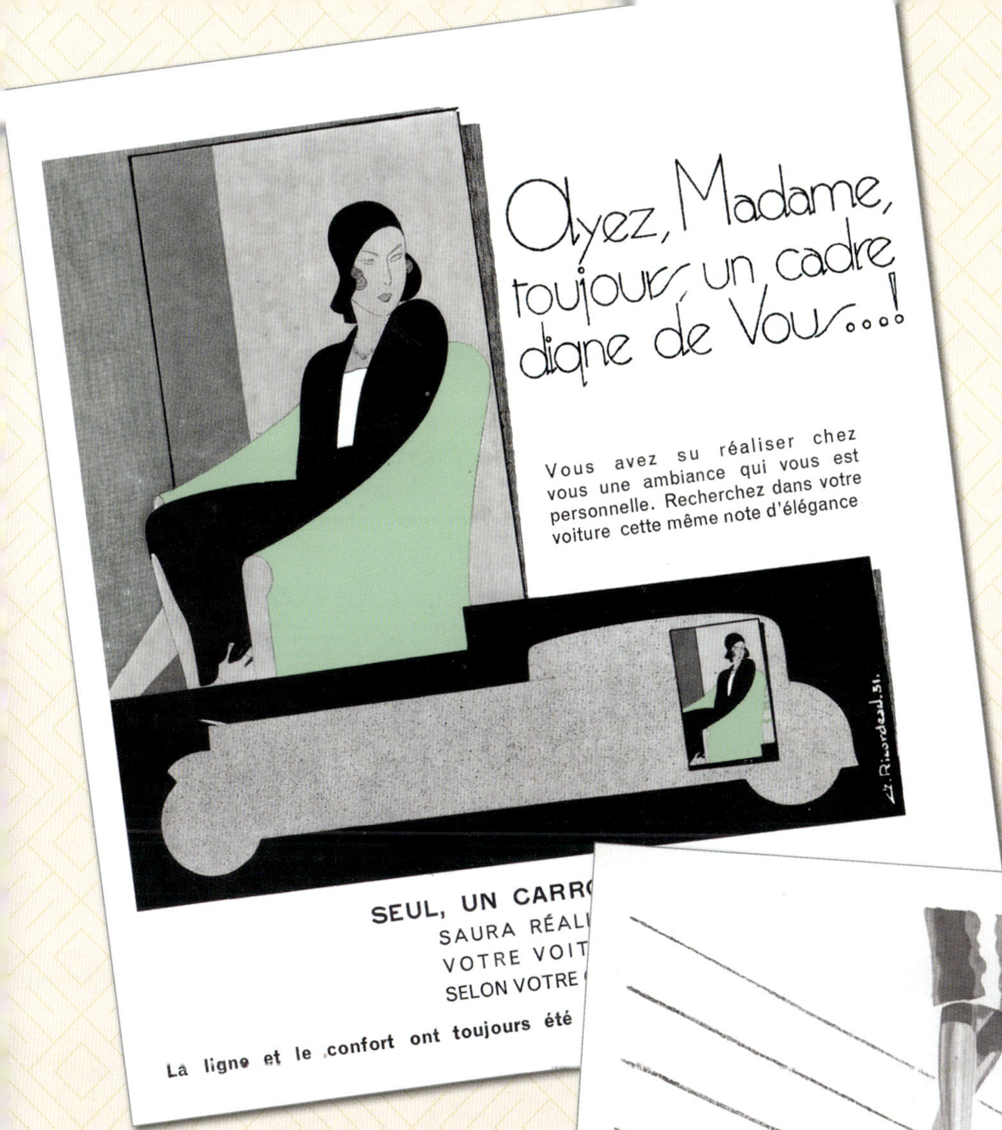


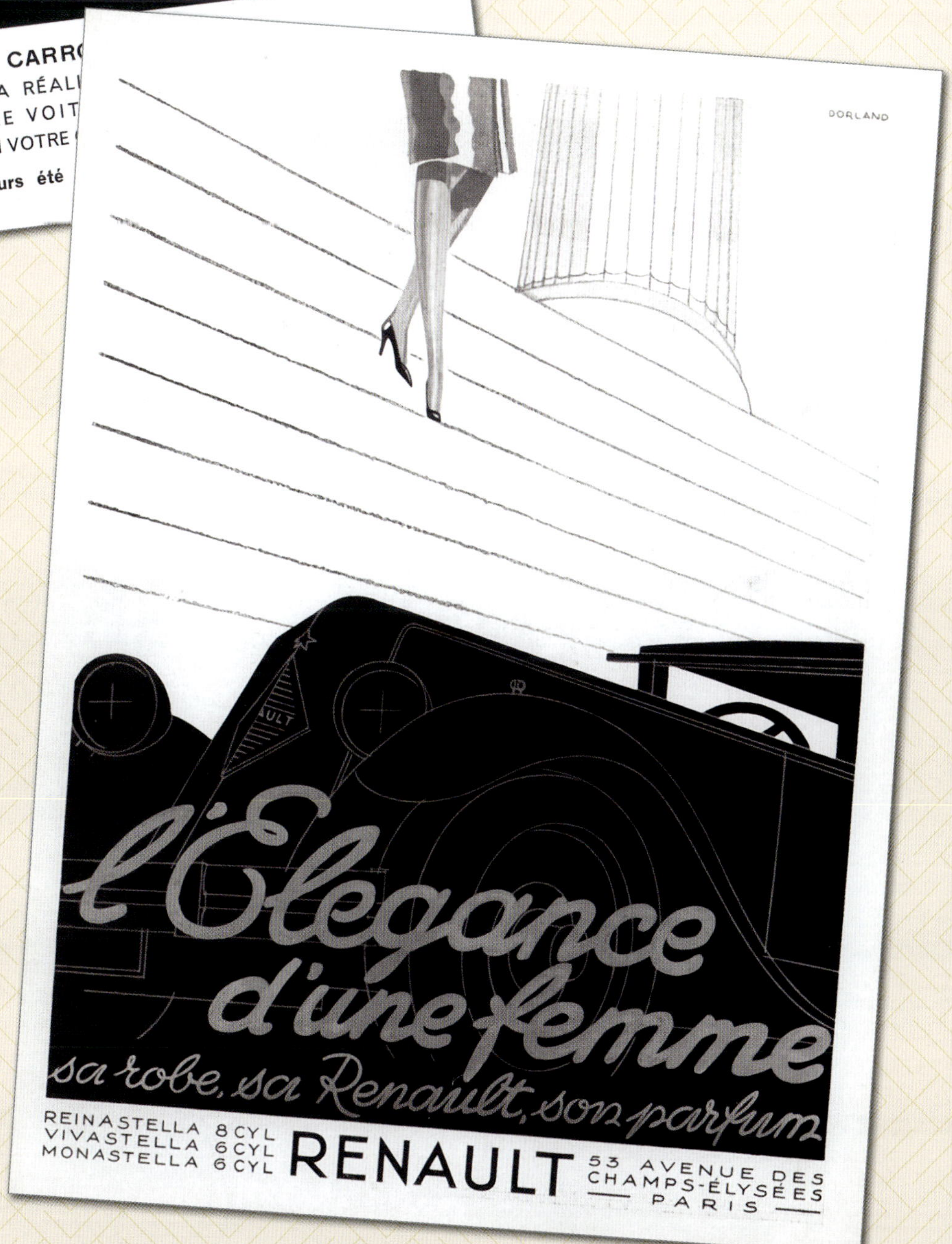


*These two stylish advertisements reveal almost abstract expressions of cars and femininity. (Vogue, August 1929)*

Lago cars. Their numerous successes were featured in advertorials and advertisements in *L'Officiel de la Mode*, with the Talbot cars sporting temporary registrations, indicating they were provided directly by the factory.

The couturiers and the automobile industry shared common values: heritage, innovation, and exclusivity were deeply embedded in both worlds. Their synergy was undeniable, and concours d'elegance events became a natural stage for their expressions of craftsmanship. Both industries exhibited the same attention to detail and precision in execution. It was not uncommon to find renowned coachbuilders, car manufacturers, and couturiers serving as judges at these prestigious events, where the task was far from simple. The concept of harmony had become increasingly refined, as André de Fouquières observed. Maggy Rouff herself declared that attention to accessories required a level of precision comparable to mathematics.

The quest for excellence and harmony among coachbuilders, car manufacturers, and couturiers epitomized the art of French refinement. These events were living masterpieces where aesthetics met unparalleled craftsmanship.

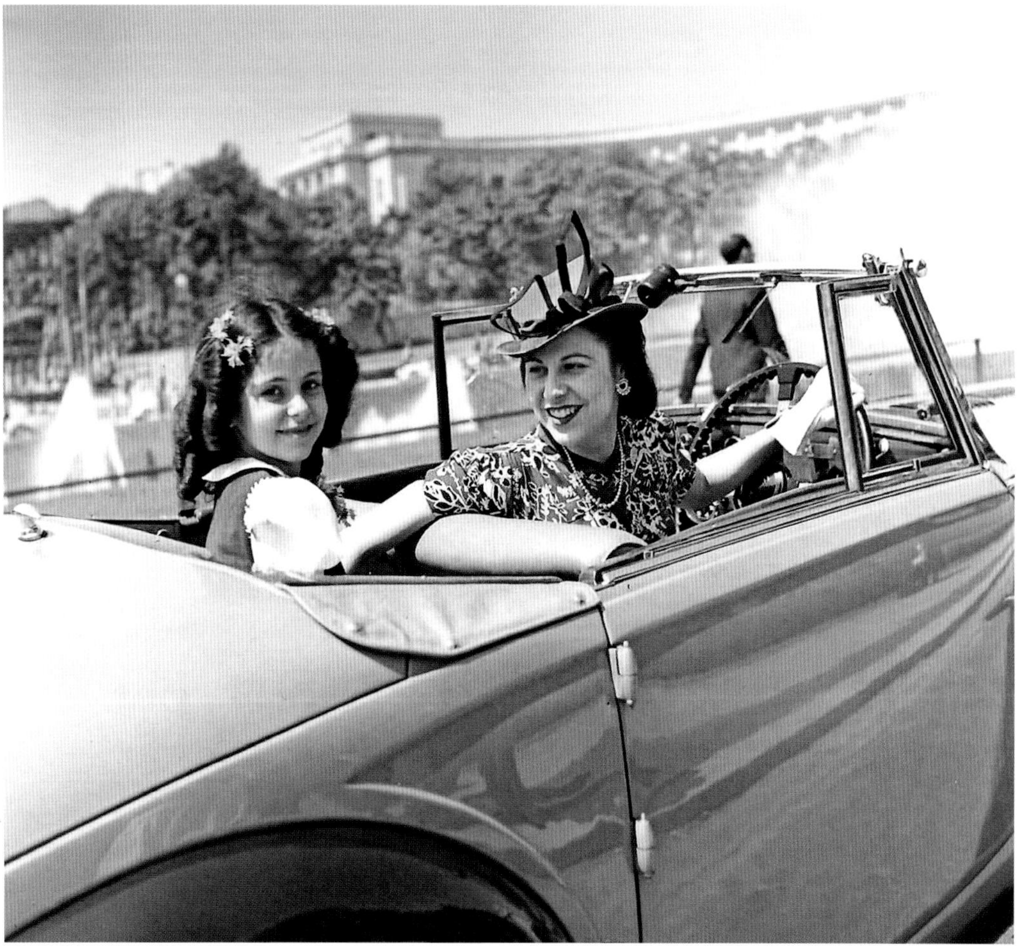

**RIGHT:** *Madame Robert Ricci and her daughter Marie-Françoise. Concours d'Elegance, June 1938, Trocadéro. (Private Collection)*

**OPPOSITE:** *An alliance of style: Nina Ricci and Talbot. (L'Officiel de la Couture et de la Mode, Leon Bénigni, May 1937)*

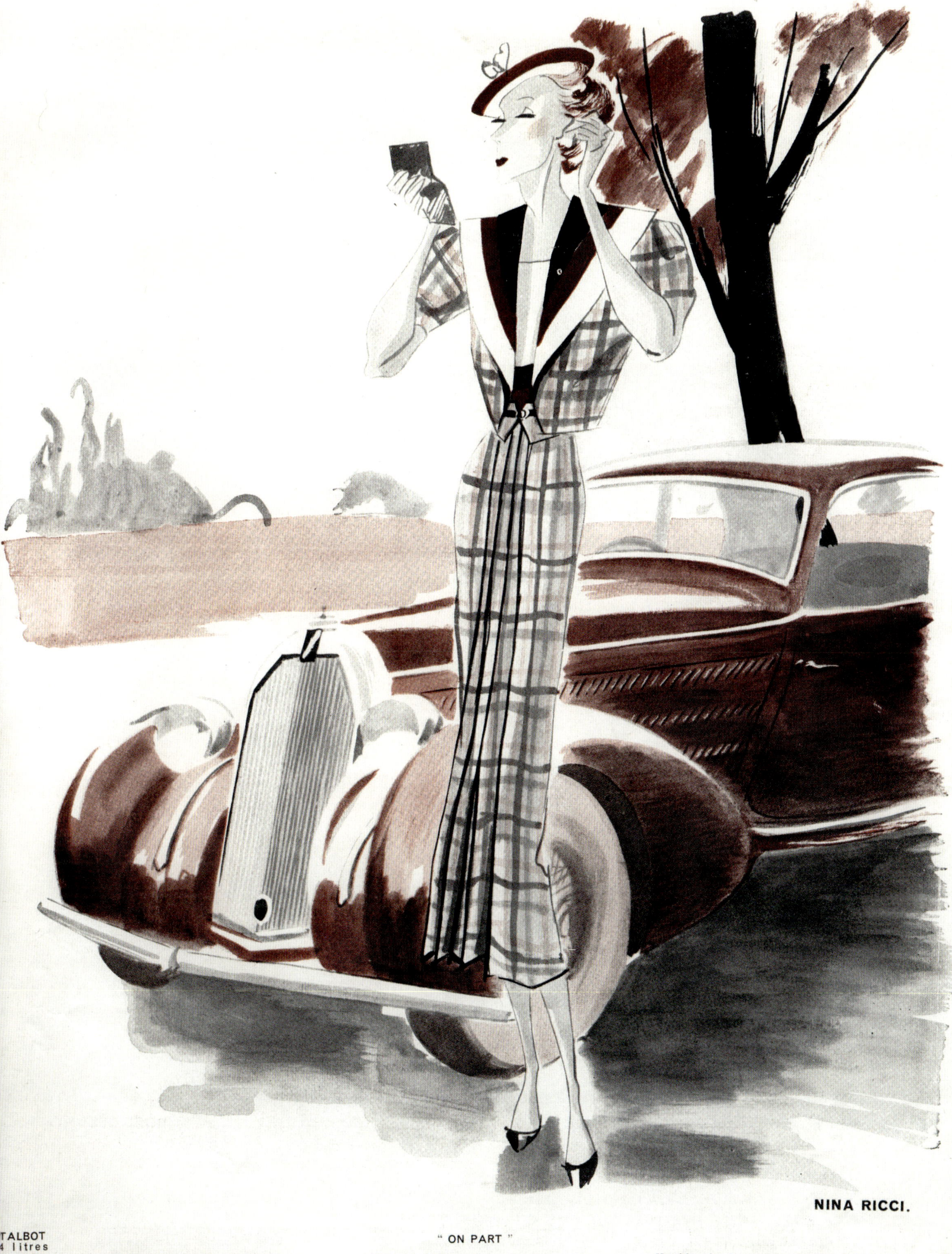

NINA RICCI.

TALBOT
4 litres

" ON PART "

*Dress in plaid woollen, with sunbathing style corsage. The little bolero removable at will.*

*Robe en lainage écossais, au corsage formant bain de soleil. Le petit boléro s'enlève à volonté.*

*Vestido de lana escocesa, en el cuerpo de forma baño de sol. El bolerito suelto.*

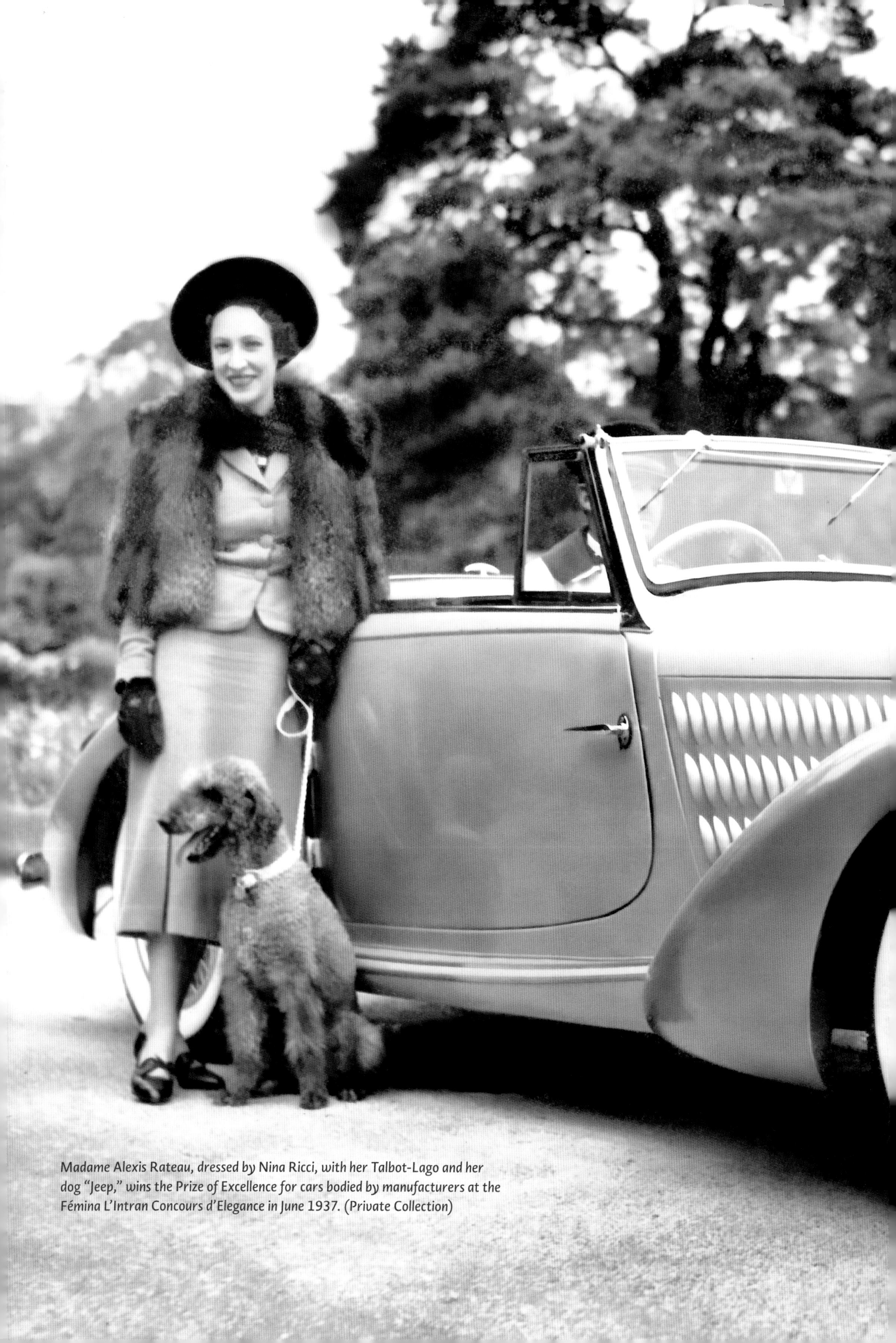

*Madame Alexis Rateau, dressed by Nina Ricci, with her Talbot-Lago and her dog "Jeep," wins the Prize of Excellence for cars bodied by manufacturers at the Fémina L'Intran Concours d'Elegance in June 1937. (Private Collection)*

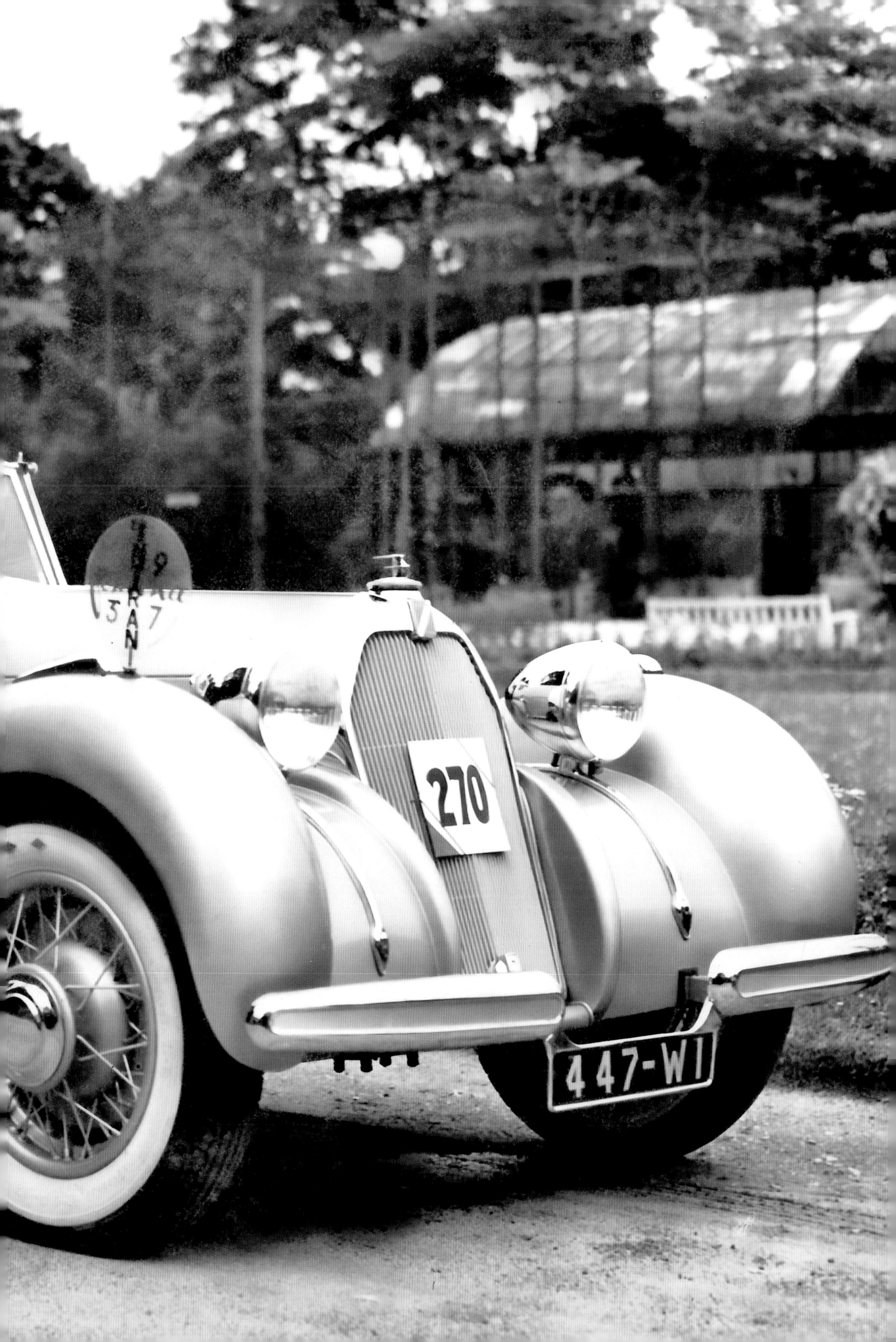
270
447-W1

# Concours d'Elegance à la française

## THE ROARING TWENTIES

## The Return of Peace Brings Women as Participants

The Great War fundamentally transformed life across many nations, and in its course, it affected the roles of women. With men away at the front, women took on traditionally "male" responsibilities, which gave them newfound independence and an expanded role in society. Driving became not only practical - essential for certain jobs - but also fashionable, stylish, and even sporty among certain circles.

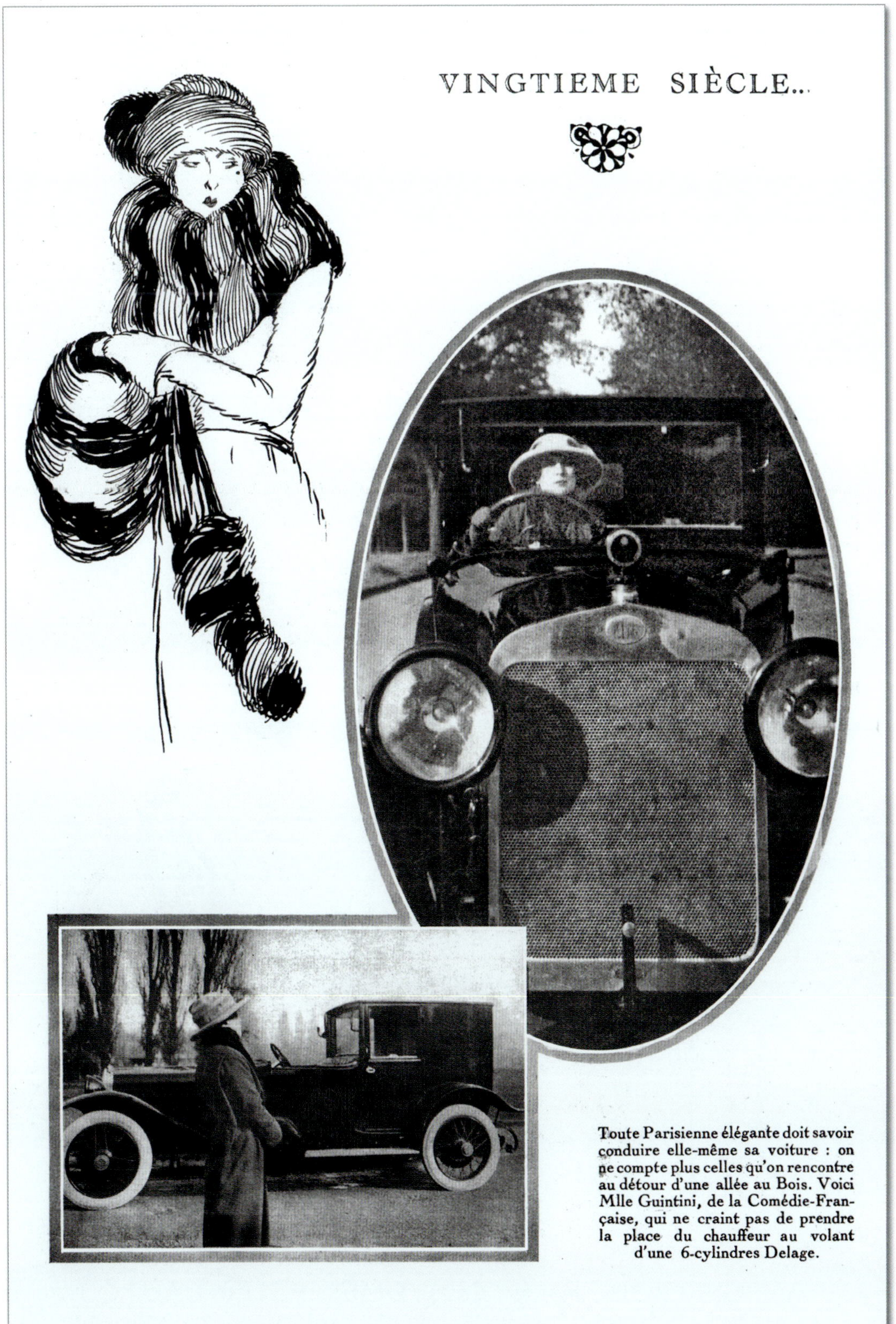

VINGTIEME SIÈCLE...

Toute Parisienne élégante doit savoir conduire elle-même sa voiture : on ne compte plus celles qu'on rencontre au détour d'une allée au Bois. Voici Mlle Guintini, de la Comédie-Française, qui ne craint pas de prendre la place du chauffeur au volant d'une 6-cylindres Delage.

*"Every elegant Parisian woman must know how to drive her car" – when a magazine like the French Fémina declares so, there is no doubt that the trend has been firmly established. (Daniel Cabart Archives)*

After a hiatus of several months of automotive competitions following the armistice of November 1918, races began to return gradually with hill climbs and speed events. However, there was hardly any mention of "concours d'elegance," as they had been conceived before 1914. In fact, even well-known coachbuilders like Kelsch could only reference past achievements, such as his victory at the 1914 concours d'elegance in Tours, in advertisements that stretched on for several years.

**RIGHT:** *Even the automobile dictates a new fashion – although it remains uncertain whether the true statement of elegance lies in driving oneself or simply being seen as a passenger.* (Daniel Cabart Archives)

**OPPOSITE TOP LEFT:** *Advertisement for Kelsch's bodywork on a 1914 Delage. While the ad highlights Kelsch's award-winning achievements in 1921, the car featured is still the one presented in Tours in 1914.* (Daniel Cabart Archives)

**OPPOSITE TOP RIGHT:** "*A young and elegant chauffeur" – or rather, chauffeuse. The Bois refers, of course, to the Bois de Boulogne, near Paris, and the avenue leading to it, the ultimate destination for those seeking to drive in style.* (Daniel Cabart Archives)

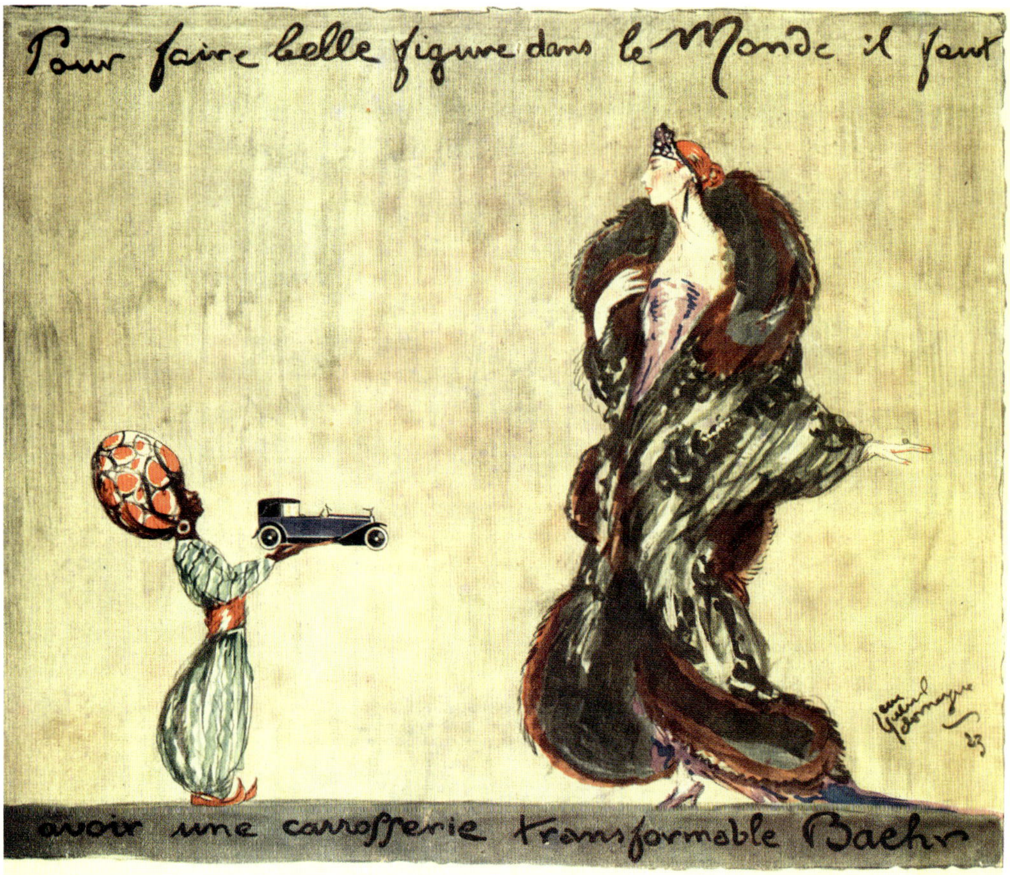


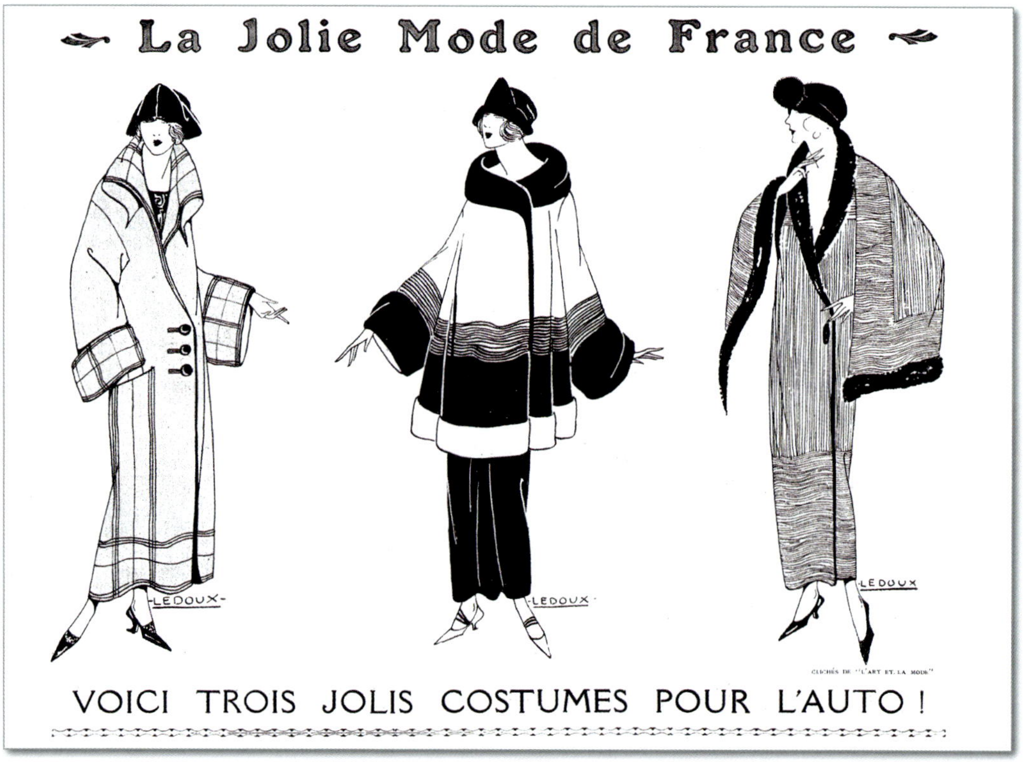

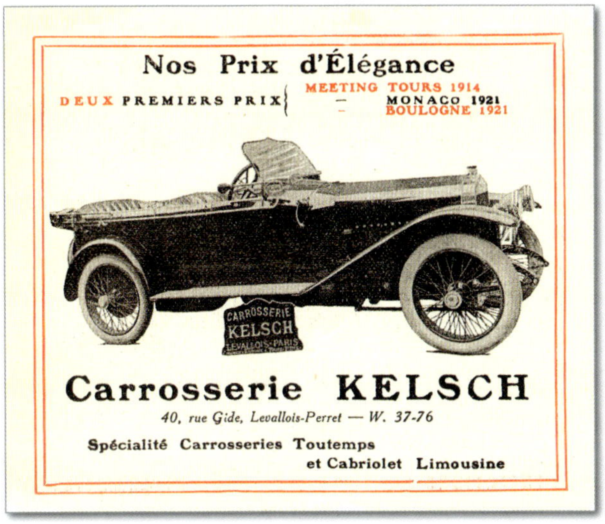


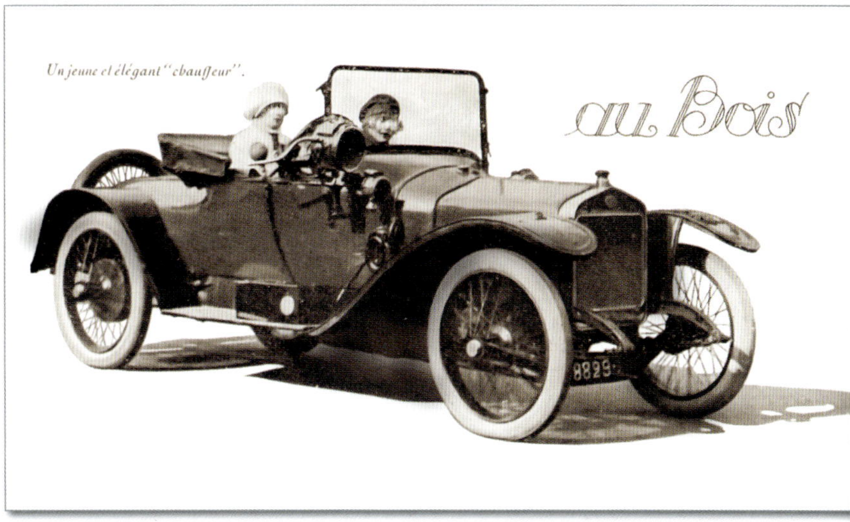


In 1919 and 1920 no notable competitions were organized in France. However, in Ostend, Belgium, a concours d'elegance was organized as early as 1919, continuing the city's tradition of competitions that dated back to the beginning of the century. To be clear, these contests were still primarily about judging car bodywork rather than the full pre-war notion of elegance.

Meanwhile, in France, the daily newspaper *L'Auto* took on the challenge of reviving events, though with a new twist in terminology. In June 1920, they introduced a fresh concept, organizing a *Concours de la bonne tenue* (Contest of Proper Attire), subtly rebranding while maintaining the essence of these competitions.

## "Good Attire Competition" *Organized by L'Auto in June 1920*

*L'Auto* introduced a unique competition in June 1920, emphasizing the presentation of the vehicle and the driver's appearance. The competition was structured into two main events: a morning session for professional drivers and an afternoon session for vehicle owners who drove themselves.

**Event for Professional Drivers** In the first session, vehicles presented by professional drivers were evaluated based on the following criteria:

1. **Condition of Mechanical Parts** – The state of the engine components under the hood.
2. **Condition of Brakes and Body Joints** – Ensuring essential safety features were well-maintained.
3. **Condition of Accessories** – Quality and maintenance of the car's additional components.
4. **Driver's Attire** – Personal presentation and attire of the driver, contributing to the overall impression.

Event for Owner-Drivers In the afternoon, vehicles driven by their owners were assessed on several qualities, highlighting both aesthetic and functional aspects:

1. **Elegance of Design** – The overall visual appeal and style of the car's lines.
2. **Comfort of the Bodywork** – The degree of comfort afforded by the car's design.
3. **Ingenuity of Fittings** – Thoughtful or creative innovations within the vehicle's interior.
4. **Arrangement of Accessories** – Both elegance and practicality were key; accessories needed to blend harmoniously with the bodywork without detracting from its design. Additionally, they were expected to serve their intended functions without obstruction. For instance, a gasoline lantern elegantly embedded in an awning may look appealing but would be problematic if it hindered proper lighting adjustments.
5. **Overall Tastefulness** – The final criterion focused on the overall harmony and sophistication of the vehicle, considering all elements as a cohesive whole.

This competition, with its detailed judging criteria, reflected the post-war interest in both utility and refined style, merging functionality with elegance in a way that celebrated the vehicle and the person behind the wheel.

This all harkens back to the similarly named competition organized by *L'Auto* in 1906, roughly fifteen years prior (see page 31 for details on the motivation behind such events: "Statement of Reasons"). When examining the purpose and structure of that earlier competition, it closely follows traditional patterns as in the report from the 1905 Monte Carlo Concours d'Elegance: "*You all know what a concours d'elegance is. About fifty well-washed cars paraded in a circle. They are judged at full speed, followed by a grand parade with banners...*"

Nearly twenty years after the first concours d'elegance en automobile, little had changed.

One notable exception, however, was on July 29, 1920, in Le Touquet; a "concours d'elegance for cars driven by ladies" was held at 3:30 p.m. in Place de l'Hermitage. Although women were now in the

## Monaco Concours d'Elegance, 1921: "A Showcase of French Coachbuilding Perfection"

(*L'Éclaireur du Dimanche Illustré*, March 20, 1921)

*"On Tuesday in Monte Carlo, it was the cars' turn to embody elegance! Meticulously polished and gleaming, with sparkling copper accents, these vehicles paraded in front of the jury seated in a grandstand on the Place du Casino. Following the morning's thorough examination, awards were presented at 2:30 in the afternoon. The level of elegance and comfort on display was so high that almost all the entries deserved first place – a testament to the impressive standard of the cars showcased.*

*"One standout was Mr. Émile Dubonnet's 1921 Hispano-Suiza Torpédo, a 32 HP, 6-cylinder model with Mühlbacher bodywork. This car, along with an identical model owned by Mr. André Dubonnet, secured two of the Grands Prix, taking first and fourth places – a remarkable achievement for the renowned Hispano-Suiza brand.*

*"Also celebrated was Mr. Delâge's latest tourer model, featuring a 6-cylinder, 20 HP engine and an elegant body crafted by Letourneur et Marchand in champagne and olive green. This car, which had been*

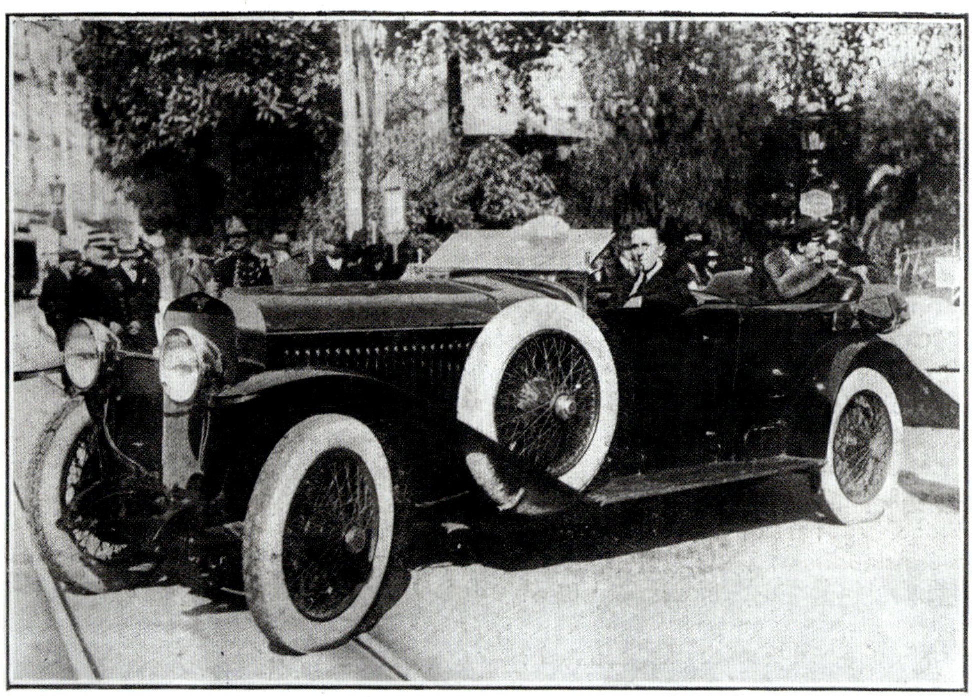

M. Emile Dubonnet, *sur Hispano-Suiza*

*Émile Dubonnet, grandson of the founder of the famous aperitif, was a pioneer in aviation. In 1921, he participated in the Monaco Concours d'Elegance with his Hispano-Suiza 32 CV Torpédo, featuring a body by Million-Guiet. His outstanding entry earned him the Grand Prix d'Honneur. (Daniel Cabart Archives)*

driver's seat, this remained largely a bodywork competition, with cars evaluated on comfort, elegance, and handling – qualities that, while refined, still emphasized the aesthetics of the vehicle rather than any new criteria introduced by the shift in driver demographics.

## 1921

Similarly, competitions were organized in Boulogne-sur-Mer as part of the Boulogne Meeting, which featured hill climbs, a speed grand prix, and a concours d'elegance event. Delage took first place – with a car "driven by a lady" – though this remained a "competition of bodywork elegance and comfort," drawing a large field of around one hundred participants.

The same format appeared in La Baule, and again in Monaco with concours d'elegance that focused on the aesthetic and comfort aspects of the vehicles, even as the presence of women drivers became more common.

*displayed at the recent London Motor Show, transported Mr. and Madame Audrain along with Madame Jacquinot and was skillfully driven by Mr. Bidal, the owner of Garage Delage. Delage made a striking impression at this concours, winning three out of six grand prizes – a feat that underscored the prestige of this French brand."*

Despite this clear celebration of exquisite chassis and coachbuilding, the spotlight remained on the craftsmanship and brand prestige rather than on "the woman" – a notable absence, even as female drivers became increasingly visible in society.

**BELOW:** *For the revival of the concours d'elegance in Monaco, the first since the end of the war, it was Letourneur et Marchand who crafted this magnificent tourer, mounted on a Delage chassis "in champagne and olive-green colors." The car won the Grand Prix d'Honneur at the Monaco Concours d'Elegance competition on March 15, 1921, awarded to its owner, Mr. Bidal.* (SBM Monaco)

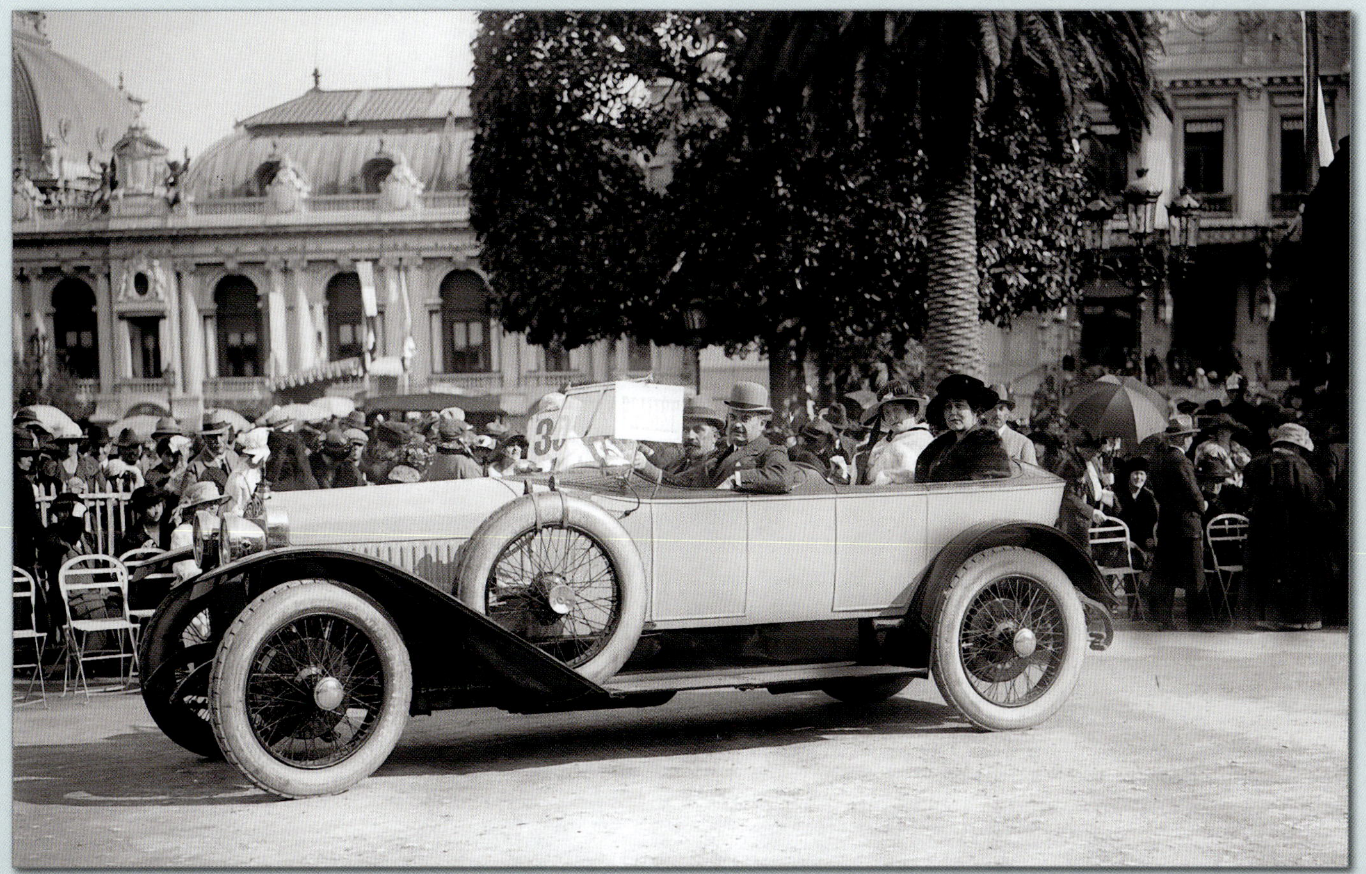

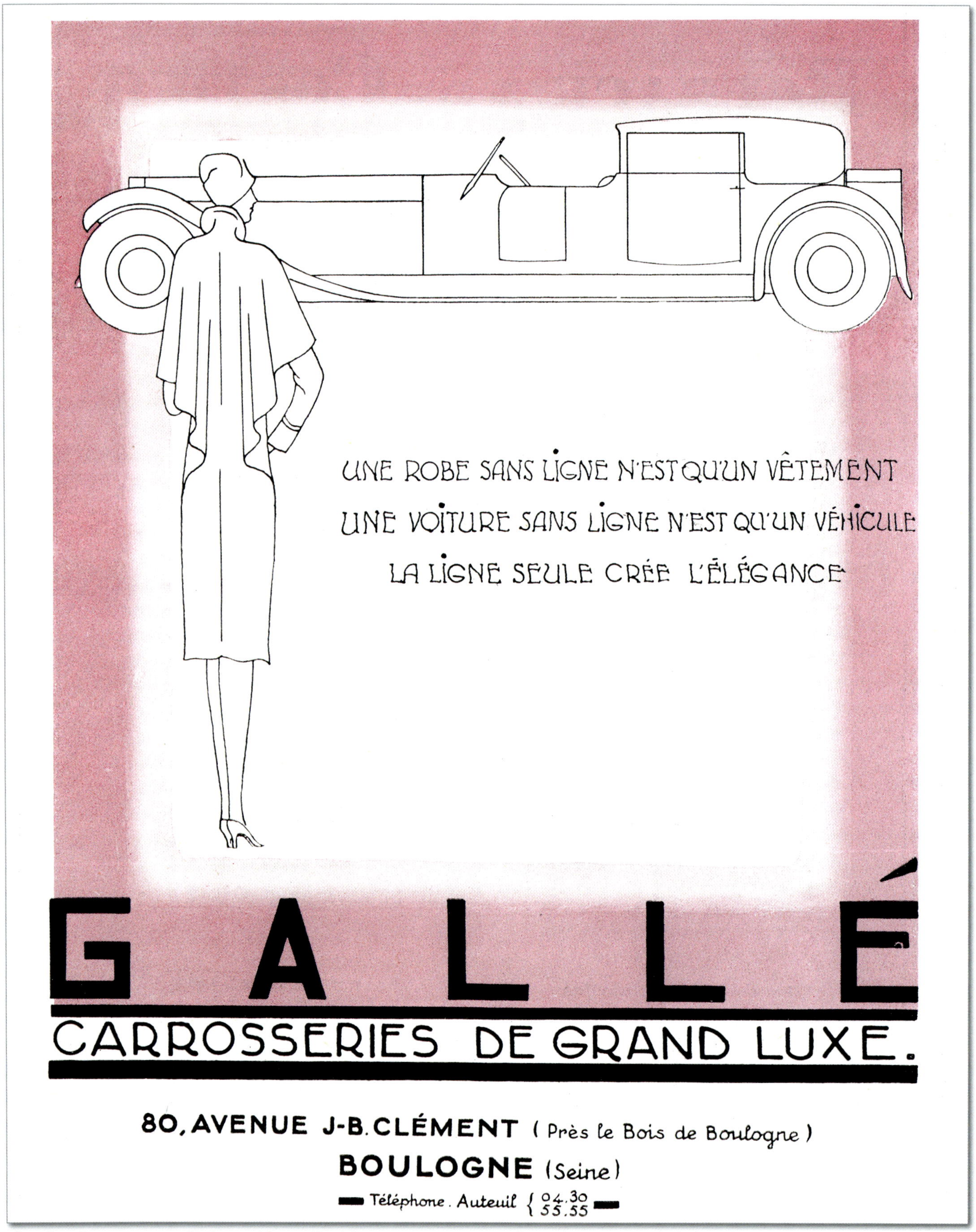


*The advertisement states: "A dress without a line is just a garment; a car without a line is just a vehicle; line alone creates elegance." Gone were the days when the word "elegance" referred to the provision and appearance of accessories. In the 1920s, the line of the bodywork, influenced by Art Deco, and the sleek new feminine fashions became the perfect embodiment of elegance. (Daniel Cabart Archives)*

## 1922: The First *L'Auto* Elegance Competition, Paris

> *"The first L'Auto Concours d'Elegance, held on June 11, 1922, in Paris, proved an extraordinary success. Organized by the sports daily L'Auto, the event attracted nearly 150 entries, with vehicles parading before the jury in a dazzling display of innovation and design. The overwhelming response from car owners, coachbuilders, and auto enthusiasts set the stage for this event to become an annual fixture.*
>
> *The event was meticulously organized, with a strong security presence ensuring smooth operations. By noon, the jury had completed their assessments, though sorting through the final scoring sheets was a time-consuming process left for later. Attendees enjoyed an impressive variety of vehicles, from sleek sidecars and nimble cyclecars to robust taxis. They showcased the ingenuity in bodywork details and the renowned elegance of French design.*
>
> *This competition, open to all from professional coachbuilders and garages to amateur car enthusiasts, offered nineteen distinct categories, each celebrating the craftsmanship of vehicle bodywork above all else. As with other concours, this was truly a competition of style and creativity, a vibrant tribute to the artistry of the French automobile industry."*
>
> (L'Auto, June 11, 1922)

## 1923

In 1923, *L'Auto* held its second elegance competition, this time with 200 cars entered – an impressive increase that highlighted the growing appeal of the event. Elegance was everywhere: in the refined lines of the bodies and the innovative details in wheel covers, bumpers, and other accessories. The entire automotive industry seized the opportunity to showcase its creations, turning the event into a vibrant display of style and ingenuity, and a prime platform for brand promotion.

### First Prize in "Closed Tourism" – A Triumph for Jean Gaborit

In the "Closed Tourism" category of the *L'Auto* elegance competition, the first prize went to a stunning creation by Jean Gaborit. This accomplishment was highlighted in *L'Auto*, June 11, 1922:

> "A MAGNIFICENT DEBUT
>
> *At just 22, Jean Gaborit, son of the deputy of Seine-et-Marne, emerged as the day's true standout. His Sizaire-Berwick, which he designed and bodied, impressed the jury and spectators alike. The car's luxury, comfort, and refined taste set a new standard, surpassing anything previously seen in this category. Although the jury's decision ultimately awarded him first prize, the enthusiastic public had already given their approval, captivated by this exceptional example of French automotive artistry."*

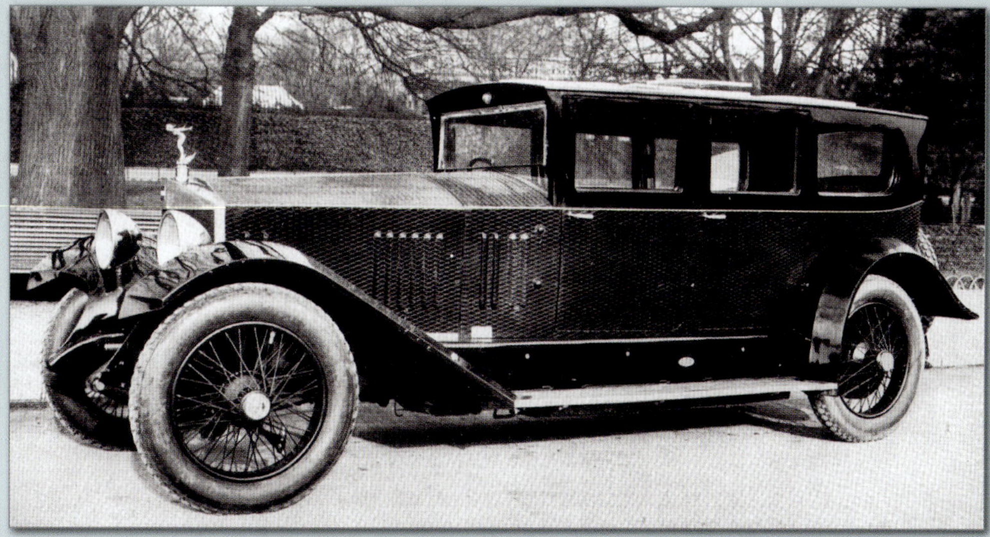

*This Sizaire-Berwick, featuring limousine bodywork by the coachbuilder Jean Gaborit, was entered in the L'Auto Concours d'Elegance in June 1922 by Félix Gaborit, Member of Parliament for Seine-et-Marne region in France, as well as the coachbuilder's father. By this time, the trend was shifting towards clean lines, directly influenced by the emerging Art Deco style. (Daniel Cabart Archives)*

*Madame Fraisse presented this cabriolet by Franay on a Sizaire-Berwick chassis at the L'Auto concours on June 5, 1925. From L'Auto, June 6, 1925:*

> "This Concours d'Elegance, organized at the Parc des Princes, was a highly successful social and artistic celebration where coachbuilders showcased marvels of elegance and ingenuity... Pretty lines.../... A number of superiorly treated sports bodies..."

*Yet, despite the event's focus on elegance, no attention was given to the presence of the poised Madame Fraisse, nor to her superb outfit, in the context of the concours d'elegance. (BnF Gallica)*

## 1924: A Step towards Female Representation

It was in 1924 that the magazine *Fémina* is believed to have organized the first women's automobile concours d'elegance, in collaboration with the daily newspaper *L'Intran*. However, historical records from that time do not provide any clear evidence of the event.

## 1925: Women at the Wheel

A significant breakthrough for female drivers came in 1925, as reported by *L'Intran* on May 13, 1925, with the announcement of an event titled "Madame at the Wheel: A Concours d'Elegance en Automobile."

Male drivers had dominated automobile competitions prior to this, leaving a noticeable void for women who were increasingly taking to the roads. This competition aimed to fill that gap, celebrating the skill and presence of female drivers. Scheduled for June 26 at 11:00 a.m. in the Bois de Boulogne, Allée des Acacias, the event invited entries in four categories:

1. Voiturettes of 5 and 6 HP (open, closed, and convertibles), driven by ladies.
2. Voiturettes of 10 HP (open, closed, or with transformable bodywork), driven by ladies.
3. Cars of less than 10 HP (open, closed, or with transformable bodywork), driven by ladies.
4. Cars of more than 10 HP, driven by a mechanic but presented by a lady.

A jury would evaluate the vehicles in each category, focusing on the elegance of the cars and the poise of the drivers. The prizes would consist of floral banners, bouquets, and diplomas awarded to those who stood out. The event emphasized that elegance and automobiles were not mutually exclusive; in fact, the elegance of the car complemented the elegance of its occupants, promising to be one of the most original events of the Parisian season.

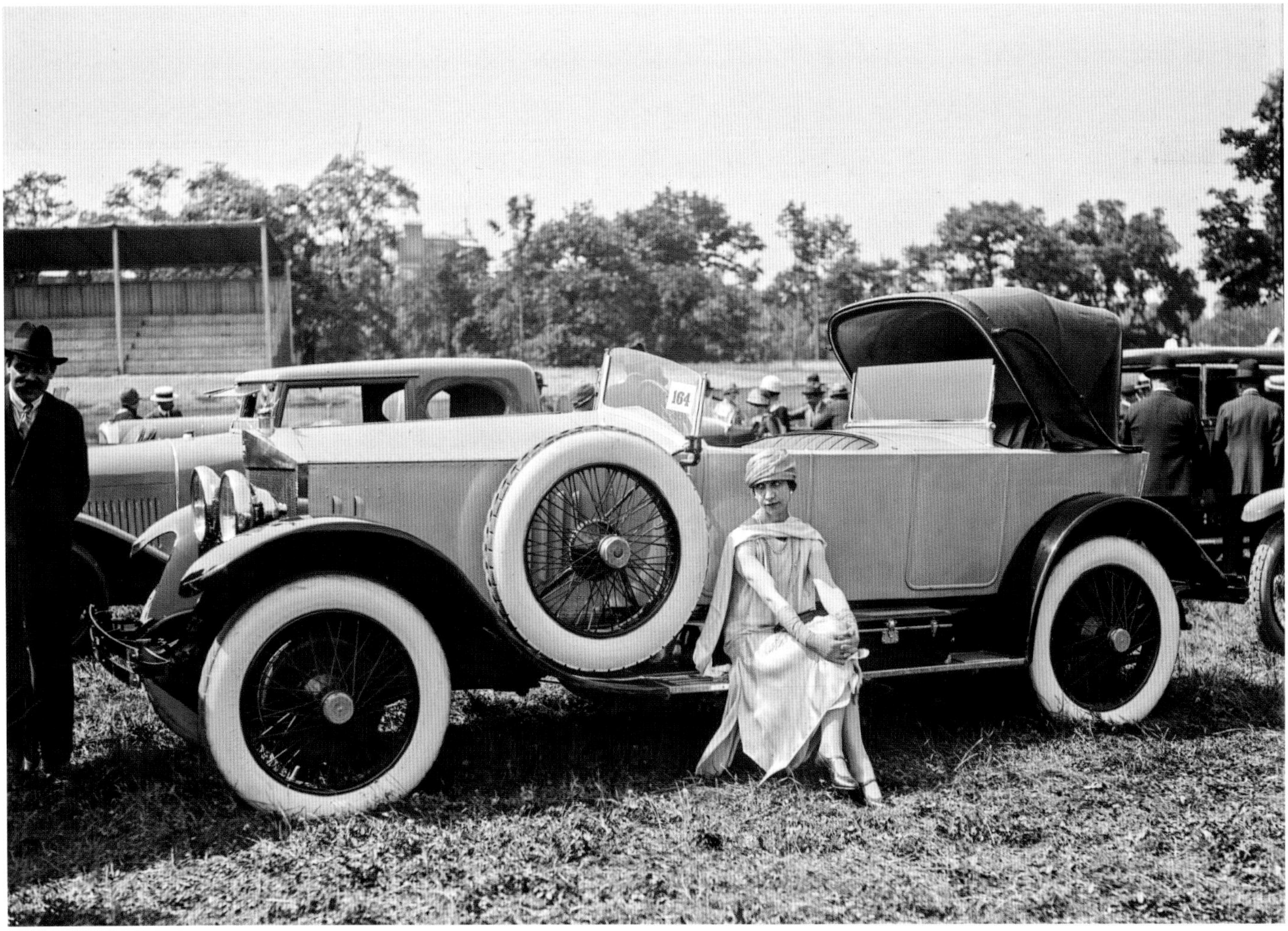

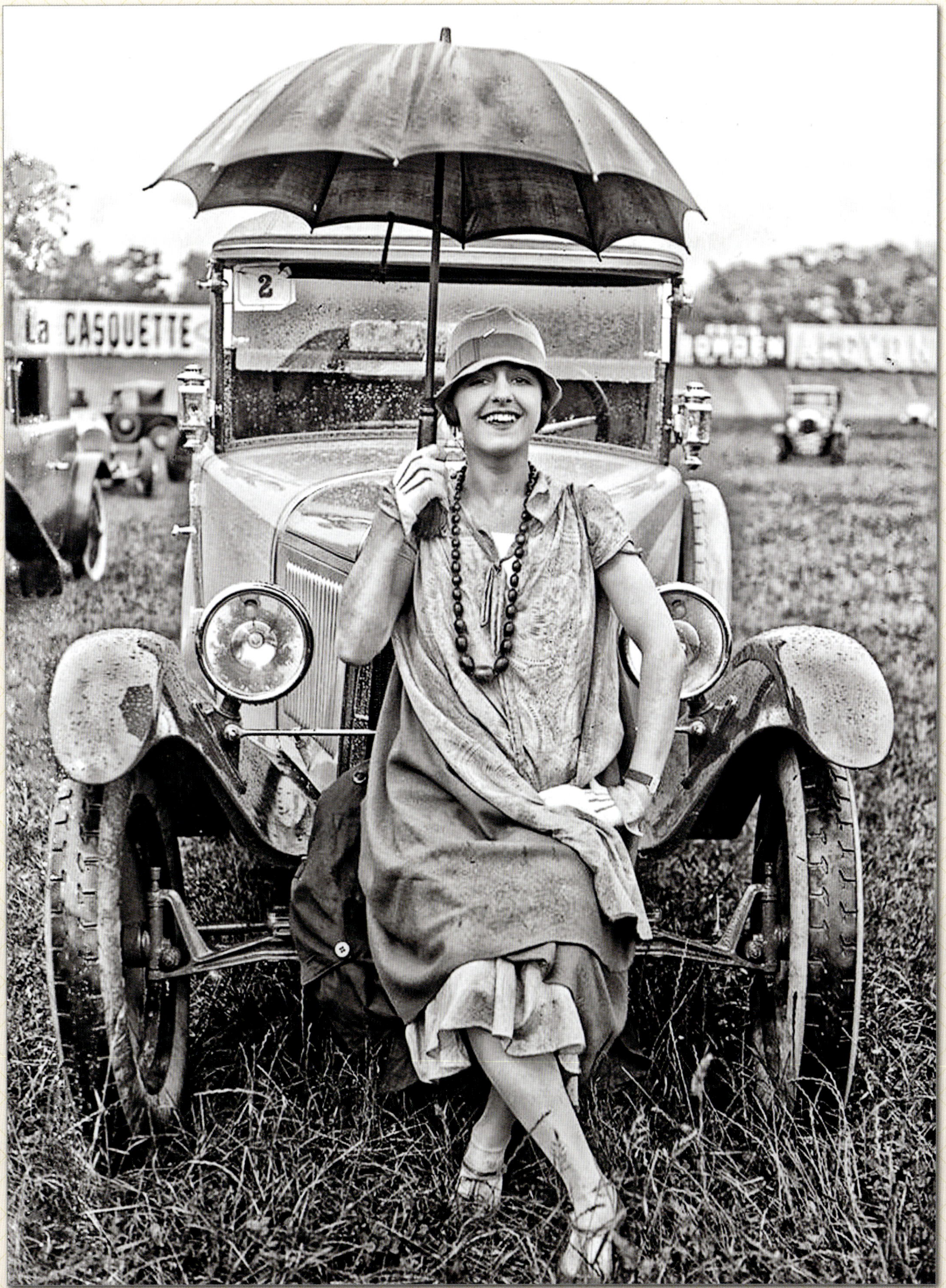


"When the jury passes, a woman poses on the cushions of the car, subtly revealing a hint of silk stockings to influence him." *Concours de L'Auto on June 5, 1925.* (BnF *Gallica*)

*The parade of automobiles during the Boulogne-sur-Mer Concours on August 28, 1925. The Delage 30 HP Type GL with Labourdette "skiff" bodywork is in front. (BnF Gallica)*

## Boulogne-sur-Mer: A Concours d'Elegance

Later that same year, Boulogne-sur-Mer hosted a concours d'elegance en automobile, marking the conclusion of the grand event known as "Boulogne Week." While the automobiles boasted undeniably elegant designs, the presence of women was sparse, making them more of an exception than the norm. Like many other competitions throughout France, the focus remained heavily on the vehicles themselves, with a strong participation from coachbuilders from Northern France, who utilized the event to enhance their visibility and publicity.

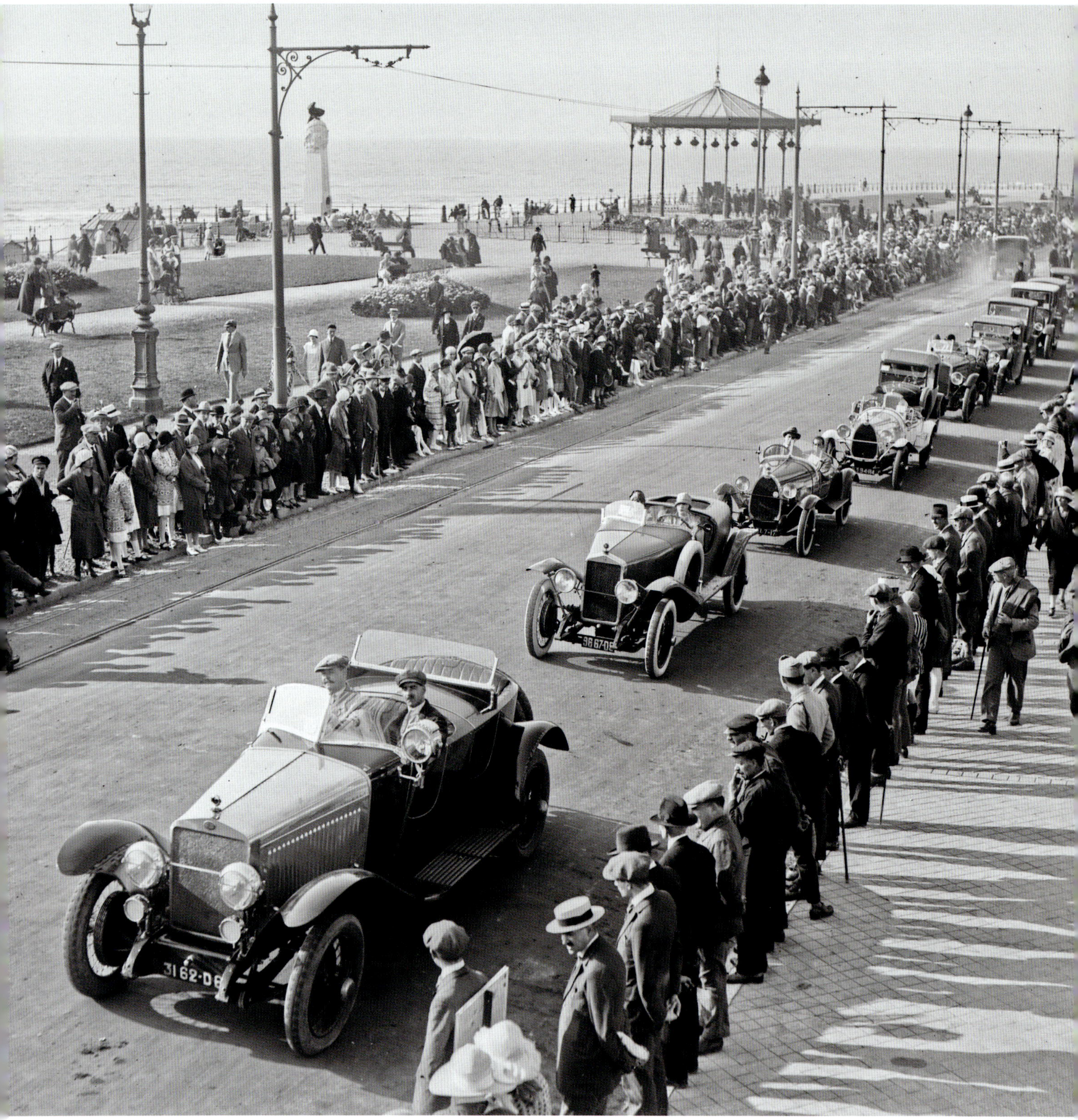

## 1926

That year, Biarritz hosted a "memorable" event which, although not identified as such in the official title, can be regarded – much like the Dinard competition of 1921 – as the true birth of a genuine Concours d'Elegance Féminine en Automobile.

The day had started in a rather "routine" way... as reported by *L'Auto*:

> *"The concours d'elegance d'automobiles organized by the Gazette de Biarritz in the prestigious setting of the Pavillon Royal took place on September 9, on a magnificent day, with complete success.*
>
> *The competitors presented a batch of superb cars that made the jury's decision difficult. This jury was installed in front of the large south door of the Pavillon Royal and it was on the entrance terrace that the Competition took place. The parade began at 4:45 p.m. and ended at 6 p.m. A large and elegant audience had settled along the track. They also showed, by their repeated applause, their admiration for the cars presented..."*

*The 1926 Biarritz competition, organized by the sophisticated fashion magazine* et La Gazette de Biarritz, *now celebrated feminine elegance just as much as the beauty of the bodywork, blending automotive and fashion refinement in a unique way.* (BnF Gallica)

Four categories of vehicles, from the cyclecar to the "big car", whose winners were rewarded with prizes ranging from the cigarette case offered by the house of Hermès, to a hood cap and a sports suit offered by the house of *Vogue*. Nothing out of the ordinary in the end...

However, a fifth category appeared following what was, as usual, a bodywork competition.

This was a "Special Prize offered for the elegance of the women occupying the car"... A first! A hat, provided by the House of Lewis, was won by Madame de Clairac in a Cadillac, and dolls were presented by Mr. Hieulle to Mlle. Boulard in a Voisin.

> *"The concours ended with a general parade of the competing cars that took place on the terrace, amidst the enthusiasm of the many assistants.*
>
> *"But it was in the evening that Fémina would once again show itself to be at the forefront of this movement and present an event that it had already organized the two previous years in Paris: the Women's Elegance Competition in automobiles...*
>
> *"After a succulent dinner, which, in ideal weather, brought together a large and brilliant audience on the terrace, the Women's Elegance Competition took place, organized by our great colleague Fémina.*
>
> *"It was at midnight that the parade of cars began, on which powerful projectors poured torrents of light. The competitors passed in their cars twice in front of the jury. The outfits were greatly admired, and the cars presented were also found to be very pretty; as for these, we noted, in passing, the car of Mr. Besançon de Wagner[1]. This time again, the jury was in great difficulty in awarding the prizes among so many graceful and richly dressed women. Finally, the prizes were awarded to Madame Bickel. Madame Girodon, already awarded at the Fémina competition, in the Bois de Boulogne in Paris, was ineligible for this competition."*
>
> (*Fémina*, September 26, 1926)

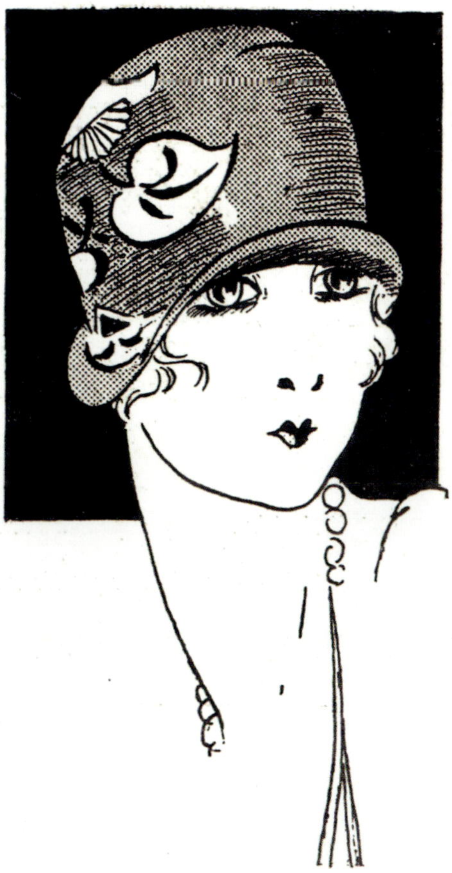

*Feutre découpé, création de LEWIS, le grand artiste de la Mode dont les Salons de la Place de la Liberté sont le rendez-vous de toutes les élégances.*

[1] *Mr. Besançon de Wagner is the husband of the most famous Paris "couturière" Maggy Rouff.*

SUR LA COTE BASQUE

# La Journée de l'Automobile au Pavillon Royal

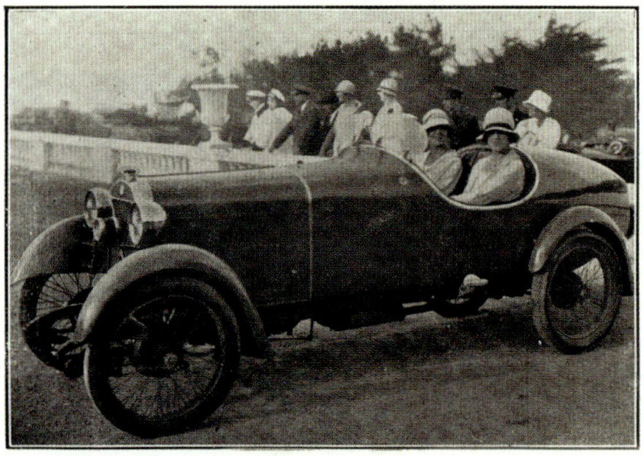

*Voitures ayant participé au Concours d'élégance automobile organisé par la " Gazette de Biarritz" au Pavillon Royal.*

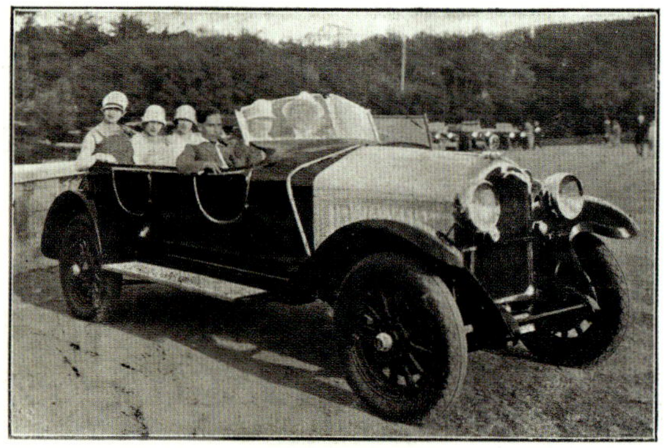

PREMIER PRIX

*Mlle de la Croix, sur Voiture Salmson.*

DEUXIÈME PRIX

*Mme Dargaignaratz sur Voiture Buick*

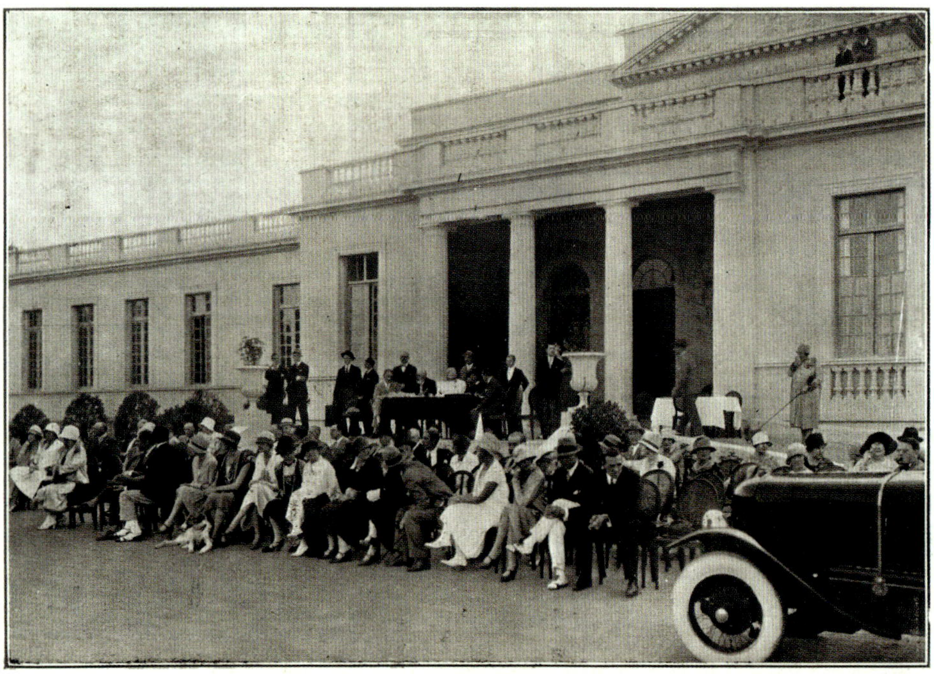

LA TABLE DU JURY ET LE PUBLIC AU CONCOURS D'ÉLÉGANCE AUTOMOBILE

*Devant l'entrée du Pavillon Royal, le Jury est installé sur le perron. Autour de la marquise de Gouy d'Arcy et de la comtesse de Rivière, nous reconnaissons : Claude Farrère, P. Ribera, M. Maurice Labrouche, le comte de Rivière, M. Metschersky, etc. Sur la terrasse, le public s'étend le long de la façade et voit défiler devant lui, les plus belles autos du monde actuellement réunies sur la Côte Basque.*

*The Judging Commitee at the entrance of Pavillon Royal. On the terrace, the onlookers watching the parade of the most beautiful cars in the world now all gathered at Biarritz for the fashionable September season.*

**THIS PAGE:** In 1926, Biarritz, fashion accessories designers began to take an interest in concours d'elegance competitions, influencing the results and adding a new dimension to the event by integrating fashion into the automotive world. (BnF Gallica)

**OPPOSITE:** What appears to be a first: cars parading at nightfall on the terrace of the Pavillon Royal. (BnF Gallica)

*Le Concours d'Elégance Féminine Automobile organisé à Minuit par " Femina "*

Studio
G. L. Manuel Frères
de Paris

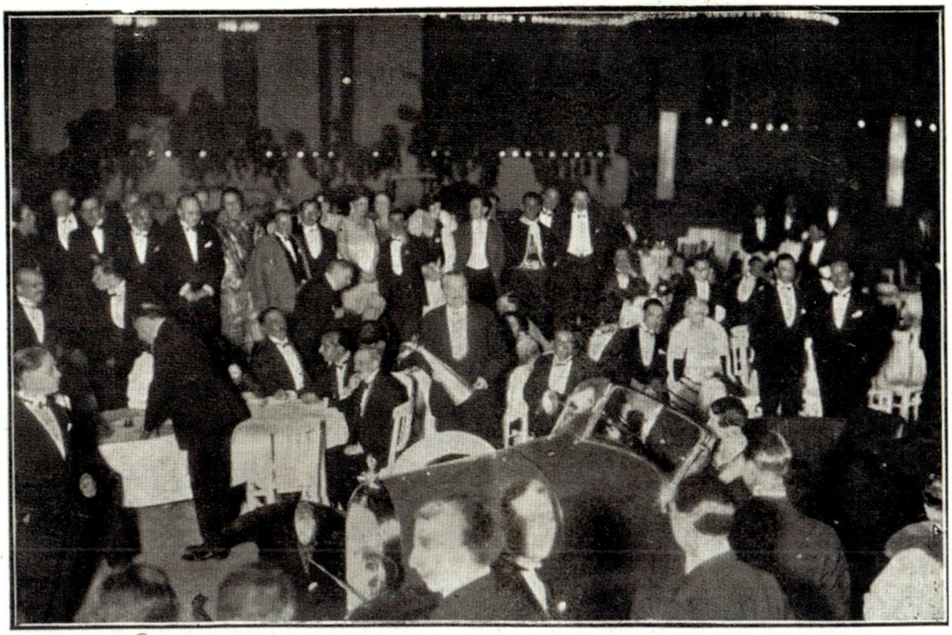

*Les Concurrents défilent devant le Jury sur la grande terrasse du Pavillon Royal, sous le feu des projecteurs.*

*The parade. The light from the projectors is directed on the various cars.*

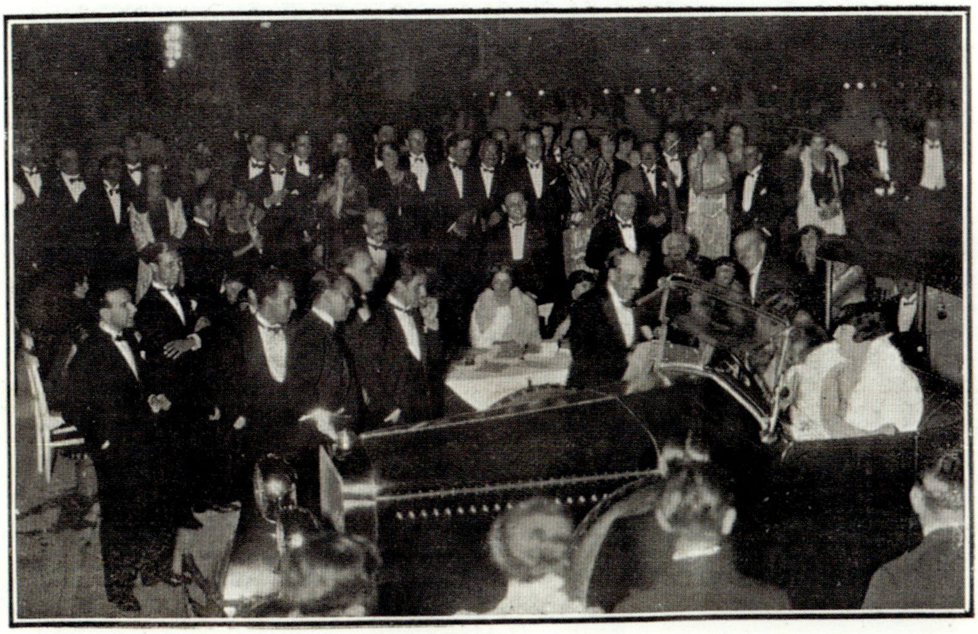

*Le Concours d'Elégance Féminine Automobile organisé à Minuit par " Femina "*

Studio
G. L. Manuel Frères
de Paris

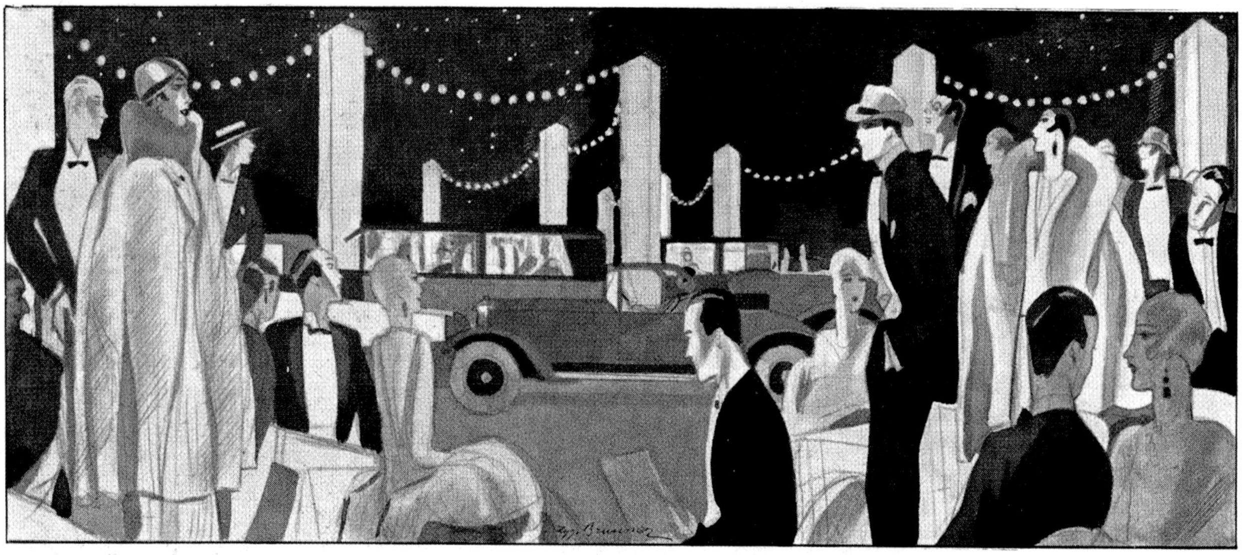

LE CONCOURS D'ÉLÉGANCE AUTOMOBILE AU PAVILLON ROYAL

*Il eut lieu à minuit, dans un cadre idéalement beau. La réunion de luxueuses voitures et de femmes admirablement vêtues, évoluant au milieu des girandoles et des pilastres lumineux, fut une vision de rare élégance.*

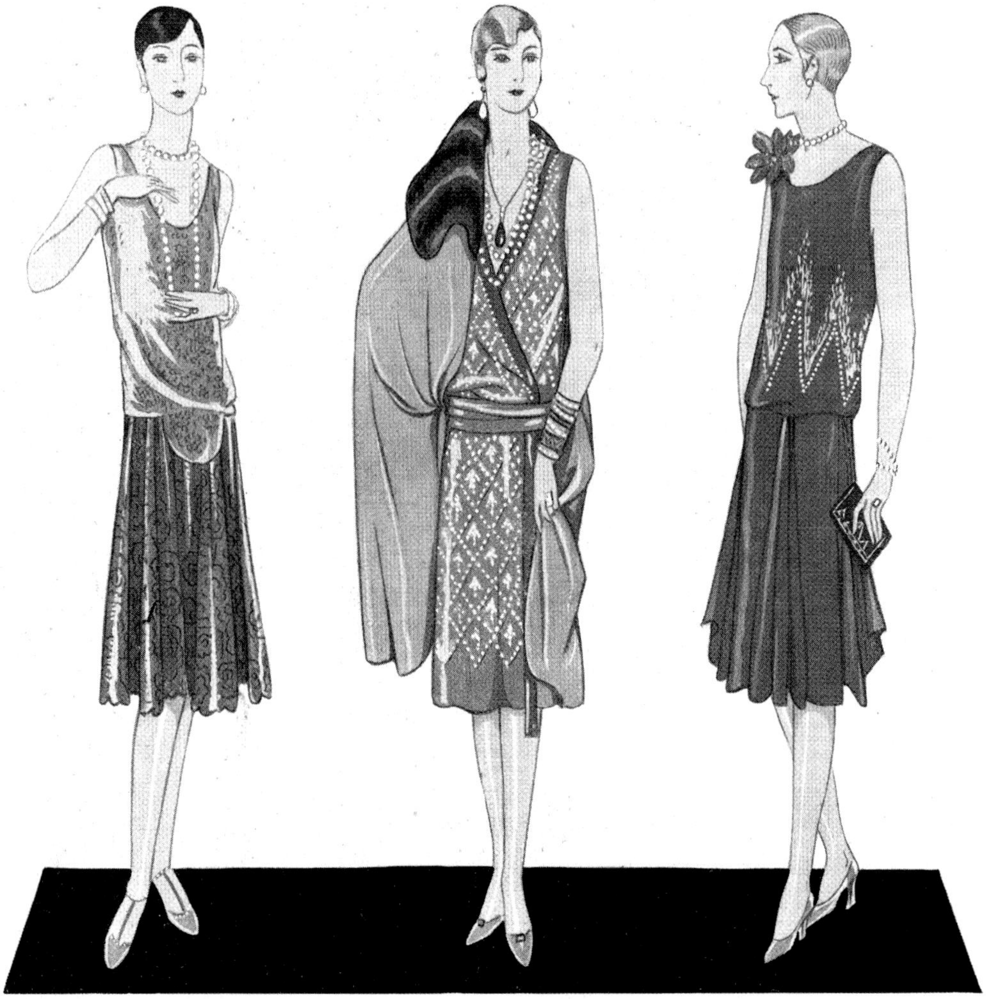

DRECOLL

*Au concours d'élégance automobile, les toilettes furent dignes des somptueuses carrosseries. Mme Maykas eut un prix : elle portait une robe dont le haut était brodé d'argent et le bas en dentelle de métal. Au bras, un bracelet formé d'un ruban de moire coupé de motifs de diamants.*

*Mme Besanzon de Wagner remporta le Grand Prix dans son merveilleux cabriolet jaune et marron. Elle était très élégante dans sa robe vert pâle brodée de verles et de strass.*

*Mme Girodon fut hors concours avec sa voiture beige et marron. Elle portait une robe de mousseline de soie mauve brodée de paillettes mauve plus foncé et d'argent. Une fleur de mousseline de soie était posée à l'encolure. Ce fut une belle soirée d'élégance recherchée et de luxe raffiné.*

Elsewhere in France, the selection and judging criteria were generally just the bodywork, including the level of innovation, line, colors, harmony, practicality and price. Even Biarritz, which had recently been one of the pioneers in the role of women and their elegance in the final ranking, did not repeat the Women's Competition.

Meanwhile, in Deauville as elsewhere in France in that same year of 1926, it seems that the concern was to satisfy the greatest number of participants who came to present their "steed." To do this, a "very judicious" way of categorizing the cars was found.

Because as the humorous text in the colored box below says... "*Every competitor, and even more so every female competitor, must have a prize, and this prize must be at least a first prize.*" The categories were multiplied, the cars separated according to their price, then according to the type of bodywork.

The cars were divided into twelve categories ranging from a value of 20,000 F to more than 150,000 F. The winners were divided between a Renault worth 20,000 F and a Hispano-Suiza worth more than 150,000 F, and also included a Delage worth 75,000 F.

And what will be judged...?

> "*The vehicles presented will be judged on the following qualities:* 1° *Elegance of the line;* 2° *Comfort;* 3° *Ingenuity of the fittings established by the bodybuilder (windshield, hood, seat, upholstery, etc.);* 4° *Arrangement and originality of the accessories (ceiling lights, bouquet holders, canteens, etc.).*
>
> *Each of these qualities will be rated according to the following formula:*
>
> *Elegance of the line,* 20. *Comfort,* 15. *Ingenuity of the fittings:* 10. *Arrangement and originality of the accessories:* 5.
>
> *The jury will be chaired by* Mr. André de Fouquières. *In each category, a Grand Prix d'Honneur and a Prix d'Honneur were awarded to the winner...*"

From 1927, Deauville repeated this until the last competition of the interwar period in July 1939 with a few variations. In this way, everyone will be sure to come back with a first prize or a *Prix d'honneur...*

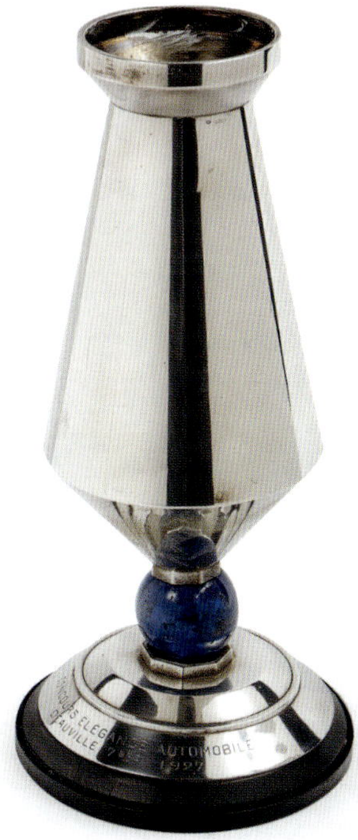

**Among the rewards, while some were modest tokens of participation in the competition, others were genuine works of art created by the greatest designers of the time, such as Puiforcat, Hermès, and others. Here is one of the cups won in Deauville in 1927. (Daniel Cabart Archives)**

## The Elegance of the Concours (*Dimanche-Auto – Le Miroir de la Route*, 1928)

In automobile concours d'elegance, every competitor must have a prize, and fantasy must not be discouraged.

From June to October, automobile concours d'elegance events flourish almost everywhere. There is no seaside resort or beach that does not have at least one.

The general principle of the organization of all concours d'elegance events, whether they take place in Paris, Deauville or in this *Divouarre-le-Troudu* immortalized by La Fouchardière, can be stated as follows:

> Every competitor, and even more so every female competitor, must have a prize, and this prize must be at least a first prize. This is the necessary and sufficient condition for bringing together an appreciable number of competitors.
>
> The application of this principle is, moreover, straightforward. We begin by creating the greatest possible number of categories: if there is only one entrant per category, the problem is immediately solved. But if, in a category there are, for example, five entrants, the following scale of rewards is applied:
>
> - *Grand Prix hors concours*
> - *Grand Prix d'Honneur*
> - *Grand Prix*
> - *Prix d'excellence*
> - First Prize

There are, of course, some disadvantages to this prize inflation. But thanks to a well-established tradition, competitors are usually satisfied with modest banners as trophies. After all, it's "honor" that's truly at stake.

## 1927: A Shift in Judging Criteria

In 1927, the automobile elegance competition in Juan-les-Pins introduced a novel approach: the public became the judge, eliminating the traditional jury. The concept was that the public, as the sovereign judge, would directly award prizes to competitors, underscoring a democratic aspect to the event. This led to a playful and competitive atmosphere, where drivers presented their vehicles, highlighting features such as the excellence of their brakes or the practical design of their mudguards. The elegance of female drivers was particularly noted, with many donning outfits that harmonized with the colors of their cars, creating a striking visual synergy.

### Fémina's Annual Competition in Paris

On June 25, 1927, the magazine *Fémina*, known for promoting female automobile concours d'elegance events since 1921, organized its annual event in Paris. This time, the competition was exclusively for female drivers, marking it as a "purely feminine" event. The setting was the picturesque Bois de Boulogne, where 200 flower-adorned automobiles paraded along the Avenue des Acacias.

The competition featured four color categories: red, green, yellow, and blue. A distinguished jury, including members of the nobility and influential figures, evaluated the cars based on their elegance, originality, and the poise of the female drivers. Each participant was rewarded with bouquets or banners, and all received diplomas the following day, emphasizing the celebratory nature of the event. Notable attendees included famous coachbuilders and directors from major automotive marques, all eager to draw inspiration for future models.

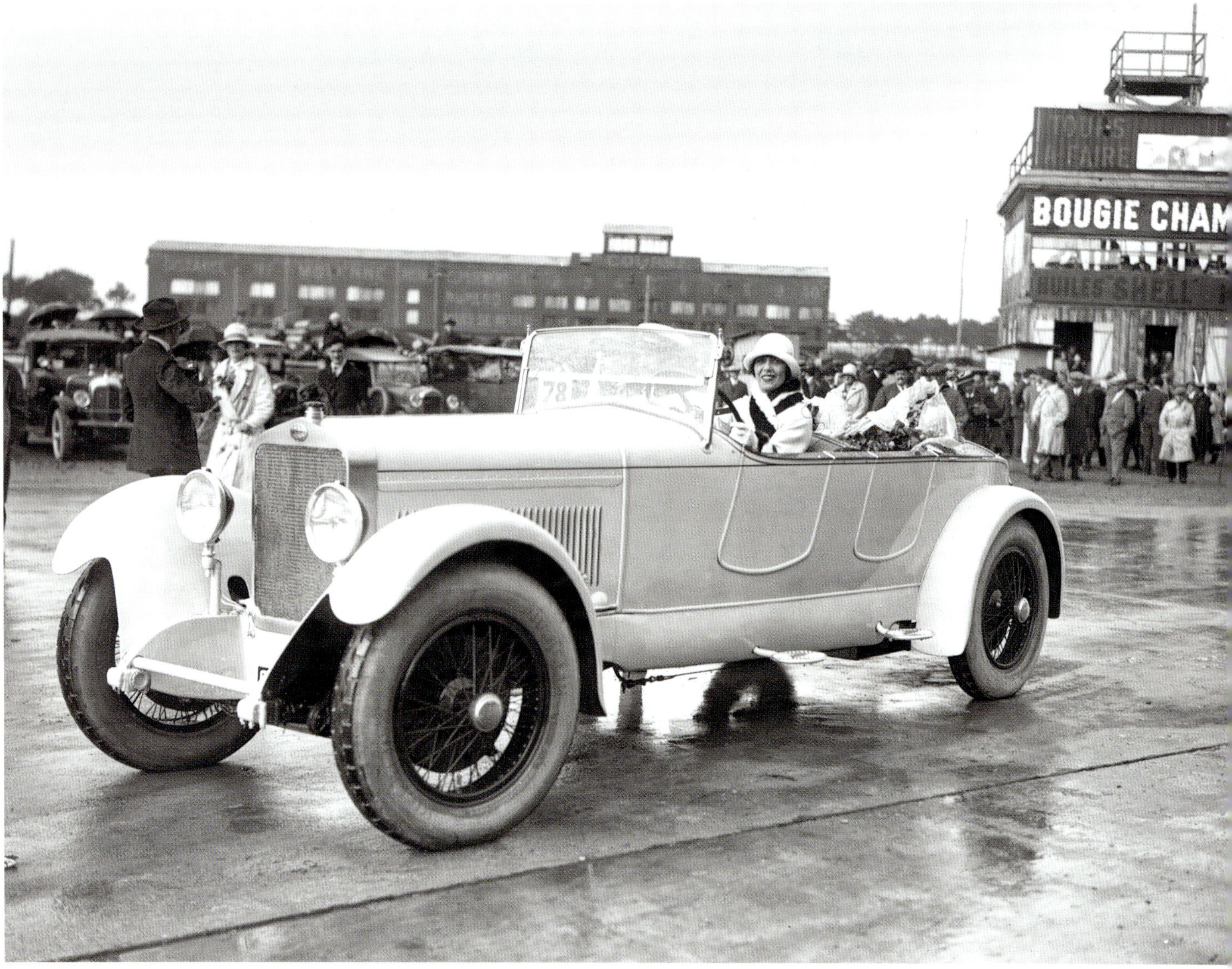


*In 1927, the popular operetta singer Mistinguett traveled across France attending elegance competitions aboard her Delage DM with Torpédo bodywork, designed by Kelsch. She won the Grand Prix d'Honneur at the elegance competition of the Journée Féminine in Montlhéry. (Daniel Cabart Archives)*

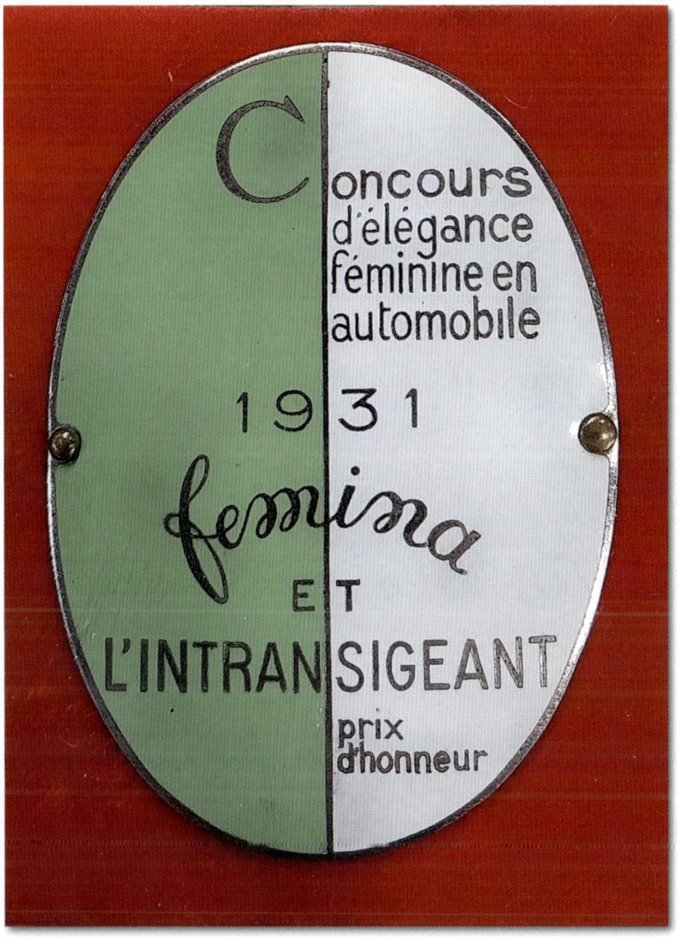


**LEFT:** *Each participant in the concours d'elegance competition managed at least one small prize such as this badge of honor shown from 1931. (Daniel Cabart Archives)*

**BELOW:** *Presented by the Viscountess de Lassus, this Ballot eight-cylinder Type RH, with coachwork by Figoni, took part in the L'Auto Concours d'Elegance competition in June 1927. (Daniel Cabart Archives)*

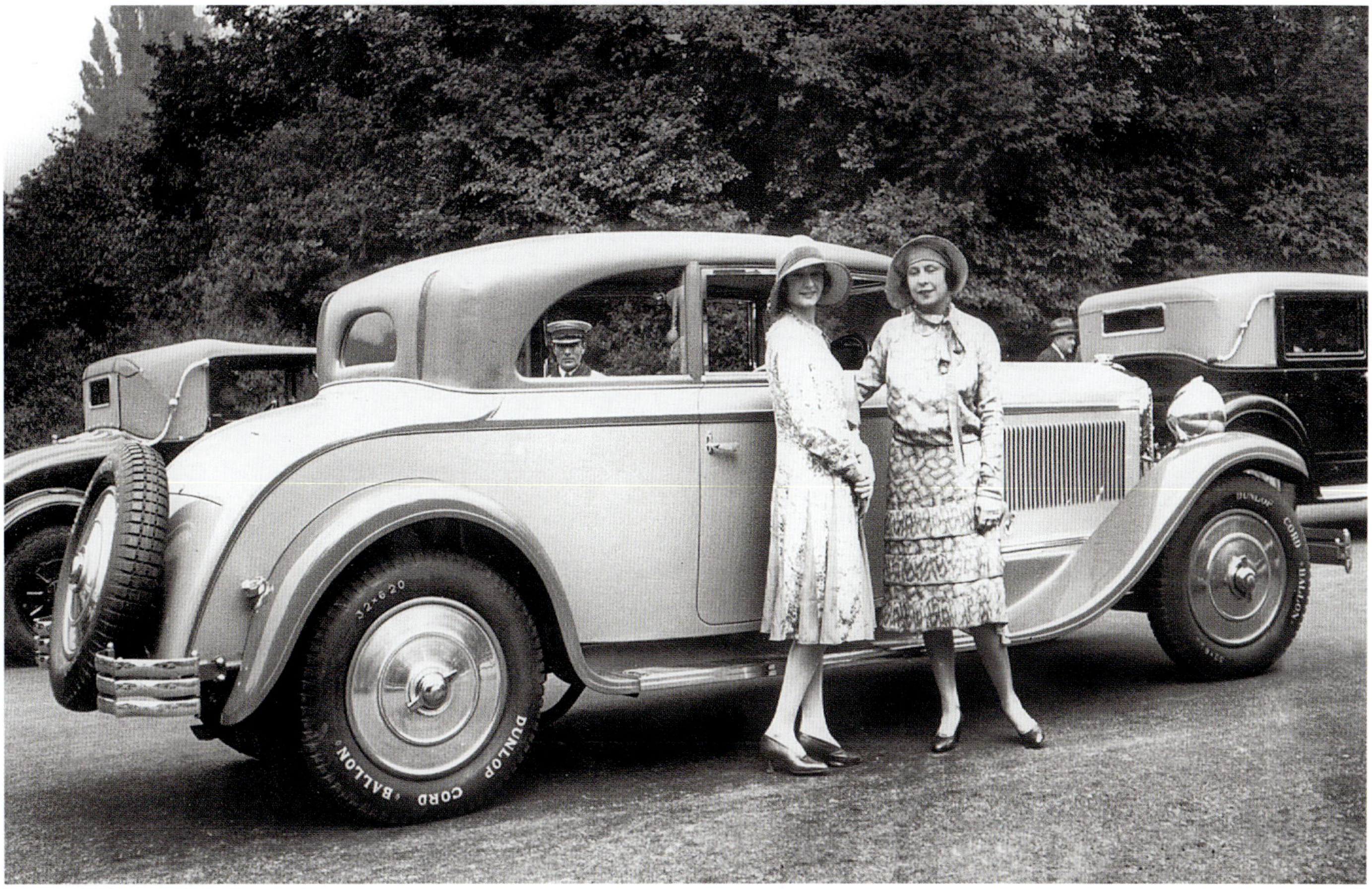

This distinctive Bugatti Type 37 coupé, owned by Count Souza Dantos, featured bodywork by Million-Guiet. It was presented by the Spanish dancer Maria Dalbacian at the L'Auto elegance competition on June 17, 1927, held at the Parc des Princes. (BnF Gallica)

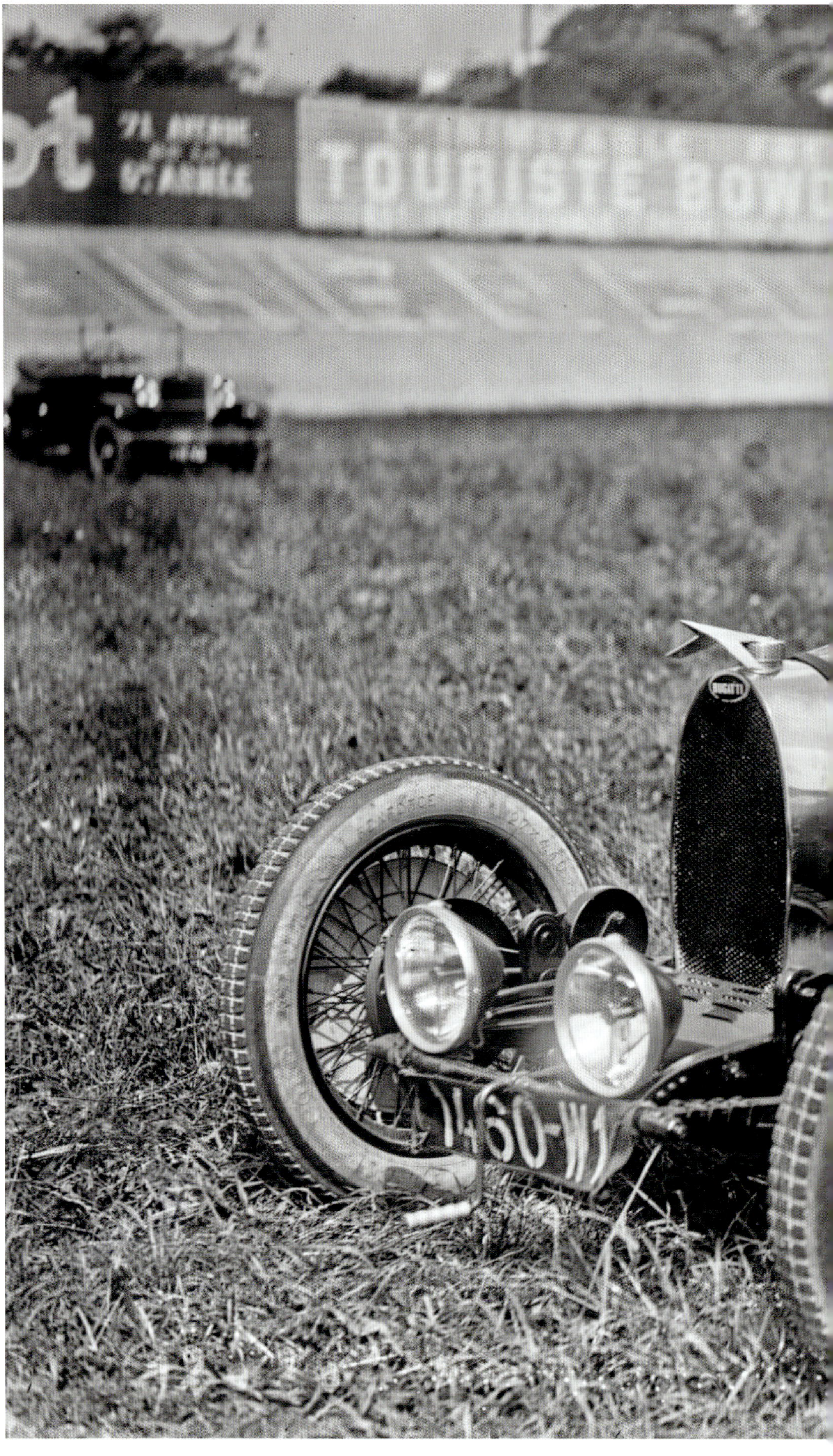

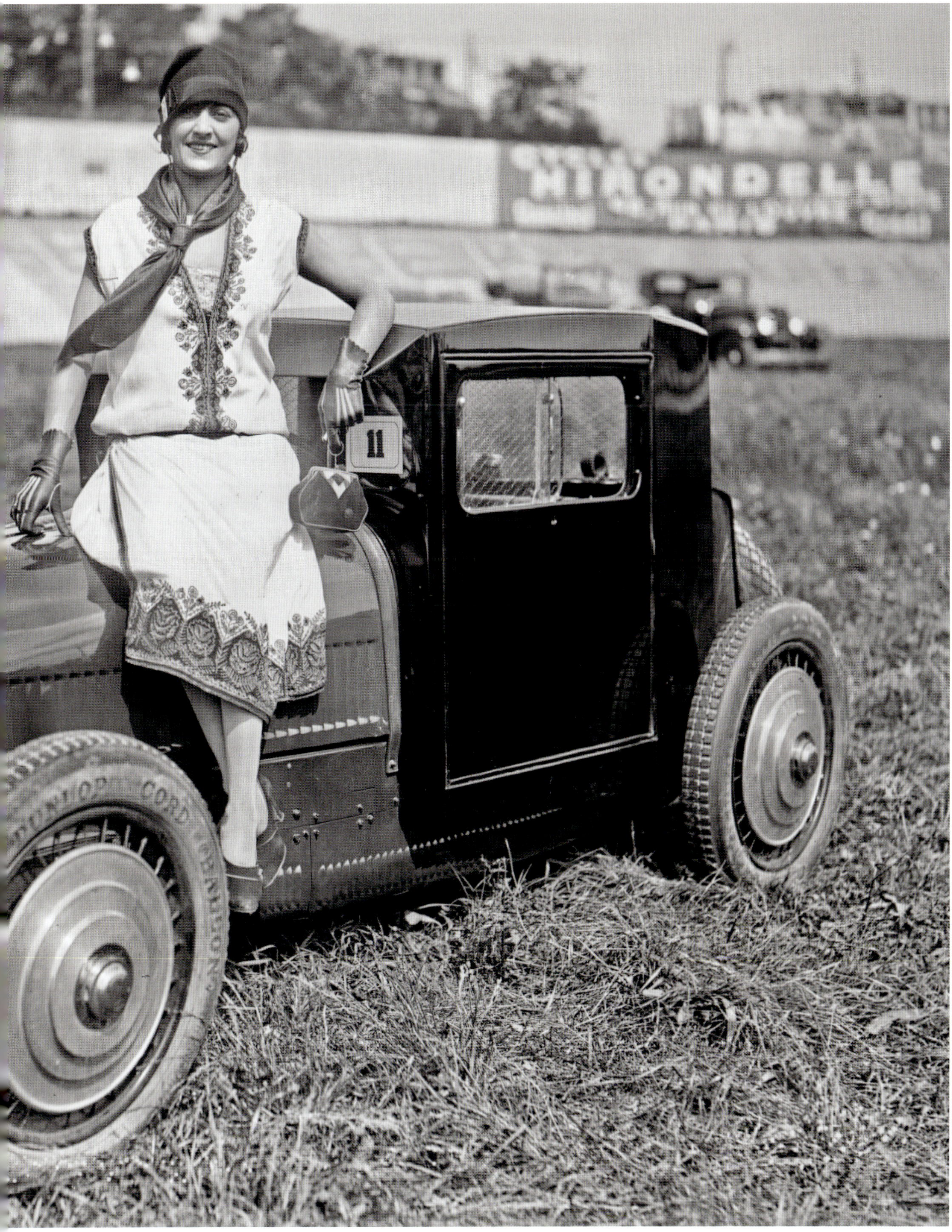
HIRONDELLE
11
DUNLOP CORD

Competitors await the jury's inspection and judging of the cars during the 1927 L'Auto concours at the Parc des Princes, near Paris. In the foreground, a Delage DM landaulet-limousine bodied by Gallé. (BnF Gallica)

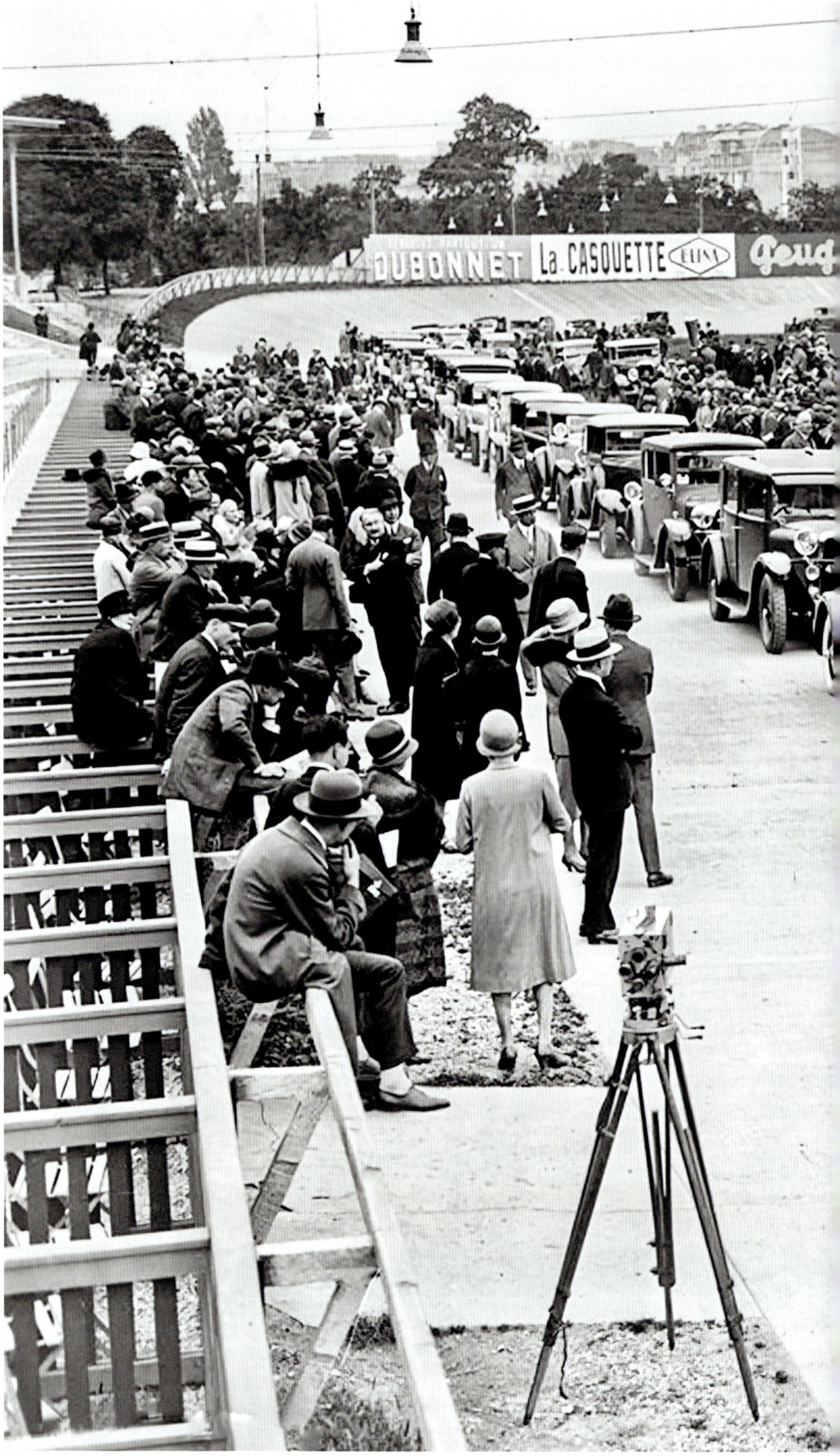

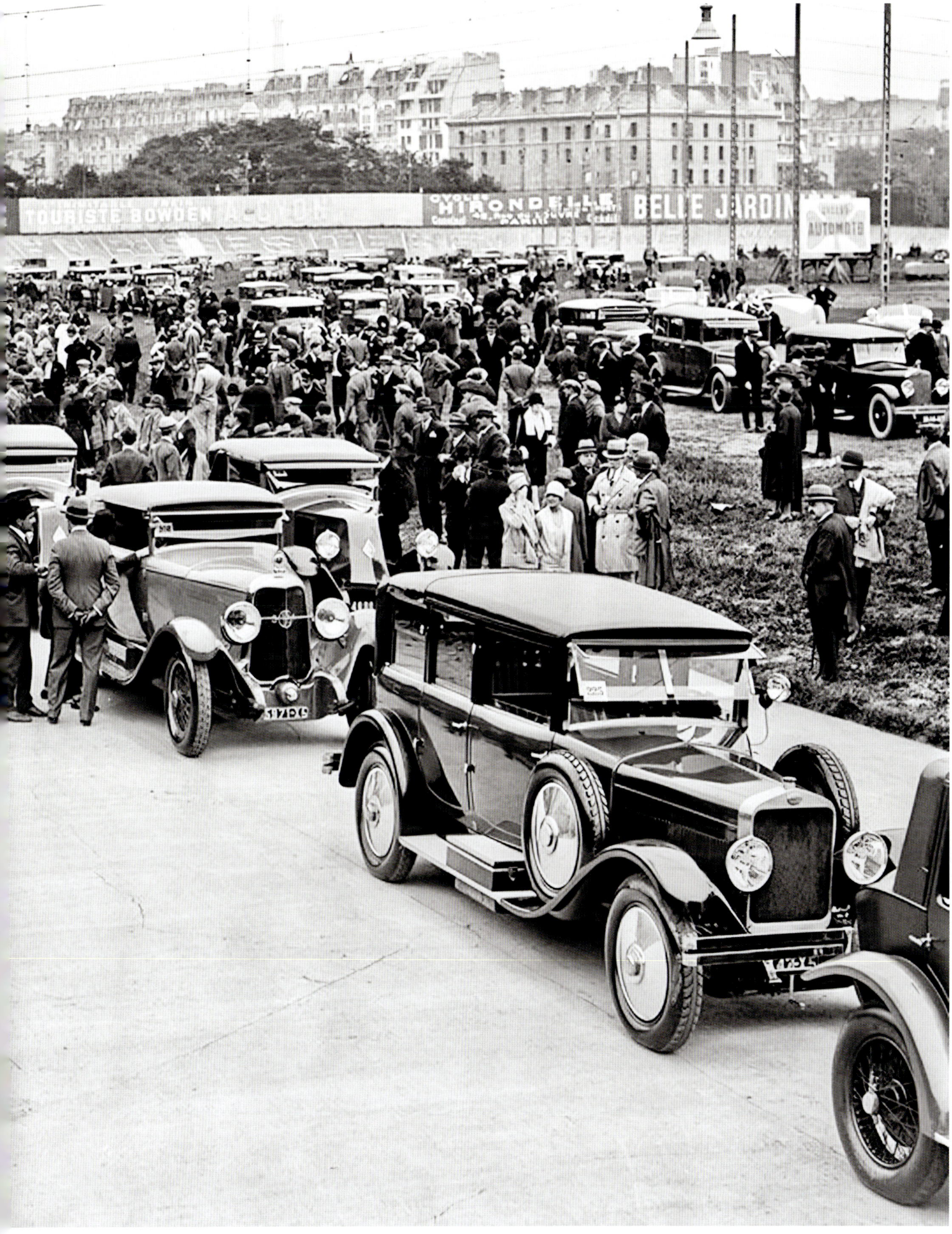
TOURISTE BOWDEN
HIRONDELLE
BELLE JARDIN
AUTOMOTO

## 1928: Harmony Between Women and Cars, Cannes

The theme of harmony between women and their automobiles continued into 1928, and was particularly highlighted during the concours d'elegance in Cannes. "*The event emphasized the intimate relationship between the aesthetics of cars and the elegance of their drivers. The competition showcased how women and their vehicles could achieve a harmonious aesthetic, with both cars and drivers presenting a coordinated visual appeal.*" (L'Intransigeant, June 25, 1927)

Commentators noted that this synergy was not just superficial; there was an intrinsic affinity between the two. Cars were said to reflect the personalities and styles of their drivers, echoing the attention both groups paid to their respective appearances. The event featured prestigious marques like Rolls-Royce, Hispano-Suiza, Packard, and Minerva, celebrating the finest in automotive craftsmanship.

The awarding of prizes was a meticulous affair, with judges deliberating over the best among a stunning array of competitors. Ultimately, banners, pennants, and plaques were awarded to the most outstanding participants, celebrating the shared elegance of both women and their cars.

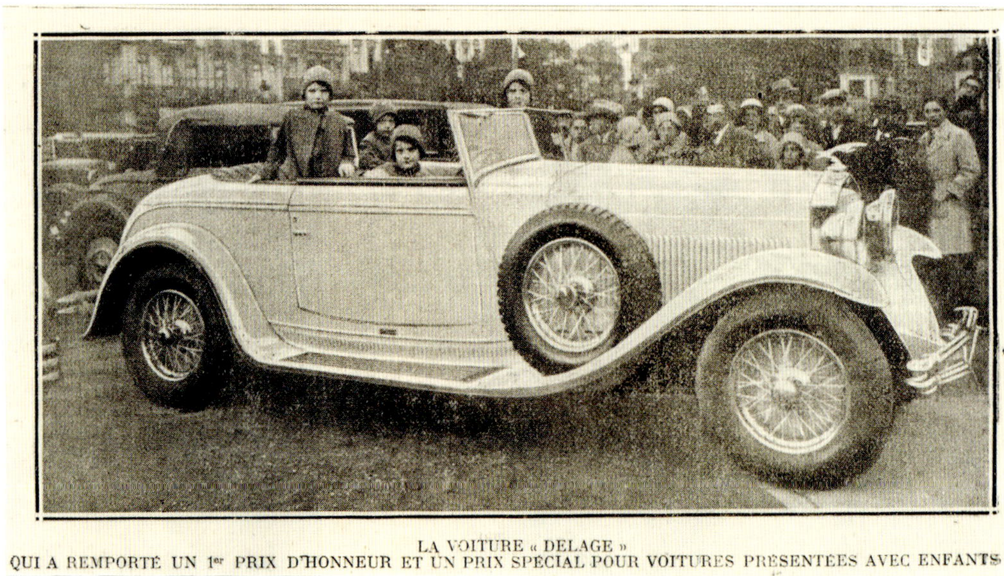
LA VOITURE « DELAGE »
QUI A REMPORTÉ UN 1er PRIX D'HONNEUR ET UN PRIX SPÉCIAL POUR VOITURES PRÉSENTÉES AVEC ENFANTS

**RIGHT:** *Now presented by children, it was another opportunity to win an award for this magnificent Delage D8 by Saoutchik. (Daniel Cabart Archives)*

**THESE TWO PAGES:** *A banner, a cup, or a plaque... Or even a simple trinket, you had to be really unlucky to leave the competition without a reward! (Daniel Cabart Archives)*

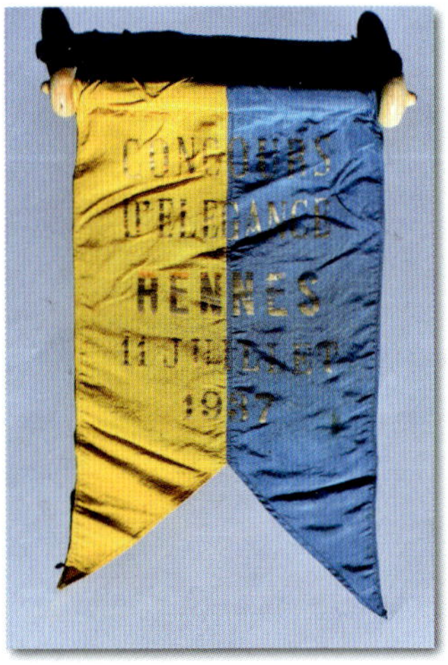

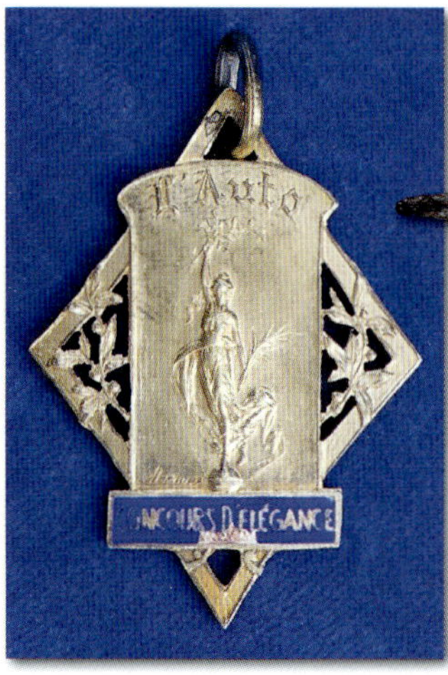


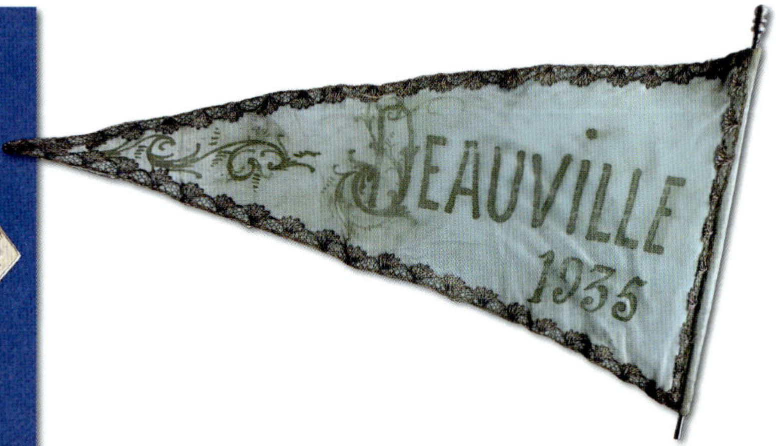


### The Triumph of Elegance in Cannes

*"The recent automobile elegance competition in Cannes was a resounding success, with around 150 cars gracefully parading along the Croisette. The participants, composed of beautifully dressed female drivers, exhibited a charming patience as they prepared for the event, adjusting their smiles and ensuring their vehicles were ready for the showcase. The cars, lined up in silent formation, took turns parading before an eager audience and a jury led by the esteemed M. André de Fouquières, known for his engaging compliments and spirited enthusiasm."* (L'Auto, April 5, 1928)

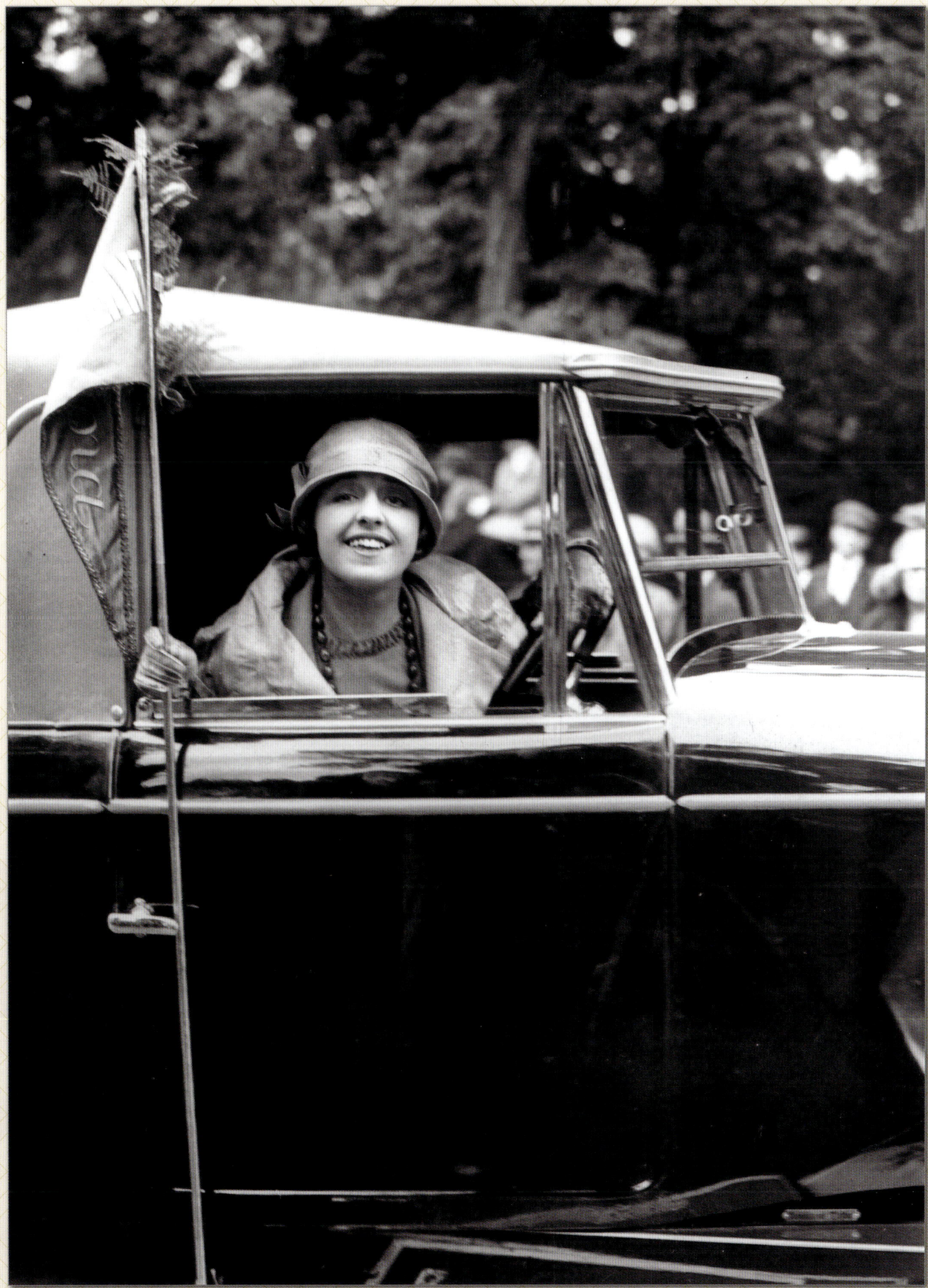

## A Shift Towards Women's Competitions

The automobile elegance competition has increasingly transformed into a women's automobile concours d'elegance competition, reflecting the growing participation and recognition of female drivers. Over the past few years, this format has begun to solidify, establishing itself as a prominent event that would persist for the following decade.

In 1929, the success of the previous year's event prompted its return, but this time with a distinct separation between the spectacle and the promotional activities of manufacturers and coachbuilders. Two competitions were organized one week apart, each with its unique focus: the daily newspaper *L'Auto* hosted a more technical examination of the cars, while the Fémina L'Intran Concours d'Elegance competition centered around the elegance of the automobiles and their female drivers.

## The Contrast Between Competitions

The press drew a clear distinction between the two events, emphasizing that while the *Fémina* competition highlighted the subjective elegance and beauty of the presenters, the *L'Auto* competition provided a more technical analysis of the vehicles themselves. This bifurcation aimed to refine the criteria for assessing the vehicles, separating the artistry of the cars from the influence of their drivers.

> *"The result was a remarkable display of automotive design, stripped of any human embellishments, where the vehicles had to showcase their inherent beauty. This emphasis on the cars alone illustrated the extraordinary craftsmanship of the coachbuilders, leaving the jury challenged in their quest to determine the best among them."*
> (L'Auto June 8, 1929)

## The Evolution of Concours d'Elegance

As the decade progressed, one significant evolution emerged: the distinction between the Concours d'Elegance Automobile and the Concours d'Elegance Féminine en Automobile. This period marked what can be considered the "golden age" of elegance competitions, with a wide range of classification and selection criteria coming to the forefront. The competitions evolved to focus on two primary aspects:

1. ***Quality of the Car:*** Competitions that emphasized the technical specifications and craftsmanship of the vehicles.
2. ***Overall Harmony:*** Events that celebrated the synergy between the car, the driver, and the driver's outfit, reinforcing a holistic view of automotive elegance.

Additionally, the presence of accessory designers and couturiers became more pronounced, as they seized the opportunity to showcase their talents within this elegant milieu. This blend of automotive and fashion excellence was particularly prominent in French competitions, where the artistry of both fields was celebrated in tandem.

## Conclusion

The automobile elegance competitions of this era not only highlighted the vehicles' aesthetic and technical merits but also celebrated the elegance of the women behind the wheel. This dual focus on car and driver established a rich tradition that continues to resonate in contemporary car shows and competitions, where elegance and style remain paramount. The efforts of the organizers, participants, and supporters of these events laid the groundwork for the enduring legacy of automotive elegance.

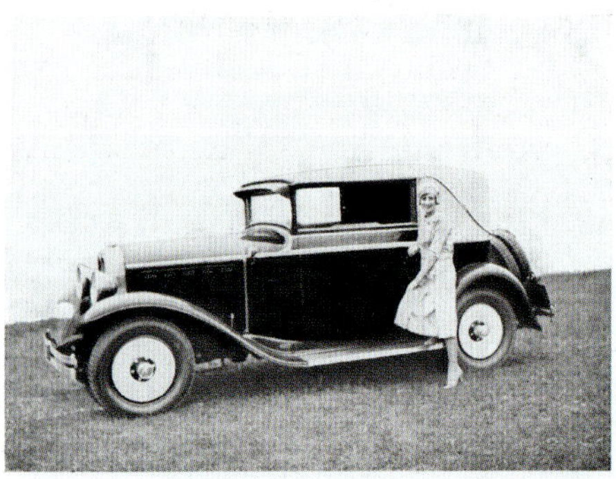

Le Grand Prix d'honneur d'Elégance du Championnat des Artistes a été gagné par Mlle Blanche Montel sur une Oakland magnifiquement carrossée par Mühlbacher et qui naturellement était émaillée au NITROLAC.

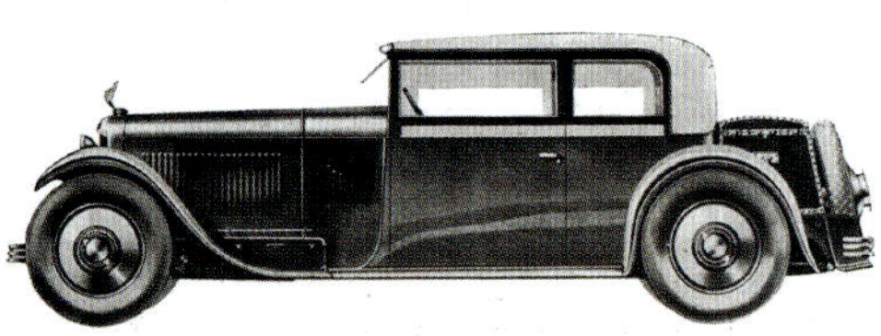

Mme Tonia Navar, de la Comédie Française, présente à ce concours une nouveauté de la carrosserie Felber, sur 8 cylindres Ballot, la conduite intérieure "Sporting Coach" naturellement émaillée au NITROLAC.

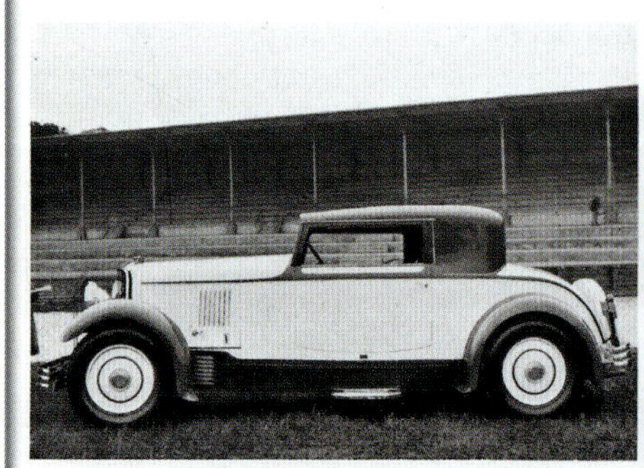

Le Trophée d'Elégance féminine Automobile du Parc des Princes a été gagné par Mme Maurice Proux sur une somptueuse 20 cv Panhard, carrosserie Maurice Proux qui naturellement était émaillée au NITROLAC.

**THIS PAGE & OPPOSITE BOTTOM LEFT:** *Renowned coachbuilders, along with their paint suppliers and major automotive brands – such as the recently introduced eight-cylinder Ballot in the middle picture above – seized the opportunity to promote themselves, often with the help of theater and film celebrities. Likewise, esteemed photographers were also present. The opposite bottom left advertisement features Dolly Davis, a happy winner of a prize in the Fémina L'Intran Concours d'Elegance. (Laurent Friry Archives)*

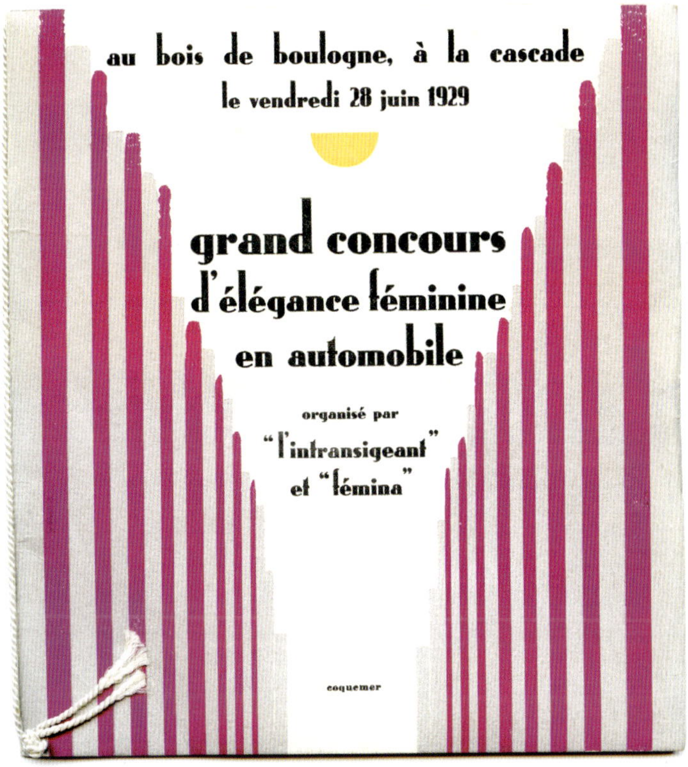


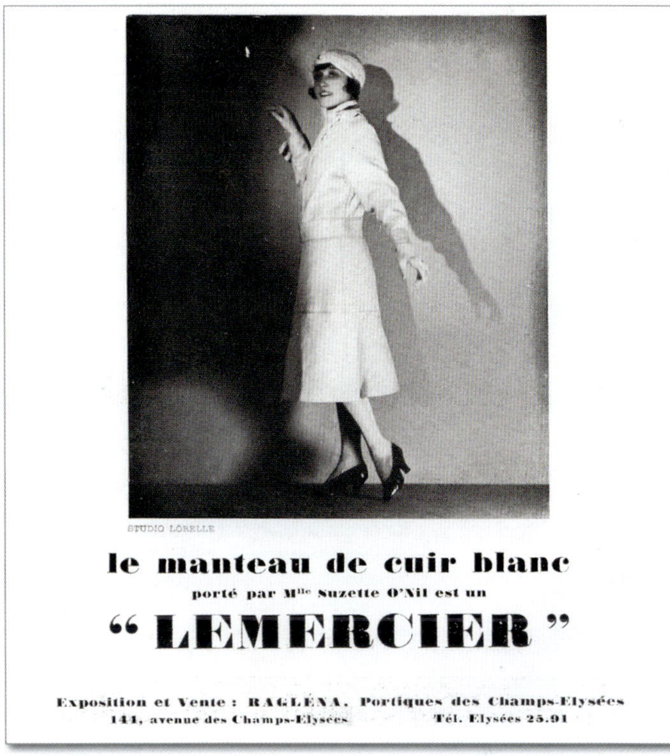


**ABOVE (BOTH):** *The program's booklet that provides the details of the event's lineup, interspersed with advertisements for clothing, fashion, and the automotive industry. On the right, an inside page features Mlle. Suzette O'Nil (later Madame Dranem, wife of the renowned singer) with a white leather coat, perfect for sporty driving. Mlle. O'Nil presented a compact Rosengart 5 CV at the Fémina L'Intran Concours d'Elegance, a car ideally suited for lady drivers in Paris. (Laurent Friry Archives)*

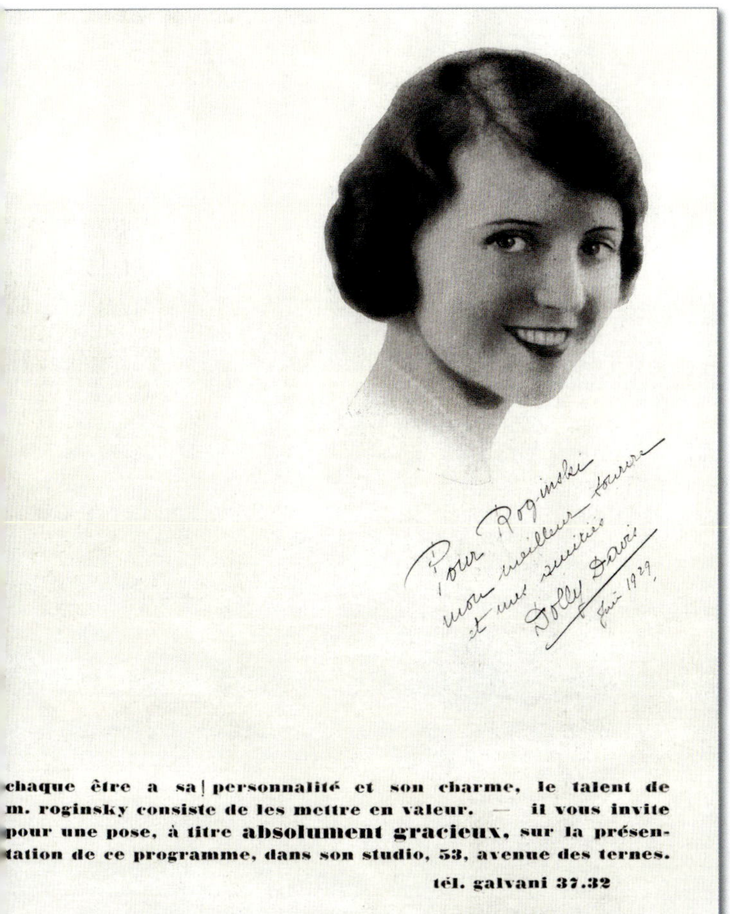


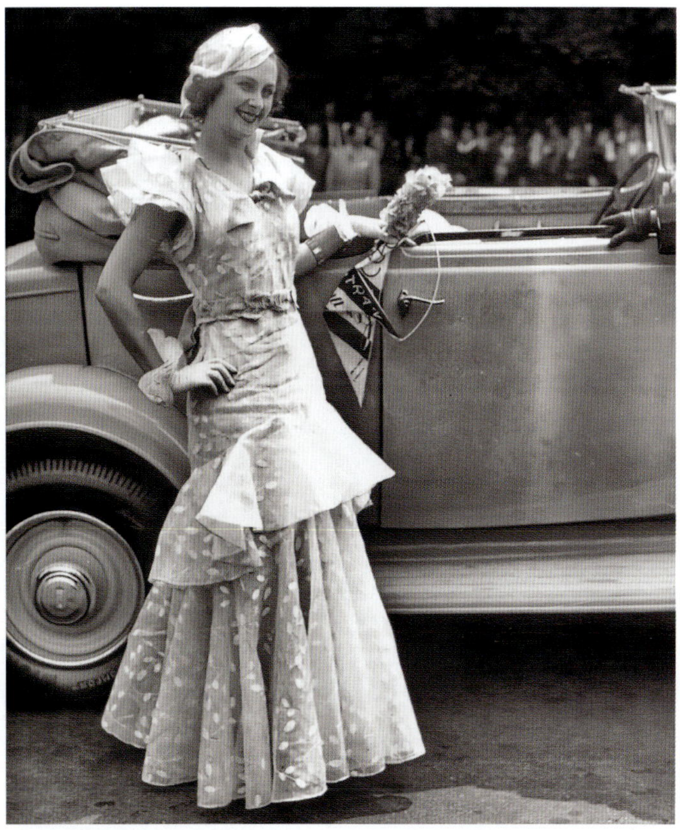

*A bouquet from the Fémina L'Intran Concours d'Elegance competition and a pennant are the very least that can be credited to this stunning presenter. (Laurent Friry Archives)*

# Concours d'Elegance at its Peak

## 1930S: THE GOLDEN AGE

## The Golden Age of Concours d'Elegance

The 1930s are often hailed as the pinnacle of concours d'elegance events, but this era didn't simply begin on January 1, 1930. Many significant trends had already started to take shape in the late 1920s, and the new decade only solidified their impact. One major development was the explosion in the number of competitions, from grand, prestigious events to smaller, local gatherings. The most distinguished contests continued to be held in traditional venues, such as spa towns, seaside resorts, and, of course, Paris. However, numerous other concours emerged throughout the country.

By August 1936, the Automobile Club de France (ACF) felt compelled to introduce regulations. They noted: "*The proliferation and significance of these events, in terms of both quantity and quality, as well as their growing influence on automobile advertising, necessitate a standardized scoring system and the classification of events by importance.*" The ACF proposed distinguishing between "national" and "regional" competitions, in addition to the numerous municipal or cantonal events that kept appearing.

A meeting at the Automobile Club de France addressed these concerns, aiming to establish universal guidelines for concours d'elegance competitions. The consensus was to differentiate clearly between technical contests, which evaluated the craftsmanship of car bodies, and women's concours d'elegance tournaments, where participants expressed their style behind the wheel. Categories would be determined using criteria such as engine power or catalogue prices. Regional events would retain their local character by allowing organizers to restrict participation to cars registered within specific regions. A standardized points-based evaluation system was also created for jury use. (*Automobile Club Féminine Bulletin, August 1, 1936*)

*The origins of the concours d'elegance competitions remain somewhat unclear, but their impact was undeniable. As one admirer poetically put it, "Who was the instigator of the concours d'elegance competitions? I don't know, but I ask that we pay homage to his merit by weaving him wreaths of cornflowers, forget-me-nots, and buttercups." These events beautifully capture the elegance, chic, and curves of both the queens of the wheel and of fashion, bringing them to life rather than freezing them in time. A perfect example is this Renault Nervastella Grand Sport city coupe, shown at the Fémina L'Intransigeant Concours d'Elegance in Paris at the Jardin d'Acclimatation on June 22, 1934. Actress Alice Field presented the car, lending the scene an air of luxury and sophistication. (Laurent Friry Archives)*

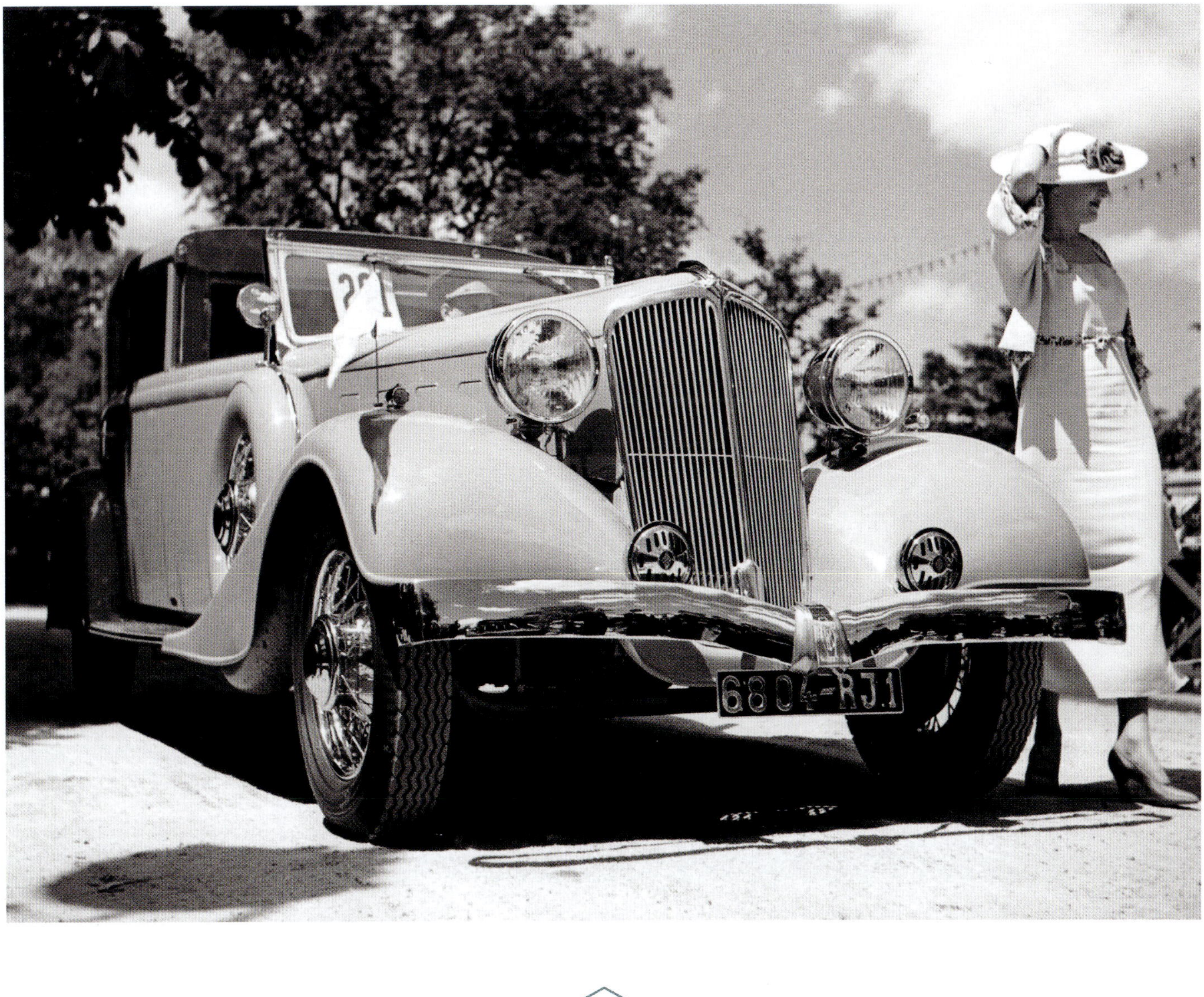

**OPPOSITE:** *Feminine and automotive elegance, everything is here. Actress Anita Page, nicknamed "the girl with the most beautiful face in Hollywood," poses for a publicity photograph in front of a 1930 Cadillac V16. (Marc Caveng)*

**BELOW:** *At the Parc des Princes, the L'Auto concours remained primarily a showcase for coachwork. Even though the Baroness de La Fontaine "played the game" of feminine elegance, it was ultimately the quality of the Grümmer construction that the all-male jury seems to be examining here. (Philippe Grümmer Archives)*

These regulations echoed the principles governing Paris's major competitions of the late 1920s, a style often described as "à la française." The groundwork for these guidelines had already been laid in previous years, such as the Biarritz event in 1926.

Two main types of contests evolved and persisted: technical coachbuilding competitions, featuring hundreds of vehicles displayed in a semi-static manner, and women's concours d'elegance tournaments. The latter saw a significant rise in female participation, with women often complementing the cars' qualities through their elegance and driving skills. The growing prominence of women in these events reflected a shift in the concours' focus.

The *L'Auto* concours, a pioneer in the field, attempted to innovate by assigning a dual jury for its 1931 event. The male jury evaluated the vehicles, while a female jury, led by the Automobile Club Féminine de France, judged the stylish female drivers.

Despite these efforts, the concours fell short of expectations of participant numbers and vehicle quality. The economic crisis severely reduced entries, and the lack of promotion worsened the situation. The June 19 final drew only a modest audience, and spectators were underwhelmed, with the event failing to spark much admiration. Notably absent were the prestigious high-end cars, such as Hispano-Suiza and Rolls-Royce, and there was little in the way of innovative designs. Instead, the event saw an influx of smaller vehicles like Rosengart, La Licorne, and Peugeot 201, with Delage standing out as the day's major success.

Faced with dwindling interest, *L'Auto* realized the need for a new approach, eventually incorporating a "tournament of feminine elegance" alongside its technical assessments. The event increasingly leaned toward spectacle rather than technical evaluations, driven by public demand.

This trend was confirmed by a competing event a week later, organized by the fashion magazine *Fémina* and the newspaper *L'Intran*. Unlike the technical nature of *L'Auto's* contest, the Concours

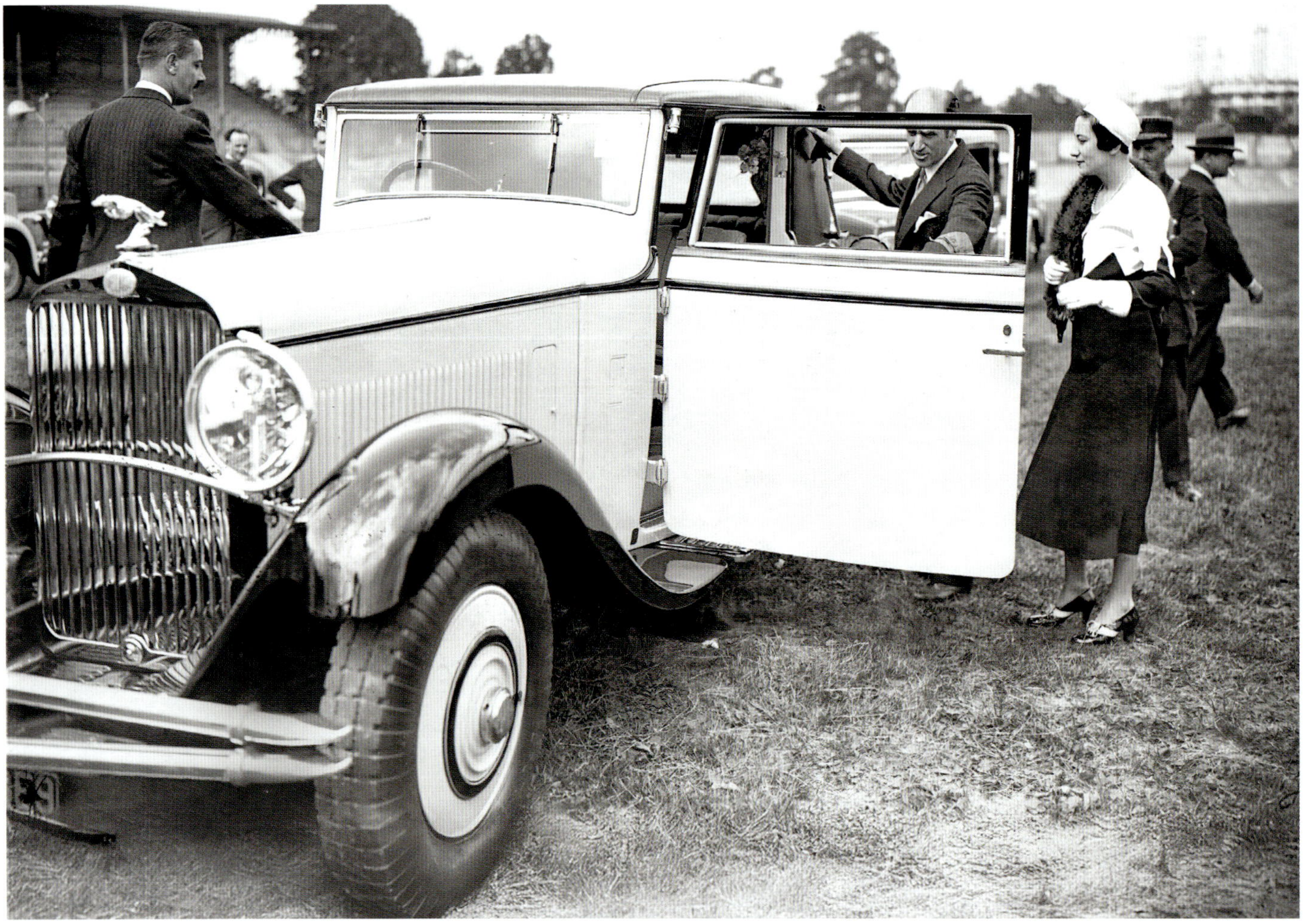

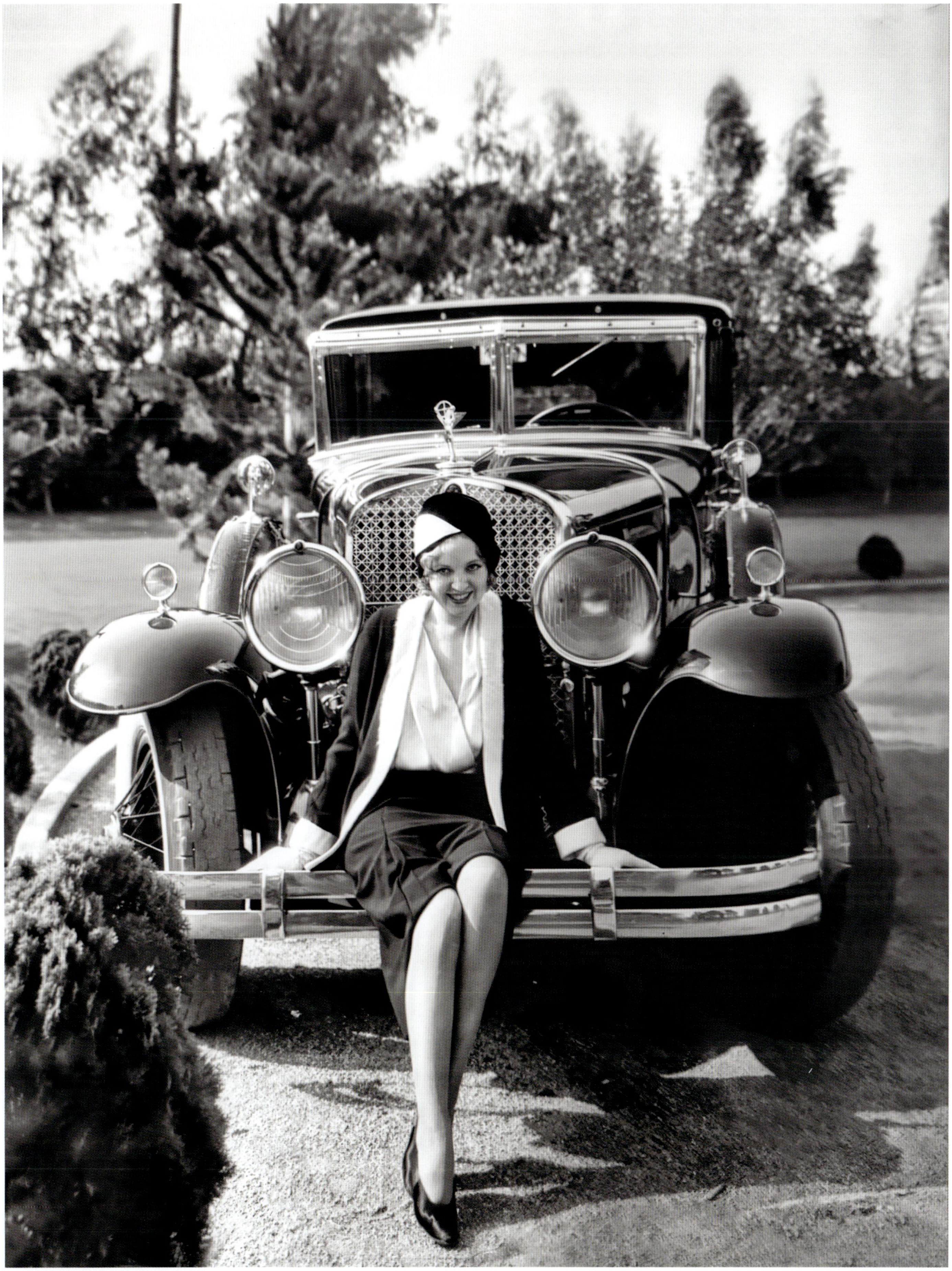

Dessin d'Hy. Fournier

Avant le Concours d'Elégance : Ultime « Raccord »

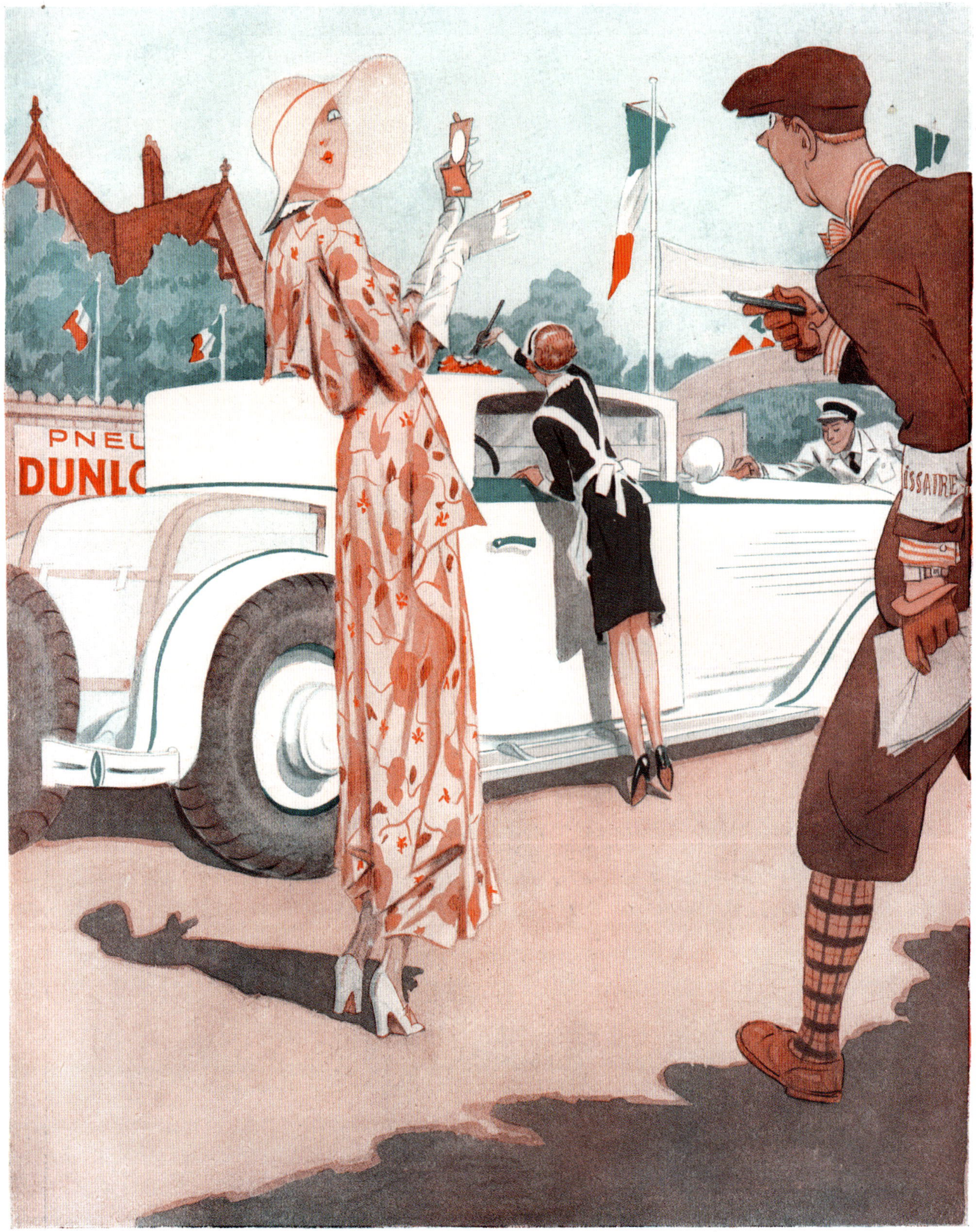


« Voyons, dépêchez-vous, Madame, le jury vous attend... »

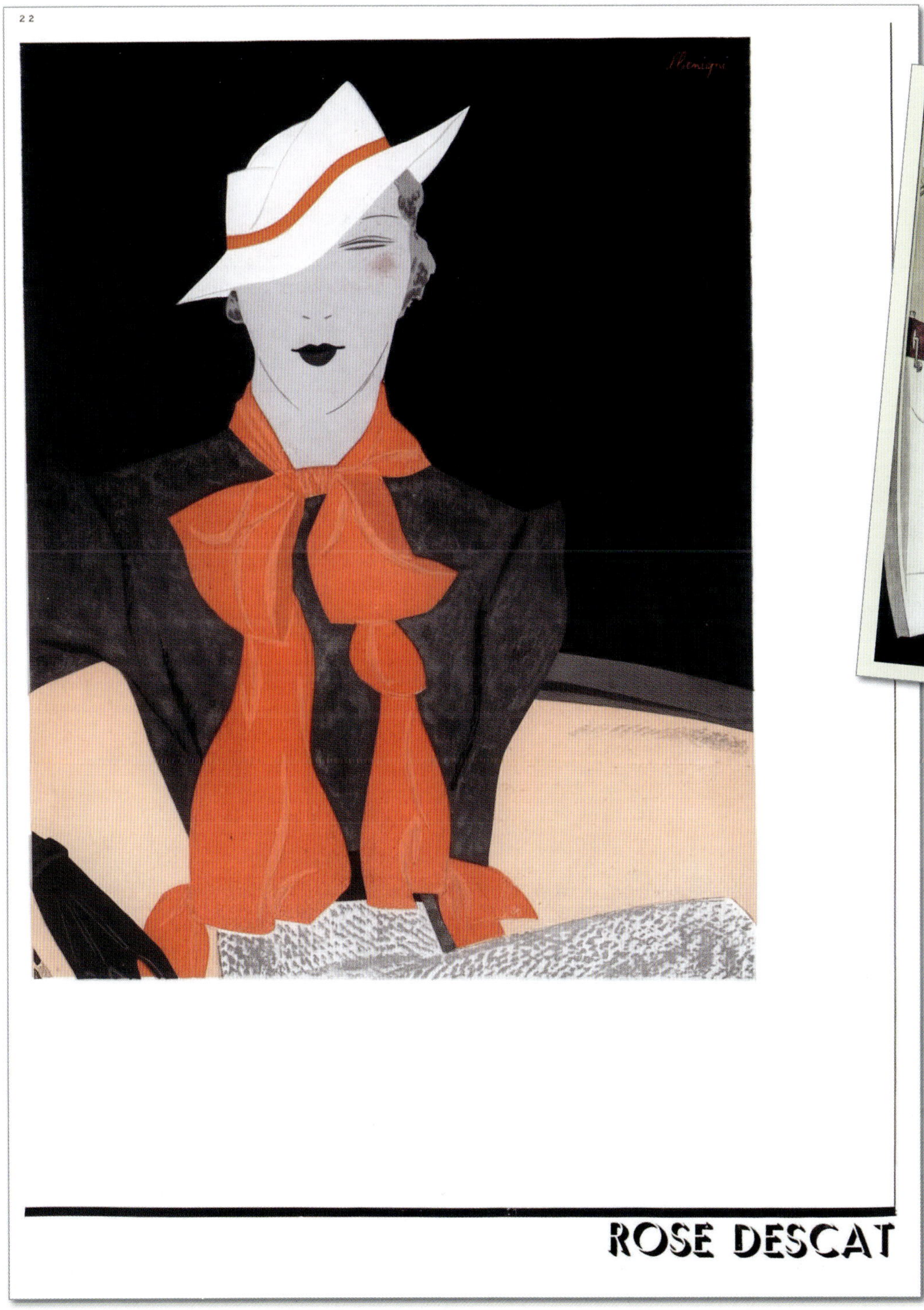


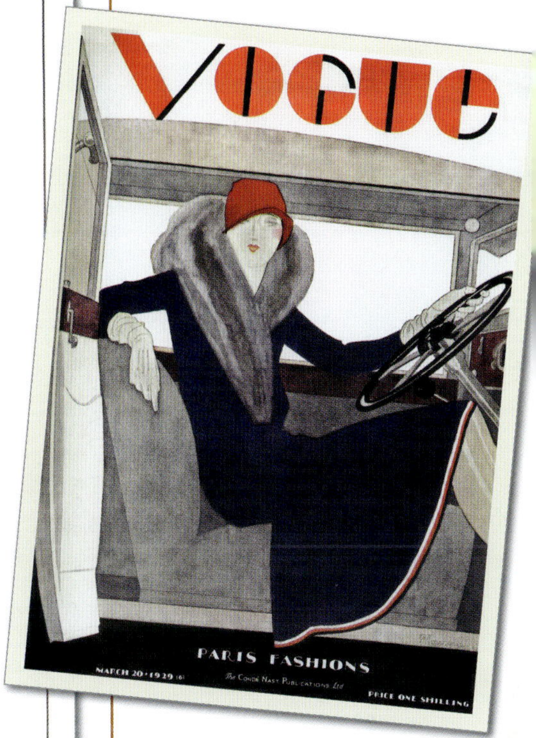


**THIS PAGE:** *Indeed, the hat became an essential accessory in feminine elegance, especially in concours d'elegance competitions. Over time, the greatest milliners – names synonymous with high fashion – found themselves increasingly receiving recognition in concours competition results, proving that style was judged not only in the lines of the car but also in the refinement of the ensemble. These concours became platforms not just for automotive craftsmanship but also for haute couture, where every detail, from flowing gowns to meticulously designed headpieces, played a role in the final impression. (Daniel Cabart Archives)*

**OPPOSITE:** *Times have changed and the lady is no longer just one part of the ensemble. She now wants to make a statement, and the impression she makes on the jury will be crucial. (Daniel Cabart Archives)*

d'Elegance Féminine en Automobile held in the Bois de Boulogne on June 26, 1931, was a resounding success. It drew 250 entries, a record, and brought together a dazzling array of car bodies, along with stylish female drivers who delighted the crowd. Competitors, dressed in stylish beach pajamas that accentuated their figures, confidently handled their cars, embodying a new era of automotive elegance.

A quintessentially Parisian gathering took place, attended by a large and distinguished crowd captivated by the elegant lines of the cars and the poise of their sophisticated drivers. The audience's enthusiasm was evident as they applauded the witty and lively commentary of Mr. André de Fouquières. Positioned at the Cascade, this fashionable assembly was also treated to a first glimpse of the latest women's fashion trends, featuring playful and intricate hats – often adorned with feathers that shaded the face or neck – worn in ways that reflected each woman's unique style.

**RIGHT:** *Once again, fascinators were an essential element of the women's collection. It would be unthinkable to appear without these accessories. These ladies are at the concours at Bois de Boulogne. (Daniel Cabart Archives)*

**OPPOSITE:** *The 1933 "Concours Fiat" in Nice was a summer event that introduced both local (or national) luxury cars and the latest range of Fiat models. (BnF Gallica)*

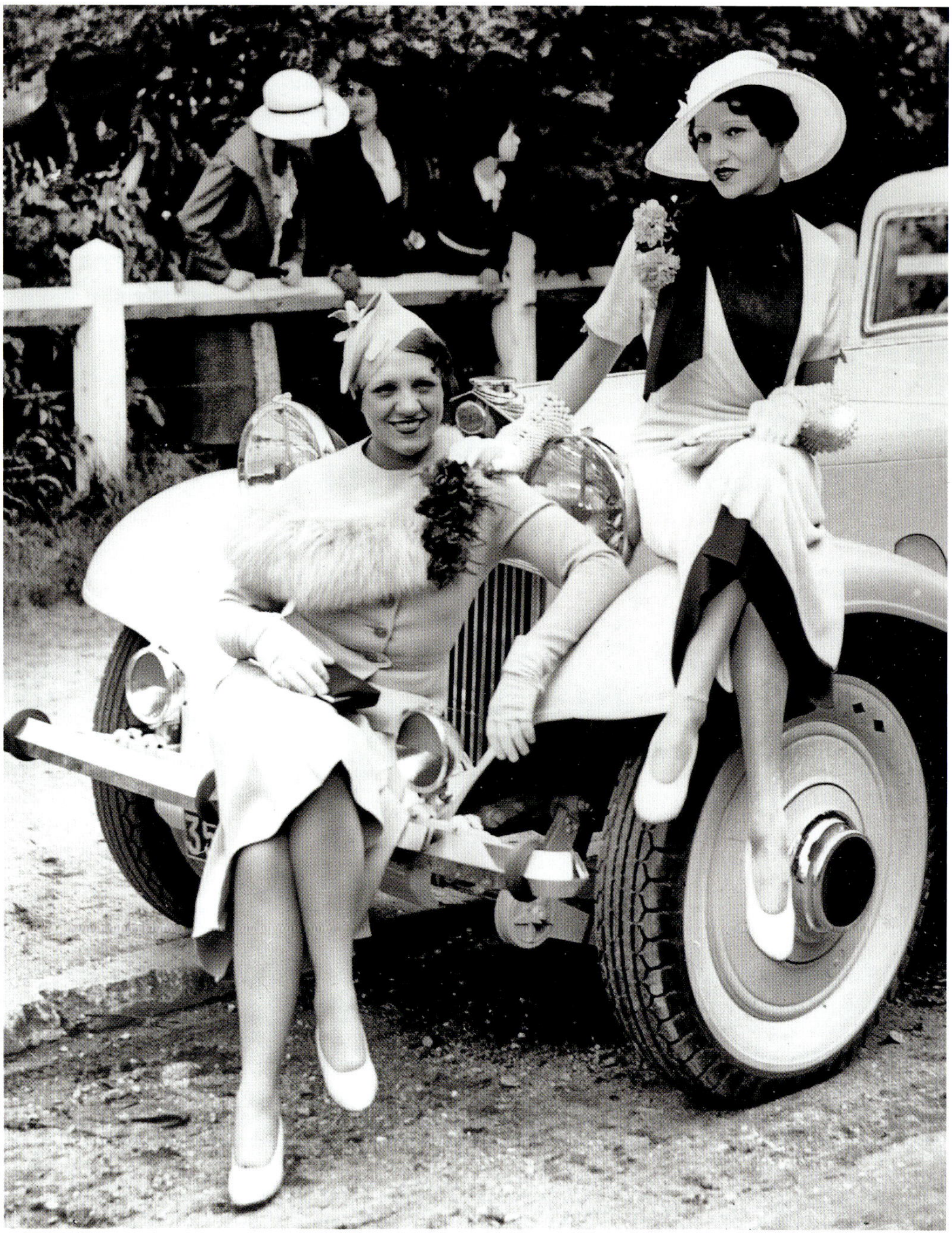

The following year, André de Fouquières, known as the "Pope of Elegance," encapsulated the newly established rules for concours d'elegance, which reinforced what had already been outlined. He described the holistic evaluation approach: "*In women's automobile concours d'elegance events, the overall impression matters, comprising several elements: the engine, the bodywork, the driver, and the outfit. Fashion plays an integral role because the outfit must harmonize with the vehicle, while even the mechanic must be impeccably dressed... No more than two colors should be used, and they must blend harmoniously – a challenge as intricate as algebra and geometry*" *(A. de Fouquières, L'Art Vivant, 1932).*

It is notable that discussions on the interplay between automotive technology, design, fashion, and elegance found a place in magazines devoted to contemporary art, underscoring the cultural significance of these competitions.

Every participant in this grand spectacle understood its potential benefits. The coachbuilders, who had been part of elegance competitions since their inception, saw opportunities for bringing attention their craftsmanship. Meanwhile, from the late 1920s onward, couturiers, milliners, accessory makers, and shoemakers embraced the chance to dress the "feminine component" of these events.

# L'ÉLÉGANCE FÉMININE EN AUTOMOBILE

## A Nice, le Concours Fiat

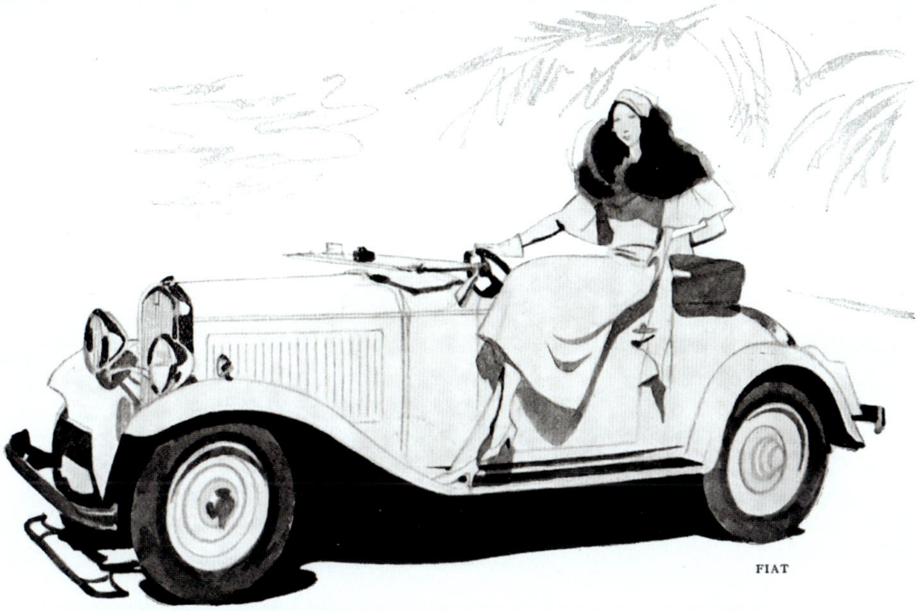


*Mme André Desgranges qui gagna le Grand Prix d'élégance Fiat était charmante, toute en blanc dans une voiture blanche.*

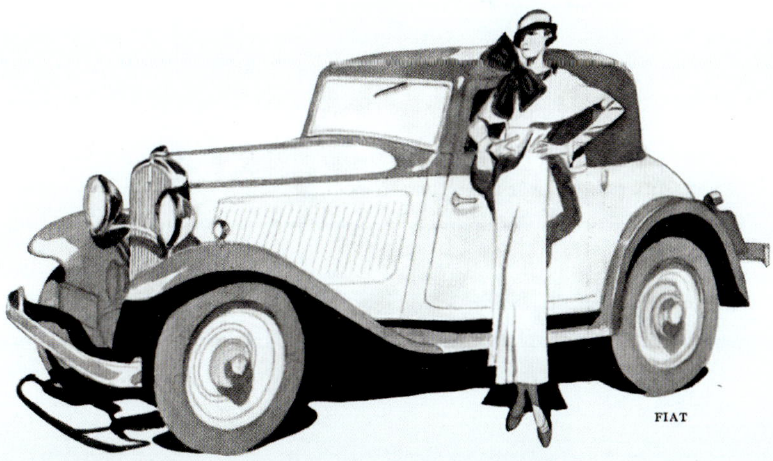


*Mlle Vera Sherwood, à qui fut décernée la Coupe d'originalité, portait une robe en angora chamois, ornée d'un nœud de velours orange.*

Il n'est pas de manifestation féminine qui offre un plus joli spectacle qu'un concours d'élégance en automobile. Choisir Nice et la promenade des Anglais pour faire défiler les concurrentes, c'était leur assurer le plus seyant des cadres. Aussi le concours Fiat, organisé par notre excellent confrère " L'Éclaireur de Nice ", a-t-il été une véritable fête d'élégance sportive. La Société anonyme pour la fabrication en France des automobiles Fiat avait confié ses couleurs à de très jolies femmes. Les charmantes conductrices rivalisèrent de goût et de recherche dans leur toilette qui était en harmonie avec les teintes de la carrosserie. Et les voitures furent très admirées pour leur ligne et pour leur confort. La tâche du jury fut agréable bien que malaisée. Le premier Grand Prix pour l'ensemble le plus parfait a été décerné à Mme André Desgranges. La coupe d'originalité a été attribuée à Mlle Vera Sherwood. La coupe sport à Mlle Monique Genin, et la coupe de " L'Éclaireur " à Mmes Laffond de Ladebat et Depret.

Au total, une belle manifestation digne en tous points du renom d'élégance dont jouit partout en France la Fiat française.

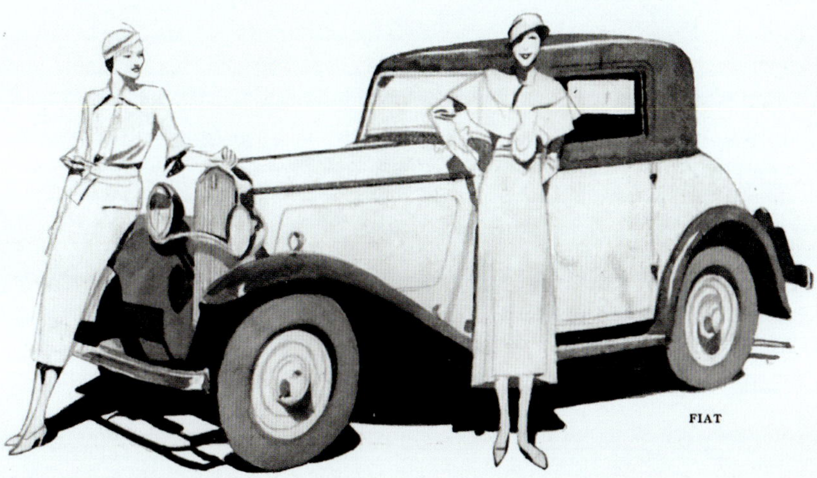


*Mmes Laffond de Ladebat et Depret obtinrent la couve de " L'Éclaireur de Nice ".*

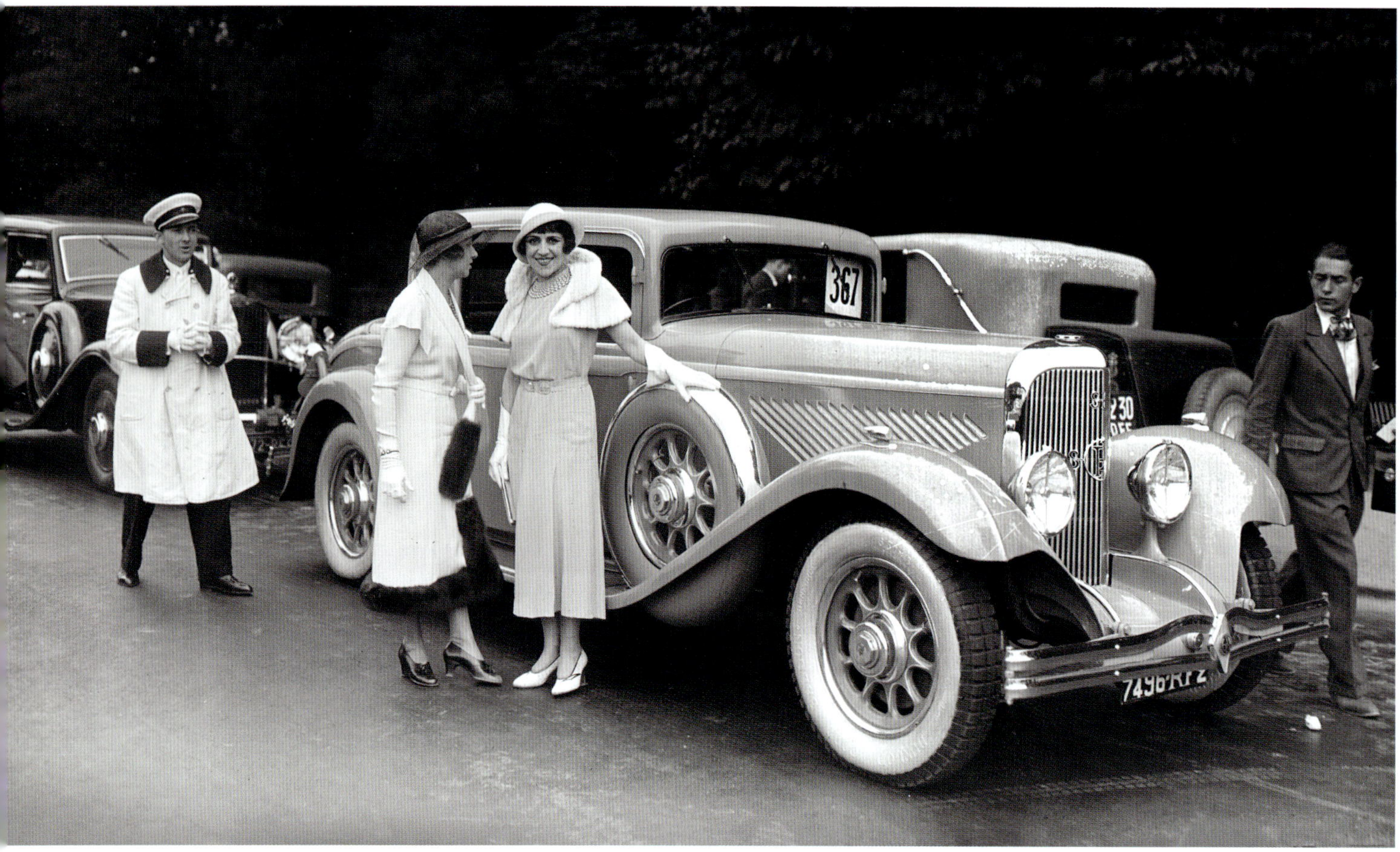

1932: The concours d'elegance is one of the most fashionable events, where spectators can ogle the celebrities – artists, actors, aristocrats, and princesses. Marie Dubas presented a Panhard at the Fémina L'Intran Concours d'Elegance competition on June 24, 1932, which was abruptly interrupted by an accident caused by a female driver who veered into the crowd, injuring several people. The chauffeur, once an occasional role, would soon become a mandatory presence at such events (see also page 147). (Laurent Friry Archives)

Car manufacturers, too, recognized the marketing potential, often developing strategies to ensure their presence at key competitions. Brands like Citroën, Amilcar, La Licorne, and Delage organized dedicated teams to represent them, while companies like Fiat went a step further, hosting exclusive concours d'elegance events for their vehicles, directly sponsored by the brand.

This tradition of feminine concours d'elegance was distinctively French, whether held in Paris or along the Côte d'Azur. By 1936-1937, across the Channel in England, a different format still prevailed. The British approach maintained the older French model from the 1920s, with a multitude of categories and only a touch of feminine elegance.

In her report for *La Carrosserie* magazine, Madame Claude Rolland described the Eastbourne competition of 1936:

> *"Automobile Elegance and the Entente Cordiale.*
>
> *Elegance competitions in England are gaining popularity and are impressively organized. These are not mere parades serving as a backdrop for social chatter, but serious contests where prizes reward the most deserving cars. Held in an exclusive atmosphere, they significantly contribute to the advancement of the coachbuilding industry. The tenth Eastbourne competition, which I had the privilege of attending, featured over 150 cars competing across 29 categories, including price-based brackets, sports cars, the most distinguished car, and categories for elegance. Judges scrutinized every detail, from the undersides of the fenders to tool placement, with exhaustive explanations provided."*

Rolland also noted the novelty of a category recognizing the harmony between car and driver ensembles. The Delage contingent, consisting of six cars in 1936, stood out, with their harmonious color schemes between vehicle and driver attire impressing even the British press (*The Motor*), which highlighted this "exotic" Continental flair.

As these competitions evolved into aesthetic and commercial spectacles – highlighting clothing and fashion-related products – coachbuilders continued to be celebrated for their artistry. Yet, the events became increasingly elaborate, often incorporating elements of high fashion and spectacle.

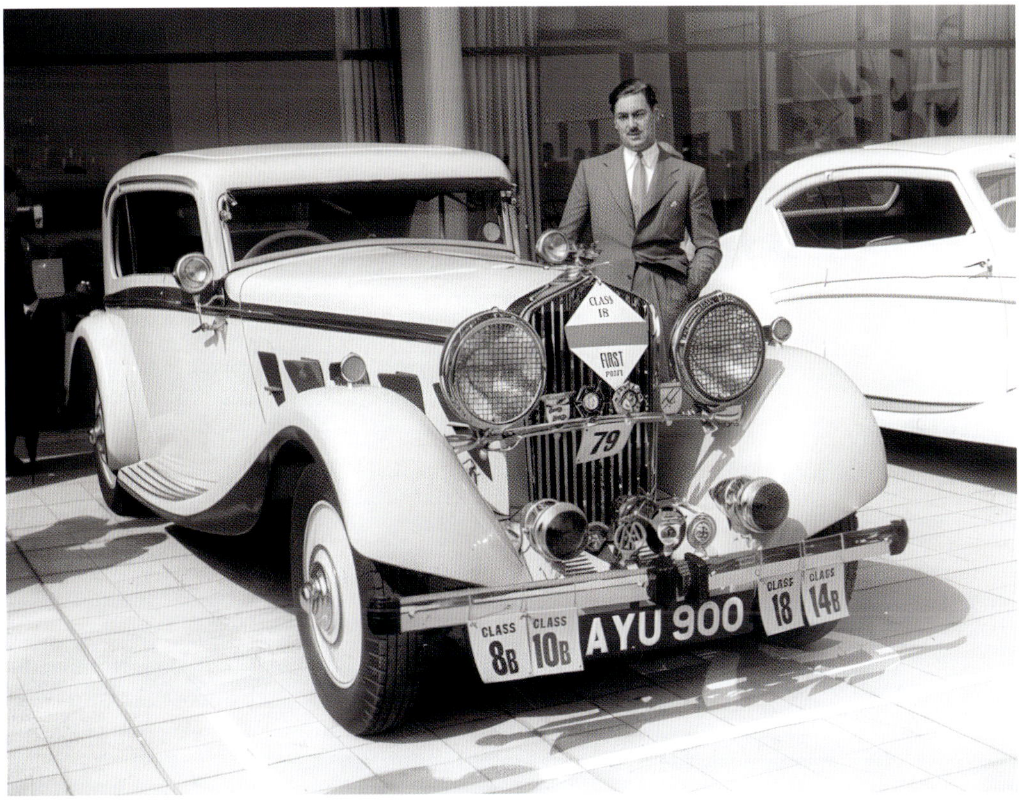


**THIS PAGE:** 1936 Bexhill Concours d'Elegance. *Obviously the feminine component is not the goal of this concours.* (Daniel Cabart Archives)

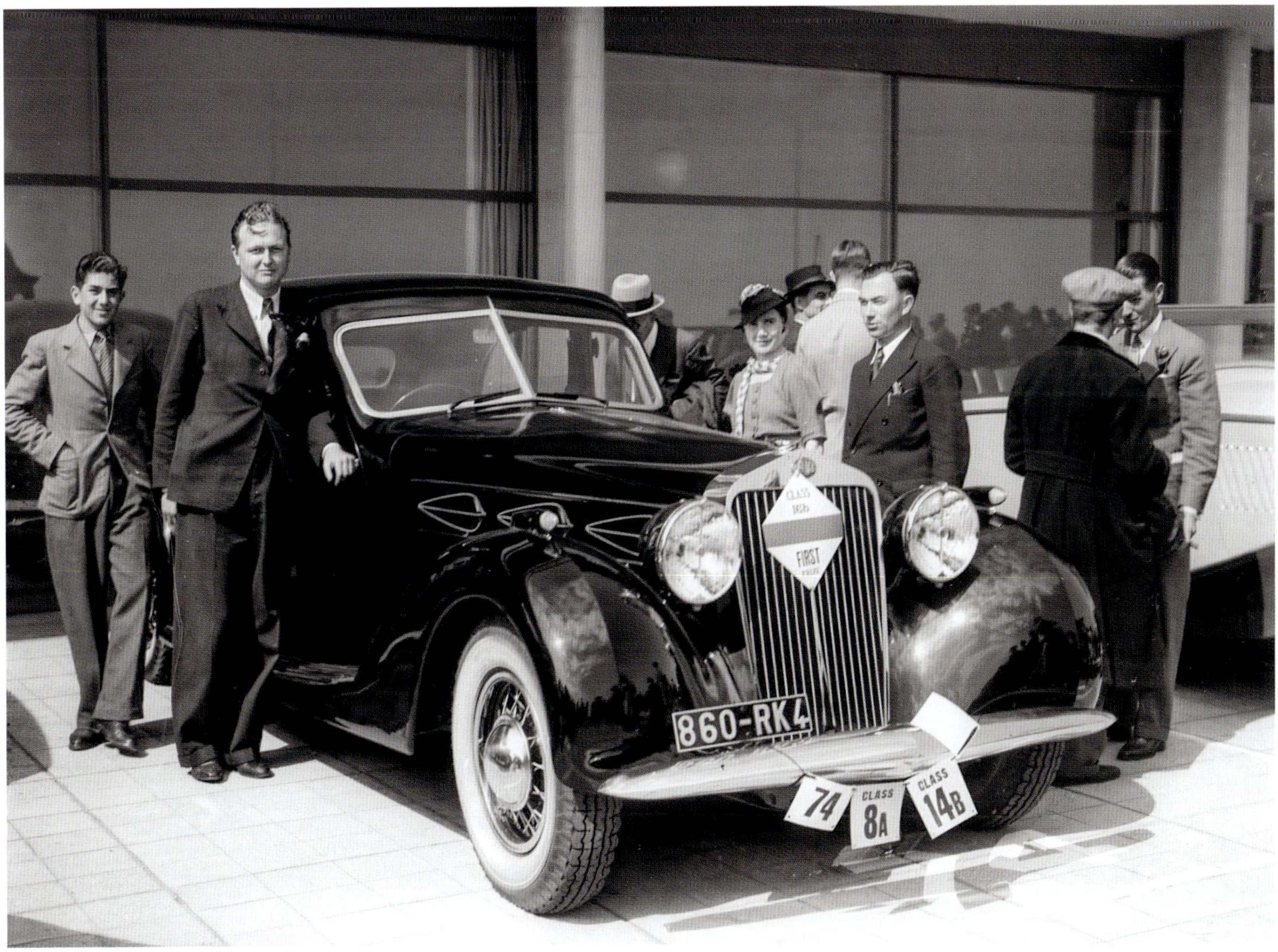

**RIGHT:** *At the 1931 Cannes competition, Renée Friedrich, daughter of the Bugatti agent from Nice, presented the new Delage 14 CV with an open-top saloon body by Chapron, paired with an outfit designed by Muller. Renée Friedrich would tragically lose her life the following year in an accident during the Paris-Vichy-Saint Raphaël rally while driving a Delage D8-S. (Daniel Cabart Archives)*

**BELOW:** *The magic of the concours at night. The Mercedes-Benz 500K Spezial Roadster with factory bodywork at the Cannes night concours on August 28, 1936. Presented by Princess Chevardchizé, the car and its crew won the 1st Grand Prix d'Honneur in the open car category. (Bernard Brulé)*

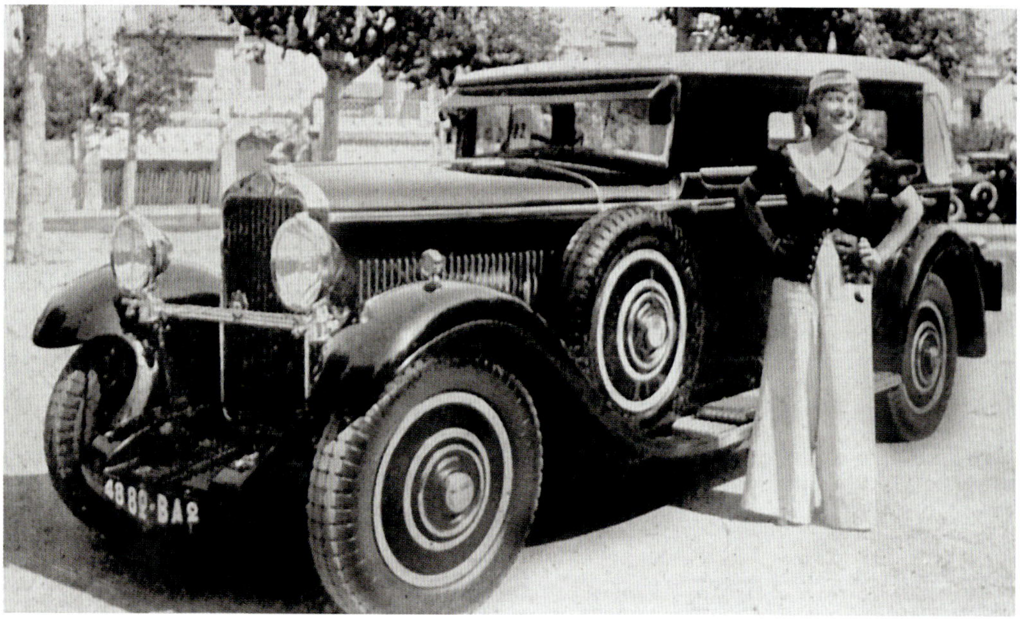


Nighttime competitions, a hallmark of the Riviera, added to this glamour. Since debuting in Biarritz in 1926 and continuing in Cannes by 1933, these nocturnal shows created a magical atmosphere. Competitions in Cannes, Nice, Juan-les-Pins, and Monaco frequently featured designers and couturiers, often honored alongside coachbuilders in the popular press.

In this dazzling world of elegance competitions, every detail, from the cars to the fashion, created a seamless blend of art, technology, and style, captivating audiences and setting trends that still inspire today.

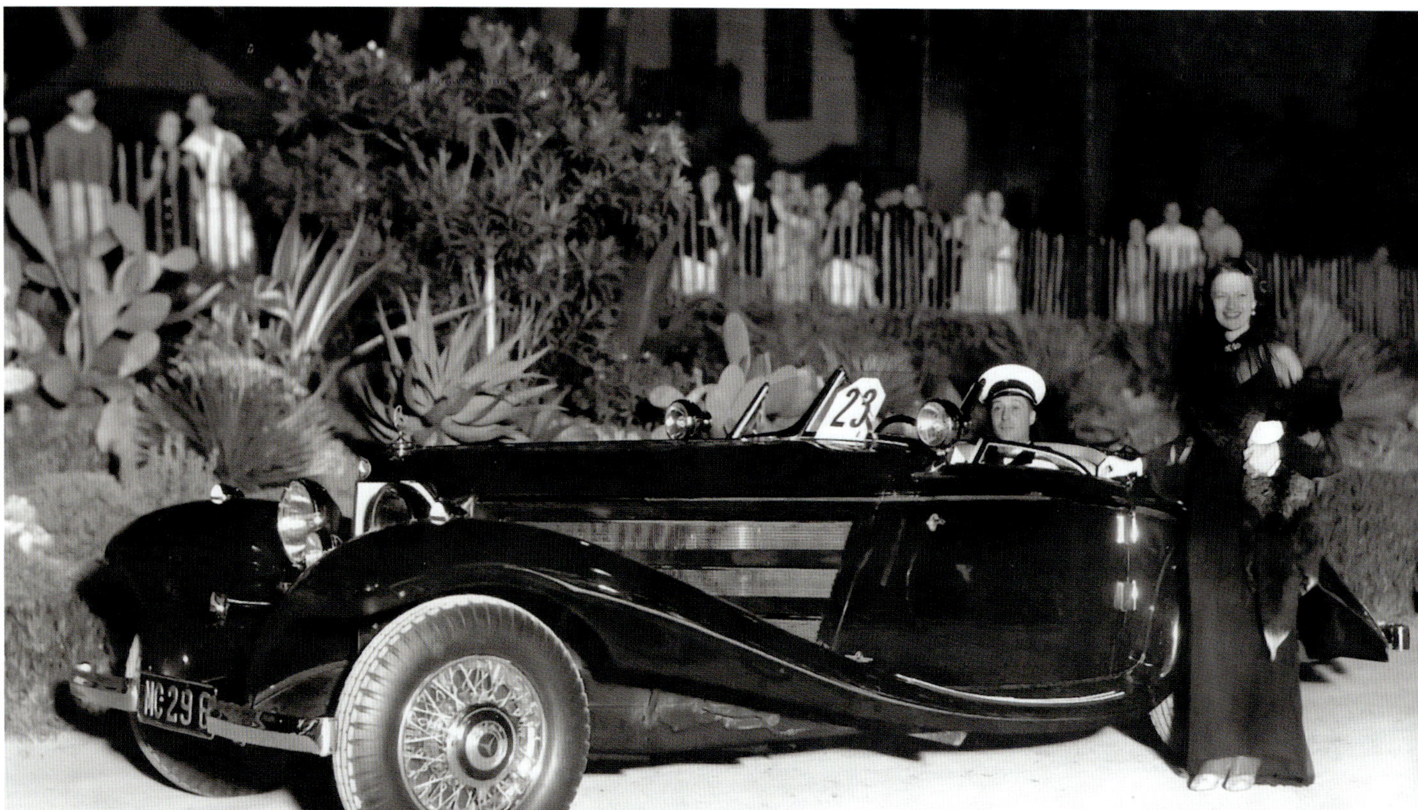

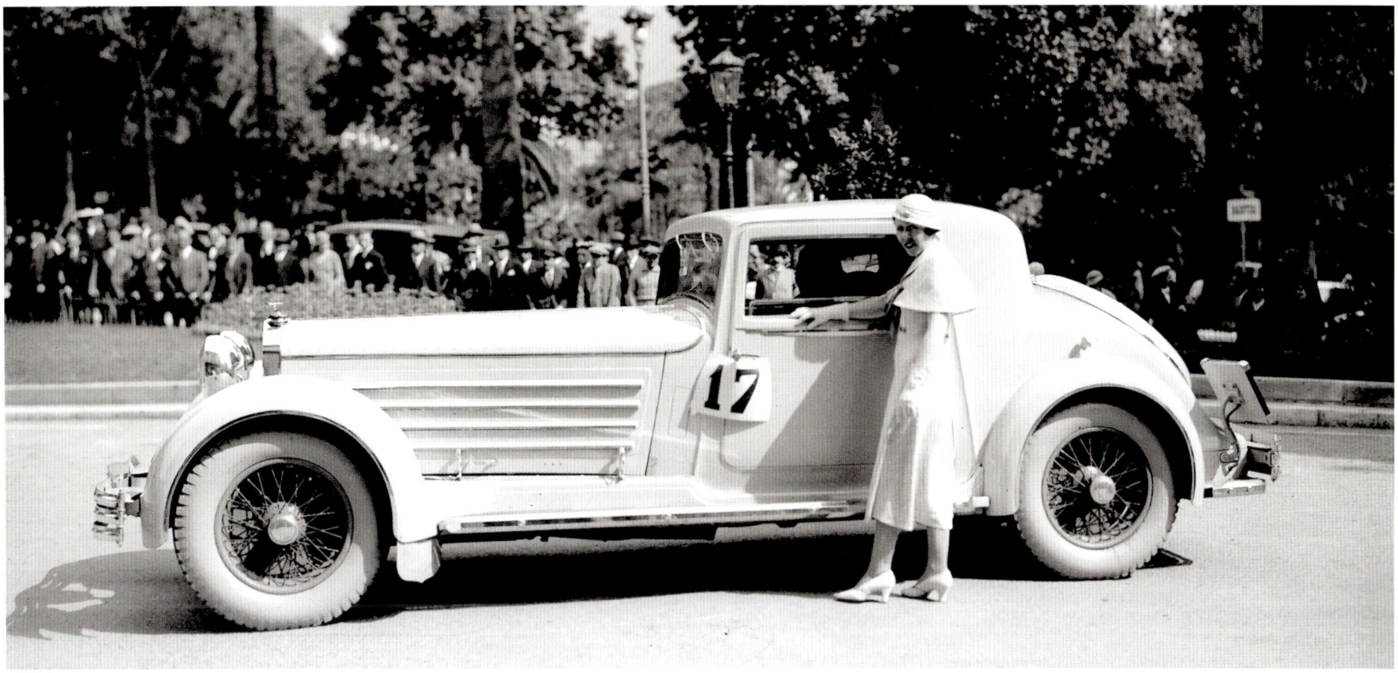


**ABOVE:** "In Cannes, as in Monte Carlo, Madame Campbell-Wood won a Grand Prix with her Delage D8-S Conduite-coupé sport, a highly unique creation by the Cannes coachbuilder Brandone." (*L'Éclaireur du Dimanche Illustré*, *June 1933*) *Local coachbuilders, such as Brandone, Bugatti, Guérard and others, consistently seized every opportunity to showcase their craftsmanship.* "Only cars valued at over 100,000 francs and delivered after January 1, 1932, were accepted at this concours – making participation quite exclusive." *L'Auto*, *June 4, 1933* (Daniel Cabart Archives)

**RIGHT:** *Competition was intense between Delage, Hispano-Suiza, and Talbot, all of whom claimed Grand Prix victories in various categories. Yet each of these marques also stood at the very pinnacle of elegance, thanks to distinctive and masterful coachwork – here exemplified by a De Villars town coupé on a Bugatti chassis.* (BnF Gallica)

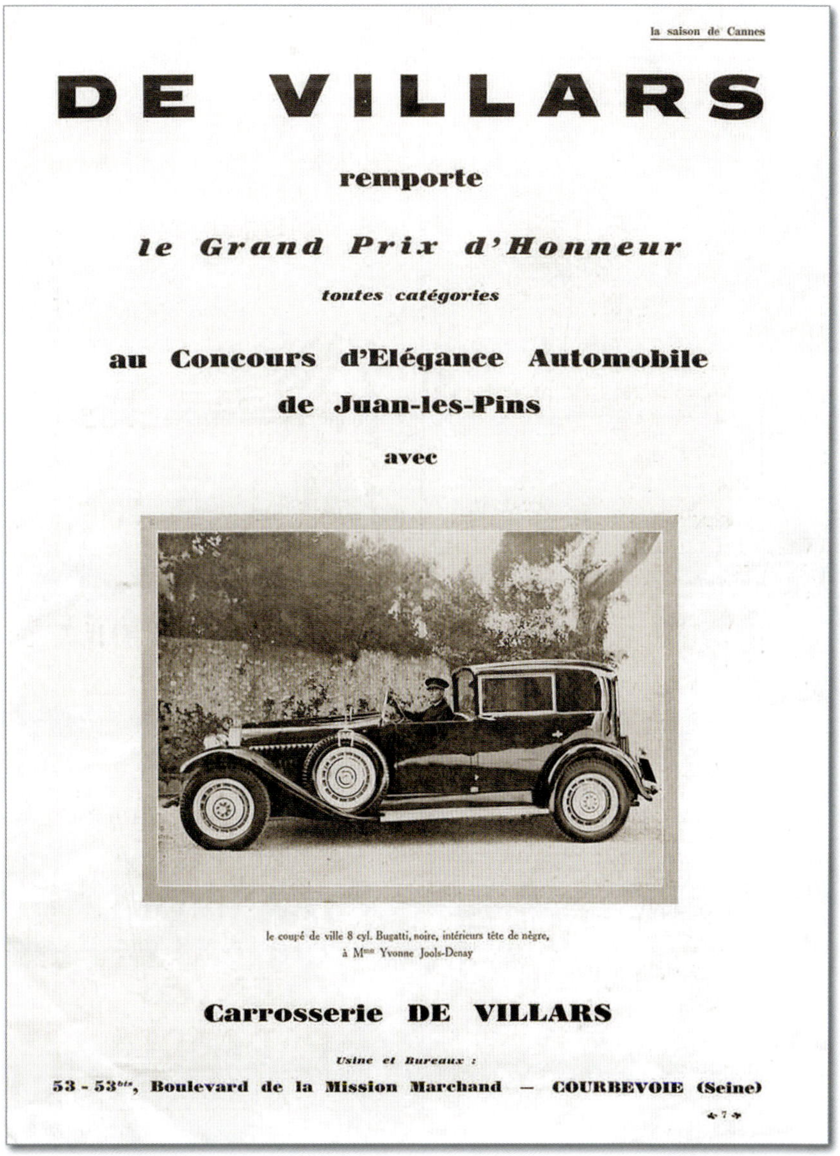

la saison de Cannes

DE VILLARS

remporte

*le Grand Prix d'Honneur*

*toutes catégories*

au Concours d'Elégance Automobile

de Juan-les-Pins

avec

le coupé de ville 8 cyl. Bugatti, noire, intérieurs tête de nègre,
à Mme Yvonne Jools-Denay

Carrosserie DE VILLARS

*Usine et Bureaux :*

53 - 53bis, Boulevard de la Mission Marchand — COURBEVOIE (Seine)

7

## Magic of the night concours (*Le journal des étrangers. Cannes, 1937*)

These elegant and luxurious companions of our seasons – our cherished automobiles – truly deserved to be honored with a glamorous nighttime celebration. Until now, they had been restricted to daytime festivities under the bright sun, and come evening, they were left waiting idly outside casinos, hotels, or restaurants. At last, justice has been served to these nocturnal beauties, as they took center stage in the first-ever evening concours d'elegance, held this past Monday with resounding success.

The scene unfolded in the enchanting gardens along the Croisette, near the Palm Beach. Palm trees were transformed into exotic spectacles, bearing loudspeakers and electric lamps like peculiar fruit. The moon lent its silvery glow, painting the cars with a soft, dreamlike light – an ideal Duco sheen, perfect in its delicacy. The atmosphere was a blend of human ingenuity and natural luminescence, with cacti and aloes, the Valley of Honor shimmering with light, and the Flowered Bridge adorned in floral splendor. The stands overflowed with eager spectators, while the jury, busy and attentive, prepared to evaluate the spectacle.

The panel of judges, consisting of Madame de la Grèze, Prince Lotfallah, Viscount Gérard de Nentque, Count Max de Pilaski, and Mr. Charvoz, made their assessments guided by a sense of soul, conscience, and refined taste. With daring flair, they even dared to rate, on a scale from 0 to 20, the individual elegance of the drivers – a bold move, as if beauty and style could ever be reduced to mere numbers!

After several rounds of careful inspections and two or three parades, the cars made one final pass over a bridge bathed in moonlight, shimmering in a silver glow. This last procession proved rewarding, as the competitors received their accolades: some left with a gleaming Cup, others with a plaque, and all with placards proudly declaring their triumphs as *Grand Prix* or *Prix d'honneur* winners.

With the celebration concluded, the cars departed triumphantly, their hard-won distinctions adorning their exteriors, a testament to their grace and splendor.

***Mr. and Madame Imbert displayed their Delage D6-70 with a "panoramic coach" body by Letourneur et Marchand at the 1936 Cannes night concours. A few months later, the same car participated in the 1937 Monte Carlo Rally, starting from Umea in Sweden and finishing a respectable 17th. (Daniel Cabart Archives)***

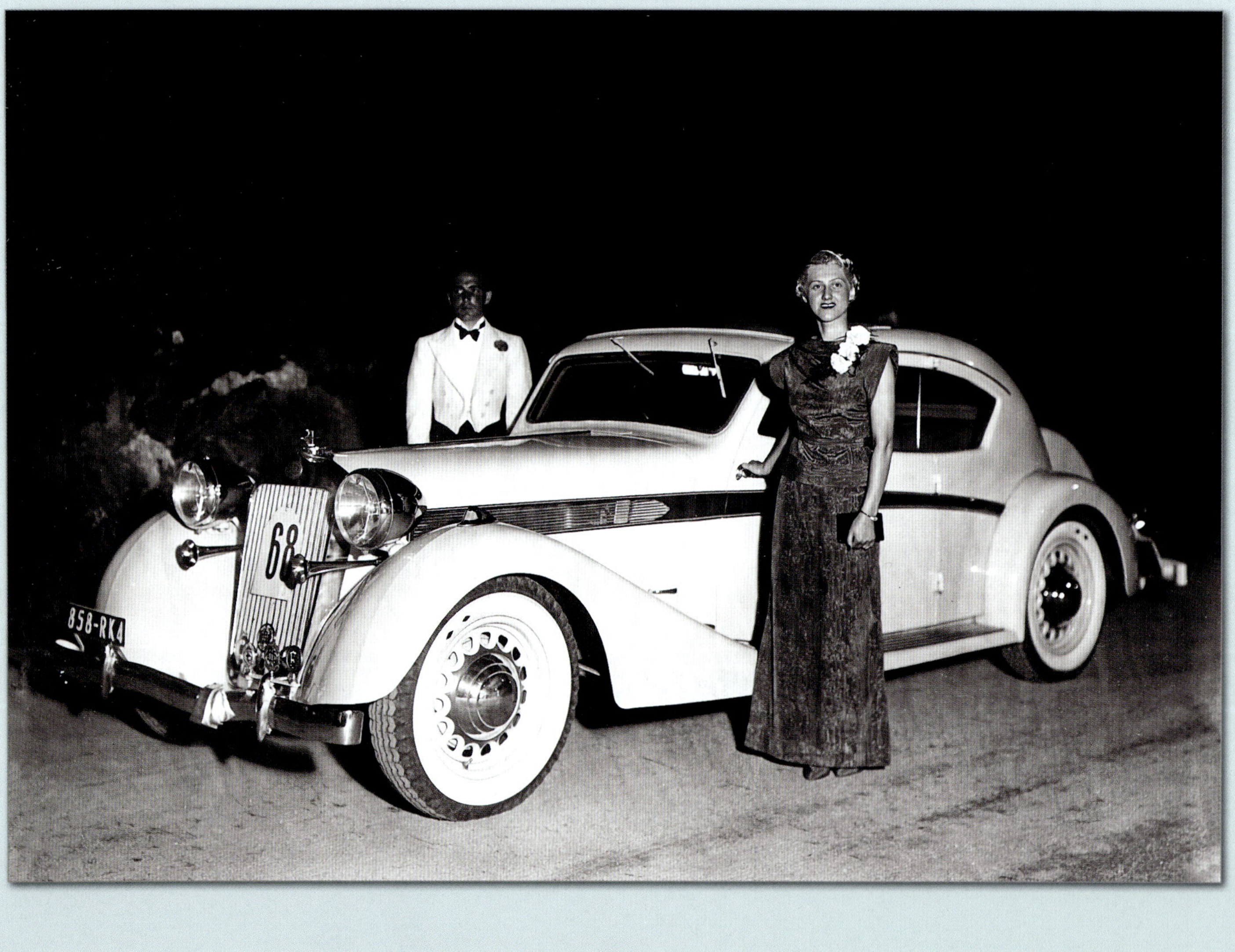

The allure of night remained irresistible. In 1938, Cannes and other Côte d'Azur resorts revisited the magic of evening elegance competitions. The magazine *La Saison de Cannes* described the memorable night of August 19 under the title "Elegances dans la nuit":

> *"For a beautiful night, it truly was an exquisite one. While there were countless beautiful evenings that summer, this one had that special something, a sparkle that set it apart. You may argue that concours d'elegance parades happen every night outside venues like Palm Beach, the Casino de Juan, or the Sporting de Monte Carlo. You could say there are always elegant young women at 'Paradise,' 'Hollywood,' or Victor's. True, Mr. Jean-Gabriel Domergue and the Mayor of Cannes aren't isolated in an ivory tower; projectors and speakers aren't rare sightings, and the Automobile Club bosses are very real. But when all these elements come together, you get 'a nocturnal and automobile feminine concours d'elegance contest.' A lengthy title, indeed, but one that perfectly captures the essence of the August 19th soiree. To complete the image, imagine the fifty most beautiful cars from the Côte d'Azur, a hundred of the loveliest women, some pedigreed dogs, and good cheer filling the air.*
>
> *This enchanting event unfolded on the Croisette, in the Jardin Albert-Ier, centered around a basin adorned with waters sparkling like diamond aigrettes. The elegant lady drivers maneuvered their cars up an inclined ramp, gracefully stepping out, circling the track, and then getting back in to drive away. Simple yet sophisticated, a true feat in its own right. It was a parade of excellence prepared long in advance, involving coachbuilders, couturiers, jewelers, hairstylists, shoemakers, and even manicurists. All this meticulous effort culminated in an ephemeral perfection – a fleeting yet radiant reward for everyone involved."*

Talbot skillfully capitalized on the popularity of concours d'elegance competitions through a strong partnership with top coachbuilders – such as Figoni et Falaschi – alongside couturier Nina Ricci and the most distinguished presenters. Here, Stella Mudge, the Maharani de Kapurthala, takes part in the Trocadéro Concours of 1938. (Daniel Cabart Archives)

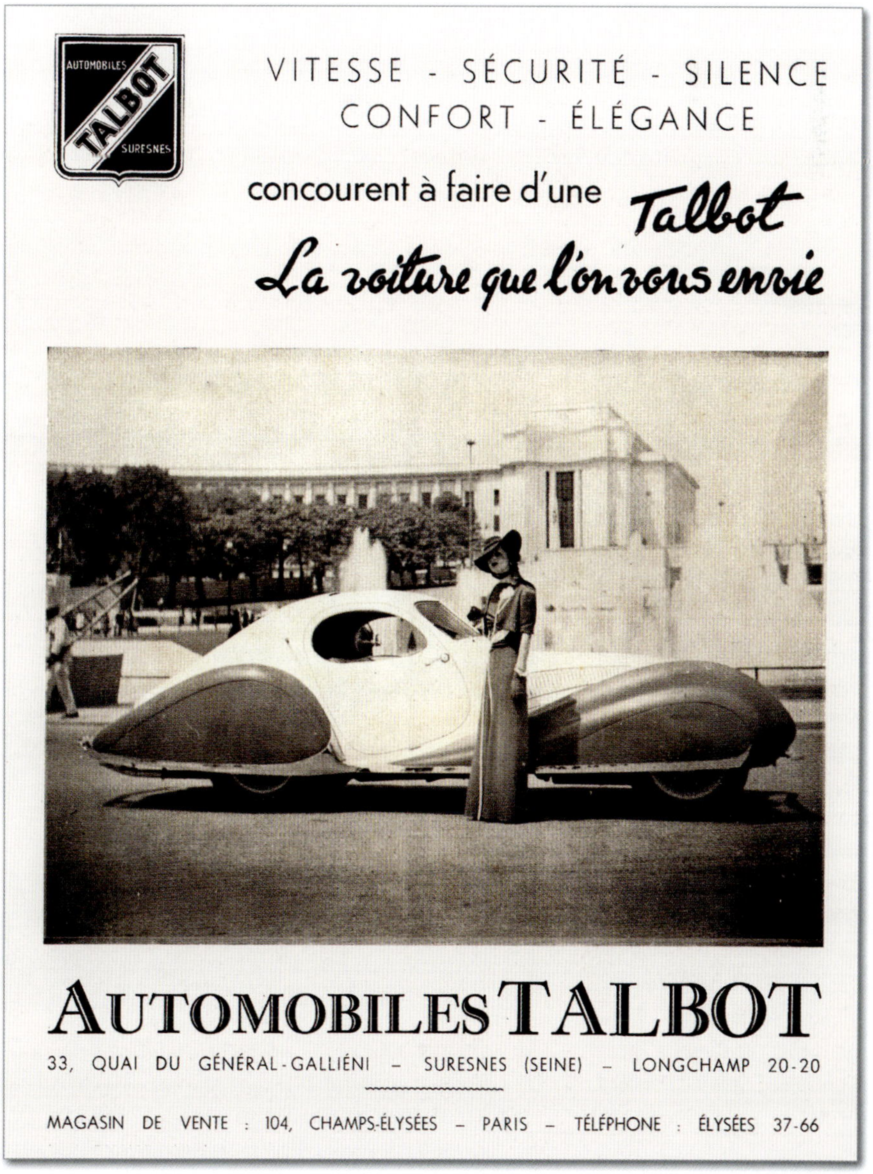


Stars and automobiles soon became an inseparable pairing. Over the years, the car owners were increasingly joined by prominent women, whose fame drew more attention than even the most beautifully sculpted automobile. Soon, celebrities from cinema, theater, and music appeared alongside the aristocratic ladies and socialites who had once dominated the scene.

> *"One by one, the competitors drove up the small path leading to the jury's stand. A gentle murmur of admiration followed their approach. Such exquisite ensembles, such refined silhouettes. The jury would have a hard time choosing the winners. Upon reaching the jury, the cars paused. The drivers emerged, warmly welcomed by Mr. André de Fouquières, who introduced them to his fellow judges with his trademark charm and humor... Here were delightful artists: Mlles de Ranwera, Marie Glory, Marie Dubas, Meg Lemonnier, Joséphine Baker, Mona Païva, Nadine Picard, Paulette Dubosc. Present were also Madame Senwiek, the Princess of Kapurthala, Madame J.P. Gaillard, Madame Hayen, and Madame Mary Costes... The parade continued, with each elegant vision more captivating than the last."*

The celebration of women and automobiles became ever more refined through the 1930s. By June 10, 1937, *La Vie Automobile* reflected on the grandeur of the recent Parisian competition in the Bois de Boulogne, noting:

> *"Automobile concours d'elegance competitions, even if they often overlook the mechanical or practical merits of the vehicles, are always a delightful spectacle. To truly succeed, four*

*elements must be perfectly combined: a stunning car, a beautiful woman, a chic outfit, and a well-groomed dog. The exact order of importance is debatable, but their harmony is essential. The ultimate aim is a seamless symphony of colors. I recall being entranced by a black and white ensemble at one such event, where the car's body, the woman's attire, and a striking Borzoi's coat created a vision of supreme elegance. It's a heavy task for the coachbuilders, designers, and even the pet groomers, all playing crucial roles in this display."*

The automobile had become one pillar among four that upheld these magnificent expressions of style.

Finally, to wrap up this glimpse into the history of the women's automobile concours d'elegance

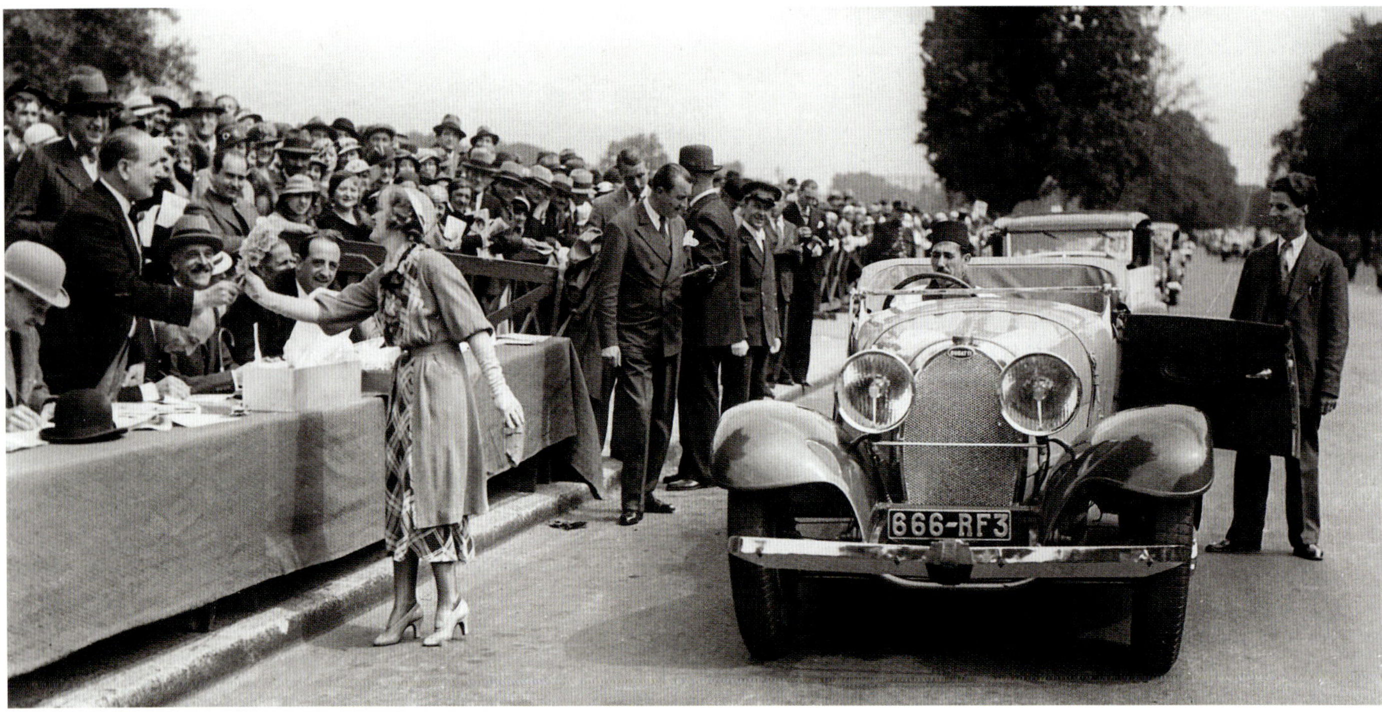


**ABOVE:** "Mr. André de Fouquières, who presented the prizes, was an attraction in himself. His presence alone would have made attending the *concours* d'elegance worthwhile, even if the event had already been flawless. For each winner, Mr. de Fouquières always had the perfect words, leaving them with a joy that came from his heartfelt speech, more so than from any plaque or banner they received," *explained actress Colette Broïdo, recalling the Fémina L'Intran Concours d'Elegance, June 23, 1933, featuring the Bugatti Type 43 Cabriolet Franay.* (Bernard Brulé)

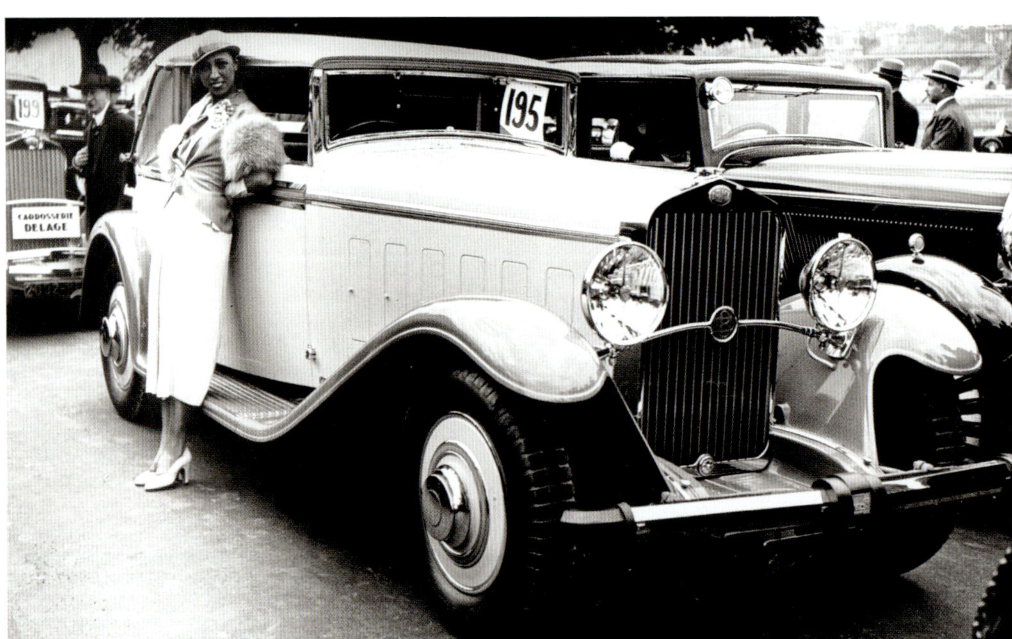


**ABOVE:** *One of the most iconic figures of the time was Joséphine Baker, a prominent celebrity and personal owner of a Delage. She not only cherished the brand but also skillfully promoted it, particularly with the introduction of the new eight-cylinder 'Queen of Concours d'Elegance' competitions. Pictured here is a D8 convertible, featuring bodywork by Letourneur et Marchand.* (Daniel Cabart Archives)

competition and its uniquely French flair, there was a report from *L'Auto* on its annual double competition, which encapsulated this union of style and sophistication.

The 1939 *L'Auto* Concours d'Elegance event in the Bois de Boulogne stands as a testament to the sophistication and artistry that characterized pre-war Parisian society. As described in the June 9th event report, this annual double competition was a celebration of style, engineering, and harmony, merging the world of automobiles with high fashion. It was more than a simple exhibition of beautiful vehicles; it was a grand meeting place for all of Parisian society – aristocrats, industrialists, and socialites alike.

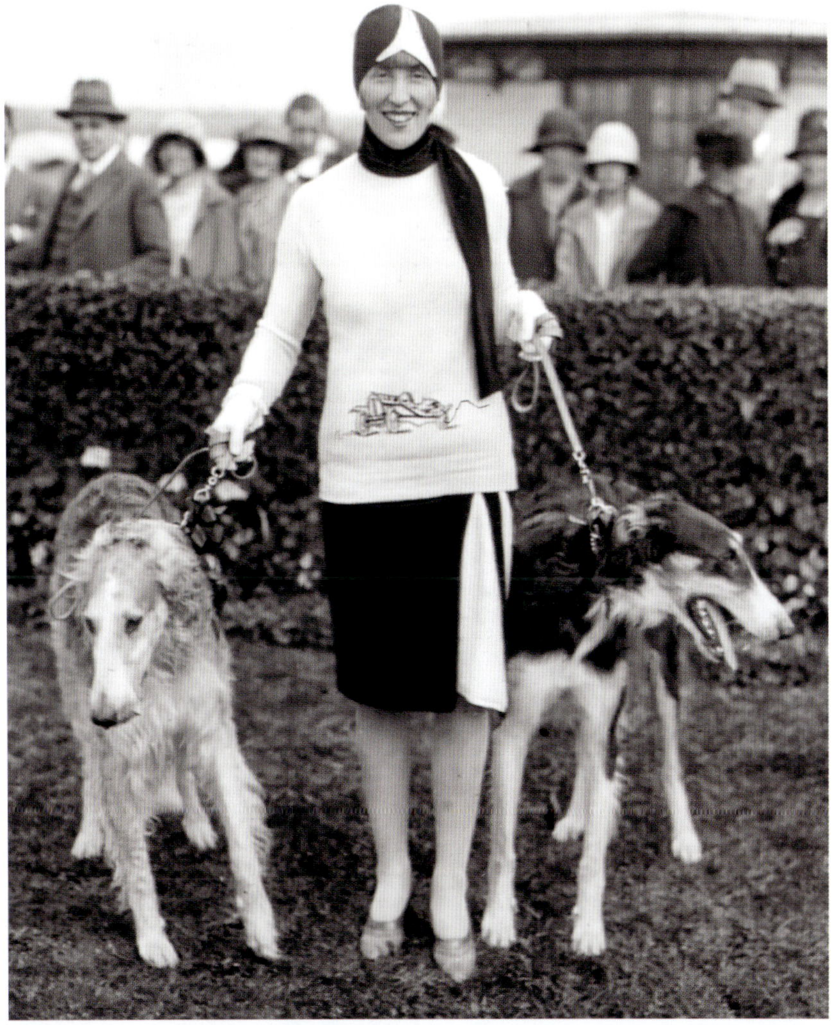

**RIGHT:** *An elegant woman, accompanied by her two dogs, was an "essential" element in any presentation aiming to stand out. These "four-legged friends" played a crucial role in winning over the jury. This particular competitor boldly embroidered her beliefs onto her attire, making a statement with both style and conviction.* (BnF Gallica)

**BELOW:** *An elegant car –* a Hispano-Suiza, *a* stylish Milord *convertible by* Binder *– and two* Borzois. *The perfect embodiment of the winning formula:* "a beautiful car, a beautiful woman (or two), a beautiful dress (or two), and a beautiful dog (or two)." (Bernard Brulé)

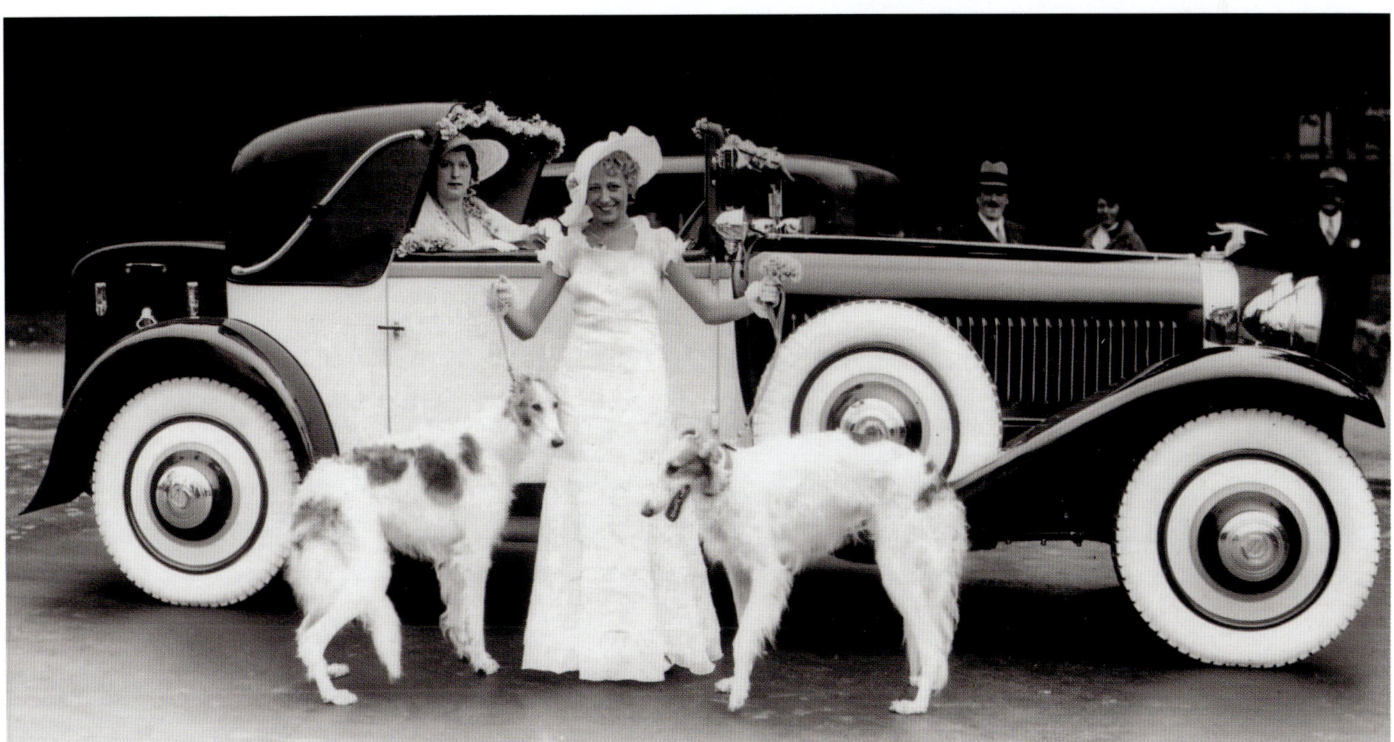

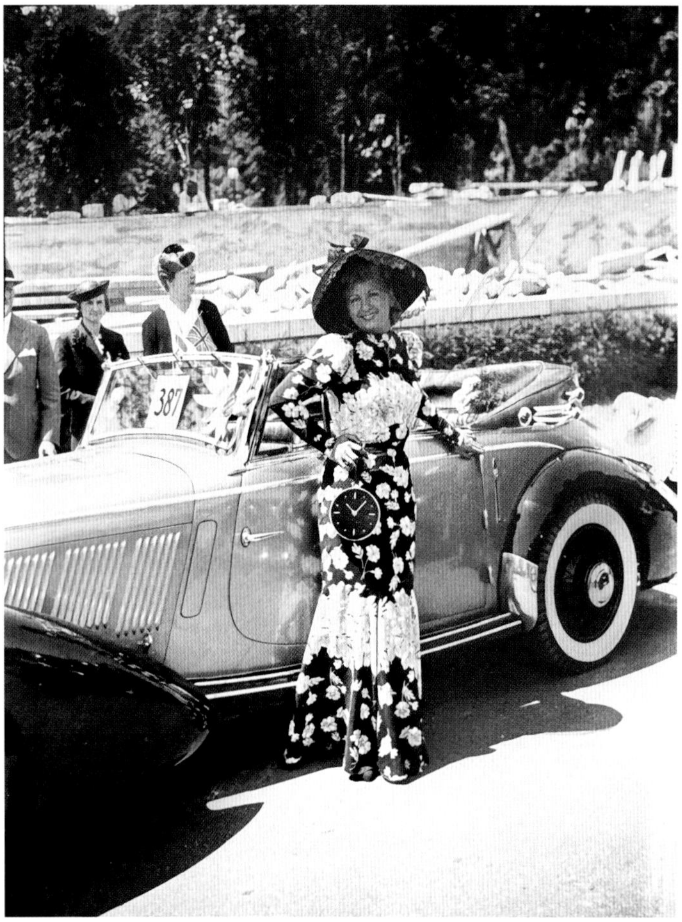
387

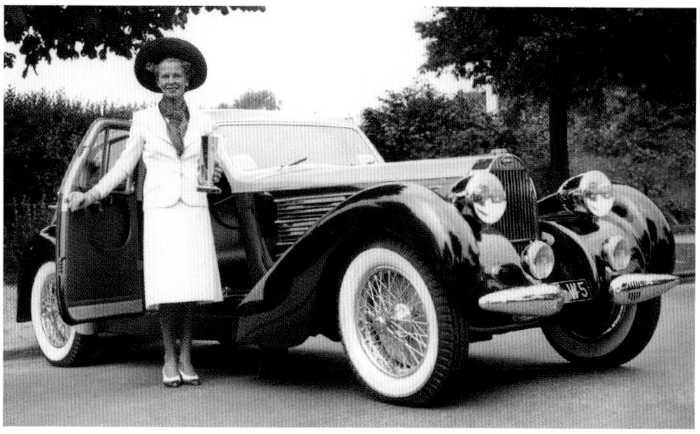

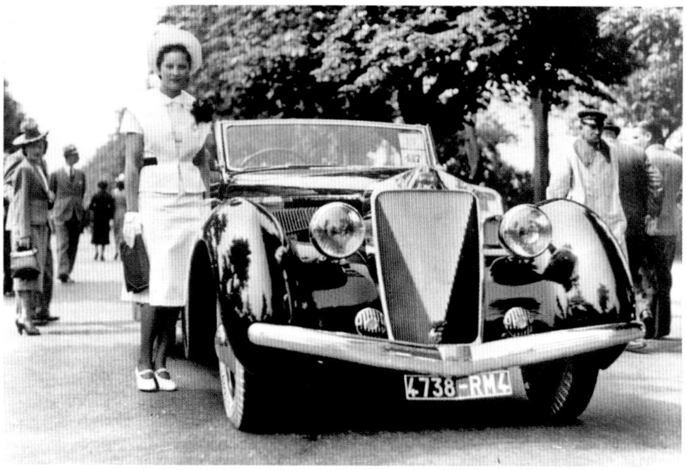
4738-RM4

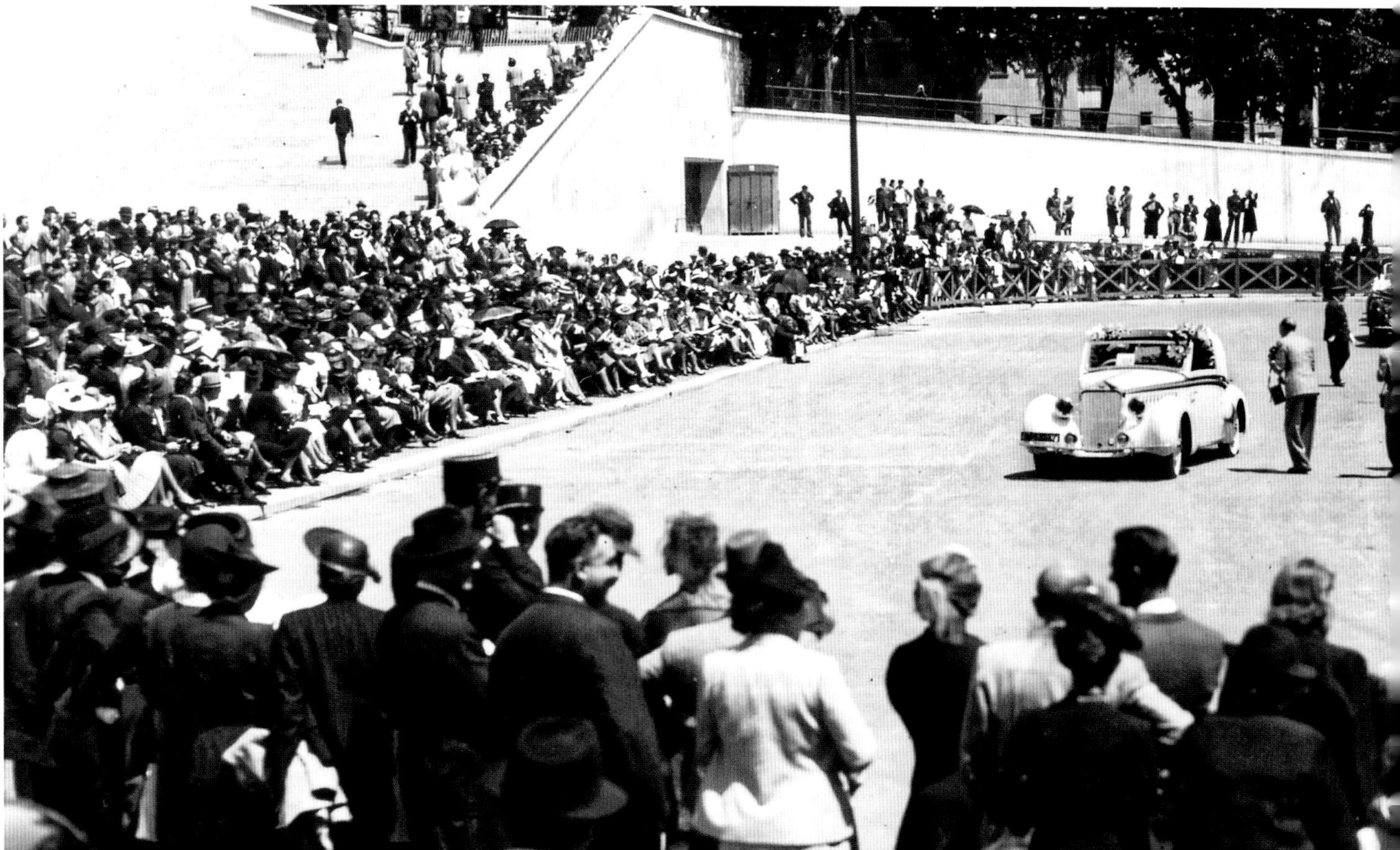

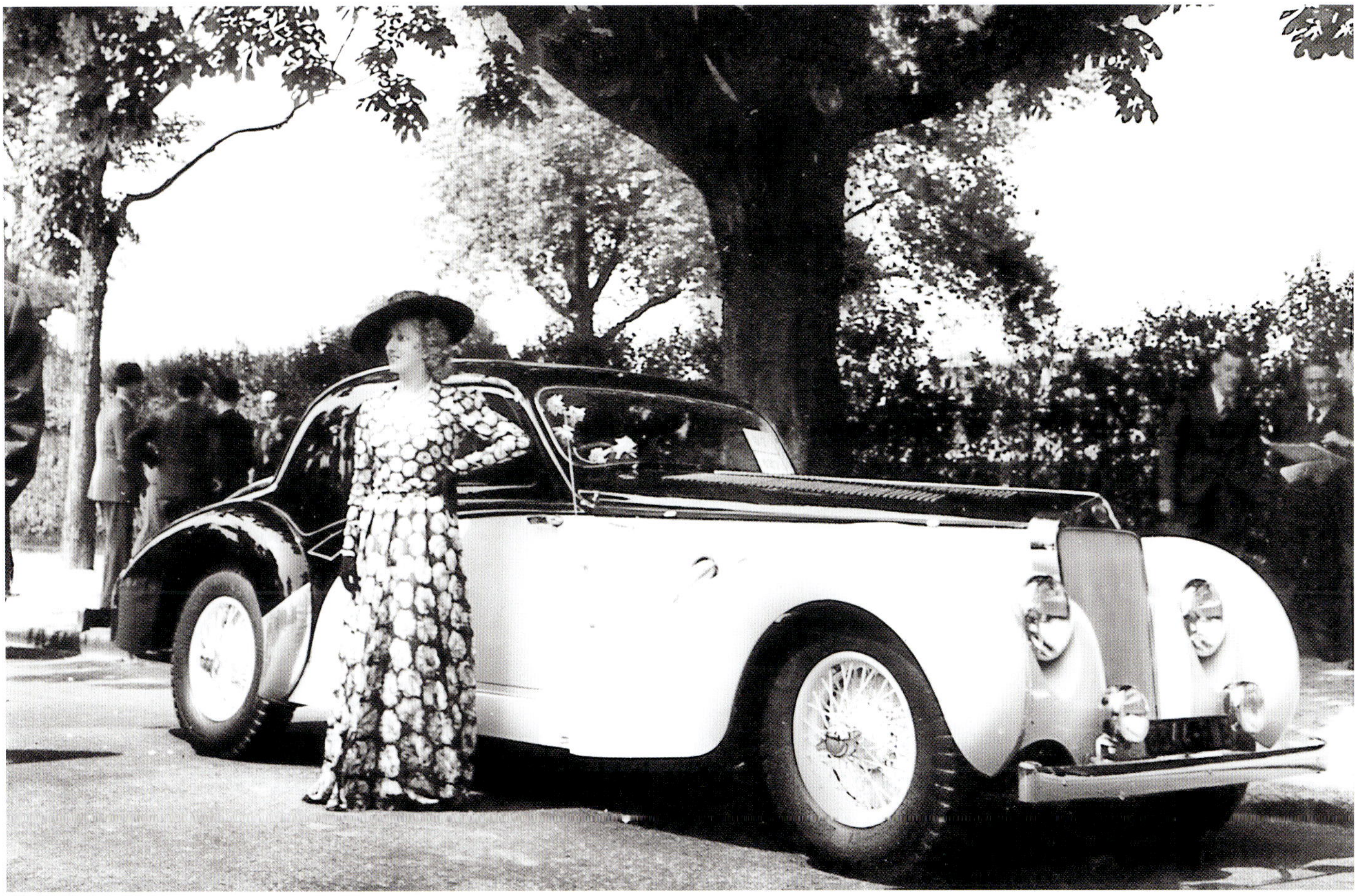

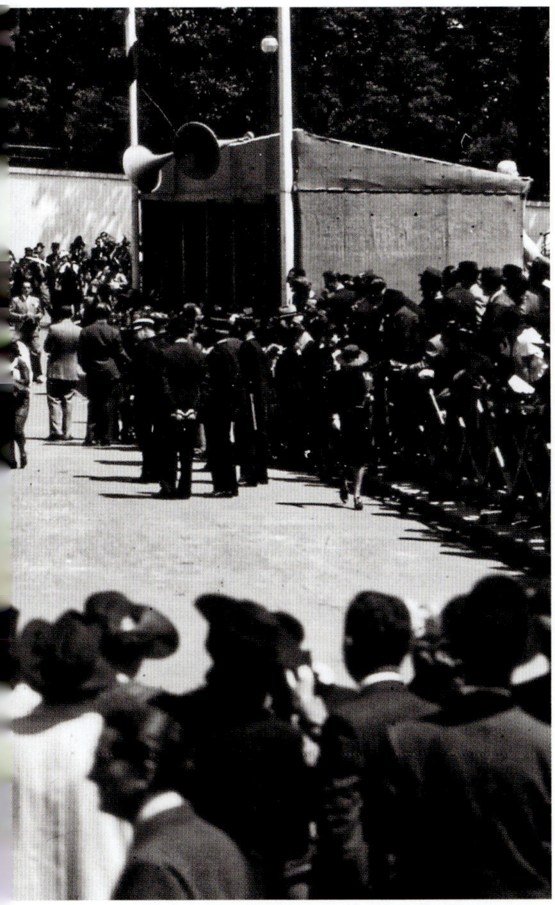

**ABOVE:** *At the L'Auto competition on June 10, 1939, Mlle. Marguerite Valmont appeared with a Delage D8-120 S with coachwork by Vanvooren, styled as a coach (or faux-cabriolet) with sheet metal. In this category, the car won the Grand Prix for cars bodied by coachbuilders. (Daniel Cabart Archives)*

**OPPOSITE TOP LEFT:** *The concours d'elegance: a platform for milliners, couturiers, and accessory designers to showcase their craft. (Marc Caveng)*

**OPPOSITE TOP RIGHT:** *At the L'Auto competition on June 10, 1939, the Countess d'Oncieu de Chaffardon presented a Bugatti Type 57 Galibier, dressed by Jean Patou in a white moiré suit. She wore a blue muslin scarf, also in matching tones, with navy felt trim and navy gloves. This elegant ensemble perfectly complemented her blue-and-white Bugatti, capturing all the available votes. She went on to win the Grand Prix for Féminine Elegance. (Laurent Friry Archives)*

OPPOSITE **MIDDLE RIGHT:** *Sonia Stevenson-Watney, niece of industrialist Walter Watney (then CEO of the newly established Delage company after 1935), was dressed by Callot Soeurs in this presentation. She was showcasing a Delage D6-70 convertible by Figoni et Falaschi, with its grille painted in "blue-white-red." (Daniel Cabart Archives)*

**LEFT:** *On June 23, 1939, "In the elegant surroundings of the Palais de Chaillot gardens, the event unfolded in a refined atmosphere of true Parisian sophistication... Spread across the neighboring terraces, offering a panoramic view of the entire competition, the crowd gathered in large numbers, undeterred by the heat of the sun, admiring the dresses and cars for two hours. The great couturiers were in the spotlight..." More than any other concours, the one organized by Fémina placed the greatest emphasis on "dressing in harmony with the colors of the car." (Daniel Cabart Archives)*

**BELOW:** On June 24, 1938, "the traditional competition for feminine elegance in automobiles, organized by *Fémina* and *L'Intransigeant*, was held in the lush setting of the new Trocadéro gardens. Beyond its beauty, the chosen location offered a clear advantage over others, as it was far better suited for the parades the competitors were required to perform." *Among the participants was a Salmson S4 DA convertible, presented by the renowned aviatrix* Maryse Hilsz. (Marc Caveng)

**OPPOSITE:** *Madame Richer-Delavau, a frequent participant in elegance competitions and rallies, is pictured here presenting her Simca 8 Sport, freshly returned from the 24 Hours of Le Mans, at the Fémina L'Intran Concours d'Elegance competition in June 1939. The car earned a "special mention" in the category of "cars driven by the competitors themselves."* (Marc Caveng)

***The event featured two main competitions:***

1. ***The Concours d'Elegance de Confort et de Technique de la Carrosserie:*** Here, vehicles were judged not only for their aesthetic appeal but also for the technical quality of their craftsmanship. The jury, composed of automotive and design experts like Charles Faroux, Henri Petit, and Jean Henri-Labourdette, assessed the ingenuity of the car fittings, the interior luxury, and the overall sophistication of the design. Each entry was expected to showcase the peak of bodywork innovation and elegant presentation.

2. ***The Trophée International d'Elegance Féminine Automobile:*** This contest focused on the elegance of the lady drivers and how their outfits were perfectly coordinated with their vehicles. The female jury evaluated the harmony between the dresses, designed by top couturiers, and the car's exterior. This event represented the essence of French chic, with couturiers and coachbuilders collaborating to create seamless ensembles that embodied refinement and style.

Under the radiant June sun, the event unfolded in a stunning setting at the Auteuil stands, with the lush greenery of the Butte Mortemart as a picturesque backdrop. It was a morning of splendor, filled with the fragrance of flowers, the gleam of polished automobiles, and the vibrancy of springtime fashion.

### *Key Highlights of the Event:*

- Countess d'Oncieu de Chaffardon dazzled in a white moire suit by Jean Patou, complemented by a blue muslin scarf and navy felt hat, perfectly matched to her blue and white Bugatti. Her ensemble captured the admiration of the crowd and the jury, demonstrating the high standards of taste that defined the competition.
- Madame Diana appeared in a playful blue silk dress with white polka dots, while Viviane Elder, the famous aviatrix, wore an eye-catching dress paired with a wide-brimmed hat, which made her disappear from public's sight.

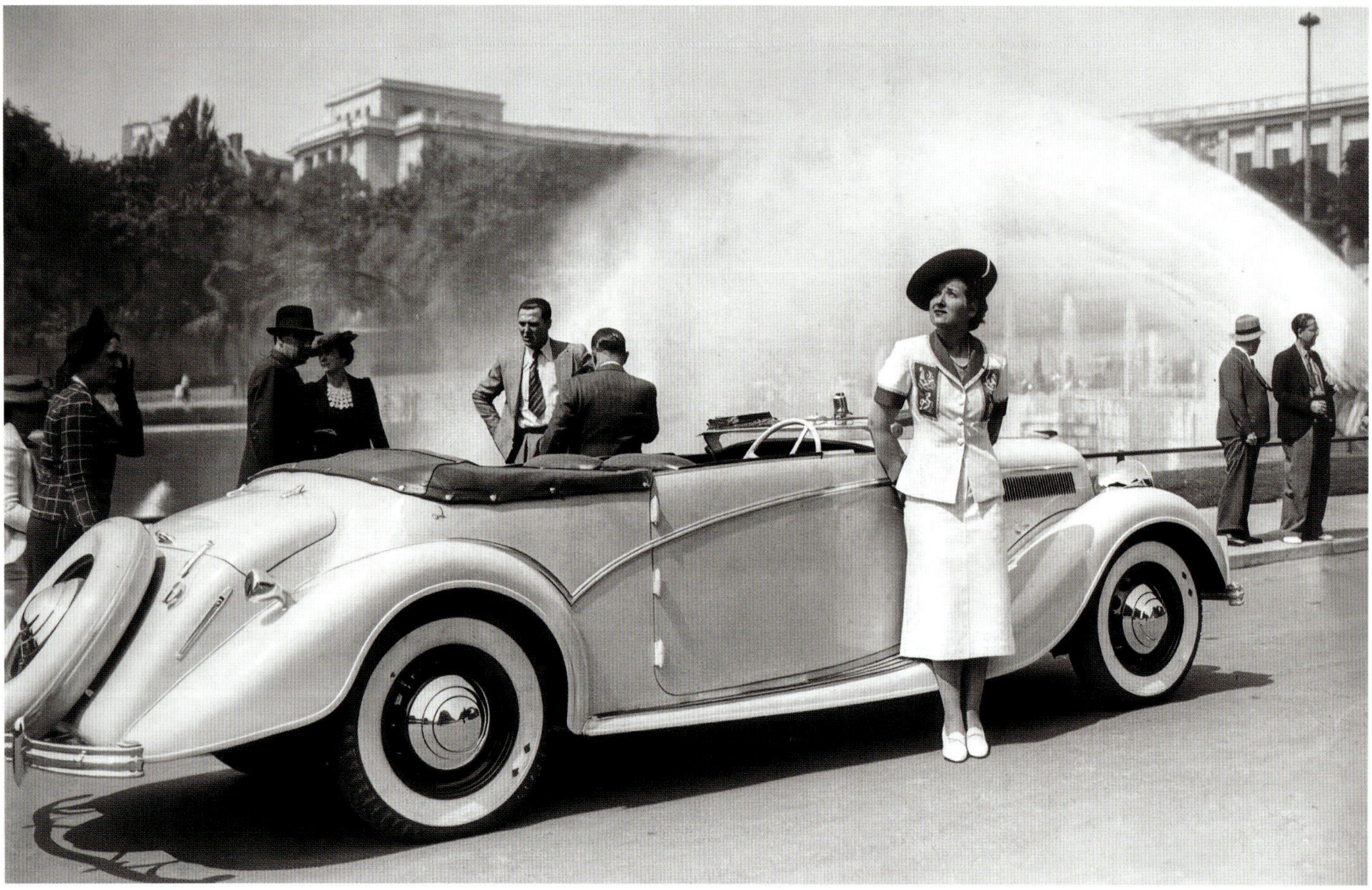

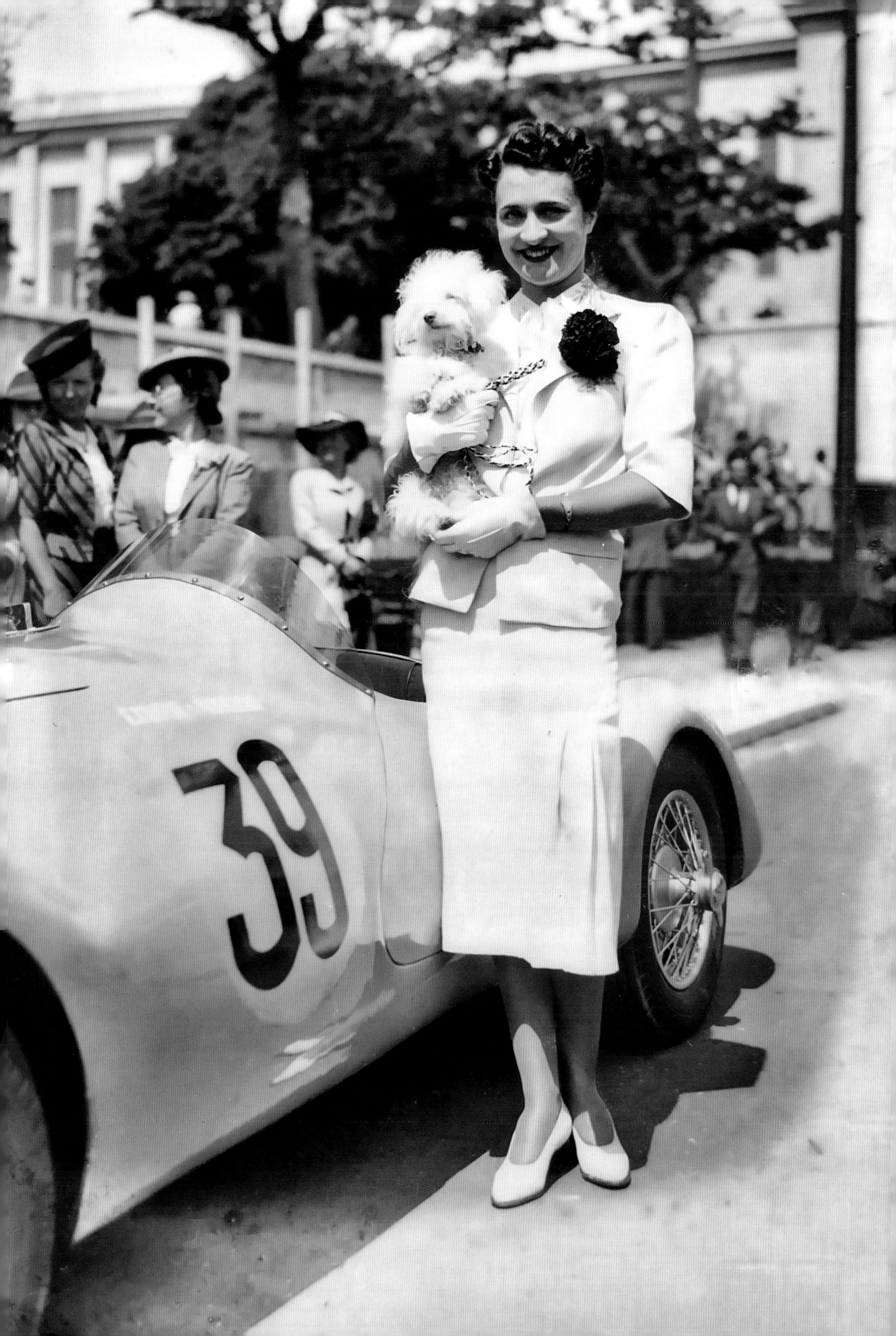
39

- Madame Marguerite Valmont turned heads in an elegant black and green outfit, and Madame Stevenson took national pride to another level by decorating her car's radiator with the tricolor flag.
- The event also featured Mlle. Moineau's charming pink and sky-blue outfit, Mlle. Sheffield's delicate pink and white print, and Mrs. Nils Boklund's striking pink and black lace ensemble.
- Gaby Fournier wore a cream linen suit, and Madame Restropo made a statement by arriving with her majestic Great Dane. Madame Robert Ricci impressed all with white silk epaulettes, while Mlle. Yolande Bemova and Mlle. Paulette Langlais brought their impeccable style with their respective prints and cream suits embroidered in black.

The entire event was a dazzling showcase of French elegance, blending automotive artistry with high couture in a way that no other country could replicate. The meticulous attention to detail, from the cars' interiors to the ensembles worn by the drivers, reflected quintessential Parisian qualities of refinement and creativity. The *L'Auto* concours, as described, was a spectacle of unparalleled sophistication, embodying the "French-style" elegance that became a hallmark of international prestige.

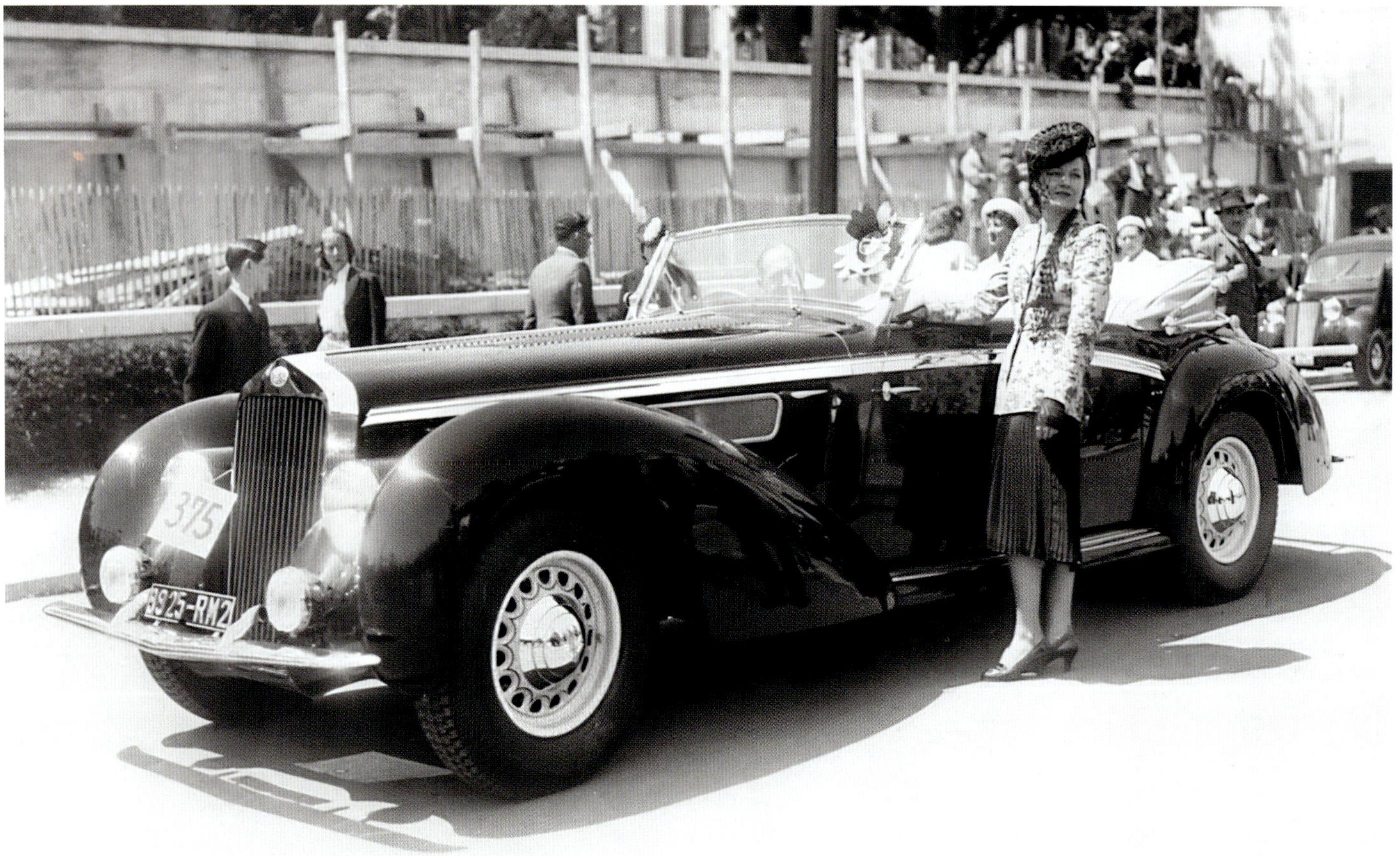


**ABOVE:** *American actress June Sheffield presented a Delage D8-120 "Deltasport" convertible by Letourneur et Marchand at the 1939 Trocadéro Competition. A few months later, the car was requisitioned by the German army and awarded to Otto Von Stülpnagel, the head of the German occupation forces in France. It would often be seen driving along Avenue Foch in Paris.* (Daniel Cabart Archives)

**OPPOSITE:** *At the competition hosted by* L'Auto *in June 1939, renowned aviator Viviane Elder displayed a Delage D8-120 convertible featuring bodywork by Letourneur et Marchand.* (Daniel Cabart Archives)

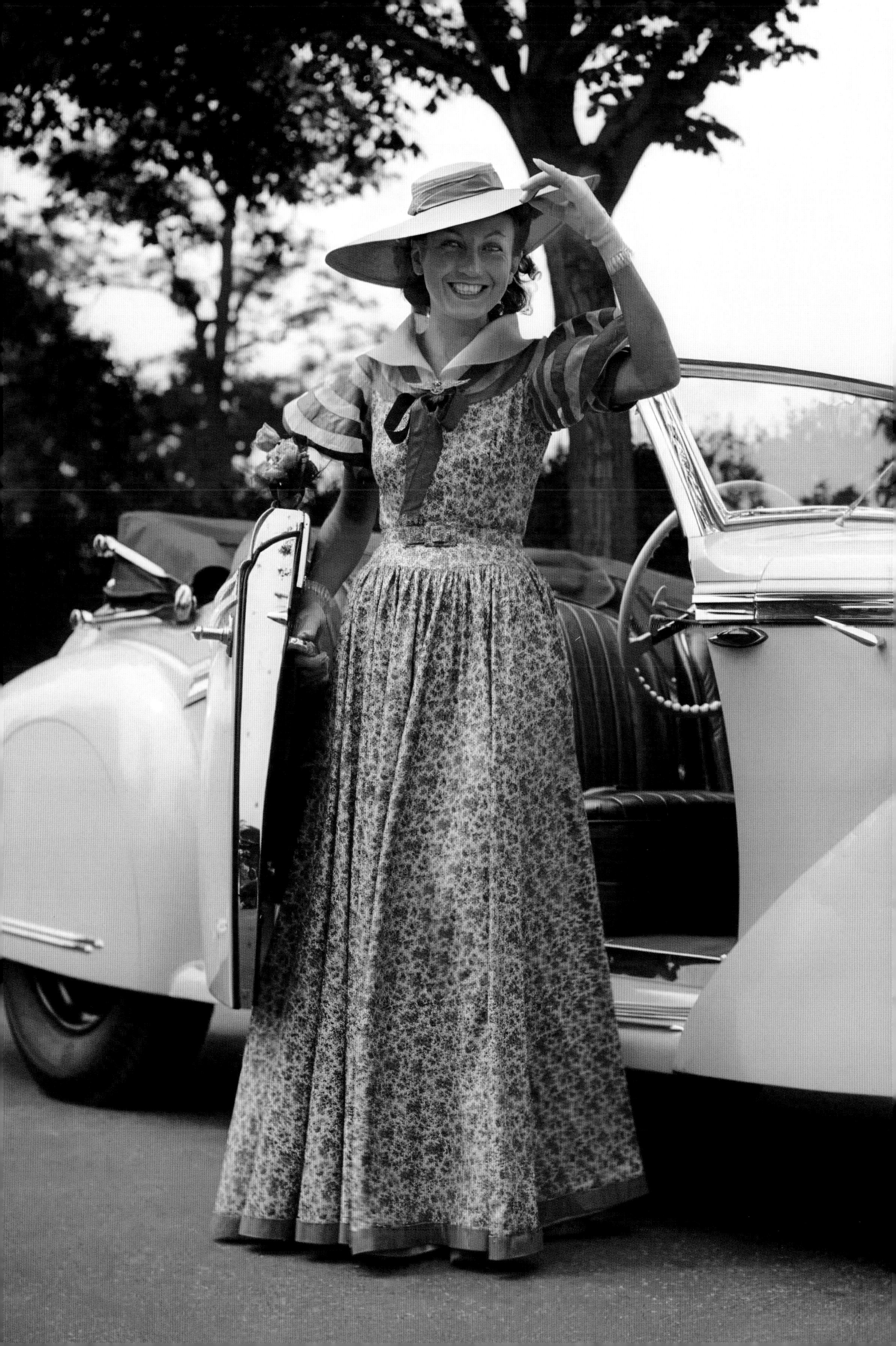

# The 'Queen of Concours d'Elegance'... The Delage D8

Bugatti and Delage – two iconic French automotive brands – share the distinction of being the only French manufacturers to win the title of World Constructors' Champion during the interwar period. However, once this milestone was achieved, each brand chose a distinct path to maintain its presence in the public eye.

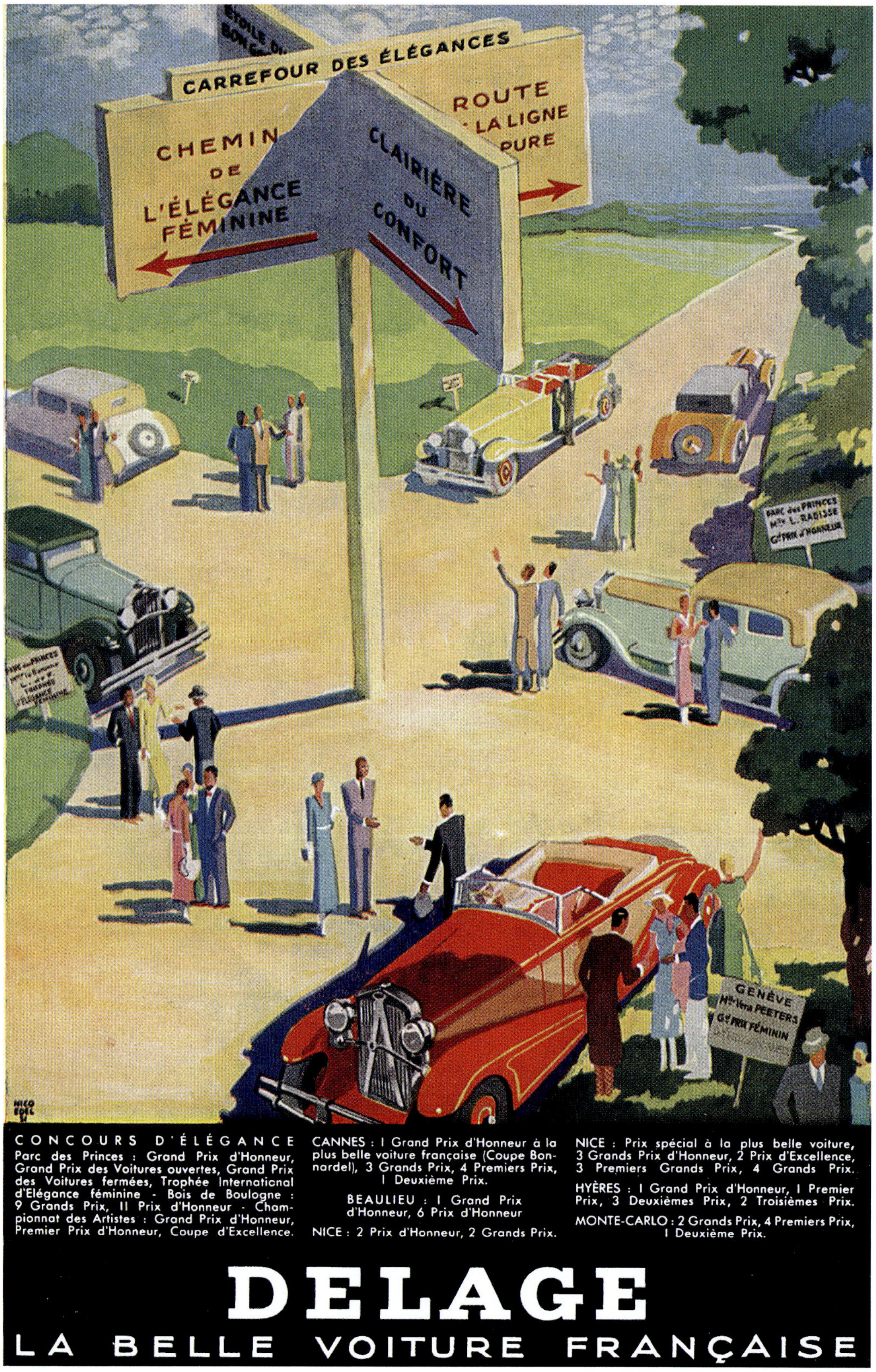

Bugatti focused on producing Grand Prix cars accessible to private drivers, ensuring its name appeared regularly in race results. In contrast, Delage shifted its focus toward unmatched manufacturing excellence and the refined elegance of its designs, appealing to connoisseurs of luxury and style.

As concours d'elegance events gained prominence, Delage recognized the importance of showcasing sophistication and craftsmanship. While the GL 30 CV, released in 1924, was undoubtedly among the finest cars available, its limited visibility in these prestigious events meant it fell short of giving Delage the widespread recognition it deserved.

Determined to shine in this arena, Delage refined its designs and elevated its presence, cementing its reputation as a symbol of elegance and prestige. The Delage, in particular, emerged as the epitome of style and engineering, earning its place as the 'Queen of Concours d'Elegance'.

Despite the elegance of the DM 17 CV, which occasionally featured in the prize lists of certain competitions, the leading and prestigious brands – such as Hispano-Suiza, Renault, and Mercedes – consistently dominated the top positions.

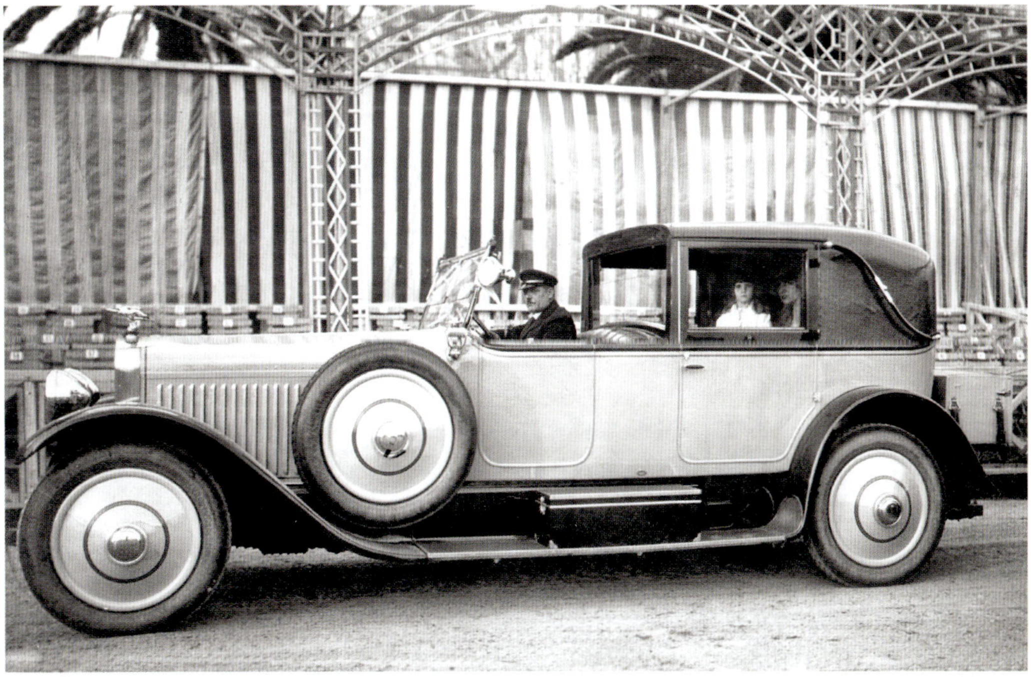

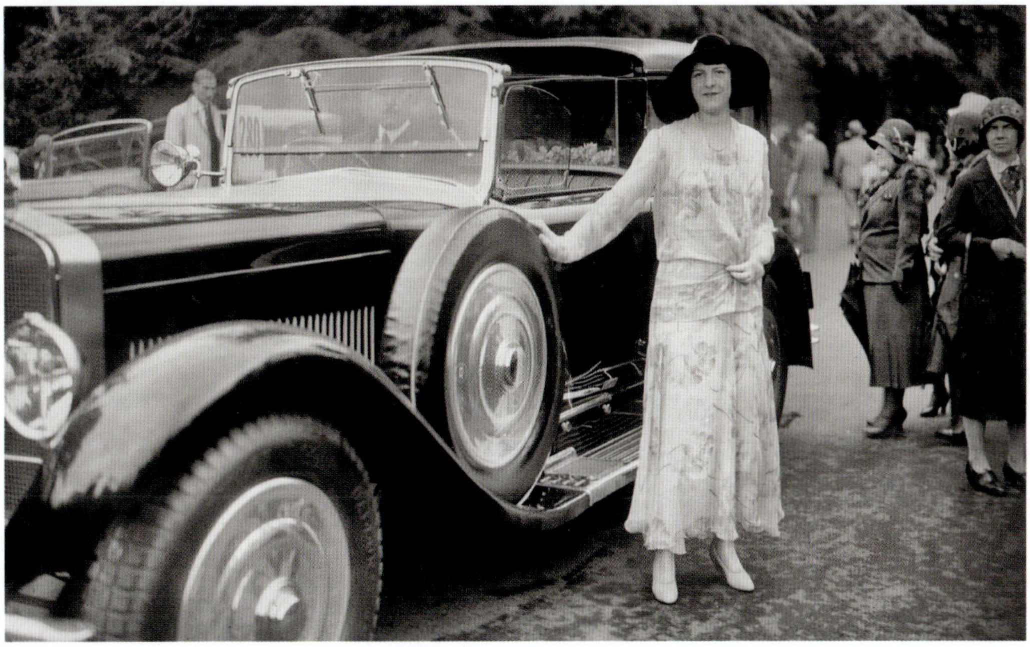

**LEFT:** *A Delage GL 30 CV, coachbuilt by Duvivier for a competition in Monaco, was often referred to as the "rival of the Hispano." Due to its limited numbers, it was unable to provide Delage with significant visibility in concours d'elegance competitions. (Daniel Cabart Archives)*

**BOTTOM:** *Alongside the GL, the stunning DM, though more modest, still exuded a quiet elegance that was often overlooked. While the Delage was known for its speed and silent performance, it had yet to be fully recognized for its elegance – an accolade it truly deserved. One notable example was the DM bodied by Saoutchik, on display at the Fémina L'Intran Concours d'Elegance competition in 1930. There, Madame Levy-Simon, striking a refined figure in her Delage 17 CV, helped Saoutchik earn a Grand Prix. (Daniel Cabart Archives)*

*Starting in 1929, with the debut of 'The Queen of Cars,' advertising, although not yet emphasizing elegance, began to showcase the D8 in one of its key domains: elegance competitions! (Daniel Cabart Archives)*

NOTRE GRANDE MARQUE FRANÇAISE

# LA D-8. DELAGE

REINE DES VOITURES

Dans tous les Concours d'élégance, dans toutes les manifestations, où s'affrontent en un pacifique tournoi, les dernières créations de nos grandes marques automobiles ; hier comme aujourd'hui. **Les DELAGE** s'inscrivent en tête du palmarès, choisies par le jury entre plusieurs centaines des plus magnifiques voitures que les grandes compétitions rassemblent chaque année en France.

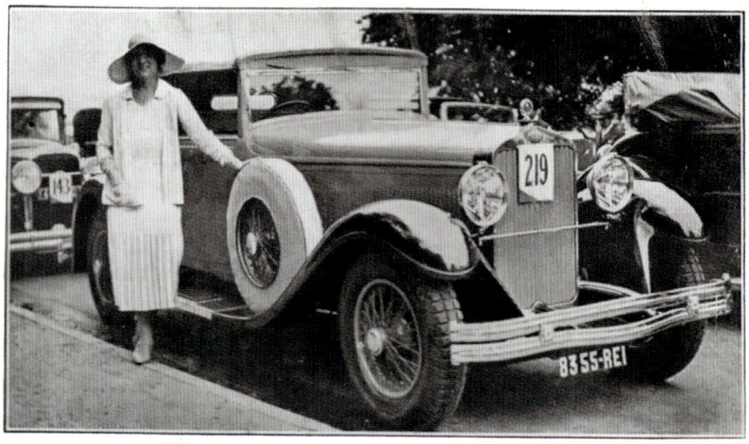


*La* **D-8** *de* Mlle DIANA [illegible] Paris.

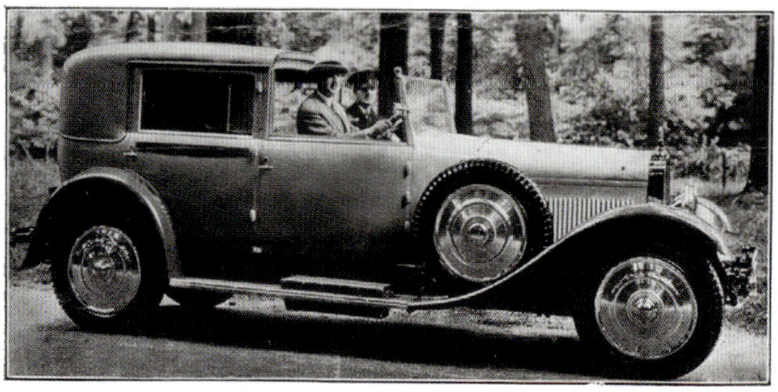

*Le célèbre chanteur-diseur* CRISTI, *dans sa* **D-6 Delage**

**La D-8. DELAGE** a, ces temps derniers, triomphé avec Mlle Andrée Vavon, de l'Opéra-Comique ; avec Mlle Diana ; avec Mlle Mistinguett ; avec Mlle Aimée Mortimer, de l'Opéra ; avec Mlle Arletty, etc.

**La D-8. DELAGE** triomphera demain sous les couleurs d'autres de nos grandes vedettes du théâtre, dont elle demeure la voiture favorite.

# LA D-8. DELAGE

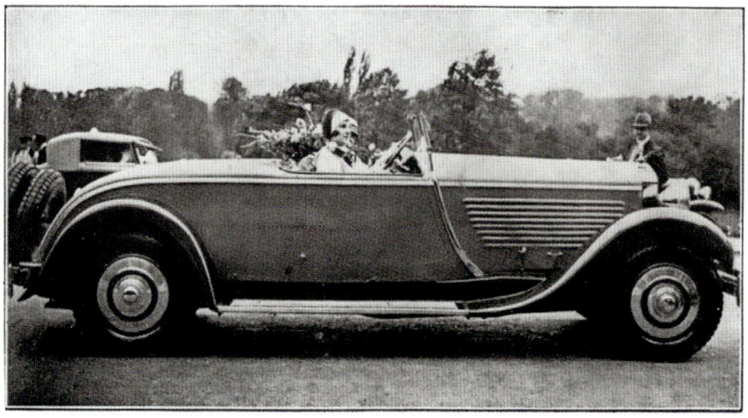

Mlle VAVON, *de l'Opéra-Comique, qui a présenté cette superbe* **D-8. Delage** *carrossée par* VAN DEN PLAS.

*Le Gérant :* G. PATRAULT. Imp. Lang, Blanchong et Cie, 30, rue du Poteau, Paris.

When Delage introduced the D8 in 1929, it marked a turning point that elevated the brand to the ranks of the greatest names in automotive history, securing its reputation for years to come. However, financial difficulties starting in 1933 eventually forced the company to scale back its presence and investments.

Critics were quick to praise the superb proportions of the D8 chassis, famously remarking that *"the Delage is the chassis that best wears a body."* This statement implied that any body designed to complement a Delage chassis would naturally achieve near-perfect proportions, as though guided by a kind of "golden ratio" established by the manufacturer.

Recognizing this remarkable versatility, many of the era's most prestigious coachbuilders eagerly embraced the D8 chassis as

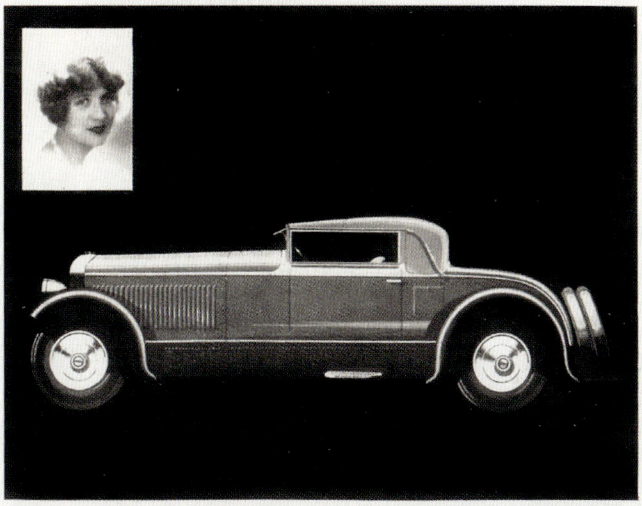

**Mademoiselle Diana, du Casino de Paris, Grand Prix d'Élégance du Championnat des Artistes, présente à ce concours une somptueuse 8 cylindres Delage carrossée par Maurice Proux et naturellement émaillée au NITROLAC.**

**RIGHT:** *The organizers never failed to announce in their programs the presence of prominent personalities from the worlds of theater and music hall. At the 1930 Fémina-L'Intran concours,* "Mademoiselle Diana won the First Grand Prize. She wore a pale pink ensemble. Her D8 was bodied by Maurice Proux and finished in a remarkable shade of 'Antibes green'." *(Laurent Friry Archives)*

**BELOW:** *Mademoiselle Mistinguett received a prestigious banner of honor for her D8, which was bodied by Gallé and finished in elegant green and brown enamel. (Daniel Cabart Archives)*

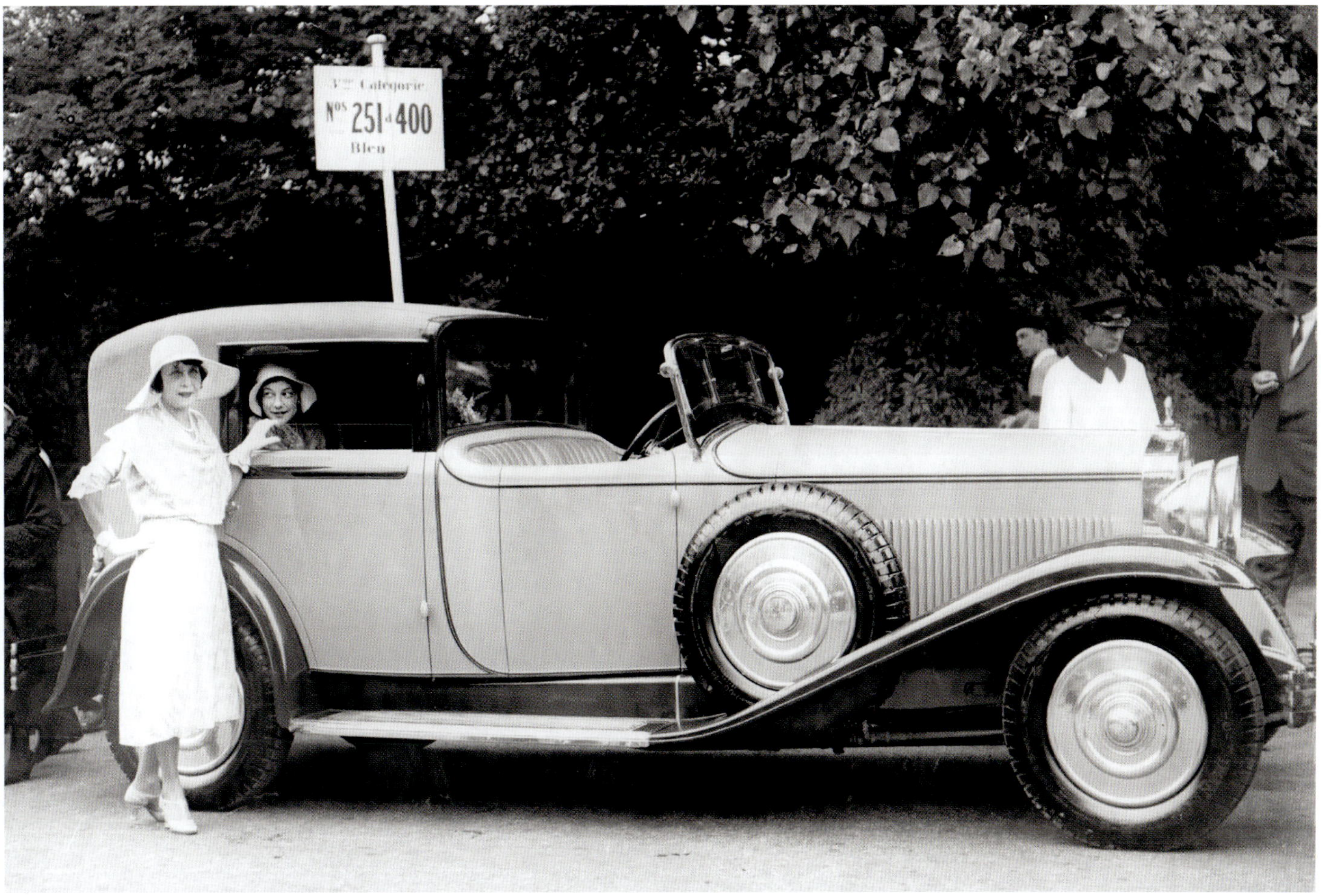

the foundation for creating exquisitely balanced and elegant bodies. Initially working with normal or long chassis, they later experimented with sportier short chassis and, ultimately, the "S" chassis – shorter, lower, and more powerful. Renowned names such as Letourneur et Marchand, Gallé, Chapron, Saoutchik, Grümmer, De Villars, and others produced true masterpieces, solidifying the D8's legacy as an icon of luxury and style.

This iconic model quickly earned the title of the "queen of elegance" or the "queen of elegance competitions."

Delage's sales department established a dedicated "team" that traveled from one competition to another, backed by the company's financial support. As a result, the same cars could be admired at prestigious events in Le Touquet, La Baule, Paris, Nice, as well as Geneva and Saint-Moritz.

Prominent figures from the entertainment world, often serving as presenters, played a key role in promoting the brand. Notable personalities included Laure Diana, Marie Glory, Blanche Montel, Joséphine Baker, Nadine Picard, and many others, further enhancing Delage's allure and reputation.

Following these triumphs, Delage agents and dealers would invariably receive a circular from the factory, urging them to prominently showcase the brand's impressive achievements.

In a move that appears to be unique in the industry, Delage even published a catalog titled "*Delage aux concours d'elegance*", exclusively dedicated to highlighting the brand's participation and remarkable successes in these prestigious events.

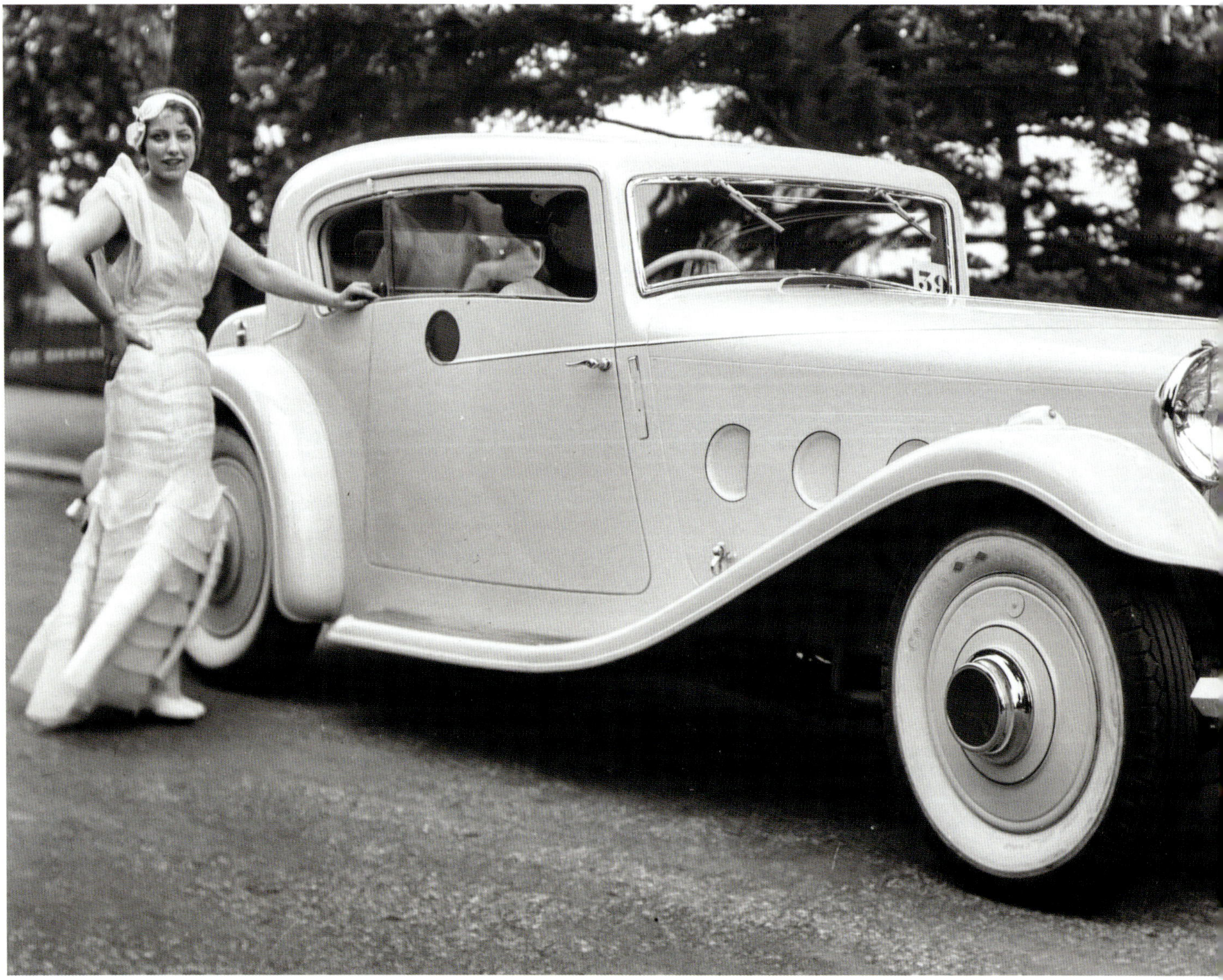

One of the most remarkable and innovative D8-S models of the 1930s: the "Yoyo"[1] by Letourneur et Marchand. Pictured here in Lausanne, presented by Josette Cortez, 'Mlle. Paris' in 1932. (Daniel Cabart Archives)

[1] *The "Yoyo" gets its nickname from the Art Deco accent line on its body. Its hallmark was a sweeping sidelight extending from the door into the rear quarter, with overlapping glass and no center pillar.*

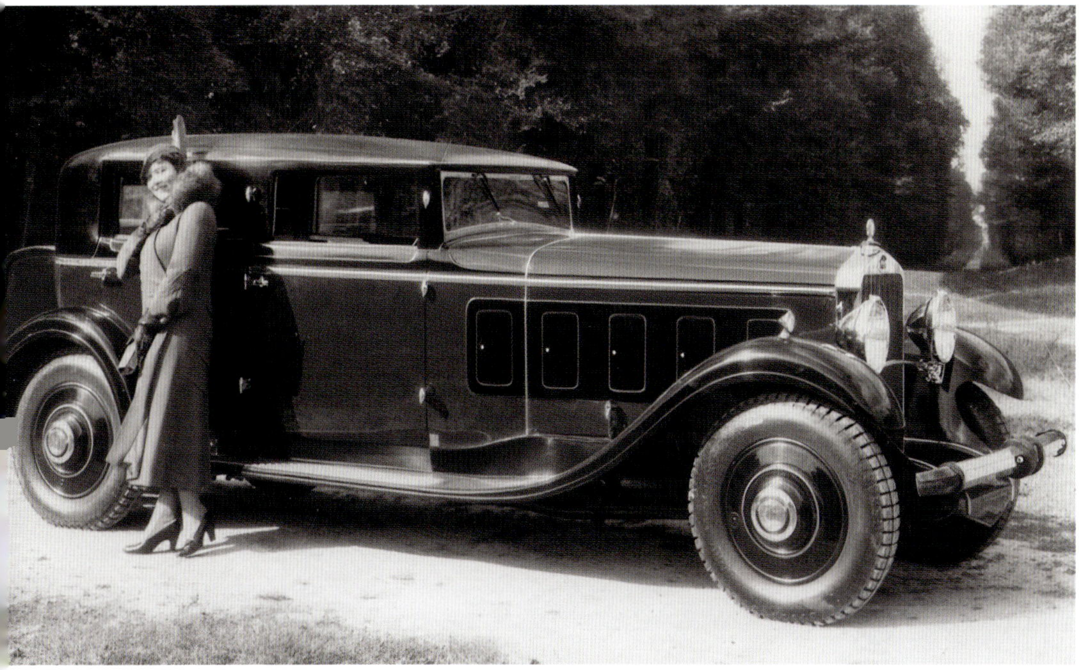

**LEFT:** *A publicity photograph of the Delage D8 with a 'Paris' saloon body by Chapron. A flawless balance of lines. (Daniel Cabart Archives)*

**BELOW MIDDLE:** *Actress Marie Glory at the Concours de L'Auto at the Parc des Princes. (Daniel Cabart Archives)*

**BOTTOM:** *Internationally acclaimed cellist Lucienne Radisse, pictured here in Le Touquet, presenting the D8-Sedan by Willy Vanden Plas. She would later perform across France and in neighboring countries. (Daniel Cabart Archives)*

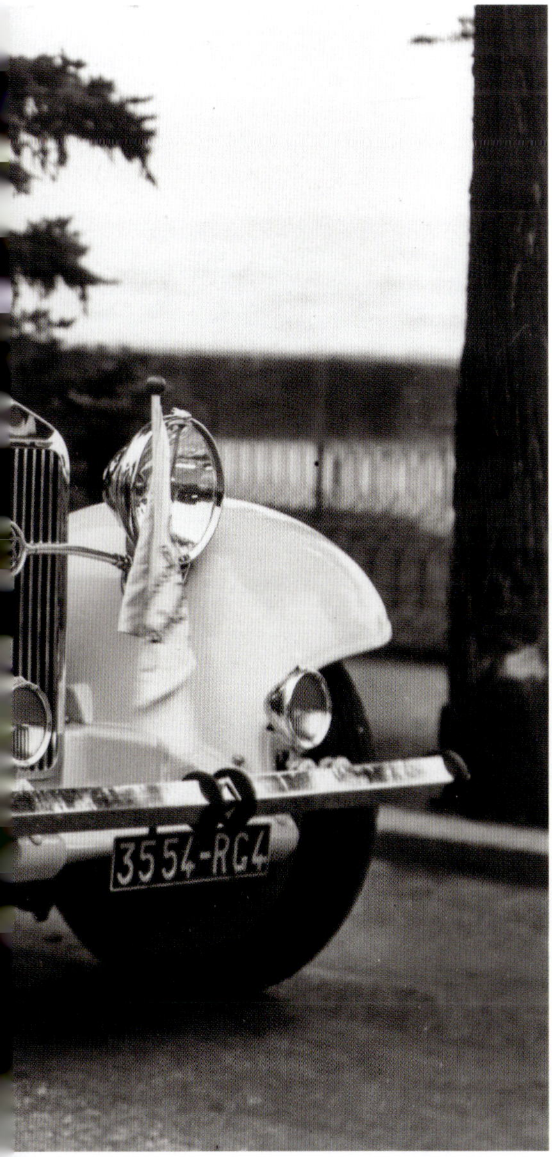


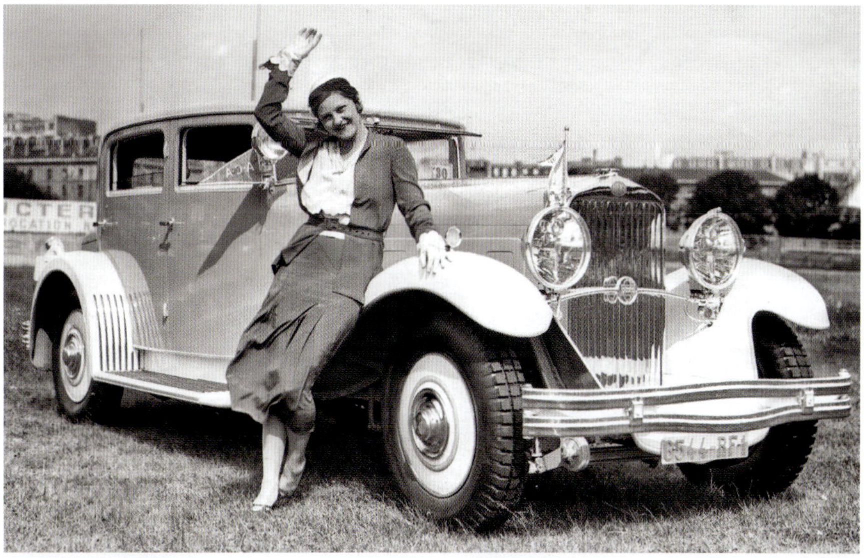

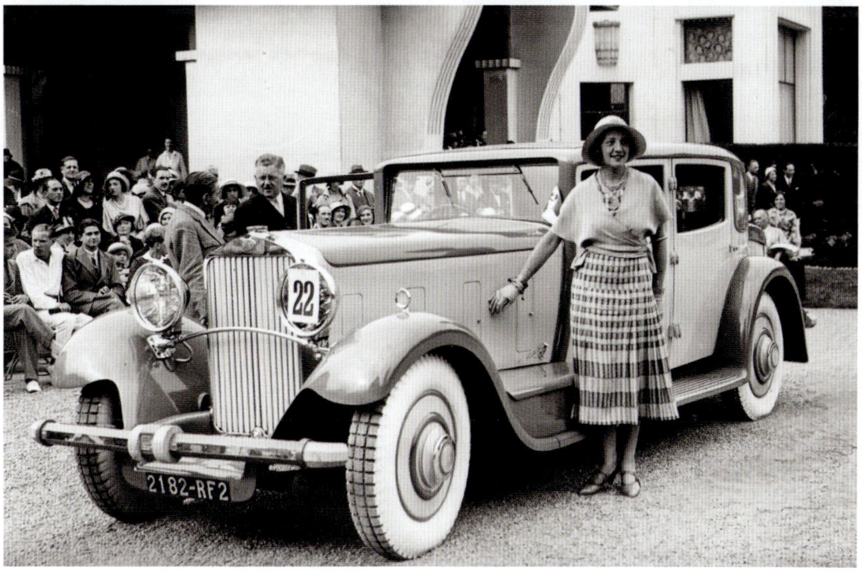

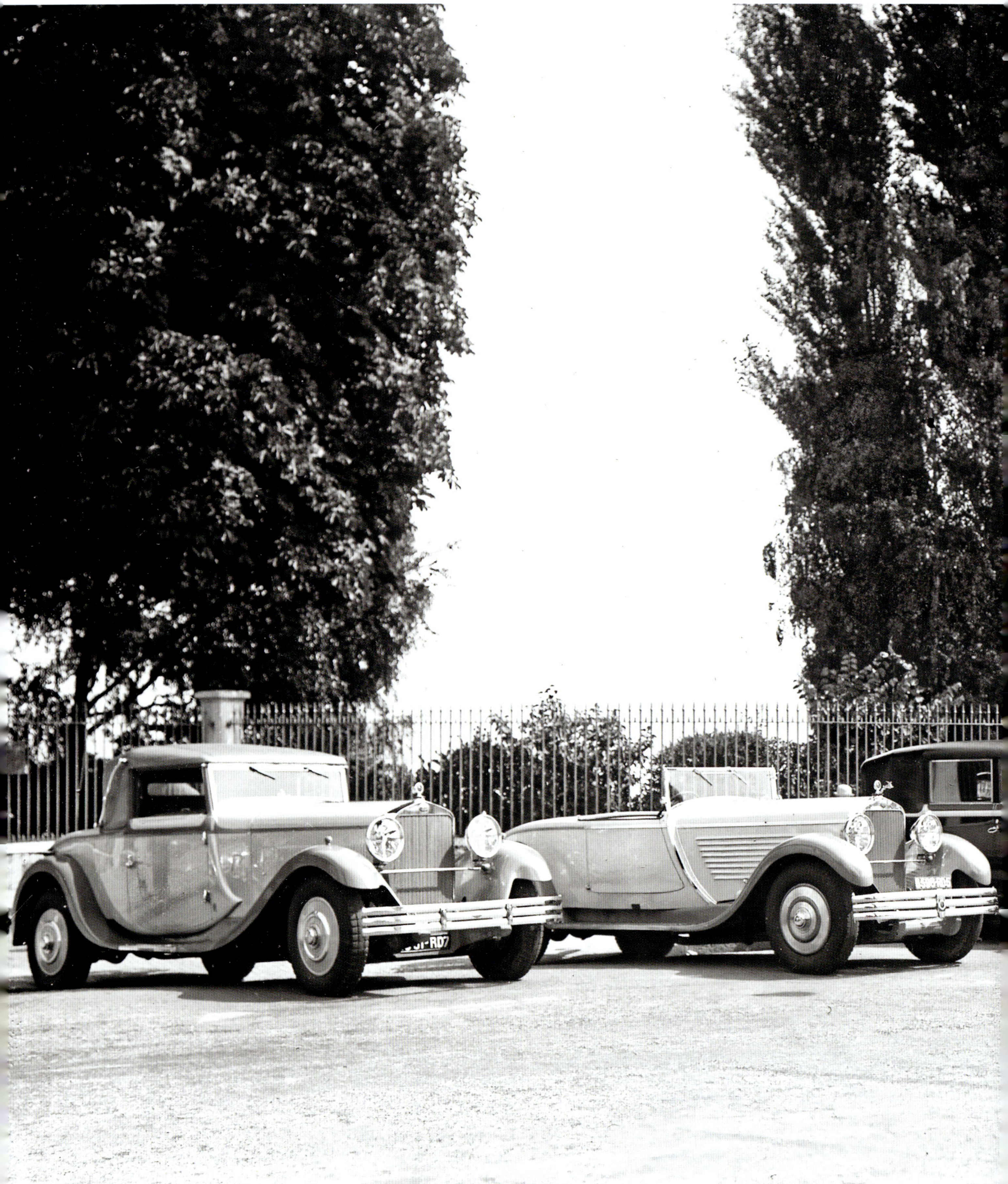

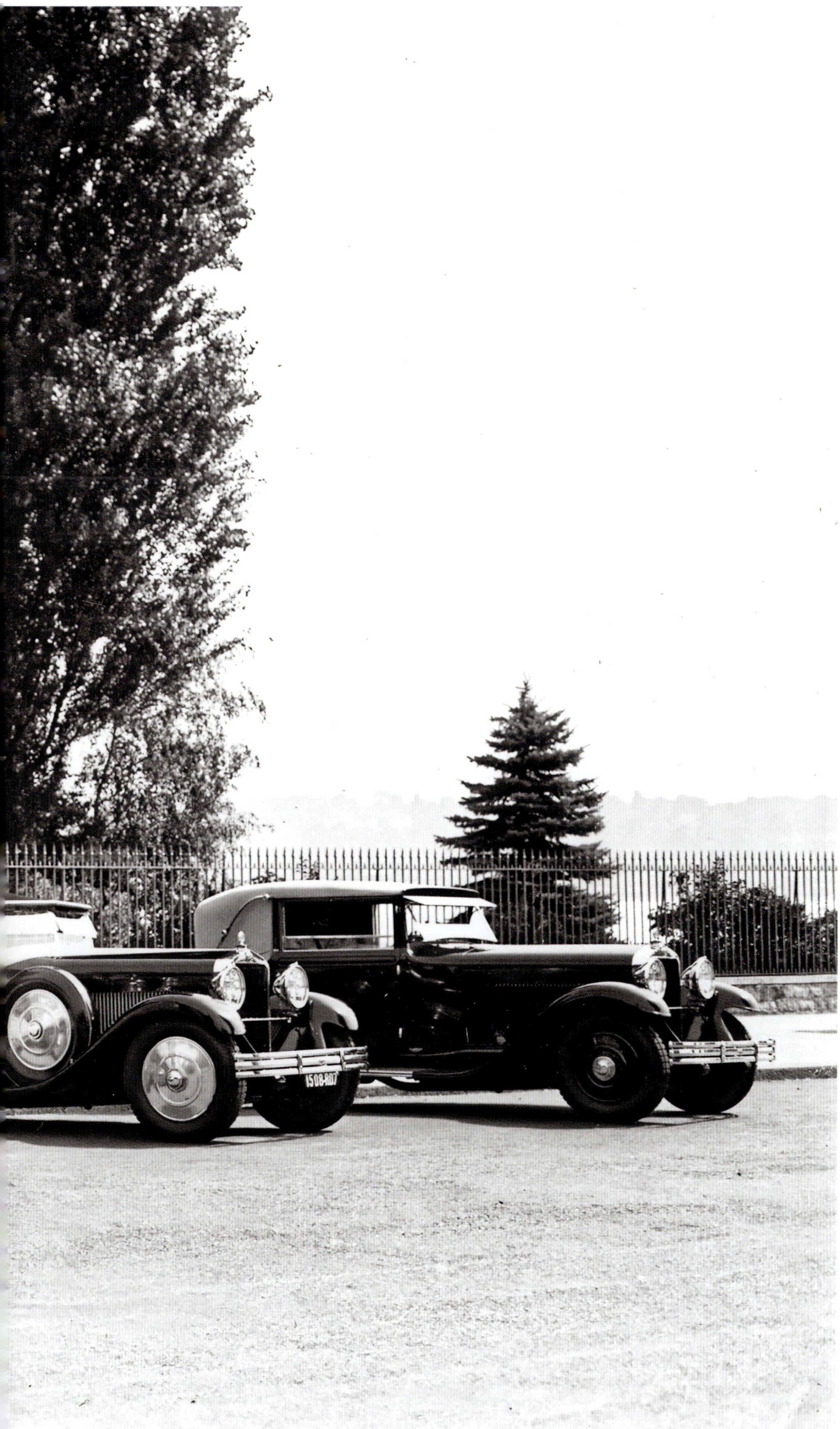

*These four Delage D8s were set to participate throughout the summer at various concours d'elegance in France and abroad. Pictured here in Geneva, from left to right: a Grümmer convertible, a Willy Vanden Plas convertible, a Saoutchik coupé-chauffeur, and a Duvivier convertible. (Daniel Cabart Archives)*

In 1931 alone, Delage took part in 39 concours d'elegance, spanning cities from Lisbon to Algiers, Nice to Geneva, and Budapest to Biarritz – further solidifying its reputation for elegance and excellence.

In 1932, at the prestigious *L'Auto* competition, no fewer than 19 Delage vehicles were honored with awards. This success was further amplified by extensive advertising coverage and reports featured in the automotive press.

The D8's successor, the D8-15, would continue to uphold the image of the "beautiful French car" and the "queen of elegance competitions," though to a lesser degree. However, Delage's financial struggles and the (temporary) closure of the company would mark the end of this glorious era.

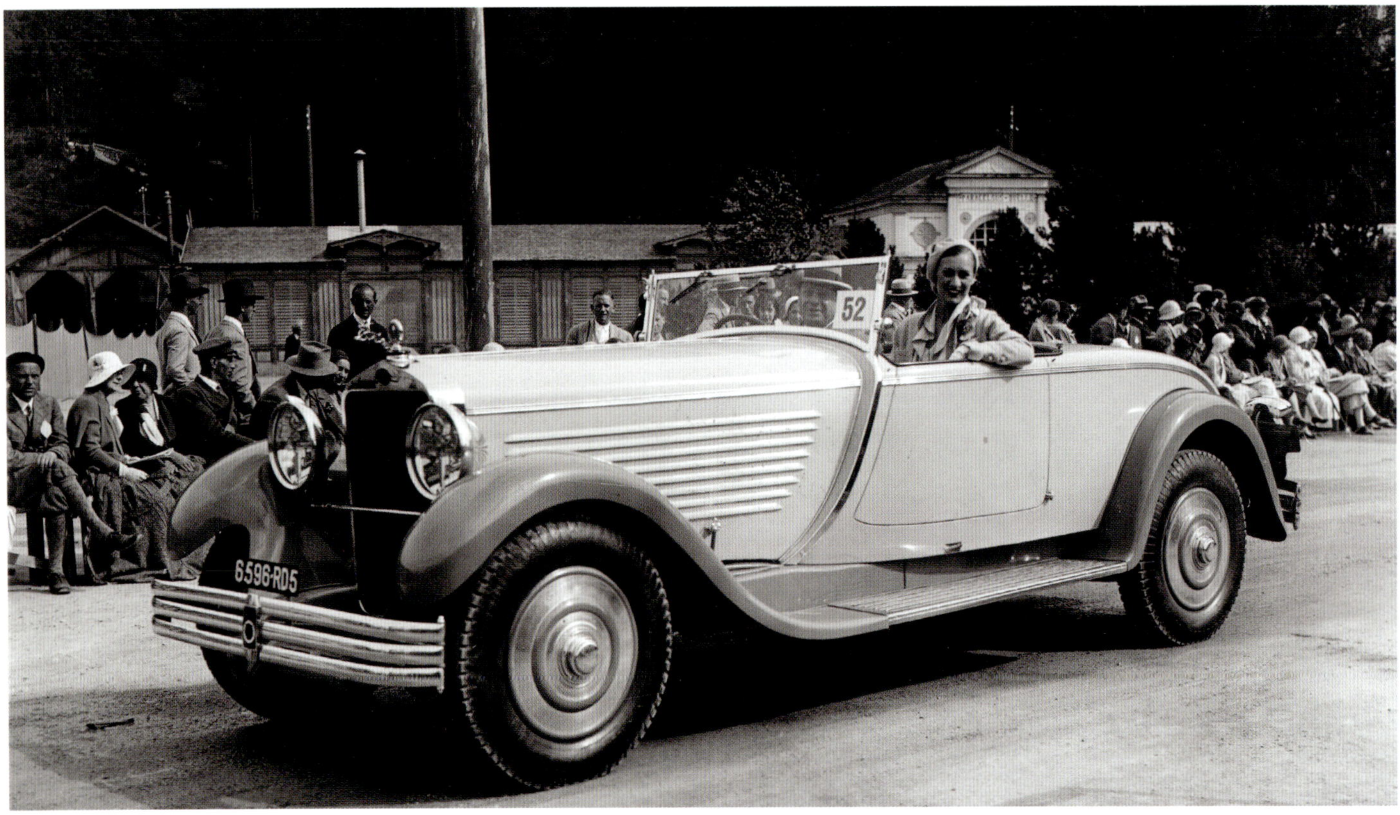


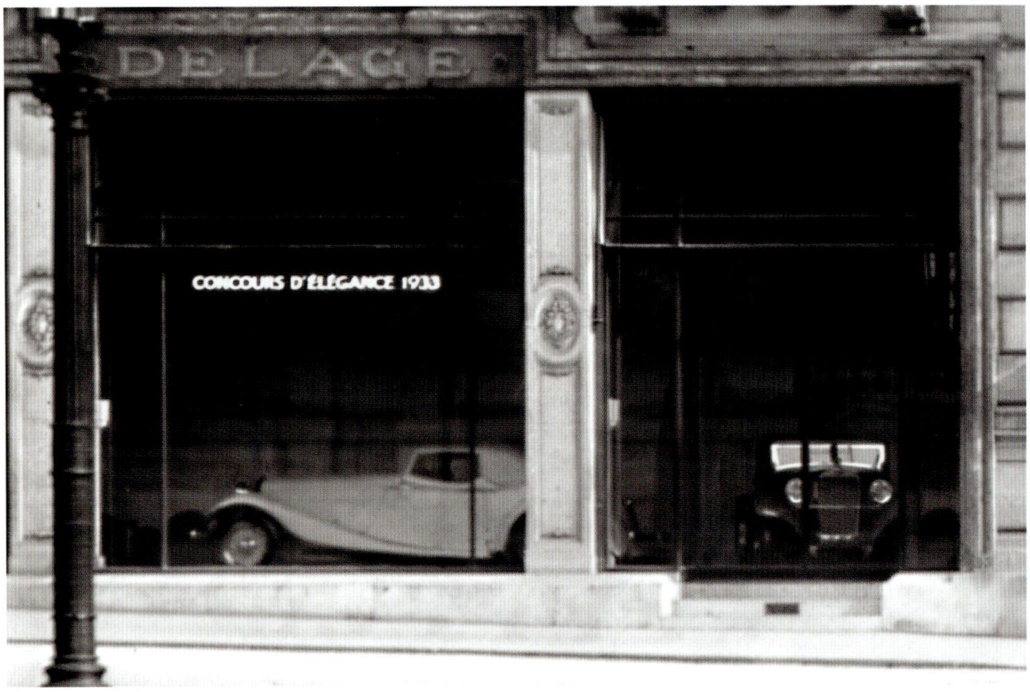


**ABOVE:** *The D8 convertible by Willy Vanden Plas, shown here in Saint-Moritz, Switzerland, and presented by Mlle. de Verneuil. (Daniel Cabart Archives)*

**RIGHT:** *In Delage's showroom on the Champs-Élysées, where Grand Prix victories were once celebrated until 1927, the display now featured laurels earned from concours d'elegance. Pictured here is a D8-S roadster by De Villars, which won Best of Show at the 2010 Pebble Beach Concours (see page 222). (Daniel Cabart Archives)*

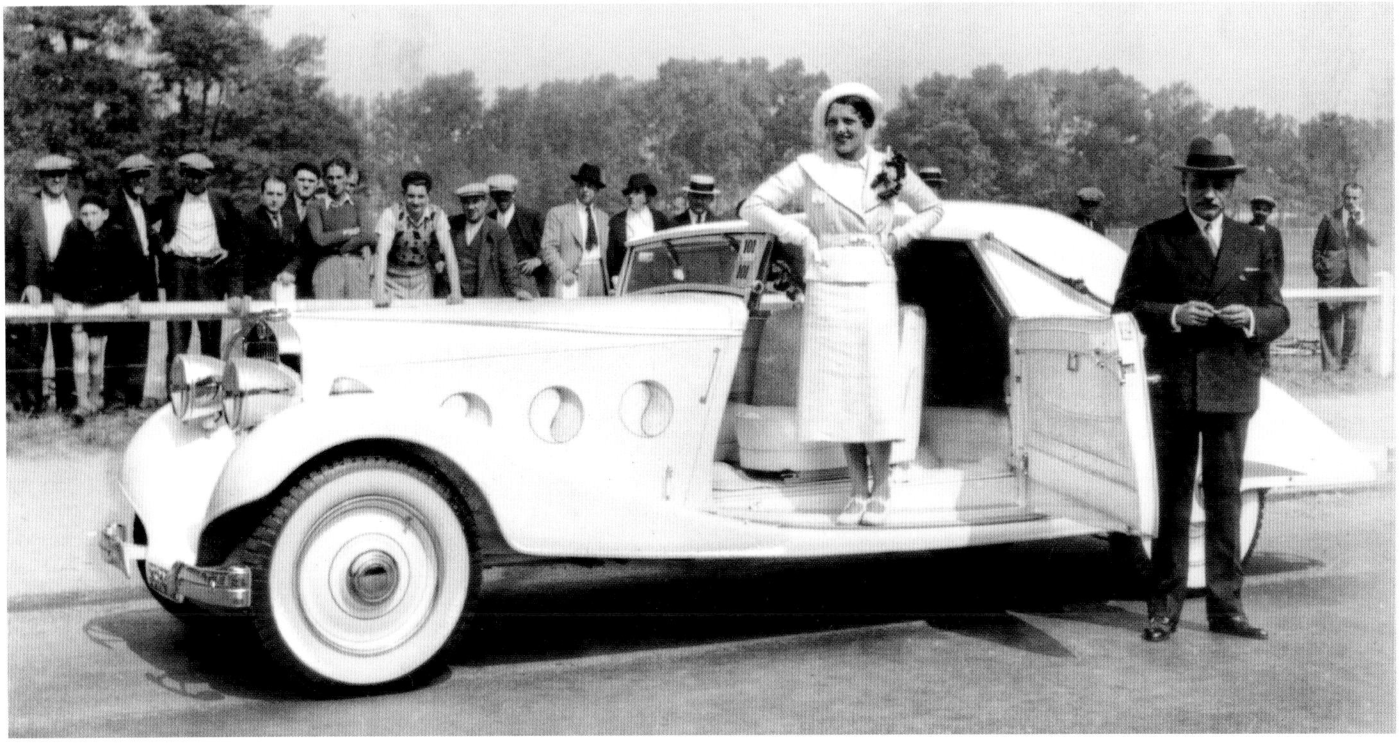

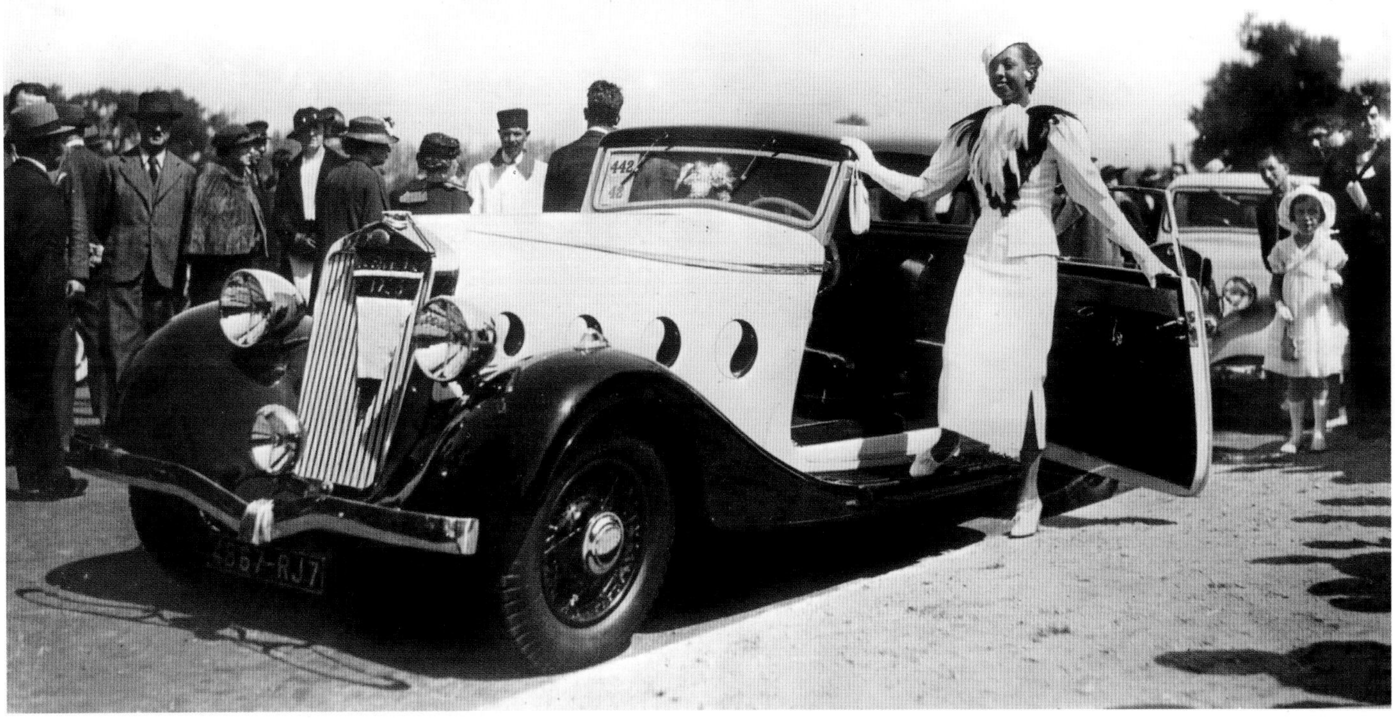


**ABOVE:** *A D8-85 with faux-cabriolet bodywork by Letourneur et Marchand, presented by its owner, Joséphine Baker, in 1935. (Daniel Cabart Archives)*

**TOP:** *Despite some remarkable successes with the D8-15 chassis, Delage struggled to retain the dominance it once held with the D8-S. Betty Spell, accompanied by Louis Delâge, presented the "Yoyo" convertible by Letourneur et Marchand in 1934. (Daniel Cabart Archives)*

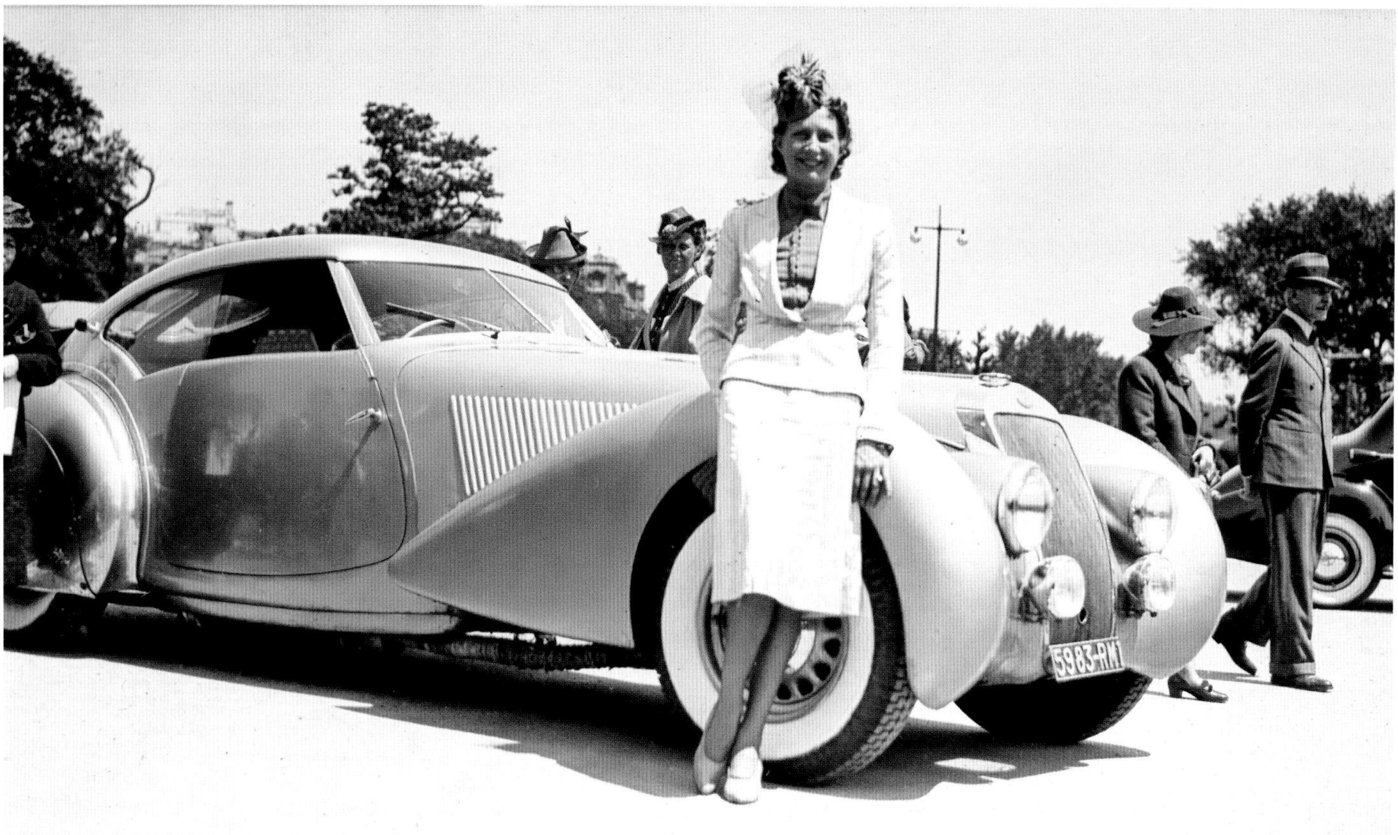
5983-RM1

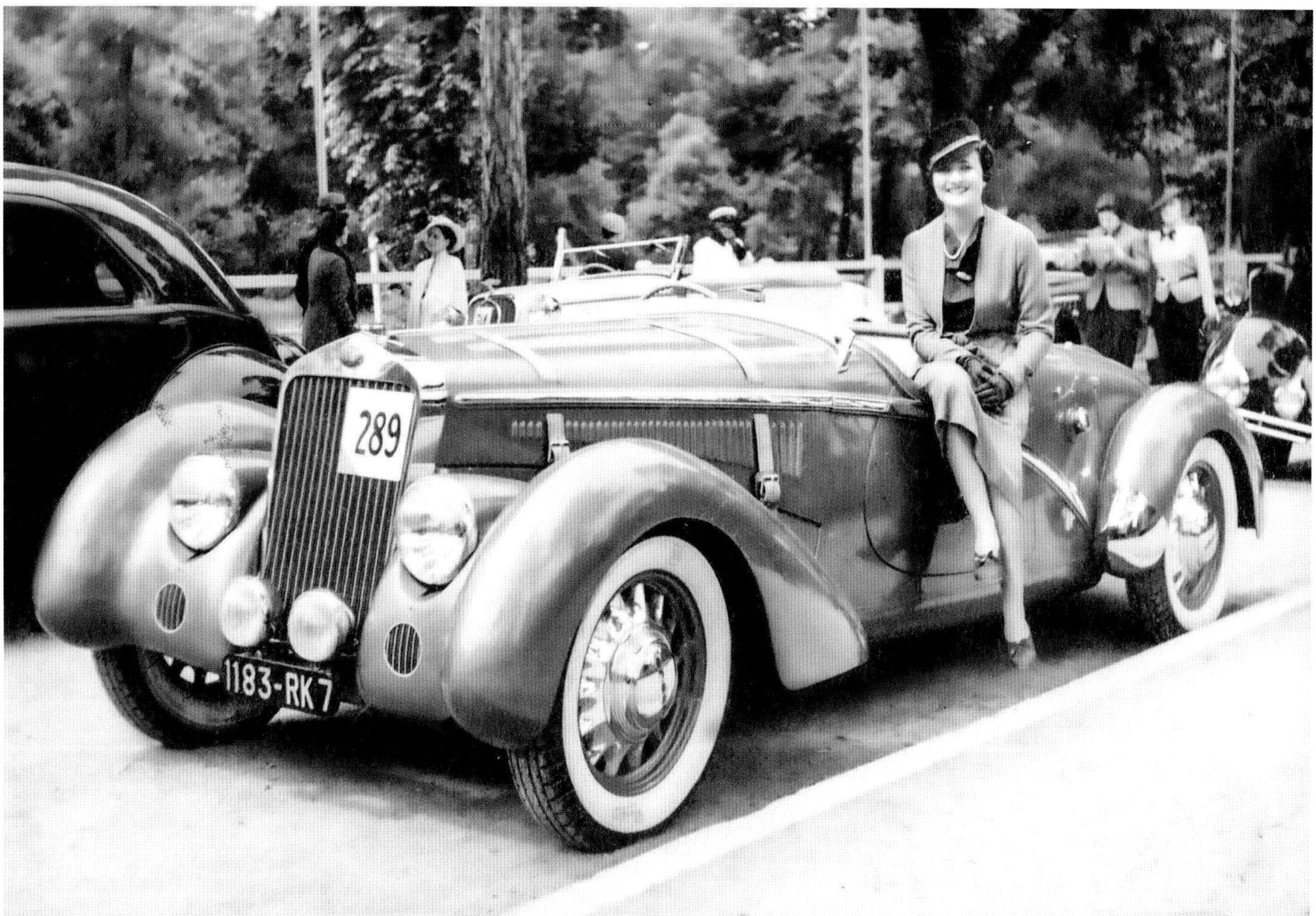
289
1183-RK 7

A second equally renowned chapter in Delage's history emerged after 1935. The Société Nouvelle des Automobiles Delage re-entered the world of elegance competitions, reigniting the brand's legacy. The new lineup, featuring 6- or 8-cylinder engines, boasted chassis with harmonious proportions. Renowned coachbuilders returned to create true masterpieces, many of which were one-of-a-kind.

Figoni et Falaschi, Pourtout, and Letourneur et Marchand helped restore Delage to the pinnacle of elegance competitions. Personalities from Parisian high society, along with figures from the worlds of entertainment and fashion, played a vital role in presenting the cars, enhancing the spectacle with their presence and exquisite outfits. A close collaboration with couturiers further emphasized the harmony of shapes and colors, showcasing the elegance of the overall design. In this domain, Delage stood at the forefront.

However, June 1939 marked the end of this era of magnificent creations. While some of these masterpieces survived the four years of war unscathed, they no longer held the same allure in the immediate post-war period. The heart of the brand had faded.

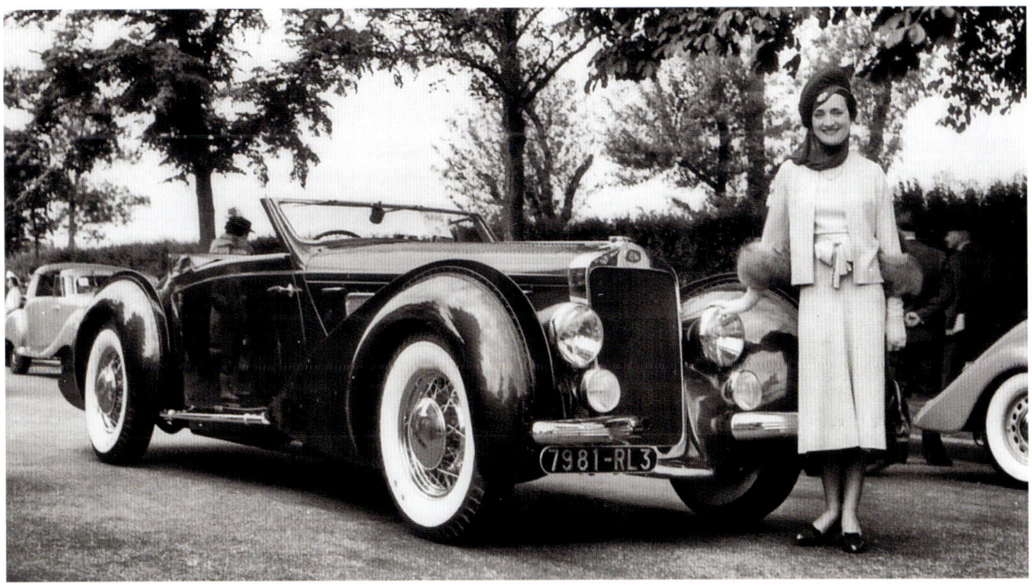


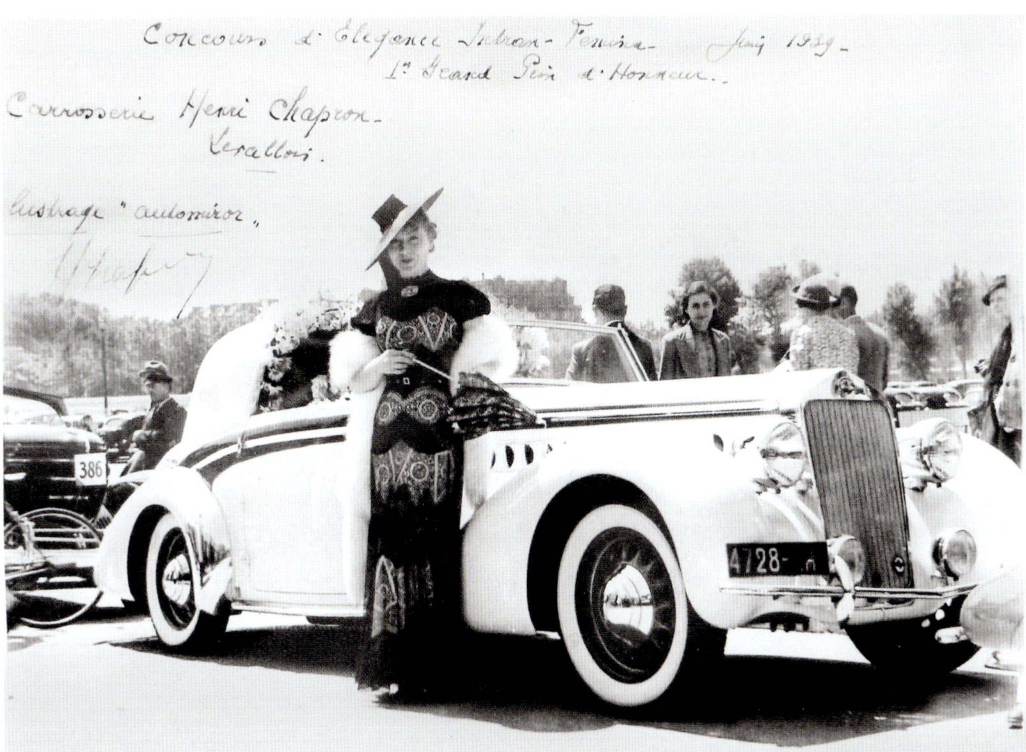


**OPPOSITE TOP:** *The D8-120 S with bodywork by Marcel Pourtout, presented here by Joan Warner at the Trocadéro Concours d'Elegance on June 23, 1939. Mademoiselle Joan Warner, known as 'the queen of the nude,' wore street clothes for the occasion! In 1935, she had been prosecuted for indecent exposure but was acquitted on the grounds that her performance took place in a private setting, not a public one. She also proved she wasn't truly nude, as she had worn a completely transparent silk slip.* (Daniel Cabart Archives)

**OPPOSITE BOTTOM:** *The D6-70 'Speciale' with roadster bodywork, a creation by Letourneur et Marchand. Both a concours d'elegance and rally car, it competed in the 1937 Moroccan Rally, driven by Madame Richer-Delavau. Its metallic green color earned it the nickname 'Grasshopper'.* (Daniel Cabart Archives)

**THIS PAGE UPPER:** *Undoubtedly one of the most remarkable Delages of the final pre-war years, the D8-120 S roadster with bodywork by De Villars. These creations by the finest coachbuilders gave rise to the claim that the eight-cylinder chassis produced the most beautiful of Delages.* (Daniel Cabart Archives)

**LEFT:** Built on a D6-75 chassis, this convertible coupé by Henri Chapron was presented in June 1939 at the Fémina L'Intran competition. (Bernard Brulé)

## L'Auto

The Concours d'Elegance de L'Auto was first held on June 10, 1922, taking its name from the eponymous sports daily. The day after its debut, the event was hailed as a "considerable and significant success" with the promise of becoming an annual tradition. "*Nearly one hundred and fifty car owners or coachbuilders responded to our call,*" the publication reported.

Initially, the competition was purely technical, showcasing "*the most varied types of bodywork, all testifying not only to great ingenuity in design but also to exceptional craftsmanship.*" Over time, however, public interest in large gatherings of several hundred vehicles began to wane, prompting the event to evolve.

The Concours d'Elegance de L'Auto transitioned into a multi-faceted event comprising several competitions with distinct objectives. These included the *Grand Prix d'Honneur de L'Auto,* awarded to a coachbuilder at the culmination of the Grand Concours International d'Elegance, de Confort et de

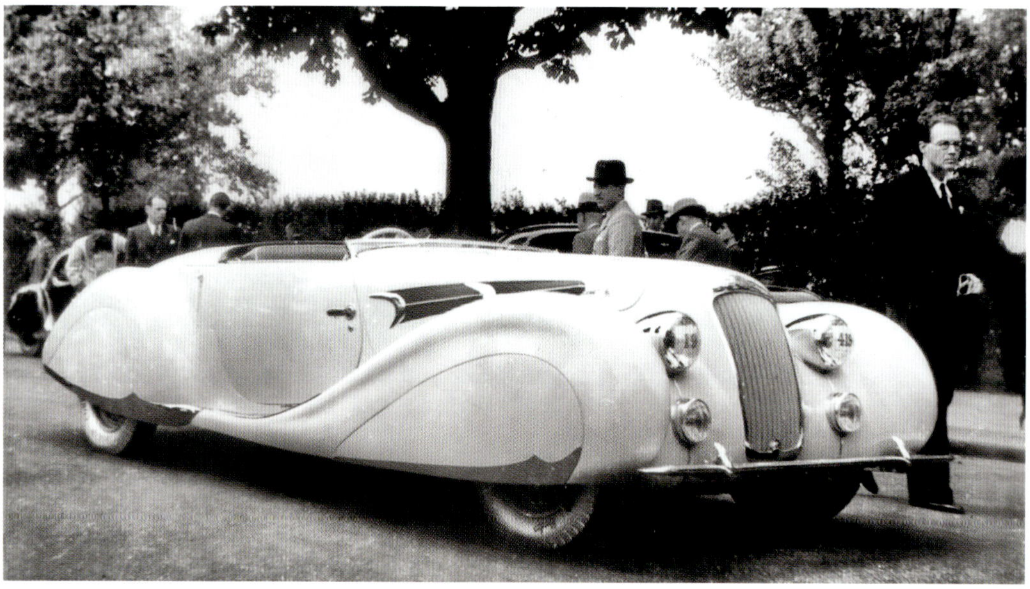

**RIGHT:** *Delage D6 3-liter roadster, bodied by Figoni et Falaschi, created for Tino Rossi, the renowned French crooner, who gifted it to his mistress, actress Mireille Balin. Pictured here on June 9, 1939, at the L'Auto concours. (Daniel Cabart Archives)*

**BELOW:** *This convertible coupé, with bodywork by De Villars, was built on a Bugatti Type 46 chassis. It was entered in the L'Auto competition on June 6, 1930. (Laurent Friry Archives)*

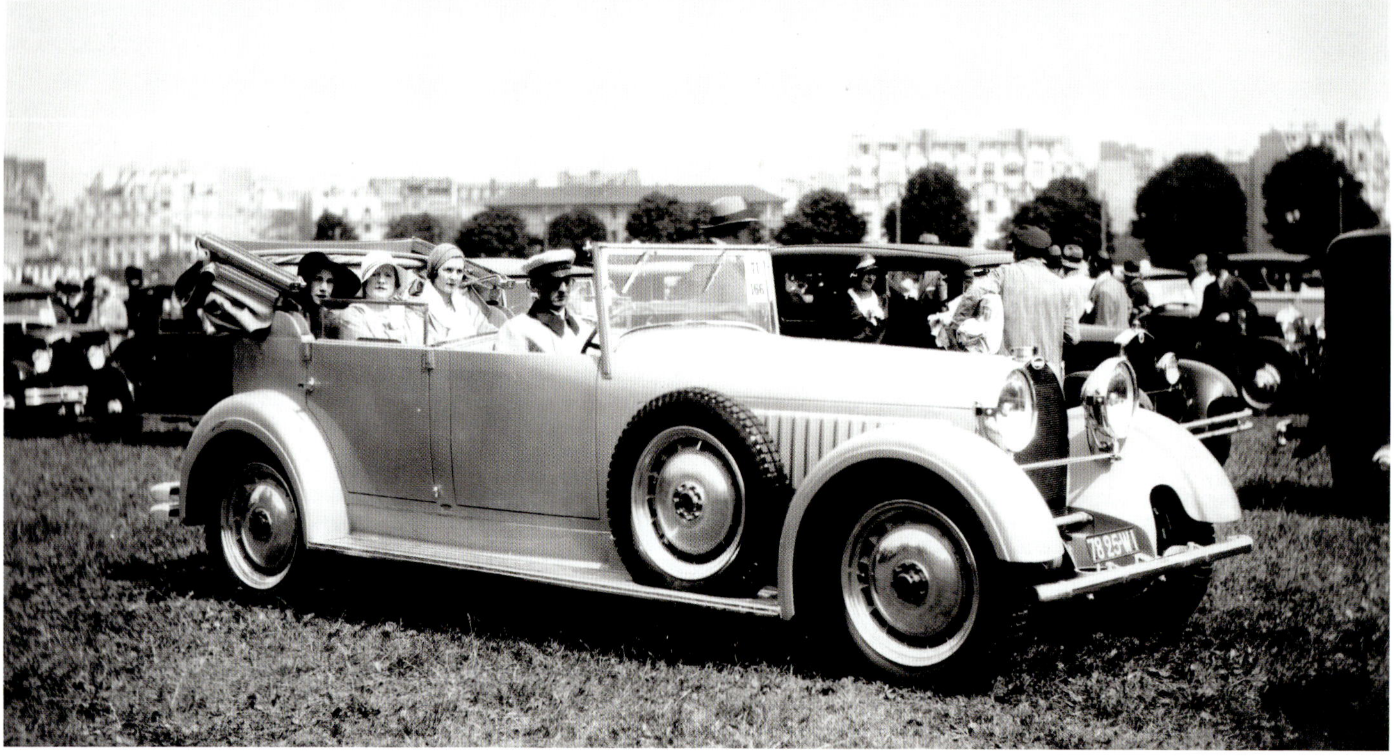

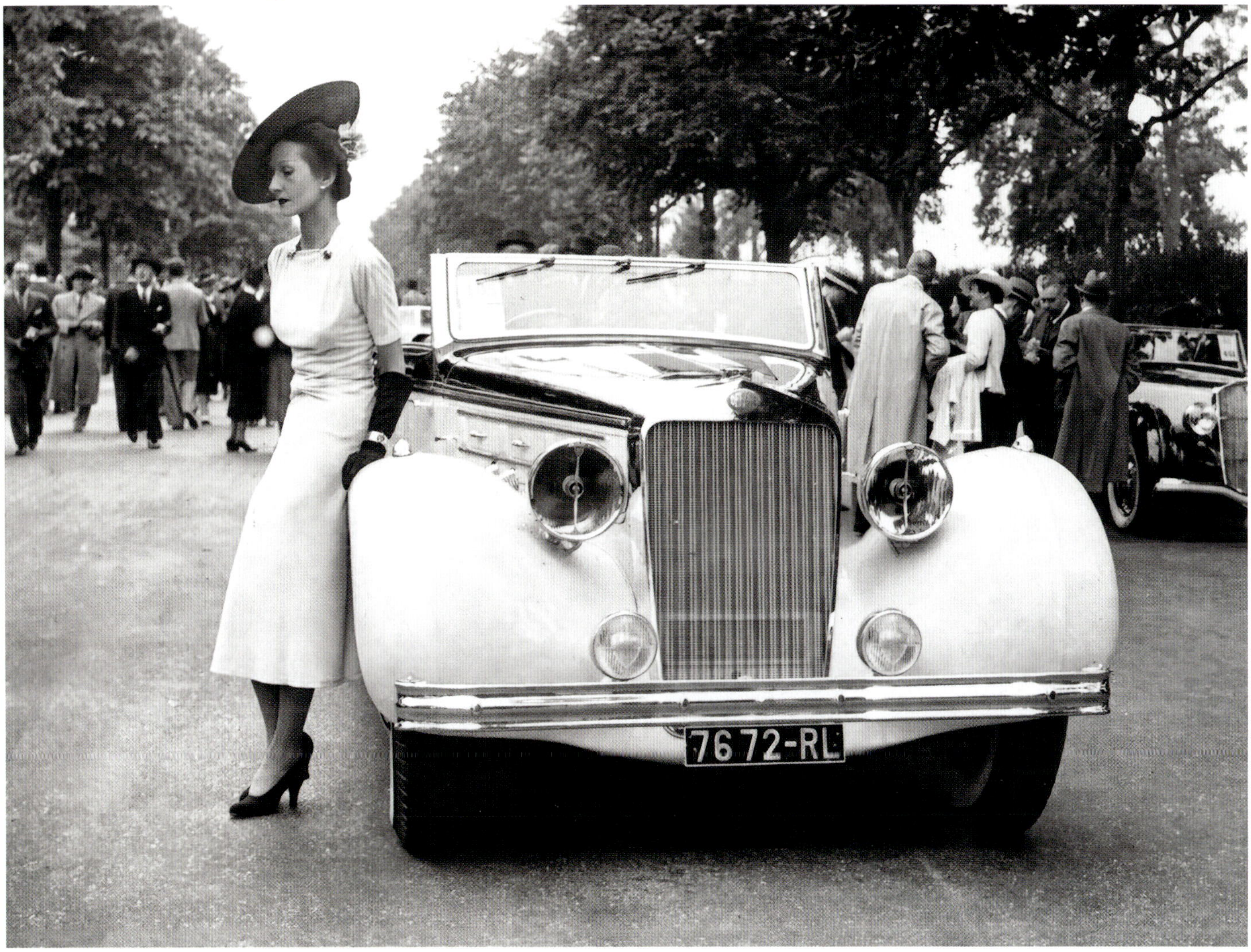


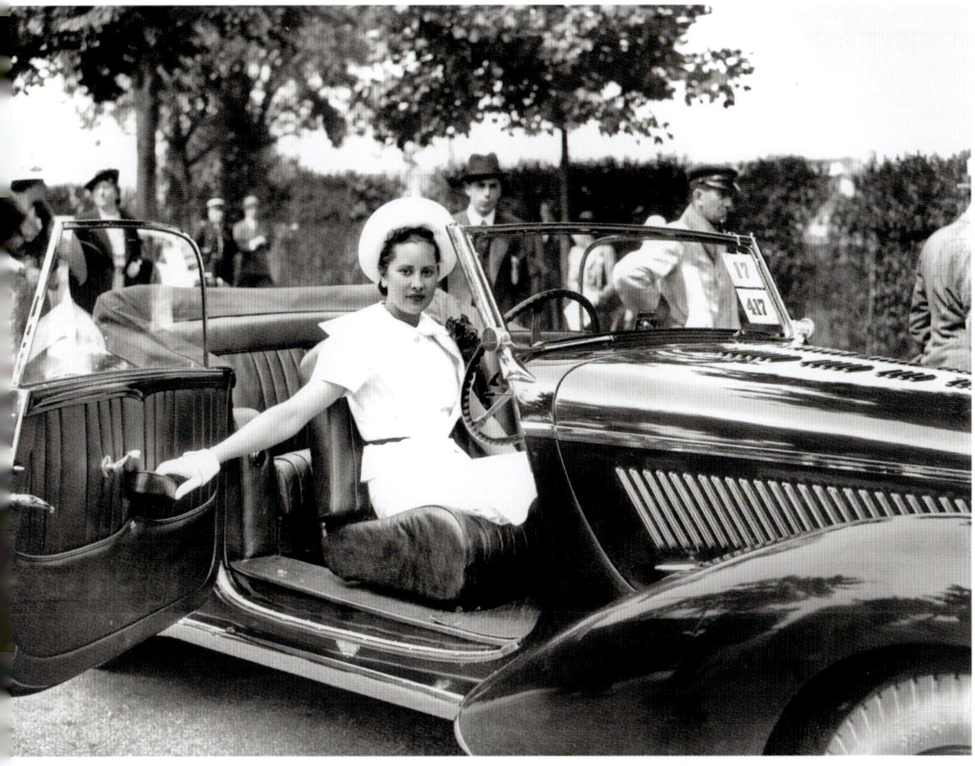

**ABOVE:** *Madame Louis Arpels (née Hélène Ostrowska, a former model for Schiaparelli, Worth, and others) presented a D8-120 convertible with a De Villars body at the L'Auto concours on June 15, 1937. This car, presented by various personalities, would later be seen in England, La Baule, and other locations.* (Daniel Cabart Archives)

**LEFT:** *A four-seat convertible by Figoni et Falaschi built on a Delage D6-75 at the competition on June 9, 1939. The car was presented by Sonia Stevenson-Watney, niece of Walter Watney, who had 'rescued' Delage in 1935. Sonia worked for the renowned couturier Callot Soeurs on Avenue Montaigne in Paris. In 1940, she joined General de Gaulle's 'Free France' in London and remained active throughout the war.* (Daniel Cabart Archives)

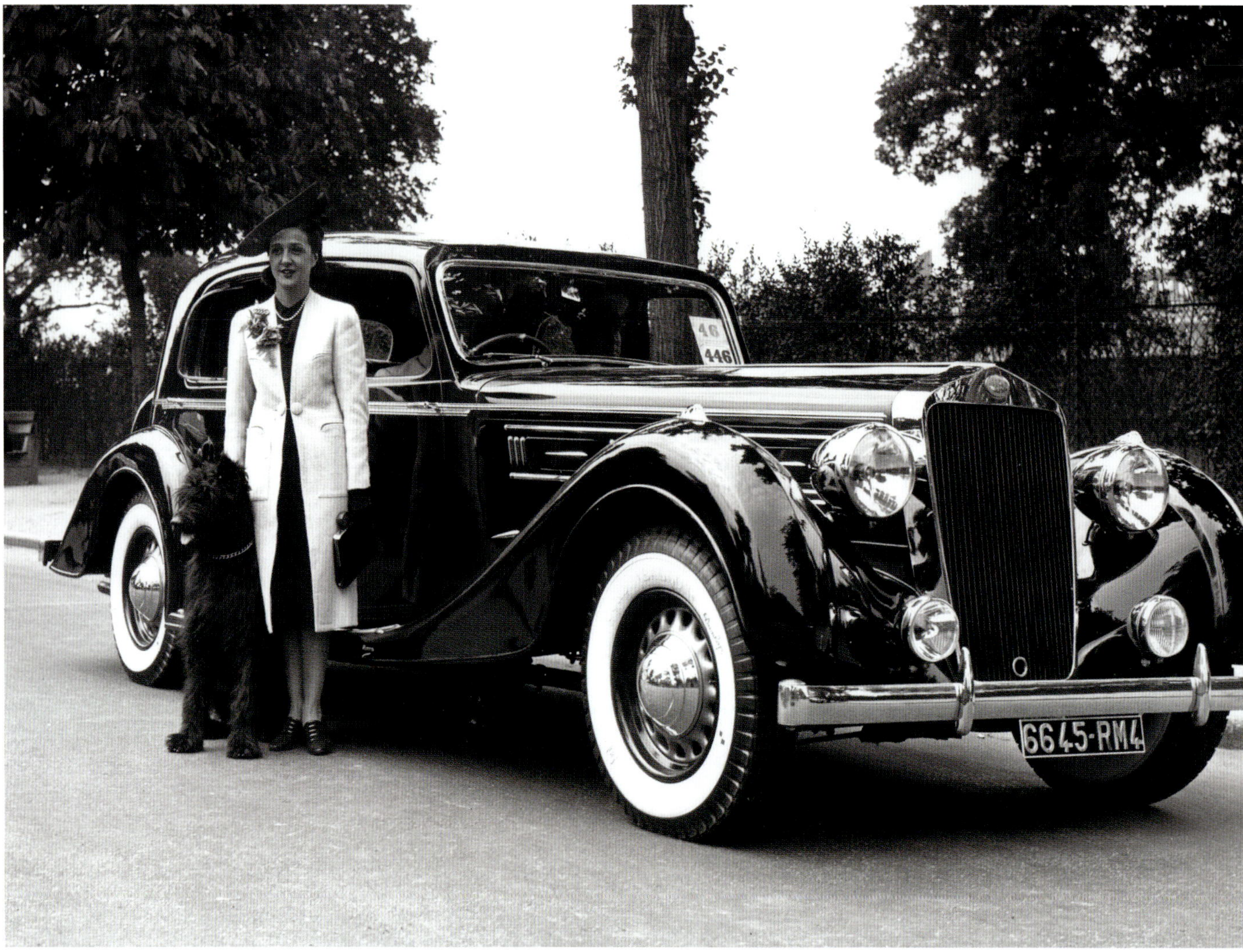


*At the same concours on June 9, 1939, Mlle. Rigaut won top honors for presenting this magnificent and classic Vanvooren sedan on a Delage D8-100. She was awarded the* Grand Prix d'Honneur de L'Auto *for a car bodied by a coachbuilder, as well as First Prize in the* International Trophy of Féminine Elegance, *recognizing the perfect harmony between her outfit and the car's bodywork. (Daniel Cabart Archives)*

Technique de la Carrosserie. The Grand Prix d'excellence de L'Auto recognized a manufacturer for their achievements. To captivate a broader audience, the Trophée International d'Elegance Féminine Automobile evaluated cars for their harmony with the attire of their drivers, a process meticulously overseen by a female jury.

The Concours d'Elegance de L'Auto became one of Paris's two premier concours events. Alongside its counterpart at *Fémina*, it embraced the allure of stardom, staging grand spectacles featuring prominent figures from the worlds of theater, variety, and later, cinema.

## Biarritz

As with many traditions, tracing the exact origins of what we now call concours d'elegance competitions is challenging, and Biarritz is no exception. However, one event stands out in the history of such gatherings. The 1926 event, organized by the Parisian publications *Fémina* (focused on fashion) and *L'Intransigeant* (a daily newspaper), introduced a scoring system that appears to have been unprecedented at the time.

First, the jury was instructed to consider the attire of the passengers, who remained seated in their vehicles. Separate from the evaluation of the cars themselves, this ranking rewarded the most elegant ensembles, with prizes sponsored by local hatters and fashion houses (also see page 83).

A second innovation occurred at midnight, when a new parade of competitors took place under the floodlights of the Pavillon Royal terraces. This time, a jury awarded special prizes based solely on the "chicness" of the passenger or driver and the overall harmony of their presentation.

The following year, the competition officially split into two distinct categories: the Concours d'Elegance

Automobiles (Automobile Elegance Competition) and the Concours d'Elegance Féminine (Feminine Elegance Competition). Over time, this event became one of the most exclusive in the Southwest region. It was subsequently organized by *La Gazette de Biarritz* and *Biarritz Illustré* at the picturesque Chambre d'Amour.

As was customary in the chicest seaside resorts, the presenters were drawn from the aristocracy, prominent members of high society, or foreign royalty. Major French manufacturers eagerly lent their latest models to these presenters for the event, allowing some vehicles to embark on a "Tour de France" of elegance – or even a tour of Europe – throughout the summer season.

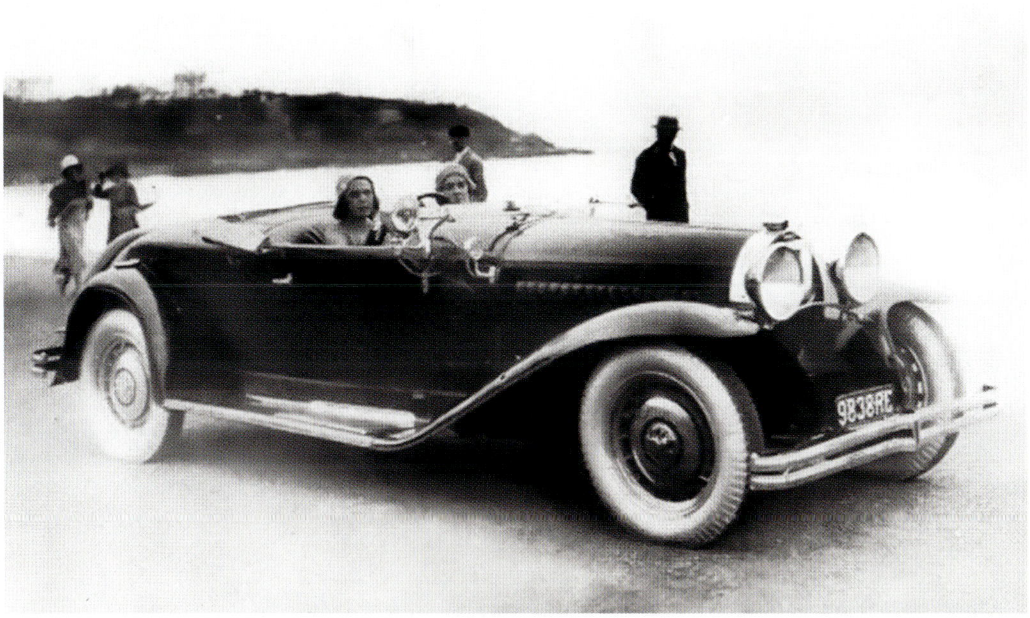

**LEFT:** *This Bugatti Type 46, chassis no. 46341, was originally built as a roadster at the factory. In July 1930, it competed in the Dinard Concours d'Elegance with Madame Gaillard, and just a few weeks later, on September 20, it made another appearance at the Biarritz Concours d'Elegance. (BnF Gallica)*

**BELOW (BOTH):** *The Biarritz Concours events rivaled their "neighbors" in the quality of their presentations, attracting the very best in the field. (BnF Gallica)*

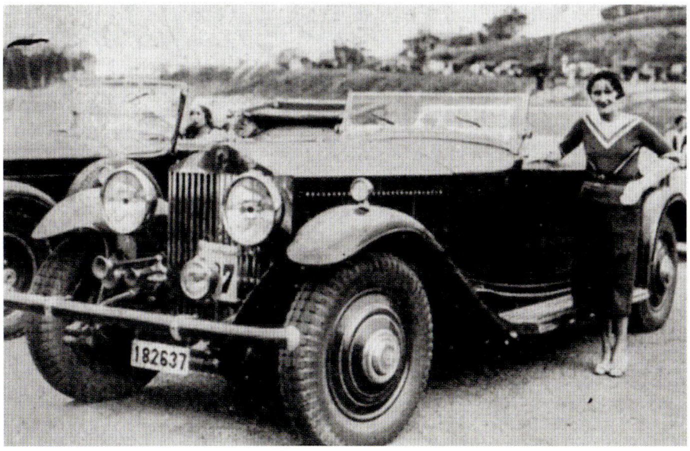

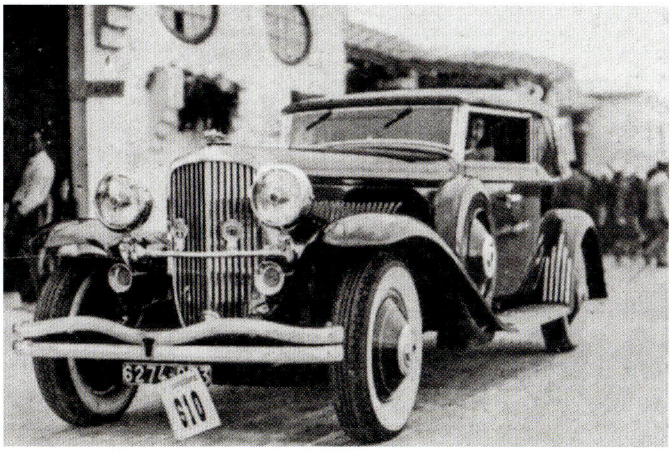

# Les Delage au Concours d'Élégance Automobile

Comme l'année dernière, les voitures Delage se sont taillé un joli succès au grand Concours d'élégance automobile de Biarritz. La voiture conduite par Mlle Radisse, la jolie artiste parisienne, qui a déjà remporté un prix au Concours d'élégance automobile de Paris, a ici triomphé à nouveau en remportant un grand prix d'honneur. Voici les quatre belles Delage qui défilèrent à la Chambre d'Amour aux applaudissements de toute l'assemblée

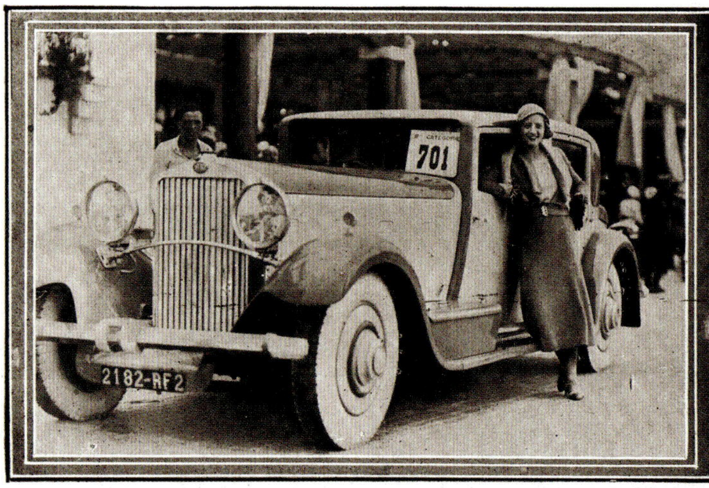


*Berline Delage 23 C. V. conduite par Mlle Radisse. Grand Prix d'honneur de sa catégorie et Prix d'élégance féminine.*

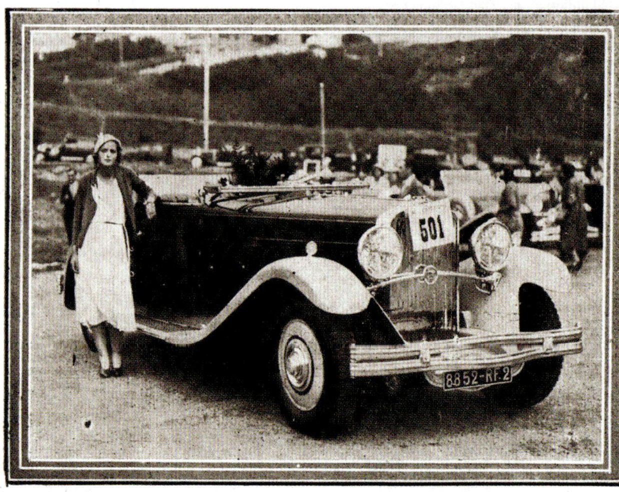


*Roadster Delage 23 C. V. conduit par Mrs Brocco Deuxième premier prix de la 6me catégorie.*

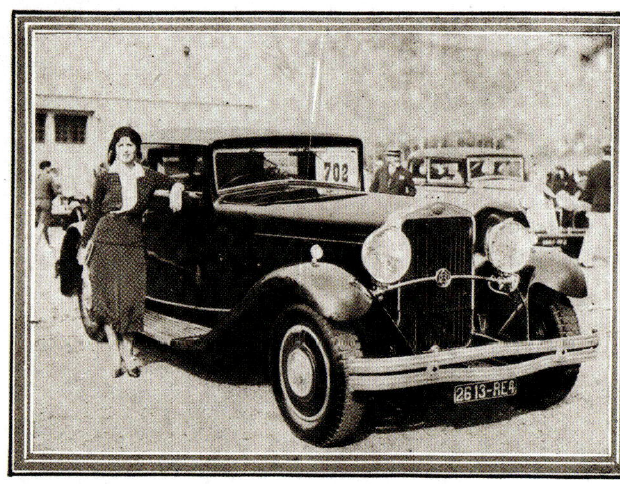


*C. I. Delage 23 C.V. conduite par Mrs Ware. Premier prix de la 8me catégorie.*

Ces superbes voitures Delage qui remportent dans chaque concours d'élégance automobile des grands prix d'honneur sont représentées à Bayonne, par Bayonne - Automobile, 3, Rue Jacques-Laffitte, à Bayonne, MM. Célestin Gambade et Fernand Forgues, gérants.

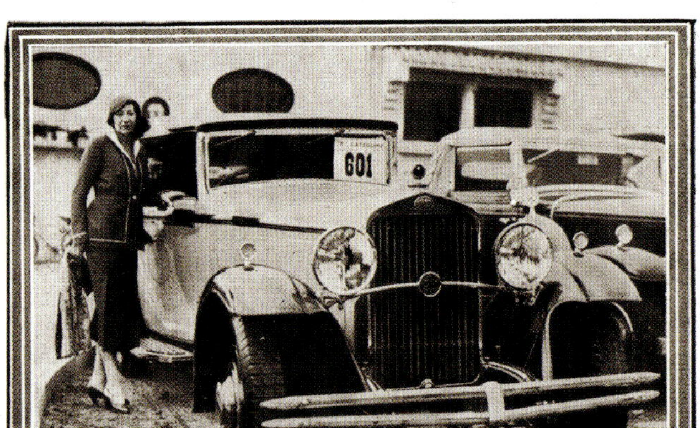


Photos Robertito

*Cabriolet Delage 23 C.V. conduit par Mme Marie de Monighetti. 4e prix.*

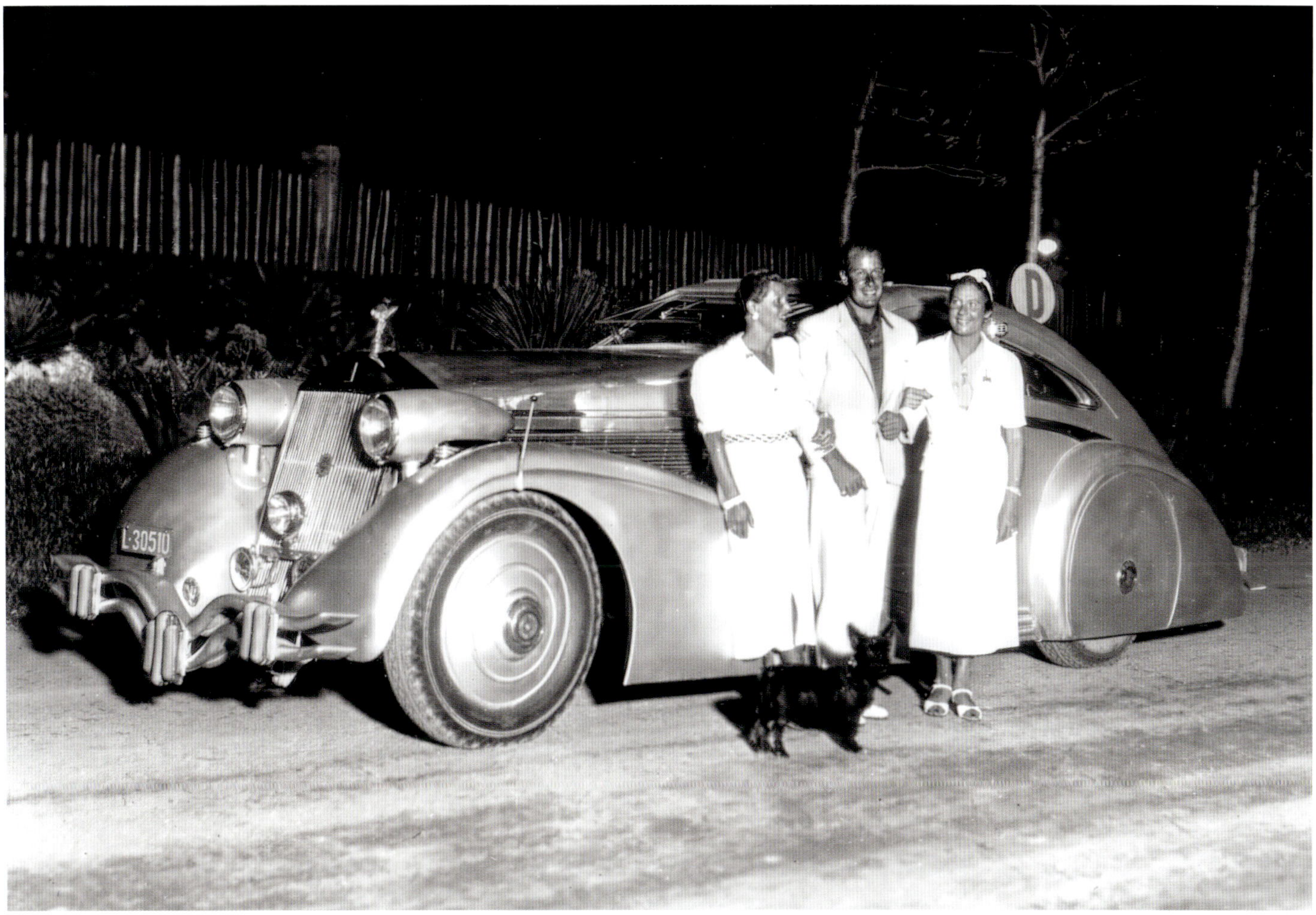

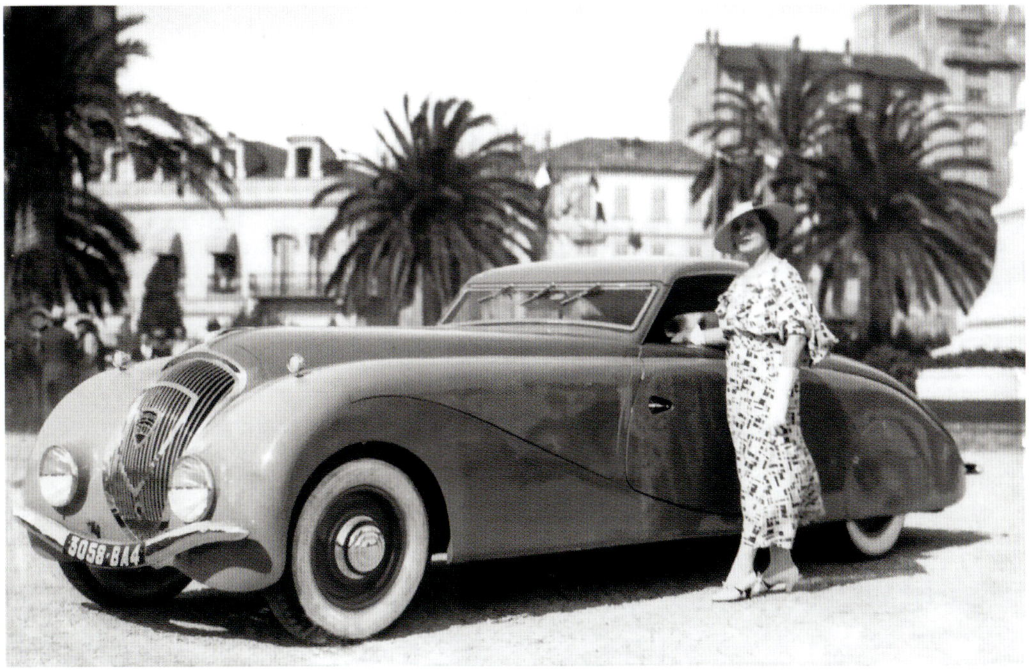

**ABOVE:** *Arguably the most baroque Rolls-Royce ever built, this Phantom I was rebodied by the Jonckheere coachbuilders of Beveren, Belgium, in 1935. Its owner, Harald Smith, presented it at the Cannes Concours d'Elegance on August 28, 1936. Today, it is a centerpiece of the Petersen Automotive Museum. (Bernard Brulé)*

**LEFT:** *As in Monte Carlo, the ultra-aerodynamic Peugeot 12 CV, a six-cylinder model with Pourtout Eclipse bodywork – hailed as the car of the future – was met with great acclaim at the Cannes Concours d'Elegance. Driven by Madame Suzanne Duman of the Opéra Comique, her dress was elegantly styled by Simone and she wore a hat from Laurence (Monte Carlo). (Bernard Brulé)*

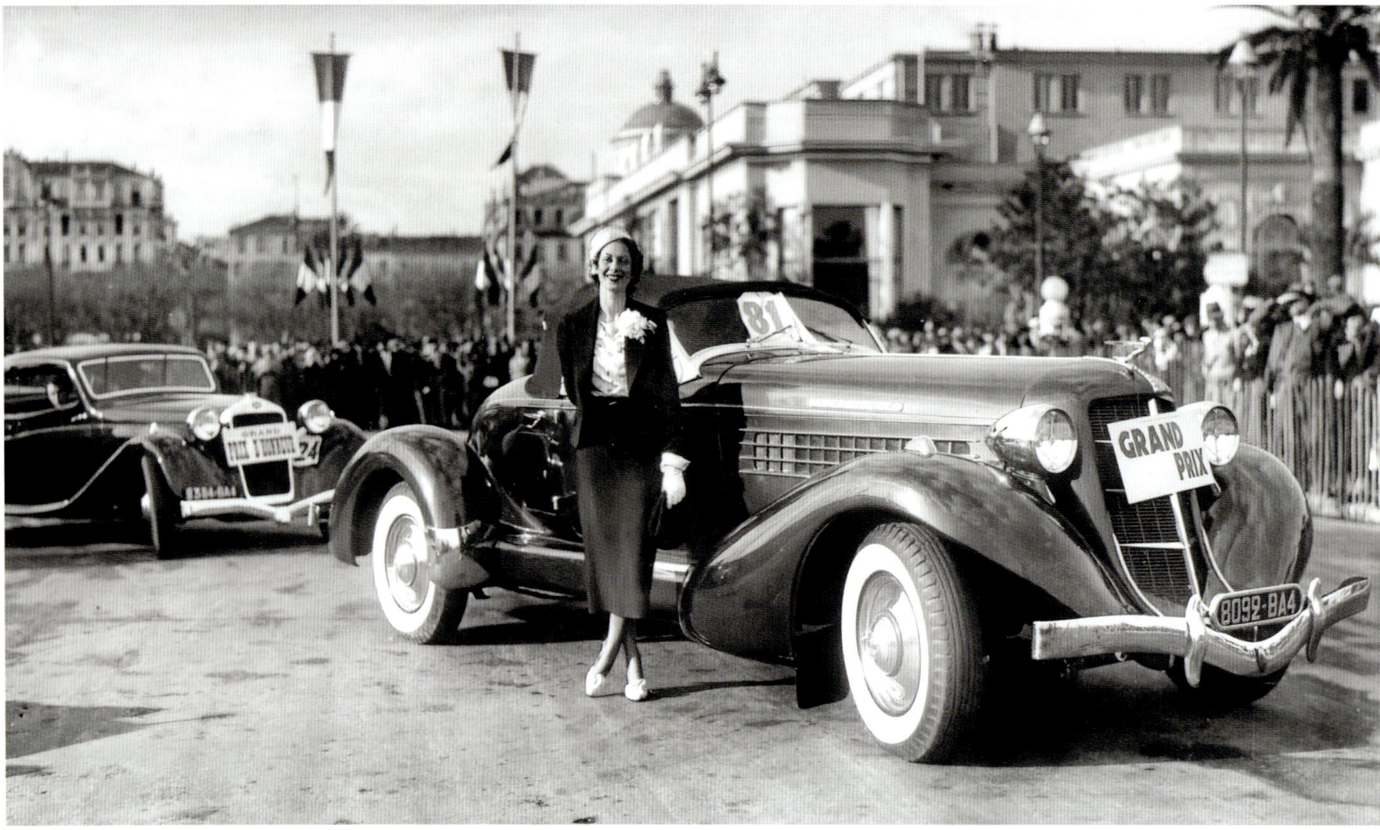


The Auburn 852 Speedster epitomized the American sports car of the 1930s. The effortlessly elegant Madame Van den Bosch appeared delighted to present this striking automobile at the Cannes Concours on March 27, 1936. (Bernard Brulé)

## Deauville

The first women's automobile concours d'elegance competition in Deauville appears to have taken place on August 26, 1926, though it is likely that similar events, such as flower parades, were held before that date. Deauville, a seaside resort often referred to as "the 21st arrondissement of Paris" due to its proximity to the capital, quickly elevated its concours d'elegance to a position of prestige. By 1939, it attracted some of the most distinguished participants – those who also excelled in Parisian concours and elegance competitions along the Côte d'Azur.

The inaugural competition was organized by Deauville's Sporting Club under the regulation of the Automobile Club de l'Ouest in the same year that Eugène Cornuché, one of Deauville's key founders, passed away. In the 1930s, the event gained further prominence under the sponsorship of newspapers such as *Excelsior* and later *Le Figaro*, with active support from the Deauville casino.

André de Fouquières, a frequent chronicler of such events, described the vibrant atmosphere of Deauville in an article for the *Automobile Club Féminine Bulletin* in July 1935: "*All the elegances gather today on the boardwalks, facing the sea. During the weekend, it is an assault of sophistication. It is the reign of stars and cosmopolitanism. At the races, polo matches, and golf tournaments, one meets both the most authentic and the most controversial princes, billionaires (what's left of them!), and women of consecrated beauty. For those weary of this artificial life, the nearby countryside offers a refreshing retreat with its blooming apple orchards.*"

Deauville became an annual rendezvous for a certain elite, much like Dinard, Le Touquet, Biarritz, or the Côte d'Azur.

The 1936 competition marked a turning point, coinciding with the introduction of France's first *congés payés* (paid holidays) under the Popular Front government. That year, the beach became a prime destination, drawing record-breaking crowds of spectators. The Delage automobile company sent its latest models, dominating the competition and claiming most of the awards.

Journalist Carlo Rim, editor-in-chief of *Vu* magazine, recounted his impressions in his autobiography, as his wife presented a new Delage at the event: "*Under a sky that had softened, an elegant crowd gathered before the Normandy Hotel. Among them were Londoners flown in at the casino's expense, mingling with Parisians who had traveled at their own cost – many among them beneficiaries of the newly introduced paid holidays. Immersed in this lively throng, at the intersection of nobility and the third estate, I could only drink in my shame, as I've always loathed such displays of worldly splendor. The loudspeaker, through the refined voice of Mr. de Fouquières, announced the names of the fortunate owners of these fine machines, which our automobile industry has every reason to celebrate. A viscountess, two renowned actresses, and three financiers' wives paraded gracefully in their sixteen-cylinder marvels, sublime Amazons of triumphant capitalism."*

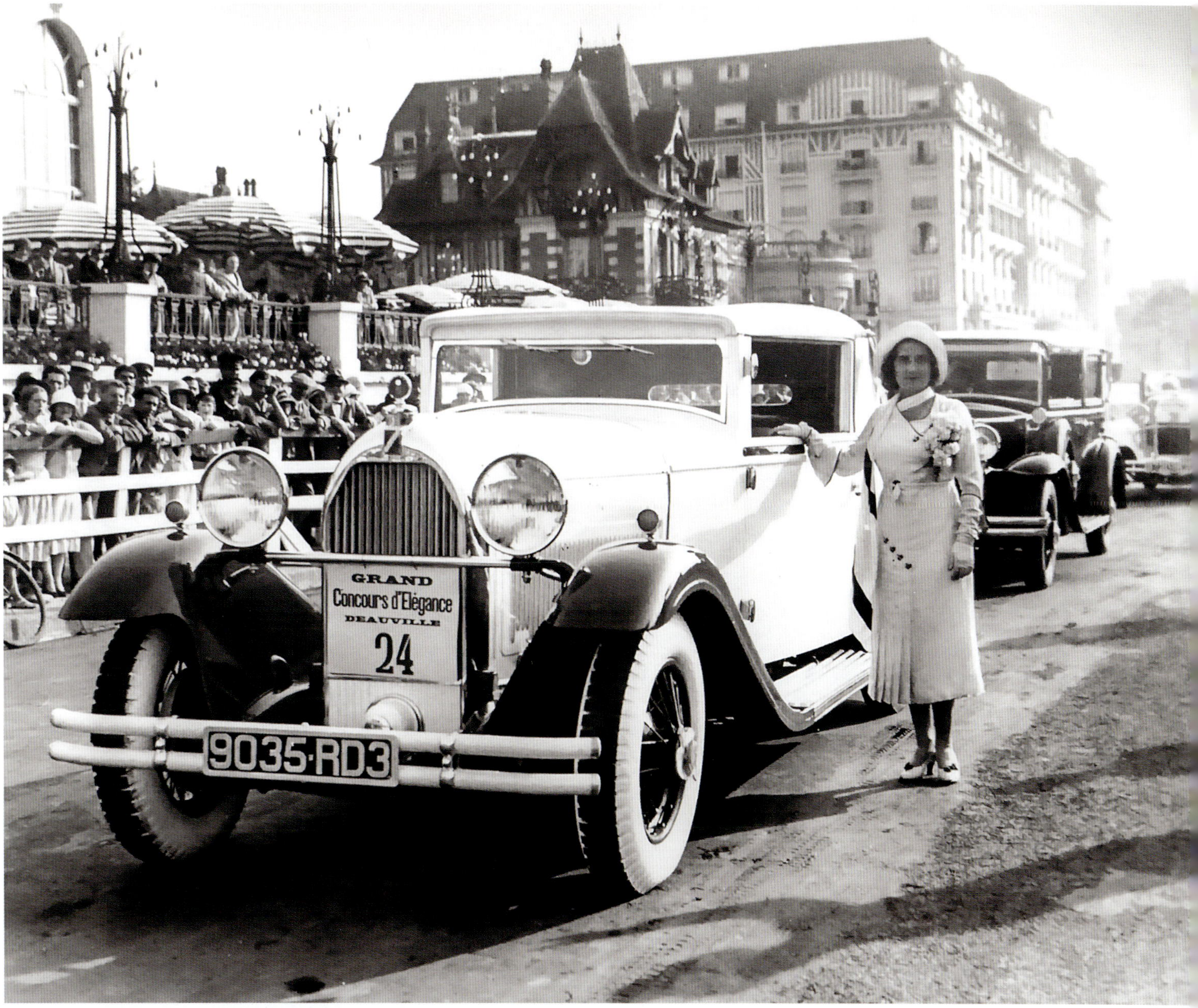


**ABOVE:** *This Talbot was presented by Mlle. Coquille at the 1930 Deauville Concours. She was the daughter of one of the directors of the Automobile Club de l'Ouest, who was also a manufacturer of automotive accessories, such as the renowned "Coquille trunk."* (Bernard Brulé)

**OVERLEAF:** *The Hispano-Suiza J12 was the epitome of French automotive prestige in the early 1930s. This coupe de ville, bodied by De Villars, took part in numerous concours events throughout the 1935 and 1936 seasons. One of its final public appearances was in Deauville on July 14, 1936, with Madame Furgerot at the wheel, where it was awarded the* Grand Prix d'Honneur. (Laurent Friry)

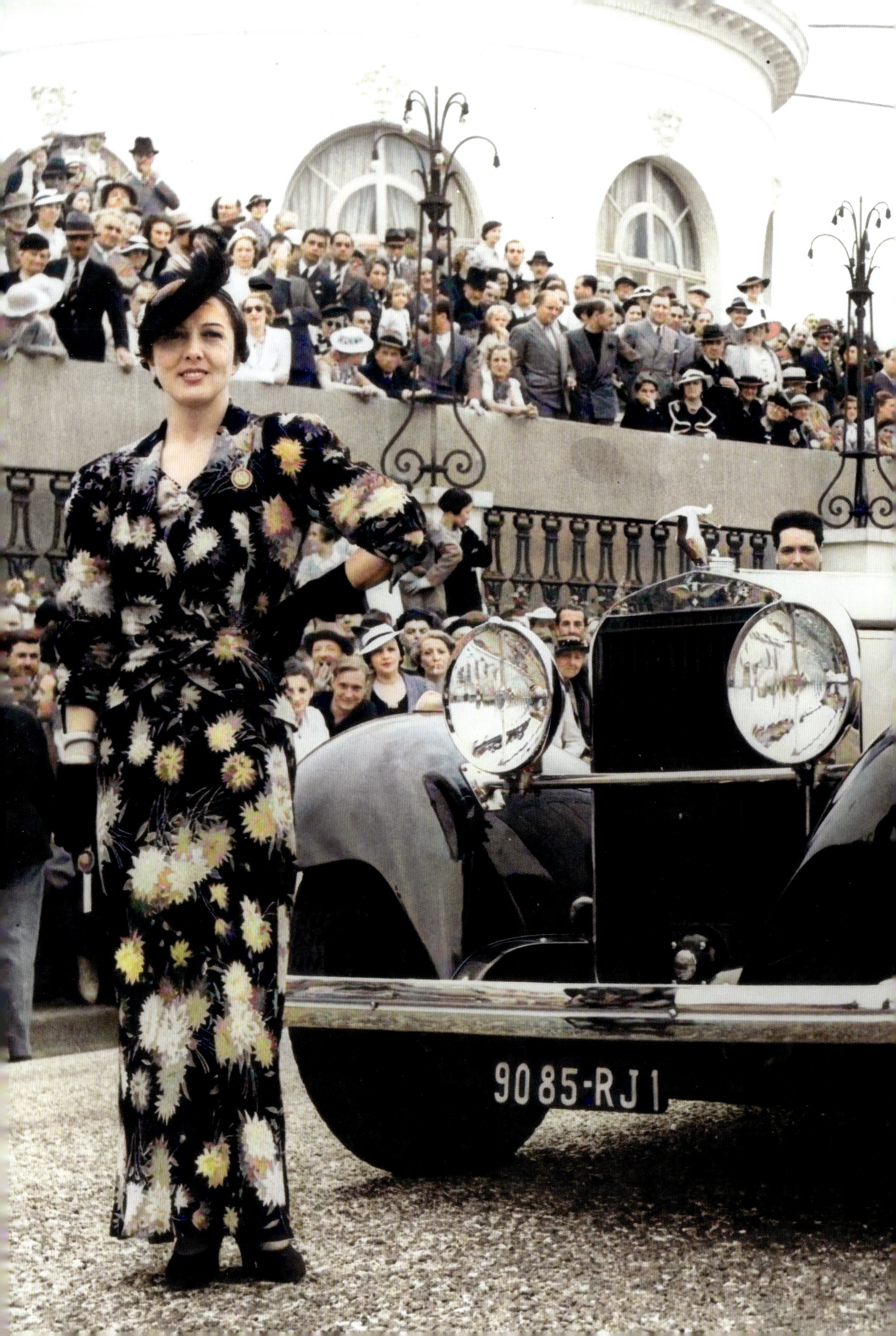
9085-RJ1

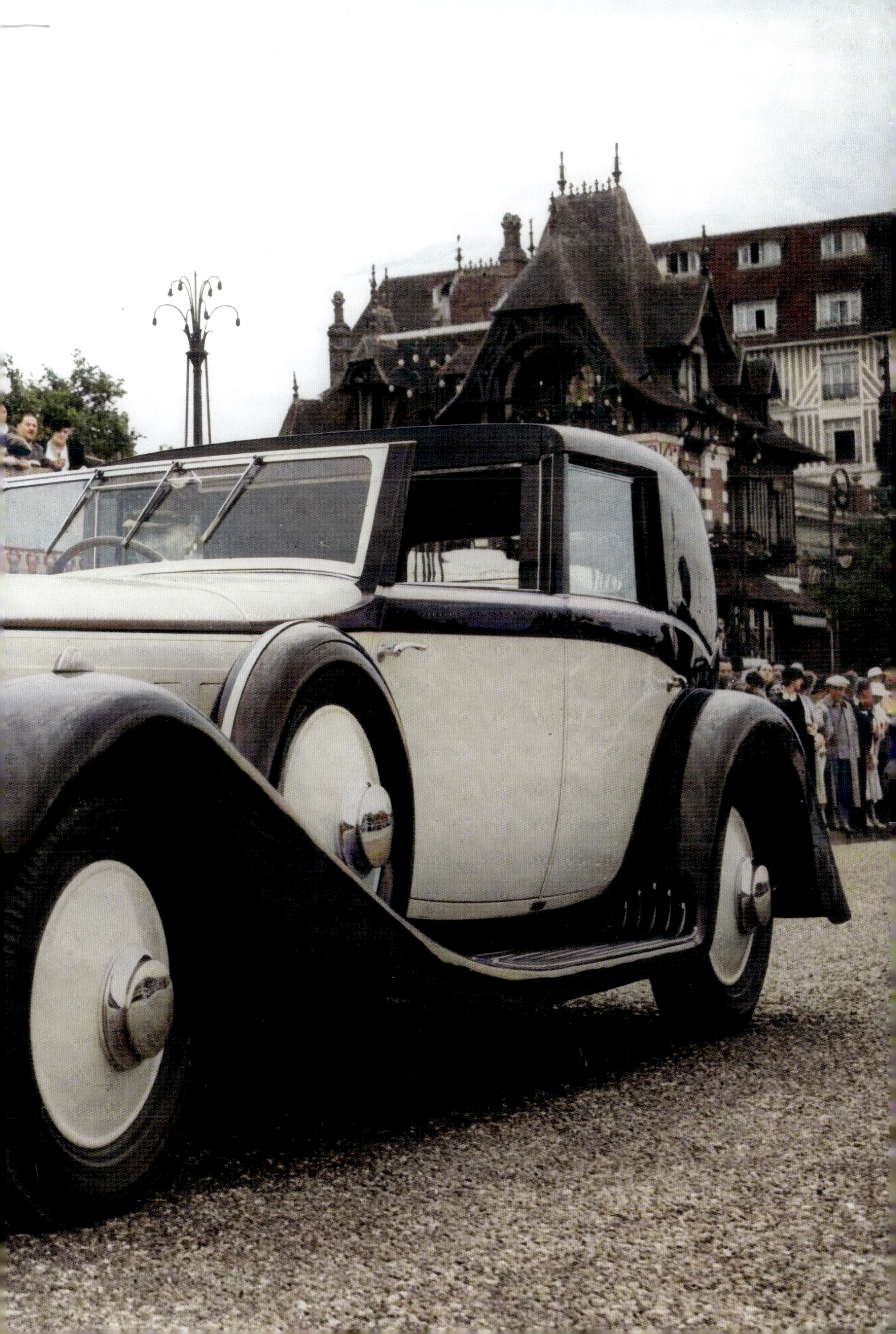

**RIGHT:** *In 1936, during its efforts to re-establish itself, Delage created a "stable" specifically for concours d'elegance. Six cars were assembled to tour across France, with a few ventured into England, securing numerous victories along the way. One of these was a brand-new D6-70 convertible by Figoni et Falaschi, owned by Madame Carlo Rim, the editor-in-chief of Vu magazine and a prominent figure in theater, literature, and cinema. (Daniel Cabart Archives)*

**BELOW:** *Madame Richer-Delavau was a member of a select Delage team in 1936. As part of this team, she participated in the Ramsgate Concours d'Elegance in England on July 9, 1938, driving a D8-100 with bodywork by Franay, finished in "pale gold metallic." The following day, she made the crossing of the Channel overnight to appear in Deauville. The "Special Award" she earned from the previous day's event in England is still visible on the grille. (Daniel Cabart Archives)*

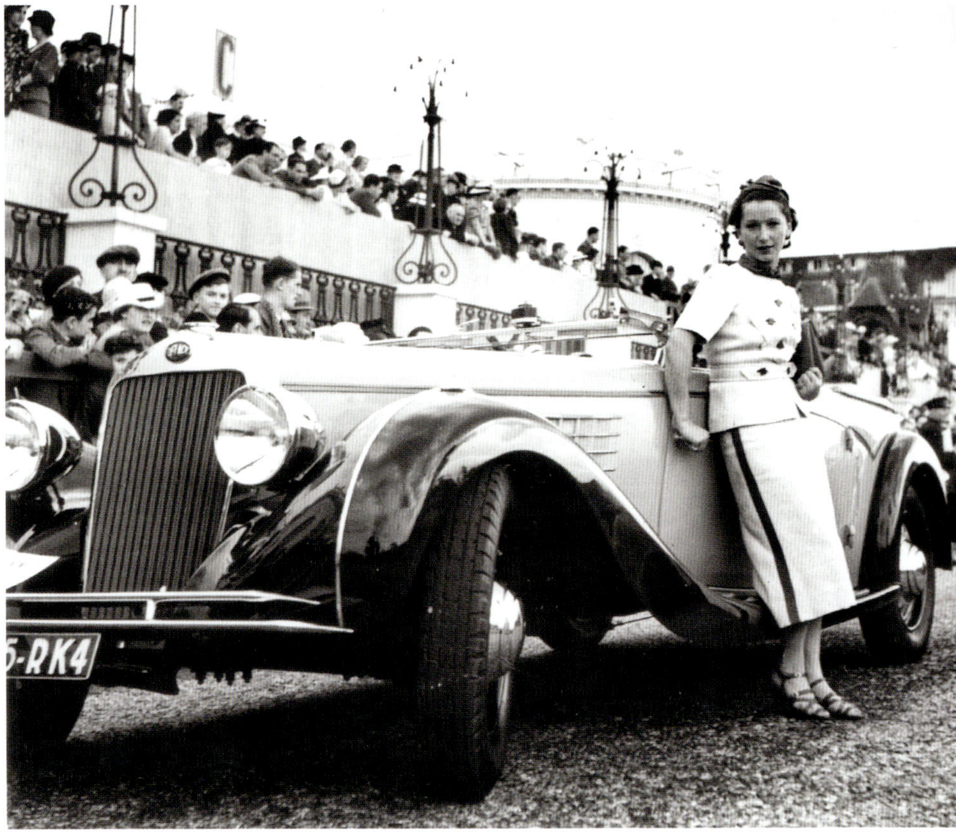


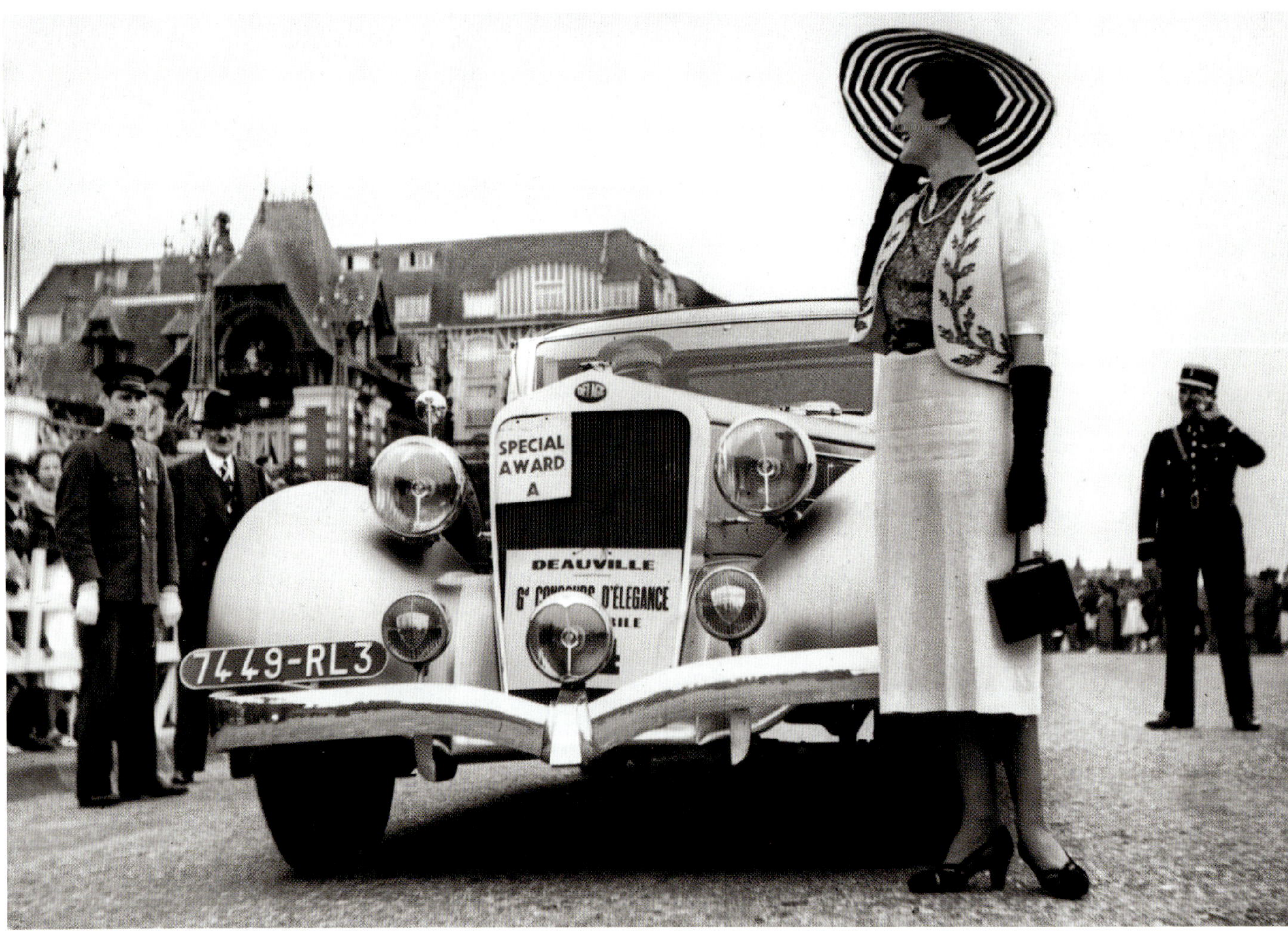

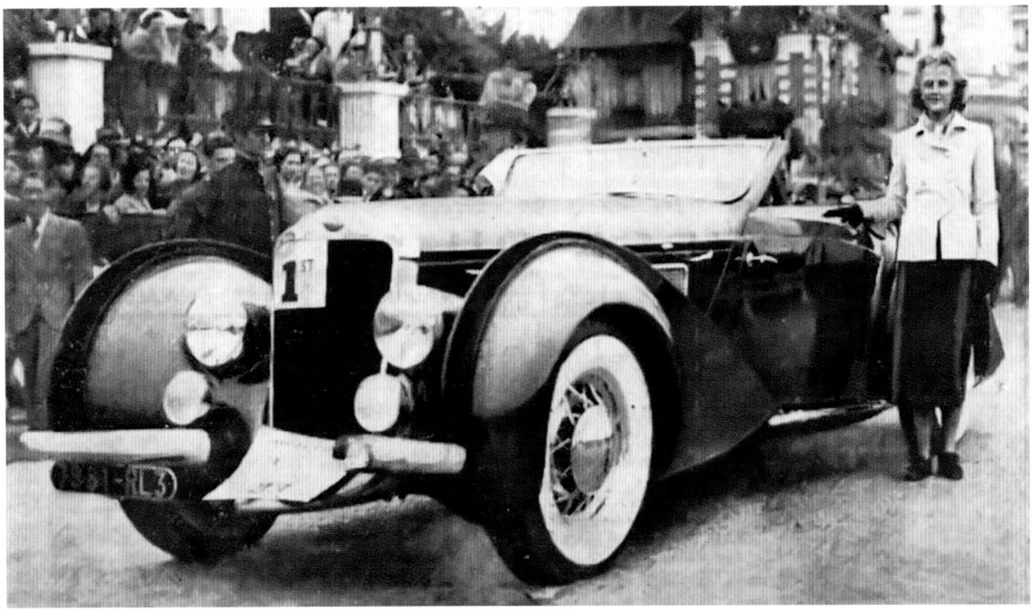

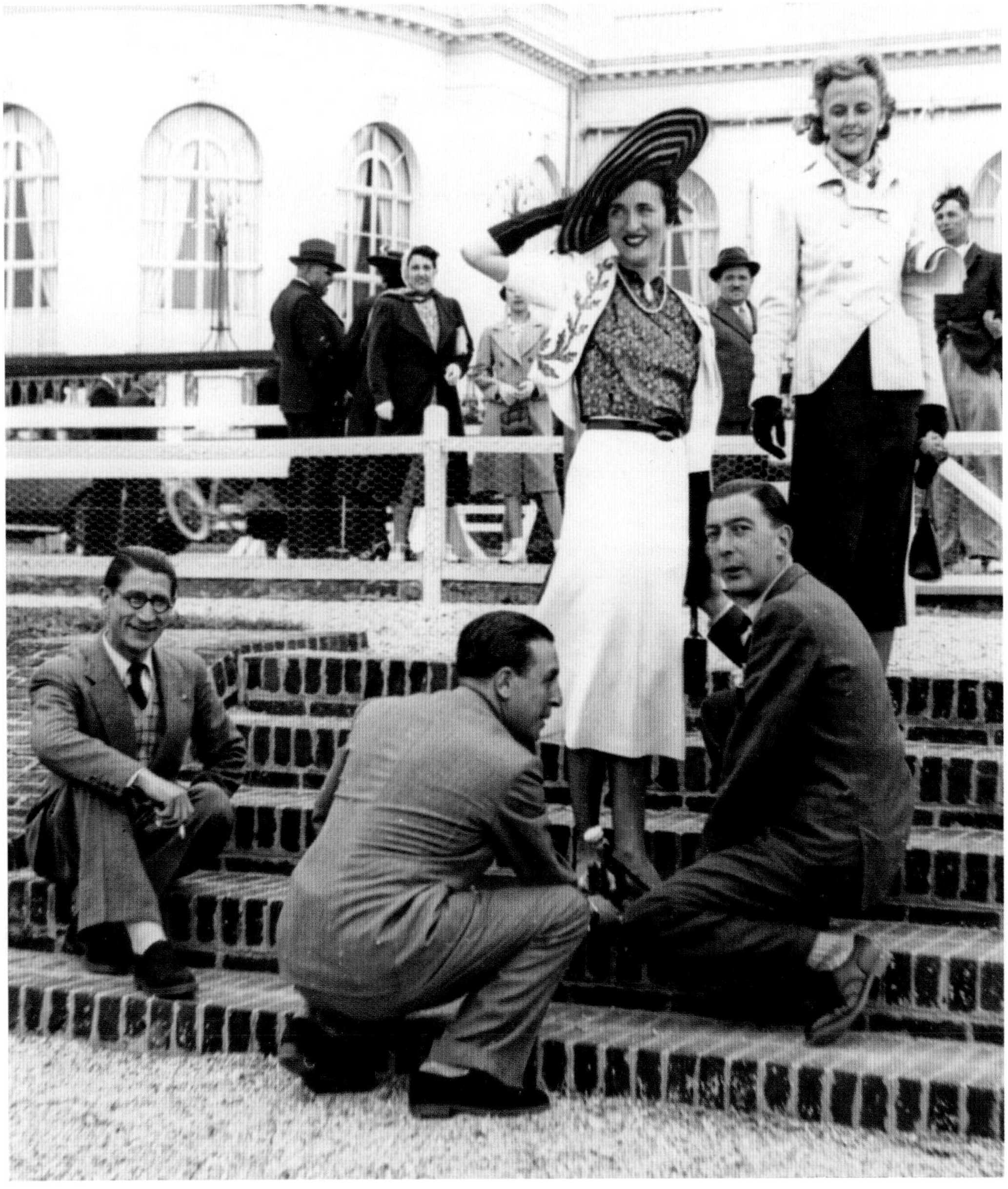

**LEFT:** *Also making the overnight journey from Ramsgate (England), Madame Claude Rolland exhibited an extraordinary De Villars-bodied roadster on a Delage D8-120 S in Deauville on July 10, 1938. The car won the Super Cup "hands down," organized by* Le Figaro. *Its victory was marked on the grille with the "1st" from the previous day's competition, highlighting the car's top ranking.* (Daniel Cabart Archives)

**BOTTOM:** Mlle. *Claude Rolland and Madame Richer-Delavau prepare for the competition. Seated on the left is Paul Pinier, the Delage engineer and creator of the renowned D6-3 Liter racing cars. The July 10, 1938* Le Figaro *stated: "The success of 'Delage' is even more deserved given the remarkable feat accomplished by the brand's executives and competitors to be present at Deauville. 'Delage' had, in fact, participated just the day before in the Ramsgate Concours d'Elegance in England, where it also claimed numerous first prizes. To be in Deauville on time, the team had to cross the Channel by boat overnight, in particularly challenging conditions due to the rough sea, and then drive from Boulogne to Deauville."* (Daniel Cabart Archives)

## Fémina L'Intran

The Fémina L'Intran Concours gained particular prominence with the 1926 Biarritz event. It was there that the competition first introduced judging based on the passenger's attire, required her to step out of the car, and, notably, held the concours at night.

The organizers then moved closer to the capital, and over the years, the *Fémina* competition, associated with the daily newspaper *L'Intransigeant*, became one of the two essential Parisian competitions, along with the *L'Auto* competition.

The concours evolved to the Fémina L'Intran Concours d'Elegance, focusing on what became its specialty: fashion and the relationship between elegance, haute couture and automobiles. The organizers recalled that this concours was "devoid of any technical character."

Like its alter-ego, the daily newspaper *L'Auto*, this concours relied on the prestige associated with stars of all kinds who came to present their cars for a day (very few stars owned these cars). Film actresses, comedians, they were very often recruited by manufacturers, by coachbuilders to add prestige to their latest creations. Not to mention fashion designers, hatters, etc. The last events before the war brought together the best in these fields. It was undoubtedly the Parisian concours that remained in people's memories as the most prestigious.

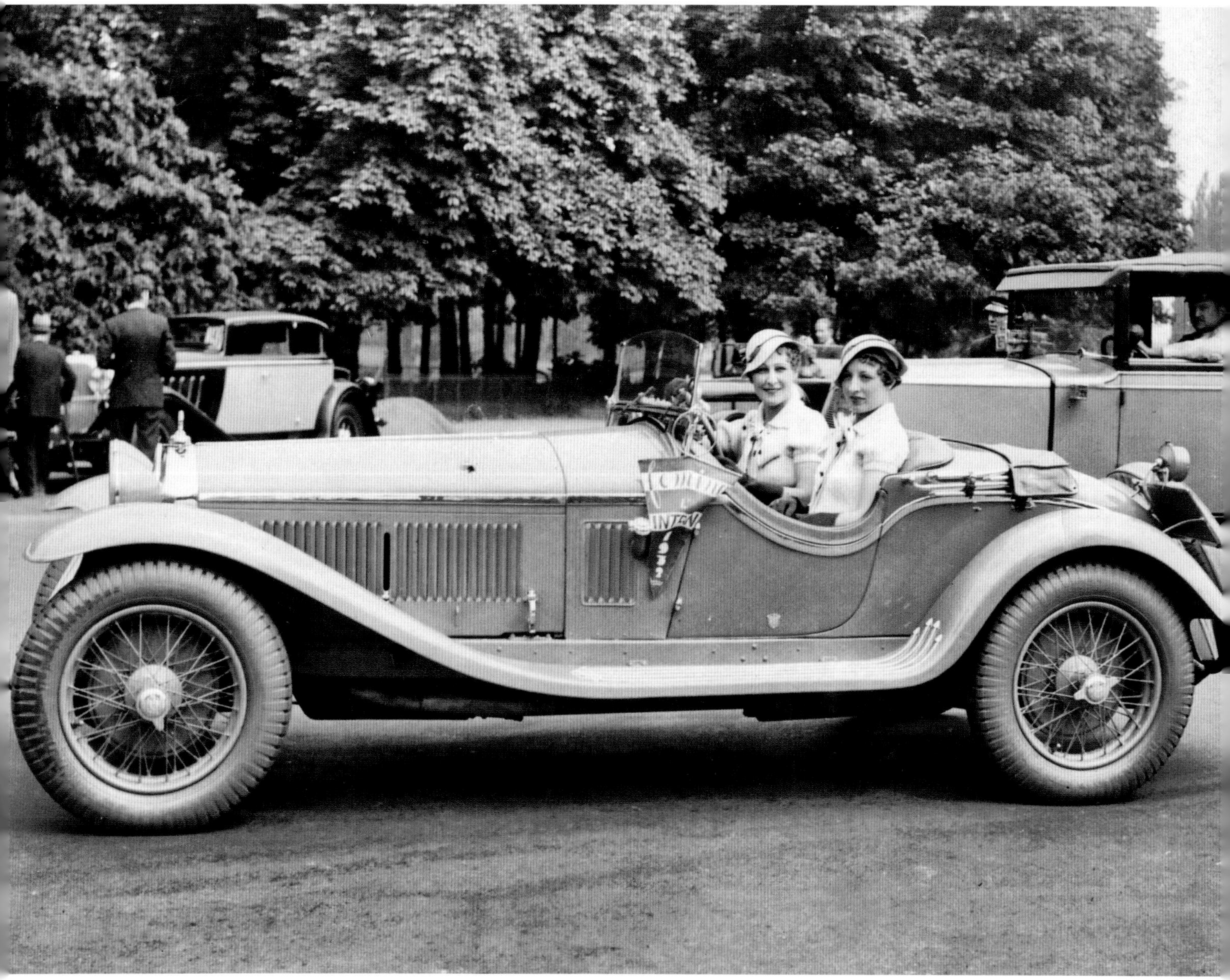

*Mary Costes and Madame de Beaufort took part with a very sporty Alfa Romeo 6C 1750 with Touring bodywork at the Fémina L'Intran Concours d'Elegance on June 24, 1932. In the background, the Panhard 35 CV with Letourneur et Marchand bodywork can also be seen. (Bernard Brulé)*

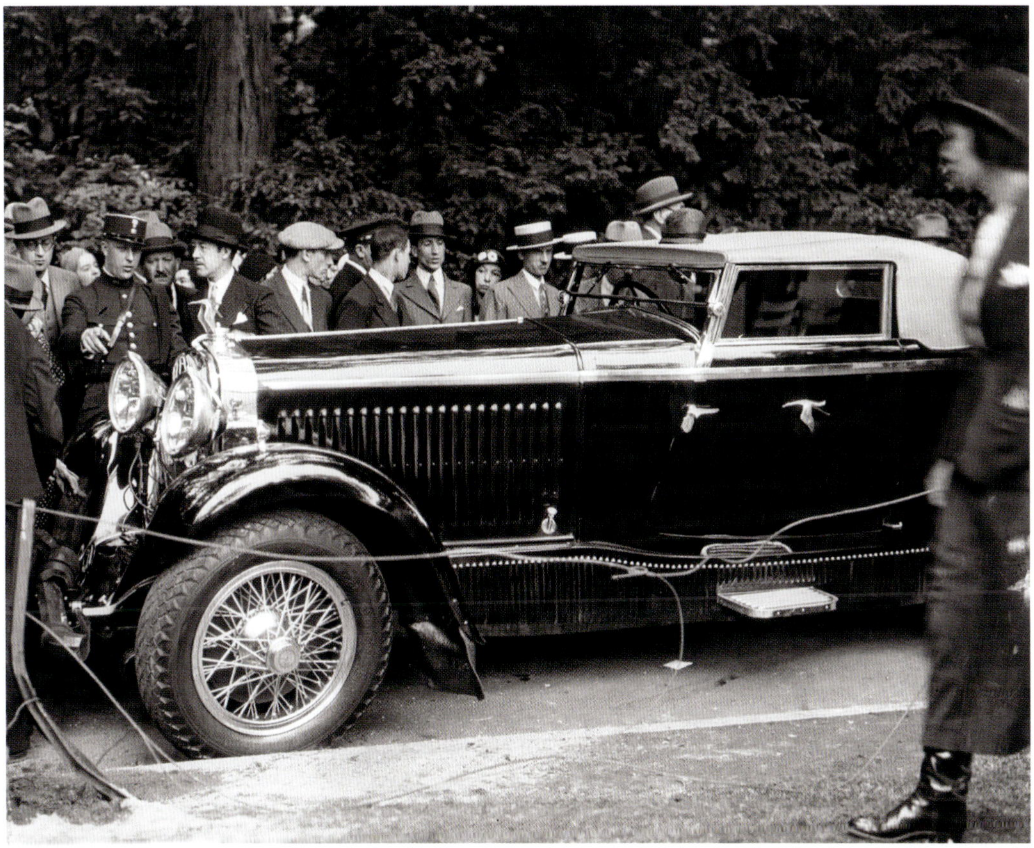

**LEFT:** *The 1932 Fémina L'Intran Concours d'Elegance was marked by a major incident. After a misstep by the driver of this Hispano-Suiza, the event was abruptly interrupted when the car veered into the crowd, injuring 14 people. This led to a ban on women driving at such events. As a result, chauffeur-driven cars became the norm, though exceptions to this rule became increasingly common.* (Laurent Friry)

**BELOW:** *Mlle. Marthe Mussine participated in the concours on June 25, 1937, presenting her La Licorne with Vutotal coachwork, crafted by Labourdette. Instead of a dog, she had her little leopard accompanying her.* (Bernard Brulé)

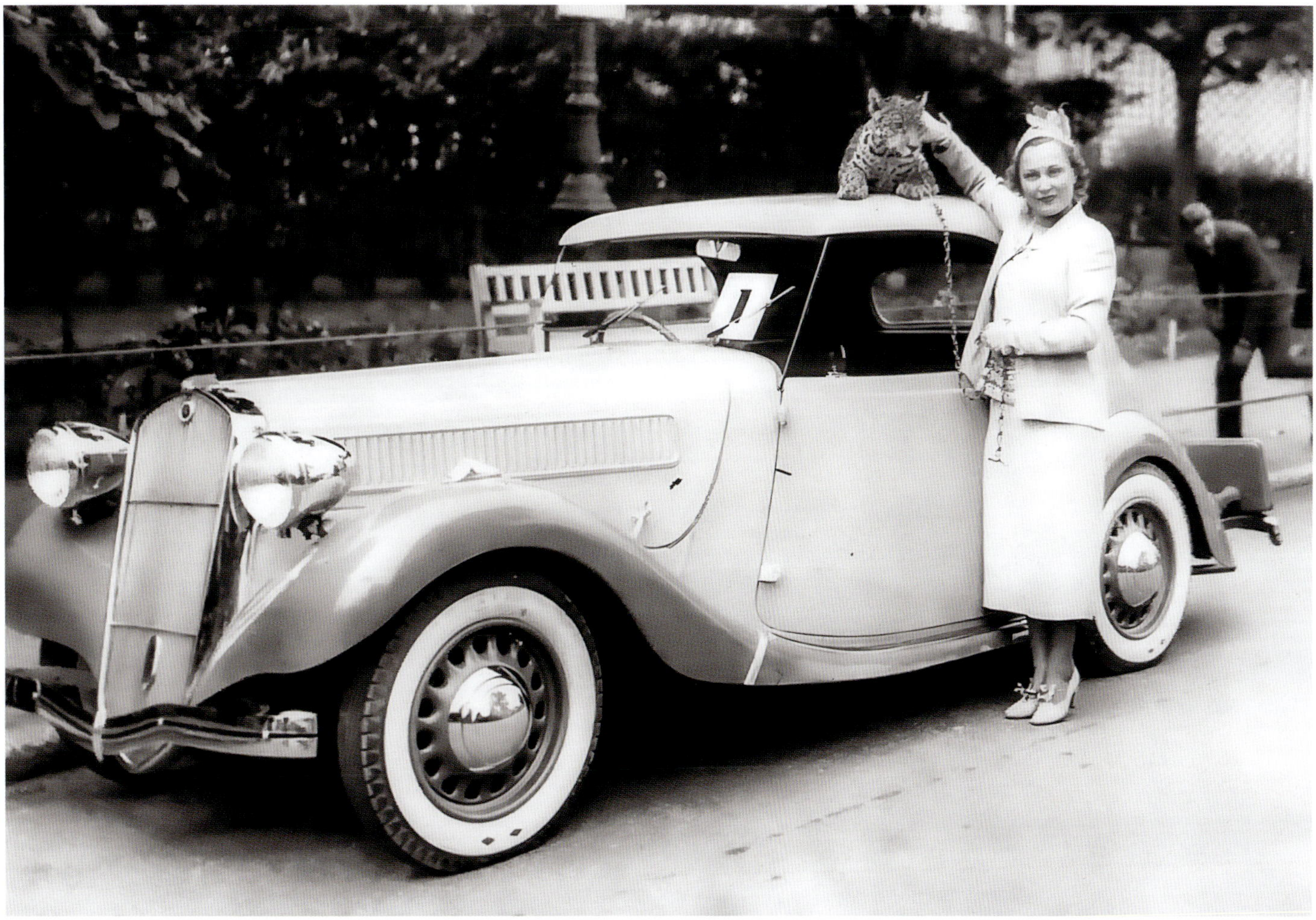

**RIGHT:** *Marthe Chenal, the renowned opera singer, presented a magnificent Renault Reinasport at the concours on June 23, 1933. The car, driven by two chauffeurs, featured coupe de ville bodywork by Henry Binder. The Renault was awarded the 1st Grand Prix d'Honneur. (Bernard Brulé)*

**BELOW:** *Swedish model Inga Lindgren, accompanied by her poodle, presented a Delahaye 135 Chapron cabriolet at the Fémina L'Intran Concours d'Elegance competition at the Trocadéro on June 23, 1939. (Laurent Friry)*

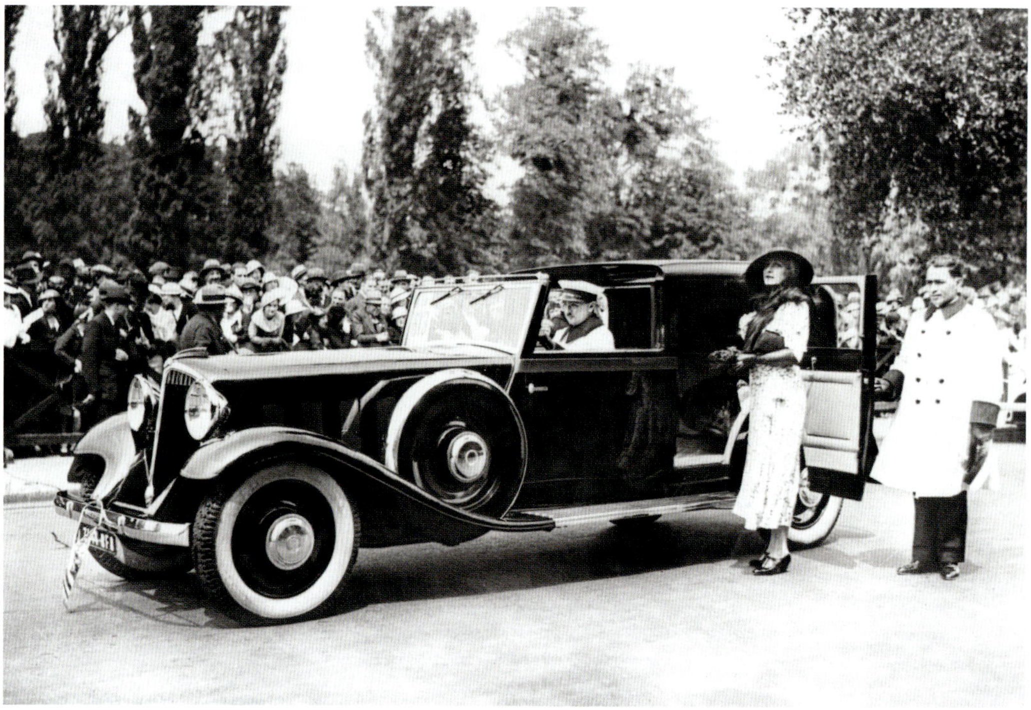

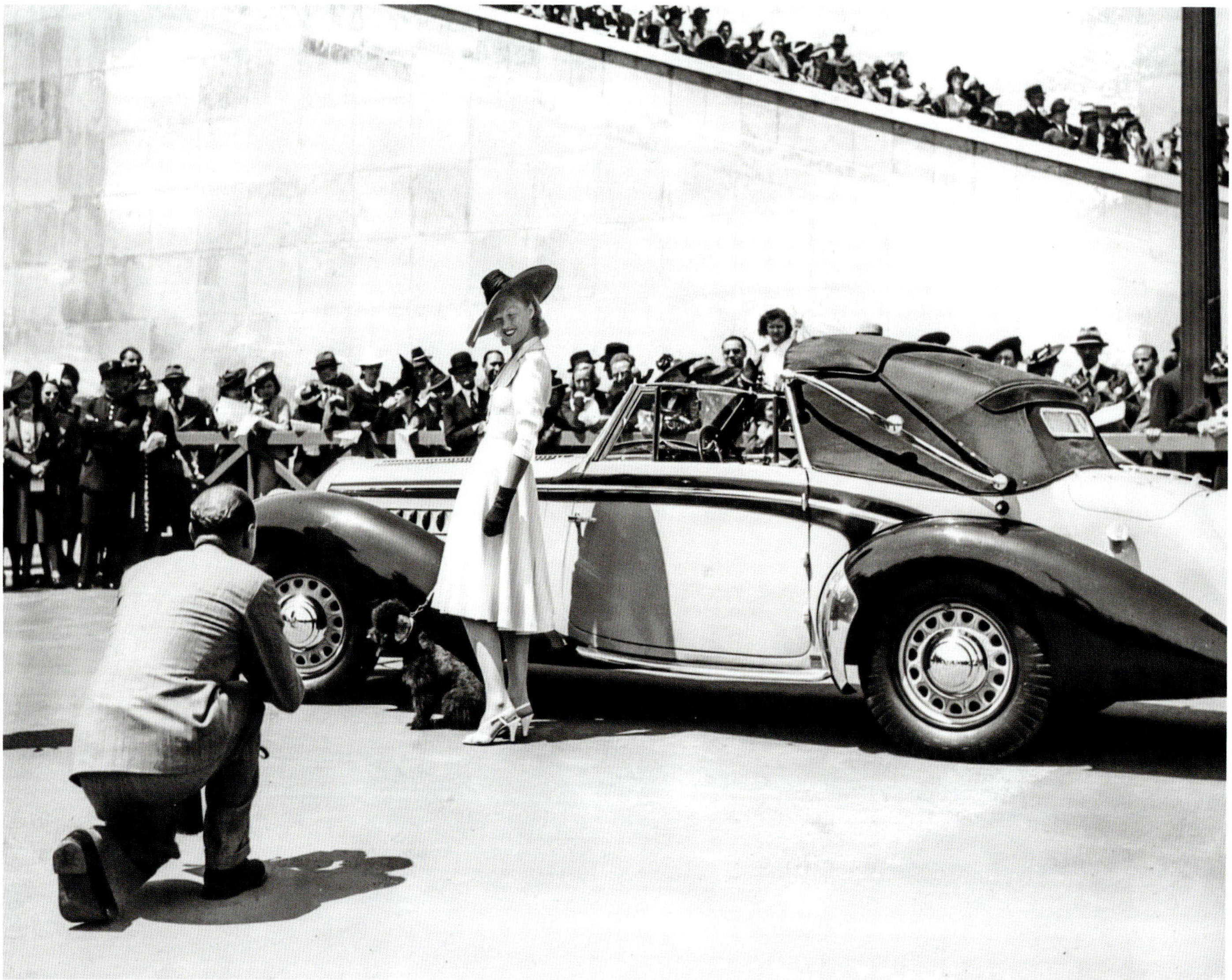

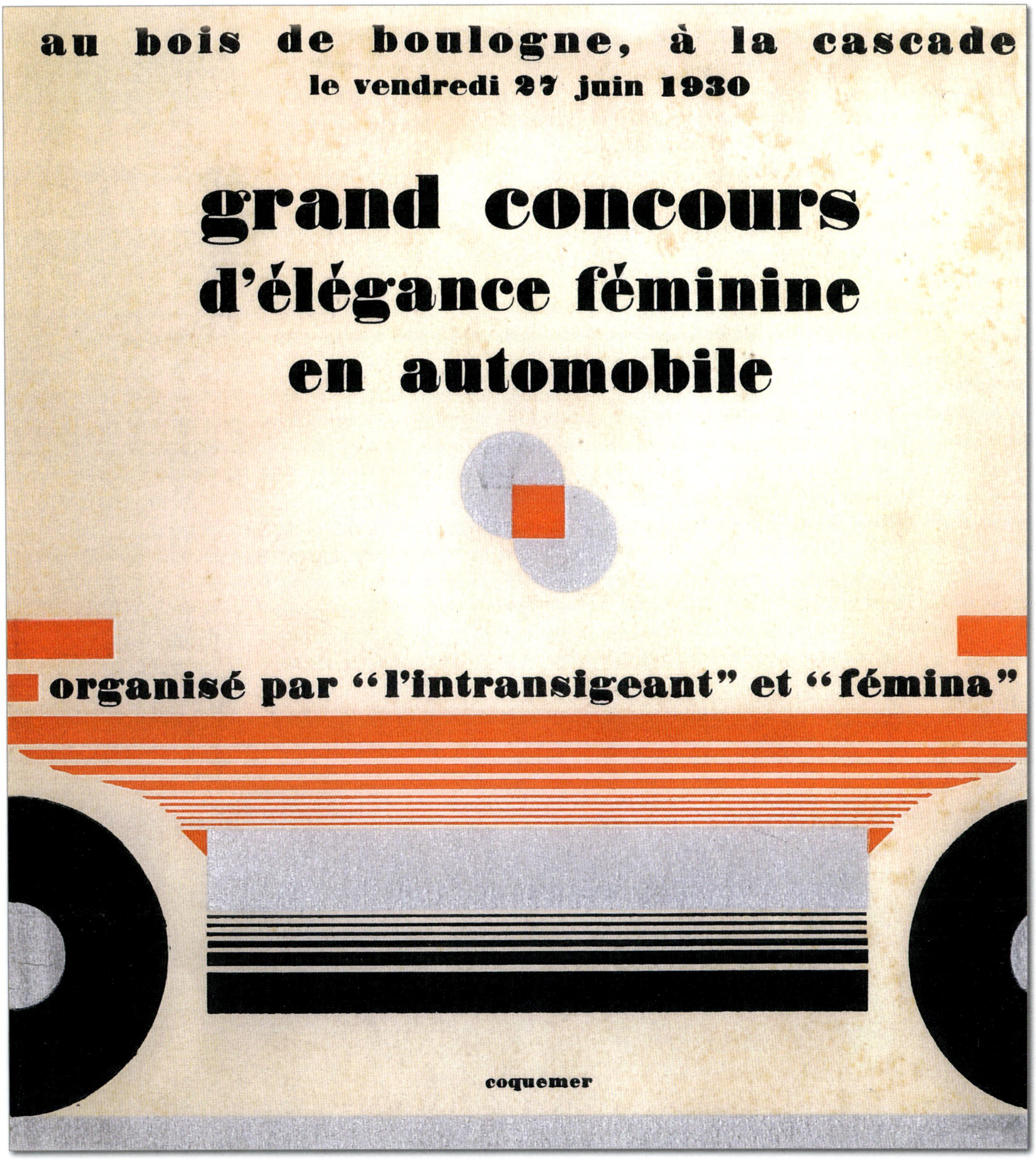


*The programs of Parisian concours d'elegance are highly coveted documents today, featuring a blend of advertisements for fashion, bodywork, jewelry, and, of course, automobiles. (Laurent Friry Archives)*

## La Baule

The story begins in 1919, when American soldiers stationed in Saint-Nazaire organized a three-day exhibition of mechanical items on the beach of La Baule to commemorate Memorial Day. This unique event featured a variety of demonstrations, including tracked vehicles, speed races between a Nieuport airplane and a Cadillac, motorcycle and car races. The event was a tremendous success, drawing thousands of spectators, as reported by the press.

The first official concours d'elegance in La Baule, however, seems to have been held on August 15, 1921. Organized by the Automobile Club de l'Ouest, the daily newspaper *L'Auto*, and magazines like *Motocyclisme* and *La Carrosserie*, this event aimed to celebrate the artistry of French automobile design. Originally referred to as a "bodywork competition," its purpose was to bolster the reputation of French automotive craftsmanship. As was common at the time, it served as a highlight within a larger event, marking the conclusion of the Angers-La Baule rally. In subsequent years, it became the grand finale of gatherings like the Meeting des Plages Guérandaises or La Semaine Automobile, which included beach speed races, gymkhana contests, and more.

François André, a visionary and influential developer, brought fresh energy to La Baule after his successes in Cannes and Deauville. In 1924, he introduced La Semaine Automobile, which featured a Grand Prix on the beach, various attractions, a tourist rally, and an elegance competition. Initially called the Grand Prix de l'Elegance or Tournoi de l'Elegance, the event soon adopted the now-famous title Concours d'Elegance en Automobile.

The first competition in 1924 took place in front of the casino, but by the 1930s, it moved to the picturesque gardens of the Tennis Club. During this *Belle Époque* era, the competition became one of the most fashionable events in La Baule. Elegantly dressed women, exuding a newfound sense of freedom, proudly drove increasingly sophisticated and luxurious cars. Often accompanied by friends, children, or pets, they embodied a perfect harmony between haute couture and automotive prestige.

The proceedings were meticulously planned: at 3pm, the vehicles assembled at the entrance of

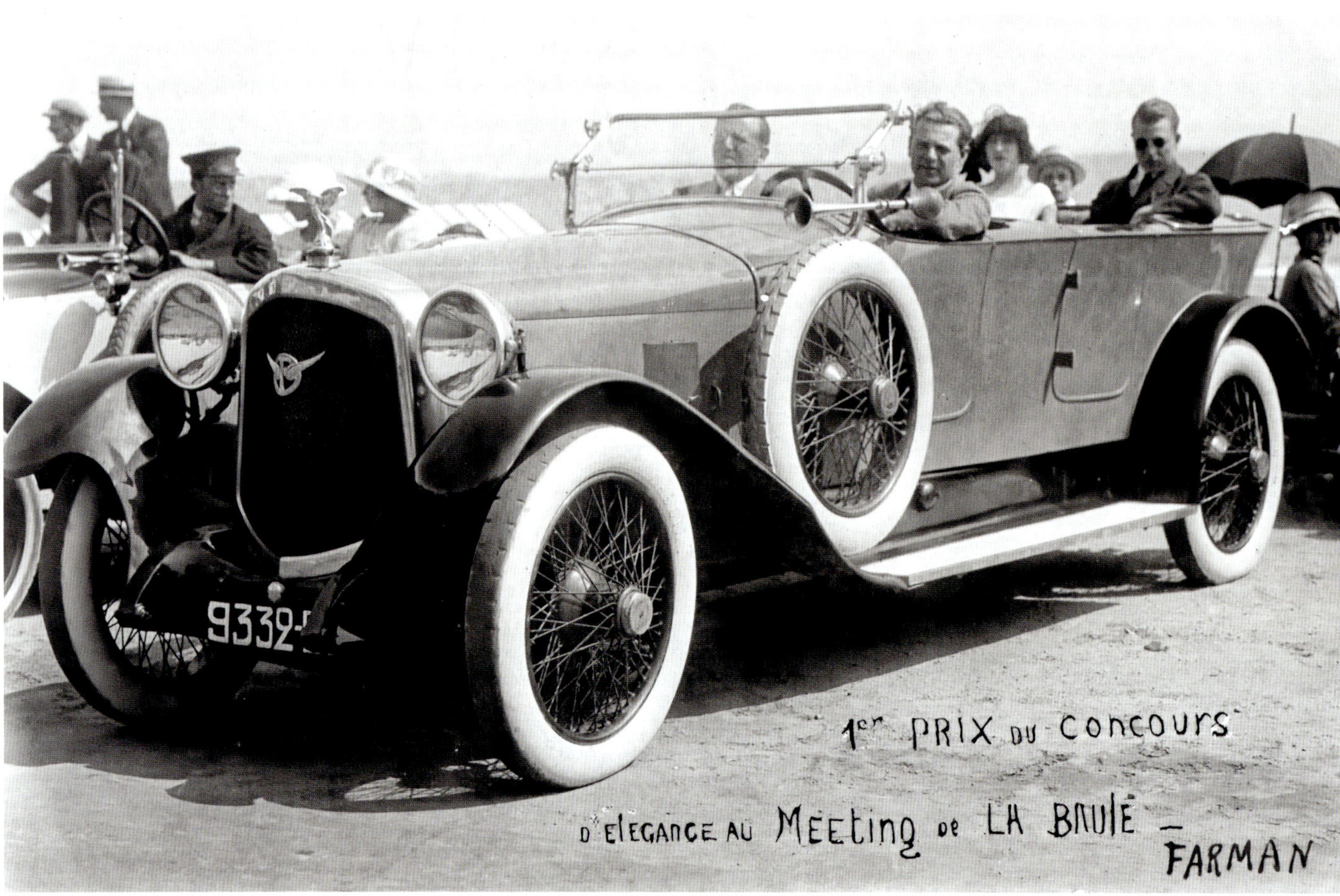


*The Farman Type A6, a factory-built tourer, was the winner of the inaugural La Baule Concours d'Elegance in 1921. Its bodywork was undoubtedly elegant, yet notably, there was no female presence alongside it. (Daniel Cabart Archives)*

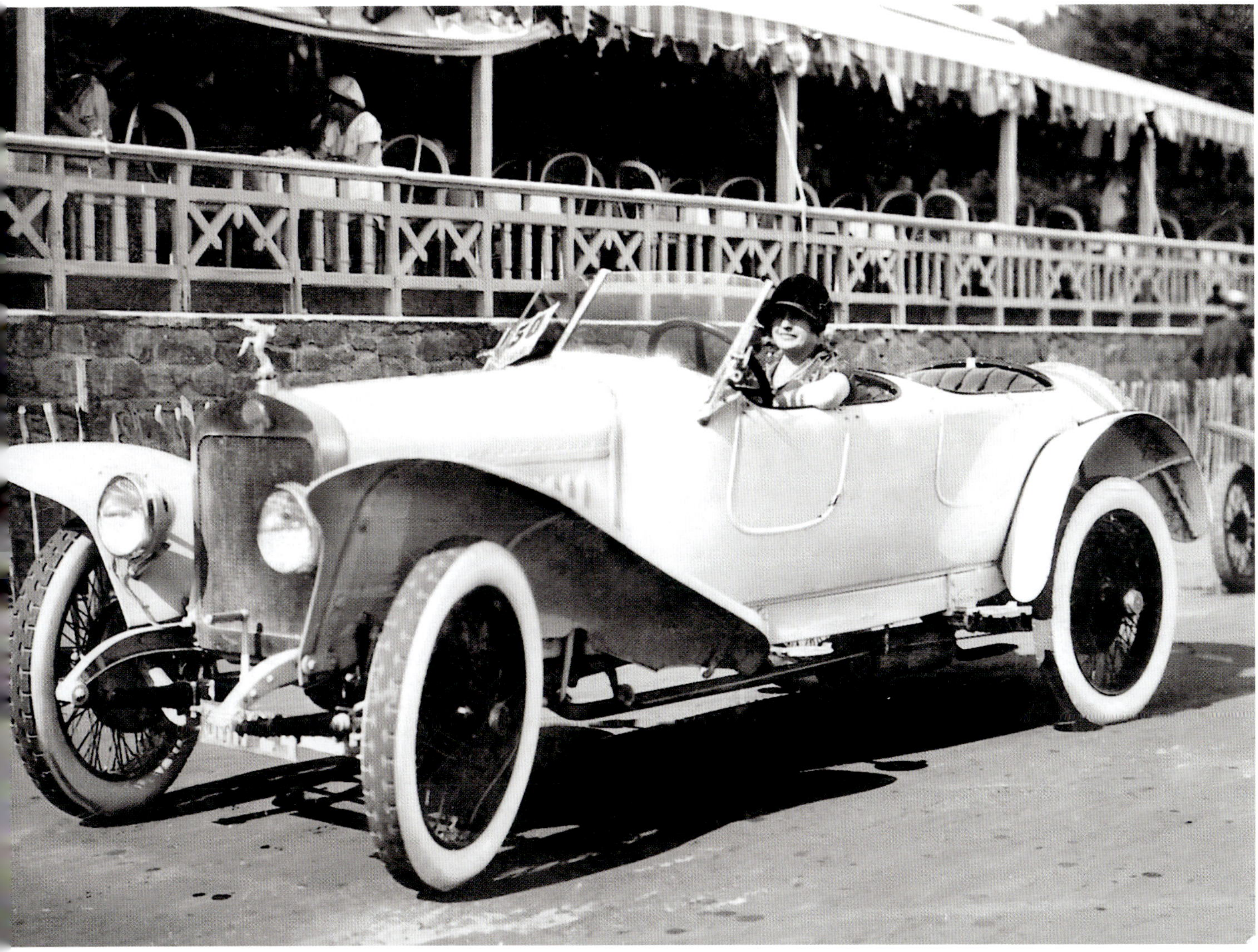

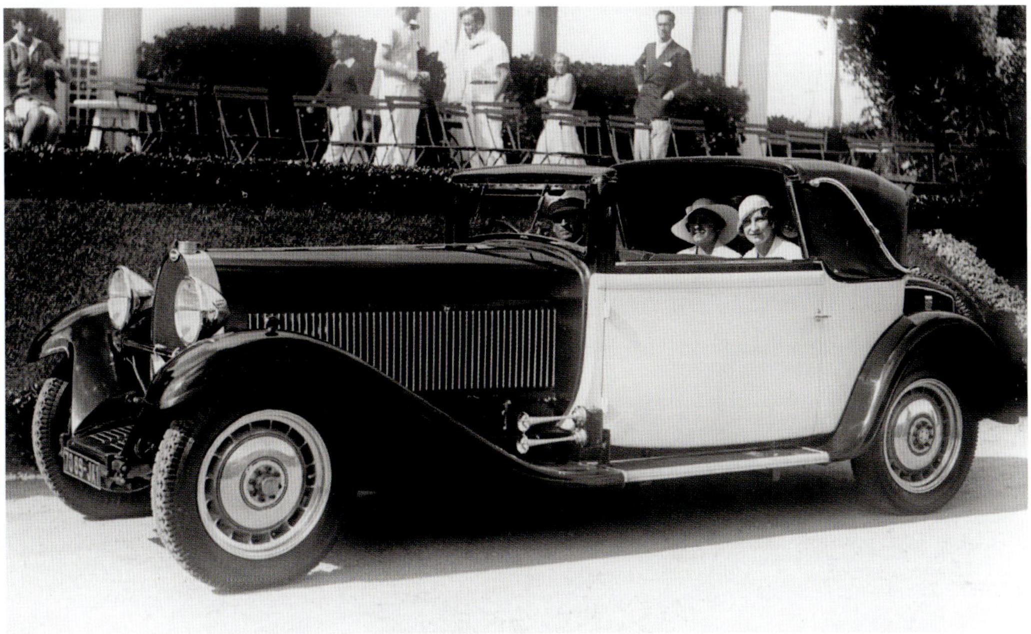

**ABOVE:** *At one of the earliest concours d'elegance held in La Baule, in front of the tennis club, Madame Guibert presented a Delage DE with Torpédo-deck bodywork by Alin-Liautard on July 23, 1924. (Daniel Cabart Archives)*

**LEFT:** *At the La Baule Concours on July 22, 1930, Madame Grosse presented a Bugatti Type 46 Vanvooren cabriolet, built under license from Weymann. (Francis Piquera)*

the tennis courts for inspection. At 4pm, the parade began, with a stop at the pergola terrace of the Clubhouse, where participants disembarked on the arm of a distinguished escort, often André de Fouquières. Each contestant introduced herself as the jury carefully evaluated her presentation. Following deliberations, the participants returned to their vehicles for the award ceremony.

Over time, the La Baule Concours, along with its counterpart in Dinard, became one of the most prestigious concours d'elegance competitions in western France. By the 1930s, major manufacturers like Renault and Delage began sending representatives - often notable women - to showcase their latest models. Due to its growing popularity, preliminary selections were conducted the day before the event. Competitors, often drawn from the local aristocracy or the entertainment world, were carefully curated, reflecting the exclusivity and grandeur of this celebrated occasion.

**BELOW:** *At the La Baule Concours on July 22, 1930, a Gardner cabriolet with bodywork by Letourneur et Marchand was presented. This design, featuring a door with a central handle, was also available on Hispano-Suiza and Delage chassis. (Francis Piquera)*

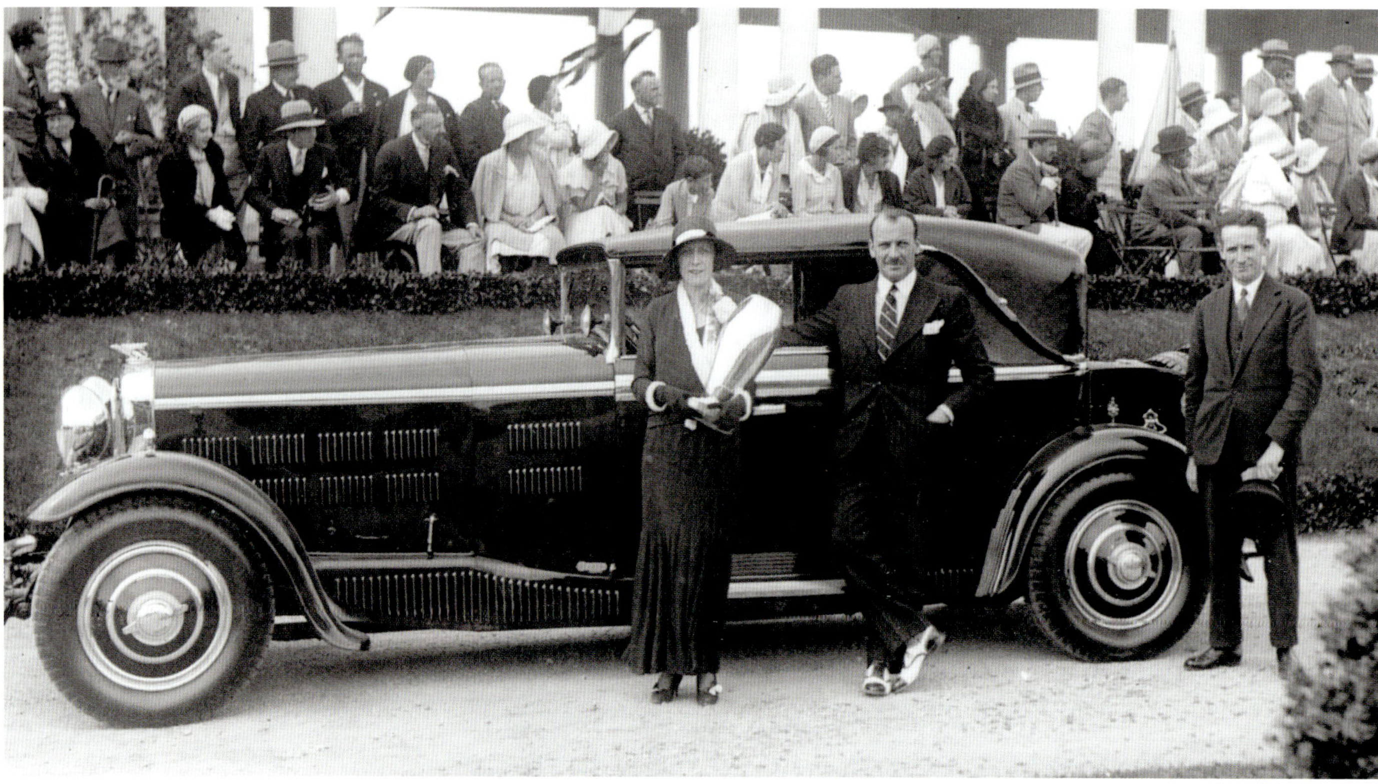

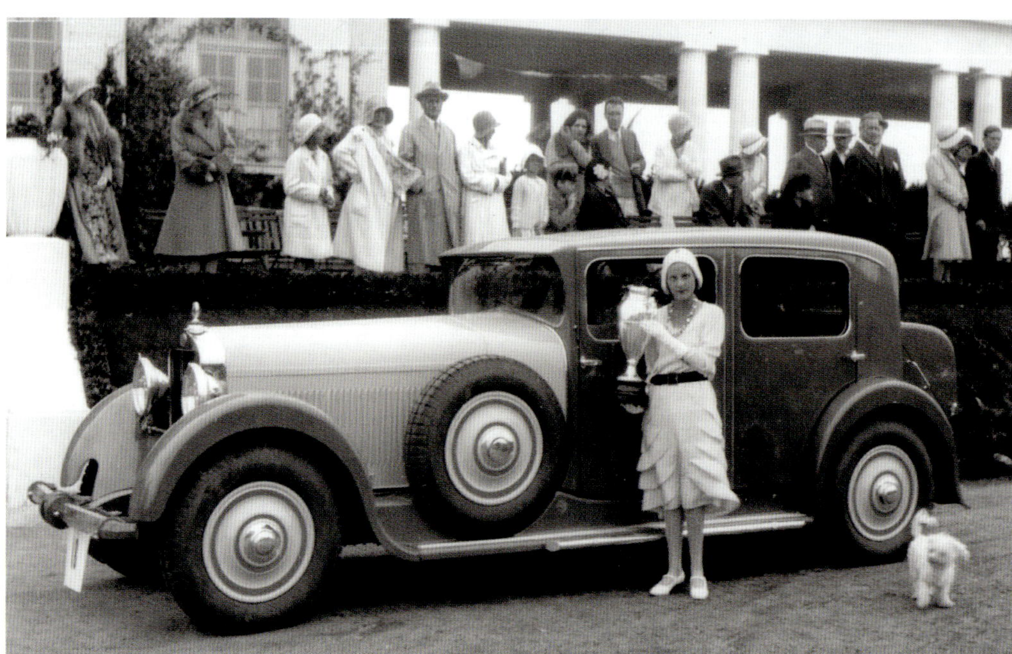

**RIGHT:** *Madame Delia Schiff-Giorgini won the Grand Prix d'excellence with her Delage D8, at the La Baule Concours on July 22, 1930. It featured a strikingly modern Toutalu aluminum sedan body by Million-Guiet, based on a Vizcaya patent. Much like the Gardner bodywork, this design could be adapted to a variety of chassis, including Hispano-Suiza, Talbot, Renault, Bugatti, and many others. (Francis Piquera)*

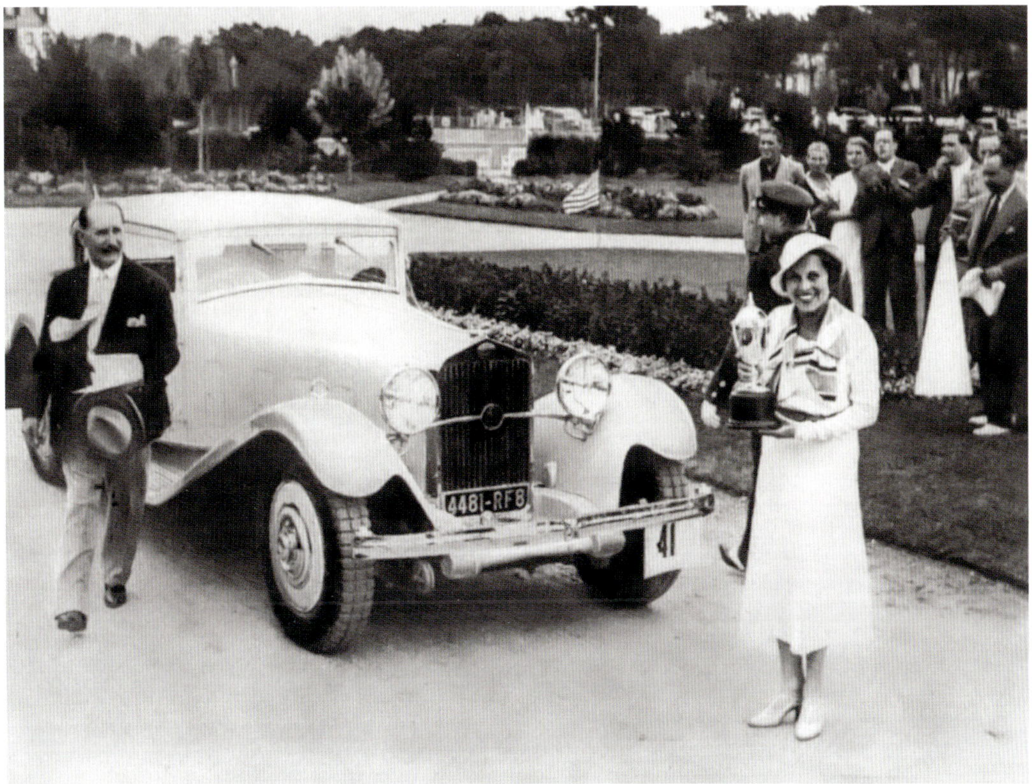

**LEFT:** *André de Fouquières presenting Mistinguett with the trophy she won while driving the remarkably sporty Delage D8-S, featuring bodywork by Letourneur et Marchand. (Gaël Archimbaud)*

**BELOW:** Until 1939, the *La Baule Concours was one of the most prestigious in France, attracting the finest entries. Among them was the Delage D8-120 with cabriolet bodywork by De Villars, a car that had previously been presented in England and Paris by Madame Louis Arpels. (Gaël Archimbaud)*

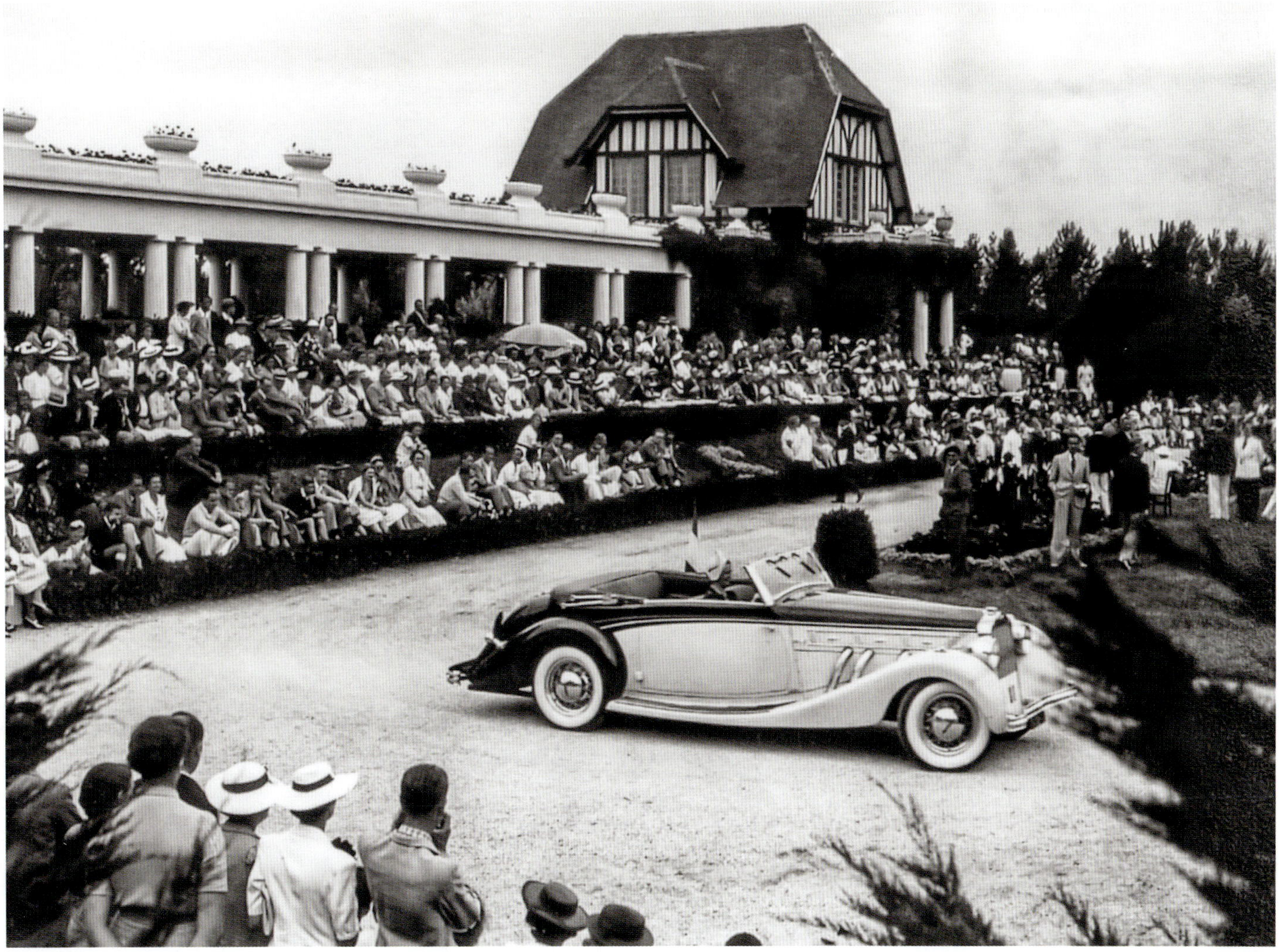

**BELOW:** *Madame Malvina Ischer's Delage D8, elegantly bodied by De Villars as a two-door sedan, won the Grand Prix in its class at the 1931 Monaco Concours. (SBM Monaco)*

**OPPOSITE:** *The proximity of Italy drew many Italian coachbuilders to Monaco, with Pinin Farina being one of the most prominent throughout the 1930s. Here, in 1931, a Lancia with sedan bodywork demonstrates the early signs of aerodynamics – demonstrating, as was increasingly essential, the perfect harmony between body design and styling. (BnF Gallica)*

## Monaco

Like the major Parisian concours and the other Riviera events, Monaco's concours d'elegance events date back to the 19th century. More accurately, we should speak of "concours in Monaco," as numerous such events were held throughout the season, organized by entities like the *International Sporting Club*, the Casino, and the *Société des Bains de Mer*.

Unlike Parisian concours, Monaco's events, like those in Nice, Cannes, and Menton, took place during the Riviera's "season," which extended from December to April. Another distinguishing feature was the large number of foreign participants, unparalleled elsewhere. Monaco was a gathering point for European aristocracy and royalty, as well as the elite of American finance and business. These events were heralded as major highlights in multilingual newspapers across the region.

The jury for these competitions was often chaired by Their Serene Highnesses the Prince of Monaco and the Hereditary Princess, who were joined in the grandstand by luminaries from the arts and, naturally, by princes, counts, and duchesses.

Initially, Monaco competed with neighboring towns like Cannes, Nice, and Menton for prominence in the early 20th century. By 1921, however, the Monaco Concours d'Elegance, organized by the International Sporting Club, had become the culminating event, serving as the grand finale that crowned winners from the preliminary competitions held in those surrounding resorts.

By the 1930s, the number of participants had declined significantly, and the selection process became more exclusive. In 1932, for example, only owners of cars valued at 100,000 francs or more, and strictly of the latest model year, were eligible to compete. Despite these stringent requirements, 65 participants registered that year. Notably, the competition attracted delegations from prominent manufacturers like Delage and renowned coachbuilders such as Pinin Farina, who traveled specifically for the event.

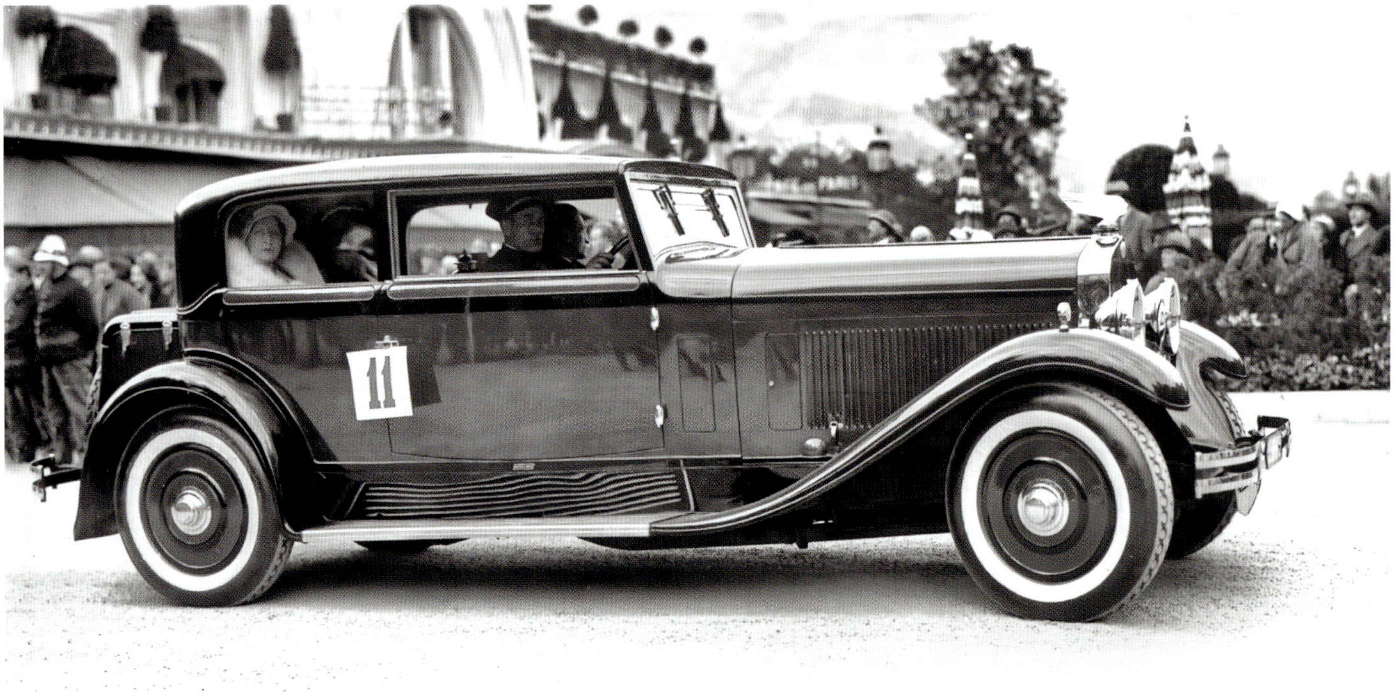

# Le XXIV^me^ Concours d'Elégance Automobile de Monte-Carlo

61 voitures engagées et cela, en dépit des conditions très dures qui exigeaient que les voitures aient coûté au moins 100.000 francs et ne soient pas d'une date de livraison antérieure au premier janvier 1931, voilà, qui en dit long sur le succès remporté par le XXIV^me^ Concours d'Elégance Automobile de Monte-Carlo, un des plus importants et des plus suivis dans les milieux de l'automobile.

Comme d'habitude, il se déroula sur la place du Casino où une foule nombreuse se pressait pour admirer les voitures et applaudir concurrentes et concurrents. Dans la tribune d'honneur où avaient pris place LL.AA.SS. le Prince de Monaco et la Princesse Héréditaire, on remarquait les membres du jury parmi lesquels on notait, le Colonel Balsan, Sir Walter de Frece, M. Noghès Antony, commissaire général; M. Alexandre Noghès, président de l'Automobile Club de Monaco; et les présidents des Automobile-Clubs de Nice, Cannes, Antibes, Juan-les-Pins et Menton.

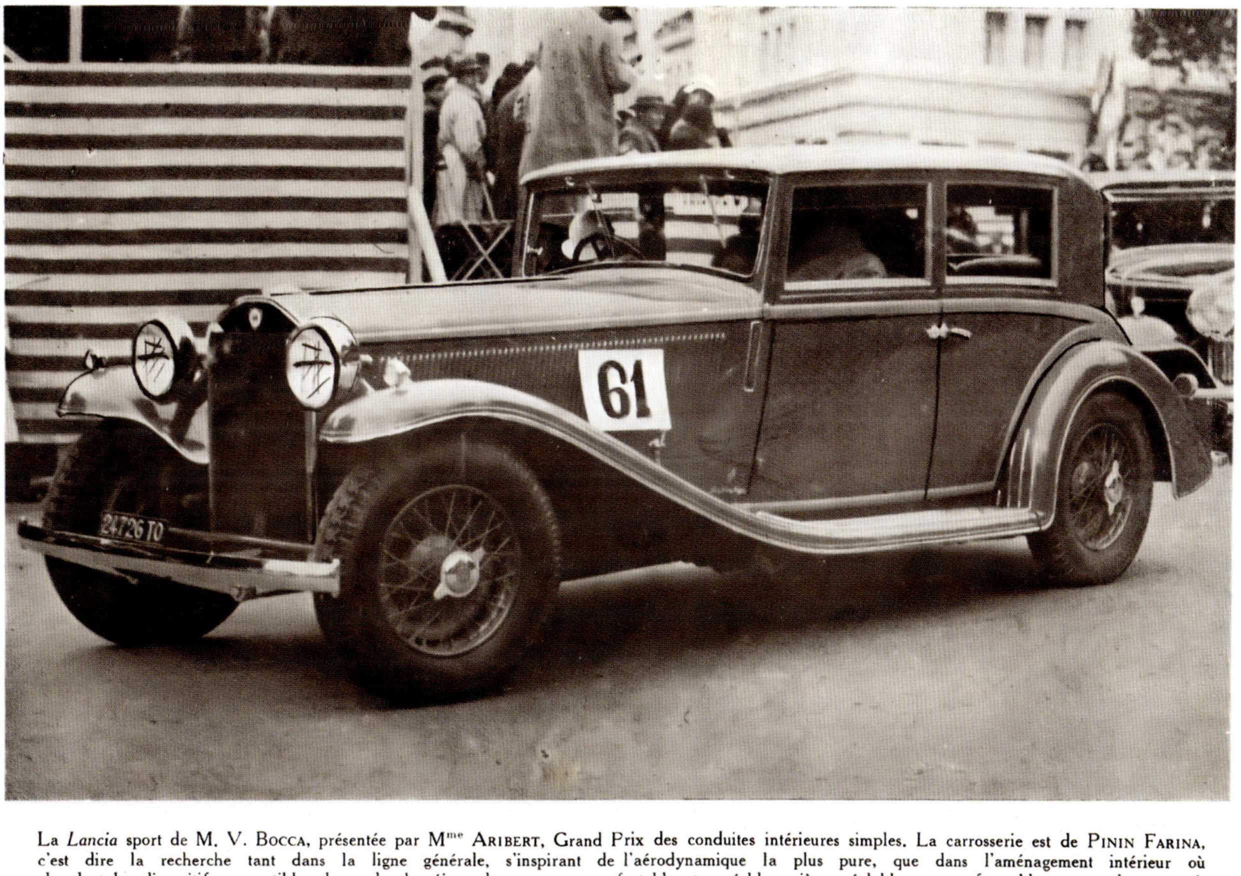


La *Lancia* sport de M. V. Bocca, présentée par M^me^ Aribert, Grand Prix des conduites intérieures simples. La carrosserie est de Pinin Farina, c'est dire la recherche tant dans la ligne générale, s'inspirant de l'aérodynamique la plus pure, que dans l'aménagement intérieur où abondent les dispositifs susceptibles de rendre le séjour des passagers confortable et agréable : sièges réglables et transformables en couchettes, petit bar, etc. ; rien n'y manque. M^me^ Aribert portait avec grâce une toilette dont la teinte s'alliait à celle de la garniture intérieure de la voiture

(Automobiles Lancia : Emmanuel Bocca, 52, rue de France, Nice).

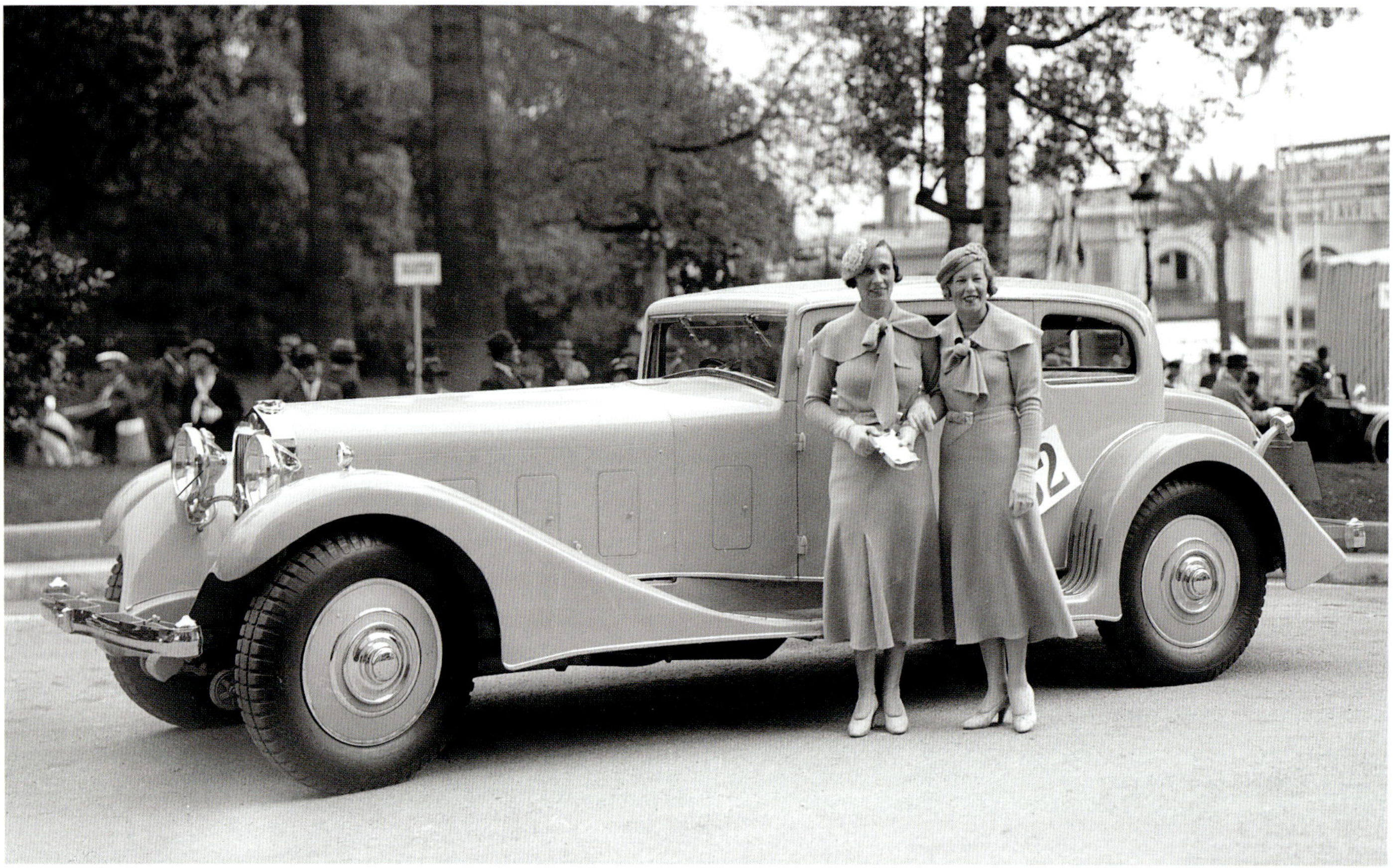

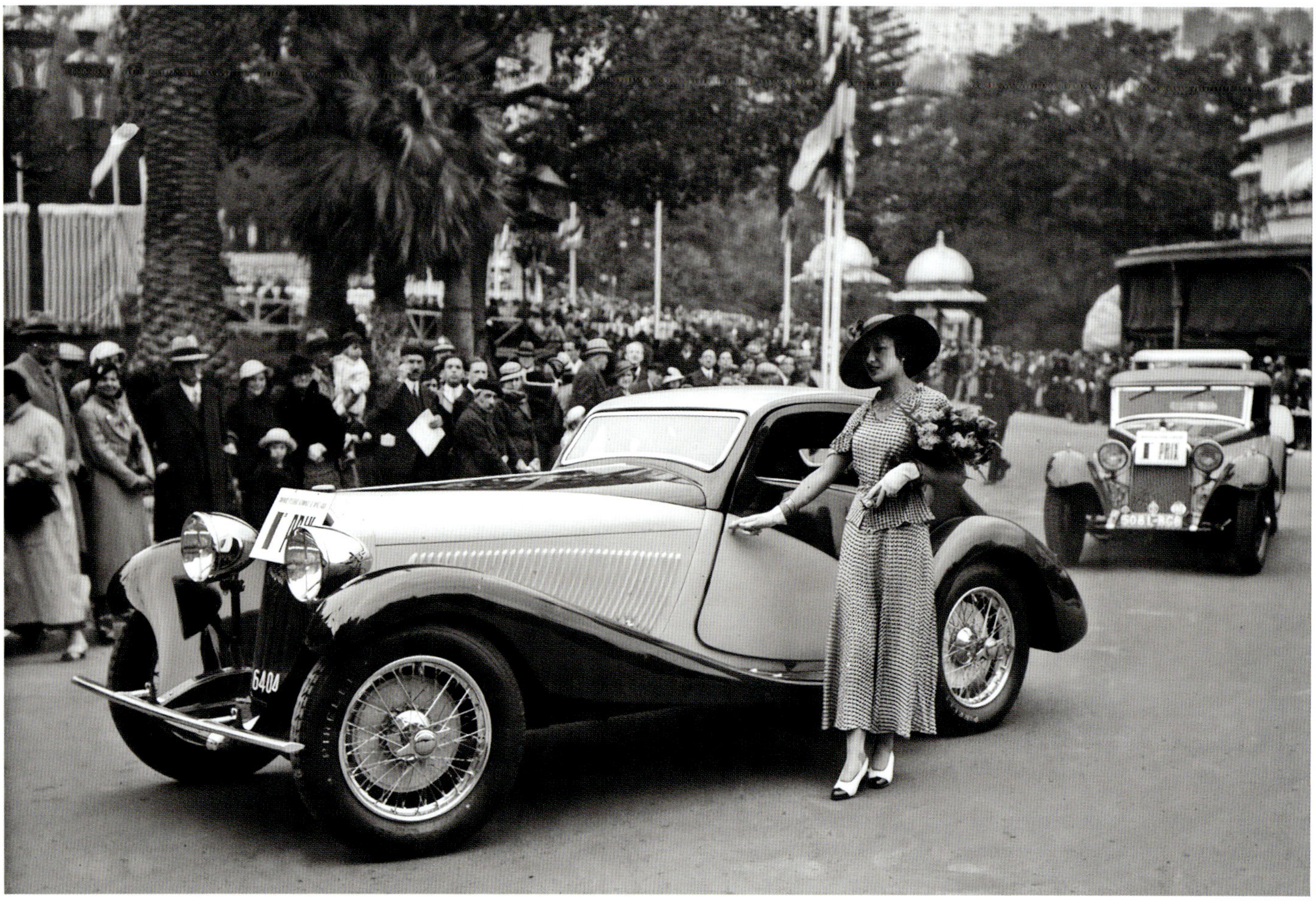
PRIX

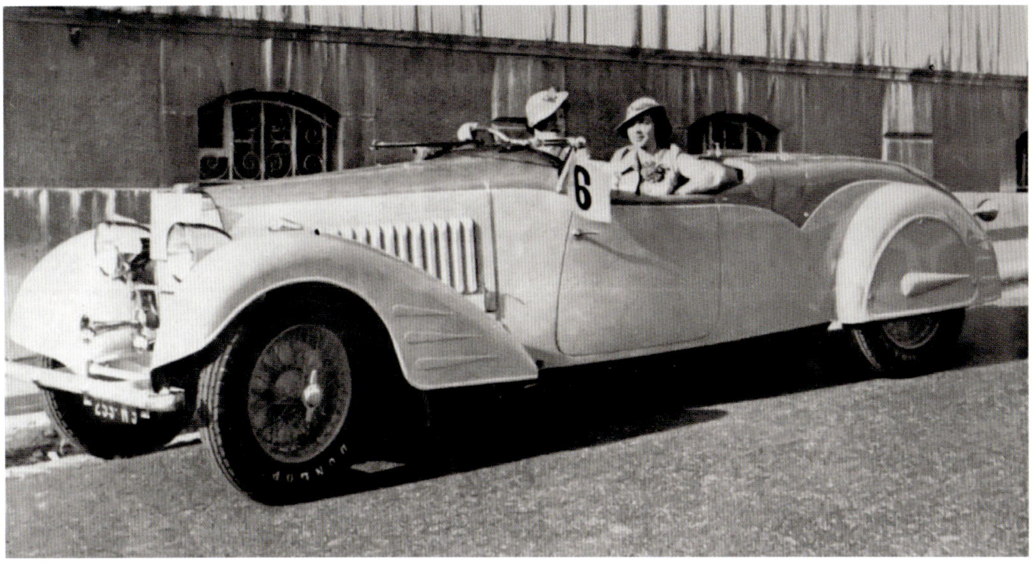

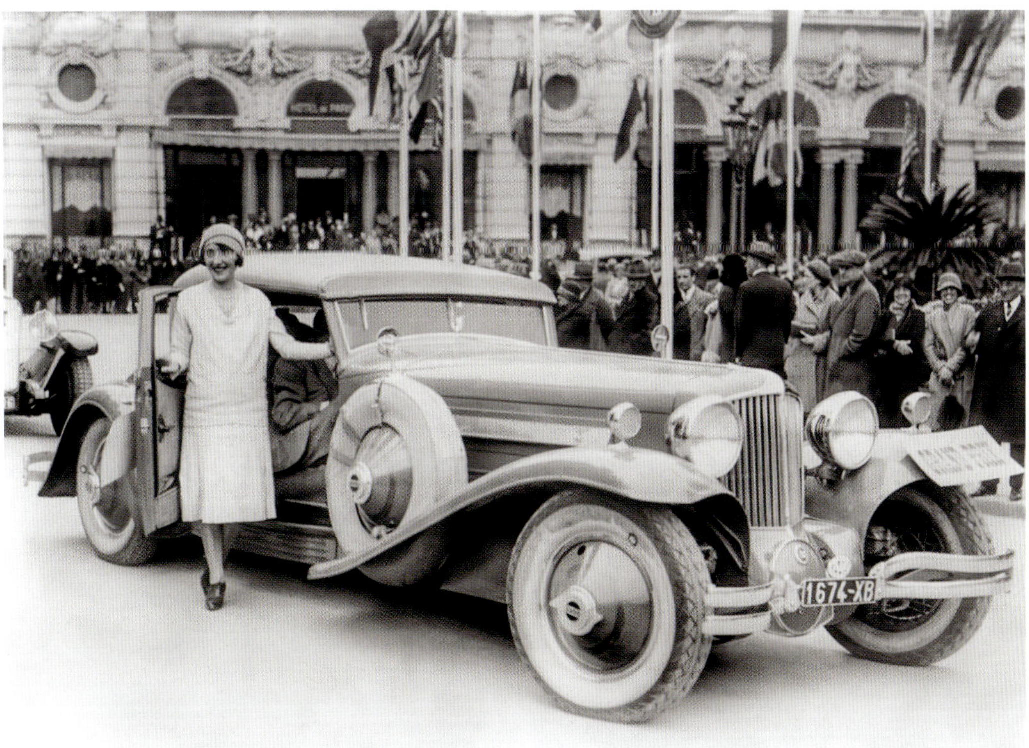

**OPPOSITE TOP:** *Mlles Meyer and Maschek, regulars at all the Côte d'Azur concours events, presented a periwinkle blue Delage D8-S with two-door sedan bodywork by Letourneur et Marchand in June 1933. Their ensembles, perfectly coordinated with the car, were designed by Jane Sylvant. (Daniel Cabart Archives)*

**OPPOSITE BOTTOM:** *This Alfa Romeo 6C 1750 berlinetta by Figoni, participated in the Monte Carlo Concours d'Elegance on April 5, 1934, presented by Mlle. Noghès. (Laurent Friry Archives)*

**TOP:** *A striking ensemble was created by this Bugatti Type 57 with its sweeping lines, accompanied by its crew, Madame Pierre de Cardaillac and Mlle. Leboucher. Their outfits by Voza-Sport harmonized perfectly with the tones of the Grand Sport Torpédo, which won the Grand Prix for roadsters at Monte Carlo and Cannes. (BnF Gallica)*

**LEFT:** *Designed by Count Alexis de Sakhnoffsky, the body of this Cord L-29 was crafted by the Hayes Body coachbuilder in Grand Rapids, Michigan. De Sakhnoffsky traveled to France in early 1930 with his car and entered the Monte Carlo Concours d'Elegance on March 22, 1930, where it won the Grand Prix. (Laurent Friry Archives)*

## Nice

The Nice competition is regarded as one of the oldest automobile elegance events in France. This claim to precedence is supported by the Automobile Club of Nice, which stated in 1939 that it had organized the first competition in 1898. As explored in the chapter on the origins of these events, while it is clear that the resorts of the French Riviera were pioneers in this arena, pinpointing the "first automobile elegance competition" remains a challenge.

Since its inception, the competition has been organized by the Automobile Club of Nice. Like similar events in Cannes, Menton, and Monte Carlo, the Nice competition has always featured a significant presence of winter visitors, particularly foreign participants, including British, Russian, and American elites.

This event arguably attracts the largest audience, with the *Promenade des Anglais* and the steps of the *Palais de la Méditerranée* providing a stunning stage for the spectacle. To accommodate a growing number of participants and to draw higher-quality vehicles, additional categories were introduced. Notably, Nice pioneered unique classifications such as the "Tournament of Elegance" and "Super-Elegance," terms that were later adopted by neighboring resorts.

As in Cannes, Juan-les-Pins, or Monaco, the Nice Concours served as a prominent showcase for local designers. The press devoted significant attention to the attire of the participants, highlighting the suppliers behind the elegant outfits. Similarly, the coverage extended to the grand hotels and residences hosting the elite social set, as well as local coachbuilders and car dealerships, which received more recognition here than at any other Riviera event.

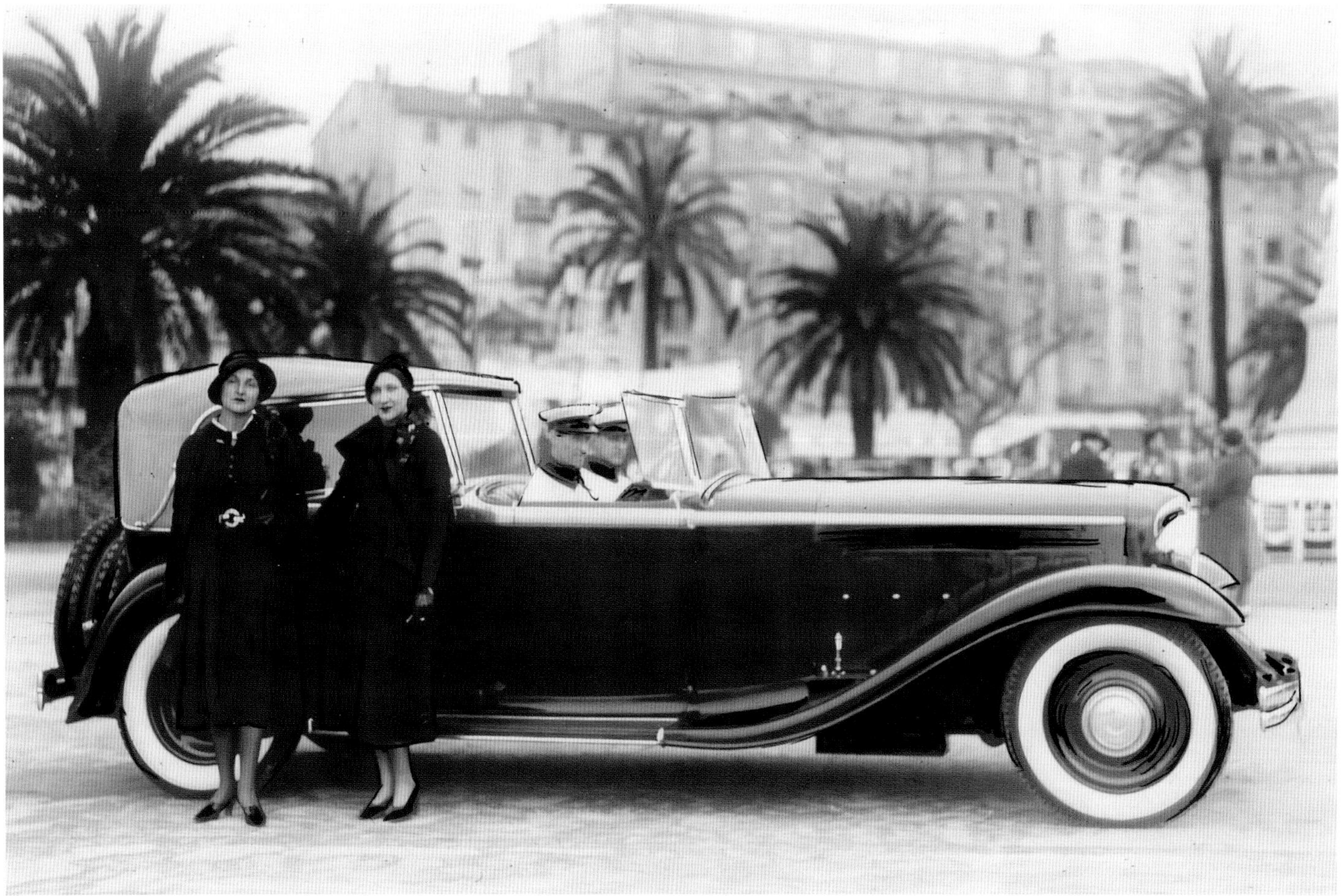

*Renault Reinastella Sport with bodywork by Kellner, was awarded the Grand Prix d'excellence at the Grand Tourney of Super-Elegance in Nice on March 31, 1932. Mlle. Dramard's clothes were styled by Gaby Mono and dressed by Worth. (Bernard Brulé)*

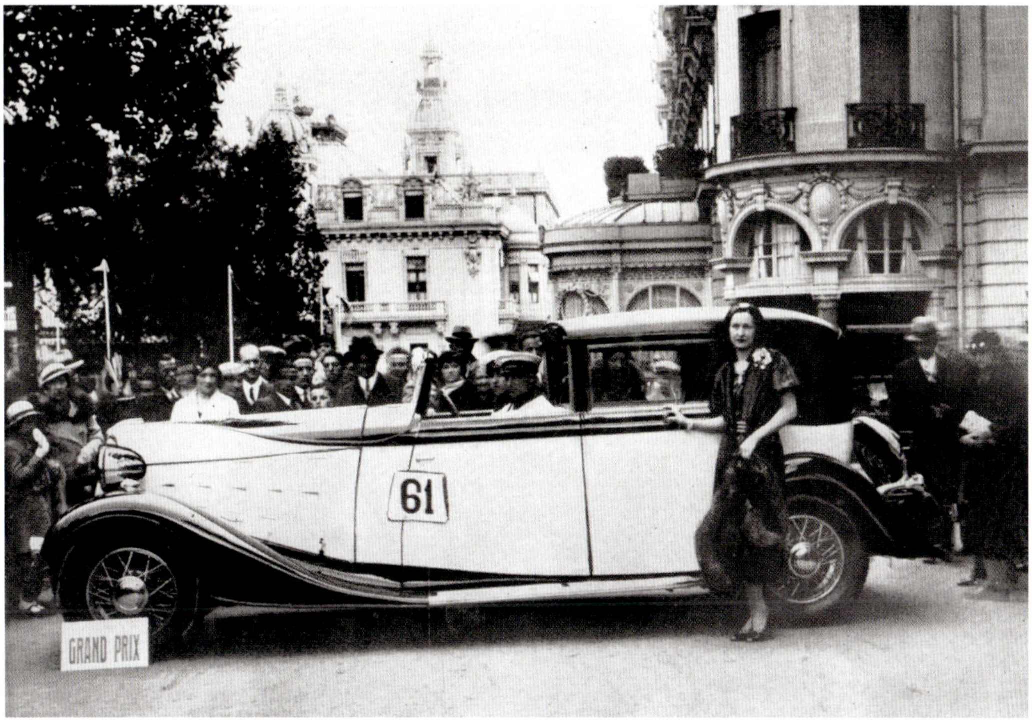


**LEFT:** *Miss France, queen of beauty and elegance, dressed by Worth, found her perfect place in this magnificent Renault with Franay bodywork; the jury awarded it the Grand Prix des Coupés. (BnF Gallica)*

**BELOW:** *Madame Holley and Madame Ras, dressed by Ras-Allard, in the Delage D8, with a coupé-de-ville body by Saoutchik, which won the first Grand Prix in its category. Everything was carefully arranged to impress the jury: two beautiful women and their outfits, a child, and two dogs. (Daniel Cabart Archives)*

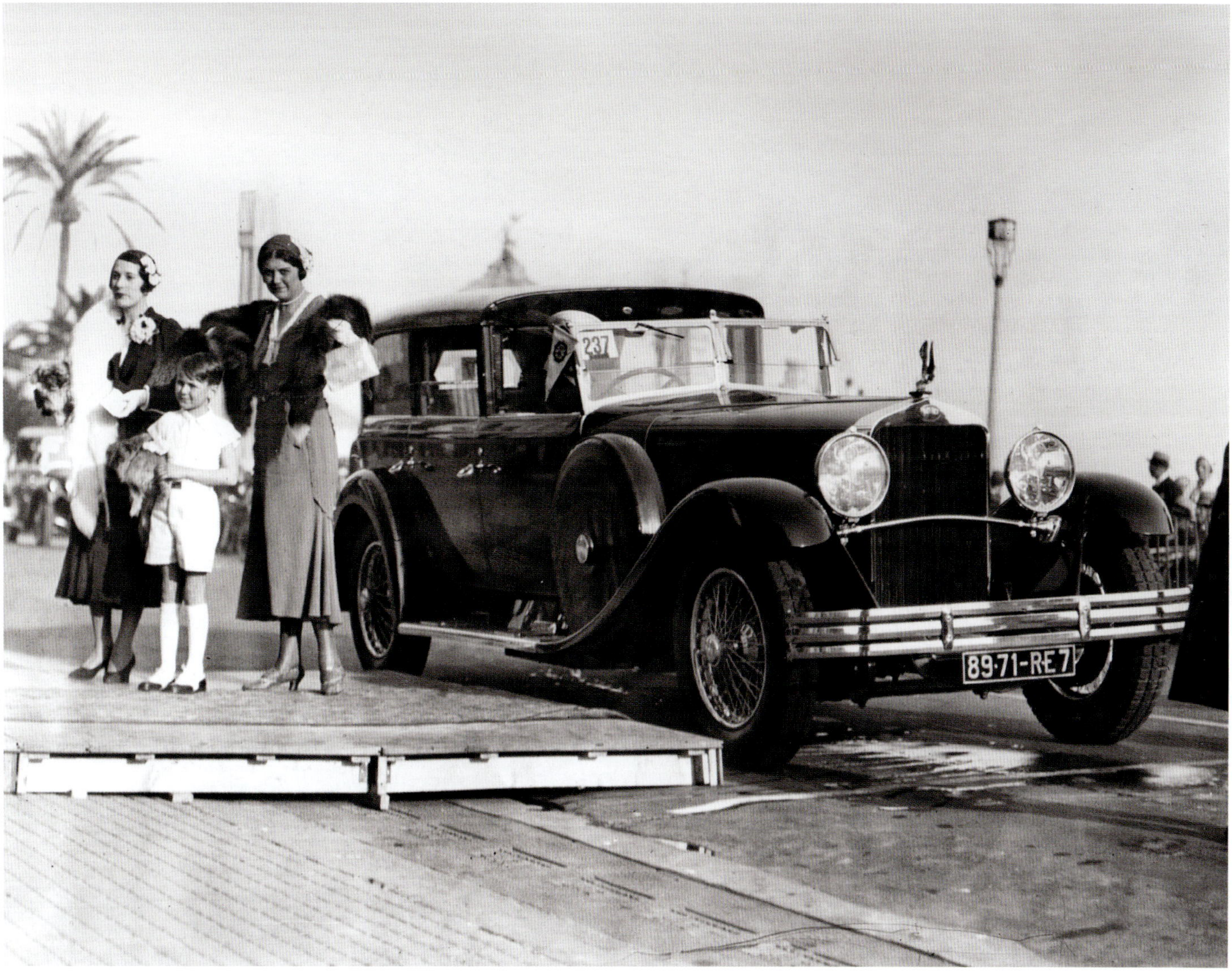

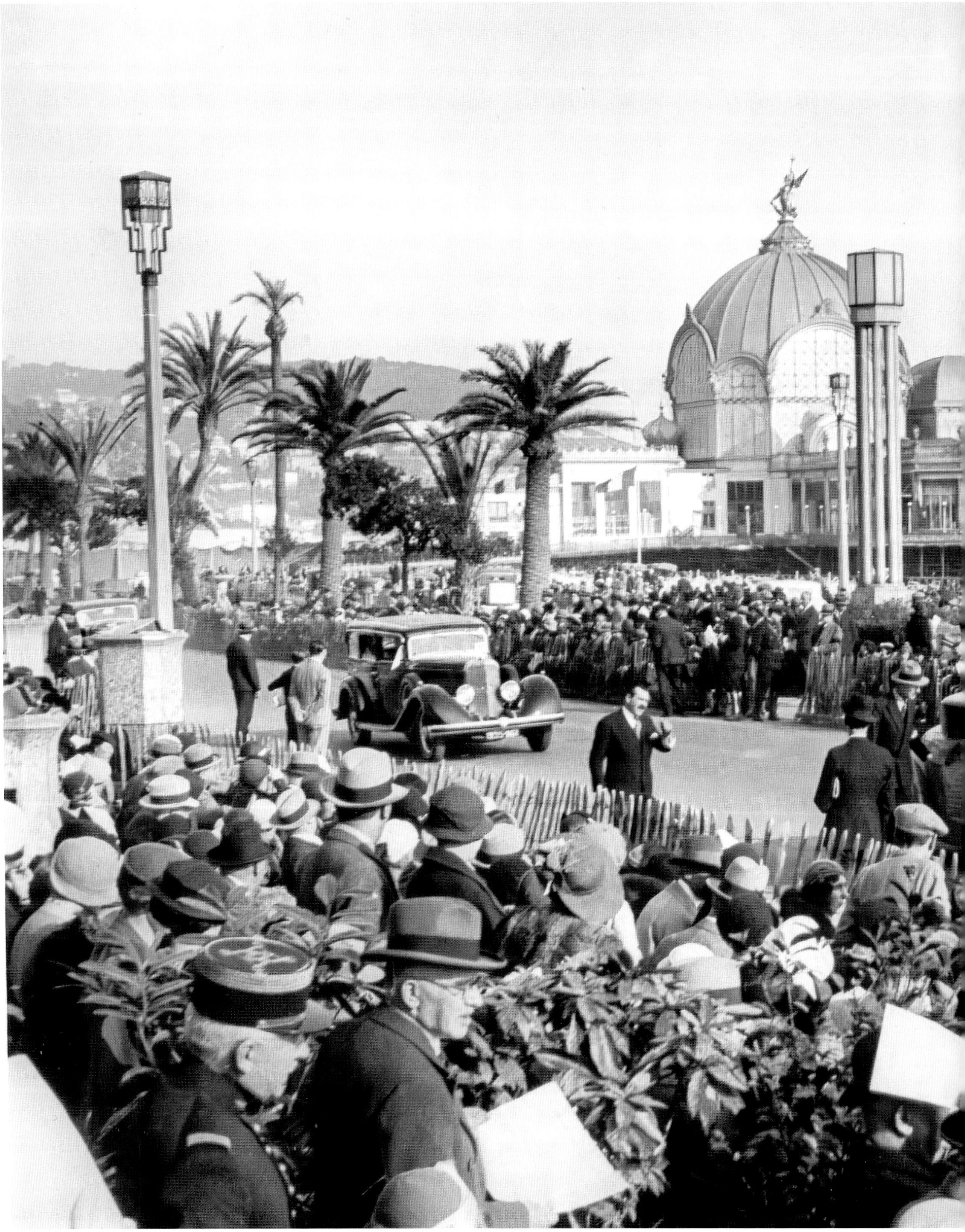

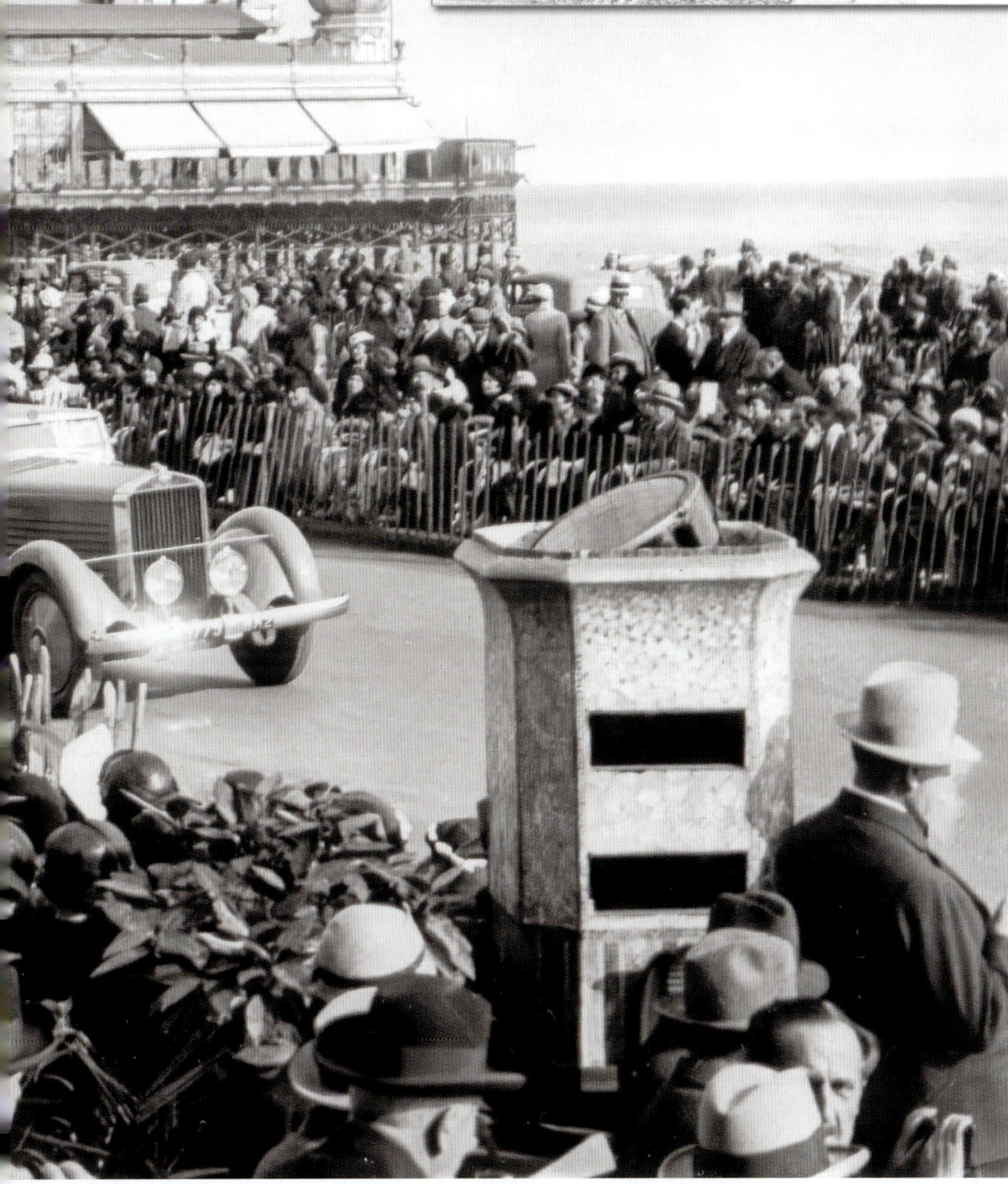

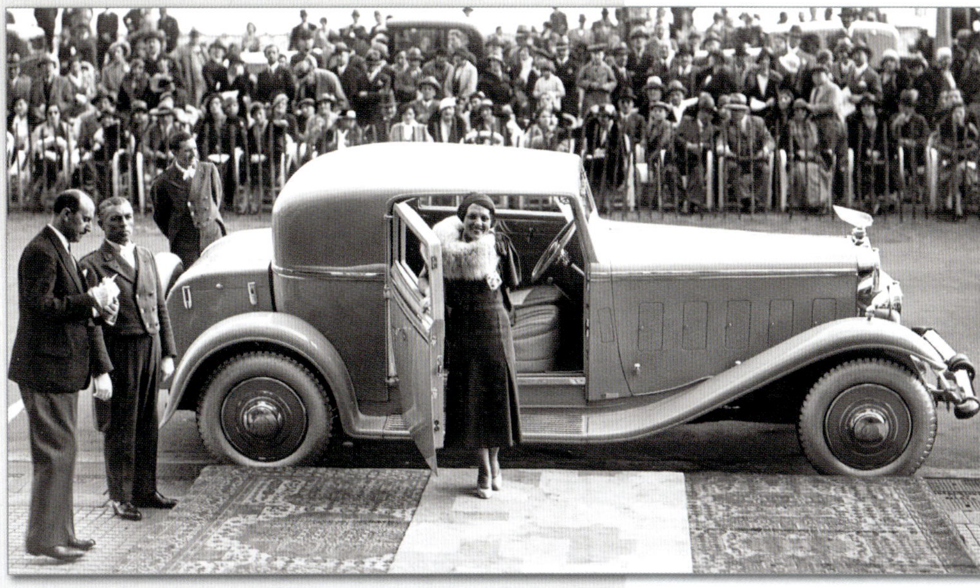

**ABOVE:** *Mlle. Betty Spell, dressed by Rosange, with her Delage D8 faux-cabriolet by Letourneur et Marchand. The Delage won the Grand Prix d'excellence in the Tourney of Super-Elegance in Nice. (Daniel Cabart Archives)*

**LEFT:** *The crowd gathered on the Promenade des Anglais during the Concours d'Elegance in Nice, admiring the stunning automobiles and the elegant atmosphere of the event. (Daniel Cabart Archives)*

# Concours d'Elegance in Europe: Turin, Sinaia, Villa d'Este

## THE REST OF EUROPE TOOK TO CONCOURS

Although the saying 'What France thinks today, Europe thinks tomorrow' may not have been coined specifically for the early days of the automobile, it could easily have been. The French were quick to pioneer the idea of adorning automobiles with flowers and organizing competitions around this theme. It wasn't long before this trend made its way across the English Channel to Britain.

In the June 21, 1902, issue of *The Autocar* magazine, there was a report on the 'Automobile Battle of Flowers' held at Earl's Court, London. S.F. Edge, a well-known entrepreneur, racing driver, and record-breaker, emerged victorious in this light-hearted competition, driving a 16 HP Napier decorated with what was likely a vibrant garland of flowers, though the image was in black and white. Incidentally, Edge was instrumental in promoting the Napier brand not only across Europe but also in India, where he participated in long-distance reliability trials and successfully sold several cars to prominent figures, including the Nizam of Hyderabad, one of the wealthiest men in the world at the time.

The concept of an 'automobile battle of flowers' was seen as an excellent promotional tool for automobiles. *The Autocar* previewed the event in its June 7, 1902, issue, noting that "*On Friday next at the Paris in London Exhibition at Earl's Court, in connection with the French Charity Fête, an automobile battle of flowers will be held. A continuous route has been planned out in the exhibition gardens, along which the procession of cars will proceed. At intervals, stands affording seats for thousands of onlookers will be placed, while bands and choruses will add to the gaiety of the meeting. Twenty-four banners will be presented to the owners of the most artistically decorated cars.*"

The June 21, 1902 issue of *The Autocar* described the event in detail, highlighting "*huge and beautiful plumes of red roses and bird's-eye at the dashboard sides, while a large horseshoe, formed entirely of red rosebuds and bird's-eye enclosing a luxuriant bunch of beautiful lilies of the valley and asparagus plume, entirely masked the bonnet. Lilies of the valley and green foliage outlined the body of the car, and lined the mudguards, while the riders, both front and back, were bowered in sweeping arches of red roses and foliage topped with a waving of superb lilies trained forward.*"

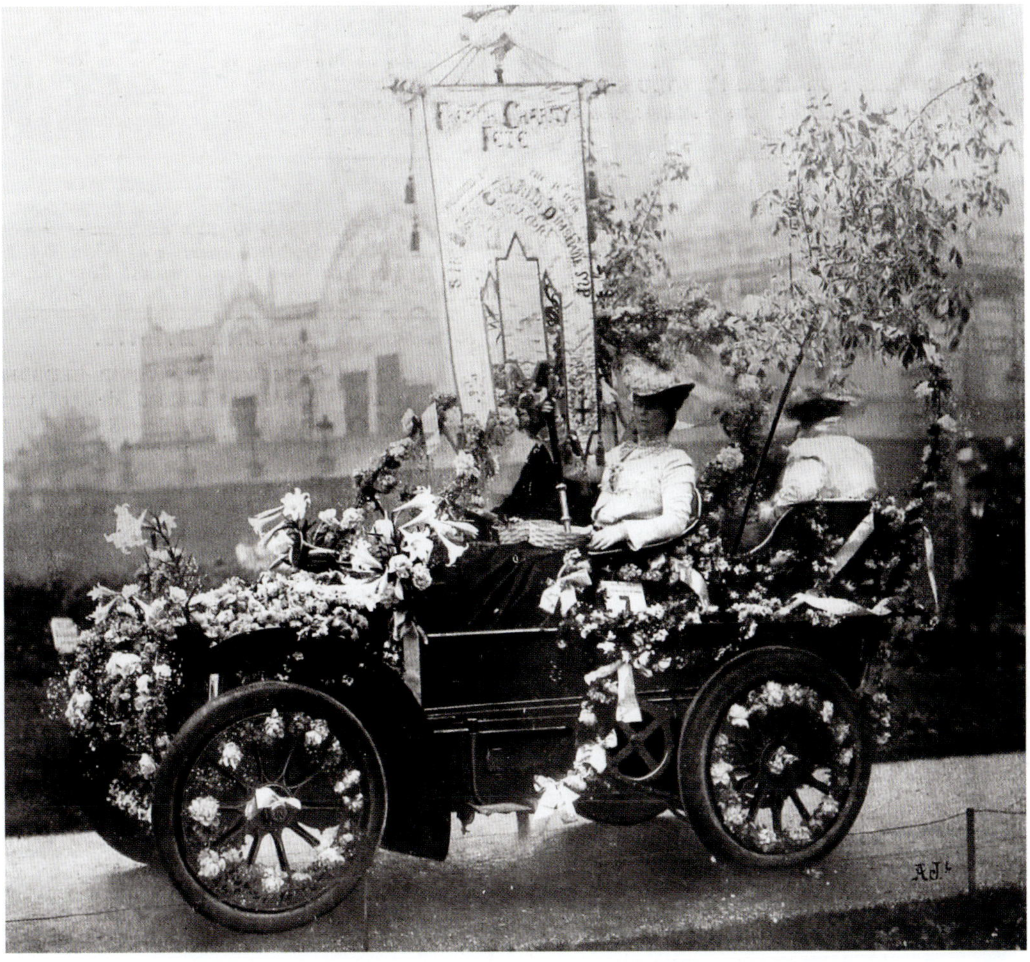

*Mr. S.F. Edge's Napier, winner of the first prize at the Battle of Flowers, Earl's Court, 1902.* "Mr. Edge's Napier, which forms our frontispiece this week, in all the pride of yellow and white and red flowers, with Gothic arches of boughs, was awarded first prize." (*The Automotor Journal, June 21, 1902*)

A similar 'battle of the flowers' also took place in Stresa, a resort town on the banks of Lake Maggiore, in Italy, on October 2, 1902, which was described as a *corso fleuri*, a flower competition. The competitors were there in cars such as French Darracqs, a couple of Panhards, a Fiat, a Mercedes, and several others.

While the 'automobile battle of flowers' was not quite a concours d'elegance à la française, neither was the event described as an 'Exposition et Concours d'Elegance' in Munich on June 9, 1907, as reported in the French magazine *Omnia*. There, the cars were judged on their equipment and features, much like the French practice of the time.

The idea spread to other countries in Europe, such as Spain. As part of the Great Automobile Week, the chic resort town of San Sebastian, in Northwest Spain, hosted a concours d'elegance on September 25, 1912, with the event "organized by the Casino of San Sebastian in honor of the Automobile Rally, which took place today."

An article in the French magazine *Omnia* described the event: "*The elegance and bodywork competition, which concluded the series of beautiful events in San Sebastian, was, like its predecessors, successful in every way. We take great pleasure in acknowledging this and once again praising the active and highly competent organizers, Messrs. Marquet and Dominguez. Through prodigious organization and impressive participation, they succeeded in bringing more than sixty cars to San Sebastian after a triumphant journey across the roads of Europe, each vehicle met with endless applause. It was, we must repeat, a truly splendid day.*"

One of the most remarkable aspects of the beauty competition for the cars in San Sebastian was the overwhelming success of the relatively unknown Belgian marque, Métallurgique. The brand secured the top three positions in the class for open-bodied tourers and claimed second and third places in the closed body class. Other notable marques represented included Berliet, Delage, Mercedes, S.C.A.P., and Anasagasti, the Argentine marque powered by a French Ballot engine, with the overall winner a Delage (likely powered by an engine made by Ballot too). The event attracted international participants, with entrants hailing from France, the United Kingdom, and beyond.

**BELOW:** *As early as June 7, 1907, Munich had a concours d'elegance, the Concours Herkomer, but this was another technical concours that judged the vehicles for their features. (Daniel Cabart Archives)*

**FAR RIGHT:** *On September 25, 1912, the Spanish resort town of San Sebastian had a concours d'elegance too but again of a technical nature. (Daniel Cabart Archives)*

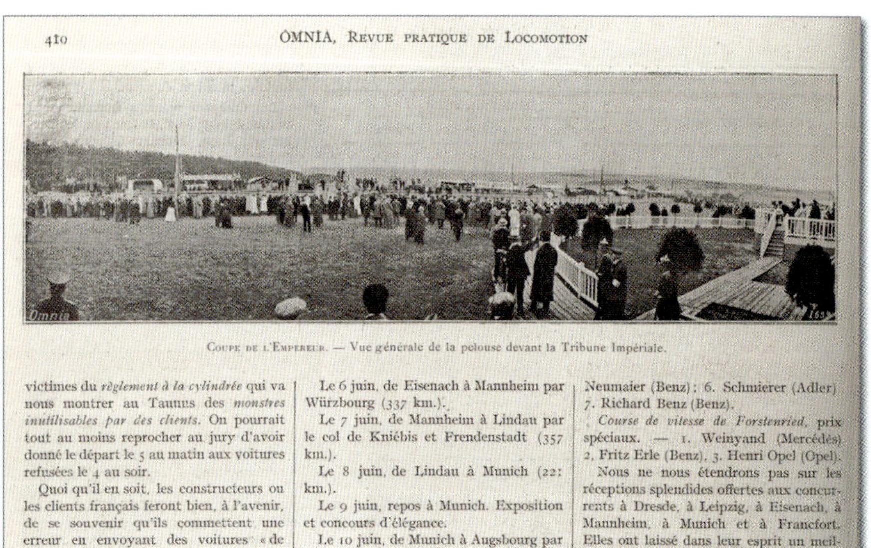

410 OMNIA, REVUE PRATIQUE DE LOCOMOTION

COUPE DE L'EMPEREUR. — Vue générale de la pelouse devant la Tribune Impériale.

victimes du *règlement à la cylindrée* qui va nous montrer au Taunus des *monstres inutilisables par des clients*. On pourrait tout au moins reprocher au jury d'avoir donné le départ le 5 au matin aux voitures refusées le 4 au soir.

Quoi qu'il en soit, les constructeurs ou les clients français feront bien, à l'avenir, de se souvenir qu'ils commettent une erreur en envoyant des voitures « de

Le 6 juin, de Eisenach à Mannheim par Würzbourg (337 km.).

Le 7 juin, de Mannheim à Lindau par le col de Kniébis et Frendenstadt (357 km.).

Le 8 juin, de Lindau à Munich (22: km.).

Le 9 juin, repos à Munich. Exposition et concours d'élégance.

Le 10 juin, de Munich à Augsbourg par

Neumaier (Benz); 6. Schmierer (Adler) 7. Richard Benz (Benz).

*Course de vitesse de Forstenried*, prix spéciaux. — 1. Weinyand (Mercédès) 2, Fritz Erle (Benz), 3. Henri Opel (Opel).

Nous ne nous étendrons pas sur les réceptions splendides offertes aux concurrents à Dresde, à Leipzig, à Eisenach, à Mannheim, à Munich et à Francfort. Elles ont laissé dans leur esprit un meil-

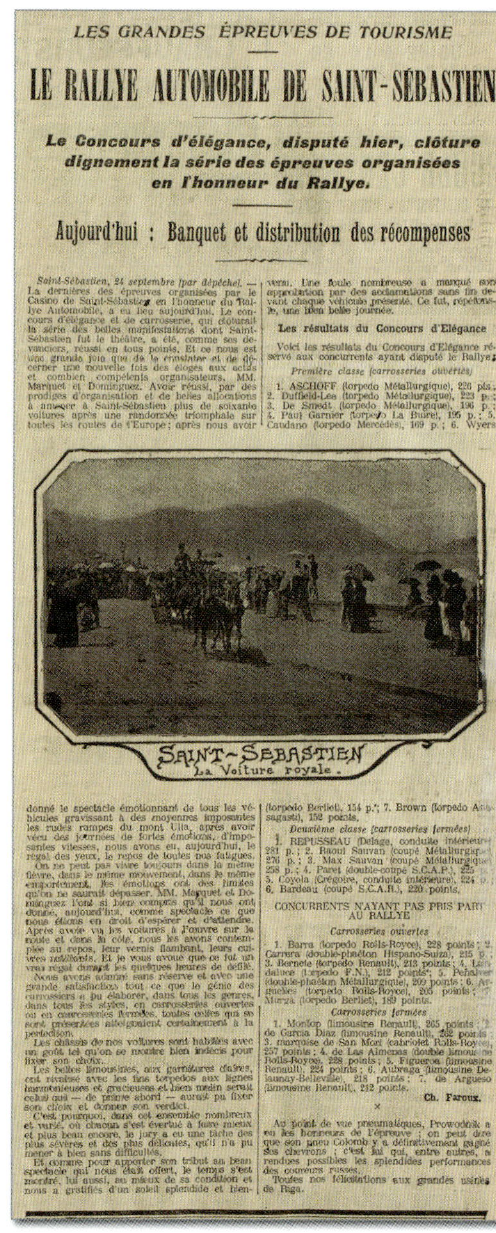

LES GRANDES ÉPREUVES DE TOURISME

LE RALLYE AUTOMOBILE DE SAINT-SÉBASTIEN

**Le Concours d'élégance, disputé hier, clôture dignement la série des épreuves organisées en l'honneur du Rallye.**

Aujourd'hui : Banquet et distribution des récompenses

*Saint-Sébastien, 24 septembre (par dépêche).* — La dernières des épreuves organisées par le Casino de Saint-Sébastien en l'honneur du Rallye Automobile, a eu lieu aujourd'hui. Le concours d'élégance et de carrosserie, qui clôturait la série des belles manifestations dont Saint-Sébastien fut le théâtre, a été, comme ses devanciers, réussi en tous points. Et ce nous est une grande joie que de le constater et de décerner une nouvelle fois des éloges aux actifs et combien compétents organisateurs, MM. Marquet et Dominguez. Avoir réussi, par des prodiges d'organisation et de belles allocations à amener à Saint-Sébastien plus de soixante voitures après une randonnée triomphale sur toutes les routes de l'Europe; après nous avoir venu. Une foule nombreuse a marqué son approbation par des acclamations sans fin devant chaque véhicule présenté. Ce fut, répétons-le, une bien belle journée.

**Les résultats du Concours d'Elégance**

Voici les résultats du Concours d'Elégance réservé aux concurrents ayant disputé le Rallye:

*Première classe (carrosseries ouvertes)*

1. ASCHOFF (torpedo Métallurgique), 226 pts.; 2. Duffield-Lee (torpedo Métallurgique), 223 p.; 3. De Smedt (torpedo Métallurgique), 196 p.; 4. Paul Garnier (torpedo La Buire), 195 p.; 5. Caudano (torpedo Mercédès), 169 p.; 6. Wyers

SAINT-SÉBASTIEN La Voiture royale.

donné le spectacle émotionnant de tous les véhicules gravissant à des moyennes imposantes les rudes rampes du mont Ulia, après avoir vécu des journées de fortes émotions, d'imposantes vitesses, nous avons eu, aujourd'hui, le régal des yeux, le repos de toutes nos fatigues.

On ne peut pas vivre toujours dans la même fièvre, dans le même mouvement, dans le même emportement. Les émotions ont des limites qu'on ne saurait dépasser. MM. Marquet et Dominguez l'ont si bien compris qu'il nous ont donné, aujourd'hui, comme spectacle ce que nous étions en droit d'espérer et d'attendre. Après avoir vu les voitures à l'œuvre sur la route et dans la côte, nous les avons contemplée au repos, leur vernis flambant, leurs cuivres rutilants. Et je vous avoue que ce fut un vrai régal durant les quelques heures de défilé.

Nous avons admiré sans réserve et avec une grande satisfaction tout ce que le génie des carrossiers a pu élaborer, dans tous les genres, dans tous les styles, en carrosseries ouvertes ou en carrosseries fermées, toutes celles qui se sont présentées atteignaient certainement à la perfection.

Les châssis de nos voitures sont habillés avec un goût tel qu'on se montre bien indécis pour fixer son choix.

Les belles limousines, aux garnitures claires, ont rivalisé avec les fins torpedos aux lignes harmonieuses et gracieuses et bien malin serait celui qui — de prime abord — aurait pu fixer son choix et donner son verdict.

C'est pourquoi, dans cet ensemble nombreux et varié, où chacun s'est évertué à faire mieux et plus beau encore, le jury a eu une tâche des plus sévères et des plus délicates, qu'il n'a pu mener à bien sans difficultés.

Et comme pour apporter son tribut au beau spectacle qui nous était offert, le temps s'est montré, lui aussi, au mieux de sa condition et nous a gratifiés d'un soleil splendide et bien-

(torpedo Berliet), 154 p.; 7. Brown (torpedo Anasagasti), 152 points.

*Deuxième classe (carrosseries fermées)*

1. REPUSSEAU (Delage, conduite intérieure) 281 p.; 2. Raoul Sauvan (coupé Métallurgique) 276 p.; 3. Max Sauvan (coupé Métallurgique) 258 p.; 4. Paret (double-coupé S.C.A.P.), 225 p.; 5. Coyola (Grégoire, conduite intérieure), 224 p.; 6. Bardeau (coupé S.C.A.R.), 220 points.

CONCURRENTS N'AYANT PAS PRIS PART AU RALLYE

*Carrosseries ouvertes*

1. Barra (torpedo Rolls-Royce), 228 points; 2. Carrera (double-phaéton Hispano-Suiza), 215 p.; 3. Bernete (torpedo Renault), 213 points; 4. Landaluce (torpedo F.N.), 212 points; 5. Peñalver (double-phaéton Métallurgique), 209 points; 6. Arguelles (torpedo Rolls-Royce), 205 points; 7. Murga (torpedo Berliet), 189 points.

*Carrosseries fermées*

1. Monlop (limousine Renault), 265 points; 2. de Garcia Diaz (limousine Renault), 262 points; 3. marquise de San Mori (cabriolet Rolls-Royce), 257 points; 4. de Las Almenas (double limousine Rolls-Royce), 228 points; 5. Figueroa (limousine Renault), 224 points; 6. Aubraga (limousine Delaunay-Belleville), 218 points; 7. de Argueso (limousine Renault), 212 points.

**Ch. Faroux.**

×

Au point de vue pneumatiques, Prowodnik a eu les honneurs de l'épreuve ; on peut dire que son pneu Colomb y a définitivement gagné ses chevrons ; c'est lui qui, entre autres, a rendues possibles les splendides performances des coureurs russes.

Toutes nos félicitations aux grandes usines de Riga.

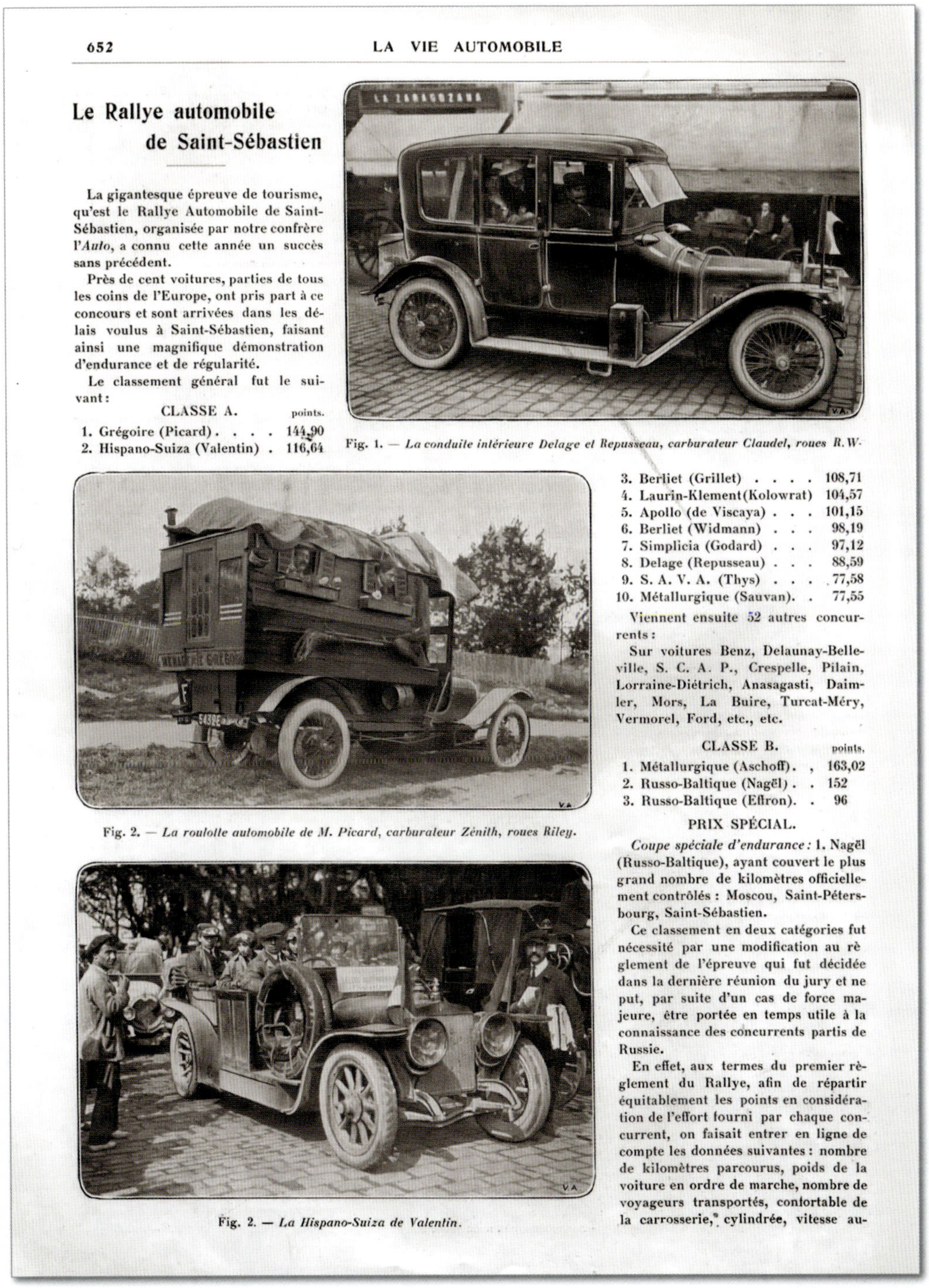

652 LA VIE AUTOMOBILE

## Le Rallye automobile de Saint-Sébastien

La gigantesque épreuve de tourisme, qu'est le Rallye Automobile de Saint-Sébastien, organisée par notre confrère l'*Auto*, a connu cette année un succès sans précédent.

Près de cent voitures, parties de tous les coins de l'Europe, ont pris part à ce concours et sont arrivées dans les délais voulus à Saint-Sébastien, faisant ainsi une magnifique démonstration d'endurance et de régularité.

Le classement général fut le suivant :

CLASSE A.

| | points. |
|---|---|
| 1. Grégoire (Picard) . . . . | 144,90 |
| 2. Hispano-Suiza (Valentin) . | 116,64 |
| 3. Berliet (Grillet) . . . . | 108,71 |
| 4. Laurin-Klement (Kolowrat) | 104,57 |
| 5. Apollo (de Viscaya) . . . | 101,15 |
| 6. Berliet (Widmann) . . . | 98,19 |
| 7. Simplicia (Godard) . . . | 97,12 |
| 8. Delage (Repusseau) . . . | 88,59 |
| 9. S. A. V. A. (Thys) . . . . | 77,58 |
| 10. Métallurgique (Sauvan). . | 77,55 |

Fig. 1. — *La conduite intérieure Delage et Repusseau, carburateur Claudel, roues R. W.*

Fig. 2. — *La roulotte automobile de M. Picard, carburateur Zénith, roues Riley.*

Fig. 2. — *La Hispano-Suiza de Valentin.*

Viennent ensuite 52 autres concurrents :

Sur voitures Benz, Delaunay-Belleville, S. C. A. P., Crespelle, Pilain, Lorraine-Diétrich, Anasagasti, Daimler, Mors, La Buire, Turcat-Méry, Vermorel, Ford, etc., etc.

CLASSE B.

| | points. |
|---|---|
| 1. Métallurgique (Aschoff). . | 163,02 |
| 2. Russo-Baltique (Nagël) . . | 152 |
| 3. Russo-Baltique (Effron). . | 96 |

PRIX SPÉCIAL.

*Coupe spéciale d'endurance :* 1. Nagël (Russo-Baltique), ayant couvert le plus grand nombre de kilomètres officiellement contrôlés : Moscou, Saint-Pétersbourg, Saint-Sébastien.

Ce classement en deux catégories fut nécessité par une modification au règlement de l'épreuve qui fut décidée dans la dernière réunion du jury et ne put, par suite d'un cas de force majeure, être portée en temps utile à la connaissance des concurrents partis de Russie.

En effet, aux termes du premier règlement du Rallye, afin de répartir équitablement les points en considération de l'effort fourni par chaque concurrent, on faisait entrer en ligne de compte les données suivantes : nombre de kilomètres parcourus, poids de la voiture en ordre de marche, nombre de voyageurs transportés, confortable de la carrosserie, cylindrée, vitesse au-

**LEFT:** *The French magazine* La Vie Automobile *featured an extensive report on the concours d'elegance at San Sebastian. (Daniel Cabart Archives)*

**BOTTOM (BOTH):** *The car that seems to have starred at San Sebastian was a French Repusseau. (Daniel Cabart Archives)*

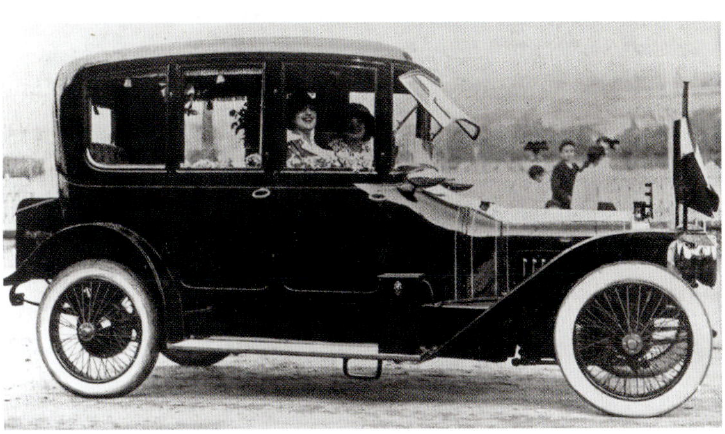

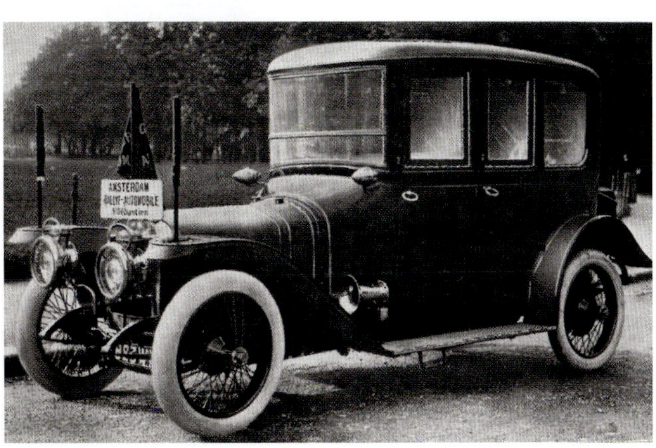

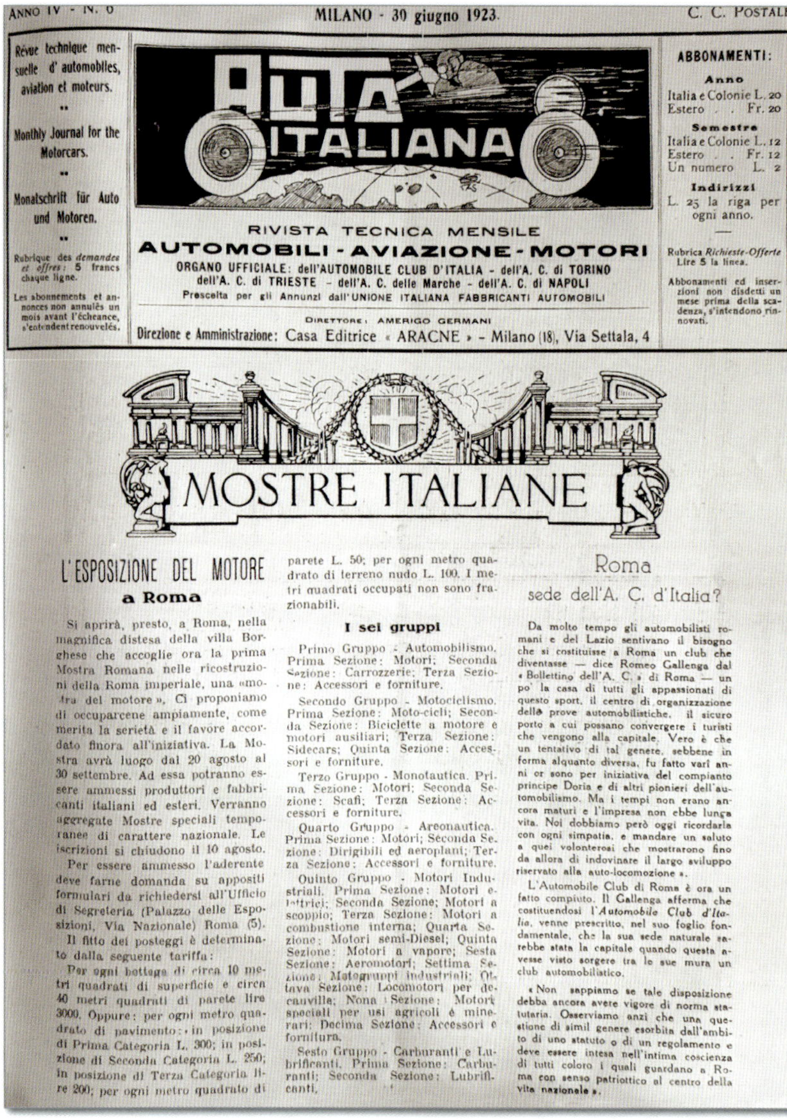

ANNO IV - N. 6 MILANO - 30 giugno 1923. C. C. POSTALE

Révue technique mensuelle d'automobiles, aviation et moteurs.

Monthly Journal for the Motorcars.

Monatschrift für Auto und Motoren.

Rubrique des *demandes et offres*: 5 francs chaque ligne.

Les abonnements et annonces non annulés un mois avant l'échéance, s'entendent renouvelés.

AUTO ITALIANA

RIVISTA TECNICA MENSILE

AUTOMOBILI - AVIAZIONE - MOTORI

ORGANO UFFICIALE: dell'AUTOMOBILE CLUB D'ITALIA - dell'A. C. di TORINO dell'A. C. di TRIESTE - dell'A. C. delle Marche - dell'A. C. di NAPOLI

Prescelta per gli Annunzi dall'UNIONE ITALIANA FABBRICANTI AUTOMOBILI

DIRETTORE: AMERIGO GERMANI

Direzione e Amministrazione: Casa Editrice « ARACNE » - Milano (18), Via Settala, 4

ABBONAMENTI:

Anno
Italia e Colonie L. 20
Estero . . Fr. 20

Semestre
Italia e Colonie L. 12
Estero . . Fr. 12
Un numero L. 2

Indirizzi
L. 25 la riga per ogni anno.

Rubrica *Richieste-Offerte* Lire 5 la linea.

Abbonamenti ed inserzioni non disdetti un mese prima della scadenza, s'intendono rinnovati.

MOSTRE ITALIANE

L'ESPOSIZIONE DEL MOTORE
a Roma

Si aprirà, presto, a Roma, nella magnifica distesa della villa Borghese che accoglie ora la prima Mostra Romana nelle ricostruzioni della Roma imperiale, una «mostra del motore». Ci proponiamo di occuparcene ampiamente, come merita la serietà e il favore accordato finora all'iniziativa. La Mostra avrà luogo dal 20 agosto al 30 settembre. Ad essa potranno essere ammessi produttori e fabbricanti italiani ed esteri. Verranno aggregate Mostre speciali temporanee di carattere nazionale. Le iscrizioni si chiudono il 10 agosto.

Per essere ammesso l'aderente deve farne domanda su appositi formulari da richiedersi all'Ufficio di Segreteria (Palazzo delle Esposizioni, Via Nazionale) Roma (5).

Il fitto dei posteggi è determinato dalla seguente tariffa:

Per ogni bottega di circa 10 metri quadrati di superficie e circa 40 metri quadrati di parete lire 3000. Oppure: per ogni metro quadrato di pavimento: in posizione di Prima Categoria L. 300; in posizione di Seconda Categoria L. 250; in posizione di Terza Categoria lire 200; per ogni metro quadrato di parete L. 50; per ogni metro quadrato di terreno nudo L. 100. I metri quadrati occupati non sono frazionabili.

I sei gruppi

Primo Gruppo - Automobilismo. Prima Sezione: Motori; Seconda Sezione: Carrozzerie; Terza Sezione: Accessori e forniture.

Secondo Gruppo - Motociclismo. Prima Sezione: Moto-cicli; Seconda Sezione: Biciclette a motore e motori ausiliari; Terza Sezione: Sidecars; Quinta Sezione: Accessori e forniture.

Terzo Gruppo - Monotautica. Prima Sezione: Motori; Seconda Sezione: Scafi; Terza Sezione: Accessori e forniture.

Quarto Gruppo - Areonautica. Prima Sezione: Motori; Seconda Sezione: Dirigibili ed aeroplani; Terza Sezione: Accessori e forniture.

Quinto Gruppo - Motori Industriali. Prima Sezione: Motori elettrici; Seconda Sezione; Motori a scoppio; Terza Sezione: Motori a combustione interna; Quarta Sezione: Motori semi-Diesel; Quinta Sezione: Motori a vapore; Sesta Sezione: Aeromotori; Settima Sezione: Motogruppi industriali; Ottava Sezione: Locomotori per decauville; Nona Sezione: Motori speciali per usi agricoli e minerari; Decima Sezione: Accessori e fornitura.

Sesto Gruppo - Carburanti e Lubrificanti. Prima Sezione: Carburanti; Seconda Sezione: Lubrificanti.

Roma
sede dell'A. C. d'Italia?

Da molto tempo gli automobilisti romani e del Lazio sentivano il bisogno che si costituisse a Roma un club che diventasse — dice Romeo Gallenga dal « Bollettino dell'A. C. » di Roma — un po' la casa di tutti gli appassionati di questo sport, il centro di organizzazione delle prove automobilistiche, il sicuro porto a cui possano convergere i turisti che vengono alla capitale. Vero è che un tentativo di tal genere, sebbene in forma alquanto diversa, fu fatto vari anni or sono per iniziativa del compianto principe Doria e di altri pionieri dell'automobilismo. Ma i tempi non erano ancora maturi e l'impresa non ebbe lunga vita. Noi dobbiamo però oggi ricordarla con ogni simpatia, e mandare un saluto a quei volonterosi che mostrarono fino da allora di indovinare il largo sviluppo riservato alla auto-locomozione ».

L'Automobile Club di Roma è ora un fatto compiuto. Il Gallenga afferma che costituendosi l'*Automobile Club d'Italia*, venne prescritto, nel suo foglio fondamentale, che la sua sede naturale sarebbe stata la capitale quando questa avesse visto sorgere tra le sue mura un club automobilistico.

« Non sappiamo se tale disposizione debba ancora avere vigore di norma statutaria. Osserviamo anzi che una questione di simil genere esorbita dall'ambito di uno statuto o di un regolamento e deve essere intesa nell'intima coscienza di tutti coloro i quali guardano a Roma con senso patriottico al centro della vita nazionale ».

**ABOVE:** *It seems that Italy's first true concours d'elegance à la française was in 1923, as reported in the June 30, 1923 issue of* Auto Italiana. *(Giovanni Bossi Archives)*

## After the Great War

After World War I, competitions that judged the appearance of cars, akin to beauty contests, became more common. For example, in its April 16, 1921, issue, *The Autocar* reported on the Midland Light Car Club's Opening Rally at Stratford-on-Avon: "*Despite Bad Weather the Midland proves a Success. Beauty contest for the cars, with prizes awarded for appearance,*" with winners including an Enfield-Allday sedan and an AC two-seater. However, there was no mention of whether ladies in haute couture costumes complemented the display.

In a Delage company press release from May 1922, there is mention of a concours d'elegance held in The Hague, where the text states: "*At the recent* concours *d'elegance at The Hague, last Sunday, three Delage cars secured the top three positions in this luxury car competition, which featured many English, American, and German marques.*" Since there are no details about the participants or what they wore, it is reasonable to assume that this event focused on the 'beauty' or 'appearance' of the cars themselves.

Italian automotive historian and historic vehicle expert Adolfo Orsi suggests that Italy may have been the first country to adopt the newer concept of concours d'elegance à la française. He notes, "One of the earliest French-style concours d'elegance may have taken place in Turin in 1923."

However, before delving into the Italians' passion for concours, it's important to recognize that other neighboring countries, such as Switzerland and Belgium, also quickly embraced this style of automotive celebration. Geneva was the one stylish city near France

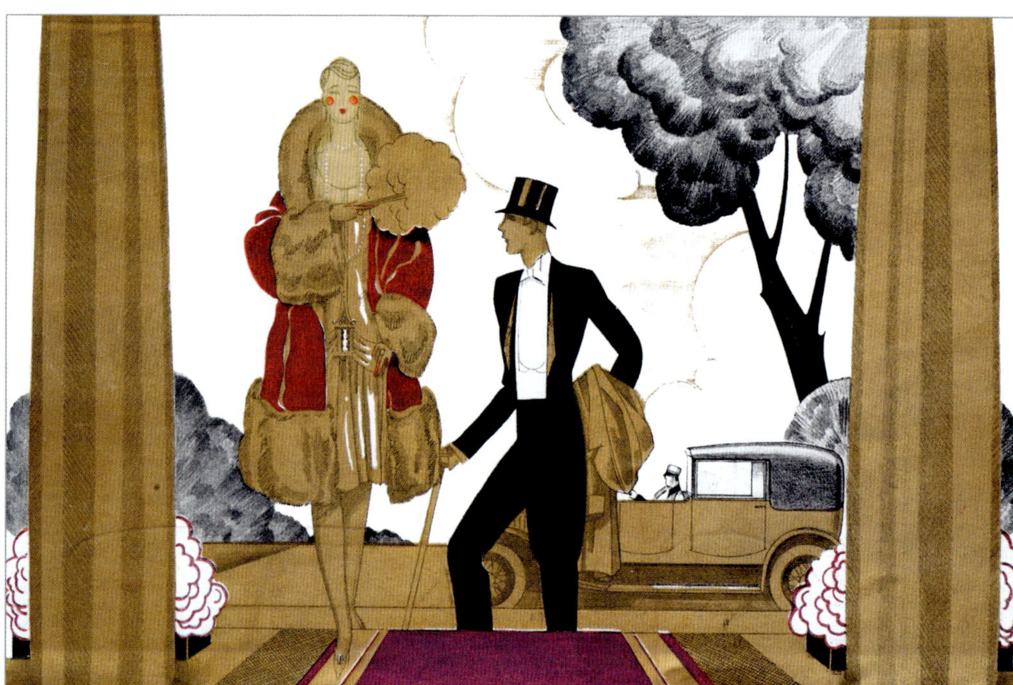

that embraced the concept of the French-style Concours d'Elegance. From 1927 onward, Geneva hosted an annual concours, though the location varied. The 1927 event was held on June 19 at the Promenade des Bastions, while the 1928 event took place on June 3 at the Parc des Eaux-Vives, which remained the favored venue for several years thereafter.

A report in the magazine *Le Temps Genève,* dated June 24, 1929, noted that the event had found the perfect setting beneath the beautiful trees of the Parc des Eaux-Vives, allowing the public to admire the latest in stylish coachwork. The event, which was an even greater success than its predecessors in 1927 and 1928, was followed by a gymkhana organized by the Automobile Club de Suisse (ACS), attracting 25 participants. The report highlighted the range of vehicles featured, from popular marques like Opel and Renault to more exclusive models such as Auburn, Packard, and several Delages. Both male and female participants were recognized as winners.

By 1931, the concours had moved to the Promenade du Lac, where it was held in 1931 and 1932. The June 18, 1932 event, organized by the *Association des Intérêts de Genève*, aimed to promote the city to international visitors and participants. A report in an (unidentified) magazine praised the event's beauty, though it lamented the lack of a larger audience. The wealth of cars on display, along with the

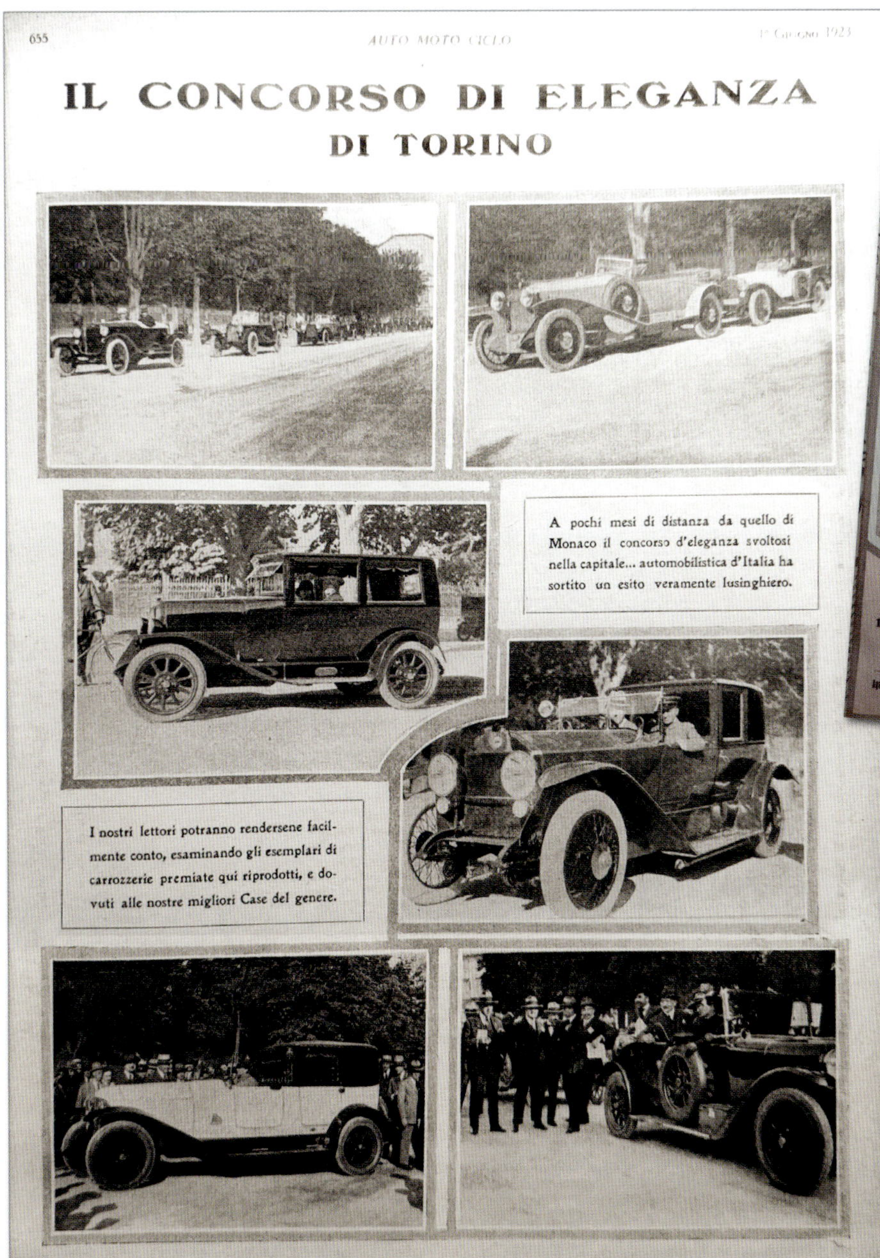

655 AUTO MOTO CICLO 1° Giugno 1923

IL CONCORSO DI ELEGANZA DI TORINO

A pochi mesi di distanza da quello di Monaco il concorso d'eleganza svoltosi nella capitale... automobilistica d'Italia ha sortito un esito veramente lusinghiero.

I nostri lettori potranno rendersene facilmente conto, esaminando gli esemplari di carrozzerie premiate qui riprodotti, e dovuti alle nostre migliori Case del genere.

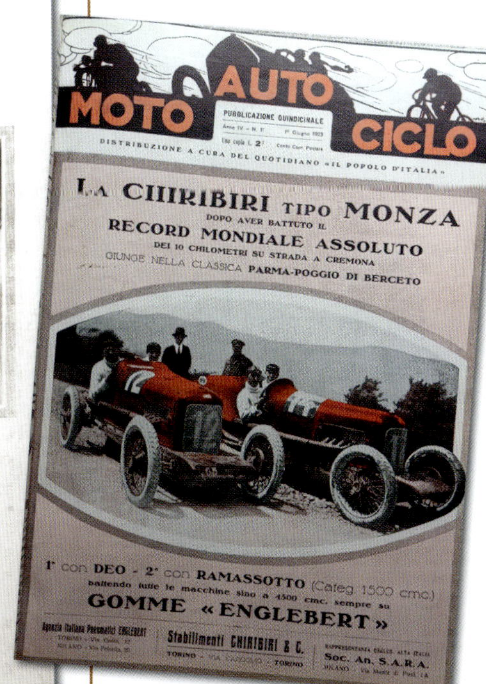


There was also a report in the July 1, 1923, issue of Auto Moto Ciclo *magazine.* (Giovanni Bossi Archives)

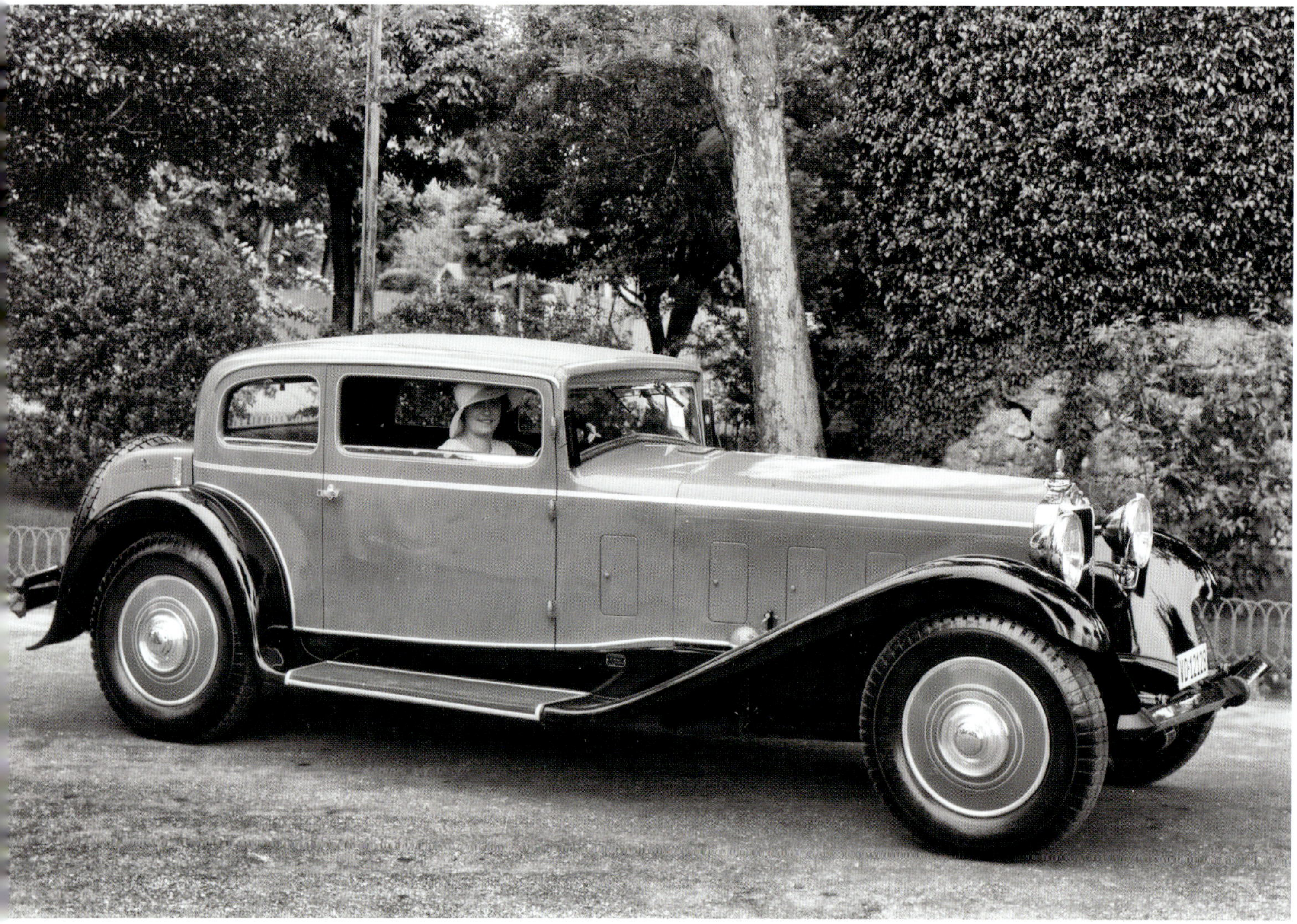

*At the Lausanne Concours on June 26, 1933, this beautiful Delage was shown – this one with factory-built (Autobineau) 2-door saloon Letourneur et Marchand coachwork and presented by Madame Terrier, wife of the local Delage dealer. (Daniel Cabart Archives)*

taste and craftsmanship of the manufacturers and coachbuilders, was met with enthusiastic applause from the attendees. Brands such as Delage, Bugatti, Renault, Nash, Chrysler, and Fiat shared the trophies awarded by the Association des Intérêts de Genève.

The Grand Prix of the concours was awarded to a Delage. A local newspaper remarked on the presence of Mademoiselles Blanche R. and Léone B. in the winning car, noting that their audacious outfits, designed by the Swiss couturier Marchal, demonstrated the unrivaled taste of the great couturiers. The results not only highlighted the car marques and models but also the coachbuilders involved, including early appearances from Pinin Farina, as well as regulars such as Letourneur et Marchand, Kellner, Gangloff, Chapron, and Castagna.

Inspired by the example set by Geneva, several fashionable Swiss towns embraced the French concept of the concours d'elegance, with the ski resort town of Saint-Moritz leading the way. On August 25, 1929, as part of the International Automobile Week in Saint-Moritz, which concluded with the Bernina hill climb, the French automaker Delage used the concours d'elegance as a platform to unveil its new Delage D8. In a press release, Delage announced that *"at the request of its Swiss general representative, Albert Goy, the company participated in the concours d'elegance with four superb and elegantly styled automobiles. These cars, equipped with the new 8-cylinder engine, were awarded silver medals."*

The press release further detailed that *"a tour of major Swiss cities was organized by Mr. Lardy, and after passing through Basel, Zurich, Bern, La Chaux-de-Fonds, Neuchâtel, and Lausanne, the four luxury Delages arrived in Geneva on Friday evening. On Saturday, the splendid vehicles – a two-seater convertible, a sky-blue roadster with Vanden Plas bodywork on a short chassis, a convertible with Gaston Grümmer bodywork, and a Saoutchik six-seater sedan with a convertible coupé limousine body – were showcased at the Kursaal garage."*

As the Saint-Moritz Concours d'Elegance became an annual event, other towns like Lucerne, Evian, and Lausanne followed suit. Lucerne's first concours d'elegance took place on August 2, 1930, and was hailed as a "brilliant success" by the local press. An Ansaldo won one category, an Alfa Romeo triumphed in another, and a variety of cars, including Bugatti, Delage, Chrysler, Mercedes-Benz, Austro-Daimler, Nash, Lincoln, Talbot, and Cord, were among the award winners.

Lausanne held its first concours delegance on June 27, 1931, and its success led to a second event on June 18, 1932. The *Grand Prix d'Honneur* was awarded to a Horch, the Grand Prix for garage owners went to a Delage, and the prize for passenger cars with standard bodywork was won by a Steyr. The 1933 Lausanne Concours d'Elegance, held on June 4, saw the *Grand Prix d'Honneur* awarded to an Isotta Fraschini. The *Prix d'honneur* for private cars went to a Delage, while the prize for garage owners was claimed by a Horch, and the *Prix d'honneur* was awarded to another Delage, presented by a certain Mlle. Josette Cortez, Miss Paris 1933. Notably, Betty Spell was among the class winners, receiving awards for presenting both a Delage and a Hispano-Suiza.

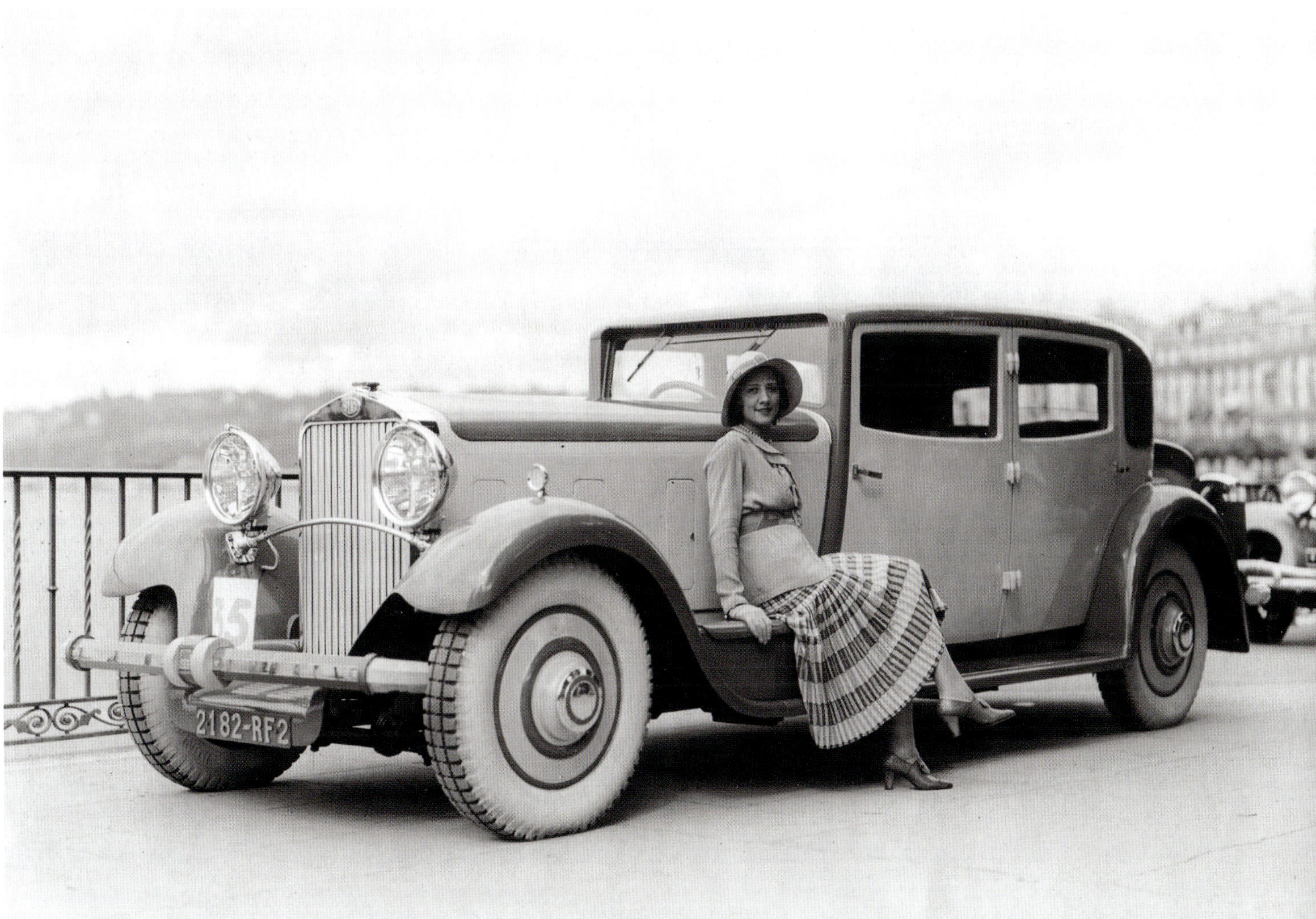


*At Geneva on July 4, 1931, Mlle. Lucienne Radisse presented this Delage D8 by Willy Vanden Plas. (Daniel Cabart Archives)*

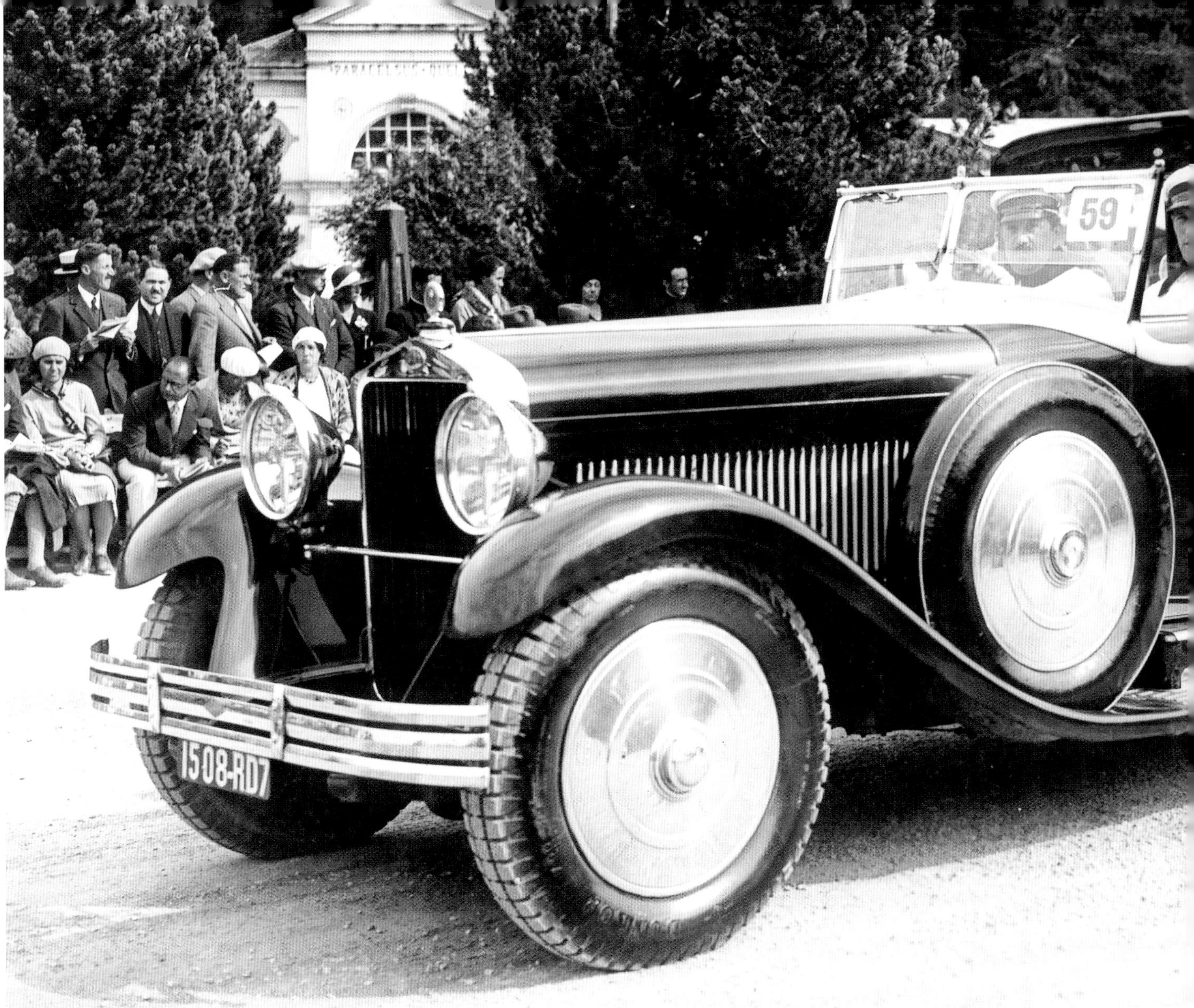
59
1508-RD7

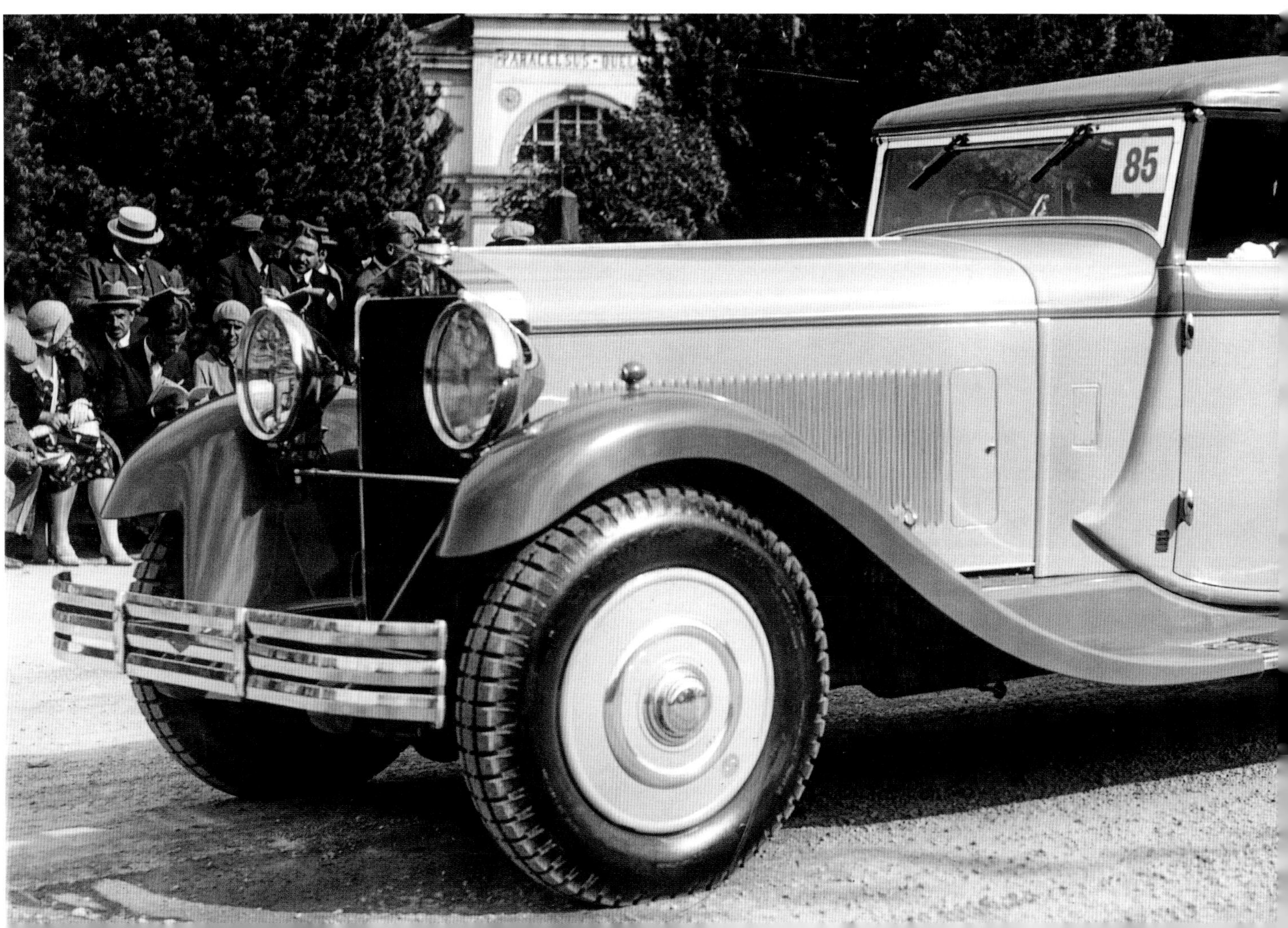
85

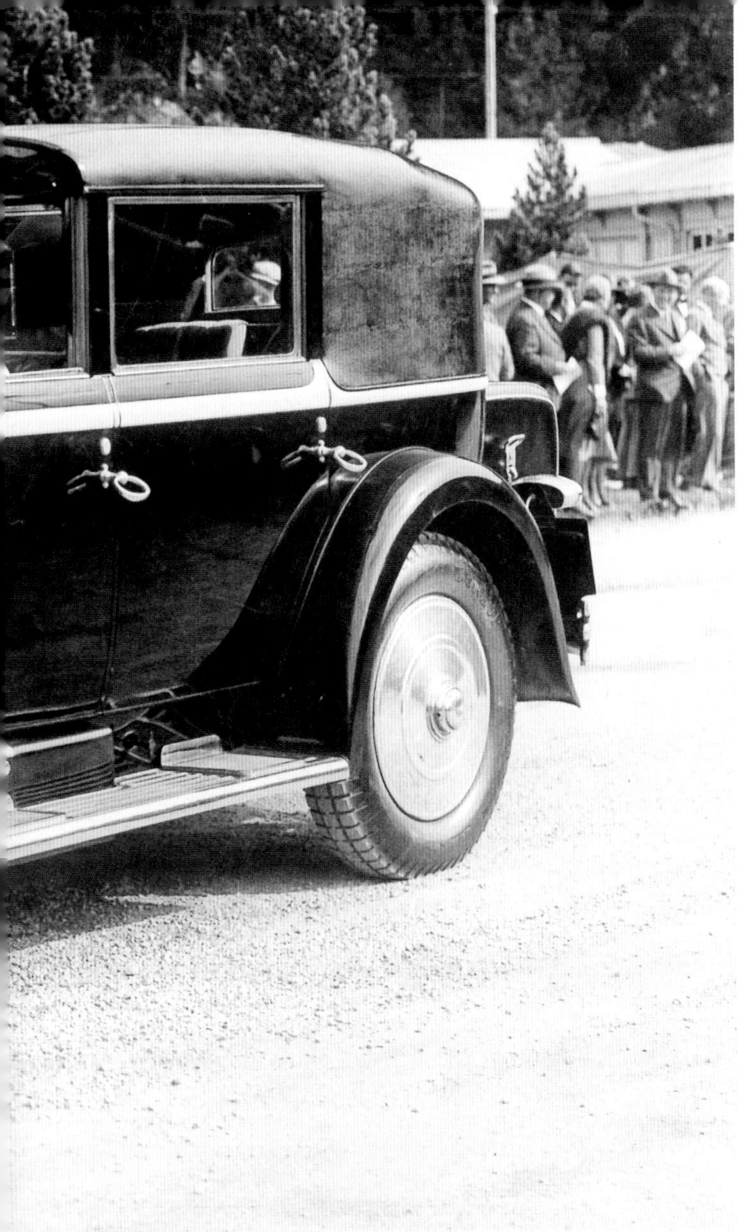

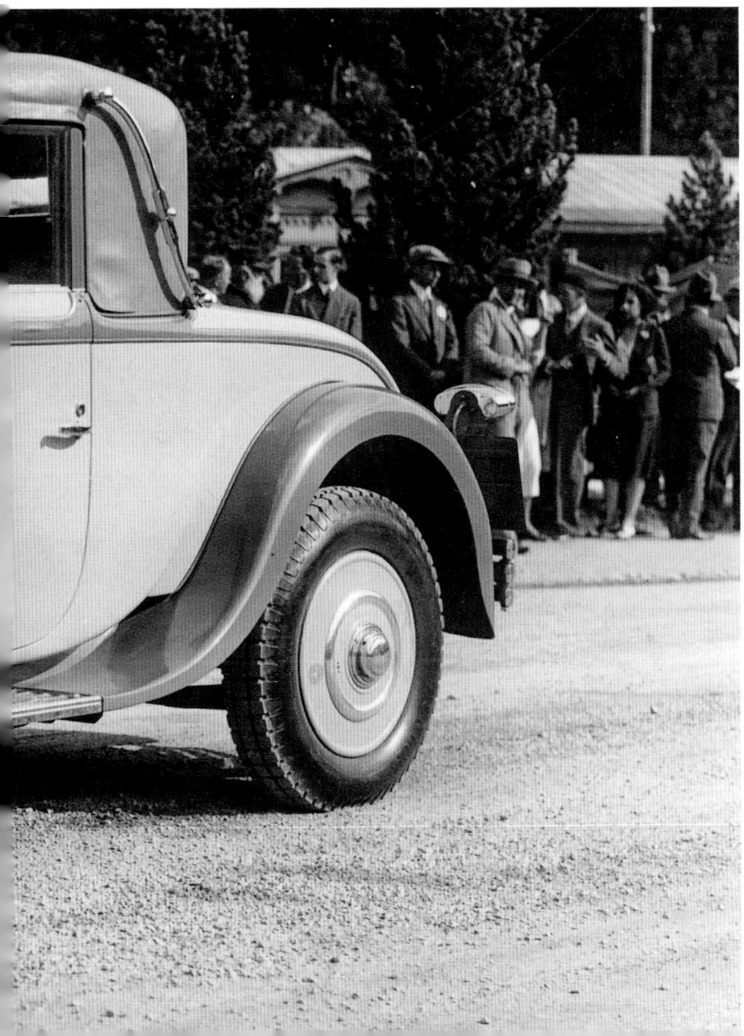

**UPPER:** *The Concours d'Elegance at Saint-Moritz was another early Swiss concours, with one taking place on August 25, 1929, where another Delage D8, one with a Saoutchik body was presented at the event. (Daniel Cabart Archives)*

**LOWER:** *For the 1930 Saint-Moritz Concours, a Delage D8-S with Saoutchik body was one of several stars of the show. (Daniel Cabart Archives)*

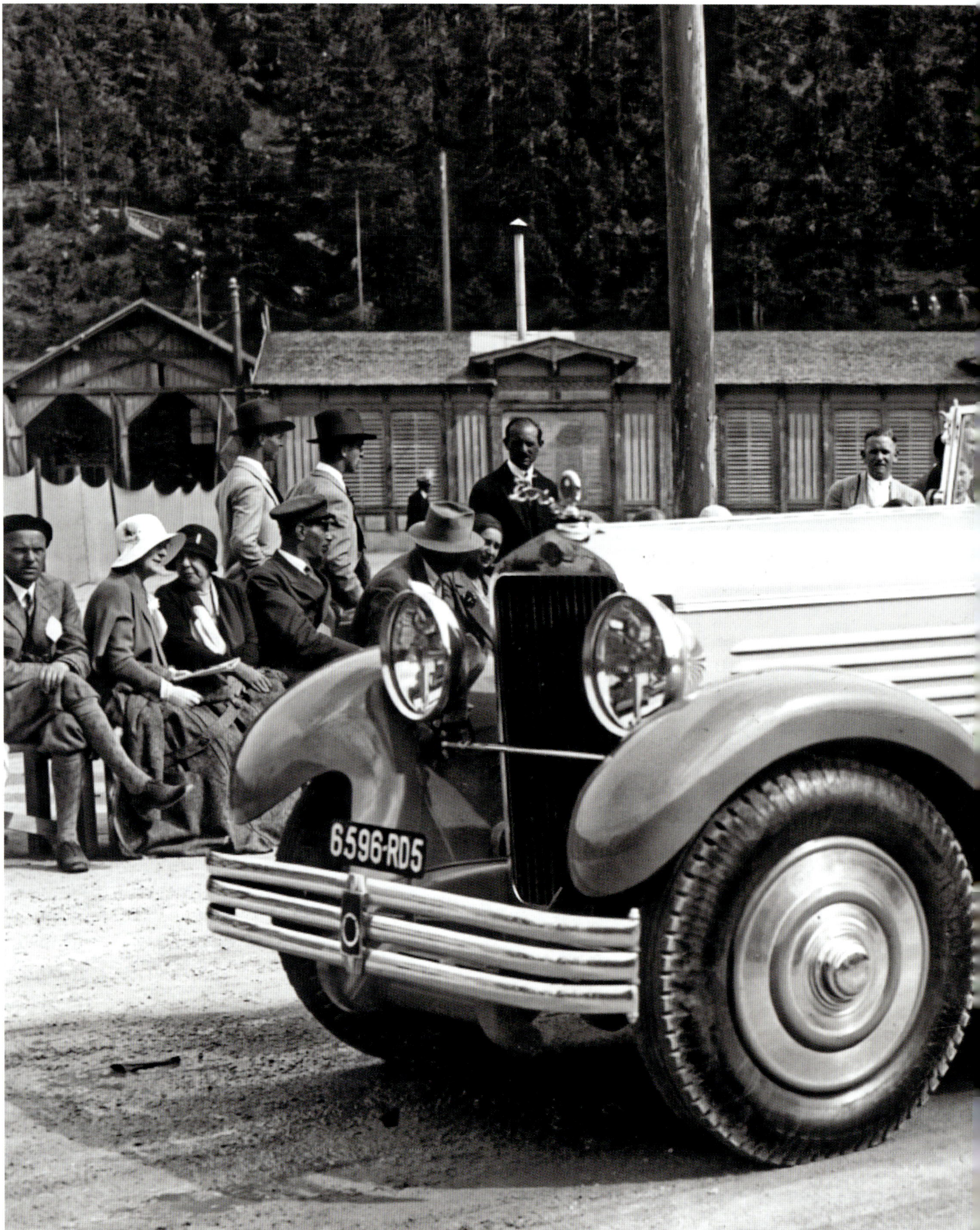
6596-RD5

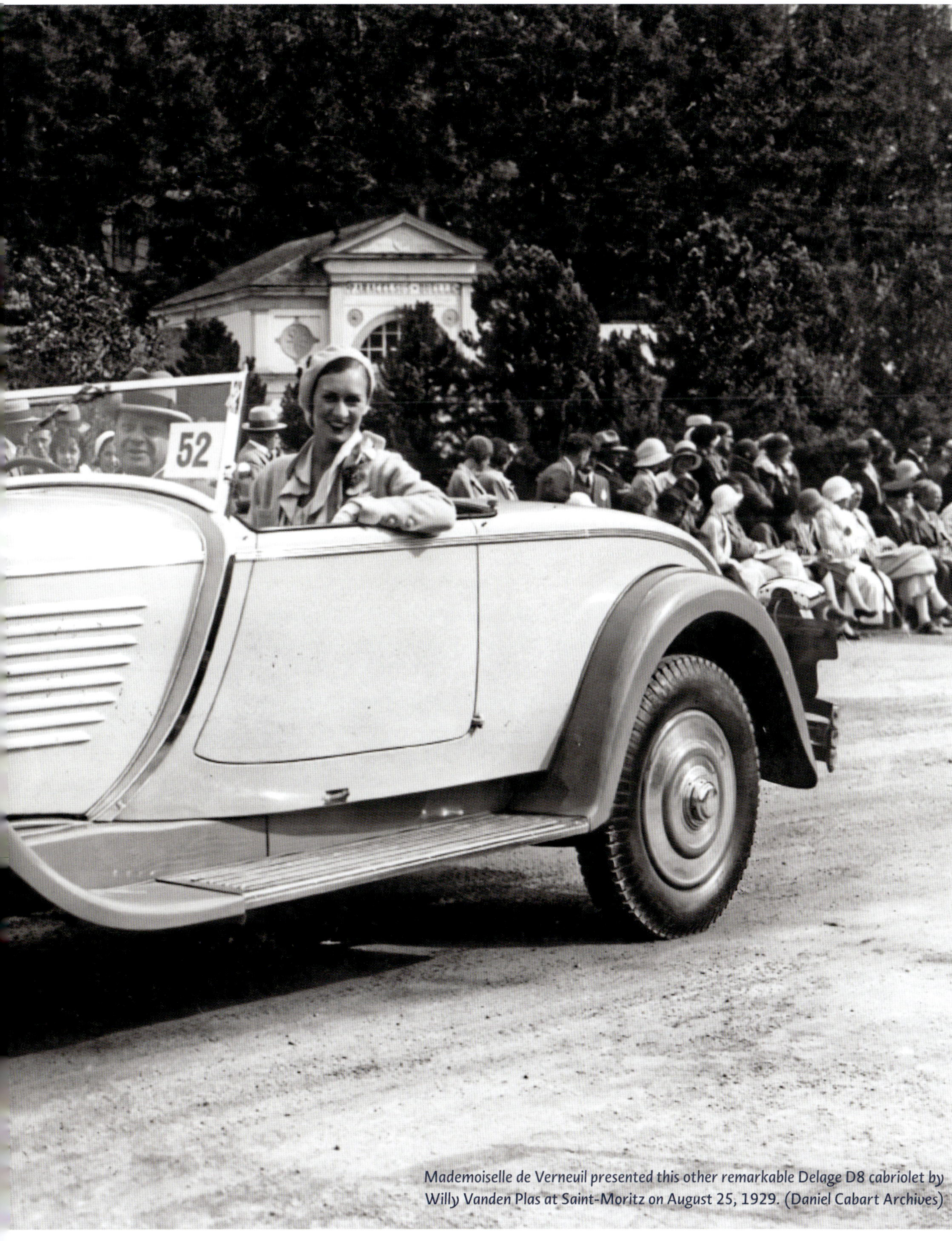


Mademoiselle de Verneuil presented this other remarkable Delage D8 cabriolet by Willy Vanden Plas at Saint-Moritz on August 25, 1929. (Daniel Cabart Archives)

**RIGHT:** *Lausanne was another popular concours location. Pictured here is a Chapron-bodied Delage for the 1930 Concours d'Elegance at Lausanne. (Daniel Cabart Archives)*

**BELOW:** *Delage's official entry at the 1930 Lausanne Concours d'Elegance was a factory-bodied car, presented by the redoubtable Betty Spell. (Daniel Cabart Archives)*

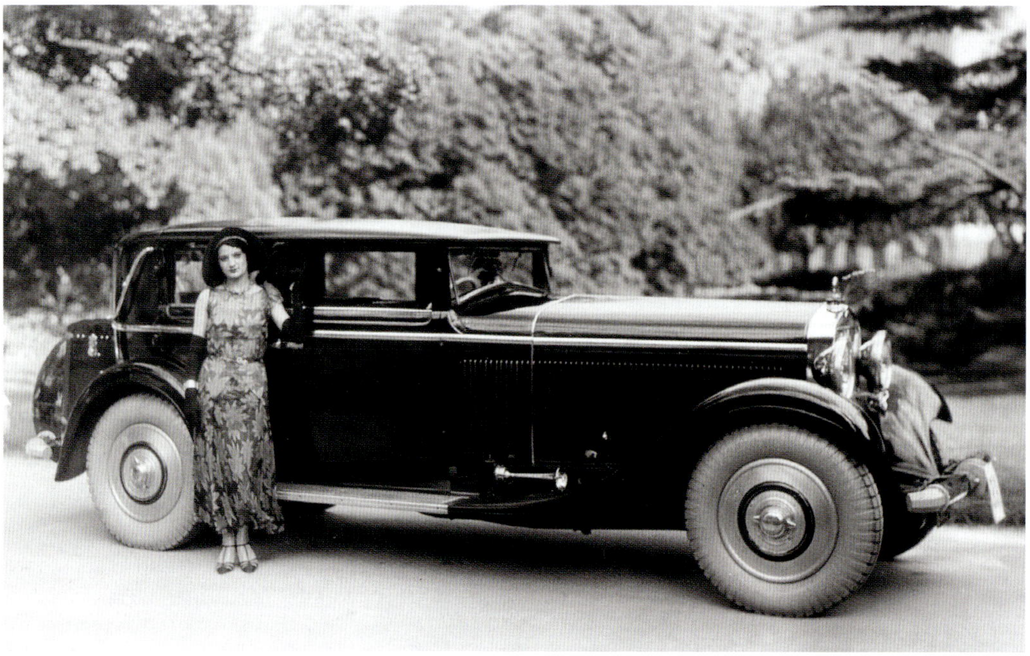

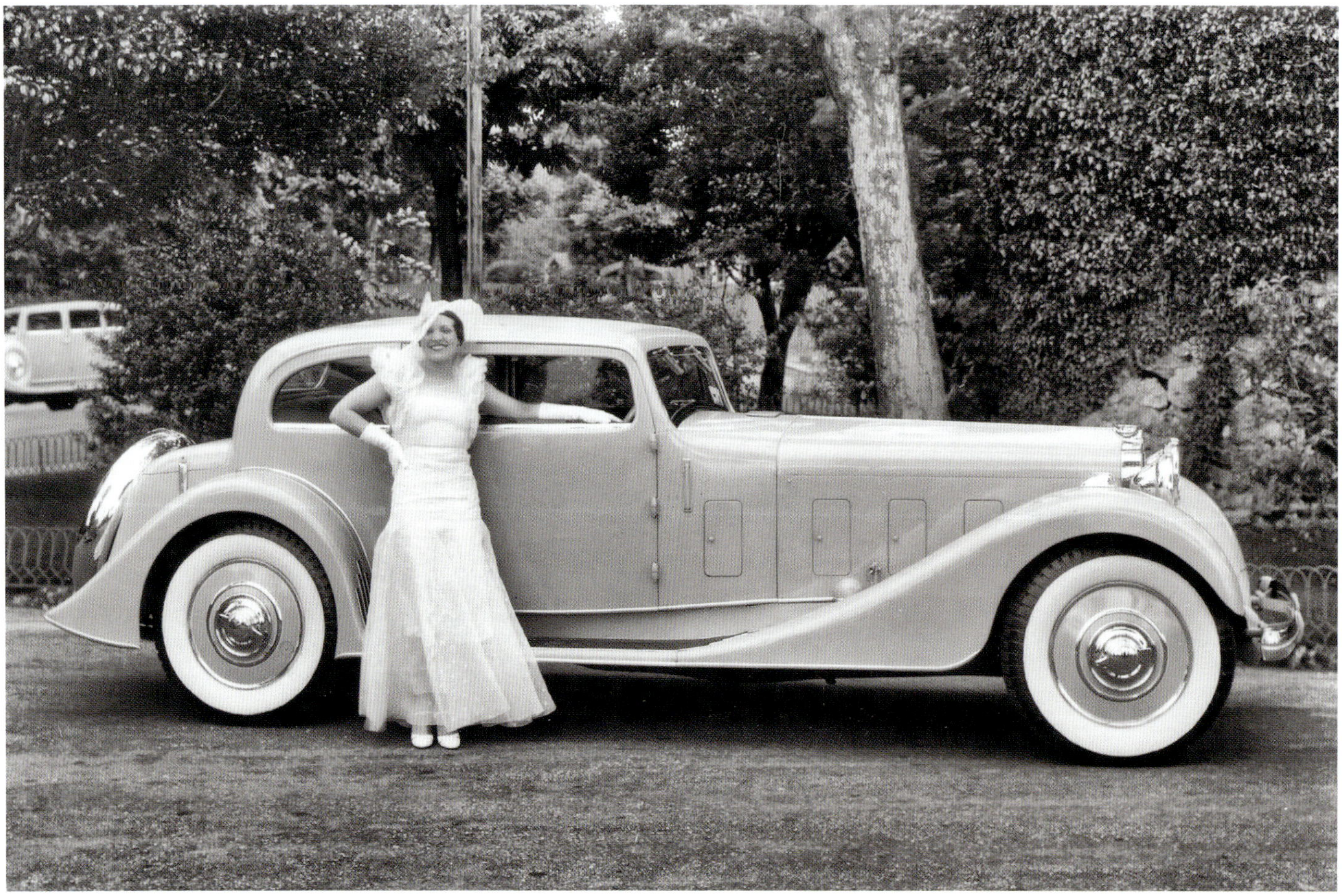

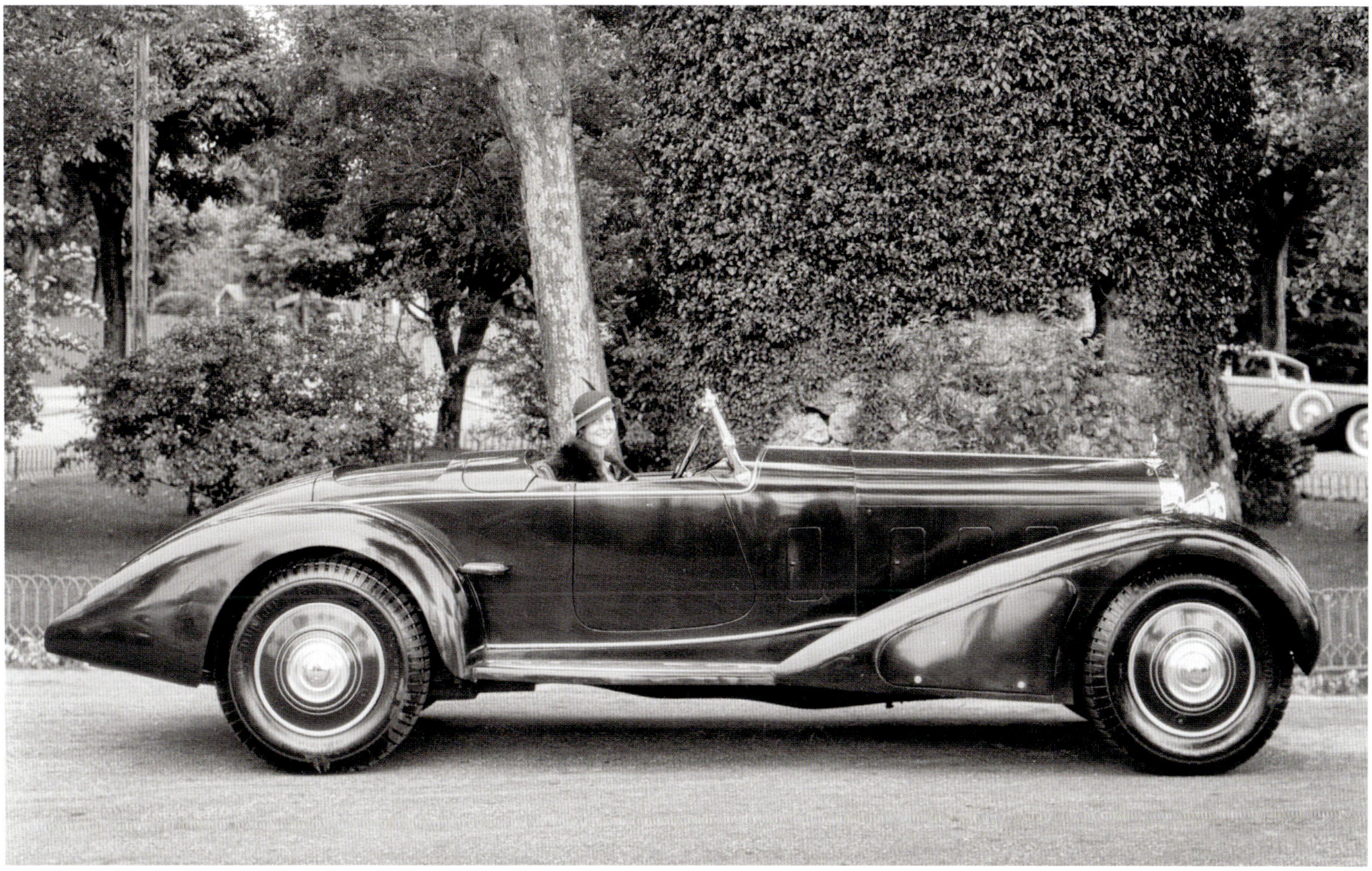

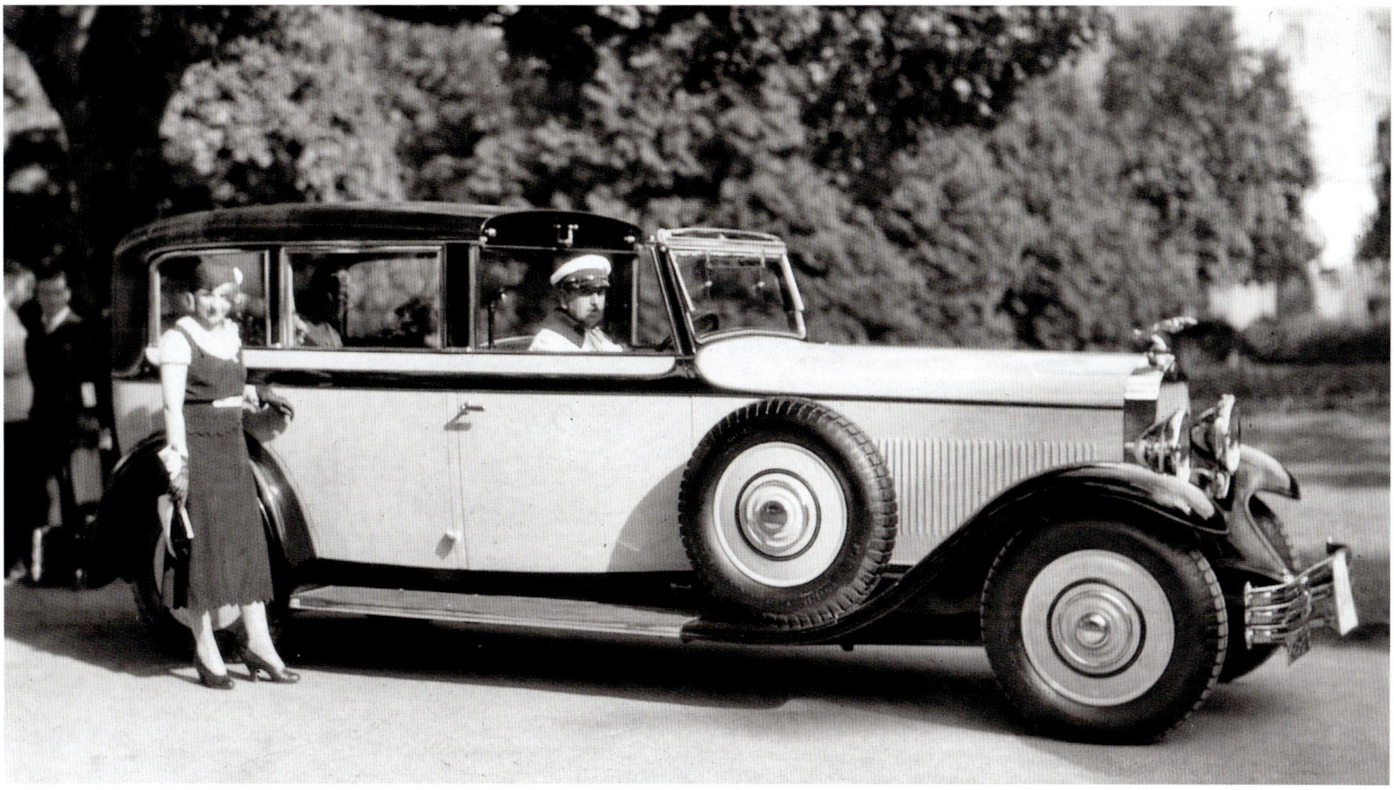

**TOP:** *Delage's dealer, Terrier Hippolythe, presented a Delage with Madame Delisle-Terrier at the wheel during the 1930 Lausanne Concours d'Elegance. (Daniel Cabart Archives)*

**ABOVE:** *A Delage at the 1931 Geneva Concours was a handsome Duvivier-bodied coupe de ville, presented by Mlle. Claude Revel (Daniel Cabart Archives)*

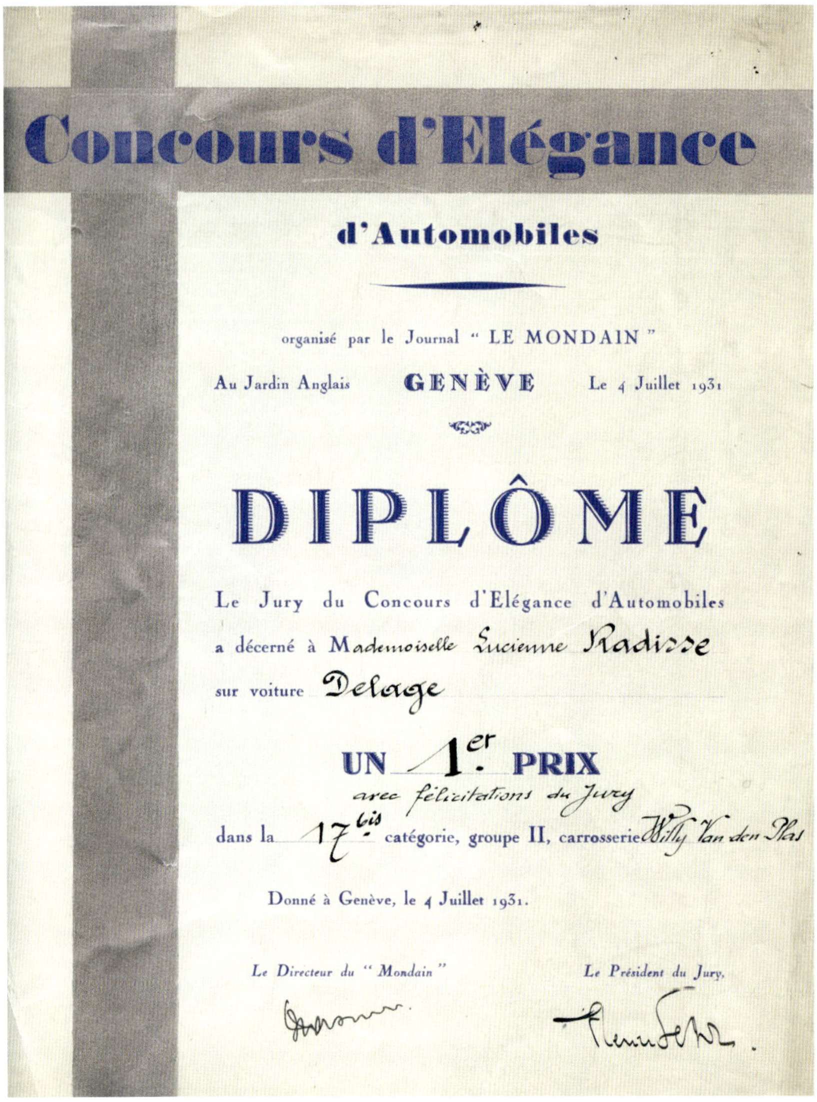

# Concours d'Elégance

d'Automobiles

organisé par le Journal " LE MONDAIN "

Au Jardin Anglais **GENÈVE** Le 4 Juillet 1931

# DIPLÔME

Le Jury du Concours d'Elégance d'Automobiles

a décerné à Mademoiselle Lucienne Radisse

sur voiture Delage

**UN** 1er **PRIX**

avec félicitations du Jury

dans la 17bis catégorie, groupe II, carrosserie Willy Van den Plas

Donné à Genève, le 4 Juillet 1931.

Le Directeur du " Mondain" — Le Président du Jury.

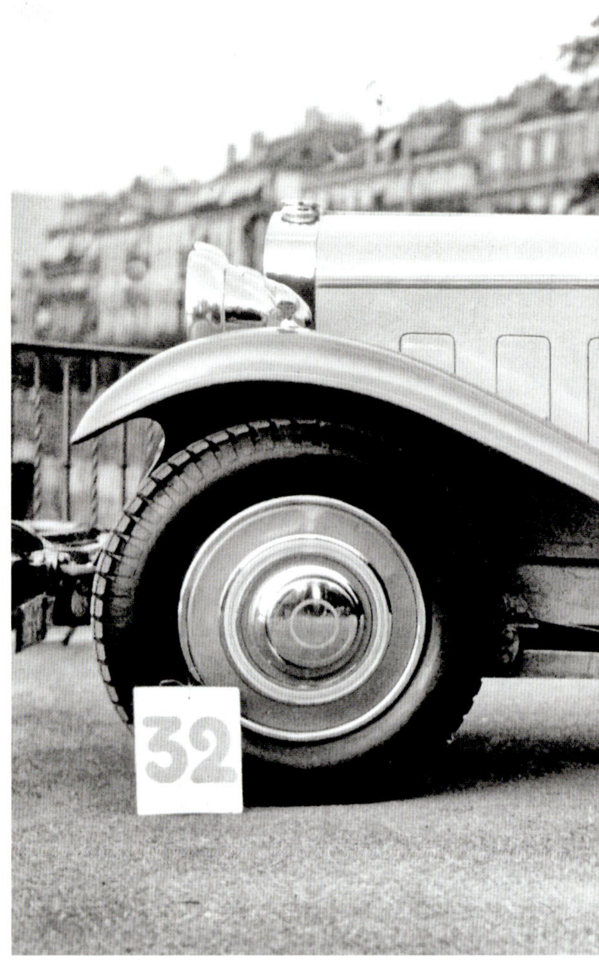


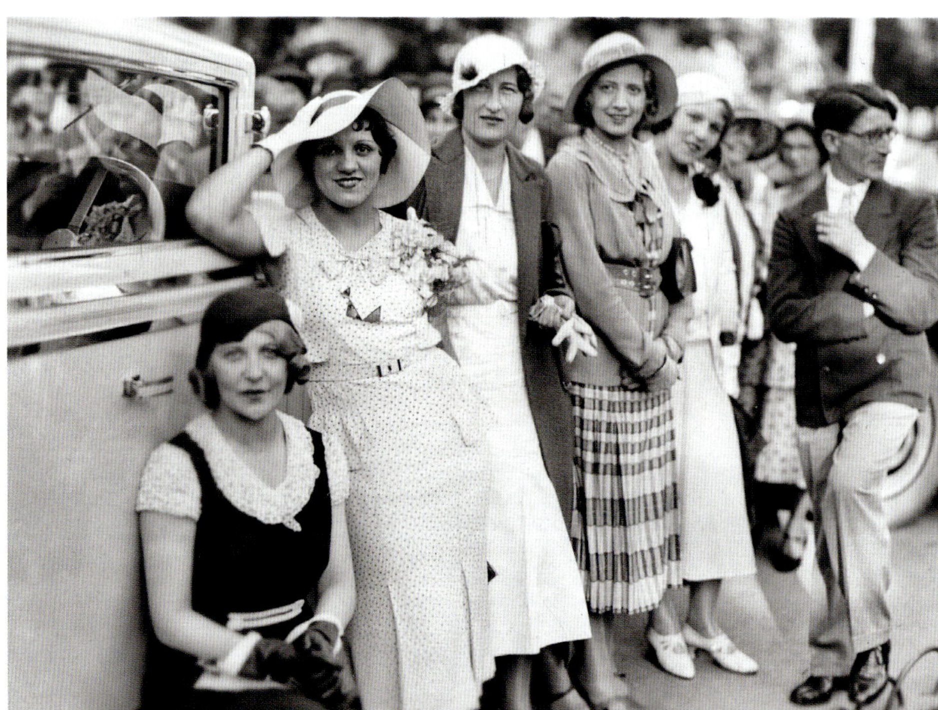

**ABOVE:** *The Swiss held their concours events regularly, such as the one in Geneva on July 4, 1931, where Lucienne Radisse presented this Vanden Plas-bodied Delage. (Daniel Cabart Archives)*

**TOP RIGHT:** *Betty Spell was at the Geneva show again, presenting another Delage by par Letourneur et Marchand. (Daniel Cabart Archives)*

**RIGHT:** *Once more in Geneva in 1931, as the ladies pose in the most stylish of dresses. (Daniel Cabart Archives)*

**OPPOSITE BOTTOM:** *At the 1932 Geneva Concours, Madame Thirion, dressed by the haute couture house Marchal, presents this other Delage D8, bodied as a D'Orsay coach by Chapron. (Daniel Cabart Archives)*

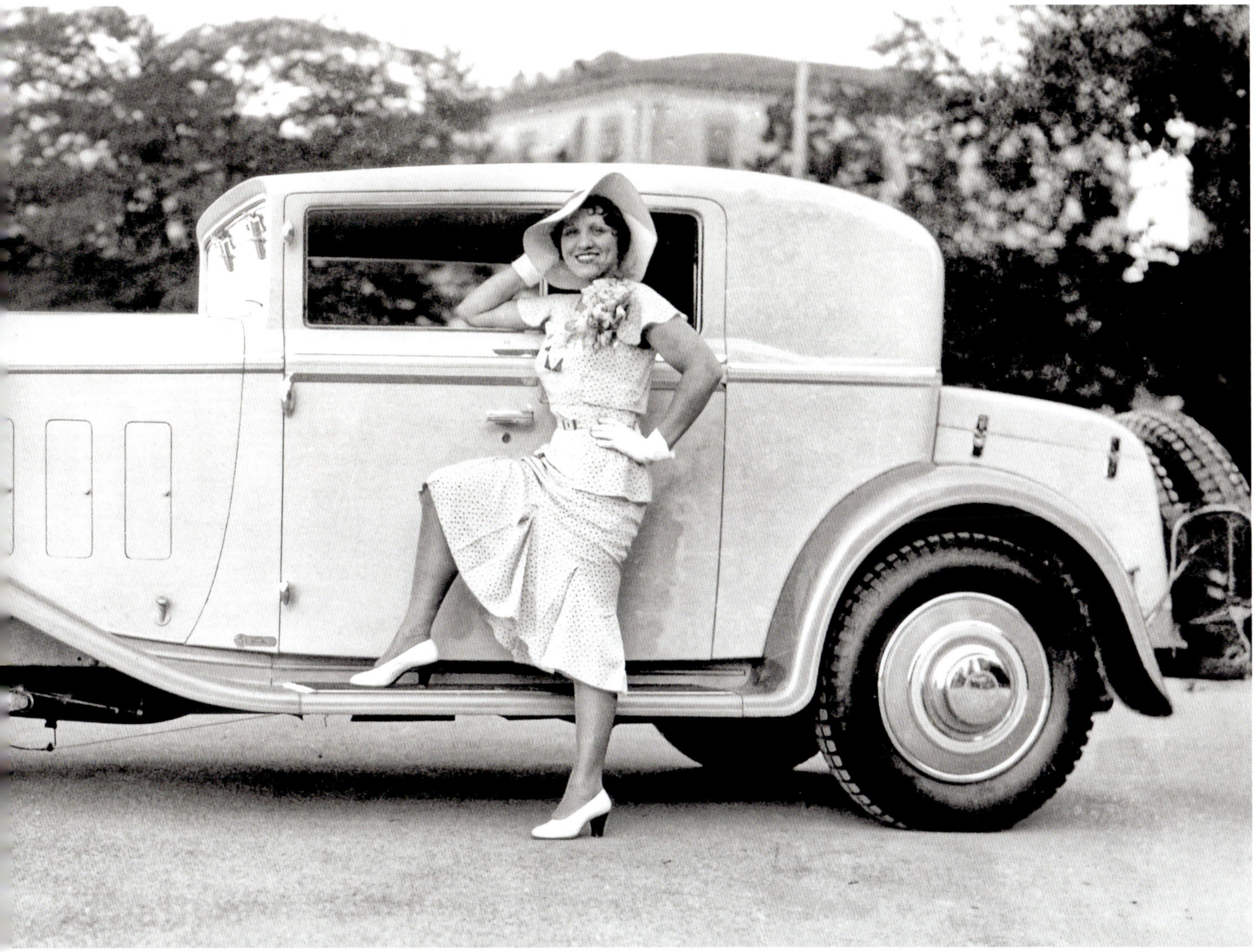

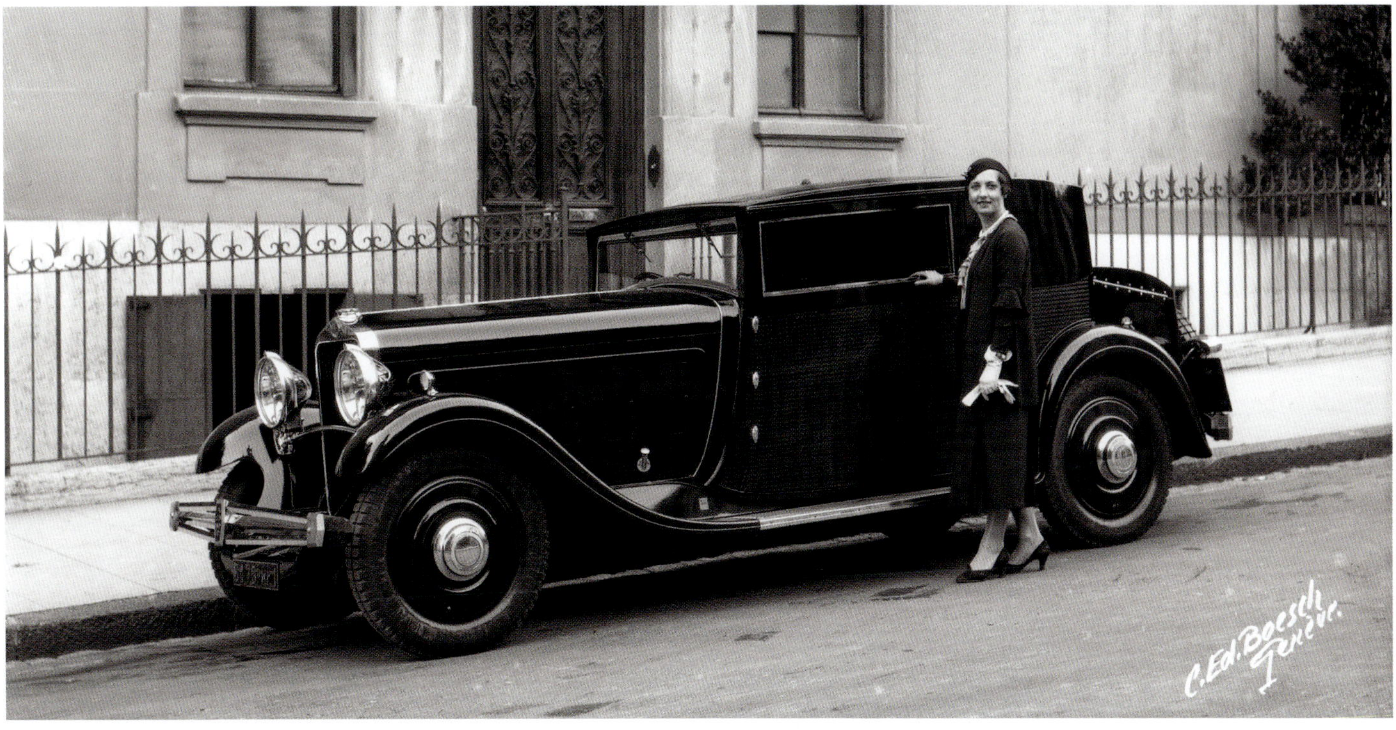
C. Ed. Boesch
Genève.

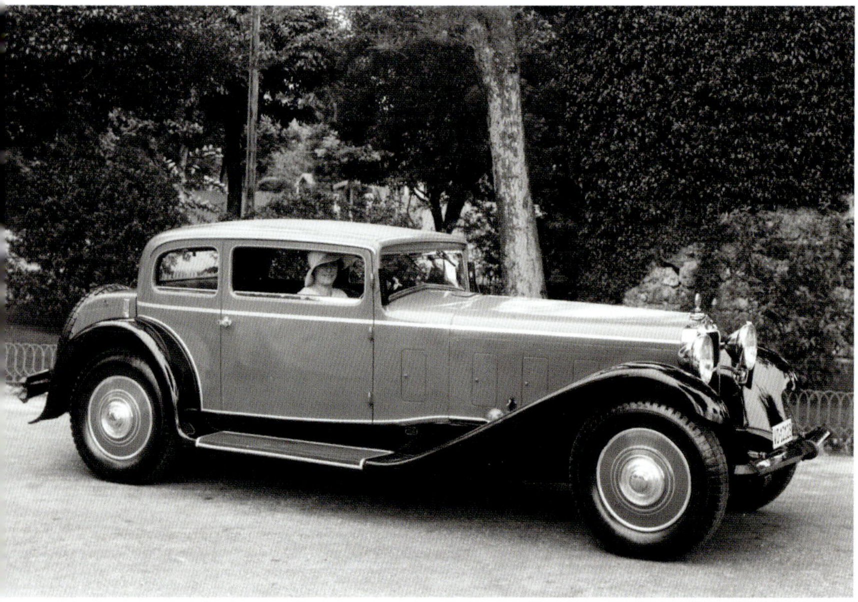

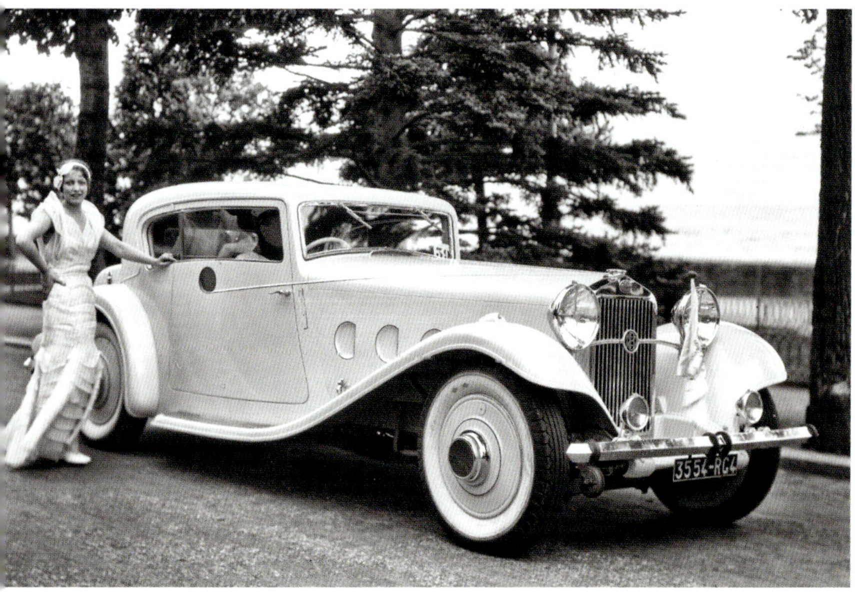

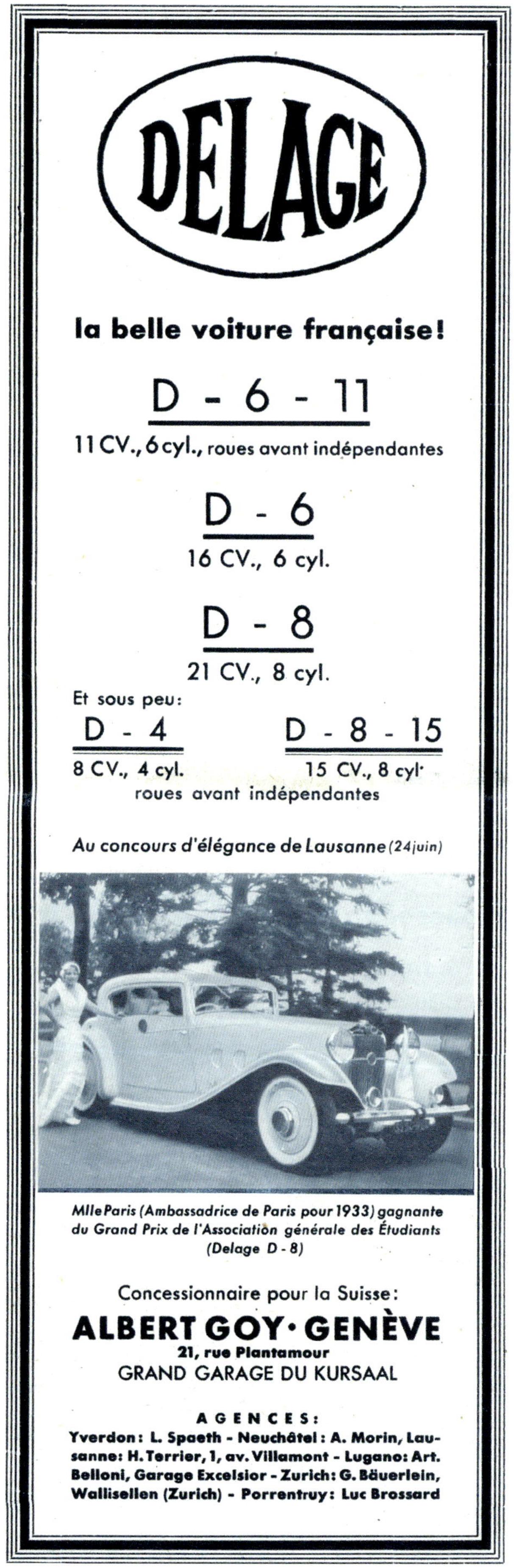


**ABOVE (BOTH):** *Lausanne remained a popular concours venue with this Delage D8-S (top) for the 1931 Concours d'Elegance at Lausanne. Mlle. Paris presented another Delage (above). (Daniel Cabart Archives)*

**RIGHT:** *Delage seems to have made the most of their endeavors at participating and winning at various concours d'elegance as can be seen from this advertisement promoting their wins. (Daniel Cabart Archives)*

## Then the English joined in

If Continental Europe seems to have embraced the French tradition of concours d'elegance, what about the stern yet resilient nation across the channel? The United Kingdom also welcomed the concept, and by the late 1920s, several concours events were attracting both participants and the public. The seaside town of Bournemouth became a trendy location for an annual concours event, with one of the earliest taking place from July 5-7, 1928.

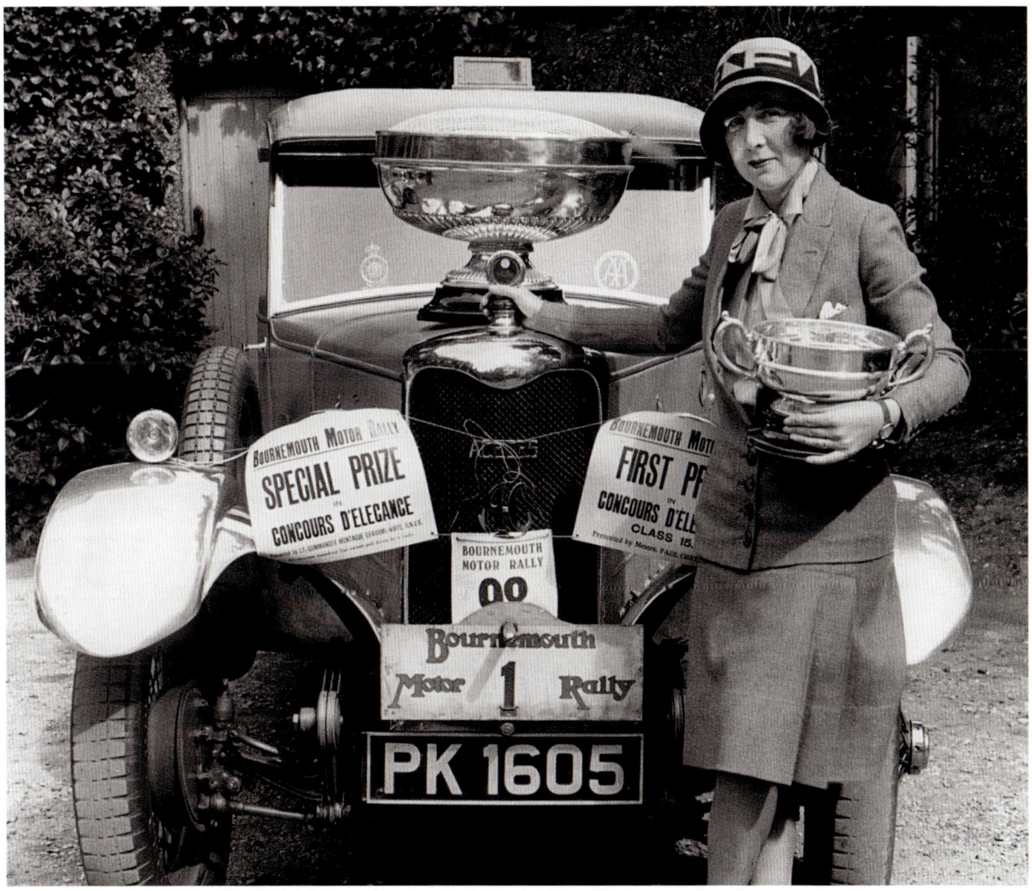


*The Honourable Mrs. Victor (Mildred Mary) Bruce with her AC Acedes at the 1928 Concours d'Elegance of the Bournemouth Motor Rally. (Alamy)*

In its June 12, 1928 issue, The Motor magazine announced, "*There has probably never been a motoring event held in this country in which so many valuable and magnificent trophies and prizes have been offered. Totaling several hundred pounds in value and presented in most cases by generous donors eager to see a Rally and* Concours d'Elegance *on a true Continental scale held in this country, the prizes consist not only of tasteful objets d'art, but of really useful articles to the motorist.*"

Among the trophies, one of the most notable was the S.F. Edge Trophy, presented by the legendary Selwyn Edge himself, who, despite being retired, remained highly active in the concours scene. *The* Motor article also mentioned a "*special prize in the* Concours d'Elegance *for the newest and smartest production of a fashionable coachbuilder.*"

Another noteworthy category was the "Old Brigade class for cars of any type manufactured before and including 1903, capable of being driven at the time of the competition."

As the concours concept gained popularity in other seaside towns like Eastbourne, Bexhill, and Ramsgate, and became increasingly popular among automotive enthusiasts, *The Autocar* magazine published an in-depth article titled "Points that Count," subtitled "How to Win Prizes at Elegance Competitions: Ideas from Past Rallies."

Written by A.G. Douglas Clease and published in the May 8, 1936 issue of *The Autocar*, the article began by stating that "*the one attribute the British owner-driver shares with Continental motorists, but apparently not to any great extent with American owners, is the pride he takes in the appearance and fittings of his car, along with the desire to add little personal touches that give the car a character of its own.*"

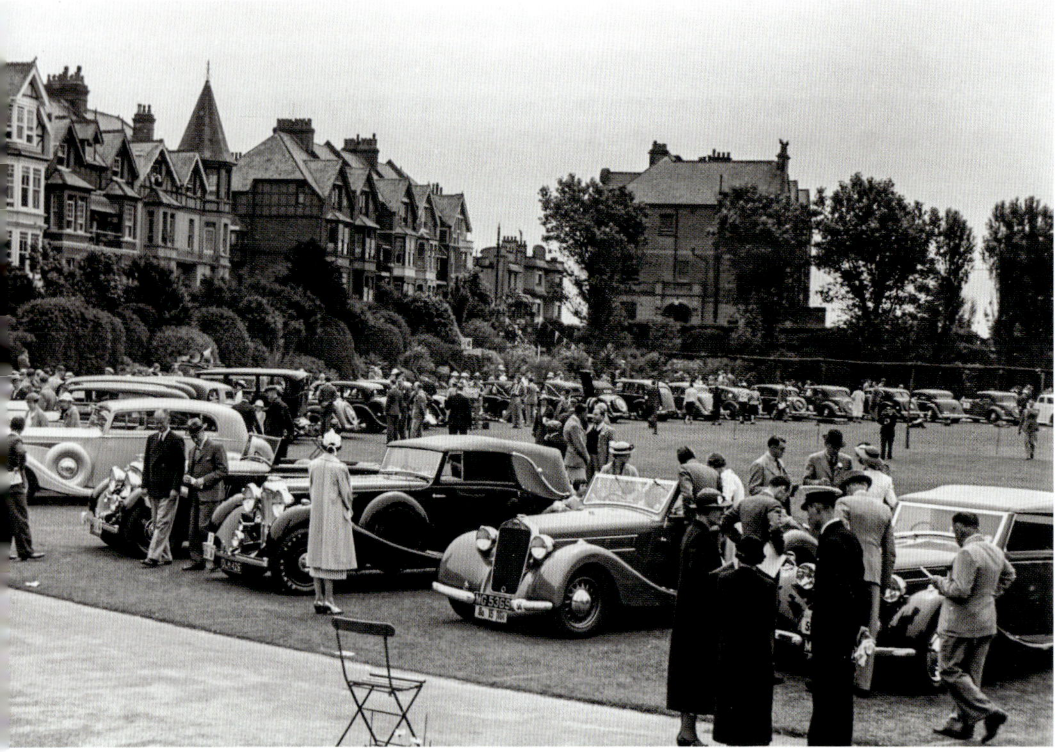

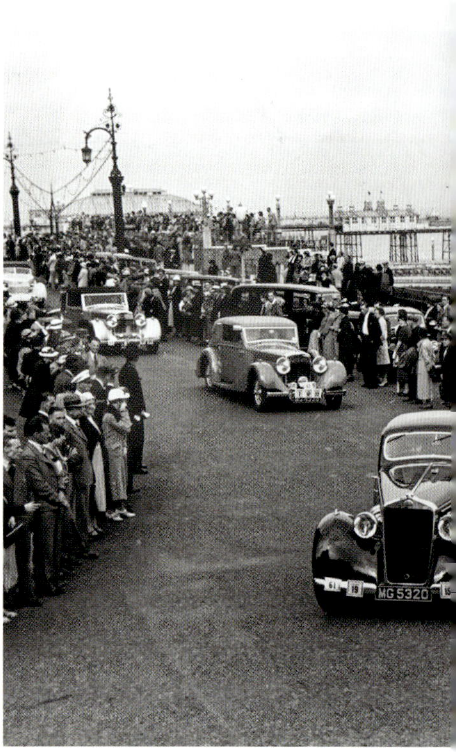


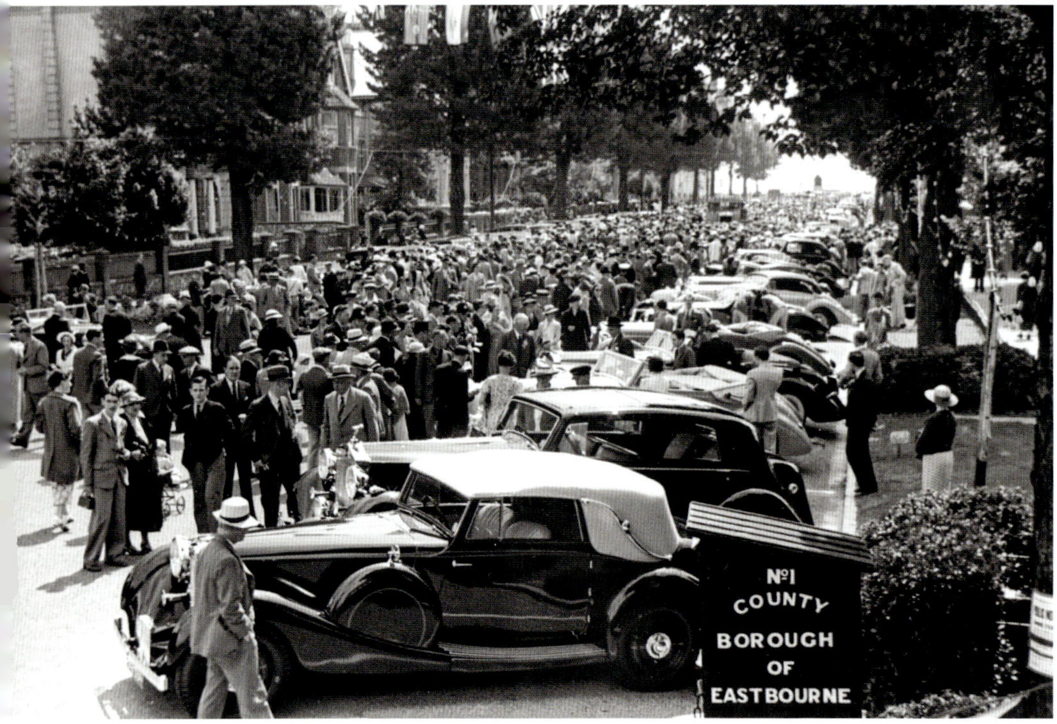


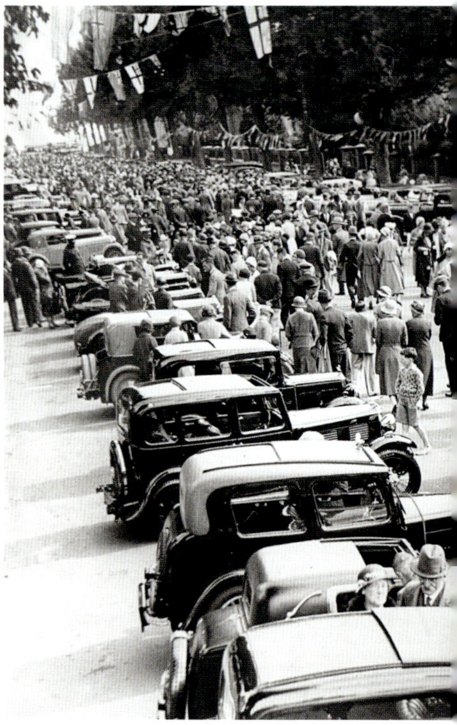

The article then offered suggestions on what should be fitted and how, noting, "*In addition to the head and side lights, a pair of Passlights is sometimes mounted symmetrically on the bumper brackets, between which a bar runs to carry the badges, while two pairs of horns are also arranged symmetrically, and a pair of spotlights is mounted on each side of the scuttle. Whether many lamps and horns are desirable is a matter of opinion, but there is no doubt that symmetrical arrangement does give a balanced appearance to the front of a car.*"

Regarding color choices and maintenance, *The Autocar* emphasized, "*Apart from the actual design, which is a matter of individual taste, there is the color scheme. Currently, fashion seems to favor a single color for the exterior, with the exception of a fine line to provide a little relief, matching the color of the upholstery.*"

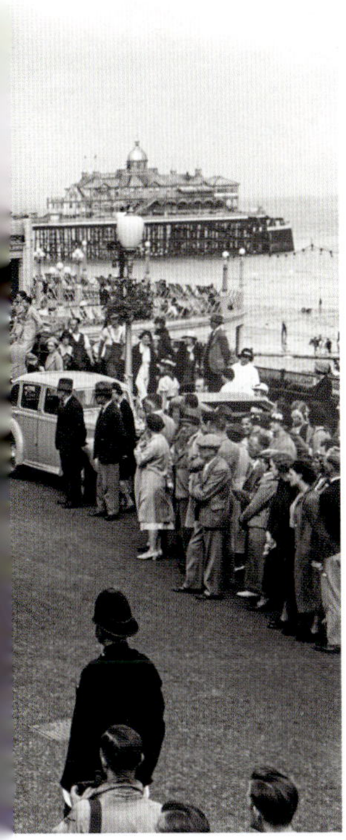

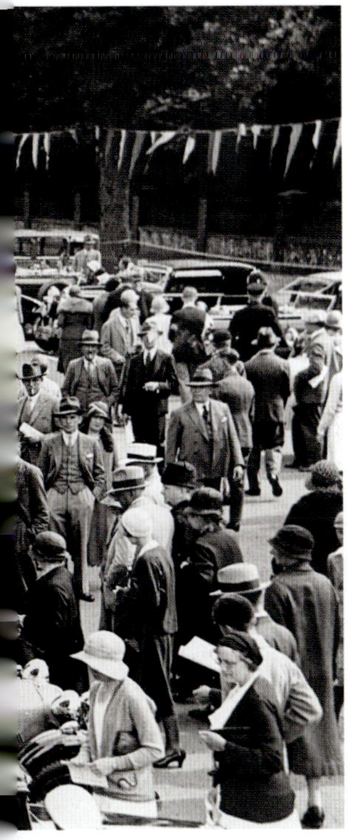

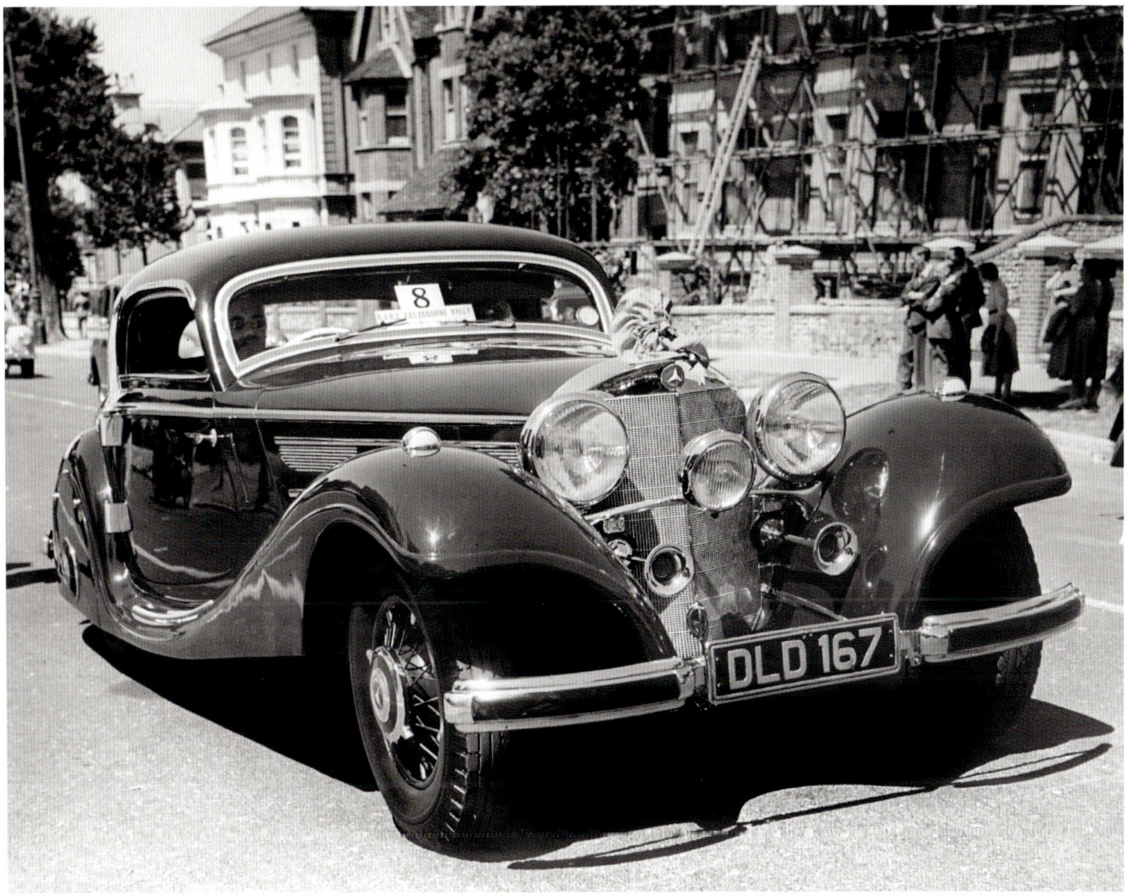


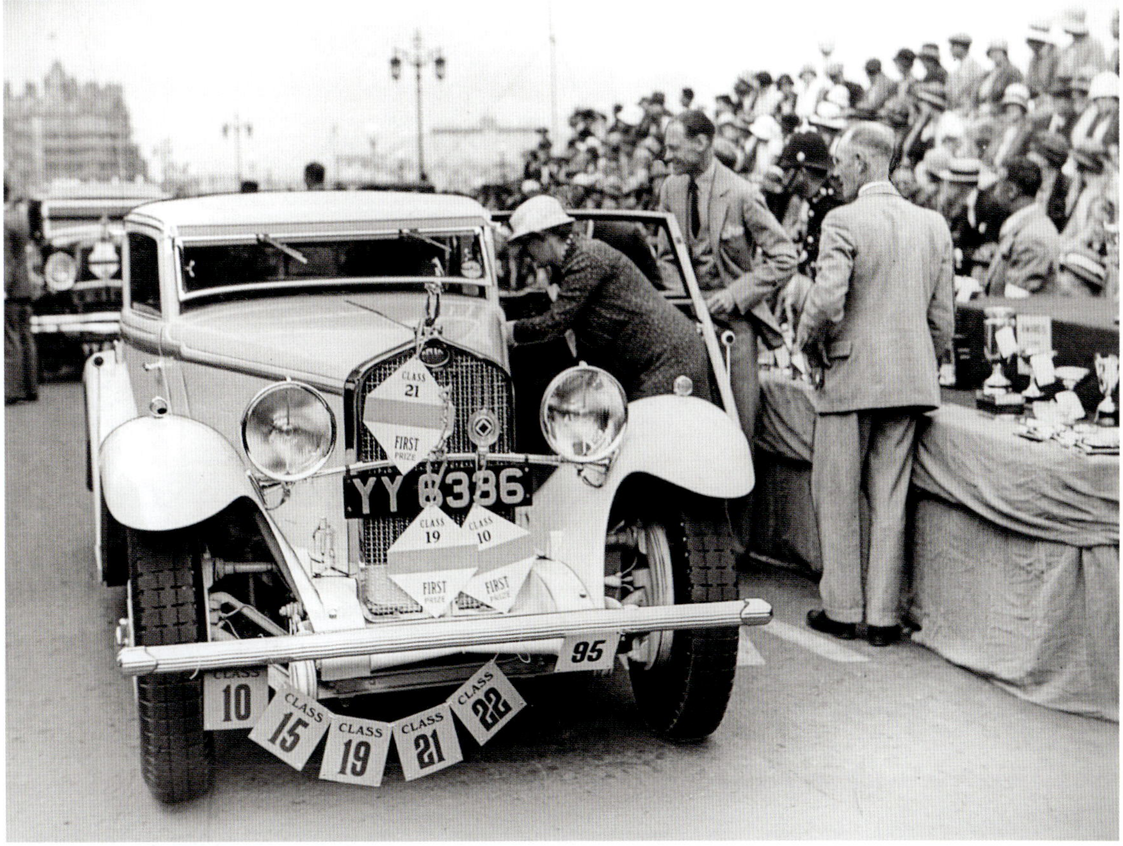


**THESE TWO PAGES:** *The English took to concours in a big way, as well, and Eastbourne was an important concours venue. This is from the 1936 concours d'elegance at Eastbourne.* (MW Archives)

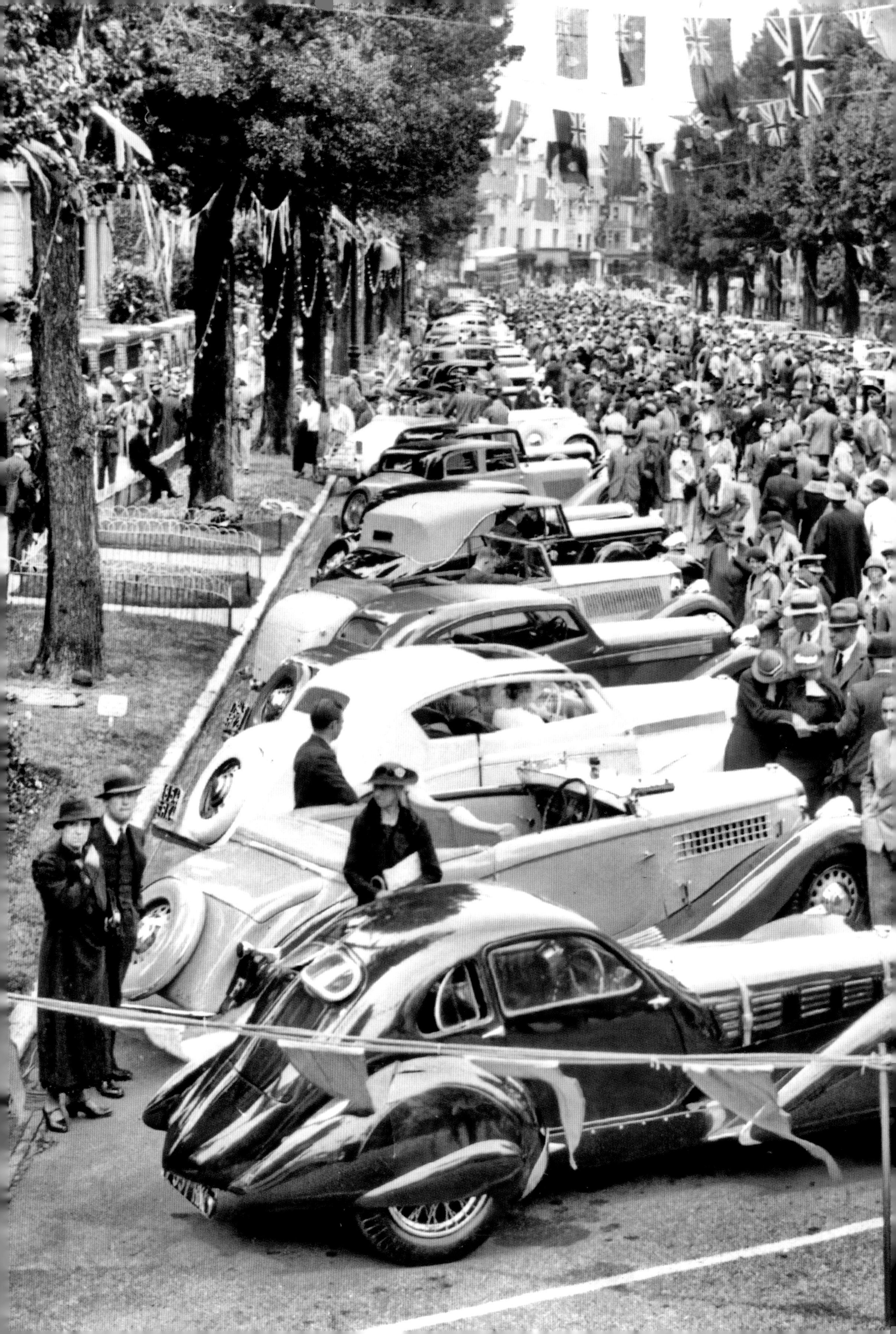

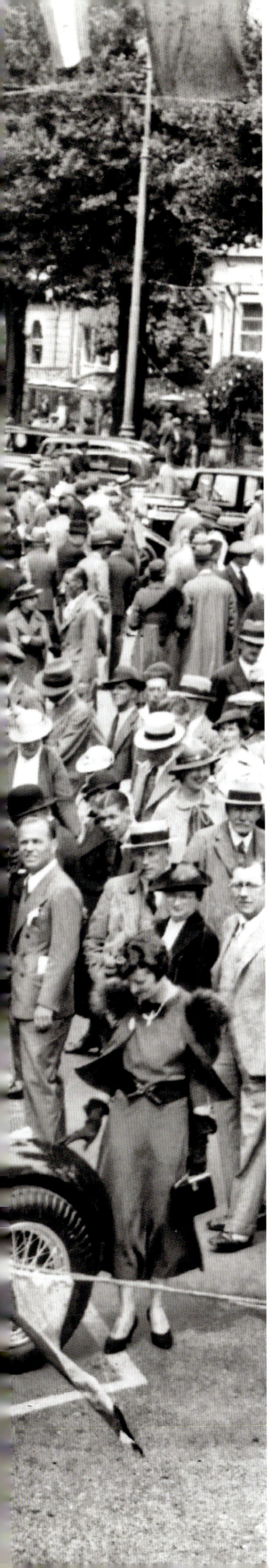

## Automotive Elegance and Anglo-French Relations

La Carrosserie, July 1936

*The Concours d'Elegance Automobile is becoming increasingly popular in England, and its organization is truly impressive. These events are far from being mere parades or light-hearted spectacles where attention drifts away from the cars themselves. On the contrary, there are serious competitions, where prizes are awarded to the most deserving vehicles, all within a refined and exclusive atmosphere. These events play an important role in promoting the prosperity and advancement of the coachbuilding industry.*

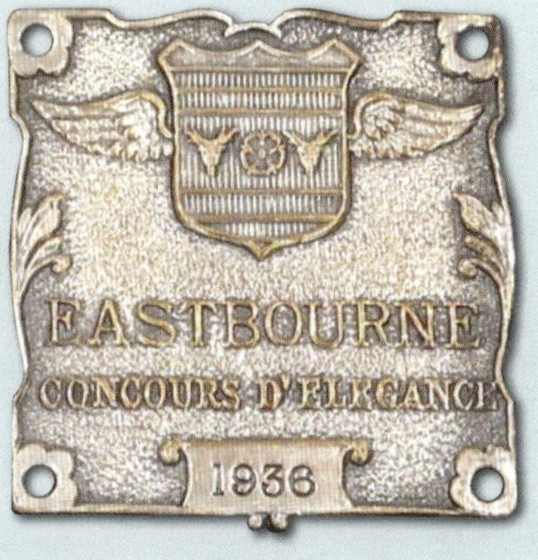


*Such was the case at this year's tenth edition of the Concours d'Elegance Automobile in Eastbourne, which we had the pleasure of attending.*

*Over 150 cars competed, divided into 29 categories: seven categories for standard cars based on catalogue prices, five for sports cars, six for the most distinguished vehicles, two for the most beautiful car presented by a lady, one for the best overall ensemble, and so on.*

*Each category was judged by two officials, who scrutinized every detail with meticulous care. One judge would examine the undersides of the fenders, while the other would assess the placement of tools, with both receiving thorough explanations about the car's construction. Policemen ensured the crowd stayed back, allowing the judges to fully focus on their evaluations.*

*Under the expert leadership of Captain Short and with the Mayoress of Eastbourne presiding, proceedings began at ten in the morning and concluded around five in the afternoon.*

*A French team, including Baroness de Fontanges, Paul Cartier, and Madame Richer-Delavau, courageously faced the discerning English judges with Delages featuring bodywork by De Villars, Letourneur et Marchand, and Figoni. Their efforts were well worth it, as these Delages earned great acclaim and honor.*

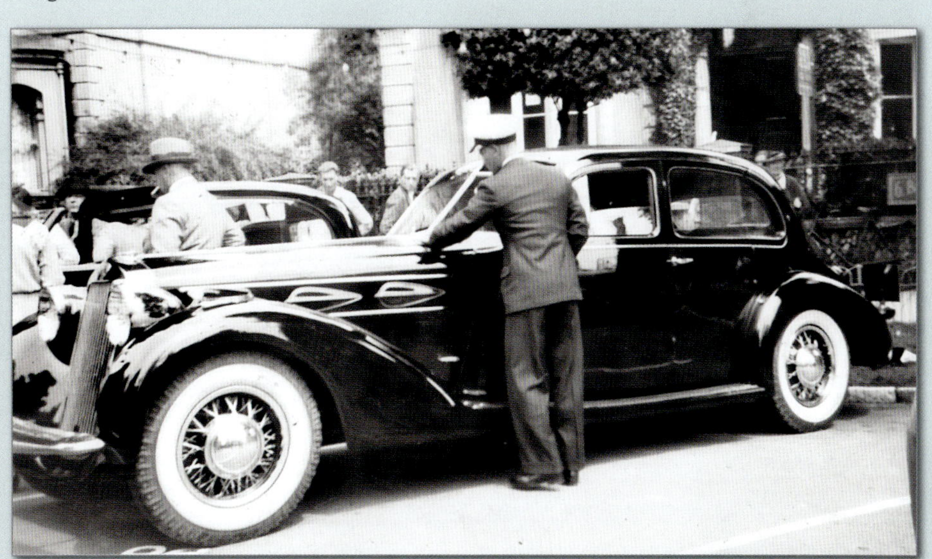

**ABOVE:** Delages seem to be everywhere – this is a D8 100 (bodied by Latourneur et Marchand) at the 1936 Eastbourne event. (Daniel Cabart Archives)

**LEFT:** *In Eastbourne, United Kingdom, in 1936, the team of six Delage cars from France was a sensation, capturing most of the trophies. The press extensively covered their victories, highlighting the harmony between the cars' colors and the outfits worn by the teams.* (Daniel Cabart Archives)

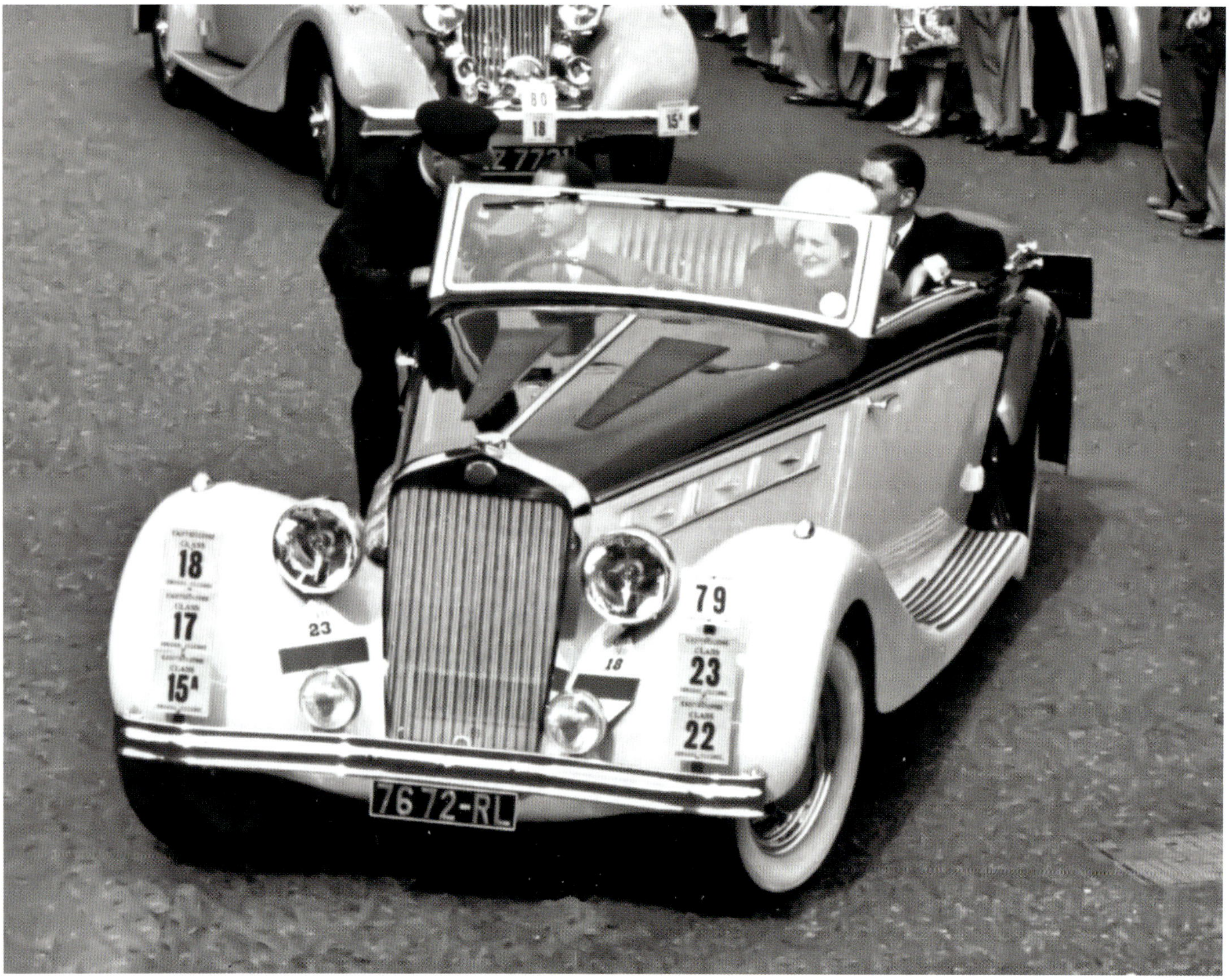

**ABOVE:** *For the 1937 concours d'elegance at Eastbourne, Madame de Laborderie presented a Delage D8-120 by De Villars.* (MW Archives)

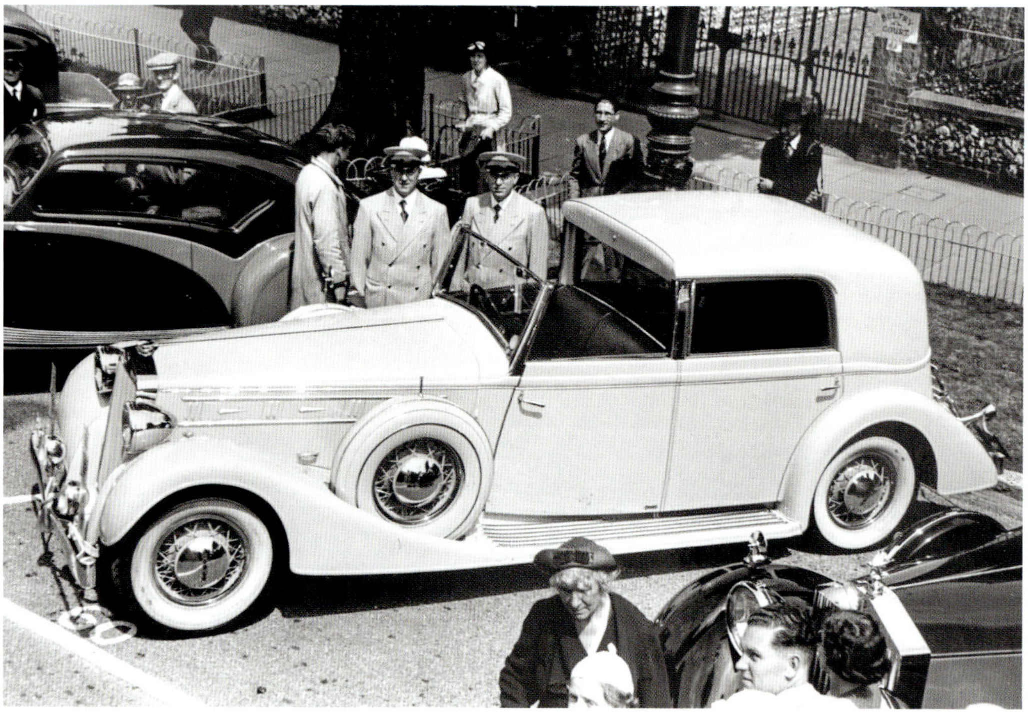

**RIGHT:** *Ramsgate was a very important concours too. This is from 1937 when a Franay-bodied Delage coupe de ville was the cynosure of all eyes.* (Classic & Sportscar)

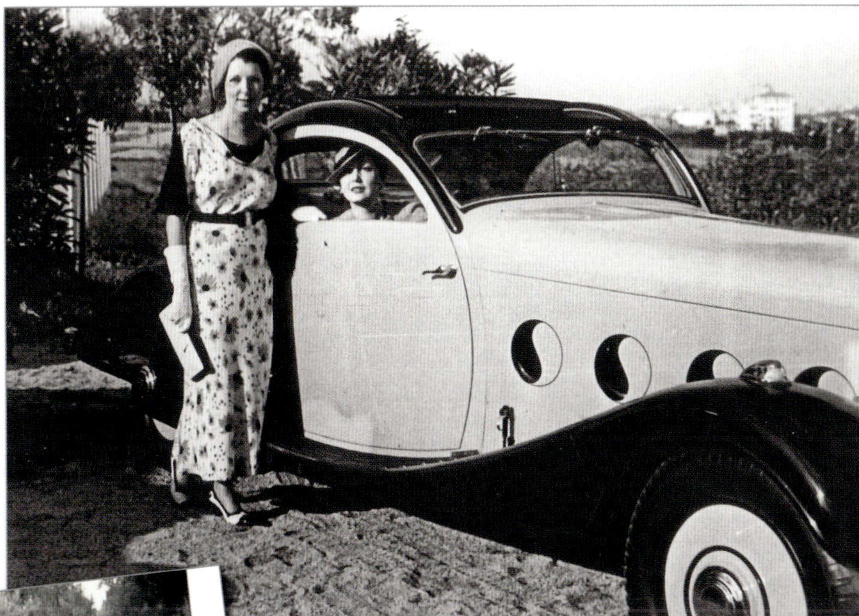

**THIS PAGE:** *Barcelona was another important concours d'elegance, and it was held every year at the Polo Jockey Club. The concours in 1934 was on June 27, and here is the famous Letourneur et Marchand-bodied Delage D8-15 "Aérosport" participating that day.* (Daniel Cabart Archives)

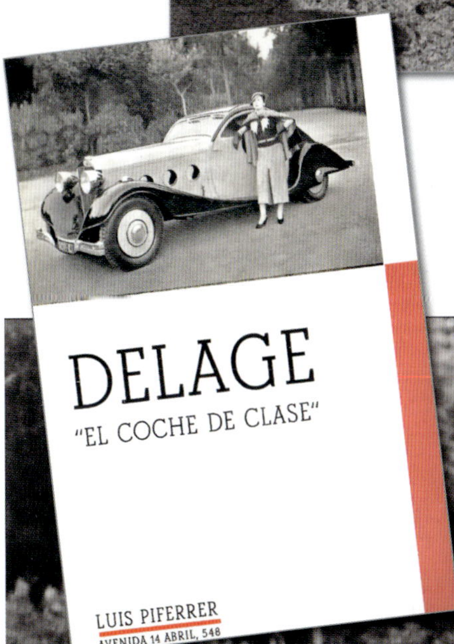


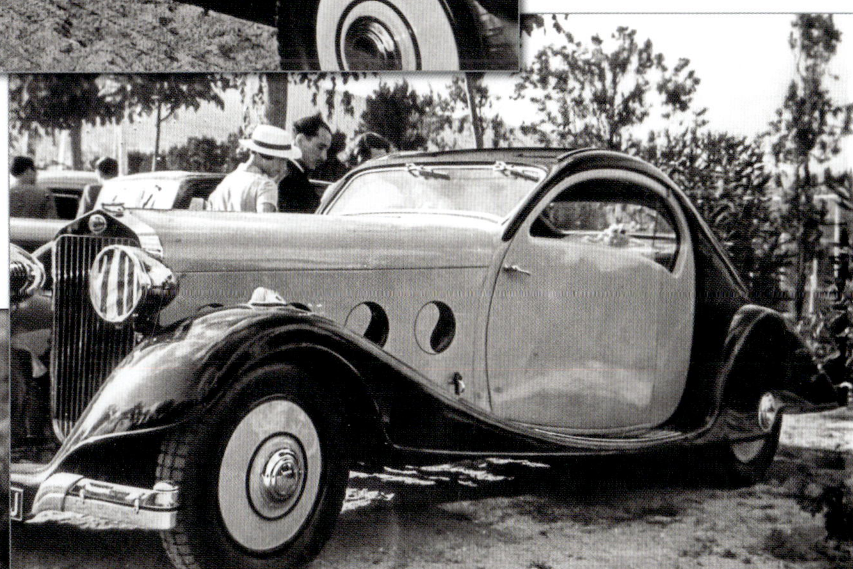

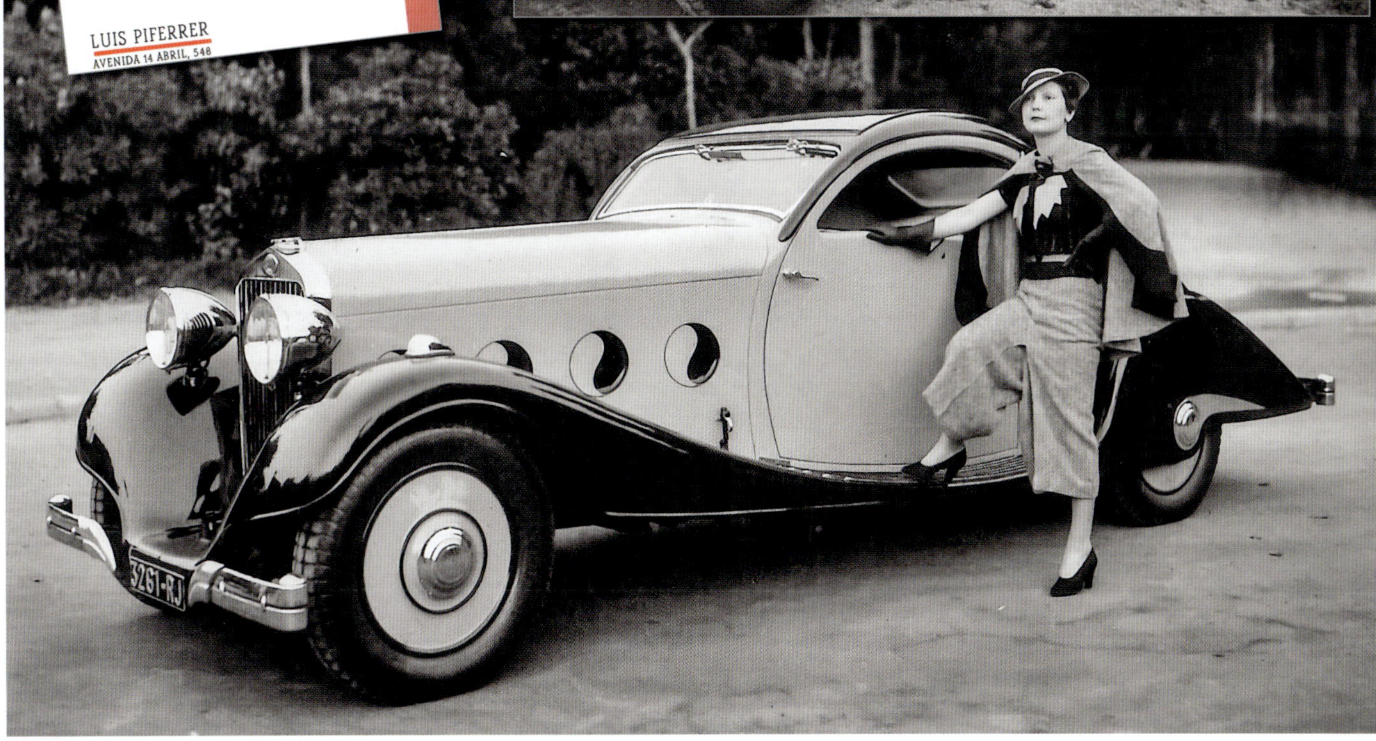

## Germany and Austria

Just as the English took concours d'elegance events seriously, so did the Germans (in Schwetzingen) and the Austrians, with the concept gaining traction in these countries by the 1930s. While Vienna initially led the way in Austria with a concours d'elegance event in 1934 – where an Austrian car, a 1934 Gräf und Stift, outshone a Citroën Rosalie 15 CV (also built under license by Gräf und Stift) – it was the historic city of Salzburg that established the concept, beginning in 1936.

An article in the September 1936 issue of *Allgemeine Automobil-Zeitung* best captured the enthusiasm of Austrian automotive enthusiasts: *"Anyone who saw Salzburg this year during the Festival Weeks, who strolled across the Residenzplatz, sat at Tomaselli's or in the 'Bazar' or in the Café Glockenspiel, felt tempted to do some tourism statistics: out of a hundred cars that rolled through the streets and parked in the squares, at least ninety were foreigners, mainly English and Czechoslovakians, but also frequently French, Belgians, Yugoslavs, Italians, Americans, Romanians, Egyptians. It was a colorful, cosmopolitan picture. Every day, starting in the early morning hours and ending late after midnight, a concours d'elegance developed in the center of the city as if by itself, which suggested the idea of organizing a real beauty contest for automobiles, which would give the many motorists from home and abroad who were present the opportunity to show off their cars in a suitably festive setting and to make them available for a comparison."*

*"The Salzburg Automobile Club put the idea into practice and found the perfect stage for this spectacular event,"* the article continued. *"It was the magnificent castle park in Kleßheim where the concours d'elegance was held. An Italian blue sky stretched over the scene, with the magnificent baroque façade of Kleßheim Castle serving as the backdrop. A select audience filled the seats."*

Despite a display of stunning cars from manufacturers like Austro-Daimler, Steyr, Tatra, Lancia, Škoda, Aero, Gloria-Triumph, Alvis, Talbot, Packard, Nash, Chevrolet, Hispano-Suiza, Oldsmobile, and Delage – which *Allgemeine Automobil-Zeitung* described as *"a rich display of beautiful cars, making it difficult for the judges to determine the most beautiful"* – the overall winner was a Mercedes-Benz 540K.

*Even Rijswijk in The Netherlands had a concours as evidenced by this advertisement for Delage – it must have been one of the last before the war, as it was on June 24, 1939. (Daniel Cabart Archives)*

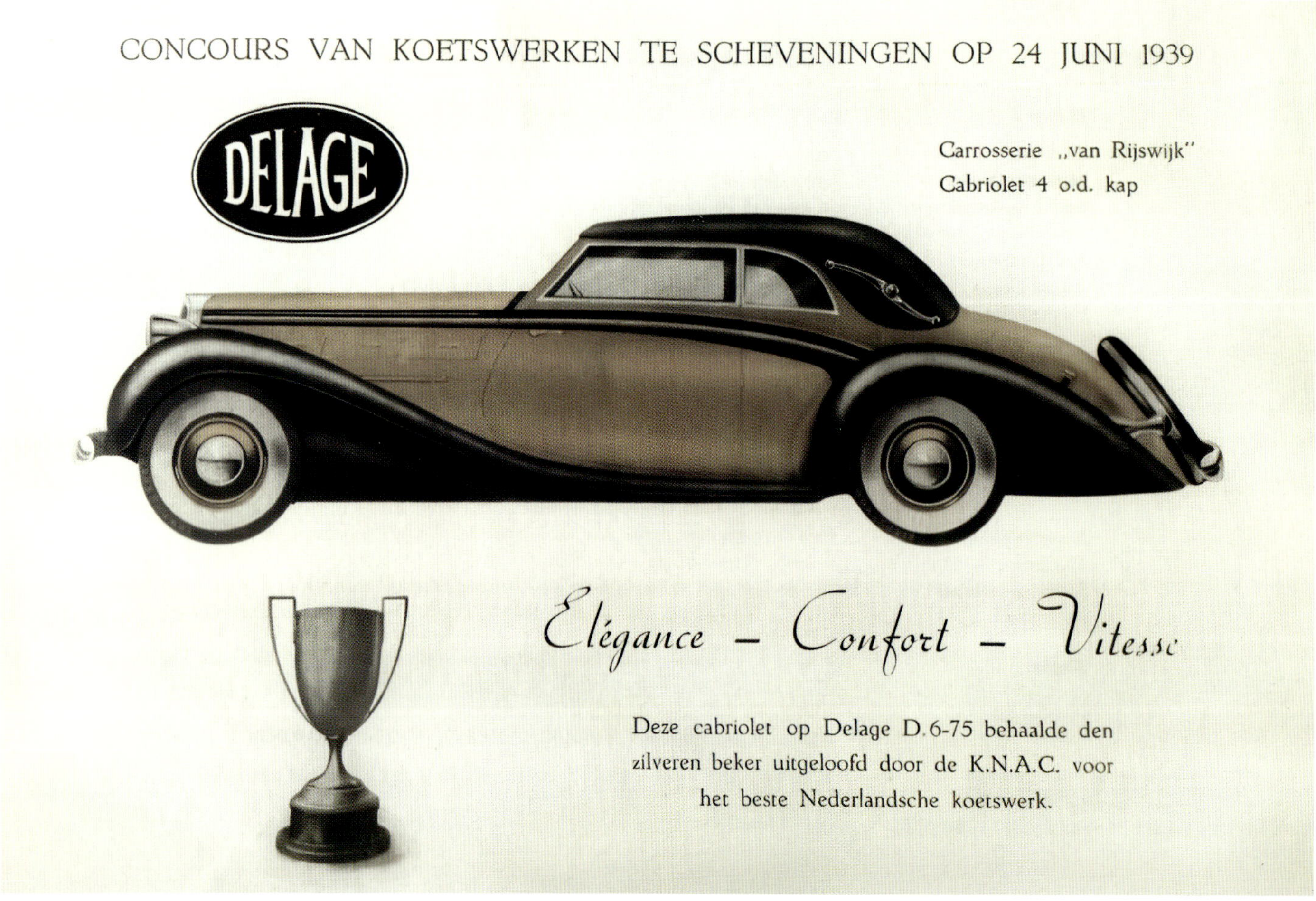

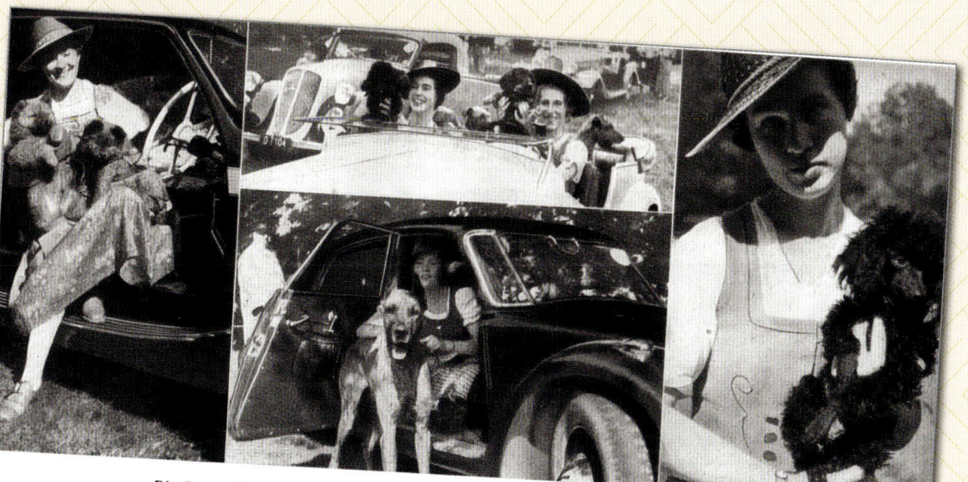

**Die Dame, ihr Wagen, ihr Hund auf dem Concours d'Elégance in Salzburg.**

Links: Mme. Mary Costes (Frankreich), auf Delage, mit ihren jungen Chow-Chow. — Mitte oben: Gräfin Dubsky (C. S. R.), auf Skoda, mit ihren Hunden. — Mitte unten: Kammersängerin Jarmila Novotna, auf Tatra-Achtzylinder, mit einer prächtigen, gesträhnten Dogge. — Rechts: Gräfin Dubsky mit dem kleinsten Pudel der Welt, genannt »Fratze.

roten Frack auf dem Festplatz erschien und gleichermaßen durch die Schönheit der Pferde wie durch die Eleganz der Reiter gefiel. Schließlich überflogen drei von der Gaisbergspitze gestartete Segelflugzeuge, mit den Piloten Dr. Aigner (Fliegergruppe Salzburg), Flugschüler Wei Chao ...

Wertungsgruppe Sport- und offene Wagen unter 2000 ccm Zylinderinhalt. 1. Preis: Goldenes Band, Start-Nr. 20, Susanne Langnese (Schweiz), Gloria-Triumph, England, 92 Punkte. 2. Preis: Silbernes Band, Start-Nr. 24, Prinz. anl. Taxis (Aigen bei Salzburg), Tatra 57, C. S. R., 76 Punkte. 3. Preis: Start-Nr. 25, Eugen Graf Dubsky (C. S. R.), Skoda, C. S. R., 72 Punkte.

Wertungsgruppe Sport- und offene Wagen über 2000 ccm Zylinderinhalt. 1. Preis: Goldenes Band, Start-Nr. 44, Annemarie Sörrensen (London), Mercedes, Deutschland, 93 Punkte. 2. Preis: Silbernes Band, Start-Nr. 39, Jill Ehlert (Argentinien), Mercedes, Deutschland, 88 Punkte. 2. a Preis: Silbernes Band, Start-Nr. 47, Mrs. Löw (U. S. A.), Oldsmobil, U. S. A., 88. Punkte. 3. Preis: Start-Nr. 46, Fr. Friedheim (Schweiz), Hispano, Schweiz, 78 Punkte.

Außerdem gelangten folgende Zusatzpreise zur Verteilung: Preis des Herrn Landeshauptmannes für den weitesten Wagen: Start-Nr. 47, Mrs. Löw (U. S. A.), Oldsmobil. Preis des amerikanischen Gesandten in Wien für den bestprämiierten Wagen englischer Erzeugung: Start-Nr. 47, Mrs. Löw (U. S. A.), Oldsmobil, U. S. A. Preis des englischen Gesandten in Wien für den bestprämiierten Wagen englischer Erzeugung: Start-Nr. 20, Susanne Langnese (Schweiz), Gloria-Triumph, England. Preis des Präsidenten österreichischer Automobil-Industrieller, Exzellenz Graf Alexander van der Straten-Ponthoz, für den bestprämiierten Wagen österreichischer Erzeugung: Start-Nr. 8, Direktor Hauser (Wien), Austro Daimler, Oesterreich. Preis des Landesverkehrsamtes Salzburg für den bestprämiierten Wagen der Wertungsgruppe IV: Start-Nr. 44, Annemarie Sörrensen (London), Mercedes-Benz (Deutschland).

Die Preisverteilung nahm das Präsidium des S. A. C. vor.

Eine gemeinsame Einfahrt der konkurrierenden Automobile in die Stadt beschloß die organisatorisch durchwegs einwandfreie Veranstaltung.









## Concours d'Elégance in Salzburg.

SALZBURG —! Kaum je zuvor trat seine internationale Geltung deutlicher in Erscheinung als in diesem Sommer, der der Stadt Mozarts einen Fremdenzustrom brachte wie noch nie. Wer heuer während der Festspielwochen Salzburg sah, wer über den Residenzplatz spazierte, bei Tomaselli oder im „Bazar“ oder im Café „Glockenspiel“ saß, der fühlte sich versucht, Fremdenverkehrsstatistik zu treiben. Von hundert Autos, die hier durch die Straßen rollten und auf den Plätzen parkten, waren bestimmt neunzig Ausländer, vor allem Engländer und Tschechoslowaken, dann aber auch sehr häufig Franzosen, Belgier, Jugoslawen, Italiener, Amerikaner, Rumänen, Aegypter ... Es war ein buntes, kosmopolitisches Bild. Tagtäglich, beginnend in den frühen Vormittagsstunden und endend spät nach Mitternacht, entwickelte sich im Zentrum der Stadt wie von selbst ein Concours d'Elégance, der den Gedanken nahelegte, wirklich einen Schönheitswettbewerb der Autos zu organisieren, der den vielen anwesenden Automobilisten und Automobilistinnen des Aus- und Inlandes Gelegenheit gäbe, ihre Wagen in einem entsprechend festlichen Rahmen zu zeigen und einem wertenden Vergleich zur Verfügung zu stellen.

Der Salzburger Automobil-Club hat den Gedanken in die Tat umgesetzt und fand für dieses sehenswerte Schauspiel auch die passende Bühne. Es war der herrliche Schloßpark zu Kleßheim, in dem der Concours d'Elégance zur Durchführung kam. Ein italienisch blauer Himmel spannte sich über die Szene, deren Hintergrund die prächtige Barockfassade des Kleßheimer Schlosses bildete. Ein erlesenes Publikum füllte die Plätze. Besonders warm begrüßt, erschien auch der Landeshauptmann Salzburgs, Doktor Rehrl, von den Vizepräsidenten des S. A. C., Kommerzialrat Klein und Kommerzialrat Schurich, empfangen. Fahnen und Blumen, Toiletten und Wagen verliehen dem Bild lebhafte Farben, die sich zu festlicher Harmonie verbanden.

Wenige Minuten nach 15 Uhr kündigten Fanfaren den Beginn der Schönheitskonkurrenz an. Es waren viele Meldungen eingelaufen. Man sah Austro Daimler, Steyr, Mercedes-Benz, Tatra, Lancia, Skoda, Aero, Gloria-Triumph, Alvis, Talbot, Packard, Nash, Chevraulet„ Hispano-Suiza, Oldsmobil, Delage — eine reiche Schau schöner Wagen, die es den Preisrichtern sehr schwer machte, die schönsten Wagen zu ermitteln.

Man hatte die Fahrzeuge in drei Wertungsgruppen geteilt: Limousinen und Innenlenker über 2000 ccm, Sport- und offene Wagen unter 2000 ccm, und schließlich Sport- und offene Wagen über 2000 ccm. Es gab demnach drei erste Preise zu gewinnen, drei „Goldene Bänder“, die den schönsten Wagen in den einzelnen Wertungsgruppen zugedacht waren. Es gab aber nicht nur eine Schau schöner Wagen zu sehen, sondern auch eine Modeschau, deren beherrschendes Motiv die Landestracht war. Und außerdem und überdies gab es noch eine Parade von Rassehunden aller Arten, so daß also das Publikum reichlich auf seine Kosten kam. Damit war das Programm noch keineswegs erschöpft. In einer Pause, in der sich das Preisrichterkollegium zurückgezogen hatte, um sein Urteil über die zum Bewerb gebrachten Autos zu bilden, wurde das Publikum durch eine malerische Gruppe von Reitern und Reiterinnen überrascht, die im

**Vom Concours d'Elégance in Salzburg.**

Oben: Direktor Eduard Hauser und Kammersängerin Rinner mit dem Austro Daimler-ADR-Achtzylinder, Sieger in der Gruppe II über 2000 ccm. Er erhielt außer dem Goldenen Band den Preis Seiner Exzellenz Alexander Graf van der Straten-Ponthoz für den bestplacierten Wagen österreichischer Erzeugung. — Links unten: Miß Annemarie Sörrensen, London, mit ihrem Mercedes-Benz, die außer dem Goldenen Band auch den Ehrenpreis des Landesverkehrsamtes Salzburg errungen hat. — Rechts unten: Baronin Susanne Langnese, Schweiz, auf Gloria-Triumph, die gleichfalls ein Goldenes Band erwarb; außerdem wurde ihr der Preis des englischen Gesandten in Wien für den bestprämiierten Wagen engl. Erzeugung zuerkannt.

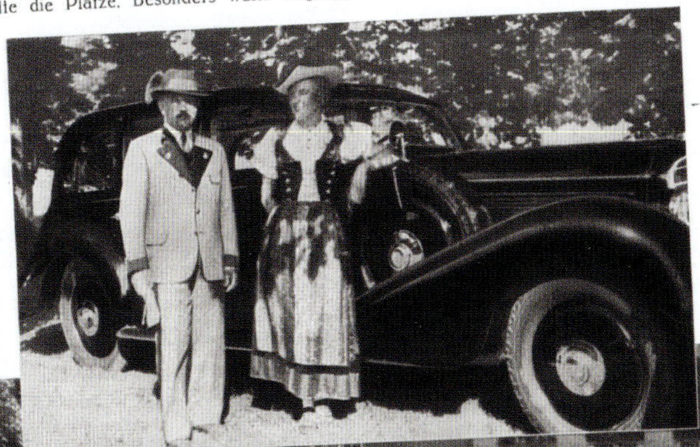

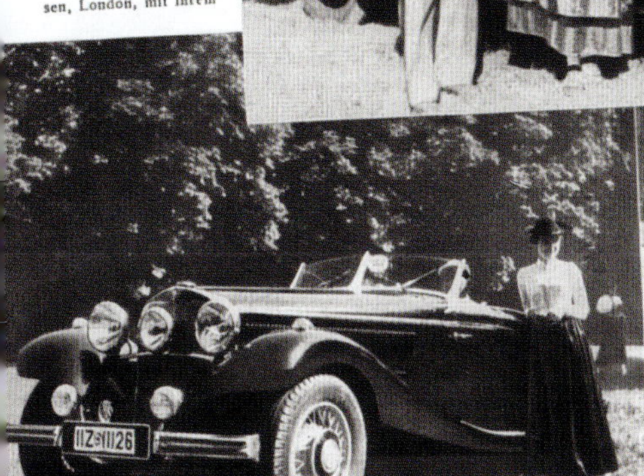



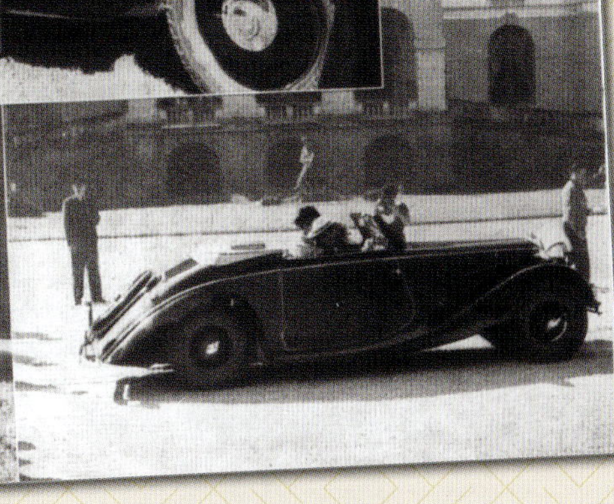

Salzburg, in Austria, was another popular Concours d'Elegance. Here is a report on the 1936 concours in the September 1936 issue of the Allgemeine Automobil-Zeitung. (Daniel Cabart Archives)

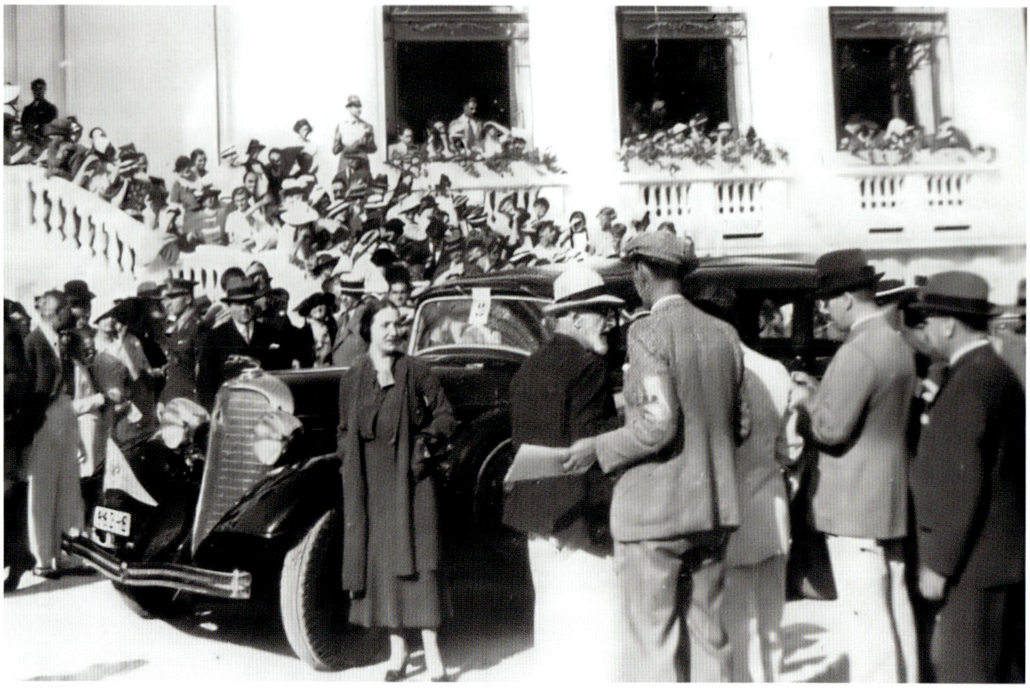

*Even Imperial Romania took to concours in a big way – this is the 1934 Concurs de Confort și Eleganță, with Madame Michelsohn in a Lincoln, which won the second place in the limousine class. (Retromobil Club Romania)*

## Even the Romanians

One country that embraced the concept of concours d'elegance in the French-style earlier than many others was imperial Romania, with the first such event taking place in the historic town of Timișoara in 1930. This event, referred to as a comfort and elegance competition (Concurs de Confort și Eleganță), featured a lineup of eleven cars, including a Mercedes-Benz, a Buick, a Chevrolet, a Czech Praga, a Horch, a Willys Knight, two Steyrs, and a Plymouth. The winning car was a Chrysler, owned by Ghica Mentiuni, with the Praga finishing as the runner-up.

Notably, Prince Nicolae, who was then serving as regent, attended this inaugural concours in Timișoara. While it's unclear whether his presence influenced the subsequent organization of the first official Elegance and Comfort Competition (Concurs de Confort și Eleganță) on August 11-12, 1934, under the patronage of King Carol II, the king's summer palace in the mountain resort of Sinaia became the epicenter of Romania's automotive celebrations during this period.

The interwar years were a golden era for the automobile in Romania. Romanian racing drivers made a name for themselves on the international stage. Petre Cristea achieved memorable victories at the Monte Carlo Rally in 1936 with a Ford V8 and at the 1939 Nürburgring in a BMW 328. Romania also hosted significant motorsport events, such as the Feleac hill climb and the International Grands Prix in Brașov (1935) and Bucharest (1936), where famed drivers like Hans Stuck competed in an Auto Union Grand Prix car. Romanian engineers, including Aurel Persu, pioneered automotive aerodynamics, and in 1936, Ford even opened an assembly plant in Bucharest to begin production of the Ford V8.

In this context, the Sinaia Elegance Contest, or Concurs de Confort și Eleganță, was first organized in 1934 by the Royal Automobile Club of Romania (ACRR), which itself was founded as far back as 1904 (then the ACR). The competition featured four classes, and three of the four winners in 1934 were women: Lilette Butculescu in a Chrysler Imperial in the tourer category, Roxana Soutza in a Delage in the closed body category, and Marie Louise Costinescu in a Chrysler in the aerodynamic cars category. The fourth winner was Ion Matak in a Rolls-Royce in the convertible category.

The success of the 1934 concours led the Royal Automobile Club of Romania to organize an Automobile Week in 1935, held from September 3-8. The week's activities included the Sinaia Rally, the Carpathian Cup, the Coastal Race, and the Elegance and Comfort Contest. The second occurrence of the Elegance and Comfort Competition (Concurs de Confort și Eleganță) took place on September 7, 1935. That evening, a grand ball was held at the Palace Hotel, where a jury selected by the ACRR awarded prizes to the most elegant and comfortable cars among the 22 participants.

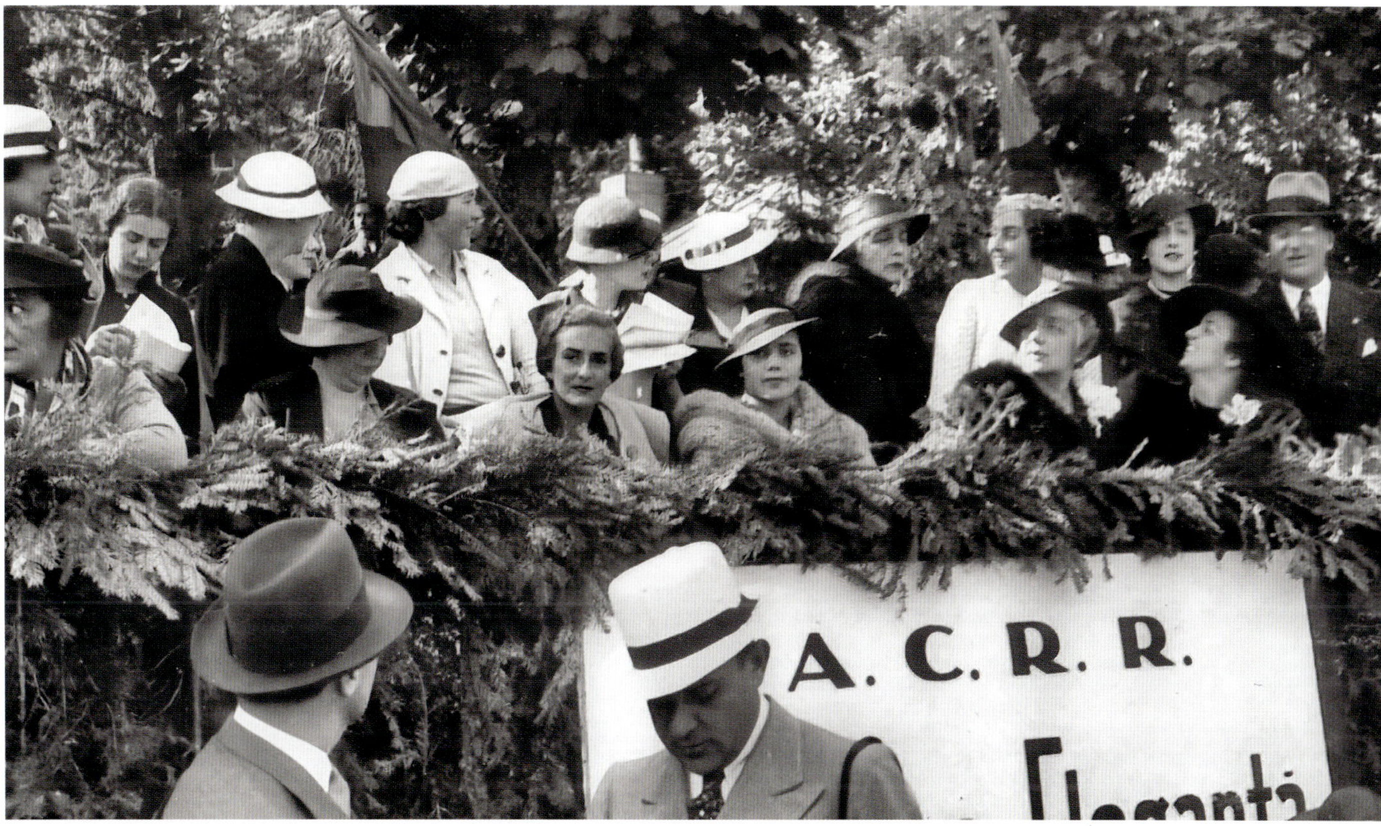


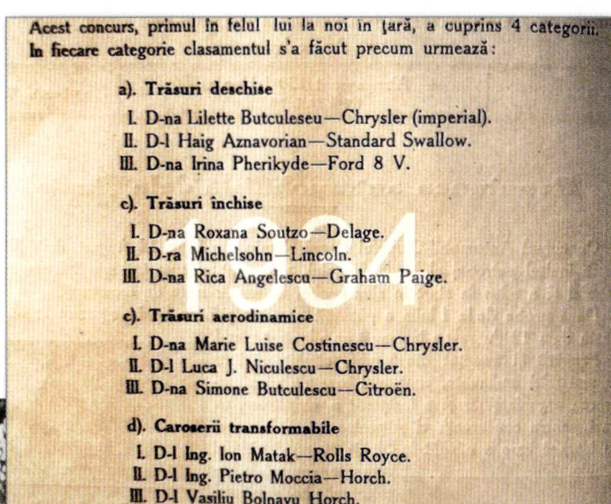
Acest concurs, primul în felul lui la noi în ţară, a cuprins 4 categorii.
In fiecare categorie clasamentul s'a făcut precum urmează:

a). **Trăsuri deschise**

I. D-na Lilette Butculeseu—Chrysler (imperial).
II. D-l Haig Aznavorian—Standard Swallow.
III. D-na Irina Pherikyde—Ford 8 V.

c). **Trăsuri închise**

I. D-na Roxana Soutzo—Delage.
II. D-ra Michelsohn—Lincoln.
III. D-na Rica Angelescu—Graham Paige.

c). **Trăsuri aerodinamice**

I. D-na Marie Luise Costinescu—Chrysler.
II. D-l Luca J. Niculescu—Chrysler.
III. D-na Simone Butculescu—Citroën.

d). **Caroserii transformabile**

I. D-l Ing. Ion Matak—Rolls Royce.
II. D-l Ing. Pietro Moccia—Horch.
III. D-l Vasiliu Bolnavu Horch.



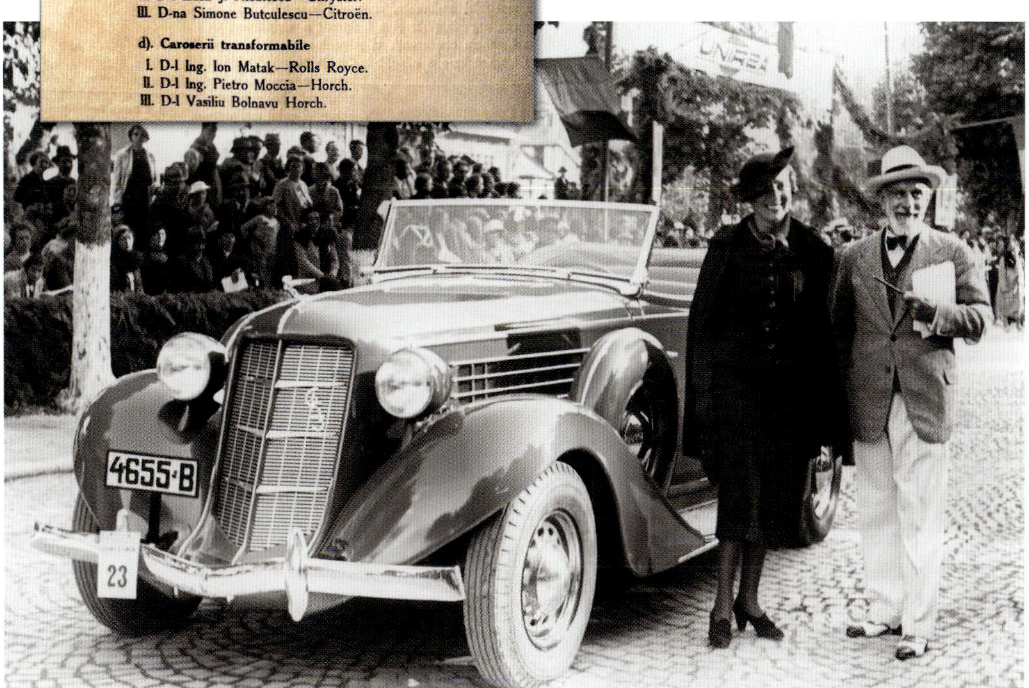


**ABOVE:** *A striking group of ladies with apparent ready access to milliners.* (Retromobil Club Romania)

**FAR LEFT INSET:** *The winners at Sinaia in 1934.* (Retromobil Club Romania)

**BOTTOM:** *Also, from the* 1934 Concurs de Confort și Eleganță, *a participant with the President of the ACRR, the Royal Automobile Club of Romania, Ion Mitilineu.* (Retromobil Club Romania)

**RIGHT (BOTH):** *The cover of the brochure and part of the entry list for the 1935 Sinaia concours. (Retromobil Club Romania)*

**BOTTOM:** *At the 1935 Sinaia concours, Marin Butculescu and his wife presented their Buick, distinguished by its special bodywork. (Retromobil Club Romania)*

**OPPOSITE:** *At the 1935 Sinaia concours, the Horch 850 Sport with Mony Magheru won Best of Show. The owner was Radu Georghe Gheorghiu, a colonel then and a general later. (Retromobil Club Romania)*

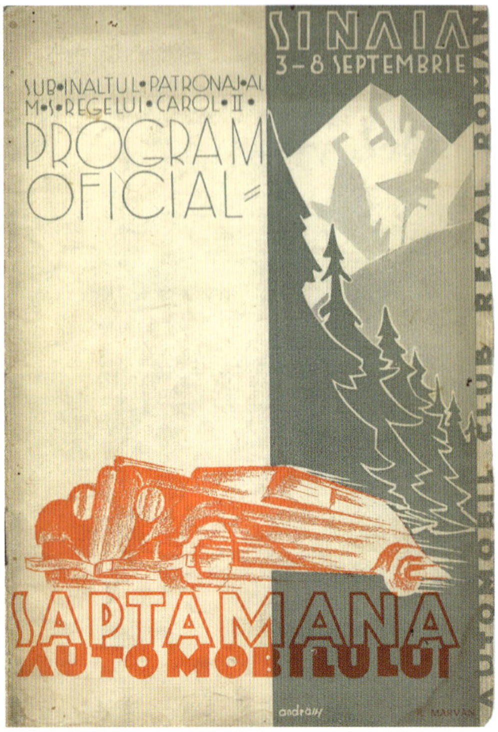


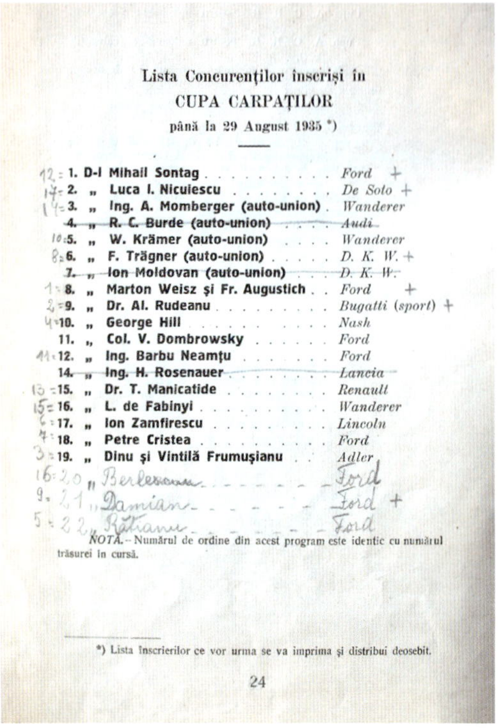

Lista Concurenţilor înscrişi în
CUPA CARPAŢILOR
până la 29 August 1935 *)

1. D-l Mihail Sontag . . . . . . . . . . *Ford*
2. „ Luca I. Niculescu . . . . . . . . *De Soto*
3. „ Ing. A. Momberger (auto-union) . *Wanderer*
4. „ R. C. Burde (auto-union) . . . . . *Audi*
5. „ W. Krämer (auto-union) . . . . *Wanderer*
6. „ F. Trägner (auto-union) . . . . . *D. K. W.*
7. „ Ion Moldovan (auto-union) . . . . *D. K. W.*
8. „ Marton Weisz şi Fr. Augustich . . *Ford*
9. „ Dr. Al. Rudeanu . . . . . . . . . *Bugatti (sport)*
10. „ George Hill . . . . . . . . . *Nash*
11. „ Col. V. Dombrowsky . . . . . . *Ford*
12. „ Ing. Barbu Neamţu . . . . . . . *Ford*
14. „ Ing. H. Rosenauer . . . . . . . . *Lancia*
15. „ Dr. T. Manicatide . . . . . . . . *Renault*
16. „ L. de Fabinyi . . . . . . . . . *Wanderer*
17. „ Ion Zamfirescu . . . . . . . . . *Lincoln*
18. „ Petre Cristea . . . . . . . . . . *Ford*
19. „ Dinu şi Vintilă Frumuşianu . . *Adler*

*NOTA.* – Numărul de ordine din acest program este identic cu numărul trăsurei în cursă.

*) Lista Inscrierilor ce vor urma se va imprima şi distribui deosebit.

24

A special ranking of elegance was created for the ladies, who were required to "harmonize their outfits, by line and color, with the car presented," according to Bogdan Coconoiu, who is the Vice President of Retromobil Club Romania, the historic vehicle federation in the country. "Prizes included a Grand Honorary Award, two gold Grand Awards of Excellence, a silver Award of Excellence, class prizes, and a special award for ladies' elegance."

In its September 12, 1935, issue, the Romanian magazine MOTOR described the event: "*The elegance contest was a splendid affair of luxury, beautiful colors, and charming bodies, complemented by a true beauty contest. It was a day of great glory for Auto Union, which won the Grand Honorary Award and the Prize of*

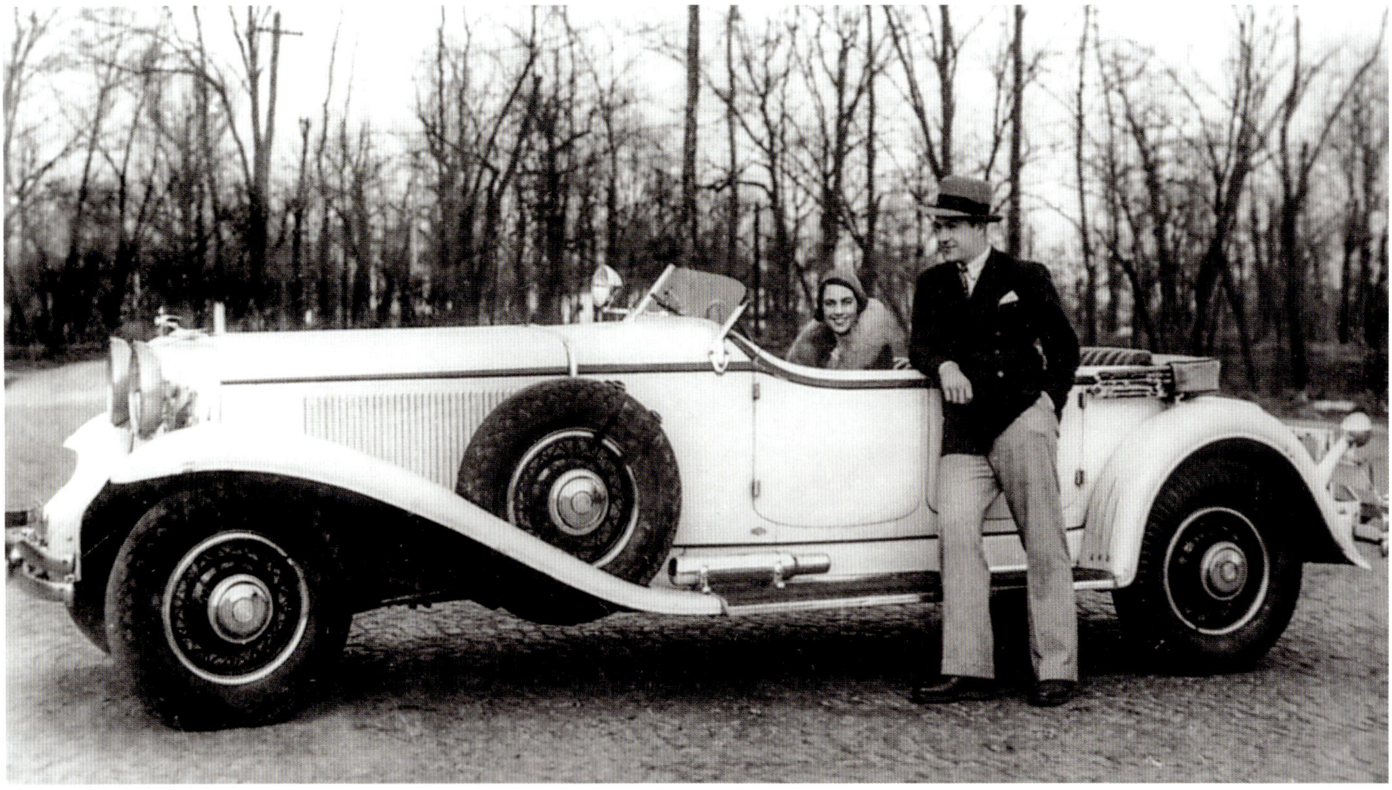

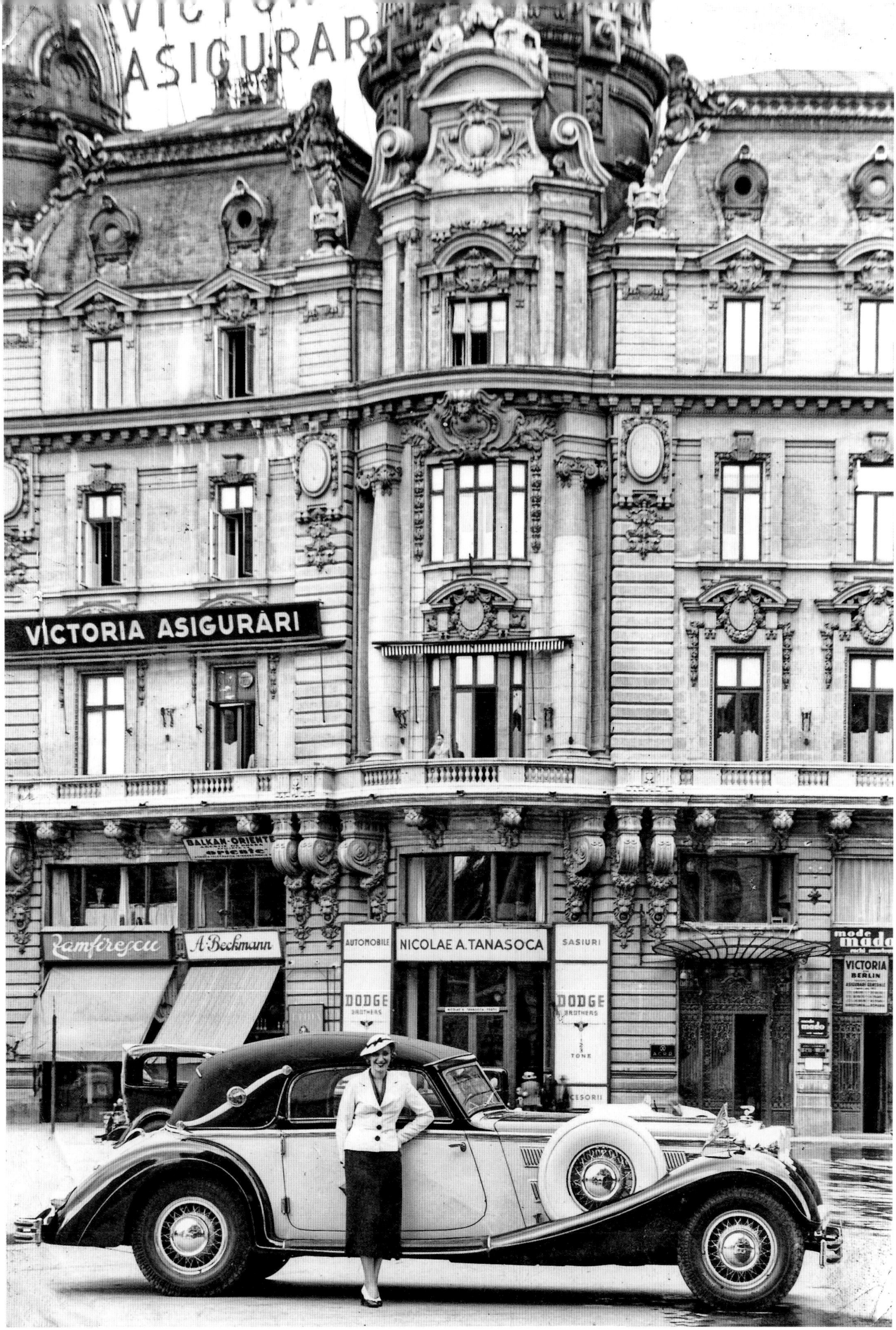
VICTORIA ASIGURARI
BALKAN-ORIENTE
Zamfirescu
A. Beckmann
AUTOMOBILE
NICOLAE A. TANASOCA
SASIURI
DODGE BROTHERS
DODGE BROTHERS
VICTORIA
BERLIN
mode mado

## THE OLDEST ITALIAN CONCORSO D'ELEGANZA

The Concorso di Eleganza per Automobili-Coppa Automobile Club Torino holds the distinction of being Italy's oldest automobile elegance competition, with its inaugural event taking place on May 21, 1923.

The debut event surpassed all expectations, prompting the Automobile Club Torino, in collaboration with Pro Torino and the A.I.A., to organize a second installment from June 11 to 15 the following year. Hosted at Valentino Park, the competition drew significant attention, even garnering the presence of His Majesty Vittorio Emanuele III, and featured 45 participating vehicles. In Category A, a Bugatti with a custom body by Gotteland & Mossotti, owned by Duke Francesco Salviati, won its class. Category B saw a Lancia Trikappa by Farina triumph, while in Category C, Farina's coachwork for the Fiat 519 owned by engineer Alcide Bona took the top spot. In Category E, an Isotta Fraschini with Castagna bodywork, owned by Baj Eggeling, was declared the winner.

After a four-year hiatus, the third event was held on November 1, 1928, featuring 160 participants split into four classes: cars up to 1500 cc with custom bodywork, vehicles over 1500 cc with custom bodywork, production cars with manufacturer-supplied bodywork, and buses or coaches designed for public or tourist transport. The event continued to thrive annually, drawing considerable crowds and a high number of participants, until it was suspended during World War II.

*(Automobile Club Torino)*

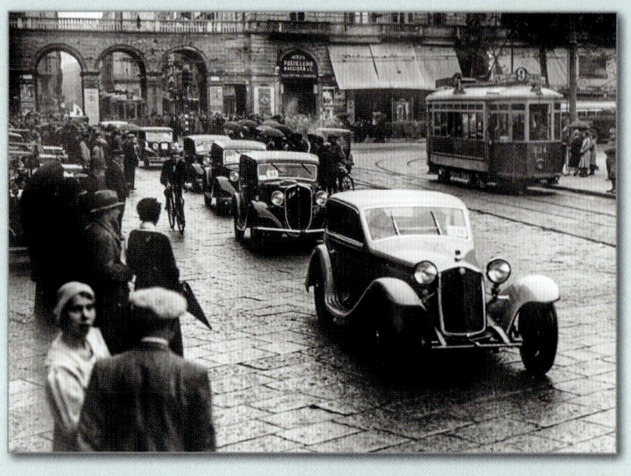

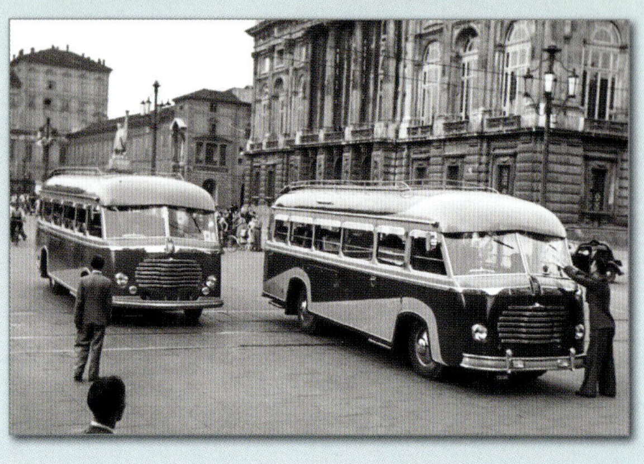

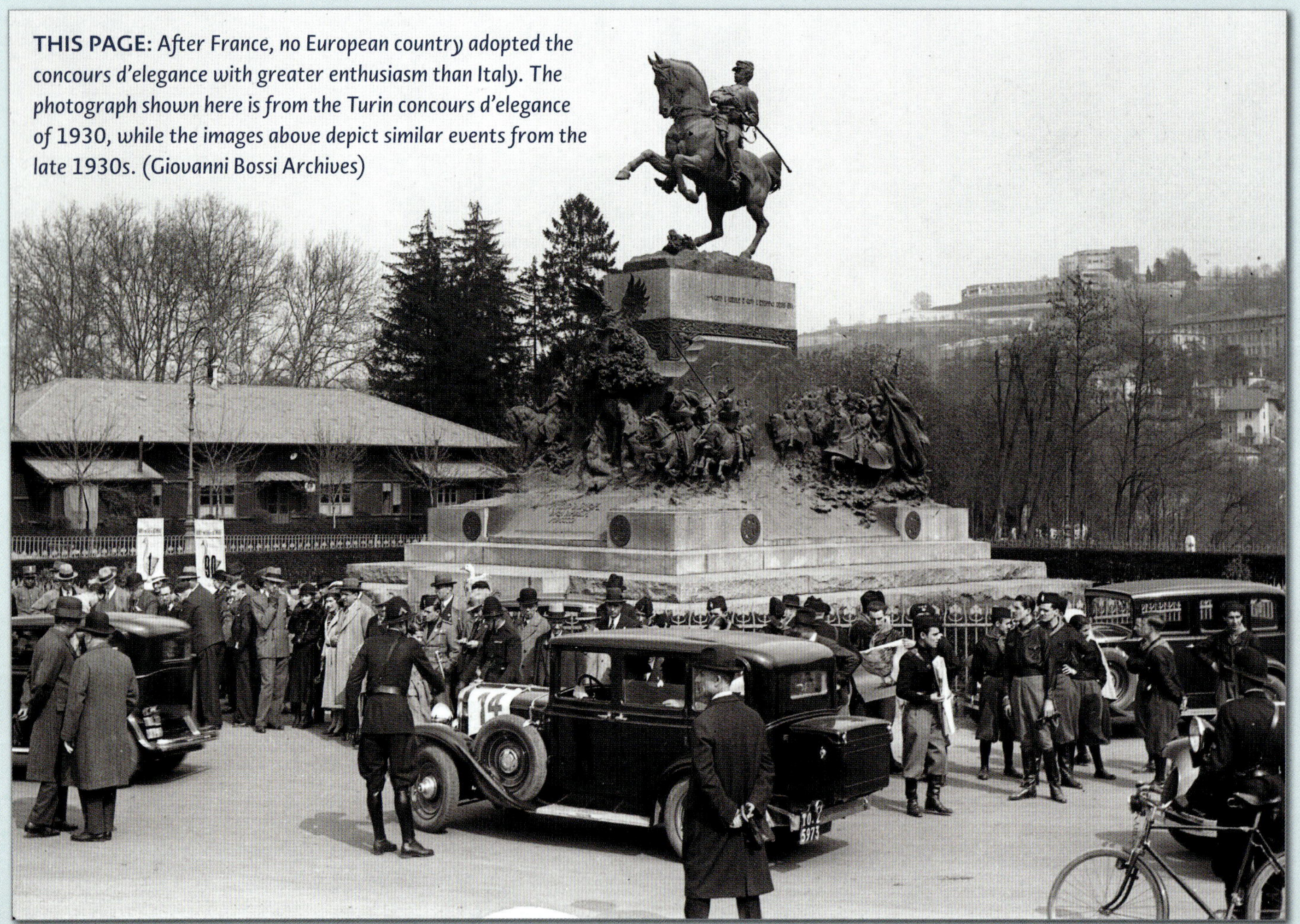

**THIS PAGE:** *After France, no European country adopted the concours d'elegance with greater enthusiasm than Italy. The photograph shown here is from the Turin concours d'elegance of 1930, while the images above depict similar events from the late 1930s. (Giovanni Bossi Archives)*

*Excellence, along with numerous other prizes."* The overall winner was a Horch 850 Sport (Horch being part of Auto Union), presented by Madame Mony Magheru.

Italy was another European country that embraced the French concept of the concours d'elegance and made it its own. As Adolfo Orsi mentioned earlier, not only did Italy debut the concept in Turin, but it quickly spread to Milan by 1926, followed by Rome in 1927, with a series of competitions that continued well into the late 1930s.

This is supported by an article from the June 30, 1923, issue of the Italian motoring magazine *Auto Italia*, shared by Italian automotive historian Giovanni Bossi. The article describes a concorso d'eleganza with the following words: "*Thanks to various technical initiatives and careful attention to tourism, a concours d'elegance event for cars was recently held. The timing could not have been better. With the support of the prefect and the mayor of Turin, a special commission diligently organized a competition that rivaled those held annually abroad, particularly in Nice and Monte Carlo.*"

The article goes on to emphasize Turin's unique ability to host such an event, stating: "*Turin is the only city in Italy capable of organizing this type of competition, as the automobile industry thrives here, and the top-rated coachbuilders have kept pace with this growth, earning global recognition.*" The event was divided into multiple categories and held in Valentino Park, a key venue for major celebrations during that time.

It concludes by noting the success of the competition: "*Just a few months after the one in Monaco, the concours d'Elegance in Italy's automobile capital was a resounding success.*"

Not only Bossi, but also historians Fabrizio Taiana and Adolfo Orsi confirm that Italy fully embraced the French-style concorso d'eleganza. Orsi highlights that events were held in 1924 in Rapallo, in 1926 in Milan and Rome, and in Cortina d'Ampezzo in 1928, which continued into the 1930s. Similar competitions were also held in Ferrara in 1929, Genoa-Nervi launched their first in 1929, continuing into the next decade. The most renowned of them all, Villa d'Este, began in 1929.

## The Beginnings of Villa d'Este

Of all the Italian concours, the Concorso d'Eleganza Villa d'Este was the one that caught the imagination of all, given its spectacular location next to the sublime Lake Como.

It was on September 1, 1929, that Lake Como hosted its first automobile elegance competition, and in no time Concorso d'Eleganza Villa d'Este became one of Italy's most significant automobile competitions, with around eighty cars judged by a panel on aspects such as elegance and practicality.

*In its first year as a concorsa eleganza at the Villa d'Este hotel in 1929 the turnout was most impressive. Here the cars are at Villa Olmo, where they were shown first, before getting them to be displayed and paraded at the more prestigious Villa d'Este. (Courtesy Carrozzeria Italiani)*

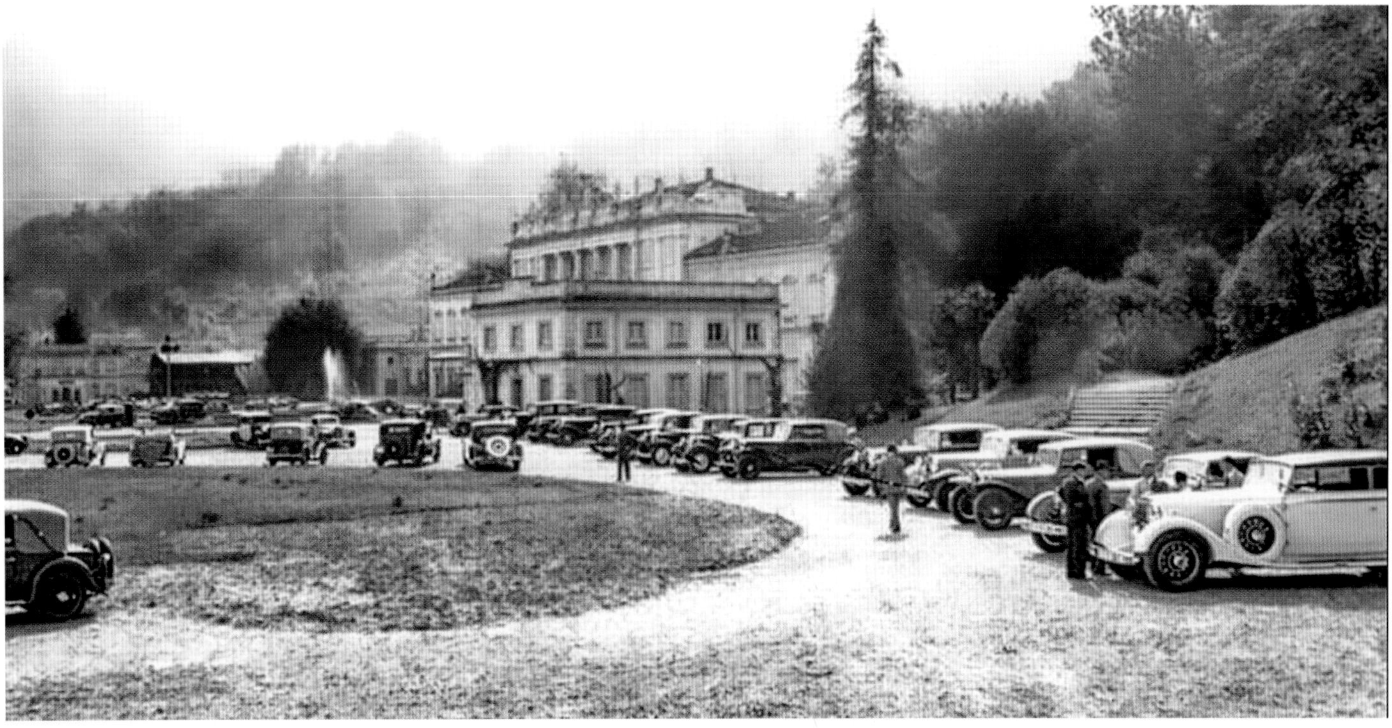

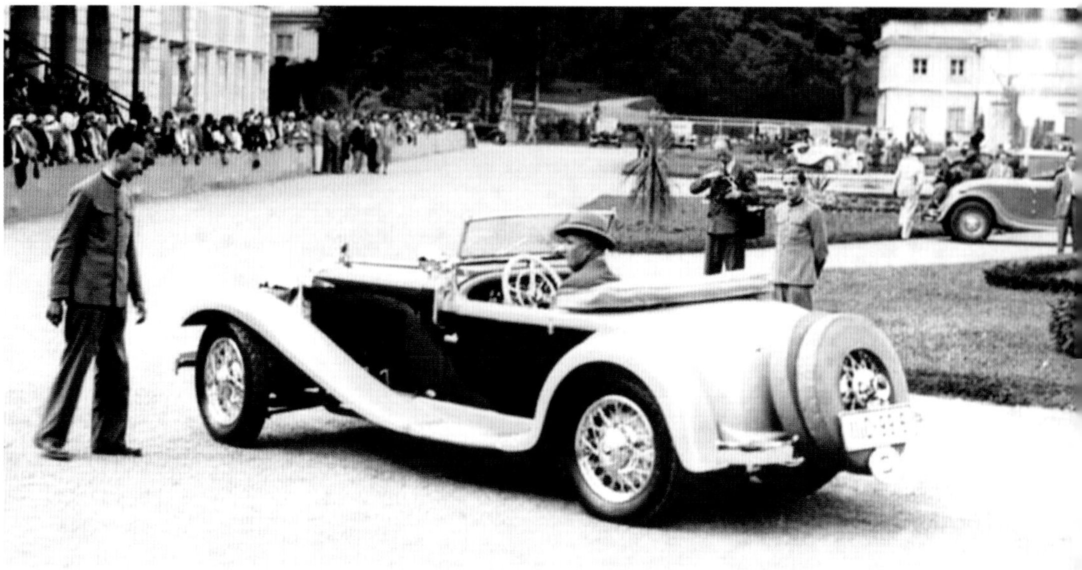

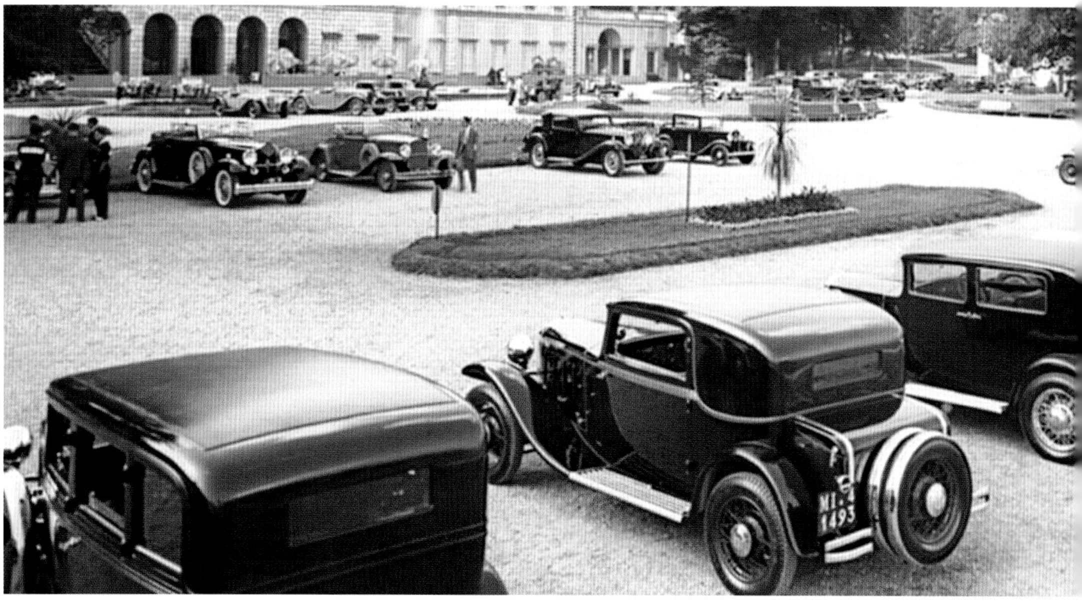

**THIS PAGE:** *Scores of mostly Italian cars recently out of the facilities of the leading Italian coachbuilders seemed to have participated in the first Concorso d'Eleganza Villa d'Este. Once again, they are at Villa Olmo, before they depart for Villa d'Este. (Courtesy Carrozzeria Italiani)*

In its inaugural year, an Isotta Fraschini, presented by Celestino Piva, won with 276 points (out of a maximum possible of 300). The competition also awarded nine cars the Targa d'Oro, including several with custom bodies by notable coachbuilders.

The second occurrence of Villa d'Este in 1930 featured 93 cars, with a Lancia Dilambda cabriolet winning the top prize. The event introduced new awards like Coppa d'onore (Cups of Honor) for first place and Gold Plaque for second, but the process was criticized for being complex and controversial, such as the inclusion of a Bianchi taxi among the winners.

Top honors went to the Alfa Romeo 8C 2300 coupé, a Maserati 2500, and a Mercedes-Benz SS for their design excellence. The competition highlighted emerging trends in automotive design, like minimal fender skirts and influences from Milan-based coachbuilders and Detroit styles.

The event gained prestige over the years, particularly with royal patronage and a name change in 1932. By 1933, the competition reflected a shift towards both custom-built and production cars, with Fiat Balillas dominating the awards. Notably, Pinin Farina's Lancia Dilambda was recognized for its aerodynamic design, foreshadowing future successes.

In 1934, the competition was grander, featuring luxury marques like Rolls-Royce and Bugatti. However, despite some beautiful entries, like a Lancia Augusta cabriolet, certain designs only received second prizes.

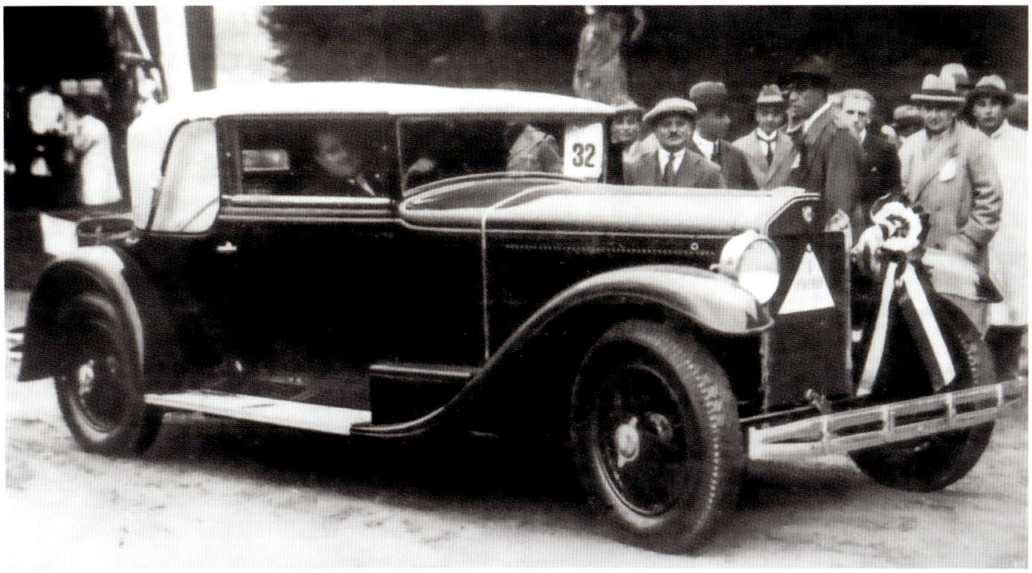

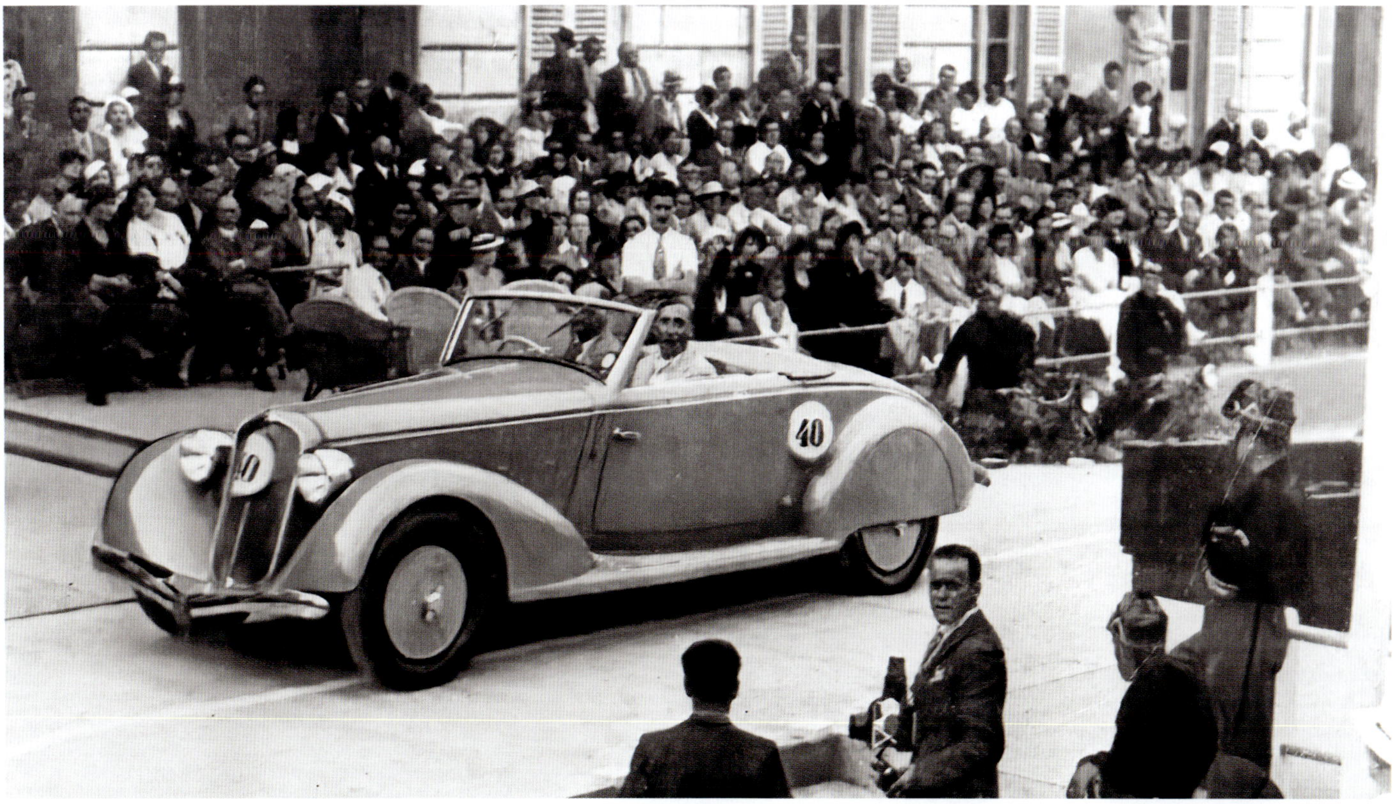

By 1935, the competition had fewer entries and focused on "worldly elegance" and "technical elegance." The Fiat 1500 series, with its unconventional design, won an award despite mixed reactions. A Swiss Martini cabriolet by Ramseier also made a memorable impact, being exhibited after the company's closure.

After a pause in 1936, the Concorso d'Eleganza Villa d'Este resumed in 1937, now under princely patronage. The event featured over 80 participants, judged by a 102-member jury, with 33 competitors receiving awards. Special awards included the Gran Premio dell'Eleganza for women and the Coppa della Provincia for "sporting elegance." The top prize went to a Lancia Aprilia coachbuilt by Pinin Farina. While some found the design eccentric, others appreciated its originality. A Graham-Paige won the Coppa del Commercio for its rational accessory design.

The 1937 event was the final Villa d'Este concours before World War II and the subsequent political shifts, which dramatically altered perspectives and lifestyles, signaling the end of an era.

**THIS PAGE:** *Villa d'Este became an occasion for coachbuilders like Pinin Farina to showcase their latest: such as this Lancia-based coupe-cabriolet in the 1930 Villa d'Este (upper) and then this sublime Alfa Romeo 8C cabriolet (directly above) which starred at the 1934 concorso. (Courtesy Pininfarina)*

# *Reviving Concours post-World War Two*

## AS PEACE RETURNED THE CONCOURS STRUGGLED

Reflecting on the late 1930s, it's clear that the golden era of the concours d'elegance was drawing to a close. On August 11, 1939, the pioneering township of Dinard hosted its concours d'elegance on a beautiful sunny day, despite the looming threat of war. That year's event featured an outstanding lineup, including a stunning Renault Viva Grand Sport, a stylish Lancia, an elegant Delage, and a refined Wolseley that had been driven over from England. The coveted *Grand Prix d'Honneur* went to Madame Meldener, elegantly dressed in blue to match her Rosengart 11CV Super Traction.

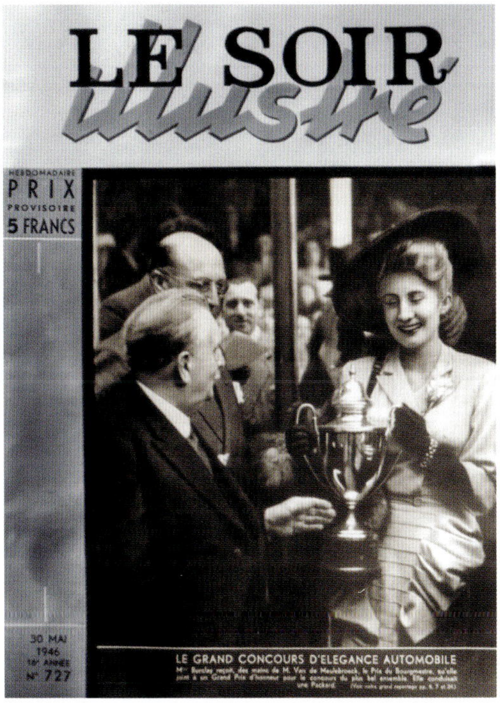


**LEFT:** *Despite the devastation of war, the good times were back early for the Belgians with a concours d'elegance in Brussels in May 1946 as reported in the issue of* Le Soir Illustré *from the same month. (Gilles Blanchet Archives)*

Just 13 days later, on August 24, France ordered partial mobilization in anticipation of Germany's invasion of Poland. By September 1, 1939, when Germany launched its attack, the world plunged into a global conflict of unprecedented scale. This war halted passenger car production in Europe, as automakers either voluntarily or under government orders shifted their focus to manufacturing vehicles and goods for the war effort.

After the war ended in 1945, American automakers were the first to resume civilian car production, while their European counterparts took longer due to the extensive rebuilding and restructuring required for many factories. Additionally, war-ravaged Europe lacked the resources to acquire cars, even as demand remained high, with countless vehicles destroyed during the conflict.

Remarkably, some concours d'elegance events in France returned as early as a year or two after World War II. In the seaside town of Deauville, the first post-war concours d'elegance was held on August 11, 1946, drawing a strong crowd of participants and spectators. The Grand Prix d'excellence that year went to Mlle. Demaillé in a Cadillac, while the *Premier Grand Prix d'Honneur* was awarded to Monsieur Eyzaguière in a Buick. The *Deuxième Grand Prix d'Honneur* was shared by a Buick, presented by Mmes Constantinesco and Francis Scott, and a Delage, displayed by Madame Fairlie-Boulay. Other notable entries included several American models – Mercury, Hudson, Nash, Packard, and a Franco-American Matford – as well as a Rolls-Royce, a Mercedes-Benz, and various French marques, reflecting the swift return of American car production.

**BELOW & OVERLEAF:** *Paris also marked the end of the war with a concours d'elegance held in 1946 at the Palais de Chaillot, beside the gardens better known as the Trocadéro. (Gilles Blanchet Archives)*

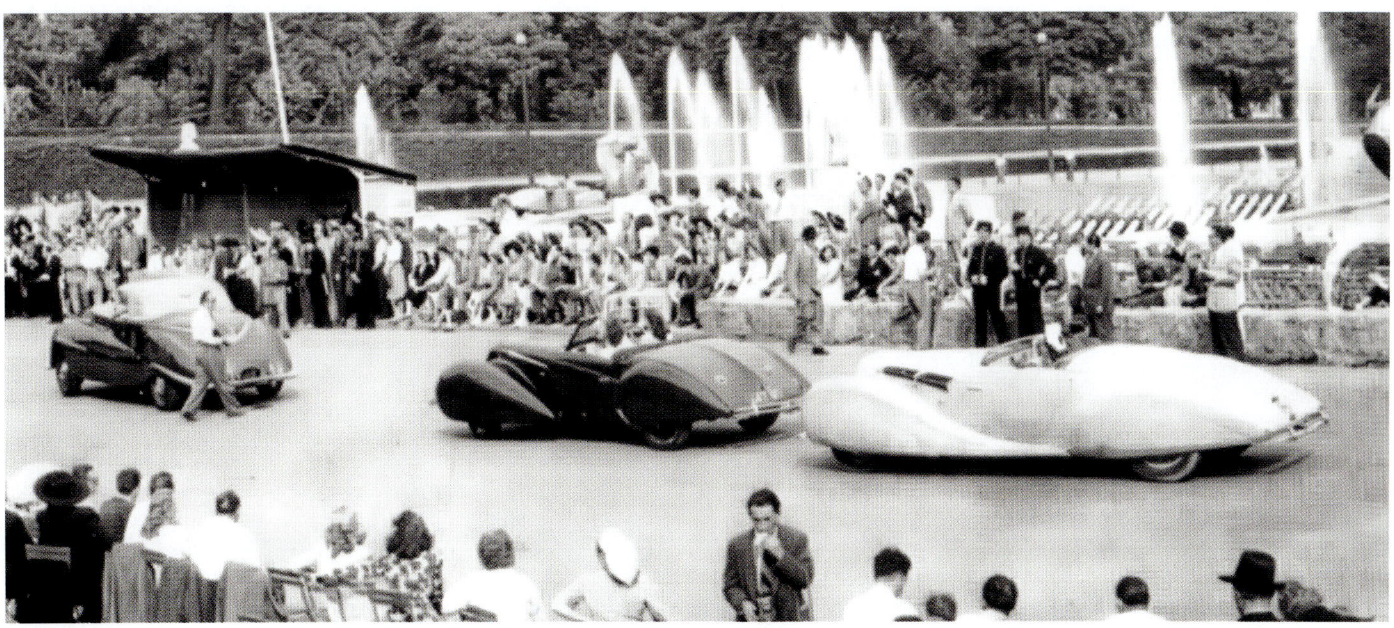

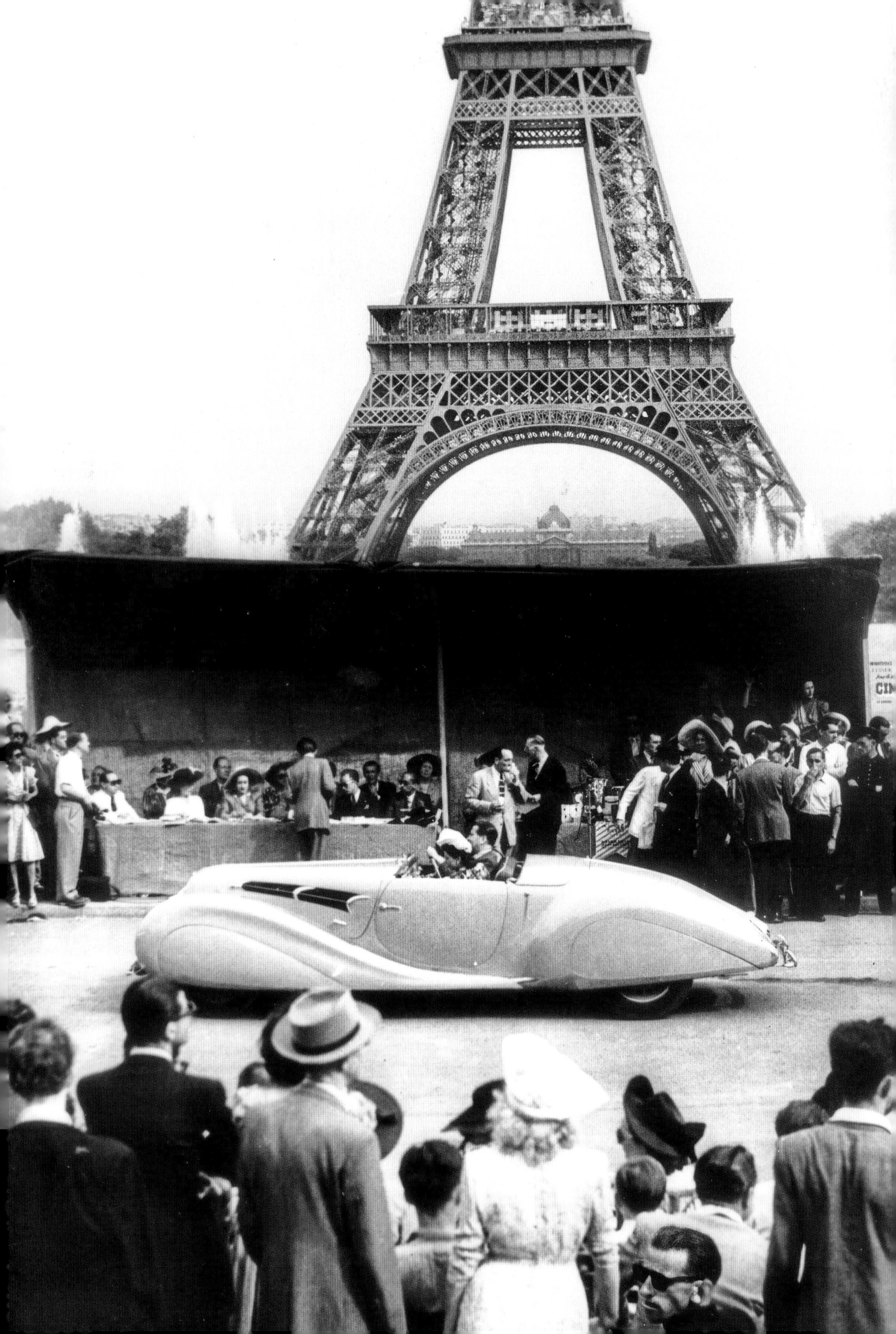

Less than a month later, Cannes held its own post-war concours d'elegance. The September 6, 1946, issue of *L'Équipe* magazine reported on the event, which took place in the gardens of the Croisette, organized by the Automobile Club de Cannes. Again, American vehicles featured prominently among the winners, including a Ford, a Nash, a Chrysler, several Buicks, an Oldsmobile, and even a Cord (which had not been in production since 1937), signaling the resilience and rapid re-entry of the American auto industry after the war. However, the Grand Prix d'excellence was shared between a Lancia, presented by Monsieur René Lang, and an Amilcar. Other awarded vehicles included several Delahayes, a couple of Talbot-Lagos, Peugeots and Citroëns, and a Hispano-Suiza.

Paris held its first post-war concours d'elegance at the Jardins de Chaillot, near the fountain at the foot of the Trocadéro, directly opposite the Eiffel Tower – a popular pre-war site for such events. Photos

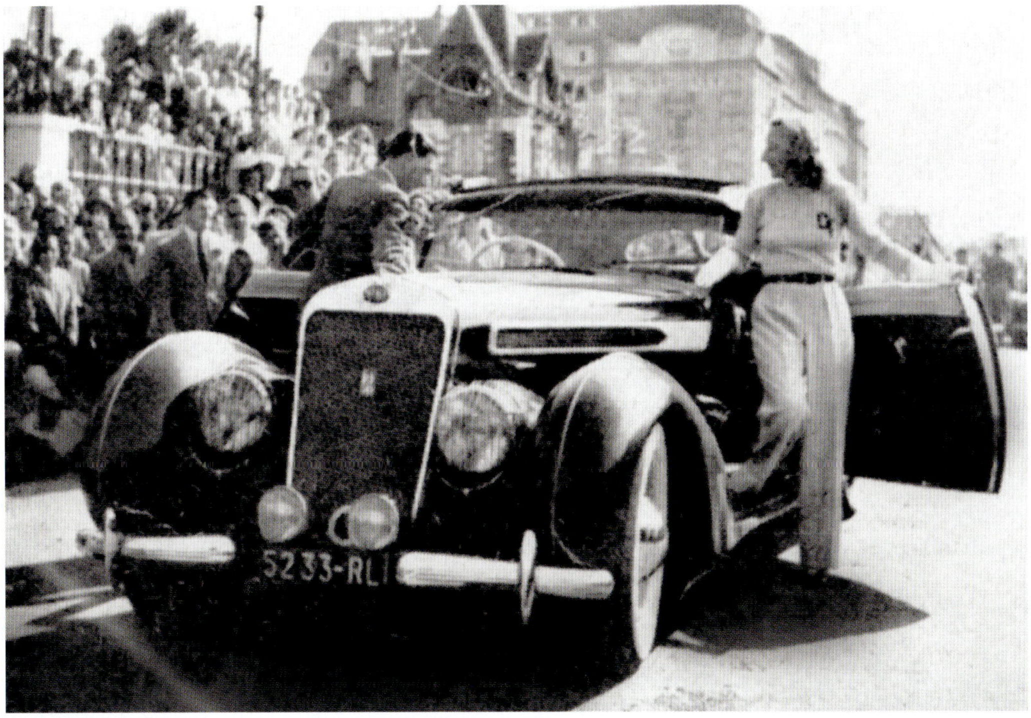

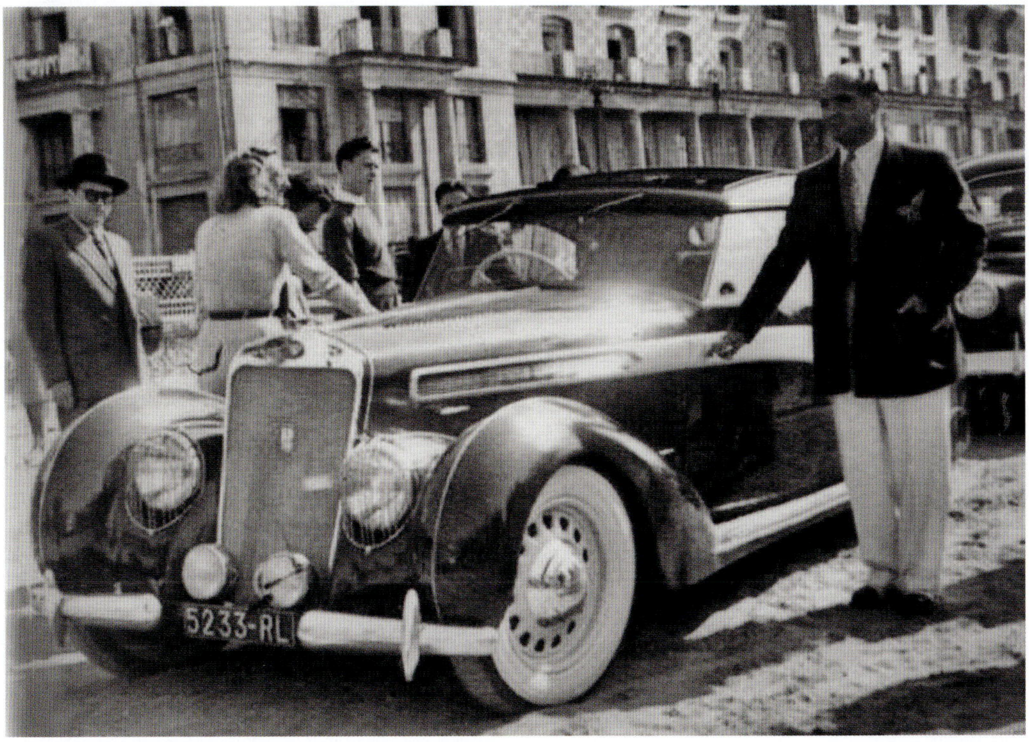

**TOP:** *Deauville got in on the act as well, staging a concours in 1946. Shown is a Delage 1938 D6-70 Vutotal by Letourneur et Marchand with a lady presenting the car. (Daniel Cabart Archives)*

**BOTTOM:** *The same Delage D6-70 Vutotal, probably with its owner in tow. (Daniel Cabart Archives)*

**OPPOSITE:** *The 1938 Delahaye 165 at the Eiffel Tower for the Trocadéro in 1946. (Daniel Cabart Archives)*

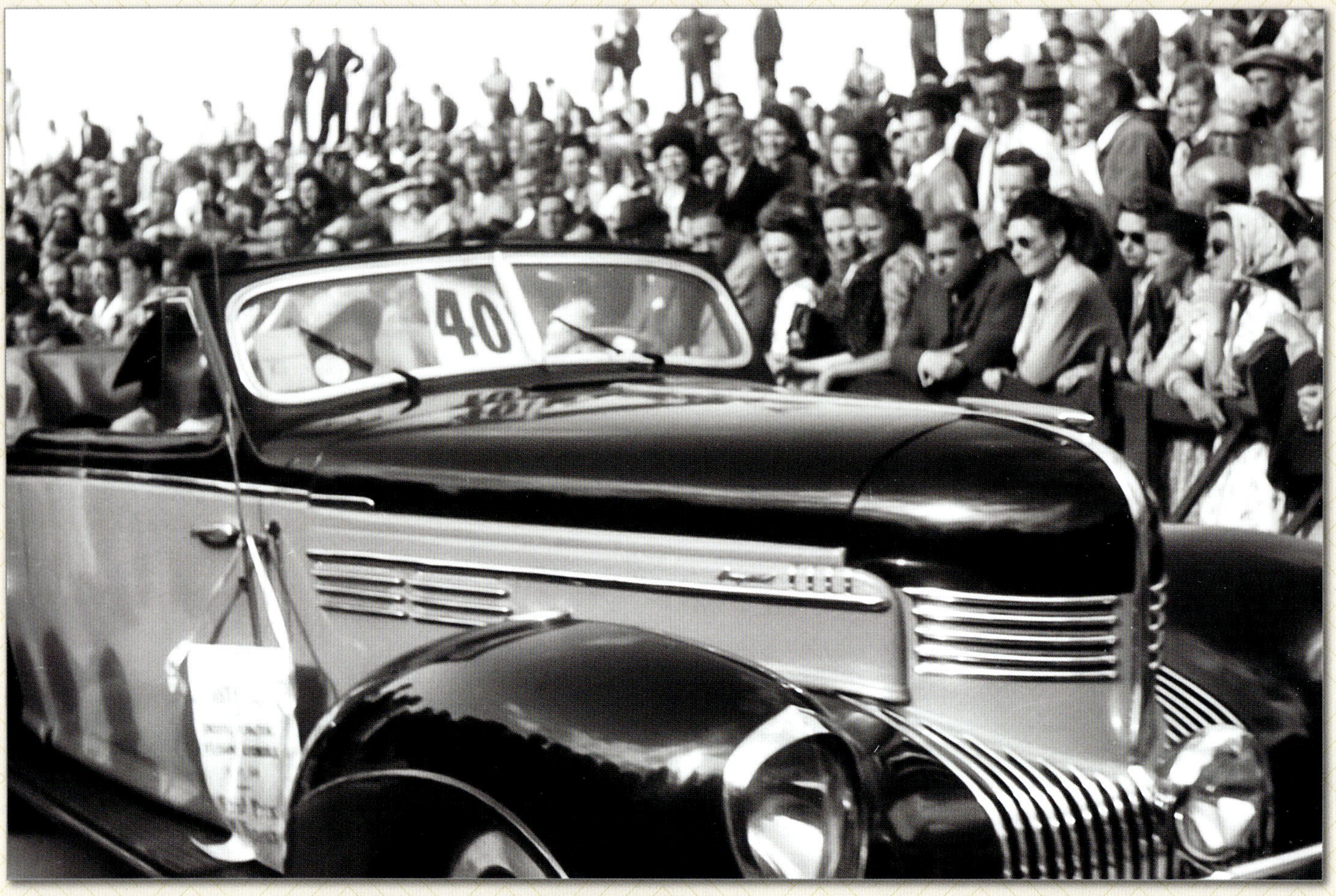
40

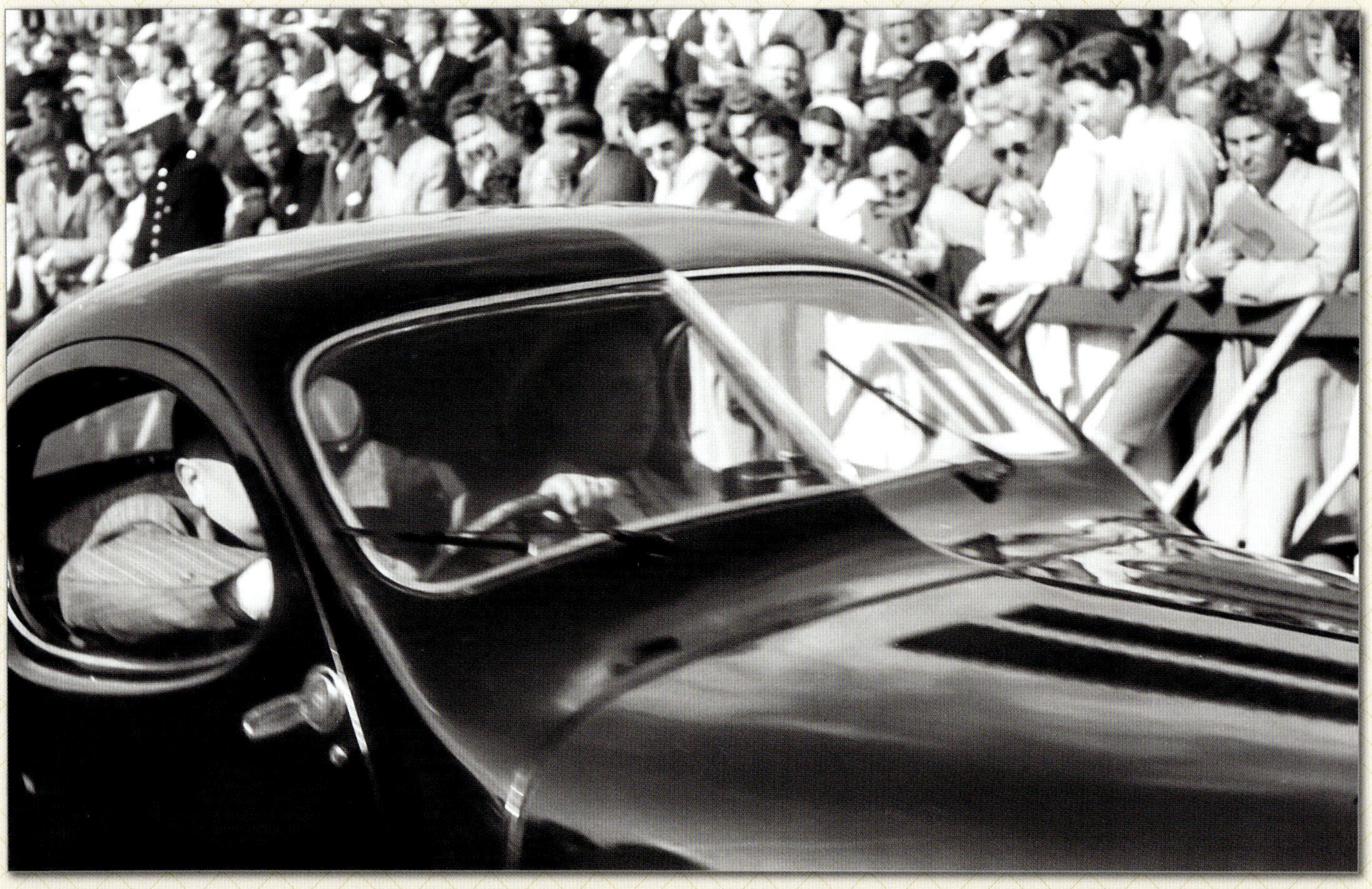

from this concours capture various cars from the immediate pre-war era, including a Delahaye 135 by Figoni et Falaschi in the "Geo Ham" style (possibly the very first in the series which was acquired by Prince Ali Khan) and another Delahaye 165 MS with a Figoni et Falaschi body or perhaps a Delage D6-70 in a similar style.

The Jardins de Chaillot was again the site for a concours d'elegance in June 1947. *L'Équipe* reported the winners, with the Grand Prix d'excellence awarded to a Delahaye with a Franay body, presented by the Comtesse de Rouget. The Grand Prix d'excellence for foreign cars went to a Buick, while the prize for a foreign car with French coachwork went to a Packard, also bodied by Franay. The first and second Grand Prix d'excellence for French cars were won by Delages, one bodied by Chapron and the other by Guilloré.

Deauville had its second post-war concours d'elegance in August 1947, where a pre-war Peugeot Darl'mat won the Grand Prix d'excellence. Among the category winners was a notable mix of American cars (Ford, Cadillac, Packard, Mercury, Buick) and French models (Delages, Delahayes, and even a Simca 8).

Across the Alps, the Concorso d'Eleganza Villa d'Este was also revived in 1947. By 1948, most major concours d'elegance in France, including La Baule, Aix-les-Bains, Enghien-les-Bains, Biarritz, Tours, and Dinard, had resumed. However, Villa d'Este skipped 1948, returning briefly in 1949 for what would be its final event for many years. This sporadic revival was mirrored by other French concours events, with only Dinard, Enghien-les-Bains, and Tours maintaining a regular presence throughout the 1950s, while others struggled to survive, and some faded entirely by the end of the decade.

**OPPOSITE (BOTH):** *Pre-war cars were back on display at the Ostend Concours d'Elegance in Belgium in 1946. Not surprisingly, a late 1930s Talbot-Lago goutte d'eau (bottom) was one of several stars at the show, along with a 1939 Chrysler Imperial (top). (Gilles Blanchet Archives)*

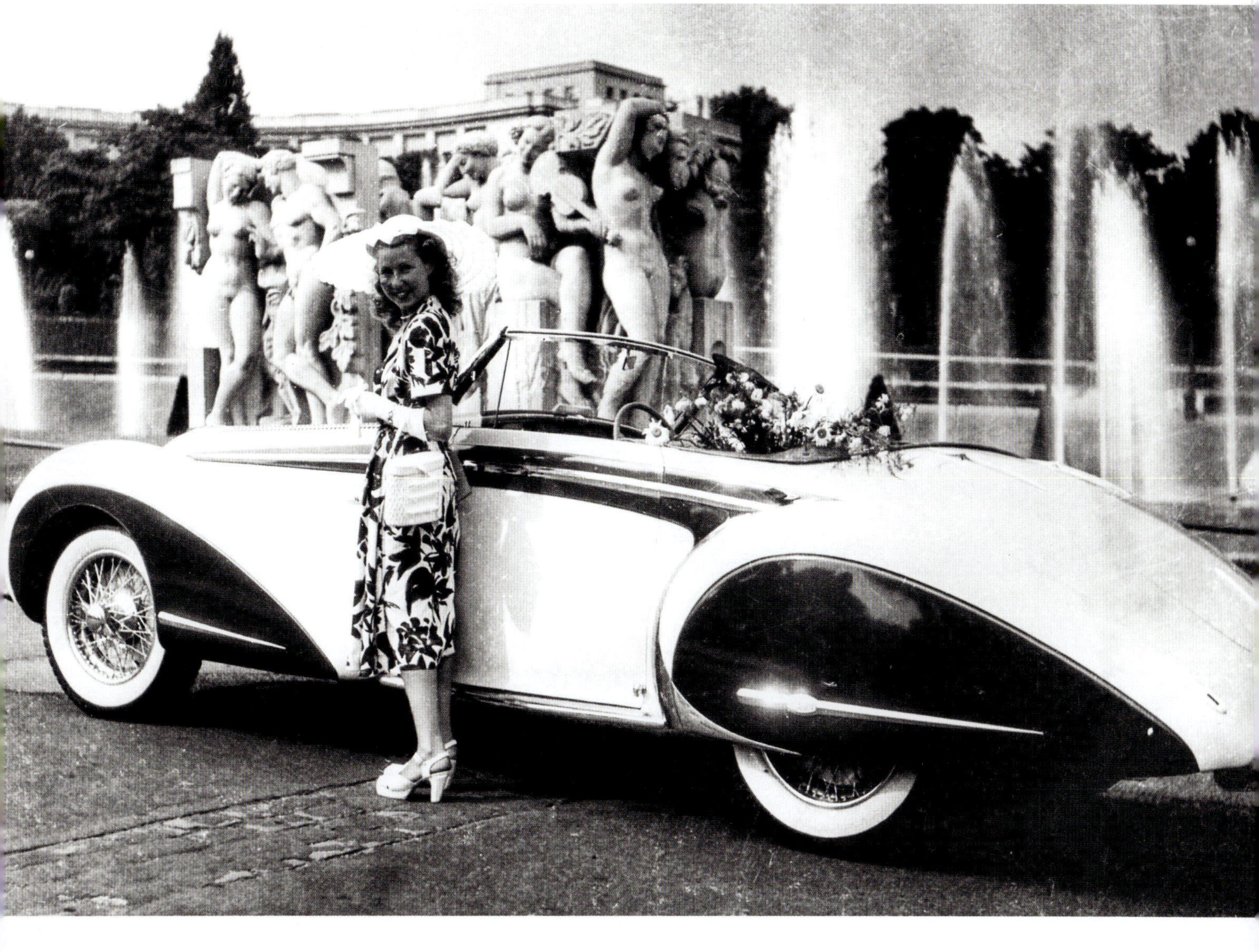

**BELOW:** *A stunning Delahaye 135 MS with Chapron cabriolet Vedette body presented by Madame Chapron on June 28, 1947, at the Palais de Chaillot concours. (Gilles Blanchet Archives)*

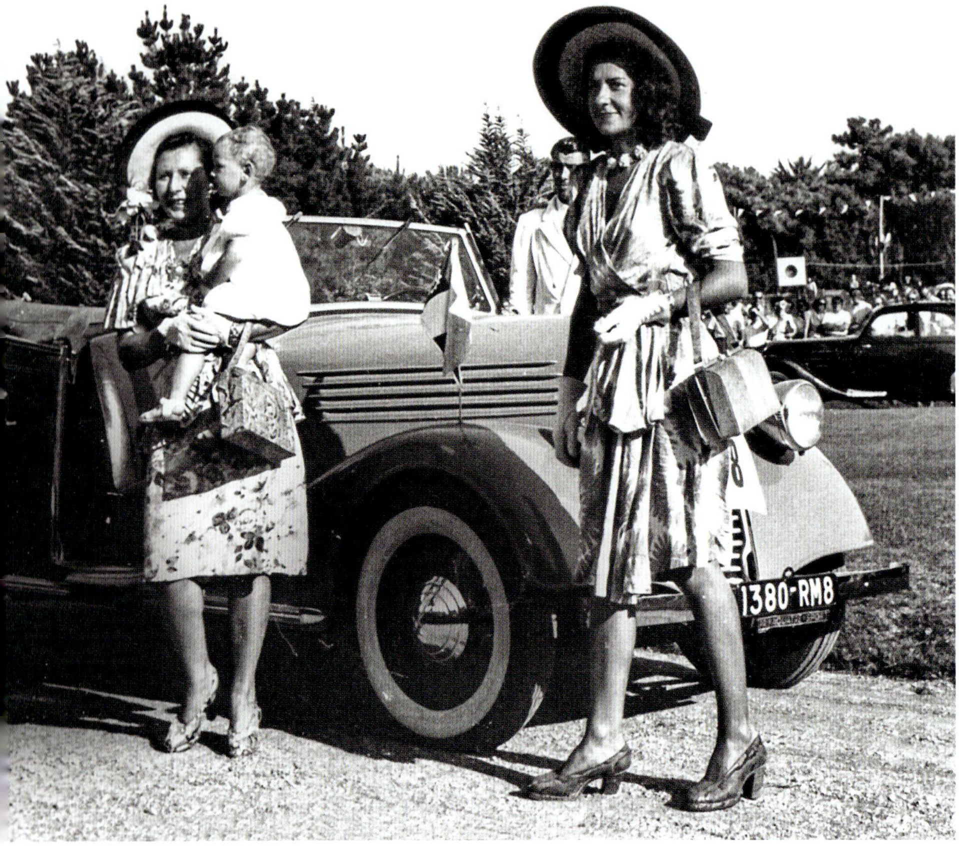


In the United Kingdom, the concours d'elegance experienced a notable revival, especially in Torquay. At this event, the Belgian Ambassador, Viscount of Thieusies, showcased his nearly new 1946 Lincoln sedan and secured first place in its class. Pre-war British cars also excelled, with a 1936 Rolls-Royce, a Bentley, and a 1939 Rover each winning their respective categories. A brand-new 1947 Armstrong-Siddeley took top honors in its class as well.

On September 6, 1947, Regent's Park in London hosted an event reminiscent of a motor show, featuring a concours d'elegance as part of the program. Less than three weeks later, on September 26, another concours was held in Tunbridge Wells, where Winston Churchill delivered a speech reflecting on nearly fifty years of motoring history. One of the standout exhibits was a 1914 Rolls-Royce 40/50 HP Silver Ghost, the oldest vehicle on display.

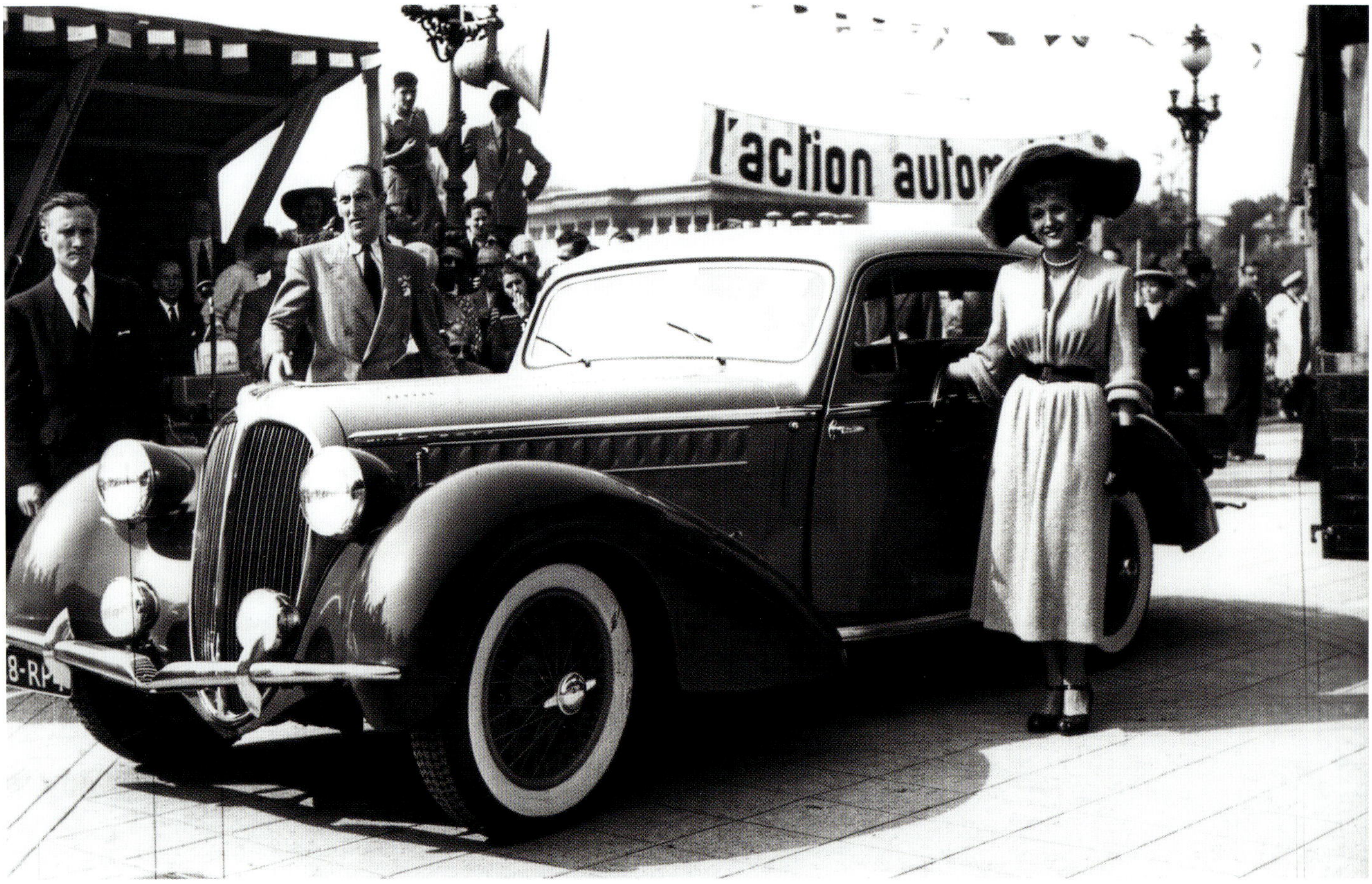

## A shift

Despite a lively atmosphere and broader audience, the concours began shifting from featuring unique coachbuilt cars to highlighting more standard production models. Prestigious cars like those from Facel Vega, Aston Martin, Lamborghini, and Porsche, as well as unique Zagato-bodied Maseratis, Lancias, Alfa Romeos, Bristols, and Chapron-bodied Citroën DS convertibles, continued to captivate attendees. Yet, the events never fully recaptured the grandeur of the 1920s and 1930s.

The entire European automobile industry had changed post-war, with the French sector particularly affected. To rejuvenate industry and the economy, the government initiated the Pons Plan, part of Jean Monnet's Modernization and Equipment Plan, led by engineer Paul-Marie Pons. With factories in ruins and production severely diminished by the war, France's state-directed economic strategy involved allocating scarce resources, particularly metals, to streamline industrial production.

Under this plan, Simca led the GFA (Générale Française Automobile), which included Delahaye (and Delage), Unic, Bernard, and Laffly. Panhard headed the UFA (Union Française Automobile) alongside truck makers Somua and Willème, while Peugeot aligned with Hotchkiss, Latil, and Saurer. Simca and Panhard were tasked with producing affordable cars, while Renault and Peugeot targeted the mid-range and Citroën the high-end market. Meanwhile, luxury manufacturers like Delahaye-Delage, Salmson, Hotchkiss, Talbot-Lago, and Bugatti were encouraged to focus on exports, as few in post-war France could afford their vehicles.

Yet these high-end French cars struggled to compete with post-war American models, which offered attractive styling, chrome details, and advanced features like automatic transmissions and electric windows at competitive prices. As a result, demand dwindled, leading to the gradual decline of the French luxury automobile industry. Bugatti faced financial difficulties by 1952, Delahaye and Delage closed by 1954, and Talbot-Lago lingered until 1960.

The downfall of the French specialist car industry also impacted coachbuilders who relied on it, reducing the demand for unique, handcrafted designs. With luxury cars losing their allure, standardized production became the norm, and concours d'elegance lost relevance. Automakers shifted to unveiling new models at large car shows, which gained prominence over elite concours events.

In France, only a few concours events like those at Dinard and Enghien-les-Bains survived into the 1960s or early 1970s. In Italy, the Rimini Concorso d'Eleganza was among the last to endure.

Looking back, the reasons are clear. Economic factors influenced the decline of concours events, as the post-war economic boom fueled demand for practical, affordable cars for suburban life and travel. Concours events, once symbols of luxury, were out of step with the economic realities of middle-class car buyers, who now dominated the market.

The 1950s and 1960s saw mass automobile production surge, led by American manufacturers like Ford, Chevrolet, and Chrysler. Cars became accessible, everyday items rather than luxury purchases, shifting the emphasis from bespoke, handcrafted cars to mass-produced models. Car culture leaned toward speed, performance, and accessibility, sidelining the elegance that had defined the concours.

Additionally, the post-war automotive industry moved toward in-house design, reducing bespoke coachbuilding. With fewer unique cars, the main attraction of the concours – showcasing exclusive luxury vehicles – faded.

Cultural shifts also played a role. The 1950s and 1960s prioritized innovation and practicality, embracing sleek, modern designs over extravagance. As youth culture rose, events rooted in aristocratic traditions appeared outdated. Concours d'elegance, symbols of an older, elite society, began to feel out of sync with mainstream interests.

Moreover, the rise of motorsport events like Formula 1 and the 24 Hours of Le Mans redirected public attention to high-performance cars, appealing to a younger audience and competing with traditional concours events. Simultaneously, major auto shows in Paris, Geneva, London and Frankfurt became the preferred venues for manufacturers to present new models, replacing the exclusive allure of concours d'elegance.

**OPPOSITE TOP:** *Even the most practical participated, like this Renault Primaquatre cabriolet sport at La Baule on August 19, 1947. (Gilles Blanchet Archives)*

**OPPOSITE BOTTOM:** *A Delahaye with Chapron bodywork for the Enghien-les-Bains Concours d'Elegance on June 12, 1948. (Gilles Blanchet Archives)*

# CONCOURS D'ÉLÉGANCE AUTOMOBILE DE DEAUVILLE 1948

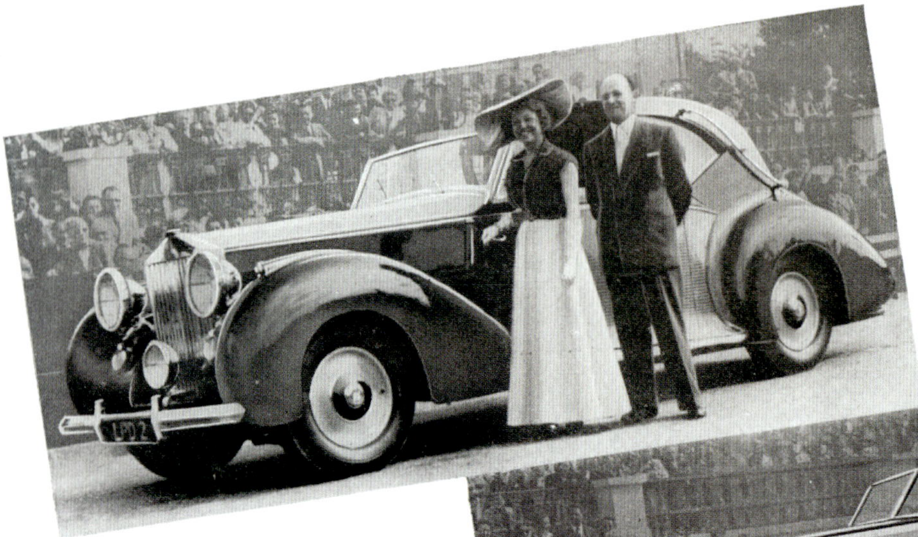

Le Grand Prix Toutes Catégories a été attribué à une Rolls-Royce carrossée par SAOUTCHIK — LAQUES SOUDÉE — présentée par son propriétaire, Sir John Gaul et par Mme Pierre Saoutchik habillée par Callixte et coiffée par Marie-Christine.

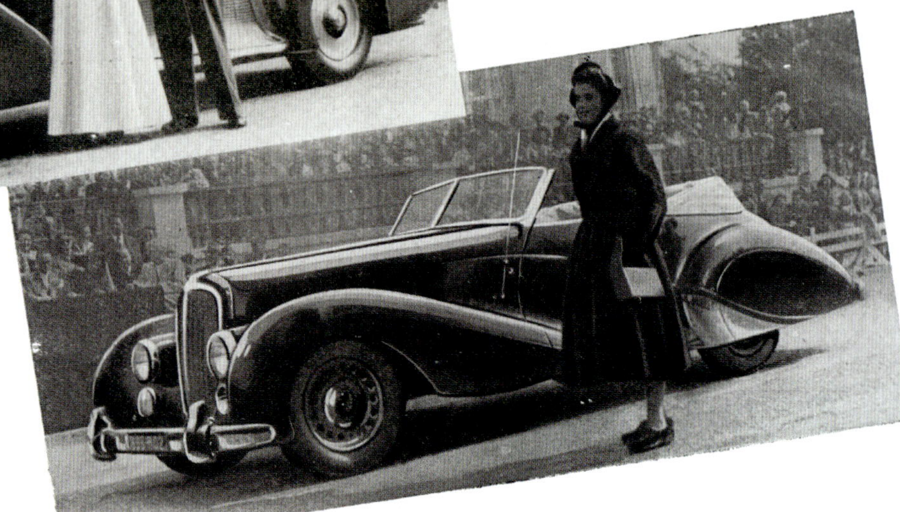

Le Grand Prix d'Honneur a été attribué à une Delahaye carrossée par SAOUTCHIK — LAQUES SOUDÉE — présentée par Mme Marc Soudée habillée par Worth.

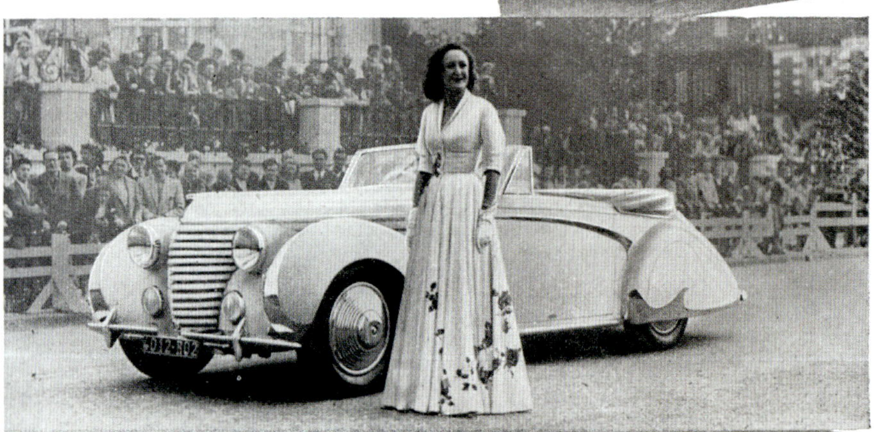

Le premier Grand Prix d'Honneur a été attribué à une Delahaye carrossée par FRANAY — LAQUES SOUDÉE — présentée par Mlle Yvette Bidermann habillée par Raphaël.

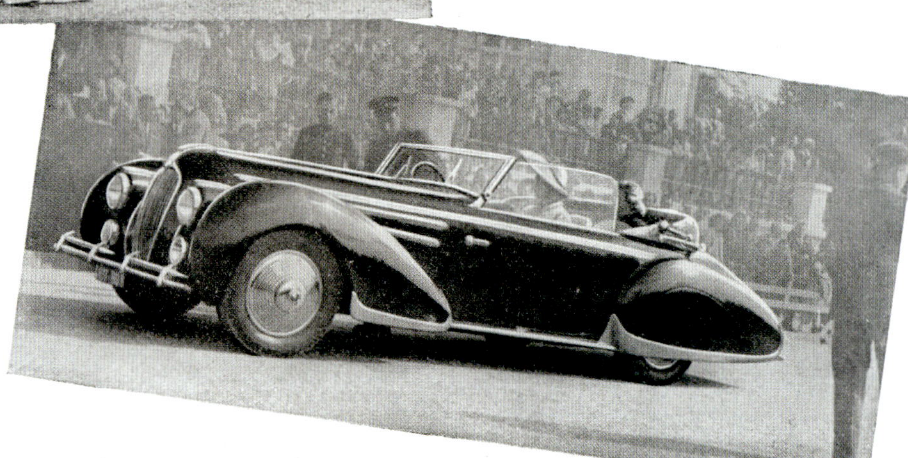

Le Grand Prix d'Excellence a été attribué à une Delahaye carrossée par FIGONI-FALASCHI — LAQUES SOUDÉE — présentée par Mme Michau habillée par Jacques Fath.

Le Concours d'Élégance Automobile de Deauville 1948, consacre la classe mondiale des Maîtres Carrossiers Français et l'incontestable supériorité des LAQUES SOUDÉE.

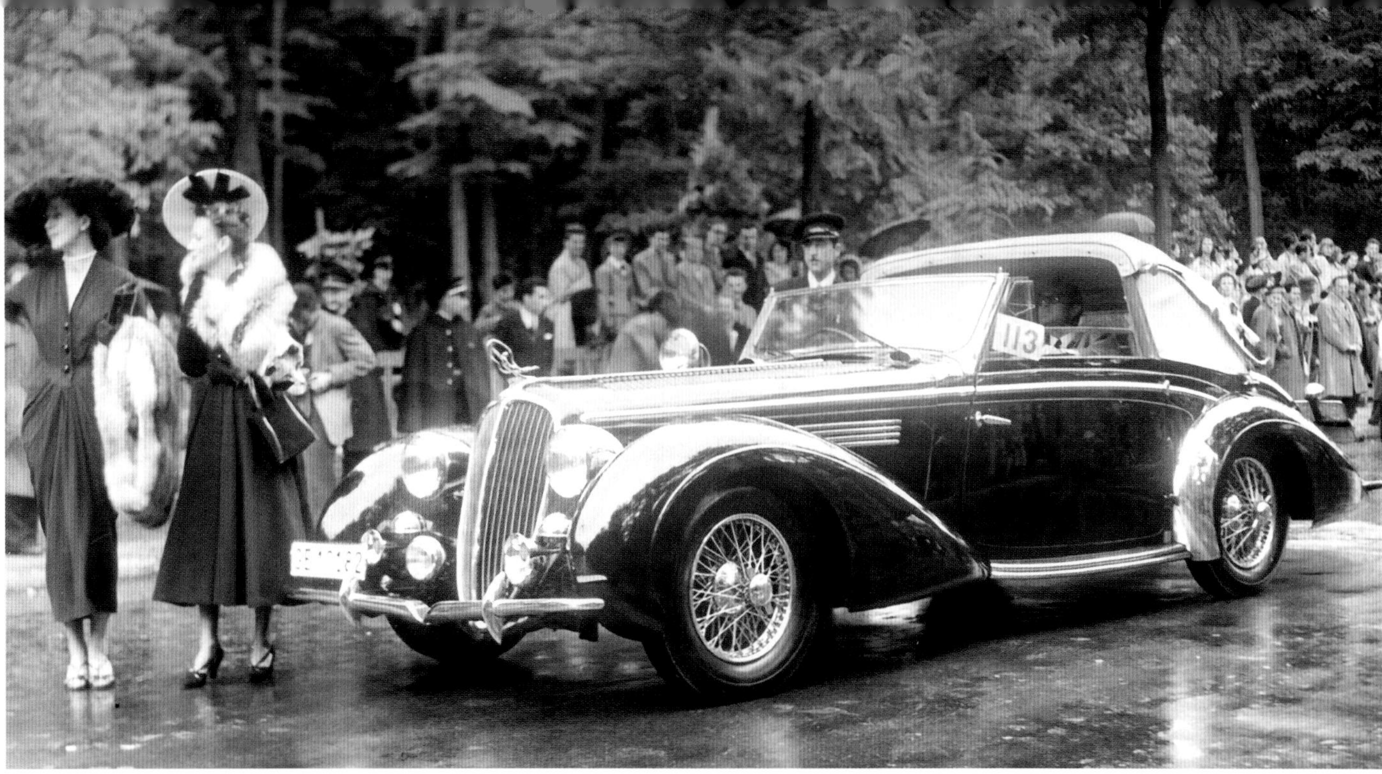


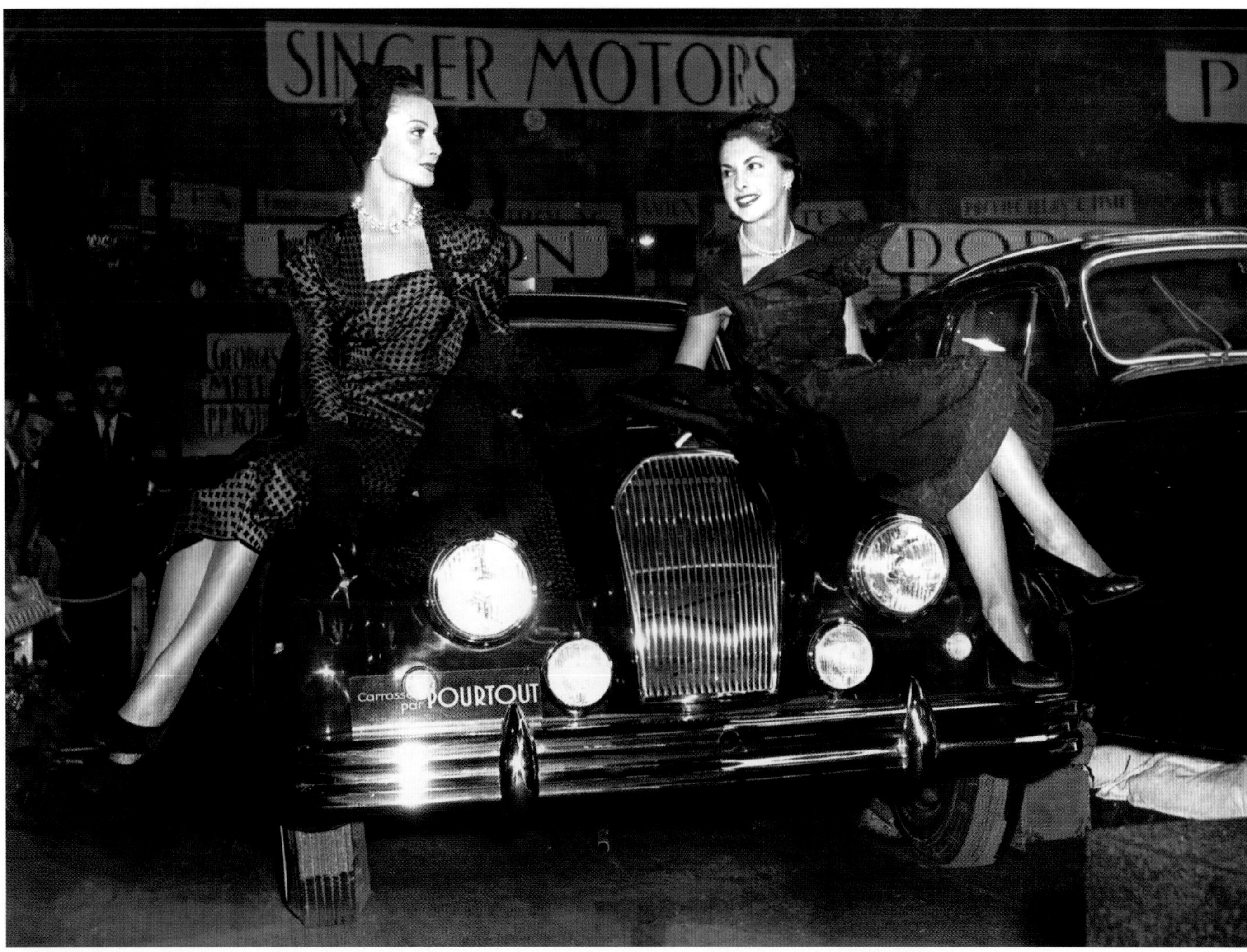


**OPPOSITE:** *A report on the 1948 Concours d'Elegance at Deauville provides a glimpse at the variety of cars presented that year.* (Gilles Blanchet Archives)

**TOP:** A 1948 Delahaye *with Chapron body at the 1948 Pavillon d'Armenonville Concours d'Elegance.* (Gilles Blanchet Archives)

**ABOVE:** *By the late 1940s and early 1950s, motor shows started to substitute for concours d'elegance events. Thus, it was not unusual to see the new* Talbot T15 Baby *– coachbuilt by* Pourtout *– at the* Paris Motor Show*, with models who were also presenting cars at Concours d'Elegance events and other salons.* (Gilles Blanchet Archives)

**THIS PAGE:** *A Delahaye cabriolet bodied by Chapron, at the 1949 Paris Motor Show. (Gilles Blanchet Archives)*

**OPPOSITE TOP LEFT:** *Another Delahaye cabriolet, also bodied by Chapron, at the 1949 Paris Motor Show with slightly different trim details. (Gilles Blanchet Archives)*

**OPPOSITE TOP RIGHT:** *Yet the style and the elegance of a concours remained as evinced by this refined lady at the Enghien-les-Bains Concours d'Elegance on June 17, 1950. (Gilles Blanchet Archives)*

**OPPOSITE MIDDLE LEFT:** *A Talbot T26 with Saoutchik body at the Gala de L'Automobile at Place Vendôme on June 7, 1950. (Gilles Blanchet Archives)*

**OPPOSITE BOTTOM:** *An image from the 1950 Tours Concours d'Elegance. (Daniel Cabart Archives)*

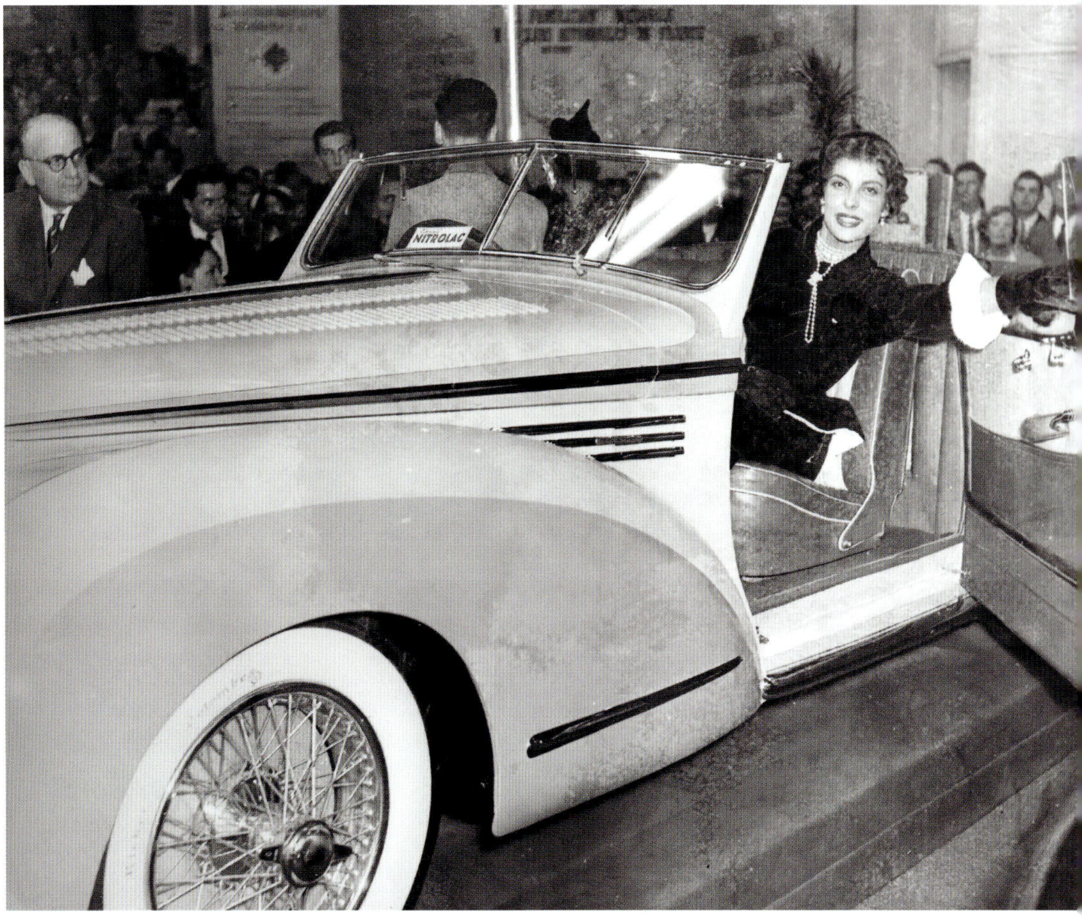


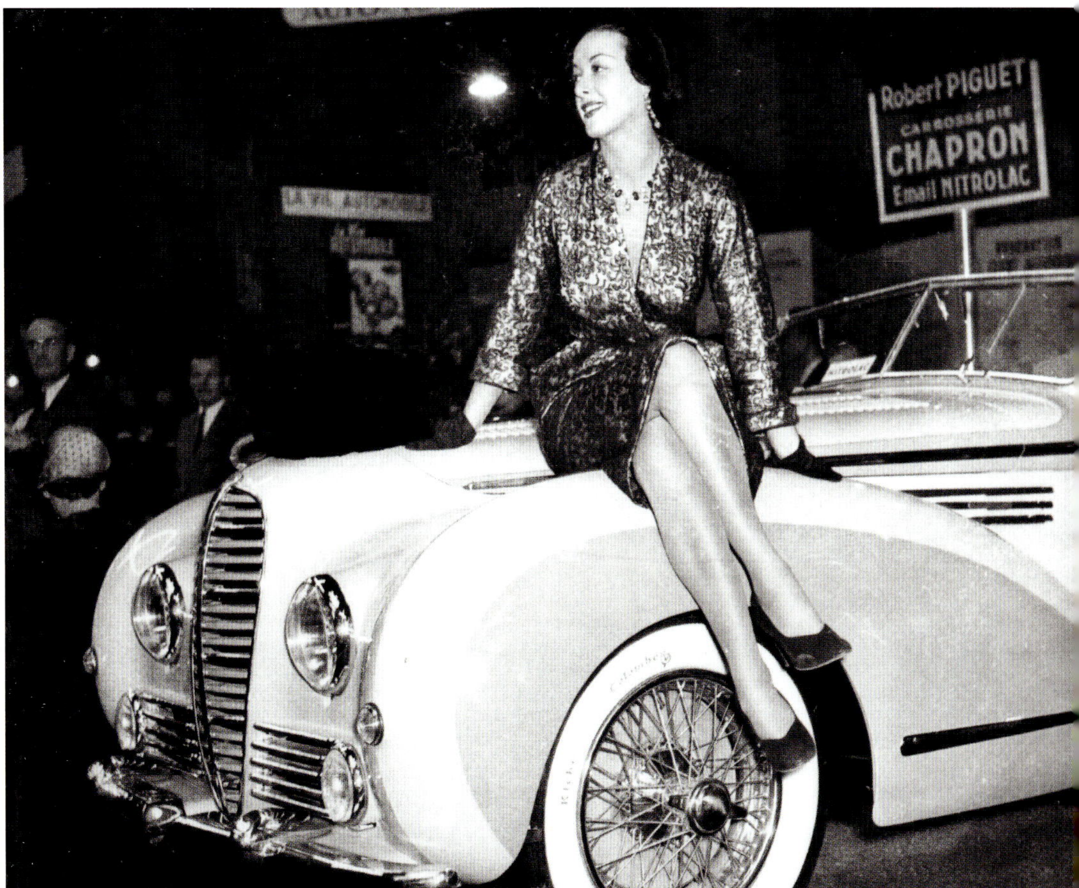

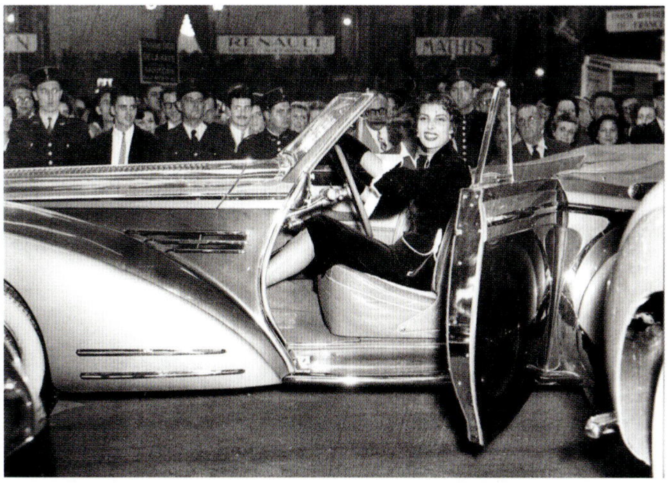
RENAULT
MATHIS

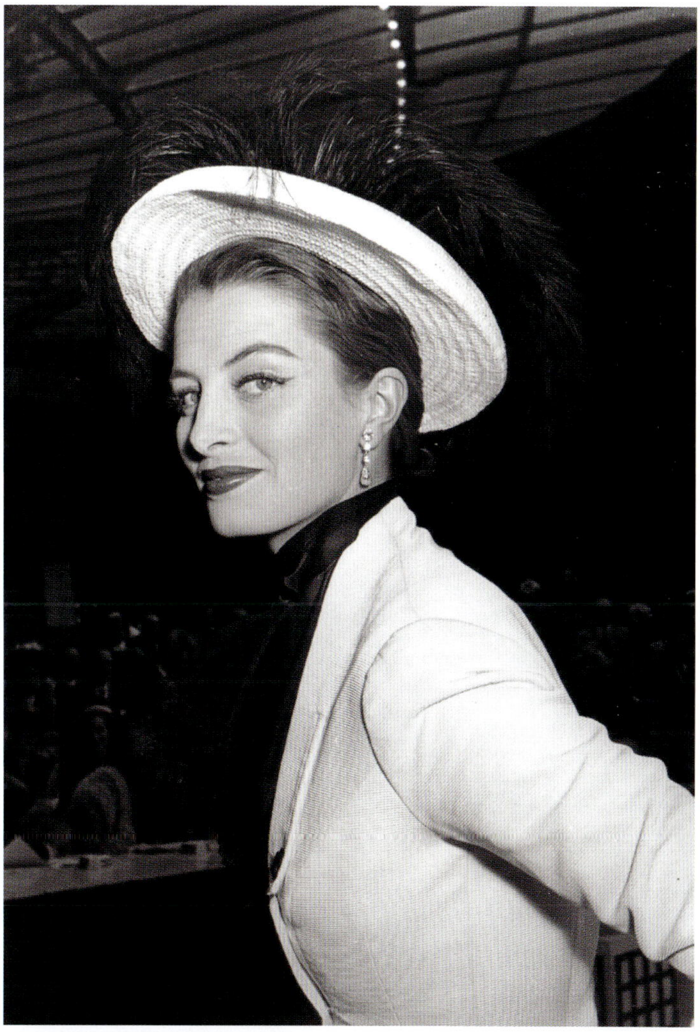

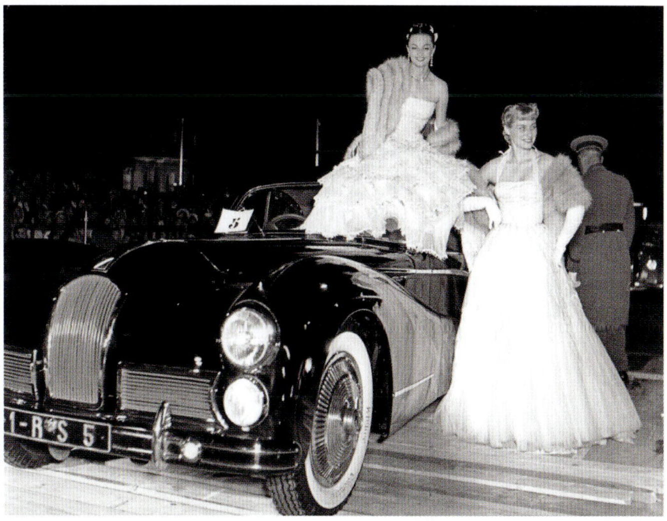

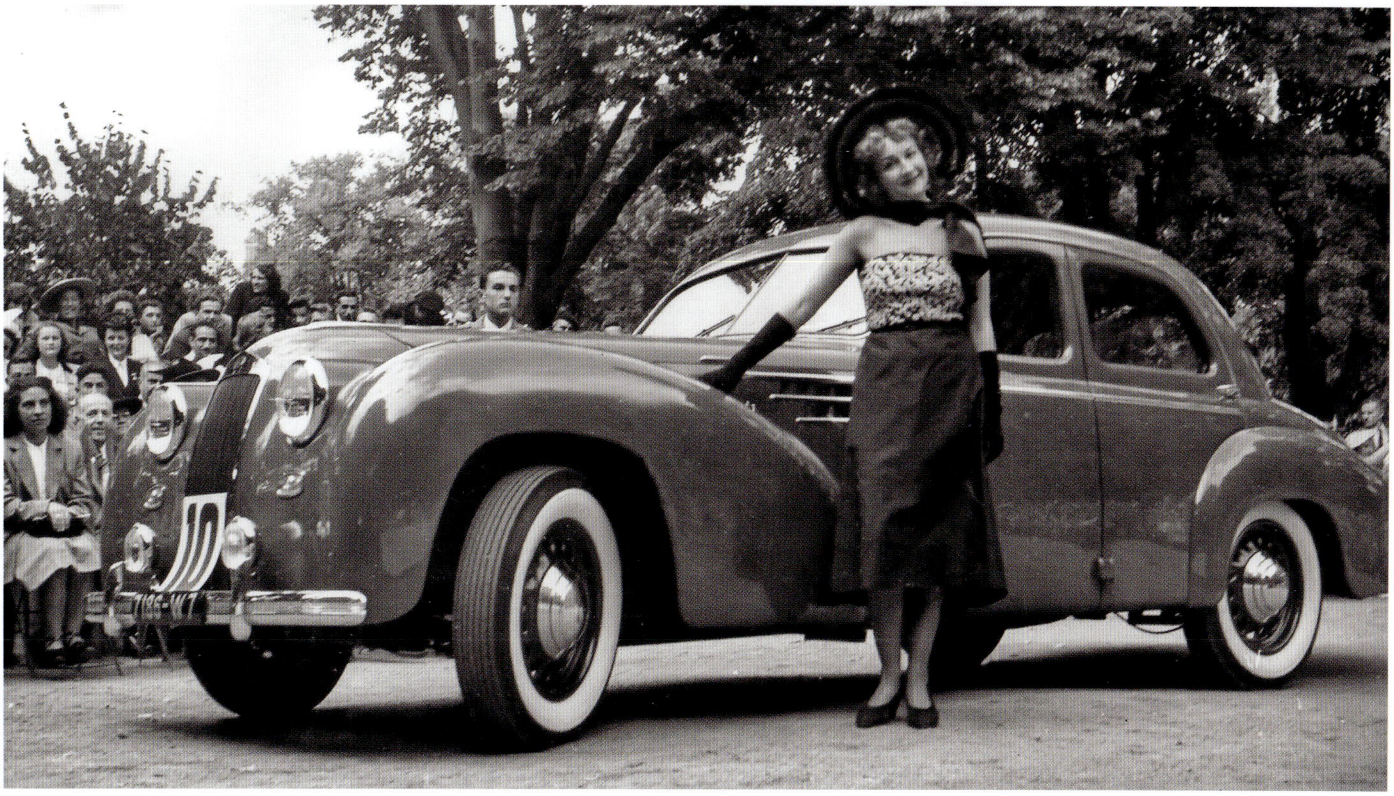

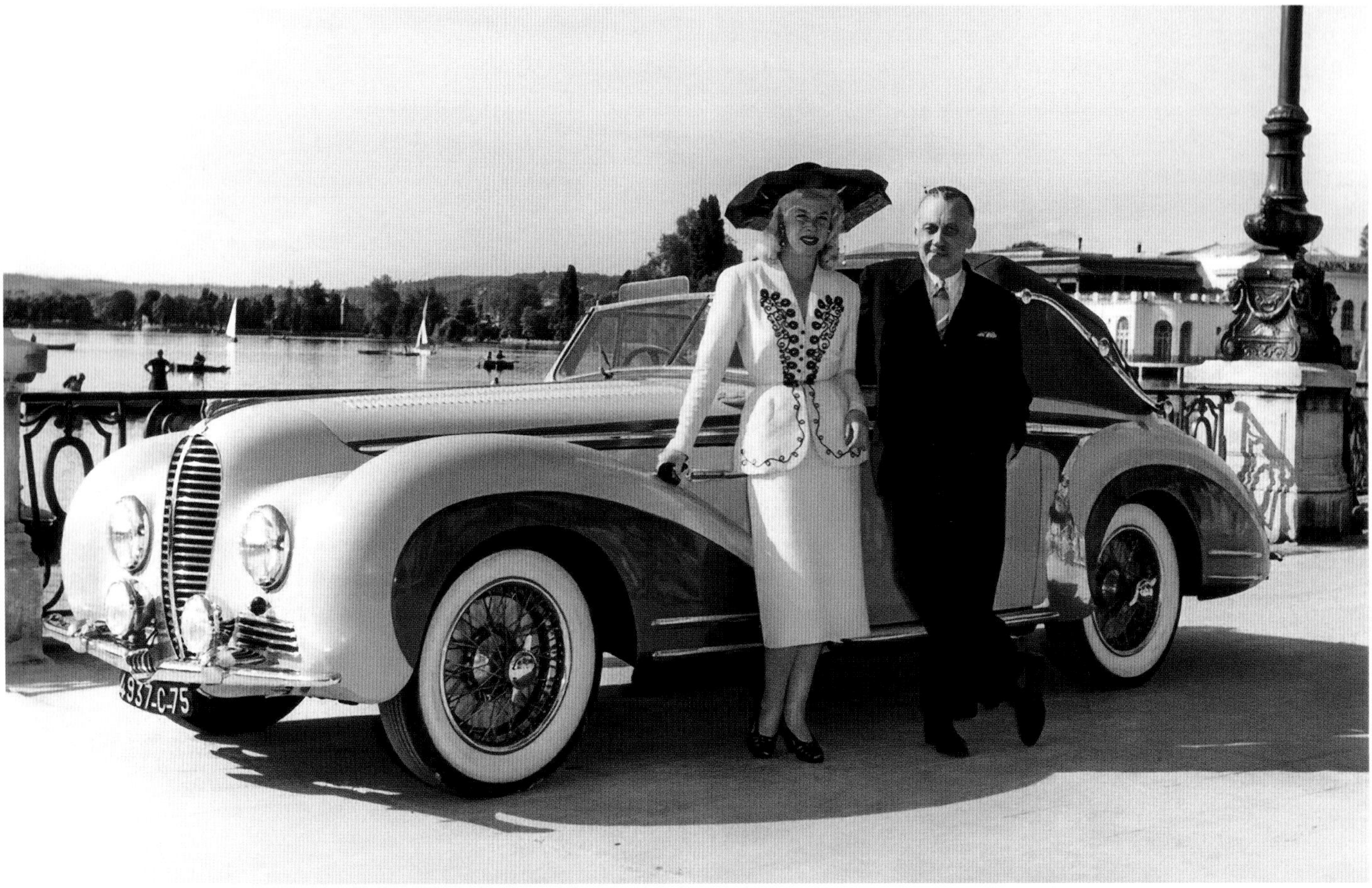
4937-C-75

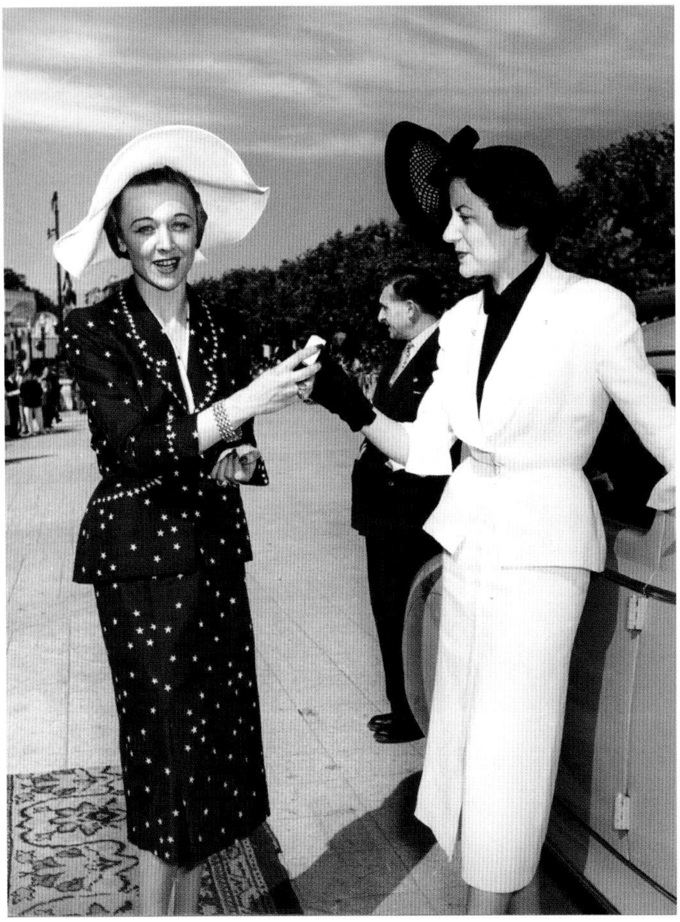

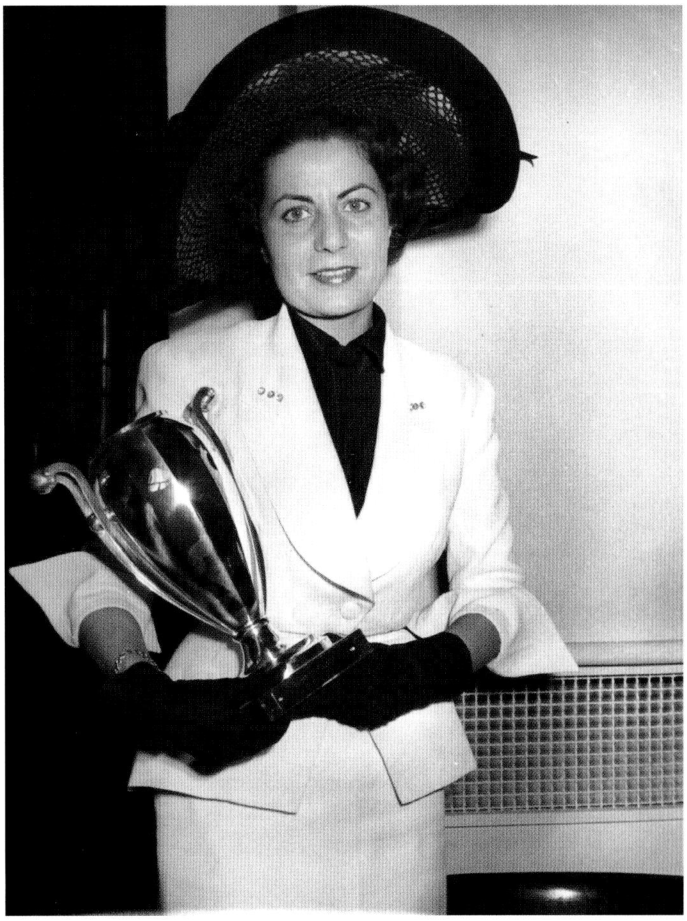

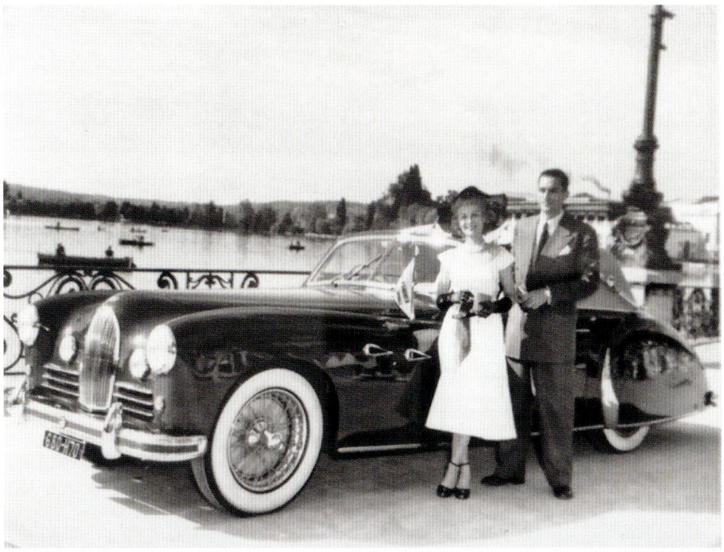

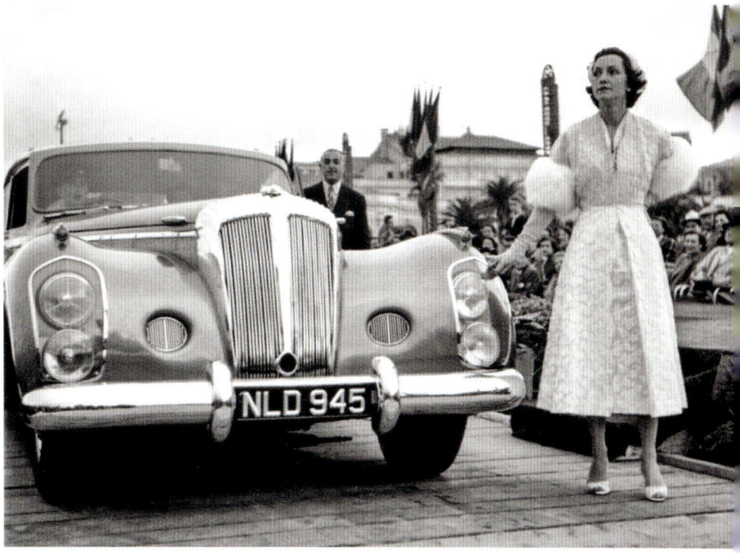


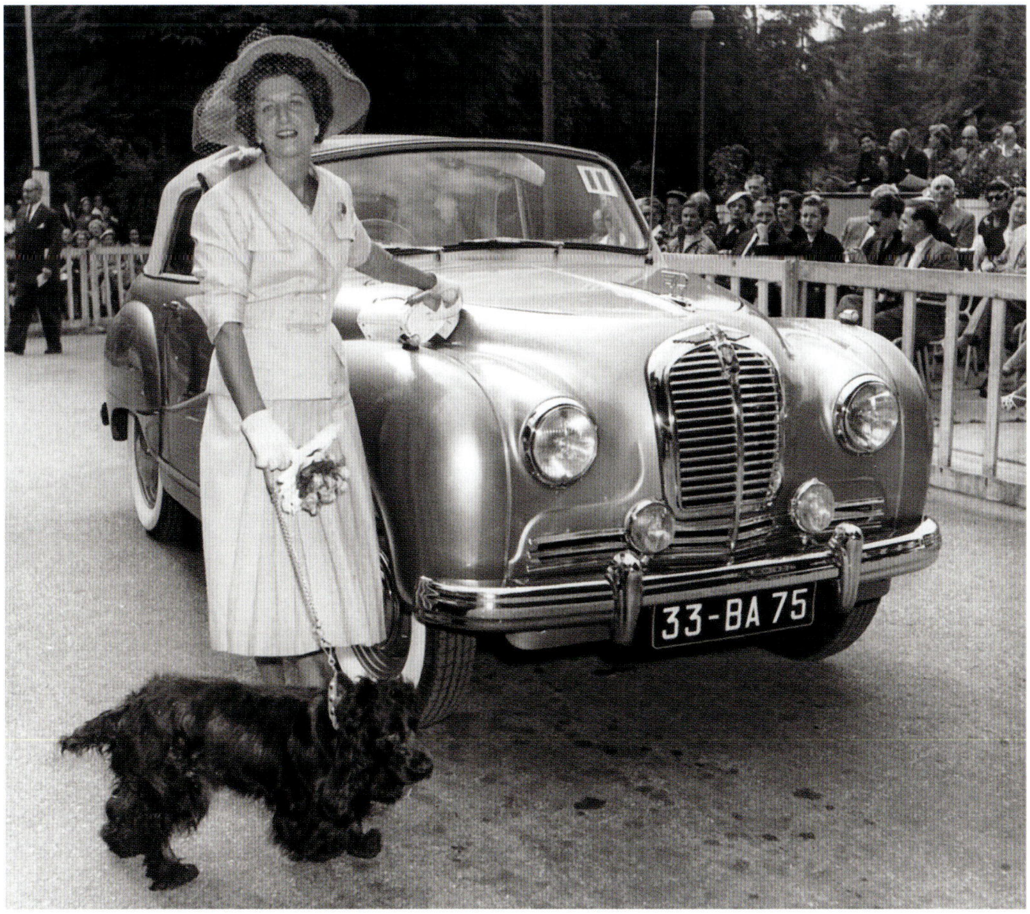


**ABOVE LEFT:** *Fabulous Saoutchik-bodied Bugatti at the 1951 Enghien-les-Bains Concours d'Elegance. (Gilles Blanchet Archives)*

**ABOVE:** *Cars like the Daimler DE36 Gold Car did the tours of the concours. Lady Docker presents the car at Enghien-les-Bains in 1952. (Gilles Blanchet Archives)*

**LEFT:** *In the post-war scenario, cars like the relatively modest Austin A40 cabriolet had a place in concours, as well. This is from an event in 1952. (Gilles Blanchet Archives)*

**OPPOSITE TOP:** *In the post-war era, the concours at Enghien-les-Bains became one of the main French concours. A Delahaye with Henri Chapron body starred at Enghien on June 16, 1951. (Gilles Blanchet Archives)*

**OPPOSITE BOTTOM (BOTH):** *Dressing up with hats and haute couture costumes remained a part of the scene during the post-war era. This is at Enghien-les-Bains on June 16, 1951. (Gilles Blanchet Archives)*

**RIGHT:** *Occasions such as the flag-off of rallies – even if a bit fractious sometimes – became another place for a concours.* At the Rallye Paris-Deauville with Maria Riquelme on July 11, 1953. (Gilles Blanchet Archives)

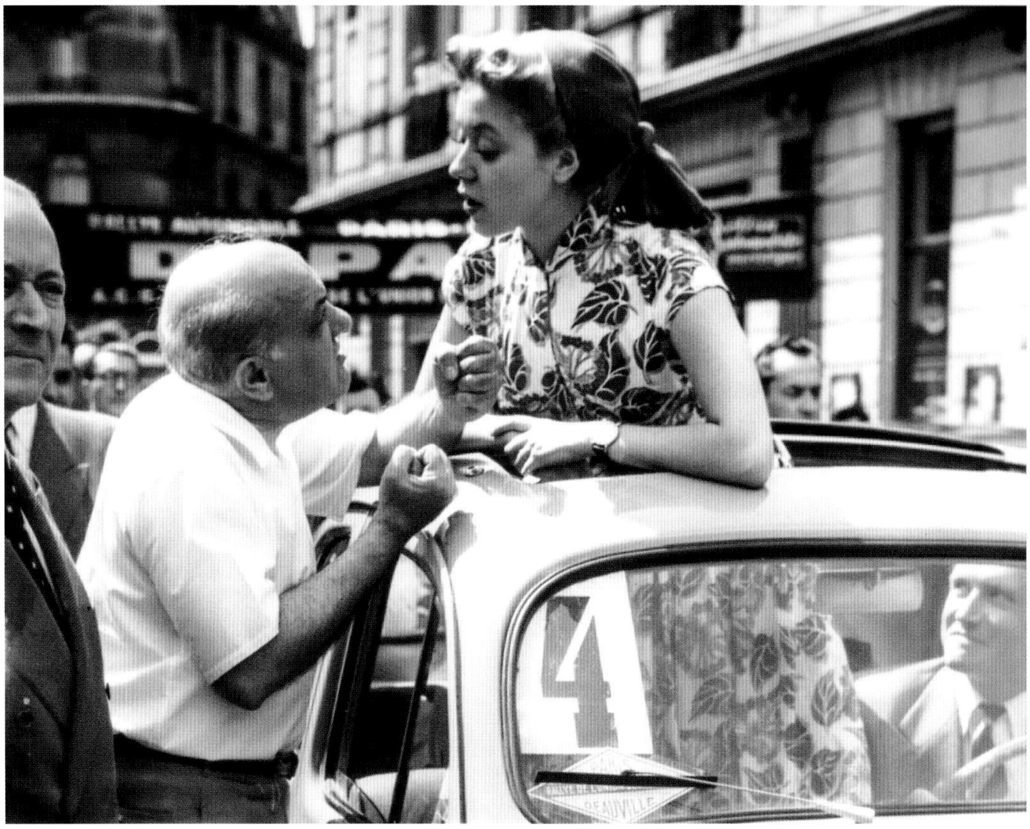

**BELOW:** It *was not uncommon to see series production American cars such as this* 1953 Kaiser Dragon *at concours in Europe.* (Gilles Blanchet Archives)

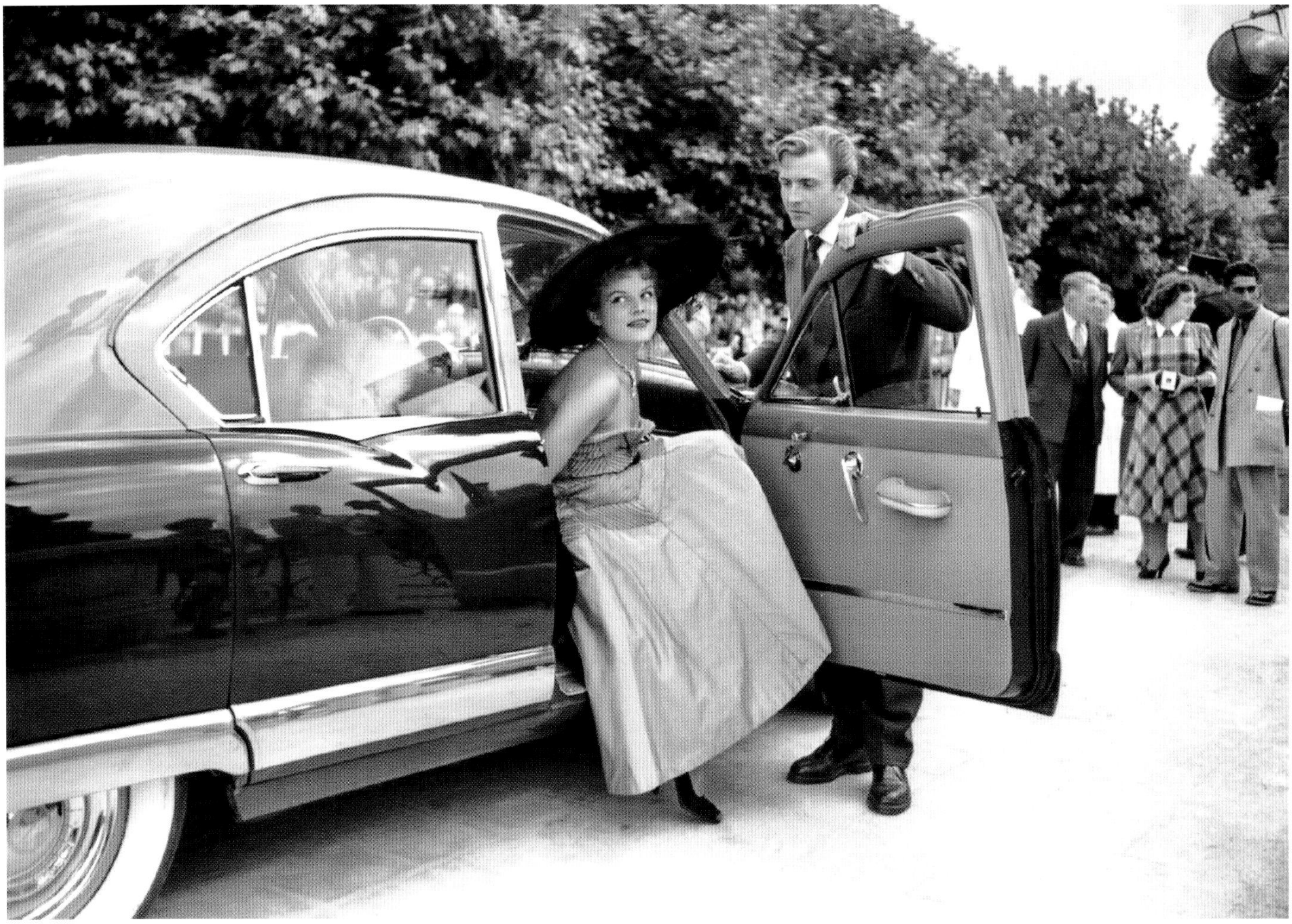

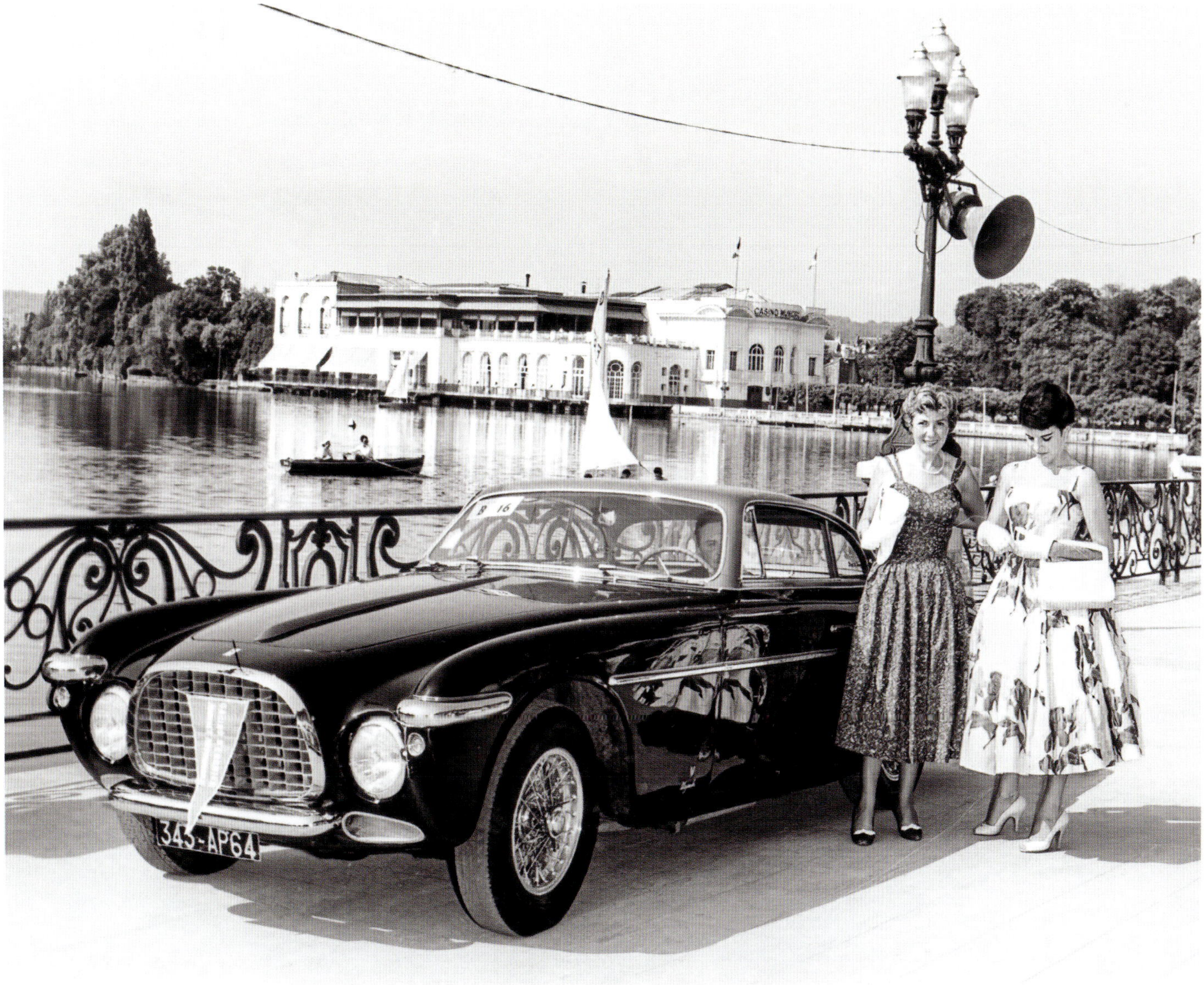


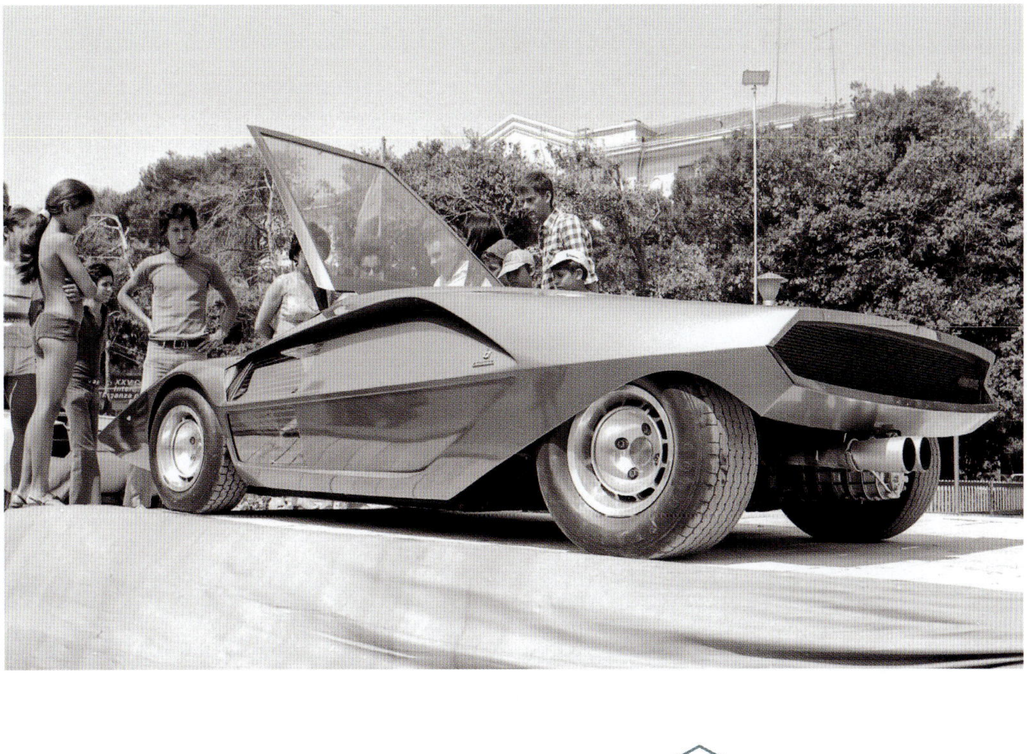

**ABOVE:** *By the second half of the 1950s, more Ferraris appeared at concours d'elegance. This Vignale-bodied Ferrari 212 Inter starred at Enghien-les-Bains on June 15, 1957. (Gilles Blanchet Archives)*

**LEFT:** *Into the 1970s just a handful of concours d'elegance were surviving in Europe. Among them was the one at Rimini on Italy's Adriatic coast. Shown is the Lancia Stratos Zero at the 1971 Rimini Concorso d'Eleganza. (Biblioteca Civica Gambalunga / Rimini Library)*

# Pebble Beach Concours d'Elegance

## THE AMERICAN WAY

After World War II, the Concours d'Elegance de Dinard, along with similar events in other French towns and cities, took a few years to resume. They returned in the early 1950s and continued for about a decade. Although the festive atmosphere and broader audience participation persisted, the events never quite recaptured the glamour and panache of the 1920s and 1930s. By the 1960s, most concours d'elegances in France had faded away.

As Europe bid farewell to these elegant events, the United States was just discovering them. Watkins Glen, near New York, hosted what might have been the first-ever American concours d'elegance, coinciding with the village's inaugural road race. Watkins Glen would later gain fame as the home of the Formula One United States Grand Prix during the 1960s and 1970s.

The first Watkins Glen road race took place on October 2, 1948, followed by another on September 17, 1949, which was won by Miles Collier, with the legendary Briggs Cunningham in a Ferrari 166 coming in second. This event possibly marked the first appearance of a Ferrari in American competition. These races sparked a series of road races across the United States, including two in California in 1950 at Palm Springs and Buchanan Field in San Francisco.

Sterling Edwards won both Californian races in his Edwards R-26 Special Sport Roadster, a lightweight

*The first Pebble Beach Concours d'Elegance was in 1950. Both modern and historic vehicles were present for the debut event. (© Julian P. Graham / Pebble Beach Concours Lagorio Archives)*

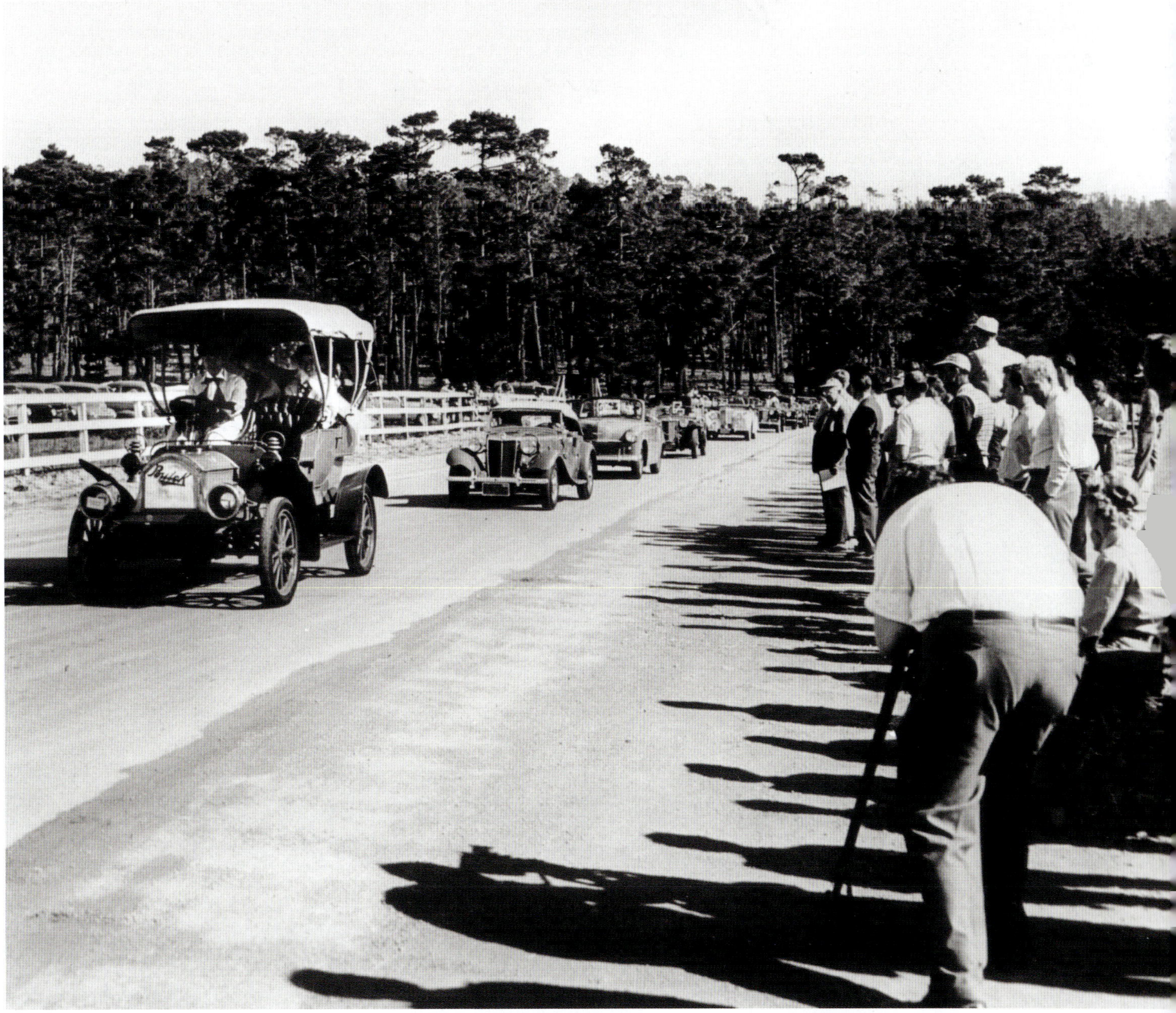

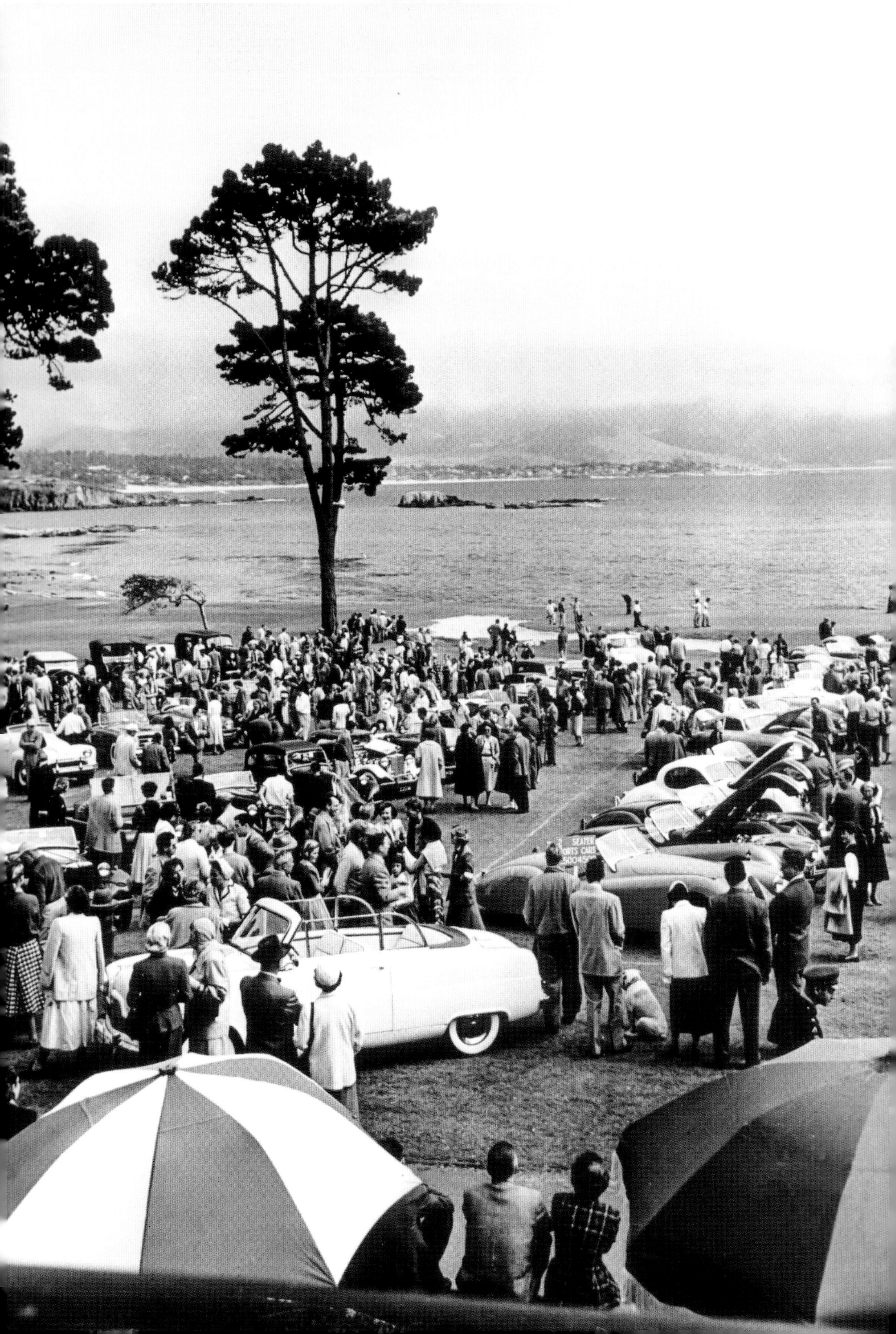

fiberglass-bodied two-seater sports car with a Lincoln engine on a shortened Ford chassis, designed by Edwards himself. He hoped to attract enough attention to secure funding for series production of the car.

Riding on his success, Edwards proposed a road race at the exclusive Pebble Beach property, owned by Del Monte Properties Company. Del Monte's president, John Boit Morse, approved the idea, and the first Pebble Beach race was held on November 5, 1950. The event featured several races, with the most prestigious being the 25-lap Pebble Beach Cup, which was won by the 23-year-old Phil Hill in a Jaguar XK120, despite Edwards finishing 14th in his R-26 Special Sport Roadster.

However, Edwards did win the inaugural Pebble Beach Concours d'Elegance with his Edwards R-26 Special Sport Roadster. Inspired by Watkins Glen's format, Pebble Beach's public relations manager, Gwenn Graham, added a stylish judged concours d'elegance to the event. Edwards' handsome, European-style convertible stood out among the entries, with around 35 cars competing for the Concours d'Elegance.

In the early years of the Pebble Beach Concours d'Elegance, new cars were entered and judged, reminiscent of the European concours before the war and into the 1950s. The 1951 event was won by a brand-new Jaguar Mk VII sedan, presented by Mrs. Charles Hornberg Jr., followed by a Jaguar XK120 Fixed Head Coupé in 1952 and 1954. The 1953 winner was a brand-new Austin-Healey 100, which was commendable given that the total number of entries that year may have topped 75.

**OPPOSITE:** *This image from the 1953 Pebble Beach Concours d'Elegance shows an interesting mix of mostly modern cars with a sprinkling of pre-war automobiles. (© Julian P. Graham / Pebble Beach Concours Lagorio Archives)*

BELOW: *1955 was the landmark year when the Best of Show went to a historic vehicle – a Pierce-Arrow 41 LeBaron Town Car from 1931 – owned by racing star Phil Hill. (© Julian P. Graham / Pebble Beach Concours Lagorio Archives)*

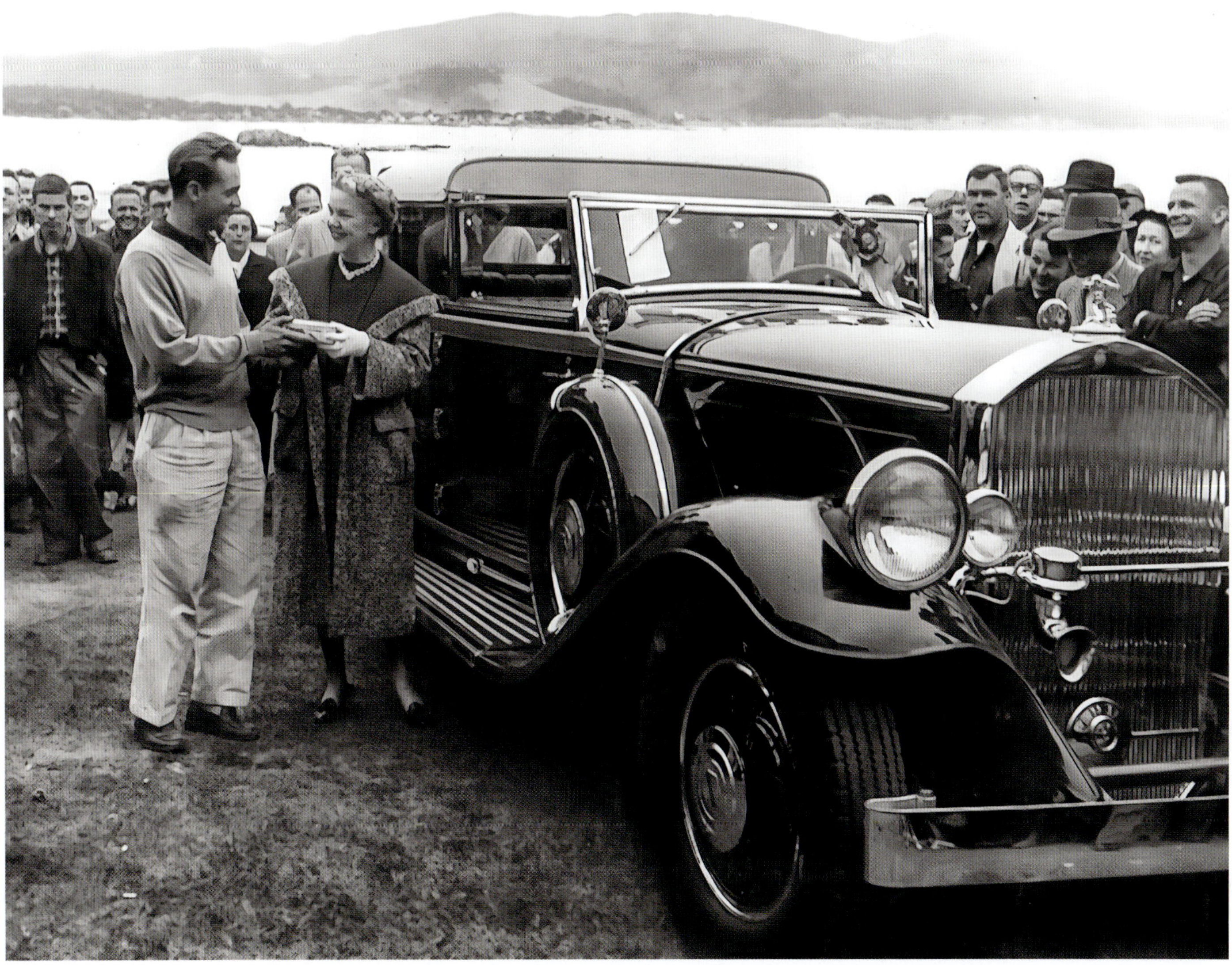

## A Landmark Year

1955 was a landmark year for the Pebble Beach Concours d'Elegance, as it marked the first time a new car did not win. Instead, the Best of Show was awarded to a beautifully restored 1931 Pierce-Arrow Model 41 LeBaron Town Cabriolet, which had been in Phil Hill's family since new. This victory, along with the runner-up being a 1934 Rolls-Royce Phantom II Continental, signaled a shift in preference towards older, pre-war cars. (Incidentally, it was an impressive double for Phil Hill, who went on to win the Pebble Beach road race too, in a Ferrari.)

This trend continued, with a 1930 Bugatti Type 37 Grand Prix winning Best of Show in 1956, marking the beginning of Bugatti's dominance. By 2024, the marque eventually won ten times, equaling the record set by Mercedes-Benz.

1956 also saw the last Pebble Beach Road Race paired with the concours, due to a fatal accident involving Ernie McAfee. This led to the event moving to the Laguna Seca circuit, which was built specifically to host the Pebble Beach Road Races, and the events were still tied together through the remainder of the 1950s. A 1937 Rolls-Royce Phantom III Mulliner Sedanca de Ville won Best of Show in 1957, beating a 1931 Cord L-29 cabriolet.

By then, other concours d'elegances had emerged in the United States, such as the Watkins Glen and the Hillsborough Concours d'Elegance, which started in 1956 in the San Francisco Bay Area. In 1958, the Greenfield Concours d'Elegance in Monterey featured as many as 300 cars.

Despite the growing popularity of concours events, Pebble Beach faced struggles and skipped the year 1960, with bad weather being the excuse. But the concours was moved to the following spring, five months later, when a 1930 Packard 740 Roadster won Best of Show. By the turn of the decade, an exciting rivalry had also developed between William Harrah and J.B. Nethercutt, with the latter's double win in 1958 and 1959. William Harrah went on to dominate Pebble Beach with wins in 1963 (1931 Pierce-Arrow 41 LeBaron Custom Club Sedan), 1964 (1935 Bugatti Type 50 Coupé Profilé), and 1966 (1931 Bugatti Type 41 Royale Binder Coupé de Ville).

By the end of the 1960s, it was J.B. Nethercutt's turn to dominate, when he captured Best of Show in 1969 with a 1934 Duesenberg J Murphy-style Dual Cowl Phaeton and in 1970 with a 1931 Daimler Double Six 50 Royal Limousine. Nethercutt would go on to win Best of Show a record six times.

*The Ferrari 166 Barchetta, which previously won the Pebble Beach Road Race participating at an early- to mid-1970s Pebble Beach Concours d'Elegance. (© William C. Brooks / Pebble Beach Concours Lagorio Archives)*

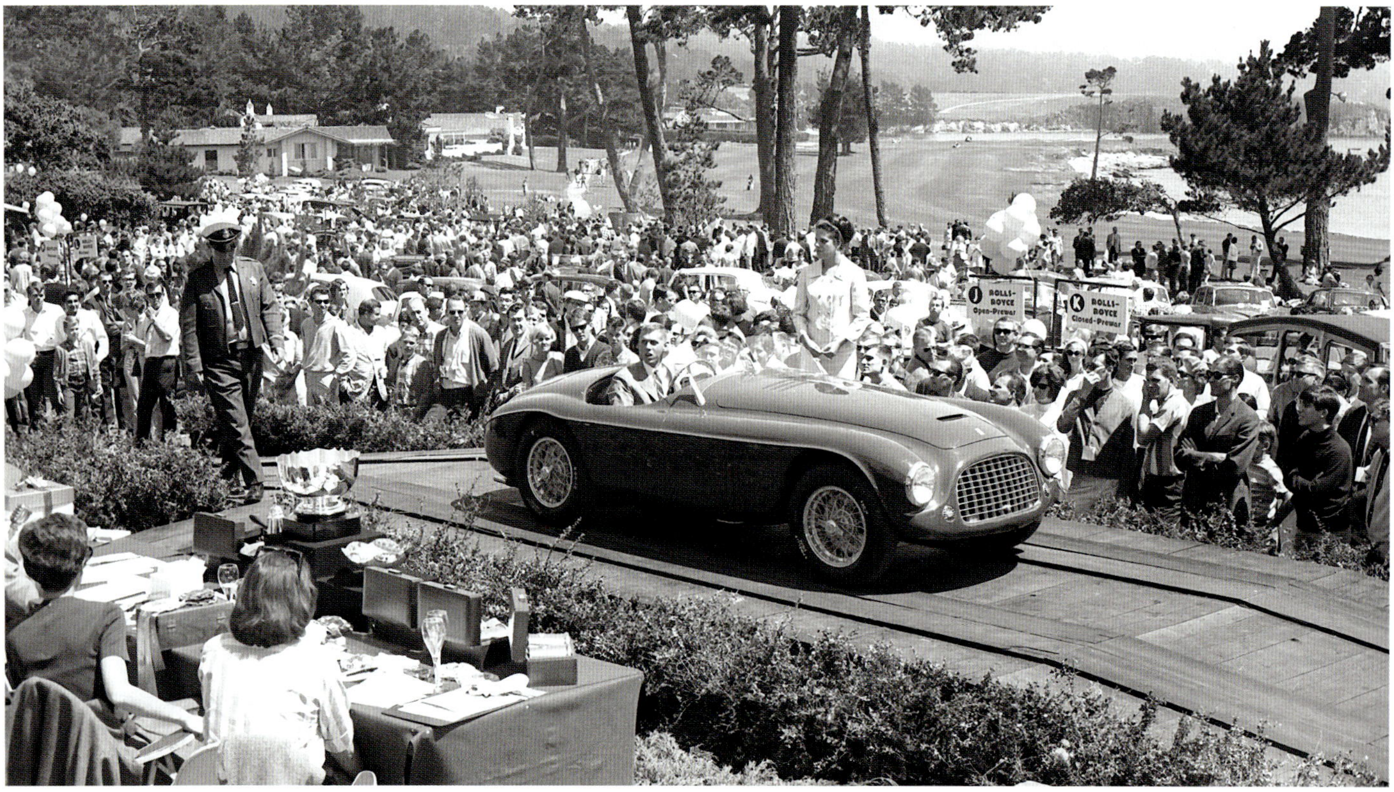

## The Heumann-Tryon Years

In 1972, Jules Heumann won with a 1922 Hispano-Suiza H6B Labourdette Torpédo. Heumann soon took on a role in organizing the Pebble Beach Concours alongside Lorin Tryon, following the passing of Lucius Beebe (a notable Rolls-Royce enthusiast) and Gwenn Graham, in 1966 and 1968 respectively. Under Tryon and Heumann's leadership, the event began to feature a more prestigious and international selection of cars, attracting enthusiasts and significant media attention.

Tryon and Heumann introduced a two-tiered judging system that prioritized recruiting class judges with specialized expertise. Under this system, cars were required to win their class before competing for Best of Show, a structure that not only became the foundation of the Pebble Beach judging process, but also influenced judging systems at many other concours events.

In refocusing the concours on classic cars, Tryon and Heumann also significantly expanded Pebble Beach's regular classes, adding new marques and models. They established an annual tradition of spotlighting special marques and boutique classes, which had previously been more occasional. Their foresight led them to introduce new classes, including one for Ferrari in the mid-1970s, and later for hot rods and microcars by the late 1990s. Additionally, Tryon and Heumann conceived and formalized the Tour d'Elegance, integrating it into the concours judging process, a lasting contribution to the event's legacy.

In 1982, Richard Adatto, an author and French car expert, began his involvement with Pebble Beach by curating an exhibit featuring over twenty Delahaye and Delage cars. The following year, an Isotta Fraschini with a stunning Castagna body won Best of Show. In 1984, a truly rare Cunningham V-7 Town Car took the top prize. However, it was the display of all six extant Bugatti Royales in 1985 and the three Alfa Romeo B.A.T.s in 1989 that truly put Pebble Beach on the map.

**BELOW:** *The golden era of the Pebble Beach Concours began in 1972 when Lorin Tryon (middle) and Jules Heumann (left) were appointed as co-chairmen. (© William C. Brooks / Pebble Beach Concours Lagorio Archives)*

**BOTTOM:** *It was in 1998 when the first Tour d'Elegance took place the day before the concours. (© Douglas Sandberg / Pebble Beach Concours d'Elegance)*

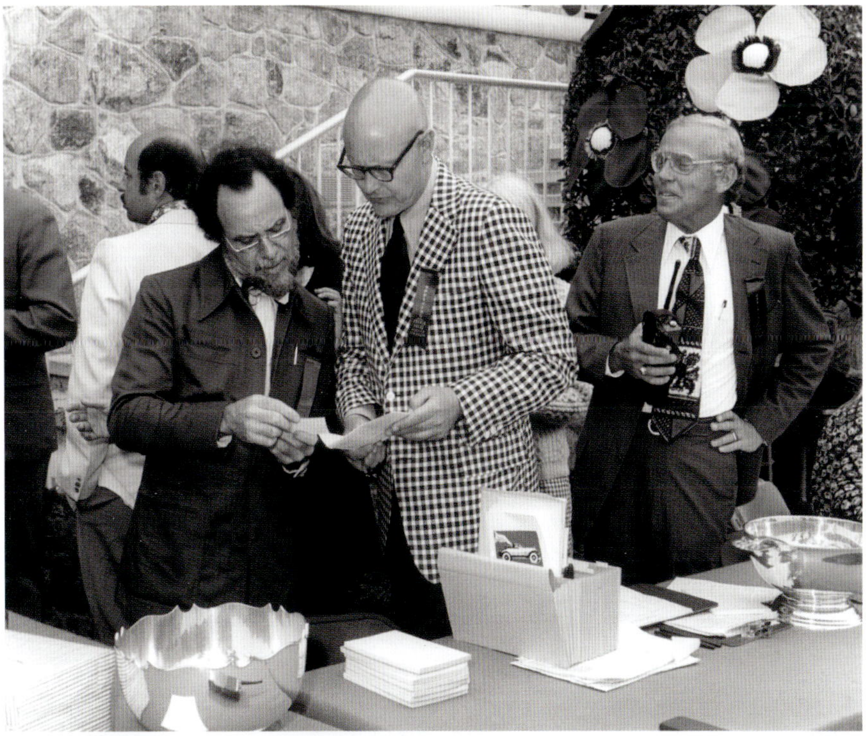

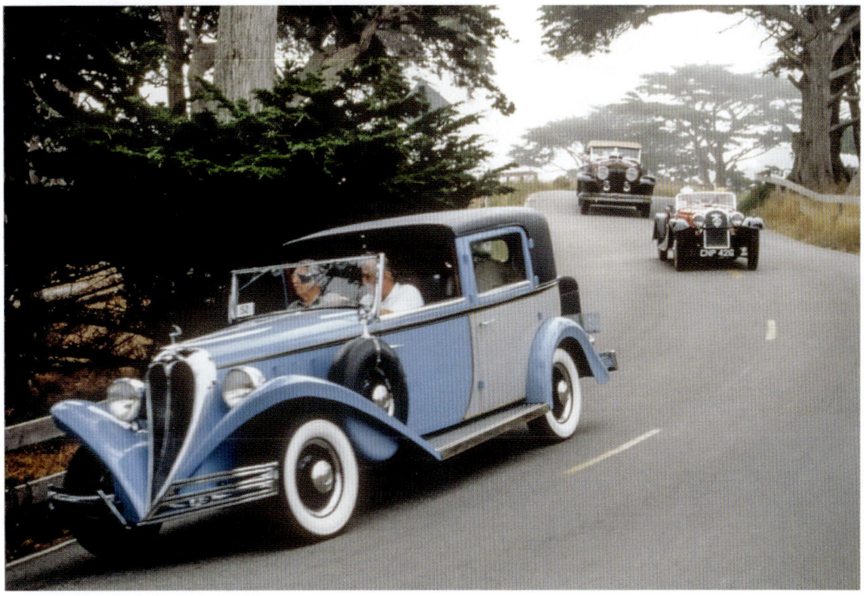

## The Sandra Button Era

In 1985, Sandra Button, a business and hospitality administration graduate, joined the Pebble Beach Company. By the end of the 1980s, Button was on the Concours Executive Committee, and she became the Executive Director by 1992. When Glenn Mounger became the new Chairman in 2000, shortly after Jules Heumann's retirement that year, Button had a say in that. Mounger, in turn, asked Button to partner with him as co-chairman in 2002. Incidentally, Mounger pushed for more recognition of preservation cars, and he also developed unique displays focused specifically on singular chassis groups. From 2005, Sandra Button became the sole Chairman of Pebble Beach Concours d'Elegance, when Mounger stepped down.

Belgian enthusiast and Monaco resident Christian Philippsen, inspired by what he had heard about

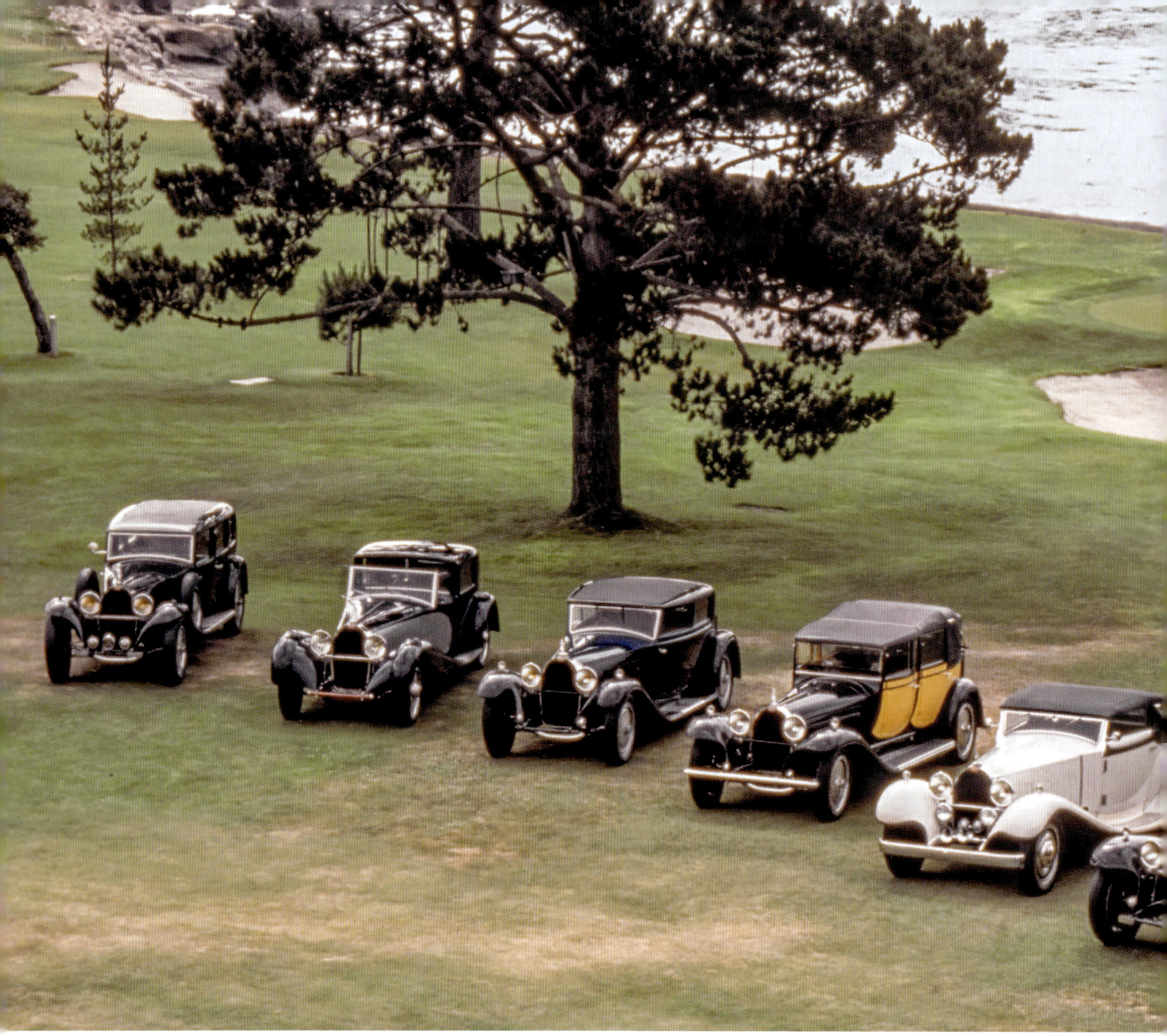

*In 1985, when Pebble Beach managed to gather the six surviving Bugatti Royales, the concours grabbed headlines worldwide. (Steven A. Gann)*

Pebble Beach, visited in the early 1980s. "I was absolutely amazed by what I saw," he said. "I thought it was too good to be true and expected to be disappointed on my return, but I was not." Philippsen noted that the concept of the concours d'elegance, originally French, no longer existed in Europe. This led him to co-found the Automobiles Classiques à Bagatelle (with the branding later changing to Automobiles Classiques et Louis Vuitton à Bagatelle, and then Louis Vuitton Classic à Bagatelle) in Paris in 1988, alongside Antoine Prunet, the editor of the prestigious French magazine *Automobiles Classiques*.

Pebble Beach not only revived the concours concept in its country of origin, France, but also inspired similar events across the United States and Europe. The Meadow Brook Concours d'Elegance began in 1979, followed by Huntington Beach Concours in 1986, and both The Amelia Concours and Greenwich Concours in 1996. Concorso d'Eleganza Villa d'Este, originally started in 1929 but dormant since 1949, was also revived in 1995.

Despite these developments, Pebble Beach remained the flagship event, leading the movement. Christian Philippsen explained that even if the event lost a bit of momentum after the early years, the duo Jules Heumann and Lorin Tryon who were called on to rescue the concours, revitalized it, focusing on classic cars. "That's how Pebble Beach became the success we know," Philippsen said. "Success breeds success and interest, leading to the revival of other concours events like Bagatelle and Villa d'Este."

Winning at Pebble Beach became a significant goal for serious car collectors worldwide. Achieving this required restoring cars to standards often surpassing their original condition, leading to a trend of

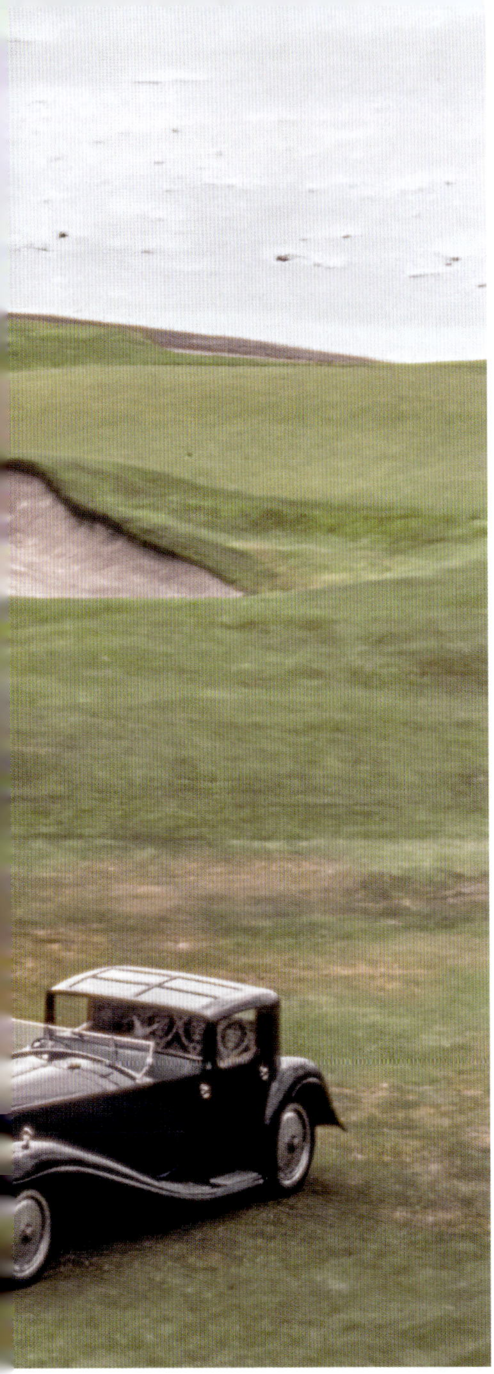

over-restoration. Philippsen explained, "When Pebble Beach became a serious event, people wanted to win Best of Show. America's car culture has always valued chrome, so the shinier the car, the better. That trend has, however, been dramatically reversed in recent years, with more attention dedicated to preservation and authenticity."

Pebble Beach also brought fame to some of the most prominent (and wealthiest) collectors of historic vehicles. During the mid-1950s to mid-1970s, collectors like William Harrah and J.B. Nethercutt vied for top honors. Later, figures such as Arturo Keller, Robert Meyer, John Mozart, Ralph Lauren, Sam and Emily Mann, William "Chip" Connor II, Charles Bronson, Robert and Anne Brockinton Lee, Jon Shirley, Jim Patterson, Peter and Merle Mullin, Bruce McCaw, Sir Michael Kadoorie, and others have dominated Pebble Beach and other leading concours events worldwide. Alongside these collectors, renowned restorers like Paul Russell, David Cooper of Cooper Technica, Stone Barn, LaVine Restorations, RX Autoworks and RM Restorations also gained recognition.

However, the issue of over-restoration became a controversial topic within the historic vehicle community. Many criticized cars restored to an overly pristine condition, labeling them as suffering from a "Pebble Beach style restoration." Sandra Button addressed this stigma: "I think we've made great progress in this regard. I think people forget that when the Pebble Beach Concours d'Elegance began, in 1950, many 'old' cars and racing greats and one-off concepts were still being scrapped in favor of all that was new. When this concours shifted its focus from new to collector

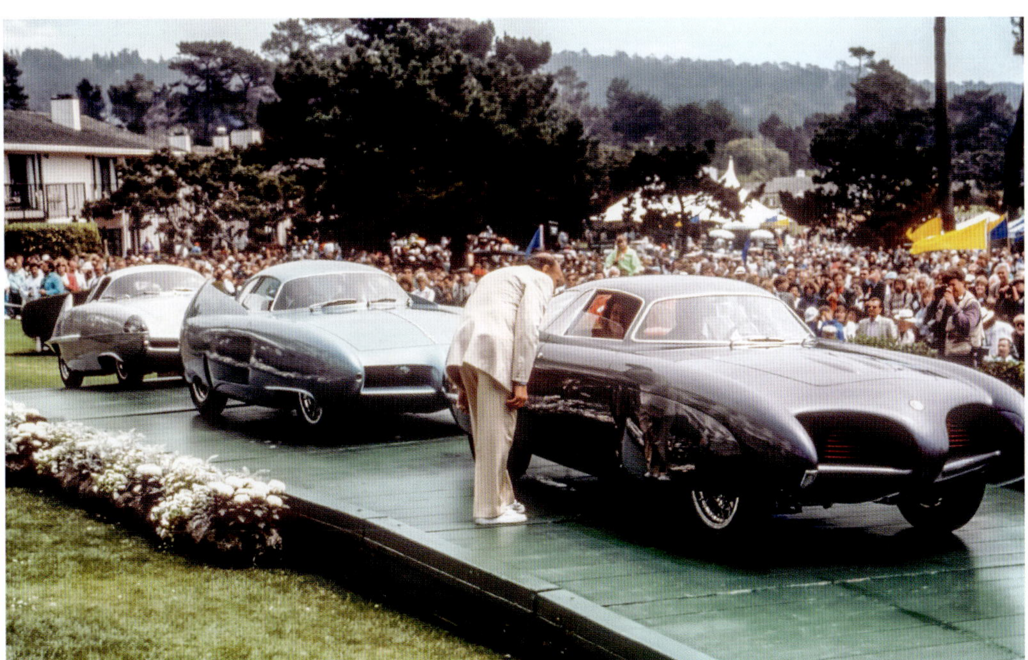

*Four years later, in 1989, the three Alfa Romeo B.A.T.s were united, and Pebble Beach managed to get Nuccio Bertone to be there too. (© Dean Batchelor, Ron Kellogg Collection)*

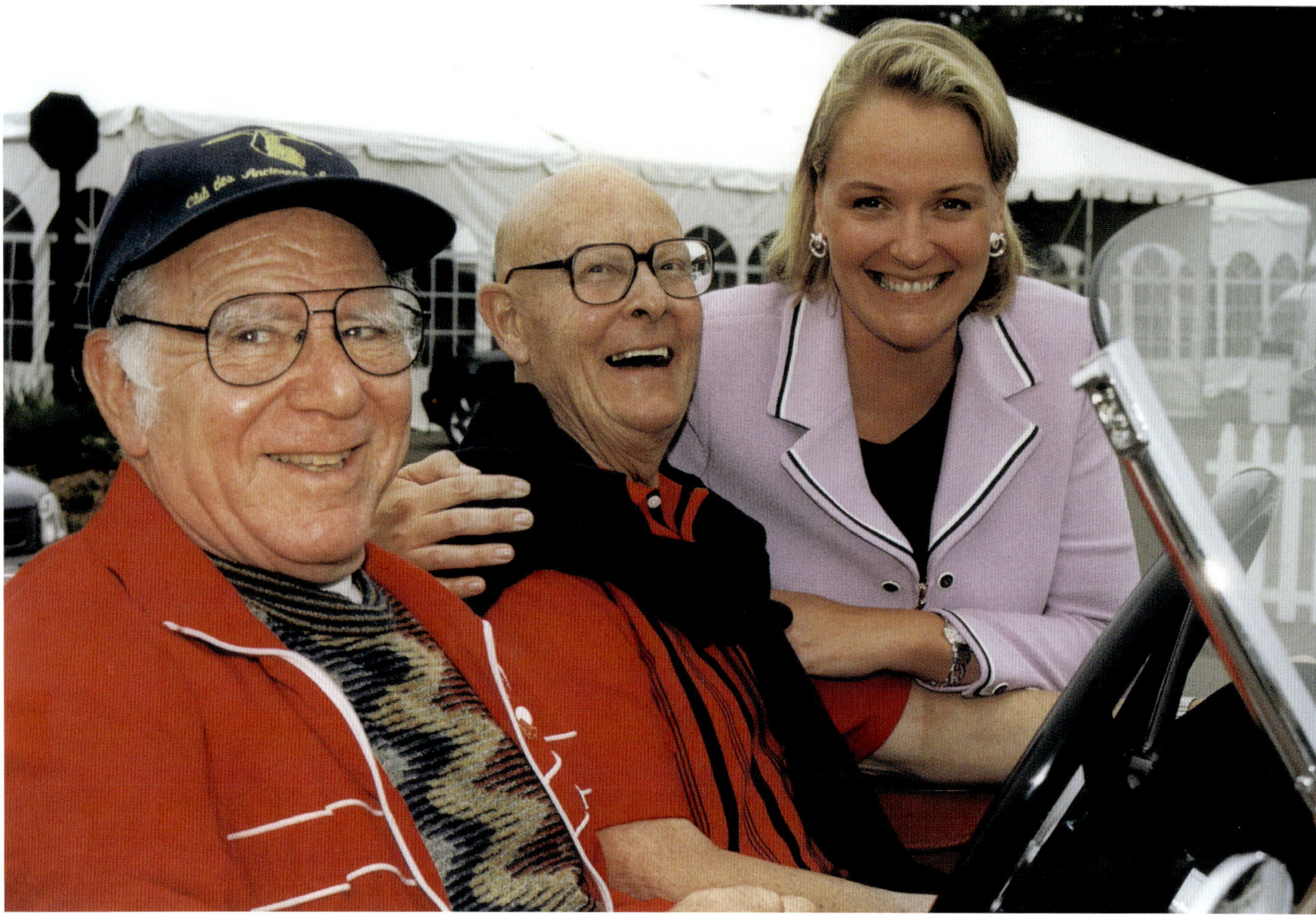

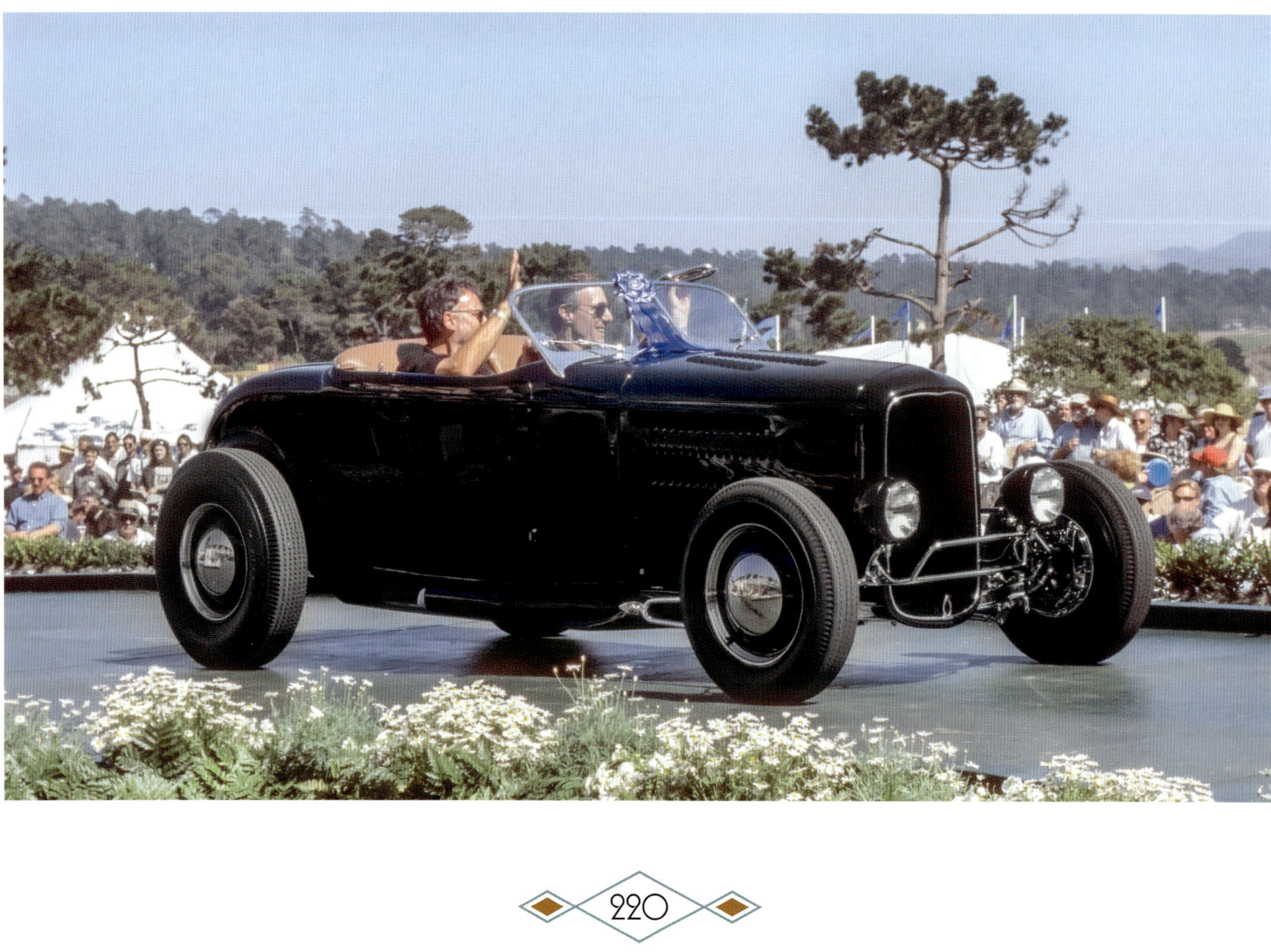

cars, and other concours soon followed suit, it gave many great old cars a reason for their continued existence. More people saved and restored and shared them. I often wonder how many great cars were saved because of this shift. Many, I'm sure."

"Inevitably, early restoration efforts were lacking in many respects; restorers did not have anywhere near the tools and resources that exist now," elaborates Button. "And yes, expectations and standards have changed. There is no doubt that increased competition led to over-restoration. But as early as the late 1970s and early 1980s, some of our key judges – Jack Passey and Phil Hill among them – began to express a desire to discourage over-restoration."

Another criticism was that many of the cars brought to Pebble Beach were essentially "trailer queens", vehicles that looked beautiful but were not intended to be, or capable of being, driven. Sandra Button countered this by pointing out: "Over the past quarter century, two important movements – the push to make certain show cars that can be driven, and the increased appreciation for originality – have also played a role in fighting over-restoration. From the start, we've had an expectation that cars on the Pebble Beach Concours should be driven. Initially, we required that they drive onto our show field and over the ramp – and several of our early entrants, such as J.B. Nethercutt, took great pride in driving their cars to and from the event. As competition heated up, the driving diminished. So, in 1998, we instituted the Tour d'Elegance to underscore the importance of driving, and participation in the Tour can directly impact the results of our class competitions and Best of Show."

## The Recognition for Preservation

One of the most significant advancements at Pebble Beach was the introduction of a class for vehicles that had remained unrestored and were in their original, well-preserved state. "It was in 1999 that we formally paired with FIVA to offer trophies for preservation," explains Sandra Button. (FIVA, incidentally, is the Fédération Internationale des Véhicules Anciens, or the International Federation of Historic Vehicles.)

"We had previously begun to showcase preservation cars, often pairing them with a restored car of a similar marque and model," elaborated Button. "In 2001, we offered our own first class for preservation. That has expanded to pre- and post-war and special marque preservation classes, and we are now working to distinguish carefully preserved cars from barn finds. In the meantime, a wonderfully preserved car was among our Best of Show nominees in 2023 – and in 2024 a beautifully preserved car, Fritz Burkard's Bugatti Type 59, won our top award."

Credit for this change should also go to veteran Italian historic vehicle expert and historian Adolfo Orsi (as well as FIVA's Technical Commission head Derek Bonzom), even if he insists that, "honestly, I don't consider myself one of the founders of the preservation movement because before me there have been other people who had already launched this message. I was in the right place at the right moment because I was called as a judge to Pebble Beach in 1997, and I judged for a couple of years normal cars and classes and I was surprised to see a lot of over-restored vehicles."

"At the time Francesco Guasti, a good friend, was the president of FIVA, and FIVA was not present in the US, and their goal was to be present there," explained Adolfo Orsi in an interview in February 2024. "So, I suggested to him to establish the FIVA Award which was given for the first time in 1999. Until then, the owners of preserved cars didn't take their cars to Pebble Beach, as they did not have any possibility to win a prize. For the first year, we had some five or six cars, the second eight, the third eleven and now we are judging more than 20 cars every year, which are in the two preservation classes of the concours, plus in all the other classes on the field."

At the same time, there have been several criticisms leveled at the Pebble Beach Concours d'Elegance, such as the tendency to favor pre-war cars over post-war ones, and that it is too elitist. Responding to the former issue which may have arisen as most of the judges are relatively elderly, Sandra Button has this to say: "Not at all – although some people might guess that was the case. Age does not predict how a judge will vote. Just over half of our ballots now go to judges in their 40s, 50s and 60s, and in recent years most of their votes have gone to pre-war cars. It is more often our older judges who cast votes for post-war cars."

The other interesting point that Button makes is: "Marque loyalty has greatly diminished – and we think that is a good thing. Several years back, I released several judges that would only vote for their

**OPPOSITE TOP:** *Sandra Button (right) posing with the stalwarts Jules Heumann (left) and Lorin Tryon (center) in the year 1998. (© Douglas Sandberg)*

**OPPOSITE BOTTOM:** *The first Hot Rod Class was introduced in 1999. (© Scott Burton, Pebble Beach Concours d'Elegance)*

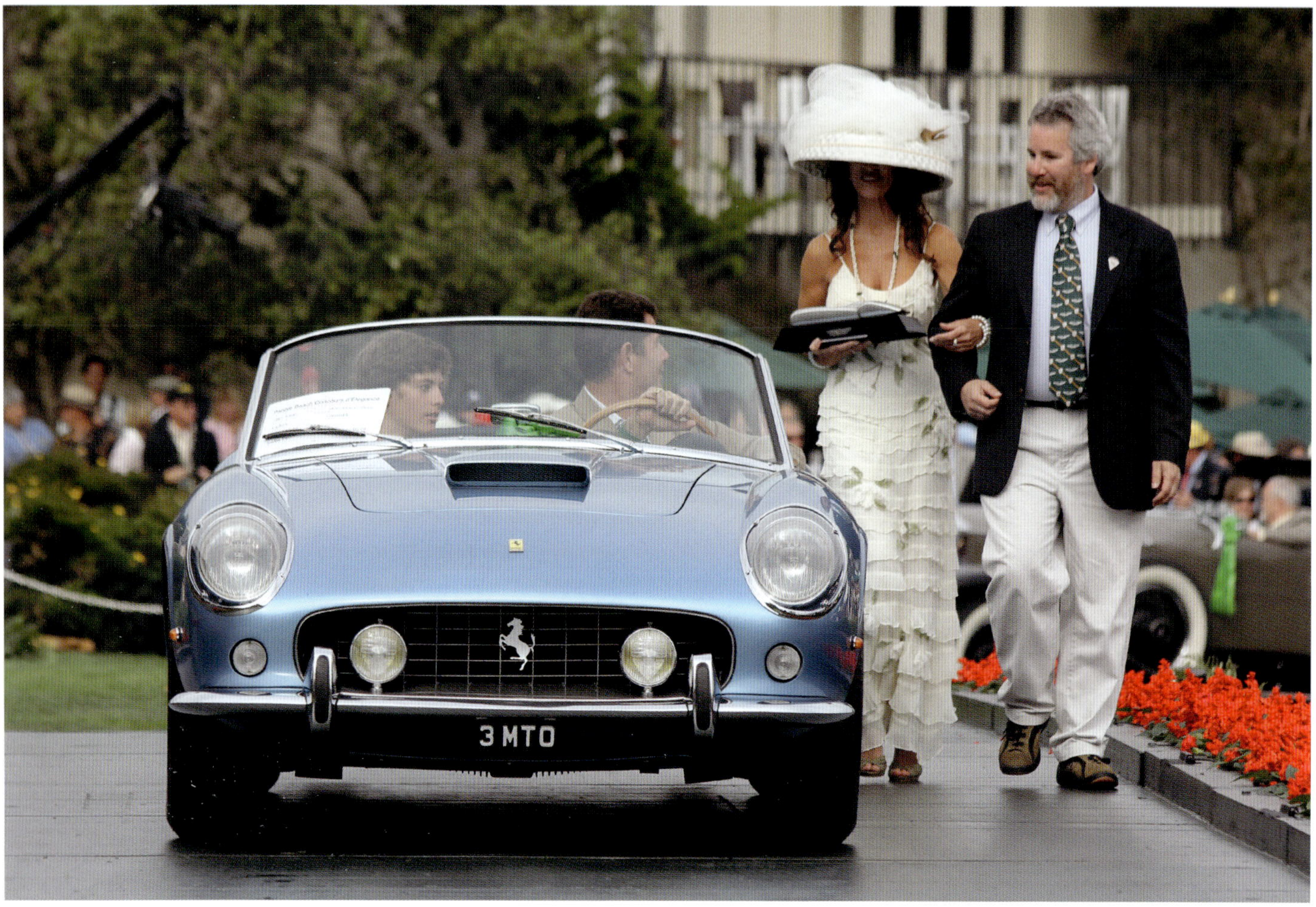


**ABOVE:** *A Ferrari 250 GT California Spyder at the 2010 Pebble Beach Concours d'Elegance. (Hugues Vanhoolandt)*

**RIGHT:** *At the 2010 Pebble Beach Concours d'Elegance an Alfa Romeo 8C 2900 A Spider regales enthusiasts. (Hugues Vanhoolandt)*

**OPPOSITE (BOTH):** *The Best of Show at the 2010 Pebble Beach Concours was a Delage D8-S De Villars roadster (shown in the Delage showroom in 1933 on page 128) owned by Jim Patterson, who poses with the car and trophy. (Hugues Vanhoolandt)*

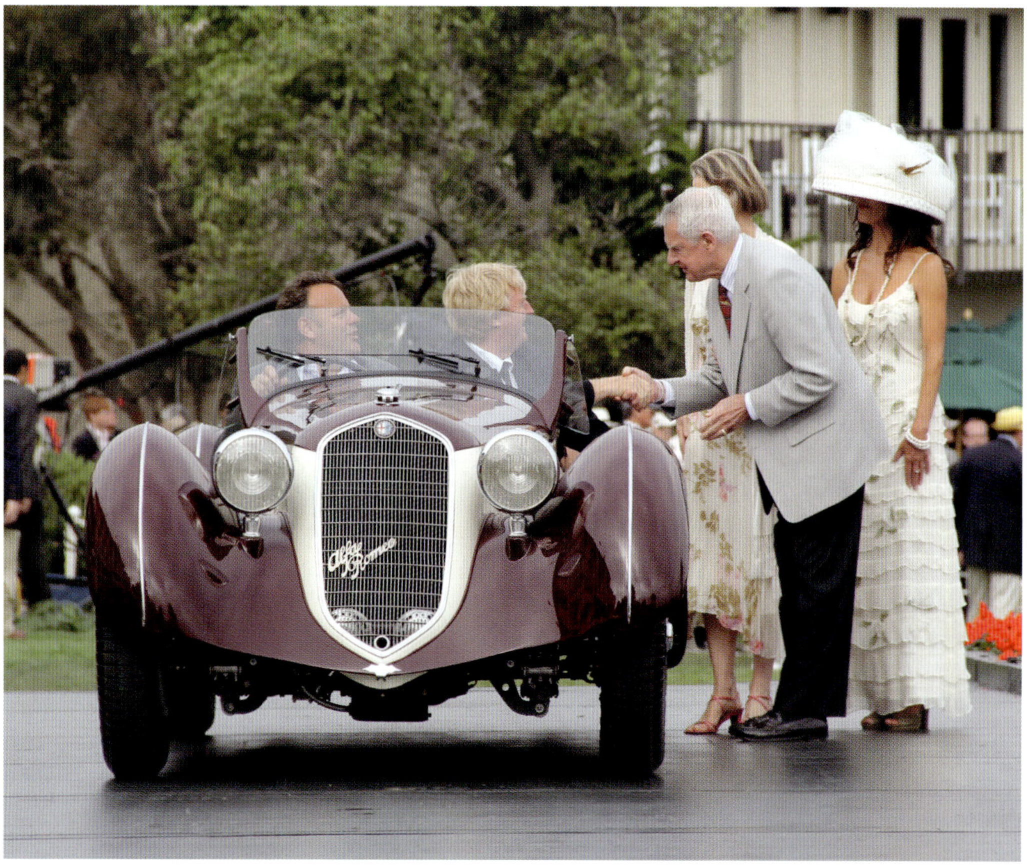

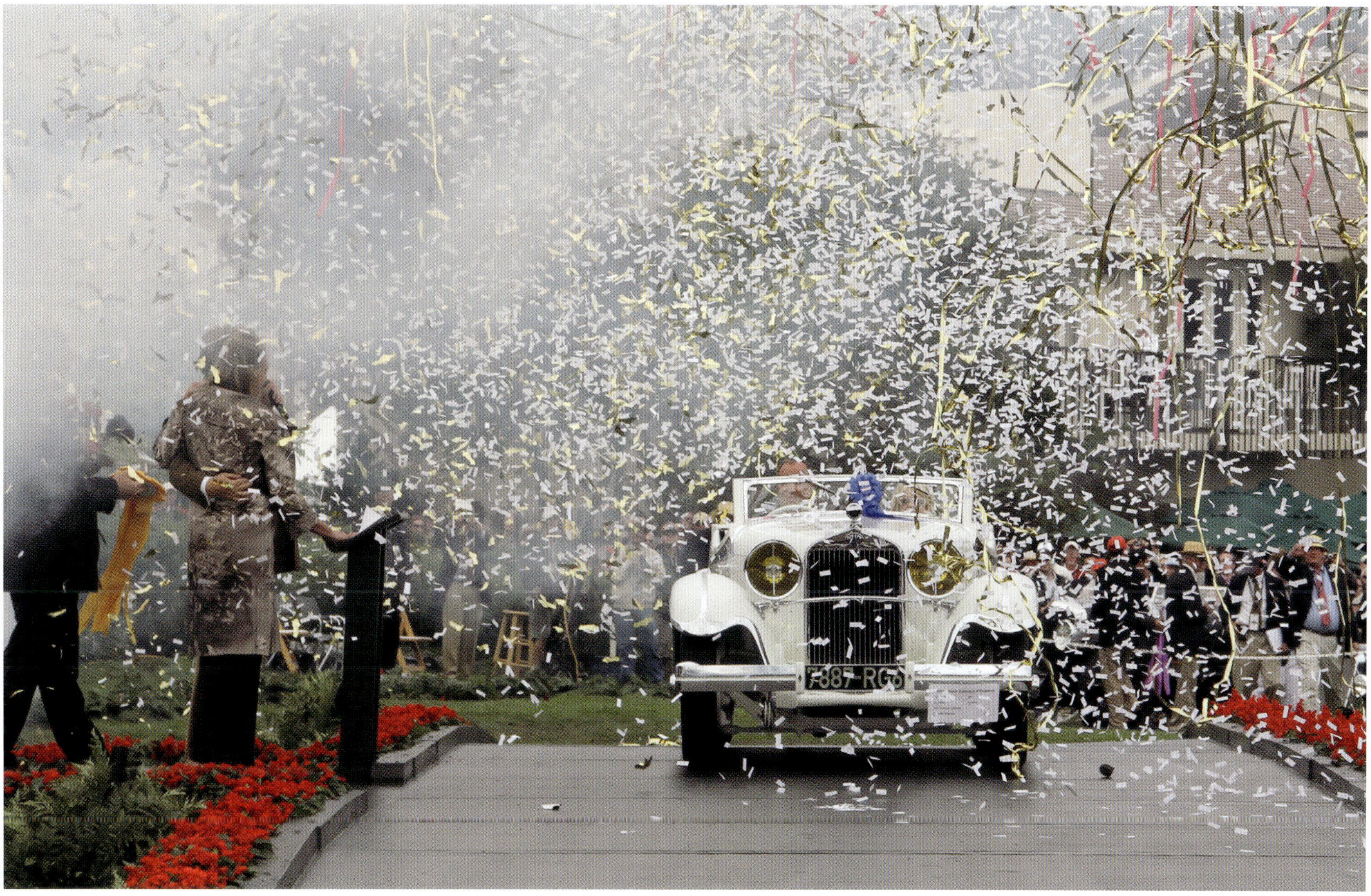

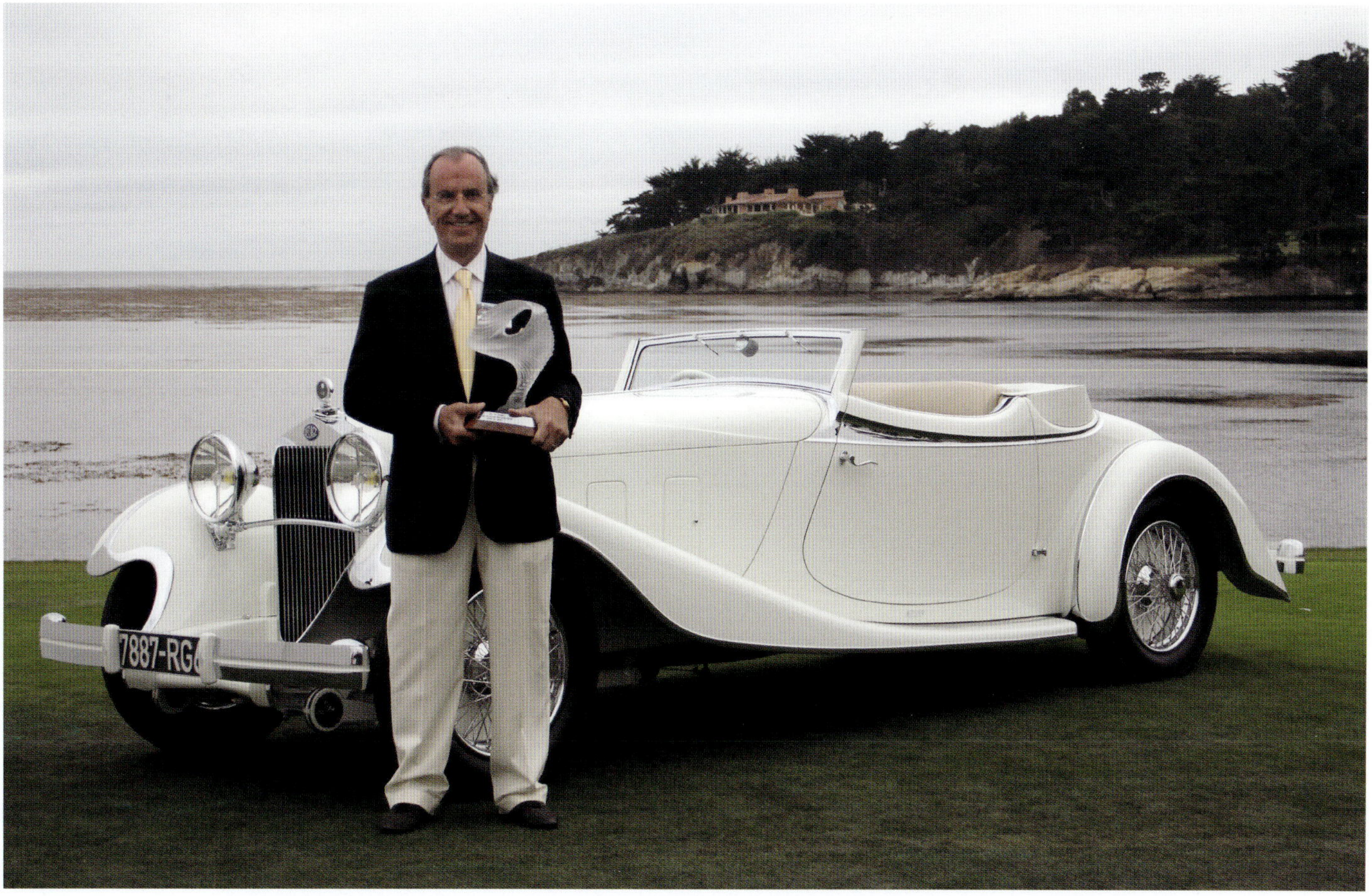
7887-RG

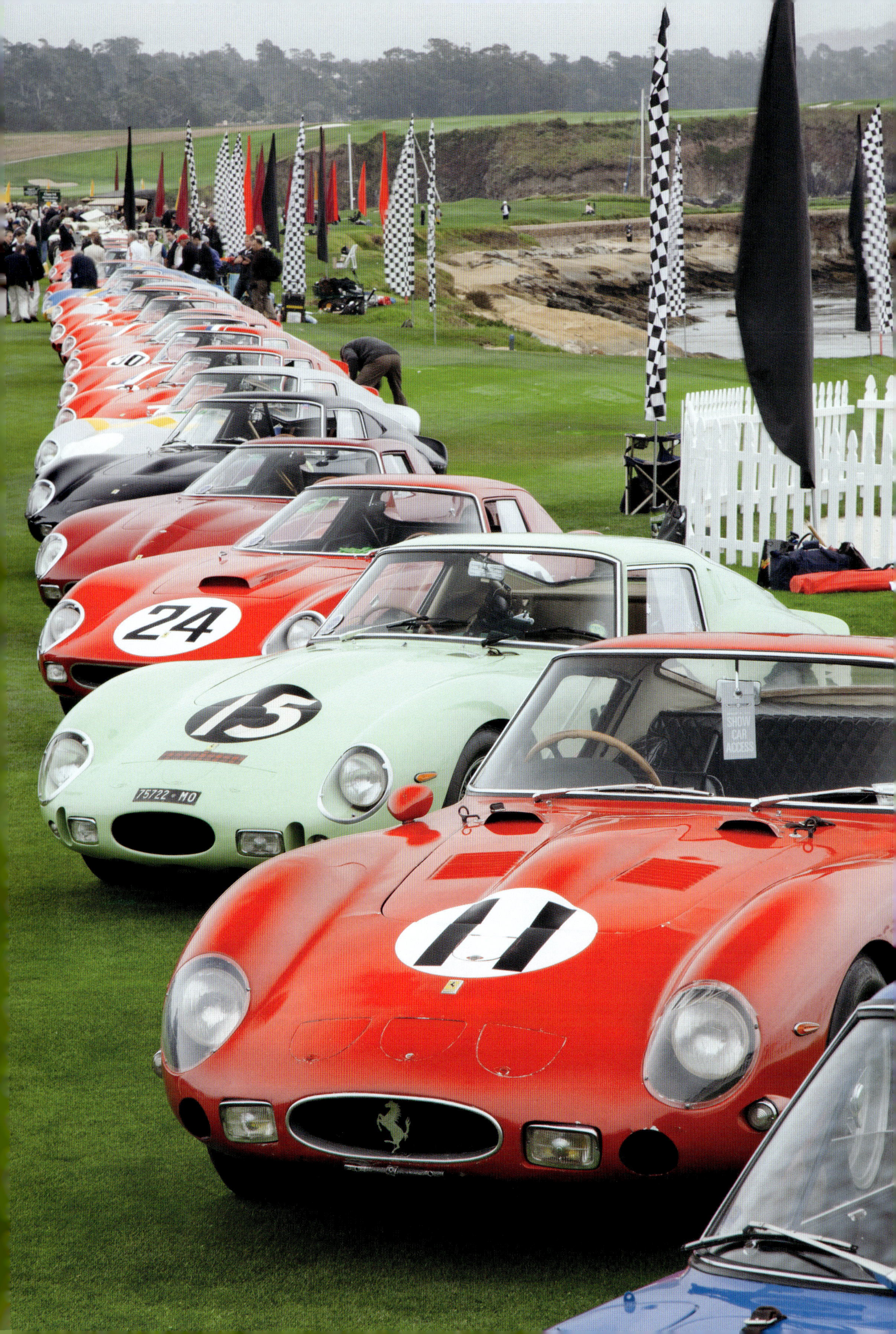
24
15
75722 • MO
SHOW
CAR
ACCESS

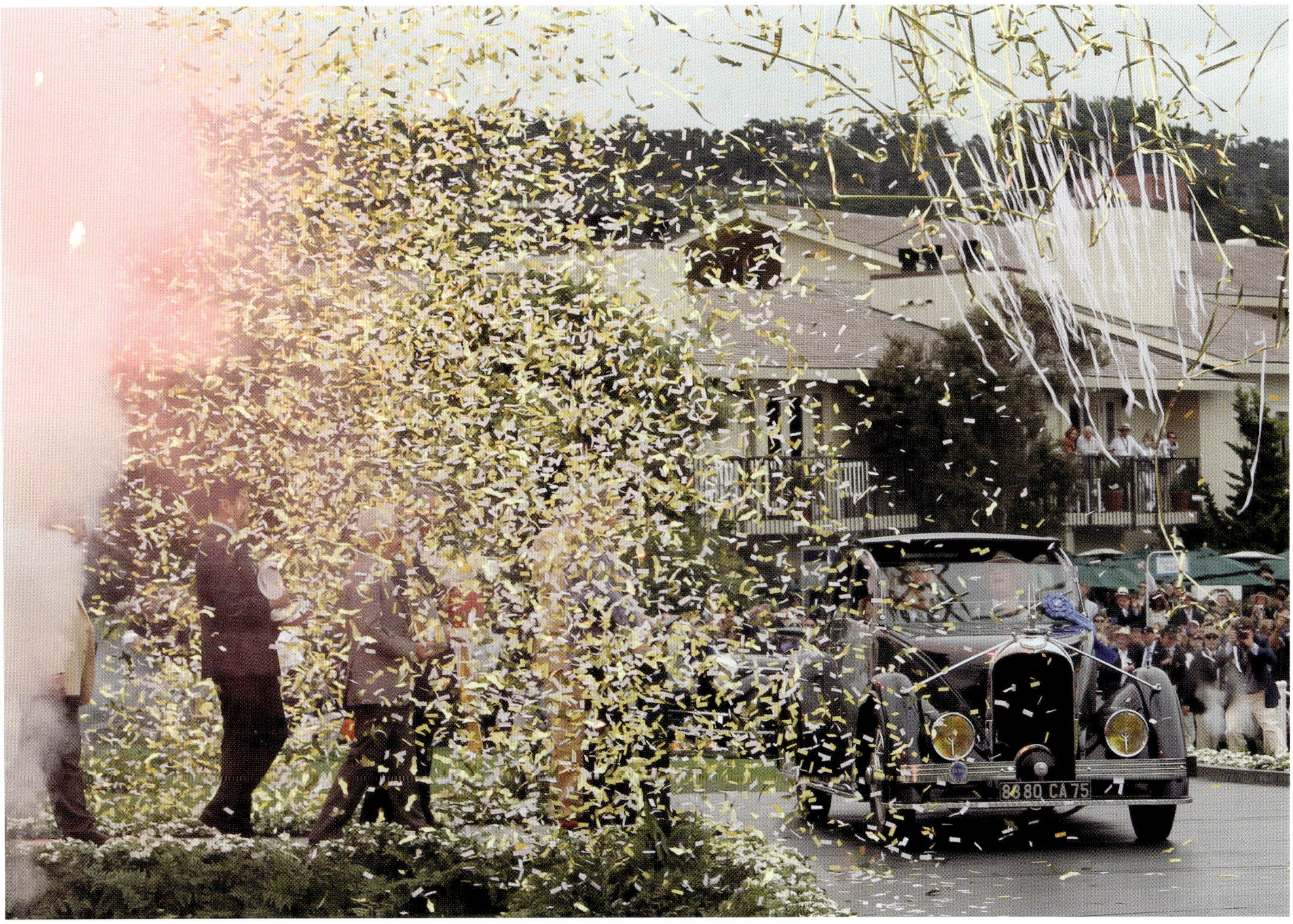

favorite marque; they refused to think 'outside of the box'. We ask judges voting for Best of Show to give equal consideration to cars from all marques and eras."

"Over the past decade, more post-war cars are getting more votes for Best of Show," insists Button. "Recent years have witnessed over a half dozen post-war cars among the nominees for that award. We are close but not yet at the tipping point for that. Also, our show field has changed dramatically over time, particularly in the past two decades. Ferrari was featured as early as the mid-1970s, but we've more recently featured other post-war marques like Lamborghini and McLaren, or post-war cars from marques like Porsche, as well as hot rods, sport customs, etc. Post-war cars generally now take up more than half of our show field. In fact, I would say the proportions are around 55:45, with about 120 post-war cars as compared to around 100 pre-war. And we are now showcasing cars created as recently as 2017 and 2023."

Most importantly, "We've also revised our judging forms to incorporate a broader definition of elegance - one that takes historic importance, race and show history into account," says Button. "Elegance is not just a matter of the eye alone; it takes the reaction of your mind and heart into account."

Also, the number of judges involved has expanded considerably: "Currently there are as many as 100 class judges and 55 honorary judges to judge 220 cars arranged across 28 classes," points out serial judge Richard Adatto. "When I started more than 30 years ago, there were barely a dozen of us judging the cars those days."

## What about the image of Pebble Beach Concours d'Elegance being too elitist?

"The Pebble Beach Concours was created by auto enthusiasts who just wanted to enjoy and share great cars," replies Sandra Button. "And to this day, we are guided by enthusiasts. That's important.

**ABOVE:** *Best of Show for 2011 Pebble Beach Concours d'Elegance was this 1934 Voisin C-25 Aerodyne of French car enthusiast Peter Mullin and his wife, Merle. (Hugues Vanhoolandt)*

**OPPOSITE:** *In 2011, Pebble Beach managed to gather most of the 32 Ferrari 250 GTOs in one place. (© Ron Kimball, Pebble Beach Concours d'Elegance)*

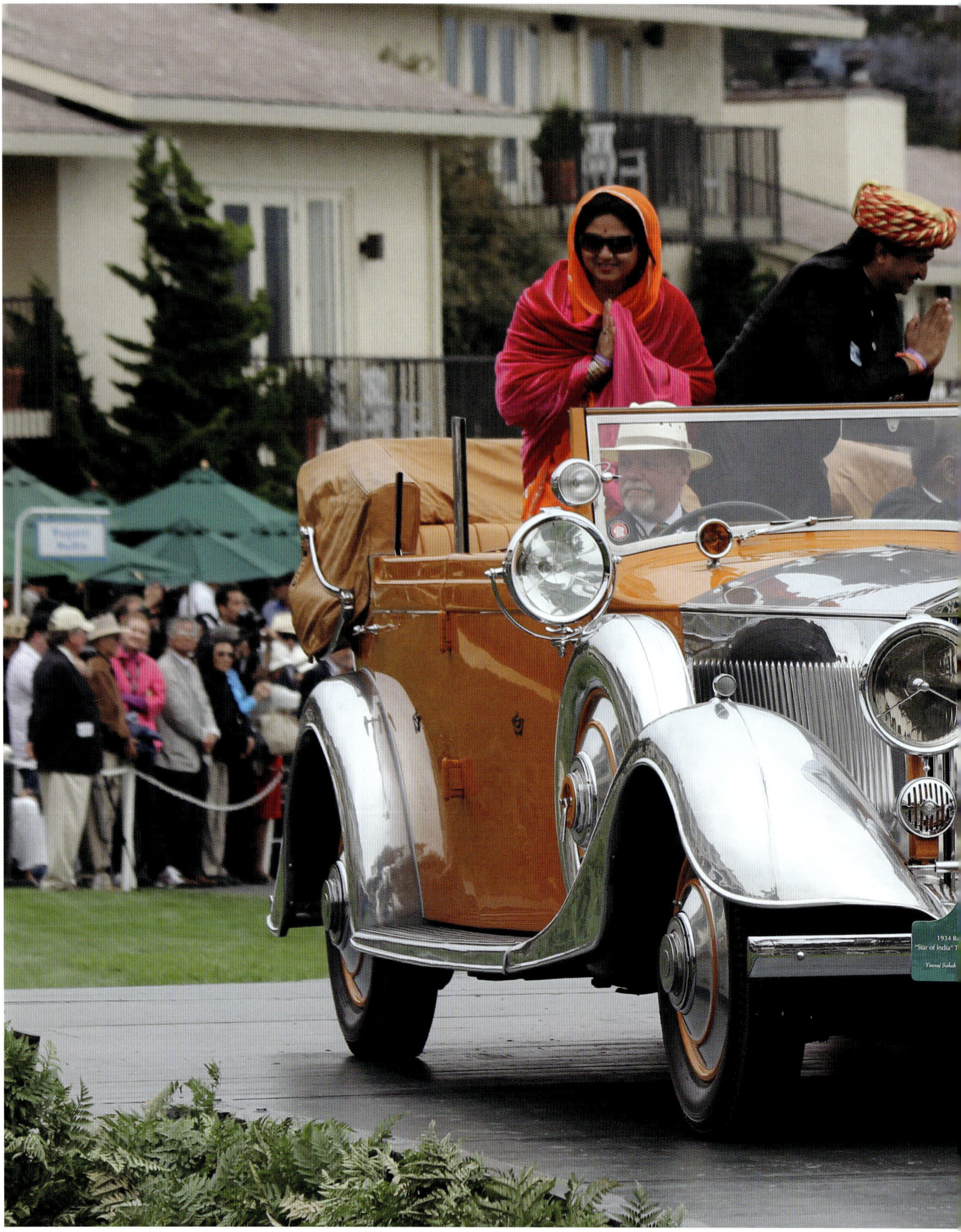

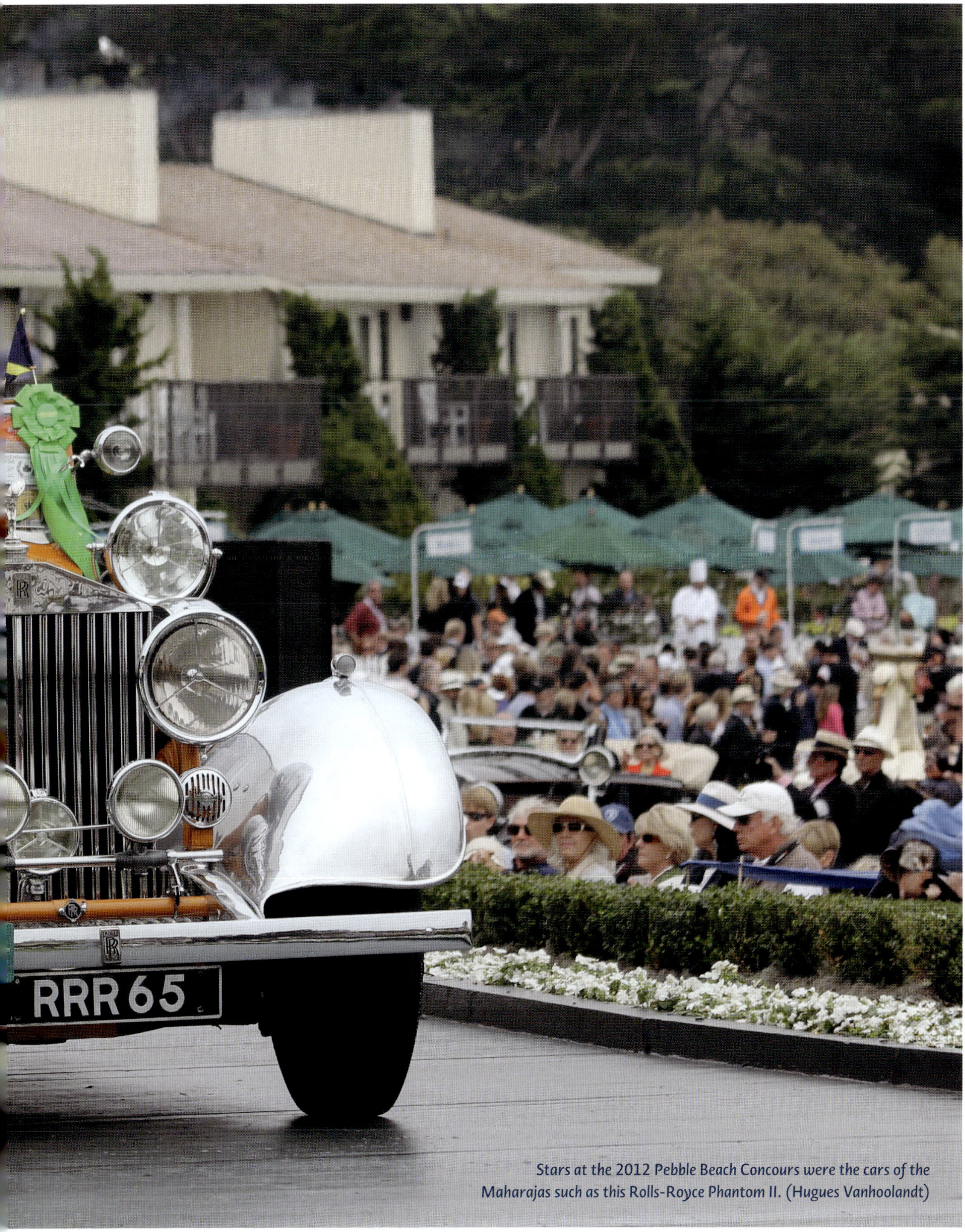


Stars at the 2012 Pebble Beach Concours were the cars of the Maharajas such as this Rolls-Royce Phantom II. (Hugues Vanhoolandt)

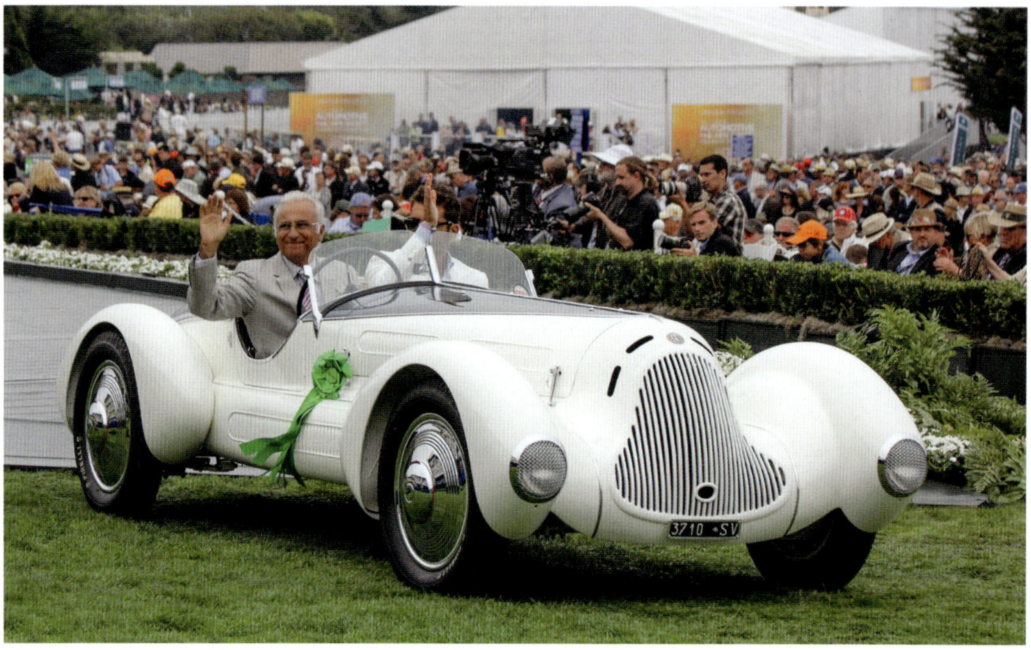

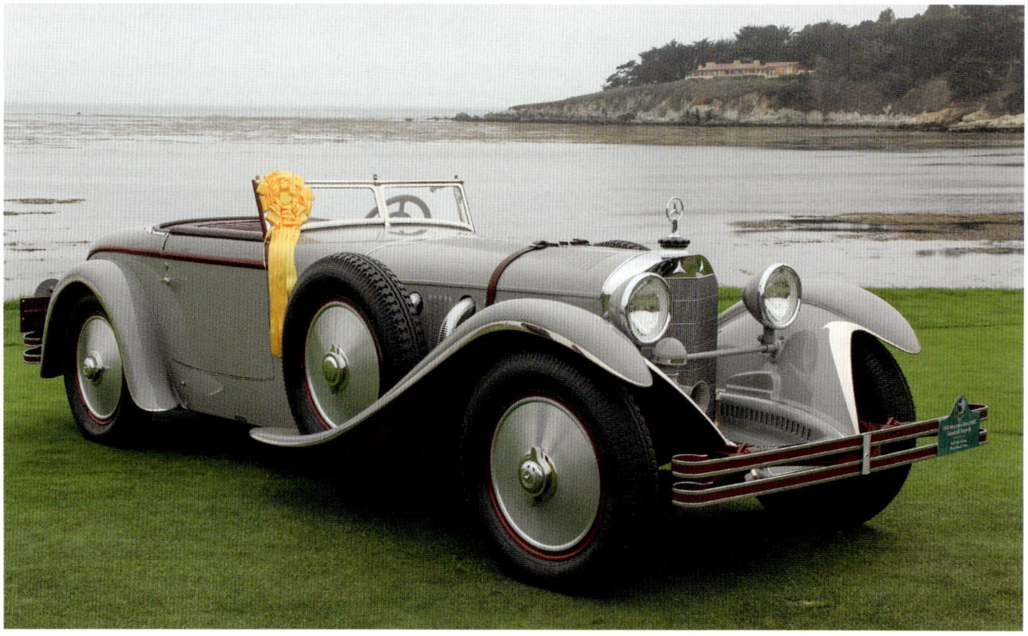

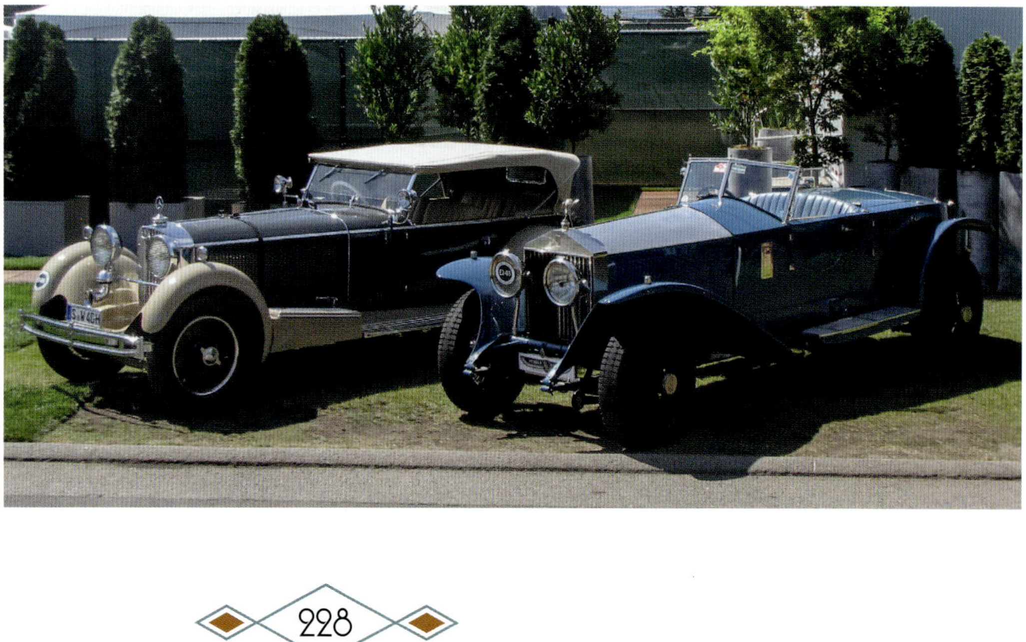

**TOP:** *One of the class winners was this 1931 Alfa Romeo 6C 1750 GS Aprile Spider of Corrado Lopresto, at the 2012 Pebble Beach Concours d'Elegance. (Hugues Vanhoolandt)*

**MIDDLE:** *With the overall Best of Show going to a unique Saoutchik-bodied Mercedes-Benz 680 S at the 2012 Pebble Beach Concours d'Elegance. (Hugues Vanhoolandt)*

**BOTTOM:** *2012 Pebble Beach Concours was also the occasion when two ex-Maharaja of Kashmir cars – the 1928 Mercedes-Benz SS roadster and the Rolls-Royce Phantom I, better known as 17EX – were reunited after eight decades. (Gautam Sen)*

We strive to celebrate the greatest cars in existence – those that exhibit technical excellence, exquisite style, and exceptional workmanship. Some were the first to include important new engineering advances, some defined elegance in their day, and some bettered all others on the track.

"Behind each of these great cars are great people who worked hard to make their automotive vision a success – and others who are striving to keep these visions alive and share them with us today. These acts of creation and caring often indeed require expertise, time, effort and money. But admiring what is great, what is elite, doesn't make a person or a thing or an event 'elitist'. Nor does celebrating the special people who have made something great. I think that what is important is that we continue to follow the lead of enthusiasts in recognizing more great cars. This involves nurturing new designers, manufacturers, restorers and enthusiasts. And then following their lead."

## What are the issues and challenges that concours d'elegance / concours d'état face?

"US concours are widely diverse in terms of focus, style and judging systems, as are European concours and those in Asia, Africa and elsewhere," explained Sandra Button. "I think all concours need to wrestle almost immediately with many things: How does our understanding of terms like preservation and restoration change as our tools and knowledge and abilities change? How do we continue to celebrate the whole range of cars as cars themselves continue to change and the range of cars expands? How will driving play a role in car events as cars change and driving also changes? How do we ensure great cars continue to be conceived and then cared for?"

*In 2014, Pebble Beach brought together an extraordinary collection of Ferrari 250 Testa Rossas. (© Ron Kimball, Pebble Beach Concours d'Elegance)*

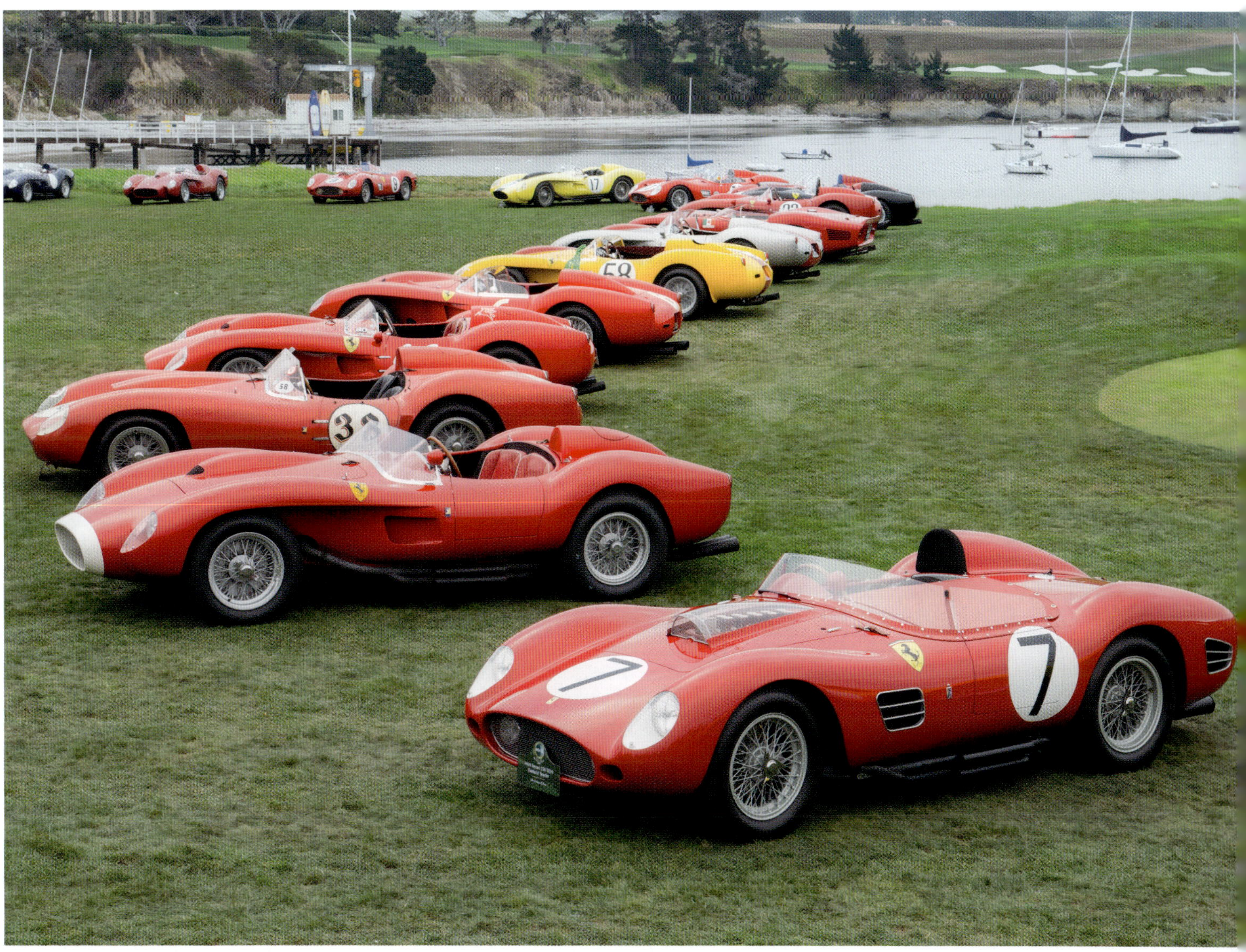

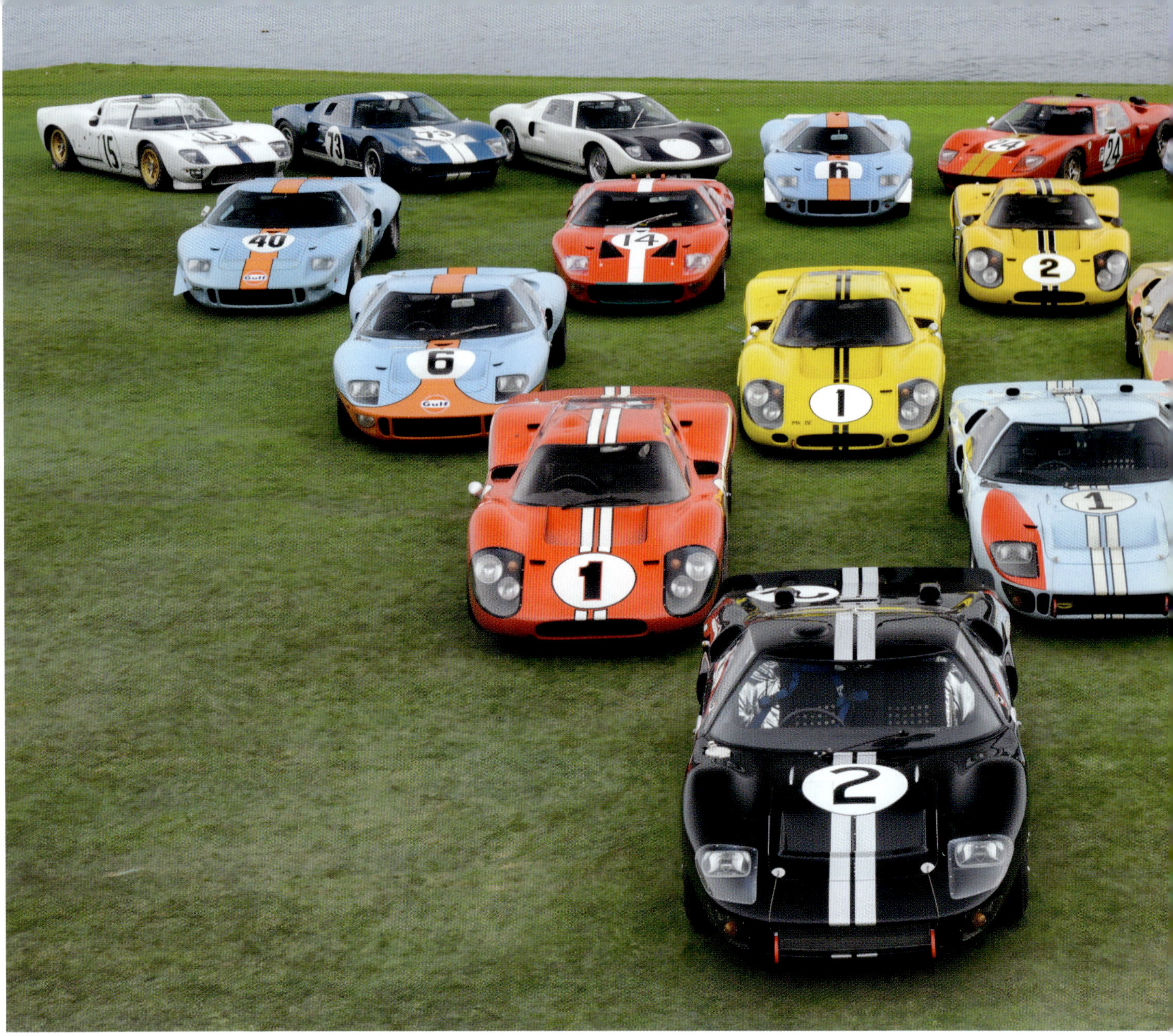
15
73
6
24
40
14
2
6
Gulf
1
1
1
2

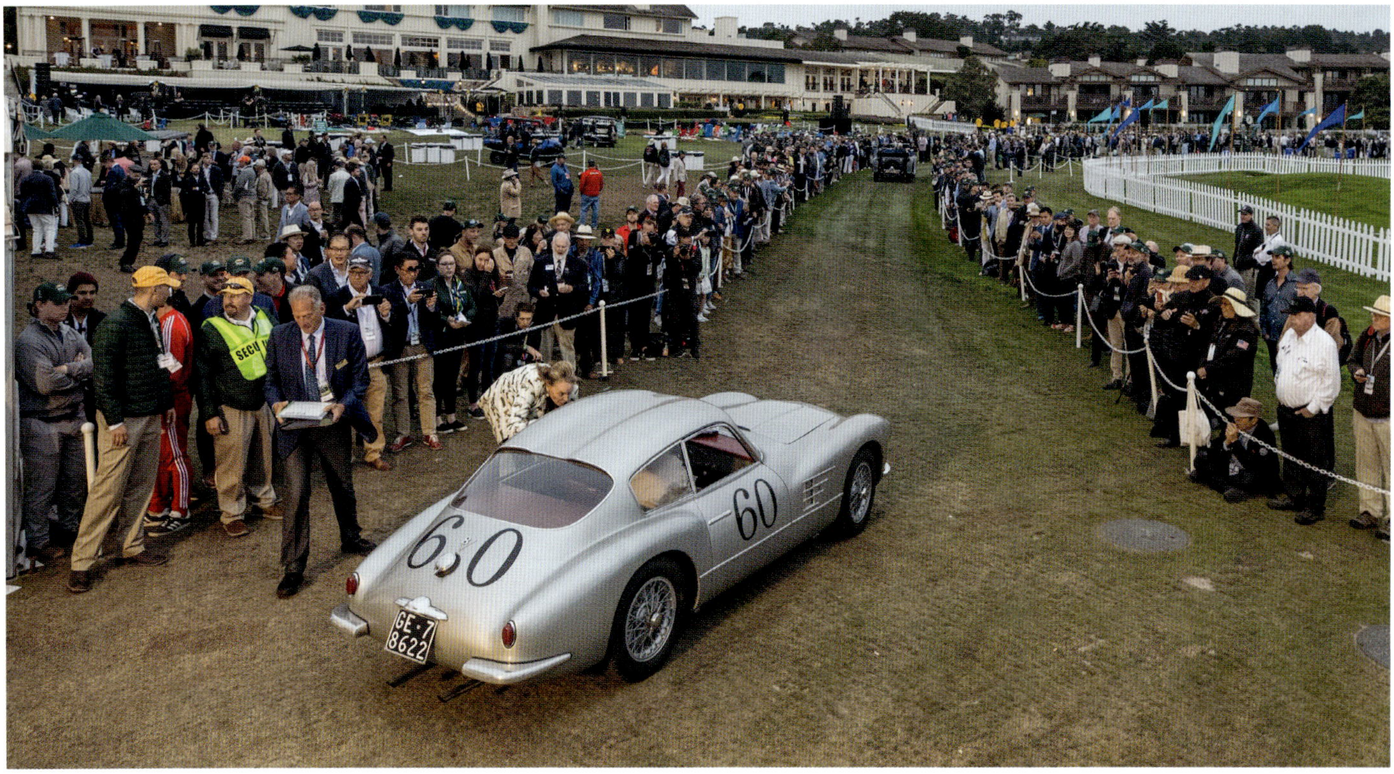
60
GE-7
8622

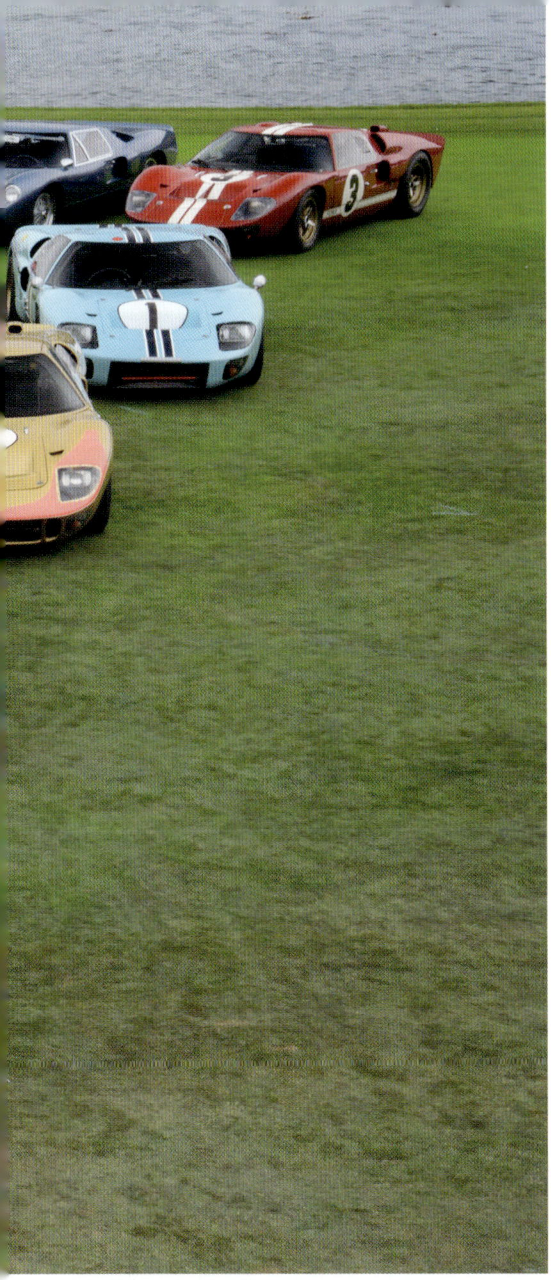

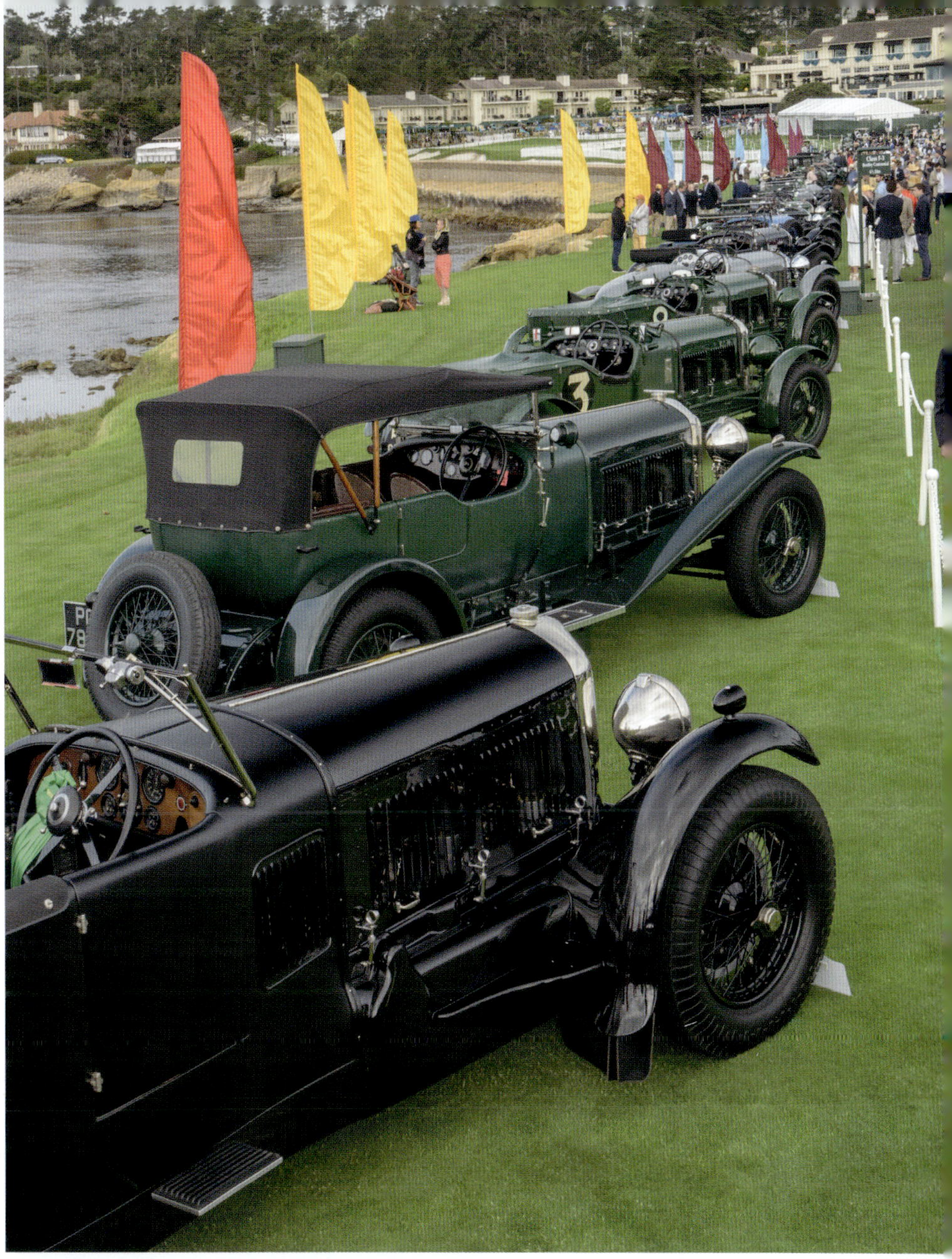

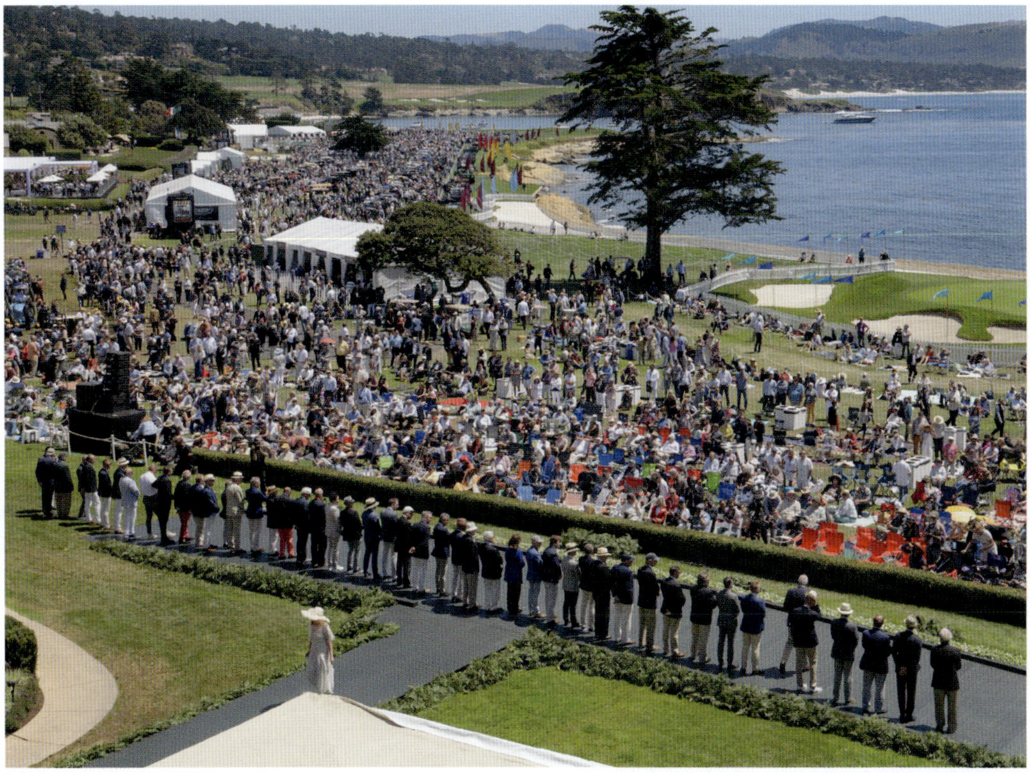

**OPPOSITE TOP:** *In 2016, every major race-winning Ford GT40 was on display at Pebble Beach. (© Ron Kimball, Pebble Beach Concours d'Elegance)*

**ABOVE:** *Racing Bentleys were assembled in 2019 to celebrate the 100th anniversary of the founding of Bentley. (© Ron Kimball, Pebble Beach Concours d'Elegance)*

**OPPOSITE BOTTOM:** *Over the years, the Dawn Patrol on Concours Sunday has become an exciting tradition. This is from 2019. (Courtesy Pebble Beach Concours d'Elegance)*

**LEFT:** *During 2019 the jury posed on the ramp for the photo session. (©Sherman Chu, Pebble Beach Concours d'Elegance)*

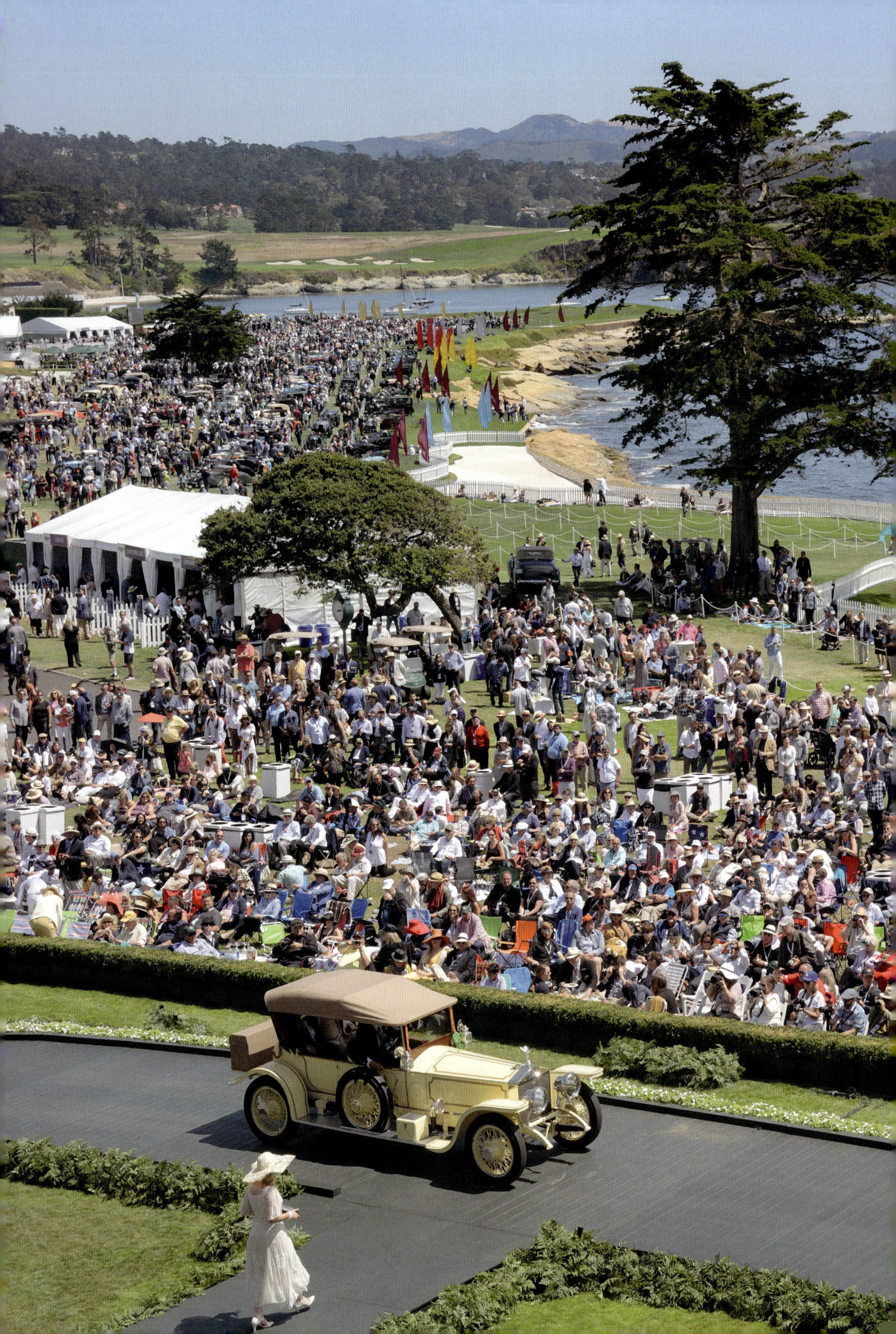

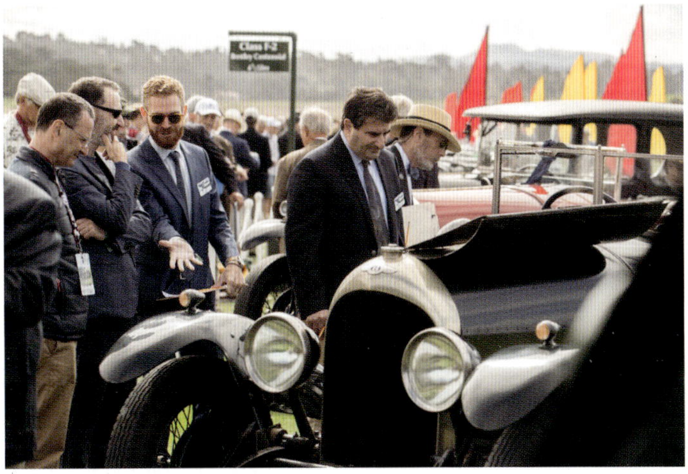

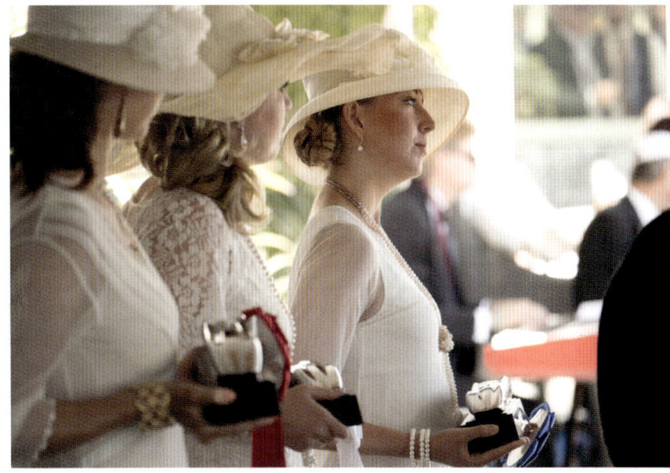

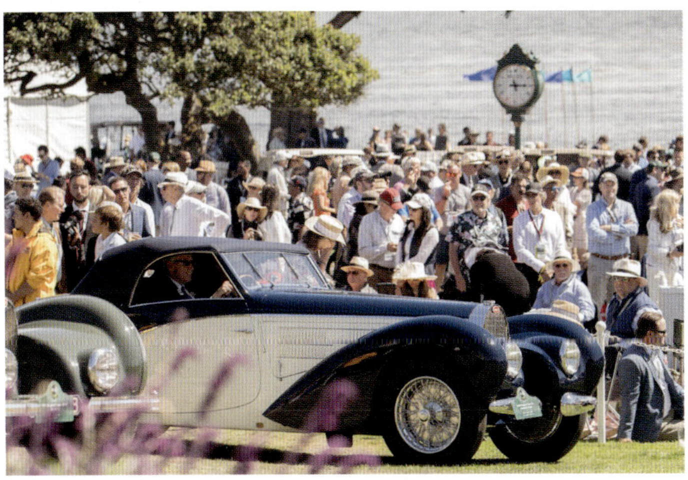

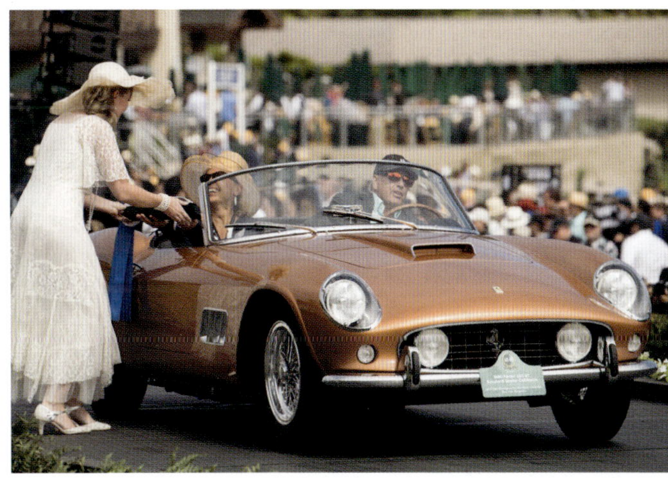

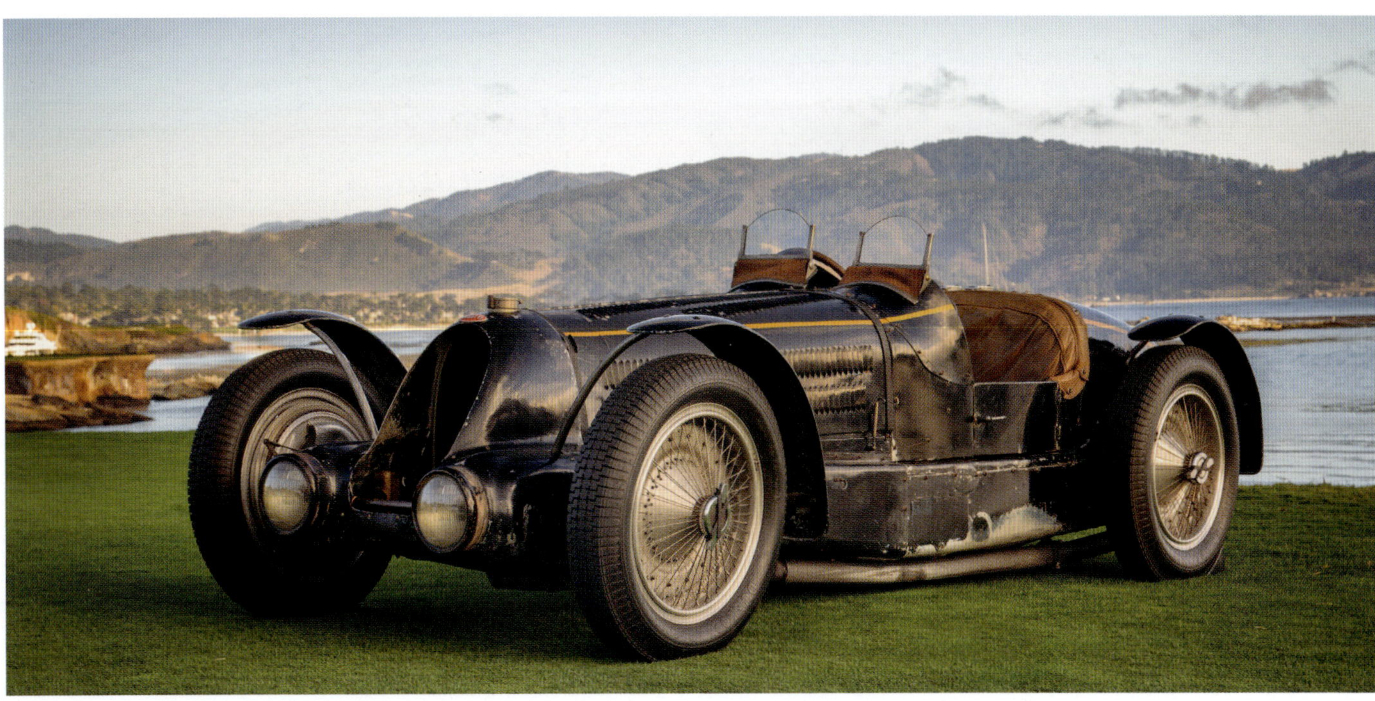

**ABOVE & OVERLEAF:** *The 1934 Bugatti Type 59, owned by Swiss collector Fritz Burkard, set a precedent by winning the Best of Show in 2024, the first time that a preserved vehicle had done so at Pebble Beach.* (© Ron Kimball, Pebble Beach Concours d'Elegance)

**TOP FOUR:** *Capturing the mood of the Pebble Beach Concours Sunday.* (©Sherman Chu, Pebble Beach Concours d'Elegance)

**OPPOSITE:** *A view of the ramp at the 2019 Pebble Beach Concours.* (© Ron Kimball, Pebble Beach Concours d'Elegance)

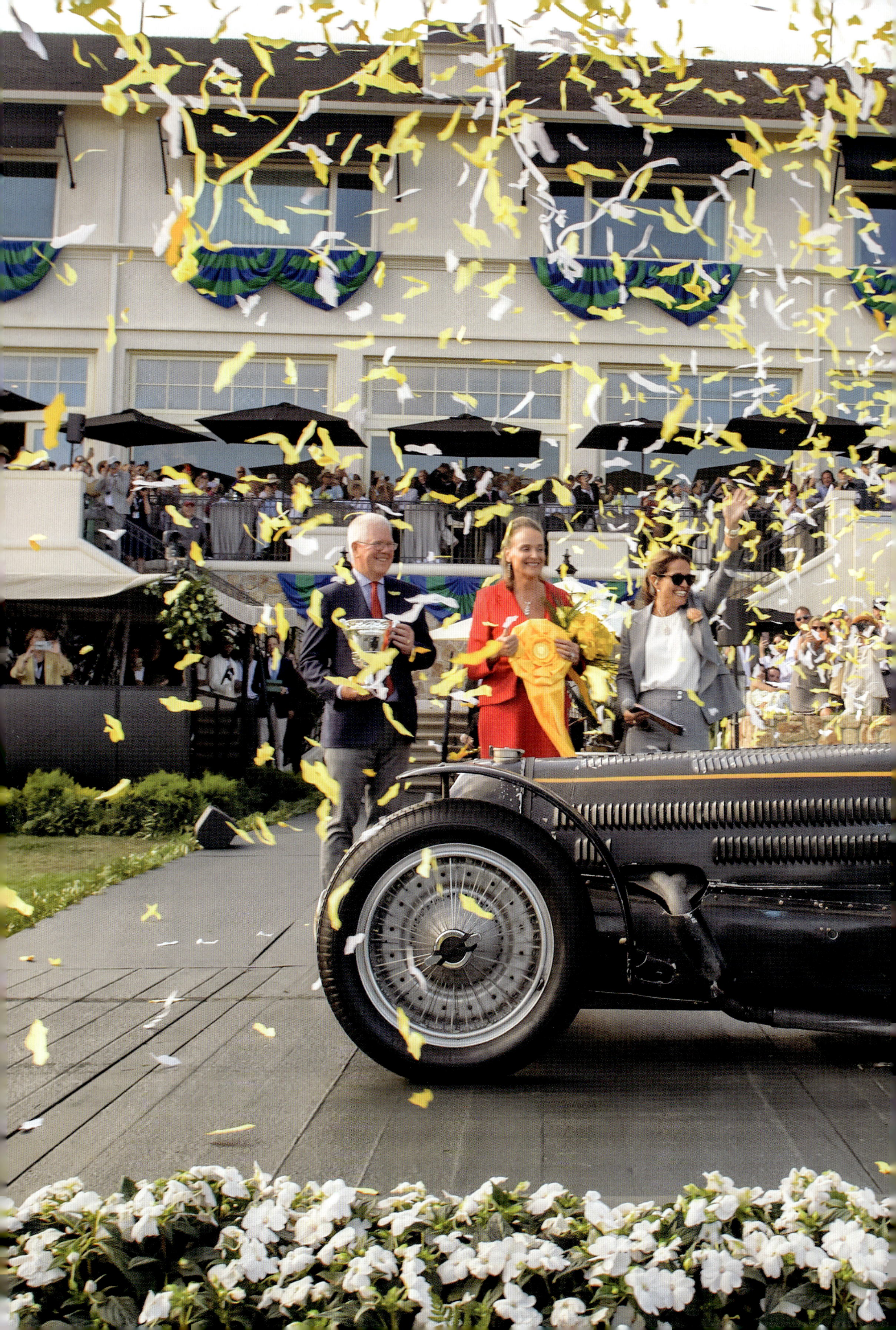

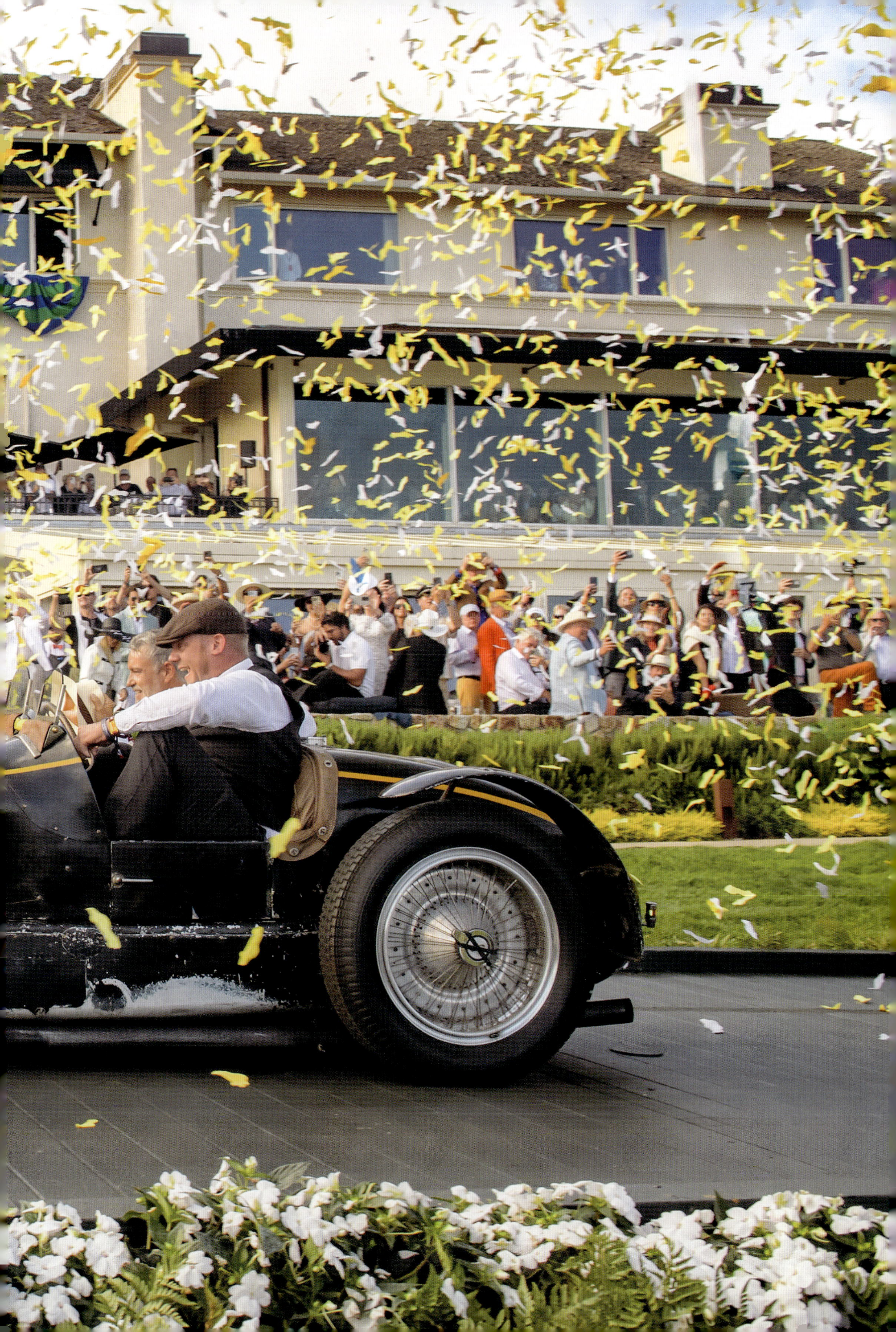

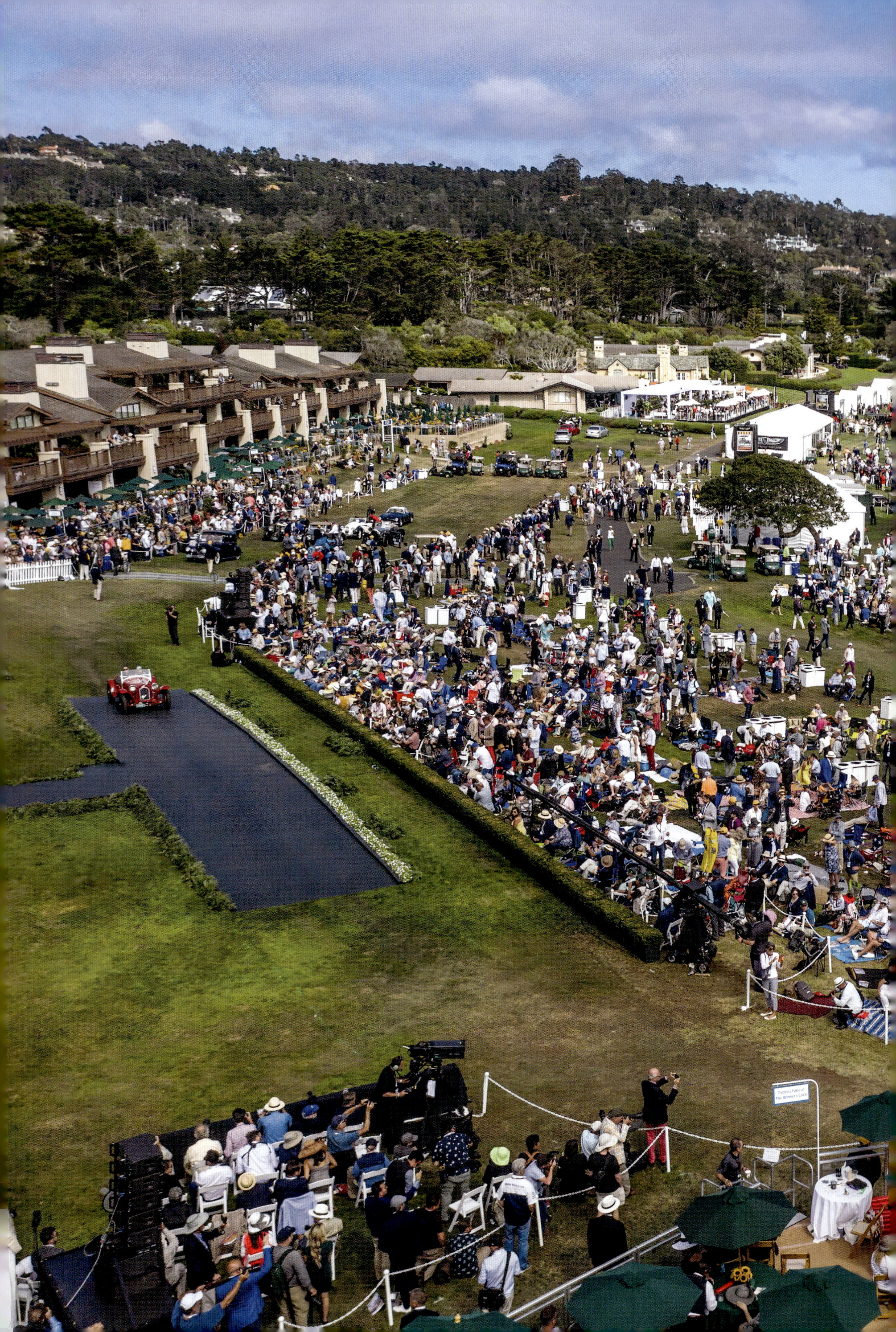

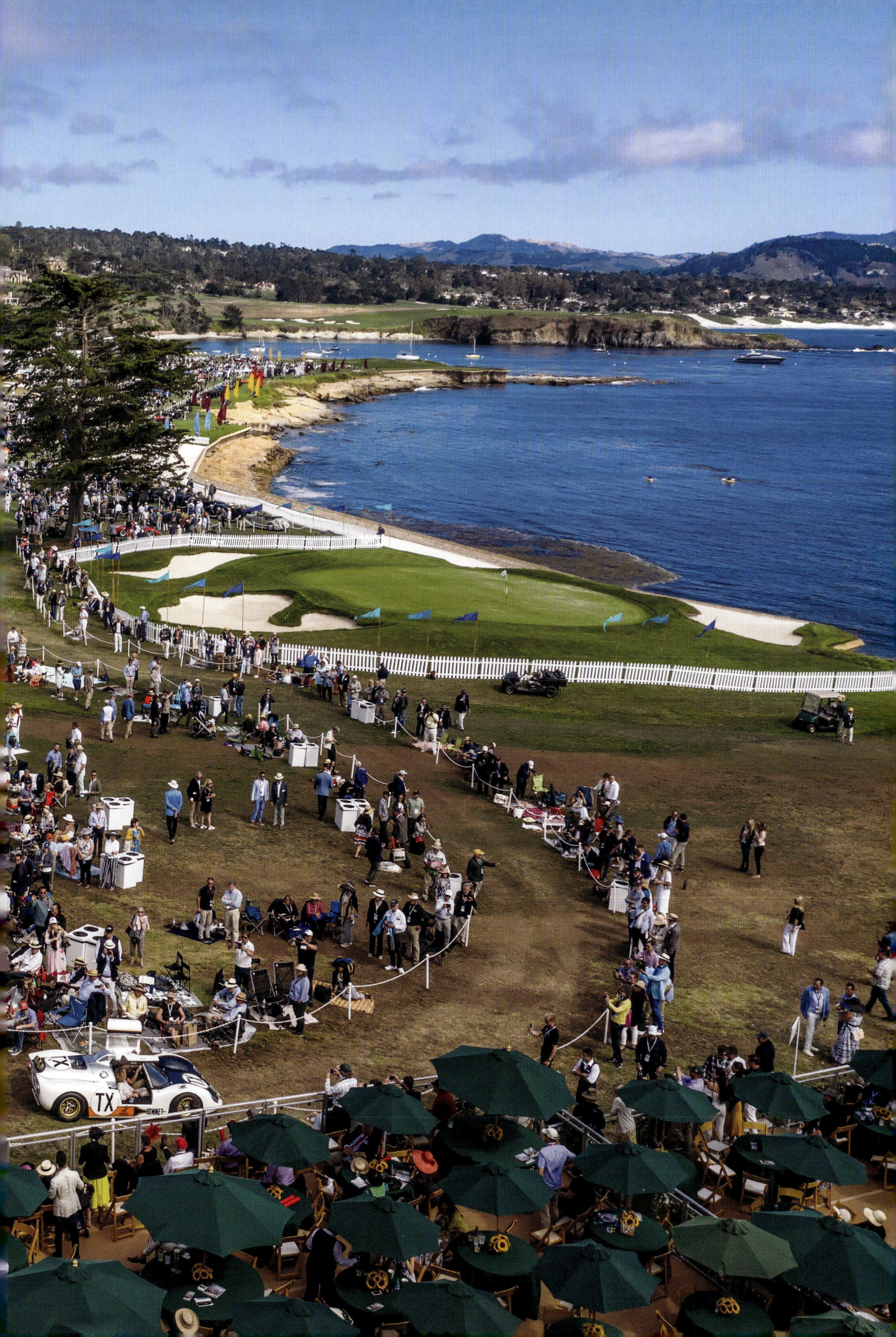
TX

# Hillsborough, Bagatelle, and the revival of Villa d'Este

## EUROPE REAWAKENS

The first Watkins Glen Grand Prix was held on October 2, 1948, on a 6.6-mile (10.6 km) course that wound through public roads. In the early years, the race passed through the city center, with spectators lining the sidewalks. However, after a tragic incident in 1952, where a car veered off the road, killing one spectator and injuring several others, the event was moved to a wooded hillside south of the town.

In 1953, the Watkins Glen Grand Prix Corporation was formed to manage the event, and a new 4.6-mile (7.4 km) course was created, utilizing existing roads. This course was used for three years until a permanent 2.3-mile (3.78 km) circuit was built on a 1.4-square-mile (2.2 km) site, incorporating parts of the old route.

Throughout these years, while the road race was the main attraction, a concours d'elegance was added to bring an element of elegance to the weekend. For the 1951 event, held on September 14, the day before the road race, a parade led by the legendary General Motors Design head Harley Earl featured his concept car, the stunning Buick Le Sabre. Earl was reportedly surprised by the enthusiastic response from the crowd. As Kip Zeiter noted in *Vette Vues Magazine*, Earl commented during the parade that the American auto industry lacked a sports car comparable to European models.

According to Zeiter, this experience inspired Harley Earl to champion the Chevrolet Corvette project. When Earl's host at Watkins Glen, Nick Fraboni, learned that Chevrolet was planning to produce the Corvette, he persuaded General Motors to send one of the first four Corvettes made to Watkins Glen for the 1953 concours d'elegance parade, held during the race weekend on September 18-19.

With more than 100 entries and a safer track located farther from town, the Watkins Glen Road Races was becoming one of the most significant races in the United States. Alongside the annual race

**LEFT:** *An early Hillsborough, California concours d'elegance was sponsored by the local High School and was organized with the help of the Sports Car Club of America (SCCA). (Rob Fisher Archives)*

**RIGHT:** *The catalogue for the third concours d'elegance at Hillsborough, from 1959. (Rob Fisher Archives)*

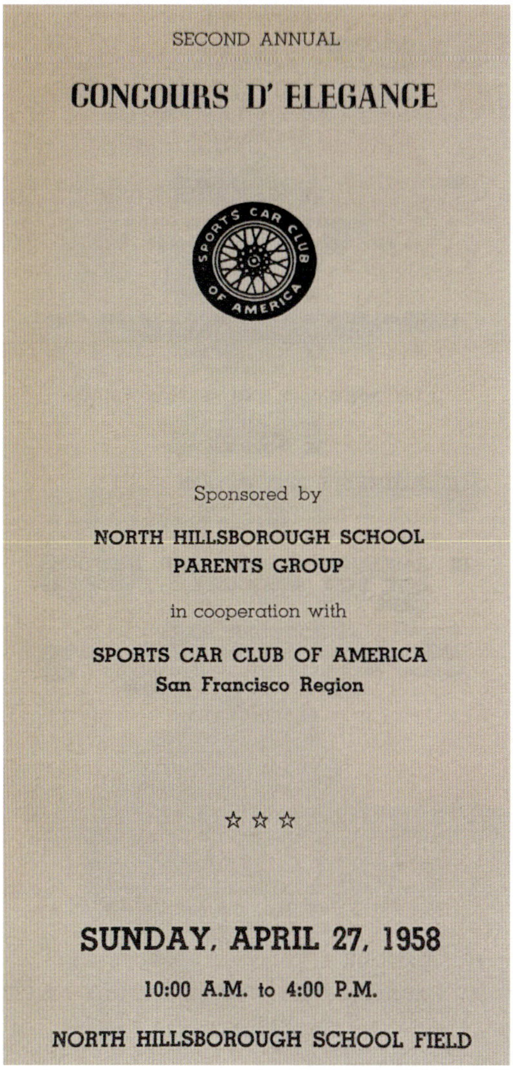


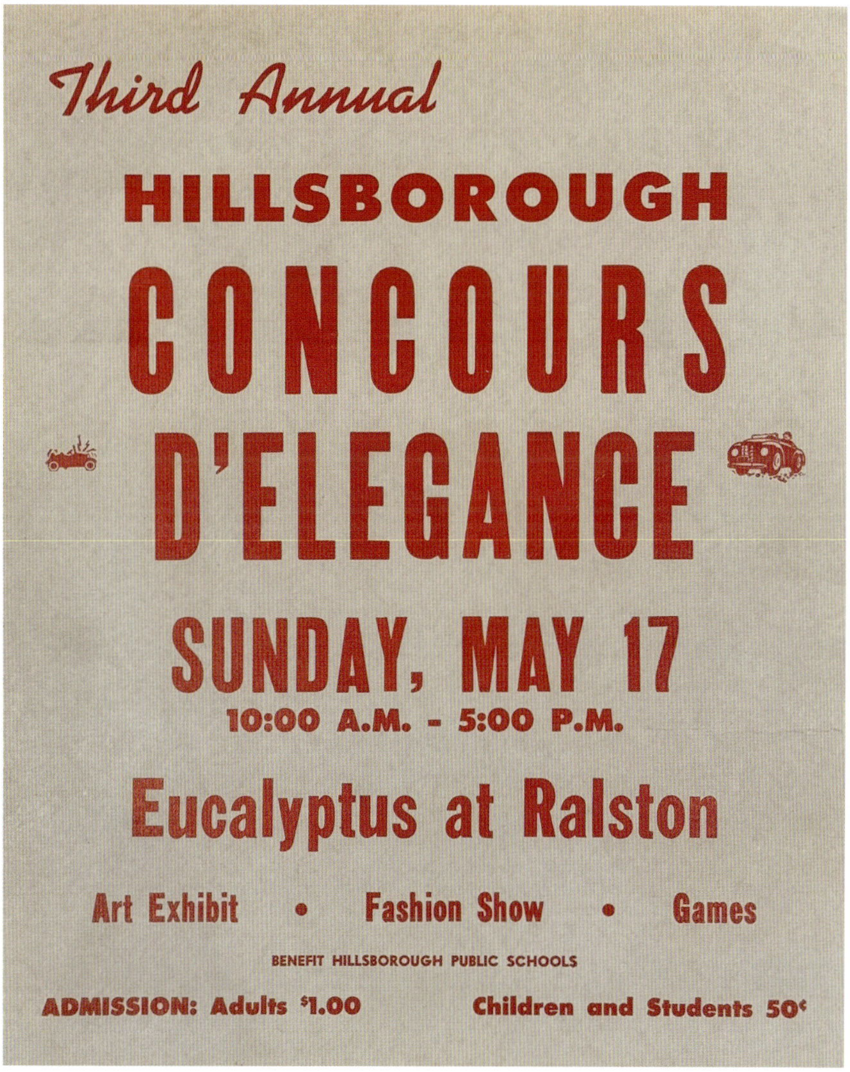

**THIS PAGE:** *The first three broad categories for the Second Annual Hillsborough Concours in 1958 were for modern cars and the last three were for historic vehicles. The organizing committee numbered many and were all volunteers.* (Rob Fisher Archives)

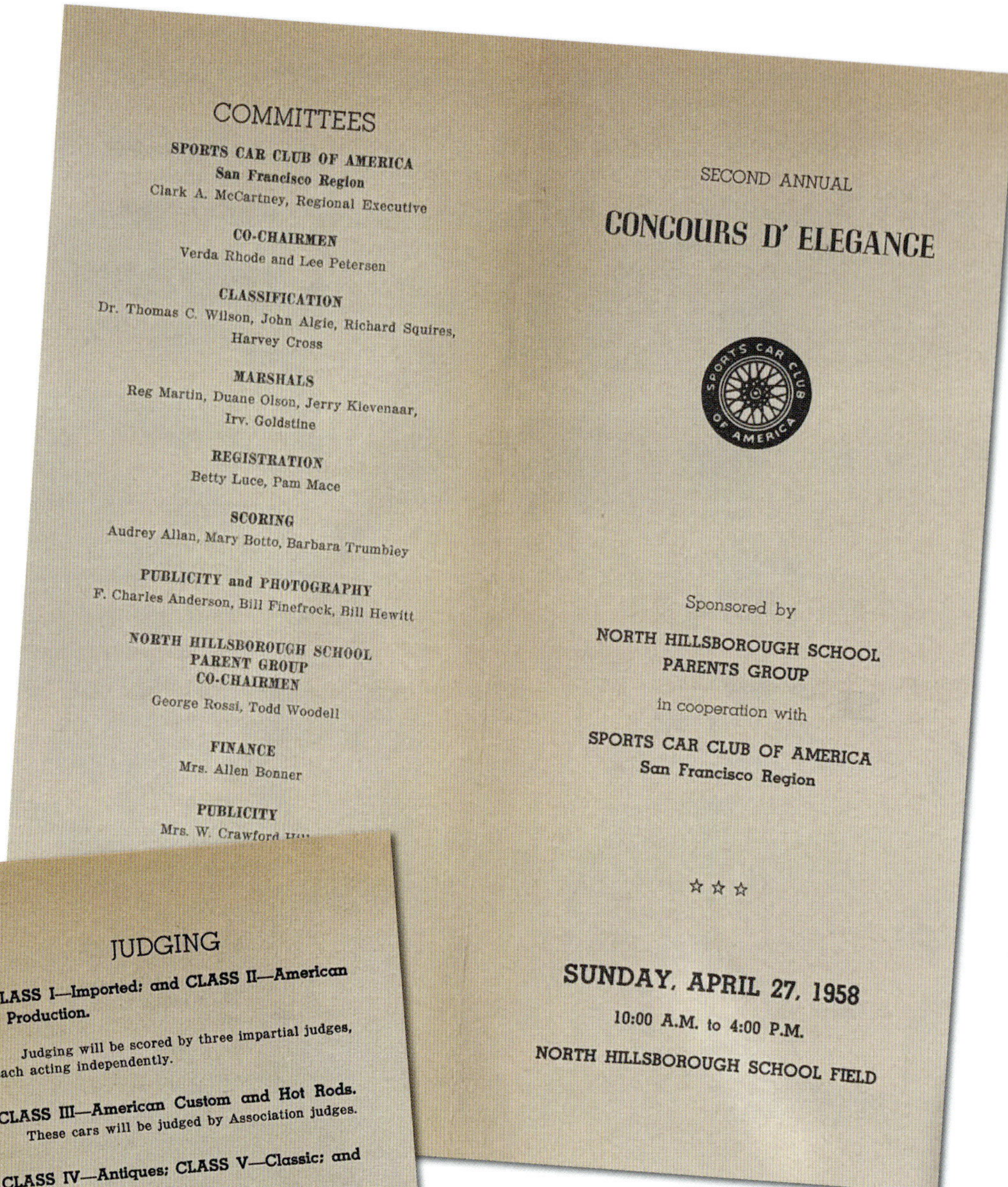

## COMMITTEES

**SPORTS CAR CLUB OF AMERICA**
**San Francisco Region**
Clark A. McCartney, Regional Executive

**CO-CHAIRMEN**
Verda Rhode and Lee Petersen

**CLASSIFICATION**
Dr. Thomas C. Wilson, John Algie, Richard Squires, Harvey Cross

**MARSHALS**
Reg Martin, Duane Olson, Jerry Klevenaar, Irv. Goldstine

**REGISTRATION**
Betty Luce, Pam Mace

**SCORING**
Audrey Allan, Mary Botto, Barbara Trumbley

**PUBLICITY and PHOTOGRAPHY**
F. Charles Anderson, Bill Finefrock, Bill Hewitt

**NORTH HILLSBOROUGH SCHOOL PARENT GROUP CO-CHAIRMEN**
George Rossi, Todd Woodell

**FINANCE**
Mrs. Allen Bonner

**PUBLICITY**
Mrs. W. Crawford H...

SECOND ANNUAL

## CONCOURS D' ELEGANCE

Sponsored by
**NORTH HILLSBOROUGH SCHOOL PARENTS GROUP**
in cooperation with
**SPORTS CAR CLUB OF AMERICA**
**San Francisco Region**

☆ ☆ ☆

**SUNDAY, APRIL 27, 1958**
10:00 A.M. to 4:00 P.M.
NORTH HILLSBOROUGH SCHOOL FIELD

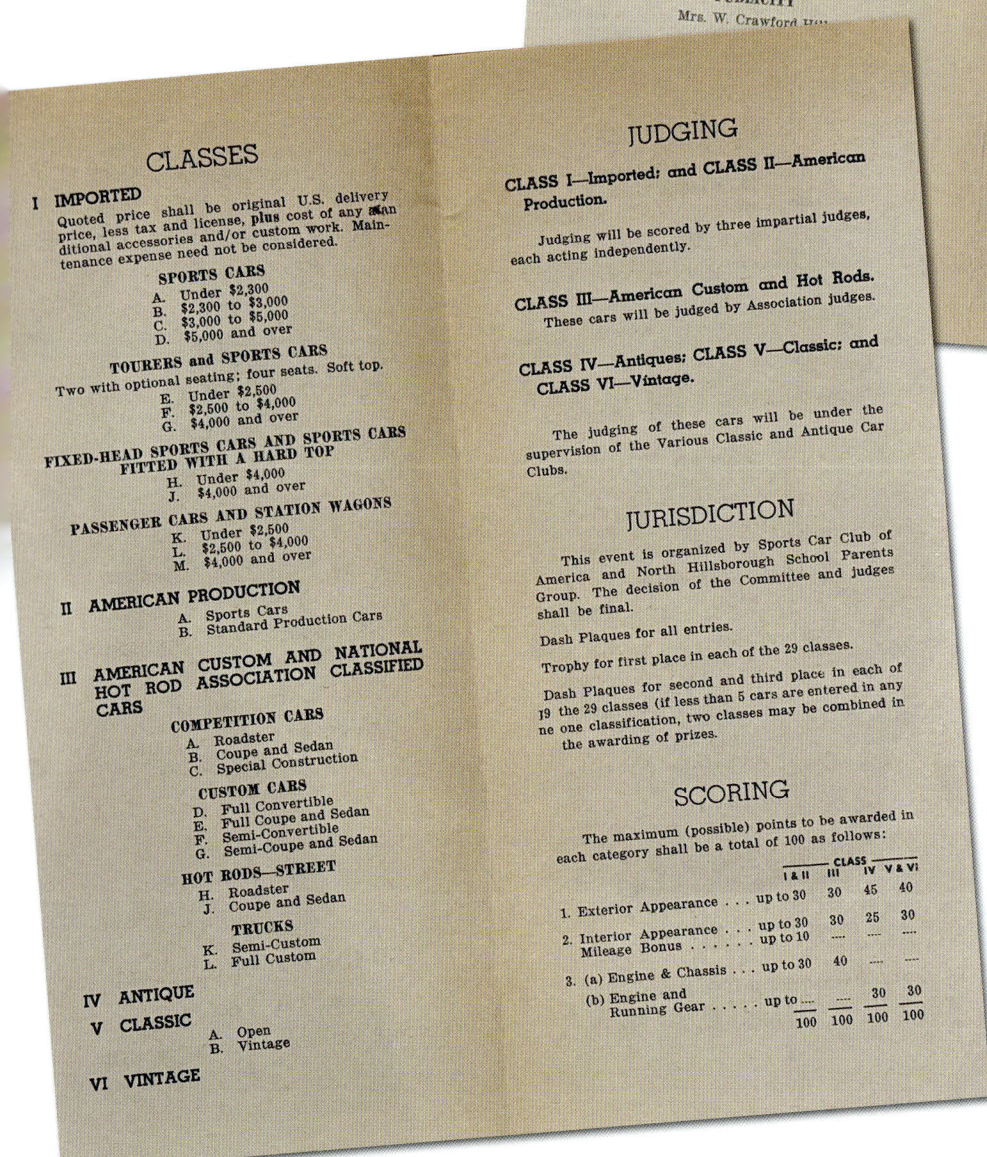

## CLASSES

**I IMPORTED**
Quoted price shall be original U.S. delivery price, less tax and license, **plus** cost of any add-ditional accessories and/or custom work. Maintenance expense need not be considered.

**SPORTS CARS**
A. Under $2,300
B. $2,300 to $3,000
C. $3,000 to $5,000
D. $5,000 and over

**TOURERS and SPORTS CARS**
Two with optional seating; four seats. Soft top.
E. Under $2,500
F. $2,500 to $4,000
G. $4,000 and over

**FIXED-HEAD SPORTS CARS AND SPORTS CARS FITTED WITH A HARD TOP**
H. Under $4,000
J. $4,000 and over

**PASSENGER CARS AND STATION WAGONS**
K. Under $2,500
L. $2,500 to $4,000
M. $4,000 and over

**II AMERICAN PRODUCTION**
A. Sports Cars
B. Standard Production Cars

**III AMERICAN CUSTOM AND NATIONAL HOT ROD ASSOCIATION CLASSIFIED CARS**

**COMPETITION CARS**
A. Roadster
B. Coupe and Sedan
C. Special Construction

**CUSTOM CARS**
D. Full Convertible
E. Full Coupe and Sedan
F. Semi-Convertible
G. Semi-Coupe and Sedan

**HOT RODS—STREET**
H. Roadster
J. Coupe and Sedan

**TRUCKS**
K. Semi-Custom
L. Full Custom

**IV ANTIQUE**

**V CLASSIC**
A. Open
B. Vintage

**VI VINTAGE**

## JUDGING

**CLASS I—Imported; and CLASS II—American Production.**

Judging will be scored by three impartial judges, each acting independently.

**CLASS III—American Custom and Hot Rods.**
These cars will be judged by Association judges.

**CLASS IV—Antiques; CLASS V—Classic; and CLASS VI—Vintage.**

The judging of these cars will be under the supervision of the Various Classic and Antique Car Clubs.

## JURISDICTION

This event is organized by Sports Car Club of America and North Hillsborough School Parents Group. The decision of the Committee and judges shall be final.

Dash Plaques for all entries.

Trophy for first place in each of the 29 classes.

Dash Plaques for second and third place in each of J9 the 29 classes (if less than 5 cars are entered in any ne one classification, two classes may be combined in the awarding of prizes.

## SCORING

The maximum (possible) points to be awarded in each category shall be a total of 100 as follows:

| | CLASS I & II | III | IV | V & VI |
|---|---|---|---|---|
| 1. Exterior Appearance . . . | up to 30 | 30 | 45 | 40 |
| 2. Interior Appearance . . . | up to 30 | 30 | 25 | 30 |
| Mileage Bonus . . . . . . | up to 10 | ---- | ---- | ---- |
| 3. (a) Engine & Chassis . . . | up to 30 | 40 | ---- | ---- |
| (b) Engine and Running Gear . . . . . | up to ---- | ---- | 30 | 30 |
| | 100 | 100 | 100 | 100 |

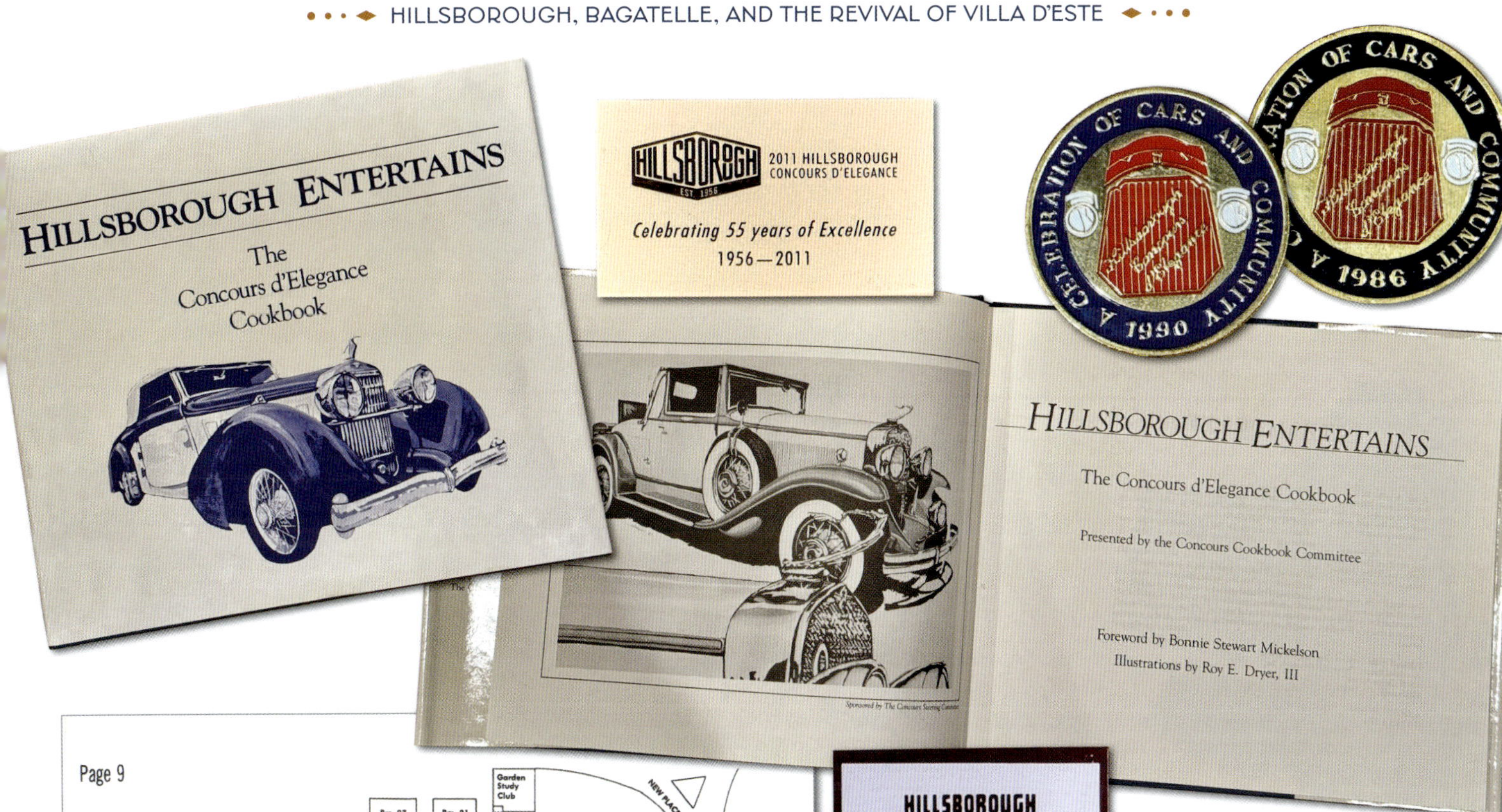


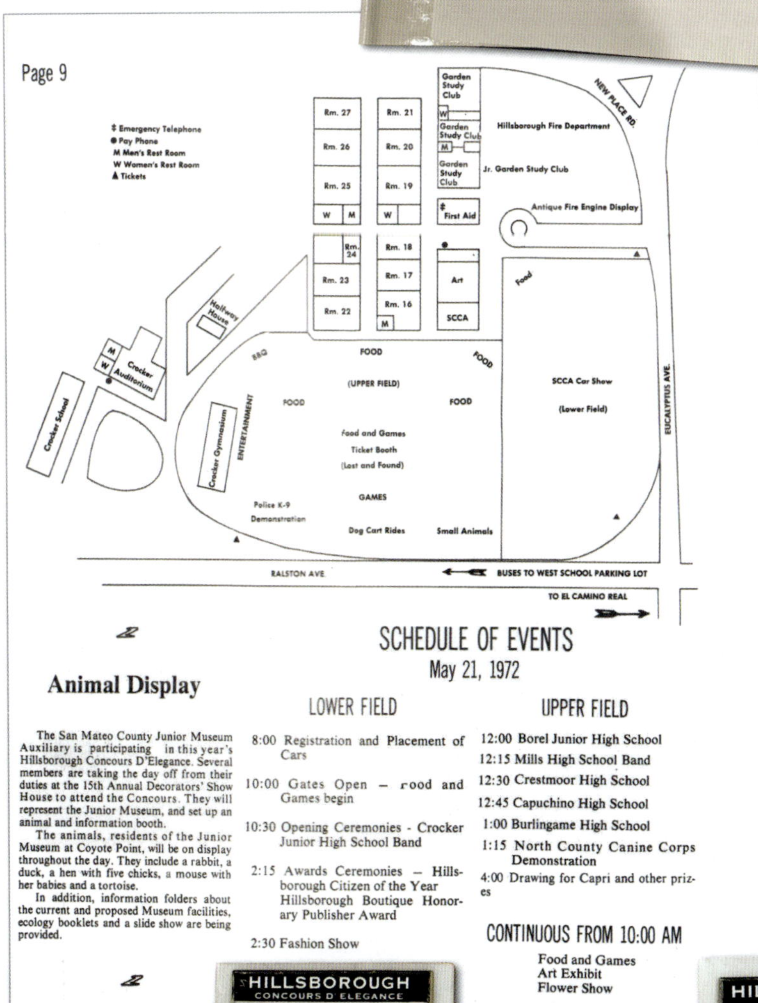

Page 9

$ Emergency Telephone
● Pay Phone
M Men's Rest Room
W Women's Rest Room
▲ Tickets

Garden Study Club
Hillsborough Fire Department
Jr. Garden Study Club
Antique Fire Engine Display
First Aid
Art
SCCA
Food
SCCA Car Show (Lower Field)
FOOD (UPPER FIELD) FOOD
Food and Games
Ticket Booth
(Lost and Found)
GAMES
Police K-9 Demonstration
Dog Cart Rides
Small Animals
Crocker Auditorium
Crocker Gymnasium
ENTERTAINMENT
Crocker School
Hallway House
NEW PLACE RD.
EUCALYPTUS AVE.
RALSTON AVE
BUSES TO WEST SCHOOL PARKING LOT
TO EL CAMINO REAL

## Animal Display

The San Mateo County Junior Museum Auxiliary is participating in this year's Hillsborough Concours D'Elegance. Several members are taking the day off from their duties at the 15th Annual Decorators' Show House to attend the Concours. They will represent the Junior Museum, and set up an animal and information booth.

The animals, residents of the Junior Museum at Coyote Point, will be on display throughout the day. They include a rabbit, a duck, a hen with five chicks, a mouse with her babies and a tortoise.

In addition, information folders about the current and proposed Museum facilities, ecology booklets and a slide show are being provided.

## SCHEDULE OF EVENTS

May 21, 1972

### LOWER FIELD

8:00 Registration and Placement of Cars

10:00 Gates Open — rood and Games begin

10:30 Opening Ceremonies - Crocker Junior High School Band

2:15 Awards Ceremonies — Hillsborough Citizen of the Year Hillsborough Boutique Honorary Publisher Award

2:30 Fashion Show

### UPPER FIELD

12:00 Borel Junior High School

12:15 Mills High School Band

12:30 Crestmoor High School

12:45 Capuchino High School

1:00 Burlingame High School

1:15 North County Canine Corps Demonstration

4:00 Drawing for Capri and other prizes

### CONTINUOUS FROM 10:00 AM

Food and Games
Art Exhibit
Flower Show

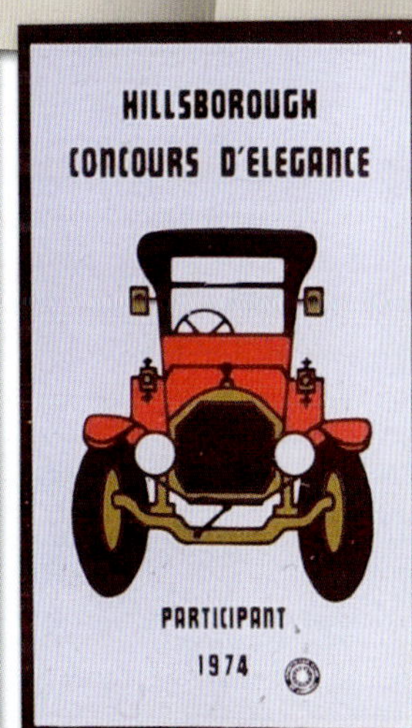


The Concours Cookbook (above) from the year 1986 featured a pre-war Hispano-Suiza on the cover.

Badges and decals have been a tradition at Hillsborough and have become highly collectible today. (Rob Fisher Archives)

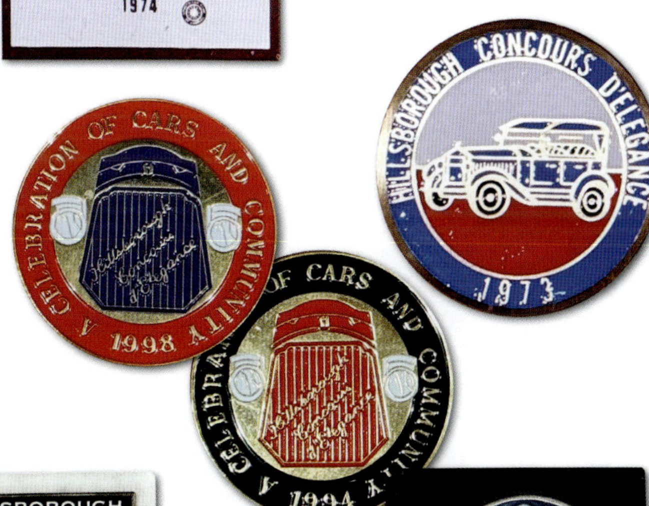


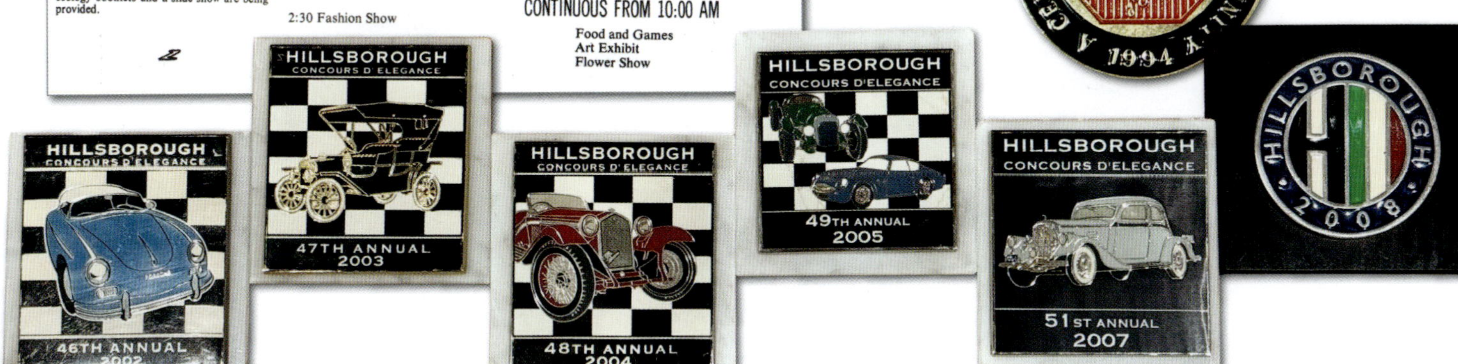

organized by the Sports Car Club of America (SCCA), the circuit hosted its first professional event in 1957, the NASCAR Grand National Division. It gained international recognition when the Formula Libre race attracted top drivers like Jack Brabham, Stirling Moss, Phil Hill, and Dan Gurney between 1958 and 1960. As racing took center stage, the concours d'elegance gradually faded away, with the last one held in 1959.

The story, though, for Pebble Beach Concours d'Elegance was different, as explained in the last chapter. With the separation of the concours d'elegance event from that of the road race, which moved to Laguna Seca, the former took on its own identity and over the years grew to be one of the most important historic vehicle events in the world.

In the spring of 1956, the small township of Hillsborough, located 17 miles (27 km) south of San Francisco and about 100 miles (160 km) north of Pebble Beach, organized a concours d'elegance for pre-war, post-war and current era cars. Like Pebble Beach, the Hillsborough Concours d'Elegance entries were to be judged on their condition, authenticity and elegance. There were over 20 judged classes with awards for first through third place. In keeping with concours tradition, a Best of Show trophy was presented.

Unlike Pebble Beach, the Hillsborough Concours d'Elegance was established as an independent fundraising event for the non-profit Hillsborough Schools Parent-Teacher Foundation. This strong, community-based focus and support have clearly contributed to its longevity. The show earned its distinction as the world's longest continually running concours d'elegance, holding events every year until the COVID-19 disruption in 2020.

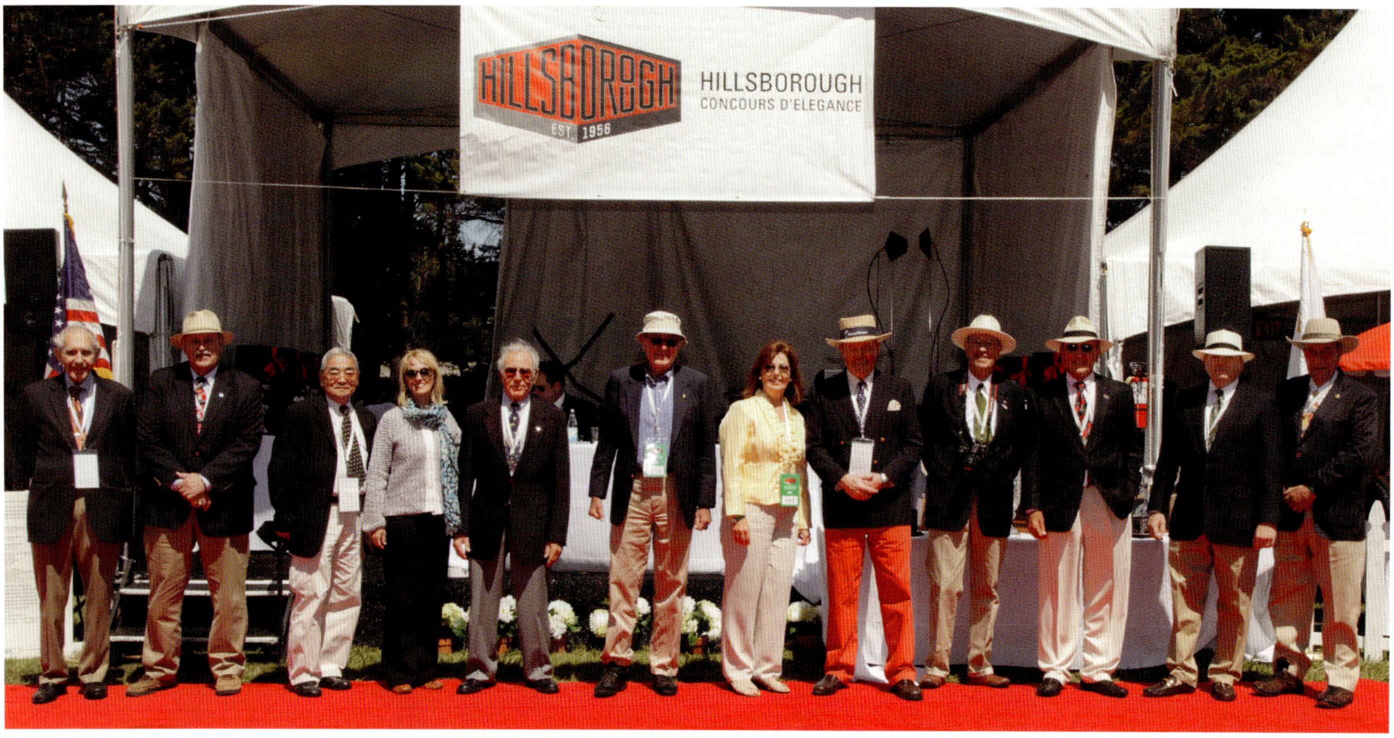


*Jules Heumann (fifth from left) was always a well-wisher, and he is seen here with Lucinda Lewis (4th), designer Tom Matano (3rd), museum director Jackie Frady (7th) and member of the jury Ed Gilbertson (6th) for the 2012 Hillsborough Concours. (Kimball Studios)*

The event resumed in 2021 and has been growing in popularity ever since. Rob Fisher, Chairman Emeritus of the Hillsborough Concours, notes that "Hillsborough began as a humble gathering on the local school fields. Fathers and mothers volunteered their time running the show. Back then, it was a three-day festival featuring a fashion show, gala, and tennis tournament with the concours as the highlight on Sunday. There was even a Car Show Cookbook with recipes from the mothers. It was a full garden party with a picnic atmosphere, where people were grilling burgers and hot dogs."

"As time passed, more classic cars were featured, but the spirit of the event, a gathering of family and friends, remained intact, always focused on raising funds for the schools," explained Fisher. "It was originally held on the school's soccer grounds, and the event grew to feature 100-120 cars."

In addition to being known for its great atmosphere, Hillsborough also became known for its display of world-class automobiles. In fact, for many years, the Best in Show winner was automatically invited to participate in the Pebble Beach concours.

Thanks to the local family and business support, Fisher mentioned the continuous involvement of the SCCA (Sports Car Club of America), which continues to provide judges – around 60 of the current 85 judges. There were local celebrity judges too, like Bing Crosby, and automotive legends Jules Heumann, Lorin Tryon, Strother McMinn and Sid Colberg, and they all played an influential role in shaping the event.

When Fisher took over the concours in 2009, he moved the show to a new venue, the magnificent Crystal Springs Golf Course, and its format was expanded. "It's much bigger now, with 225 cars across 23-24 classes, including marque-specific classes for marques like Rolls-Royce, Bentley, and Ferrari. The SCCA still provides judges each year. Initially, the concours was held in May, but we moved it later in the year for better weather. It's now held on the third weekend of June, which works best for our entrants and attendees."

Fisher believes the key to Hillsborough's enduring success, since its founding in 1956, is its charitable focus and dedication to providing a fantastic experience for its attendees and entrants: "What makes Hillsborough work is that it's a charity event, not a commercial one. Commercial events struggle to survive, which is why all the other concours in the San Francisco Bay Area have come and gone. We have exceptional collectors who continue to embrace our show as a venue for displaying the world's finest examples of automotive history in a comfortable atmosphere."

Despite the challenges of 2020, with 64 consecutive years of events and 68 events since its inception in 1956, Hillsborough remains the longest continually running concours d'elegance in the world, securing its place as one of the oldest and most enduring automotive shows.

**BELOW & OVERLEAF:** *After Rob Fisher took over the running of the Hillsborough Concours the venue was shifted to Crystal Springs Golf Course. This is from 2012. (Kimball Studios)*

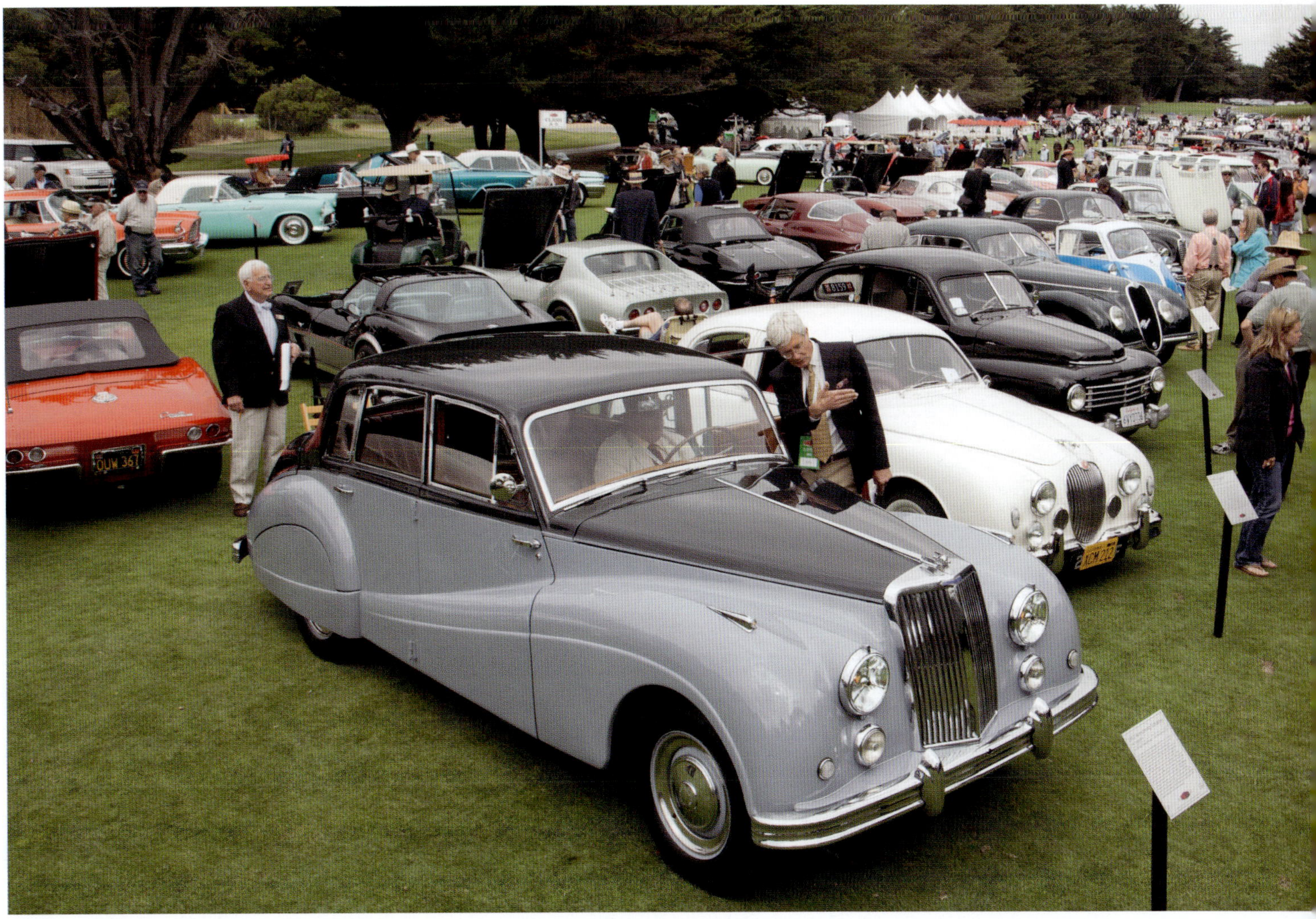

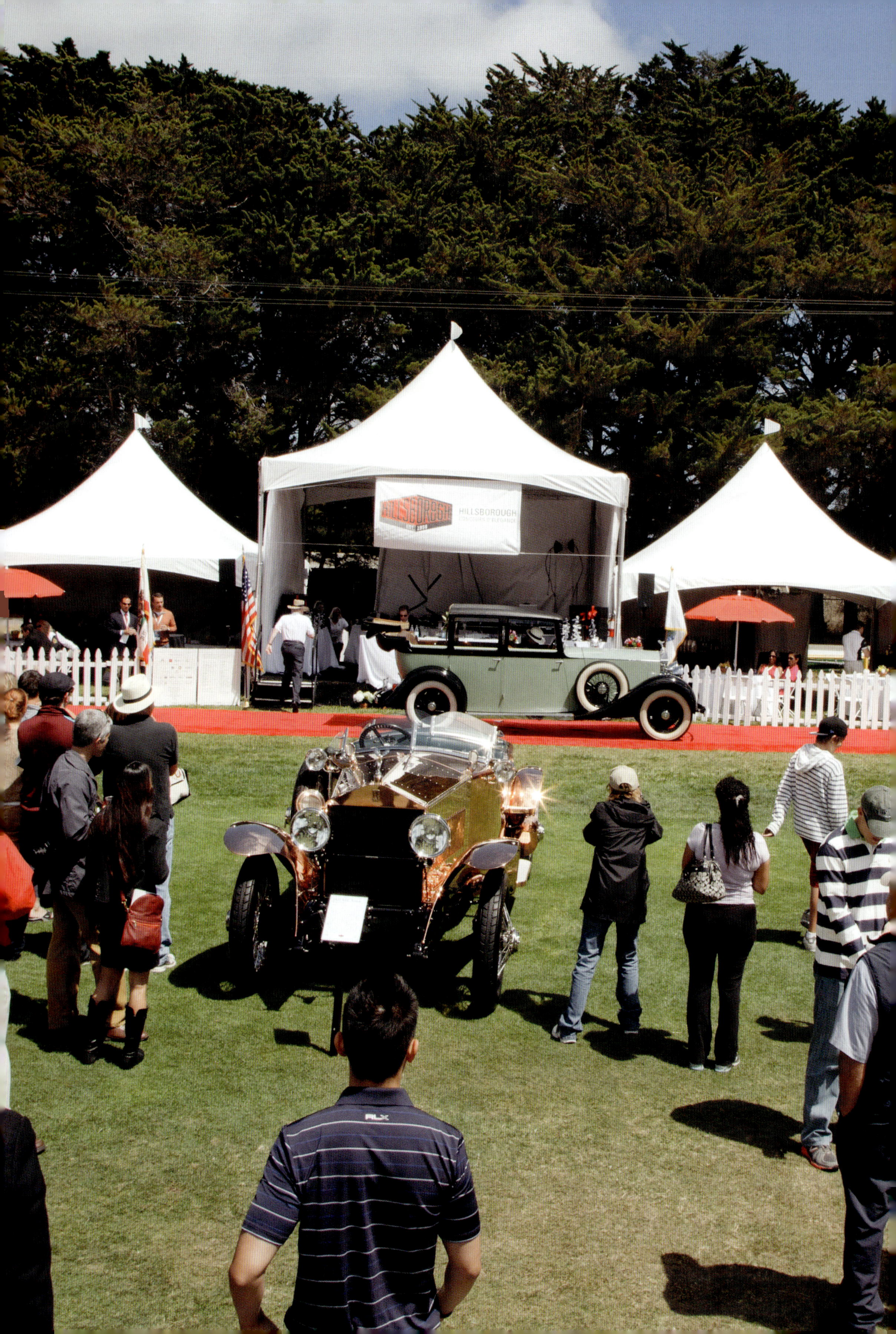
HILLSBOROUGH
HILLSBOROUGH

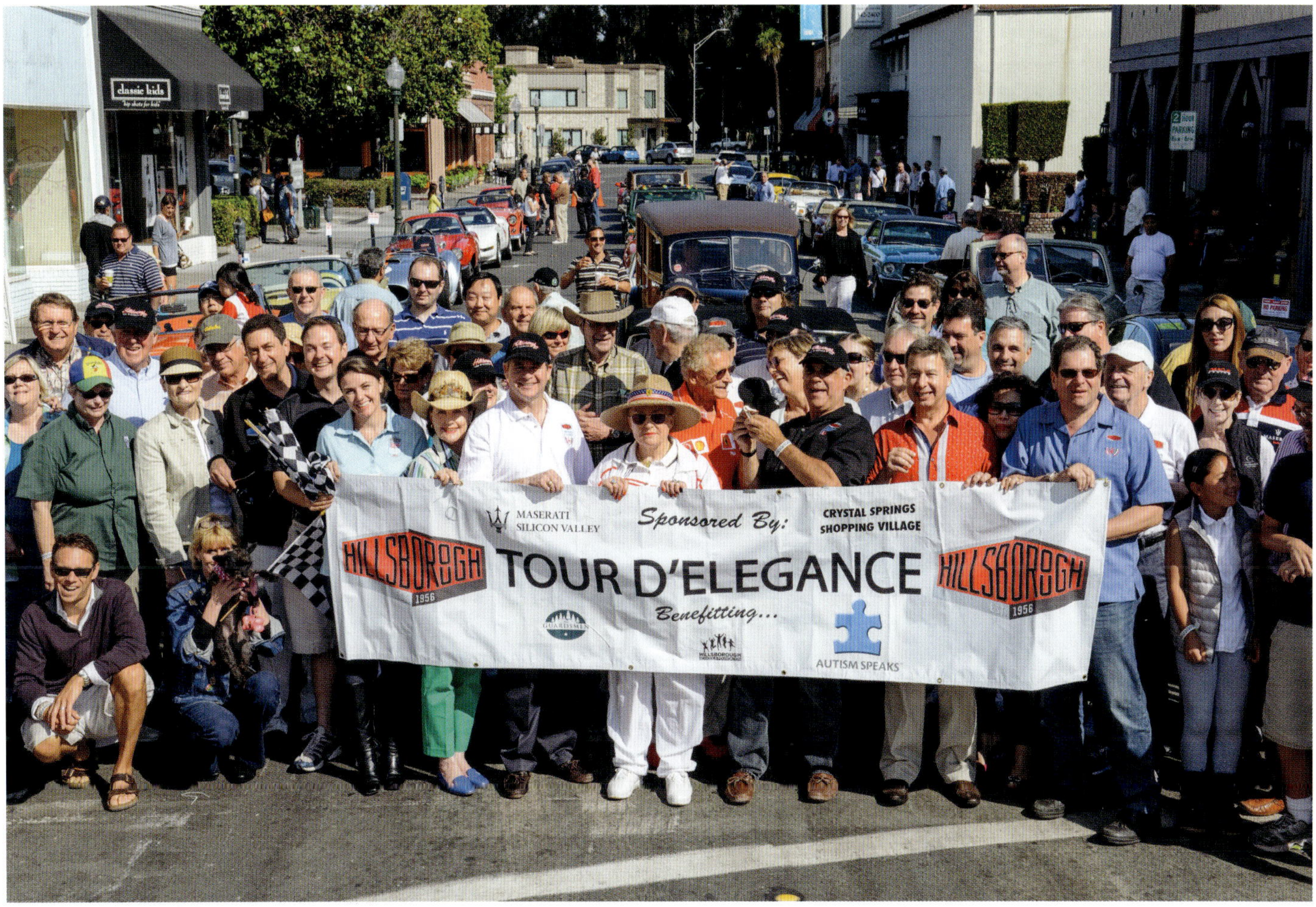


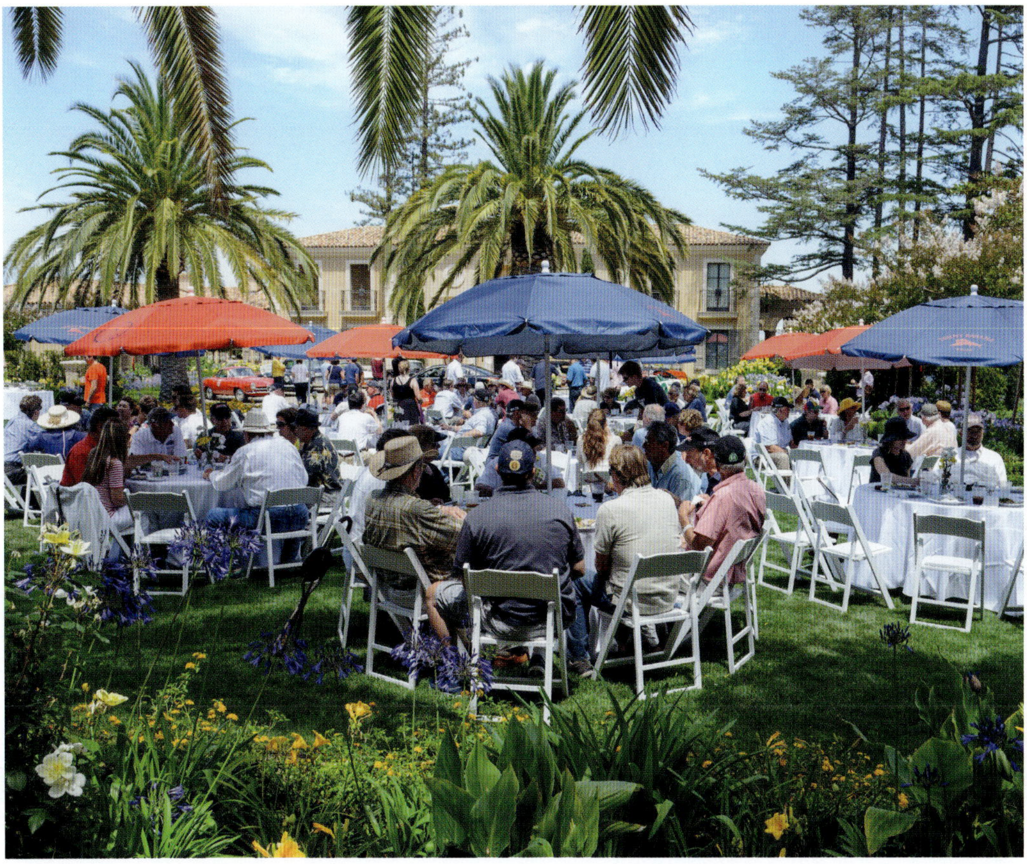

*Hillsborough also has a Tour d'Elegance. This is from 2014. And at the end of the drive a sumptuous meal awaits all participants. Also, from 2014. (Kimball Studios)*

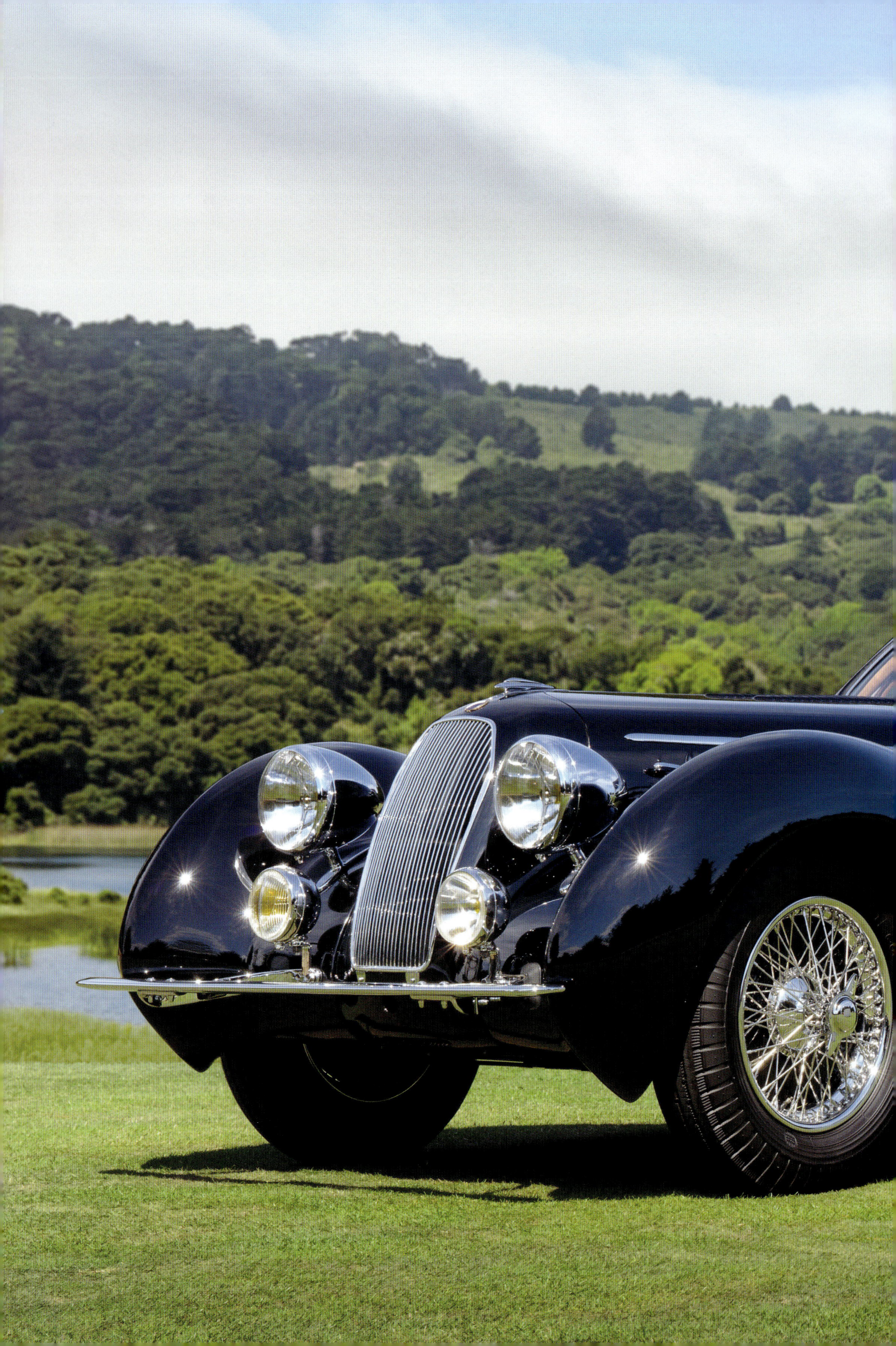

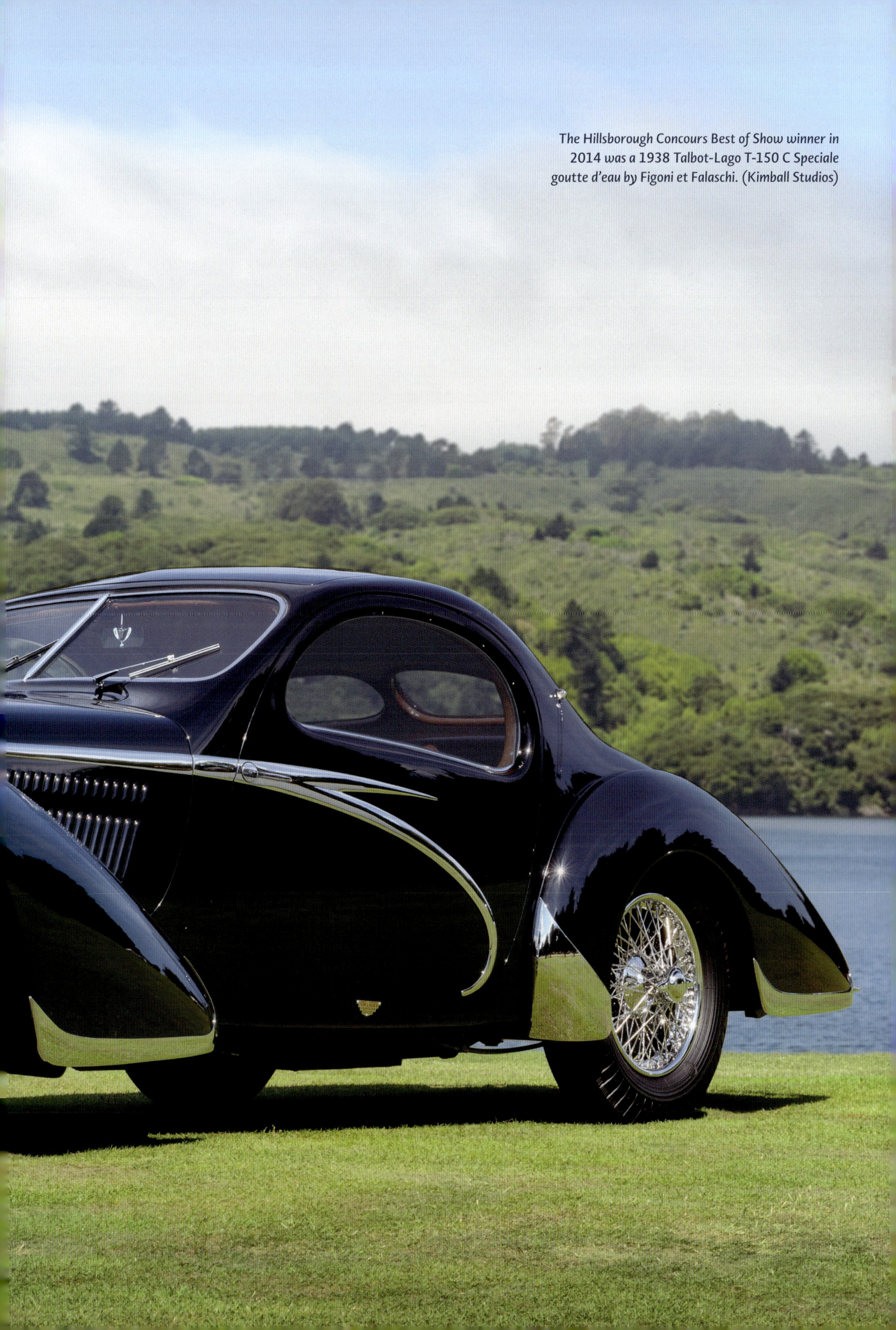

The Hillsborough Concours Best of Show winner in 2014 was a 1938 Talbot-Lago T-150 C Speciale goutte d'eau by Figoni et Falaschi. (Kimball Studios)

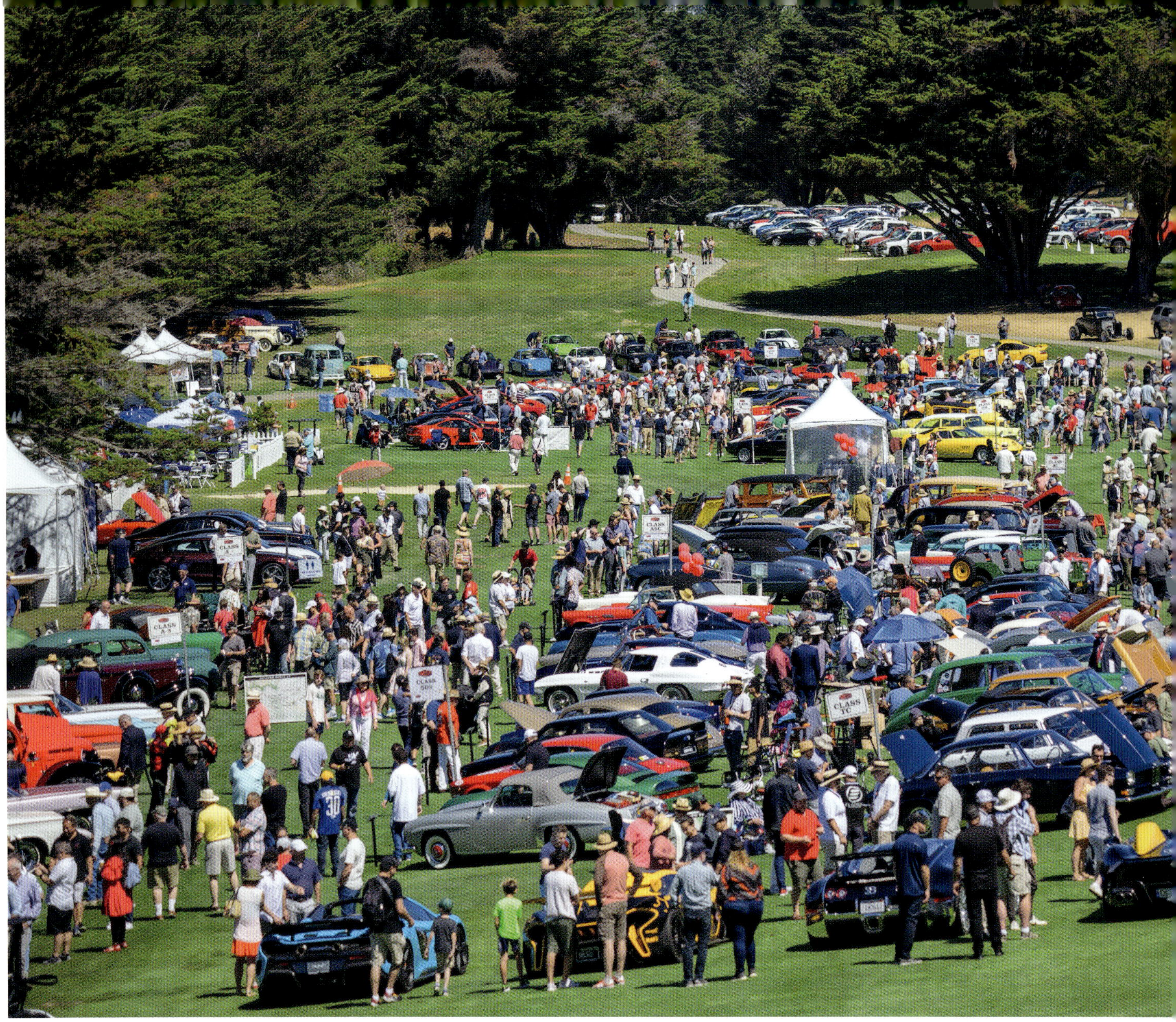

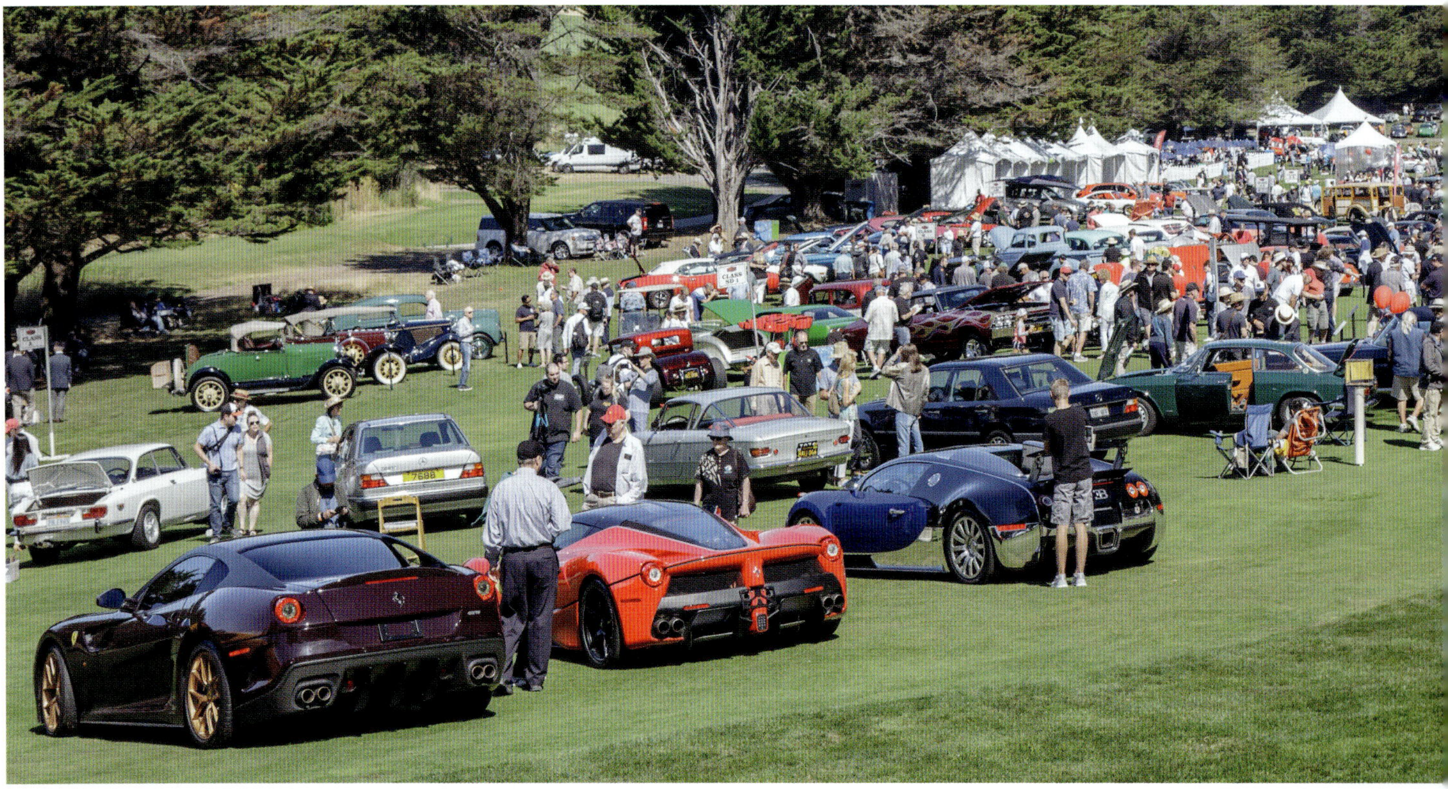

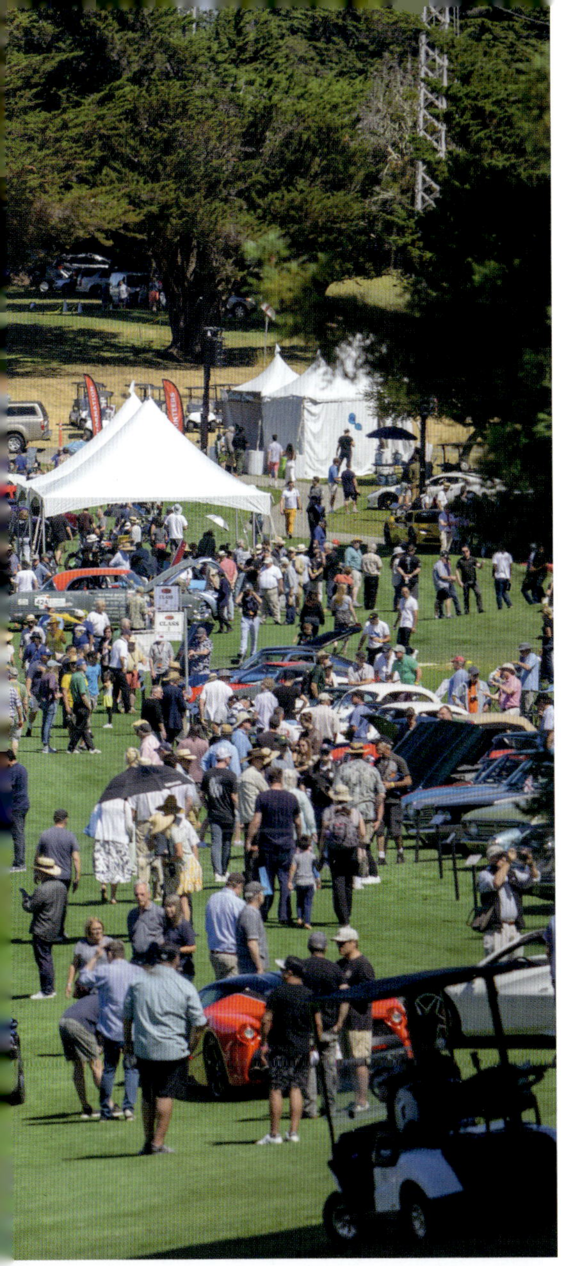

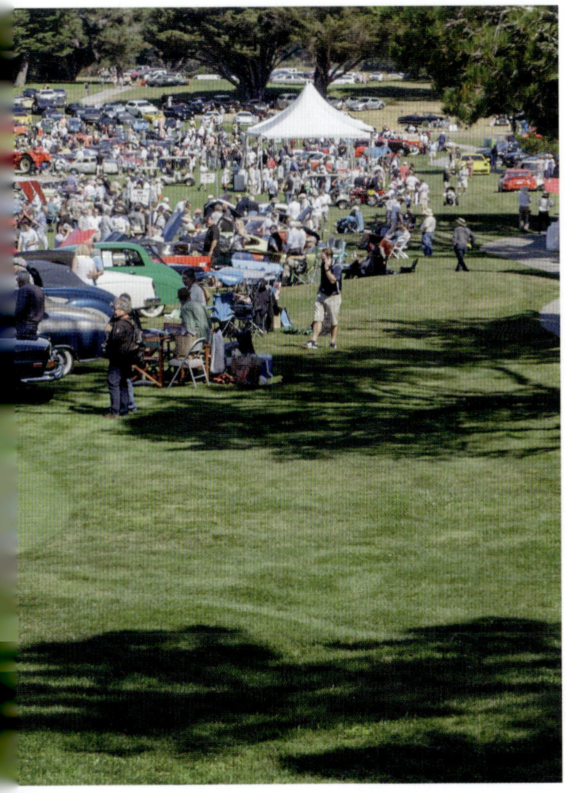

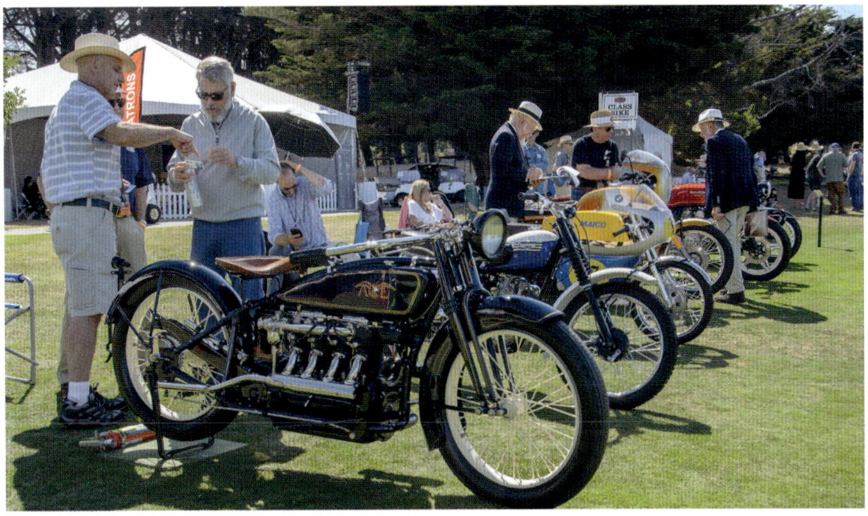

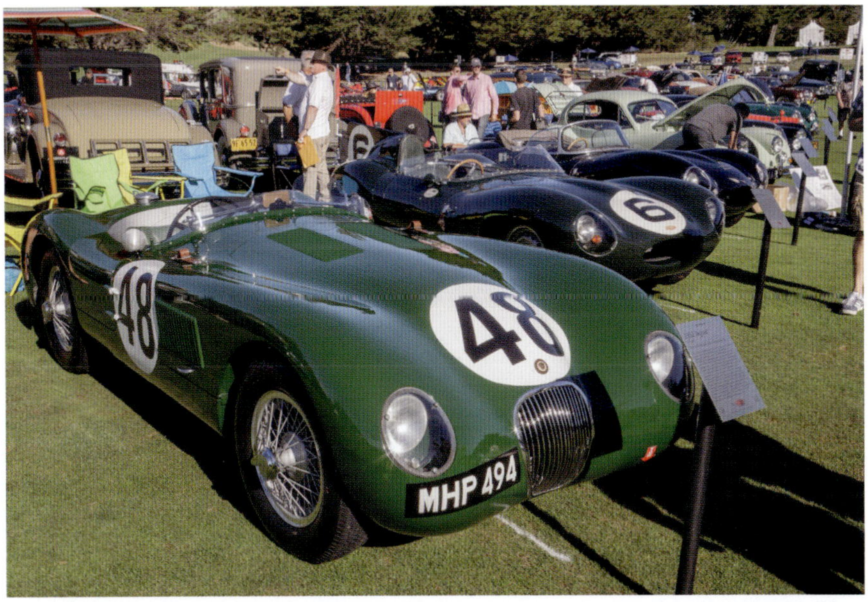


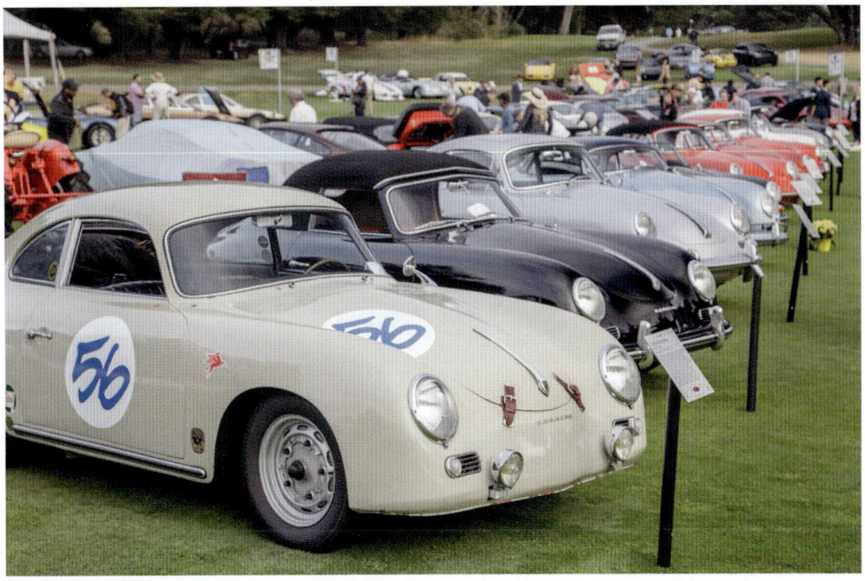


**THESE TWO PAGES:** *Despite being a non-profit volunteer organization Hillsborough has been progressing by leaps and bounds as can be seen from these images from 2017.* (Kimball Studios)

**RIGHT:** *Rarities of all kinds abound at the 2021 Hillsborough Concours. This BMW 315/1 Sport Roadster owned by Karra Canum was the cynosure of all eyes. (Kimball Studios)*

**BELOW:** *Another car that caught the eye was this 1915 Ford Model T Speedster owned by Ed Archer. What may catch the eyes of many are the rented spectators in the background. (Kimball Studios)*

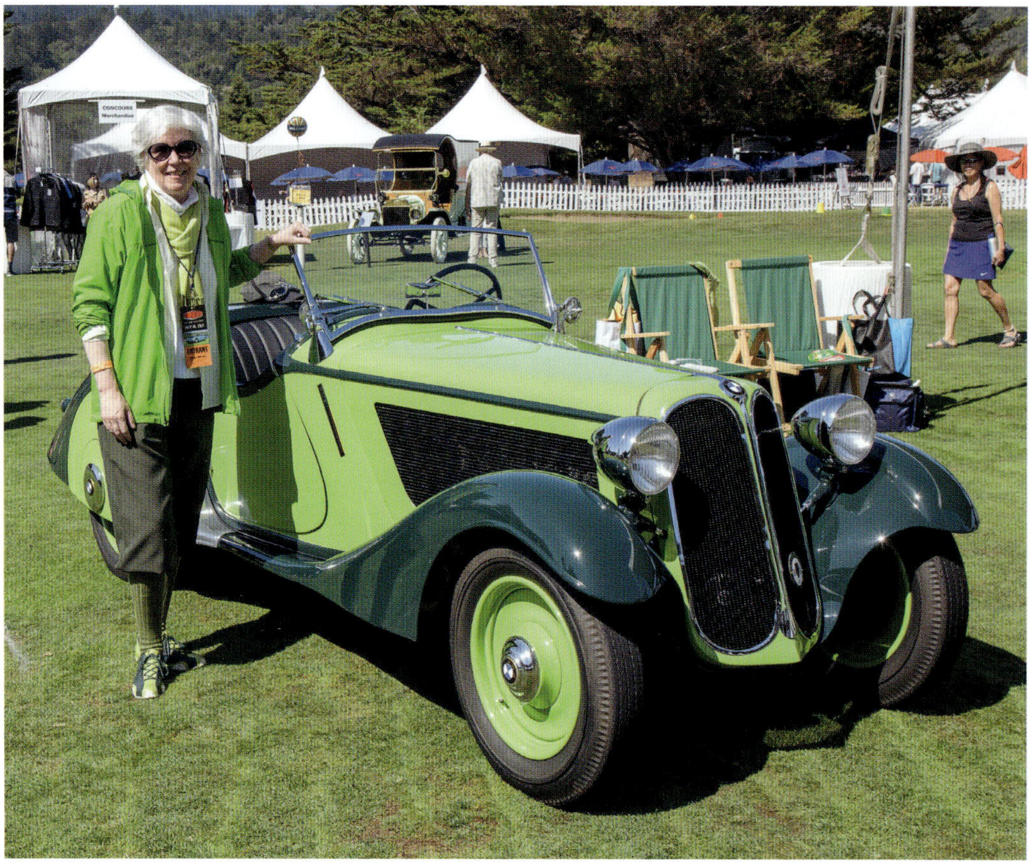

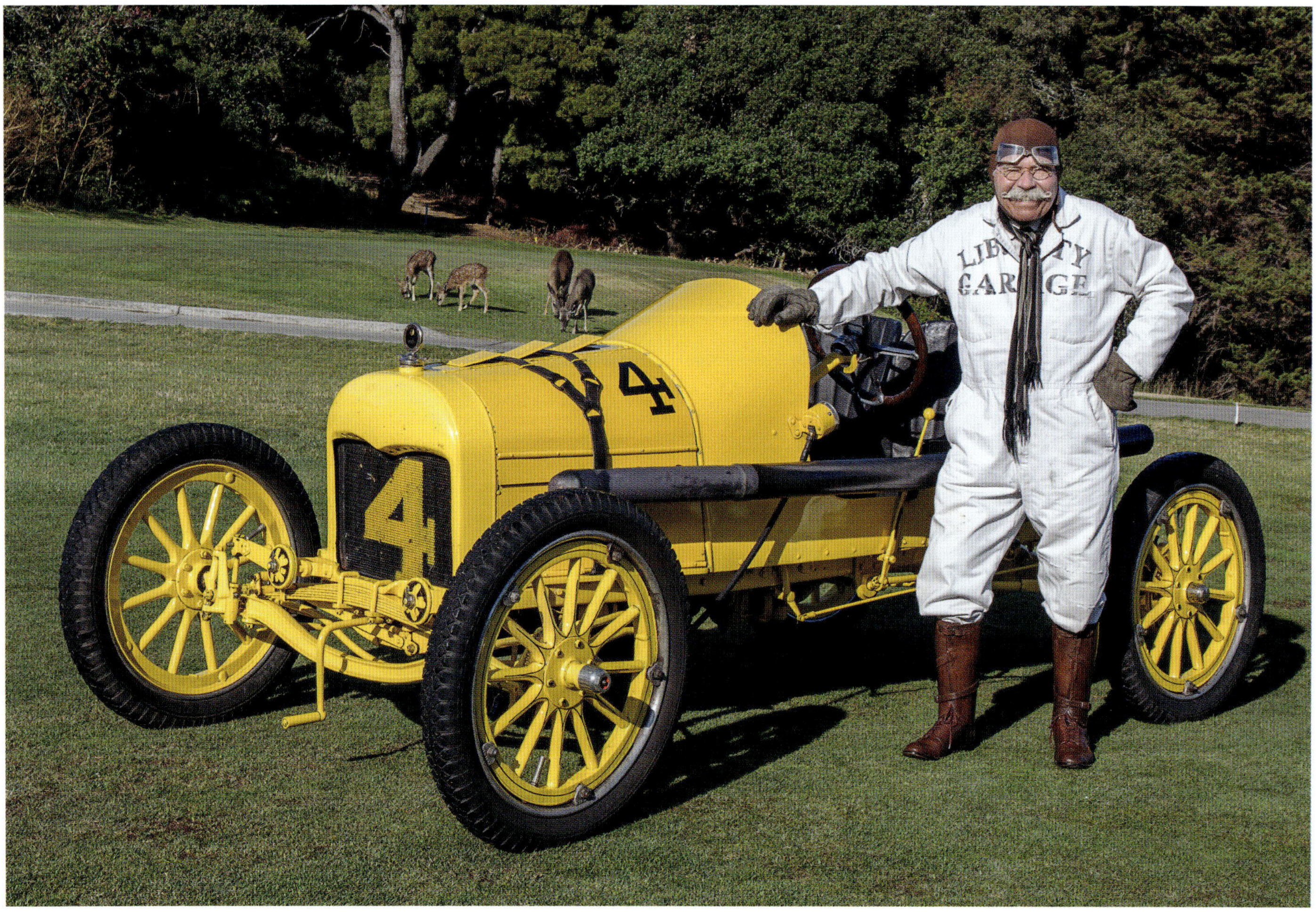

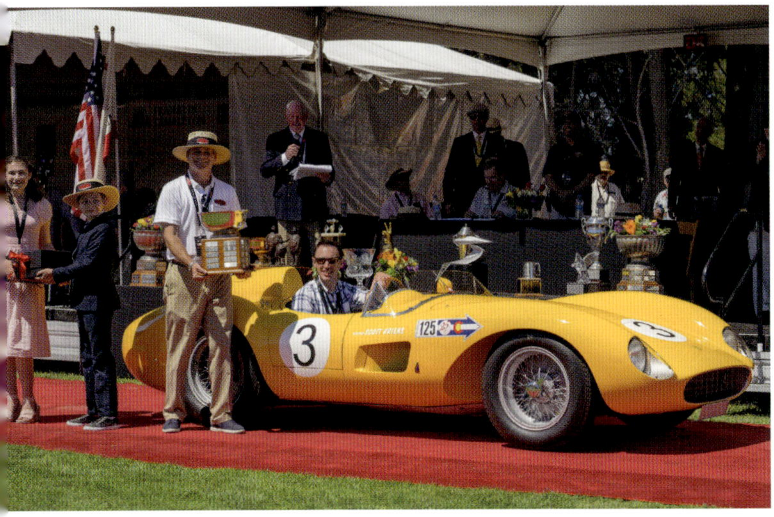

*The Best of Show for 2022 was a 1959 Ferrari 500 TRC owned by Scott Kriens. In this image, Chairman Rob Fisher also presented it with the Children's Choice Award. (Kimball Studios)*

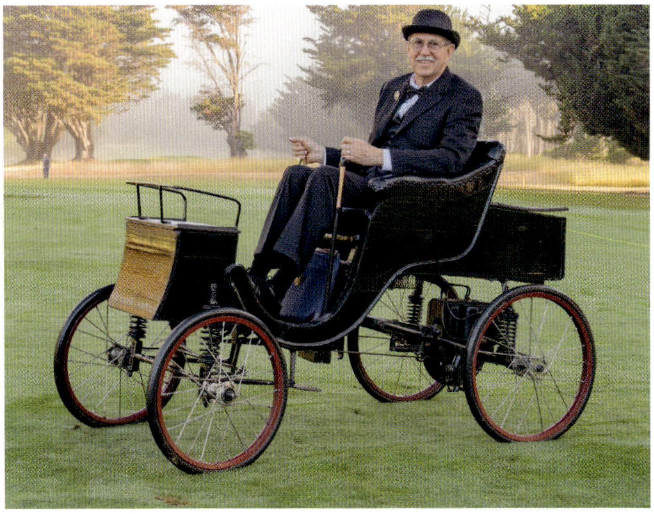

*In 2022, there was this 1896 Riker Electric Roadster driven by descendant Richard Riker. (Kimball Studios)*

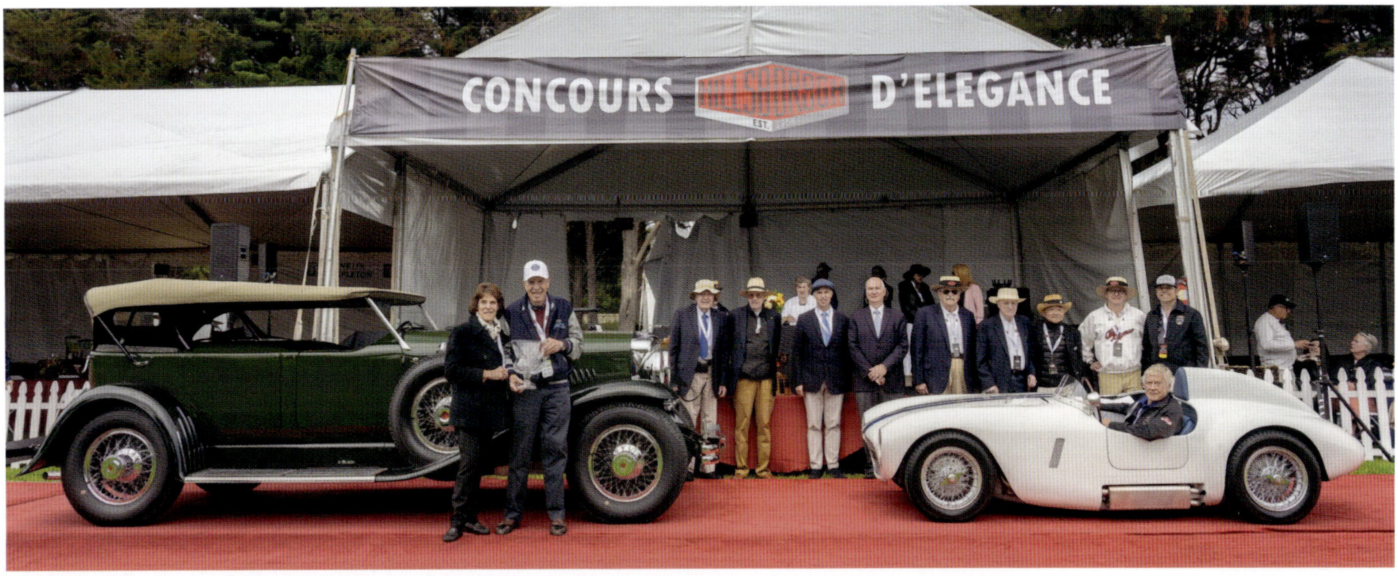


*The Best of Show for 2023 was Valerie and Aaron Weiss' 1929 Duesenberg Model J for pre-war cars and a 1955 O.S.C.A. Sports Racing Special owned by Phil White for the sports car category. (Kimball Studios)*

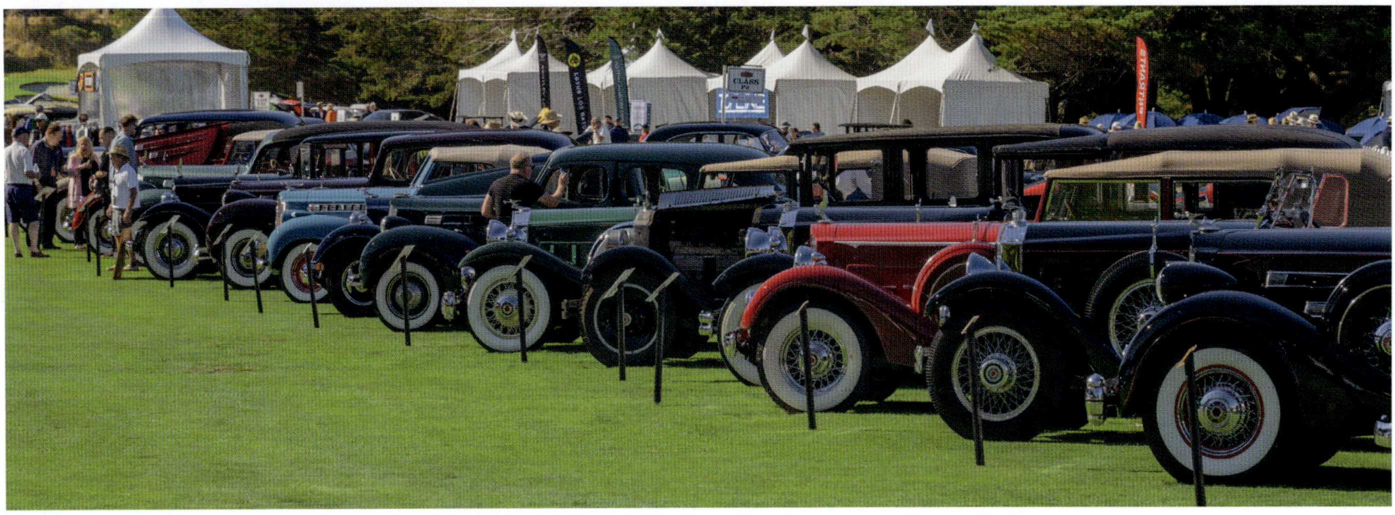

*The field at the Hillsborough Concours in 2024. (Kimball Studios)*

## As the Old Continent goes into decline

It is interesting to note that while European concours d'elegance à la française were gradually declining in the 1950s and 1960s, with perhaps the last being the Rimini Concorso d'Eleganza (featuring modern concepts and cars) continuing until around 1971, the New World was discovering the concept of concours d'elegance but for historic vehicles. This led to the emergence of a uniquely American style of concours d'elegance, something the French would later regard more as a concours d'état.

By the 1970s, several new American-style concours d'elegance events began to emerge, with the Forest Grove Concours d'Elegance, founded in 1973, becoming one of the most prominent and long-standing. Hosted in the concours d'état format (see pages 354-371), it has grown into one of the largest gatherings of automobiles, featuring as many as 325 vehicles competing in over 50 categories between 2017 and 2019. Forest Grove is located 25 miles (40 km) west of Portland, Oregon.

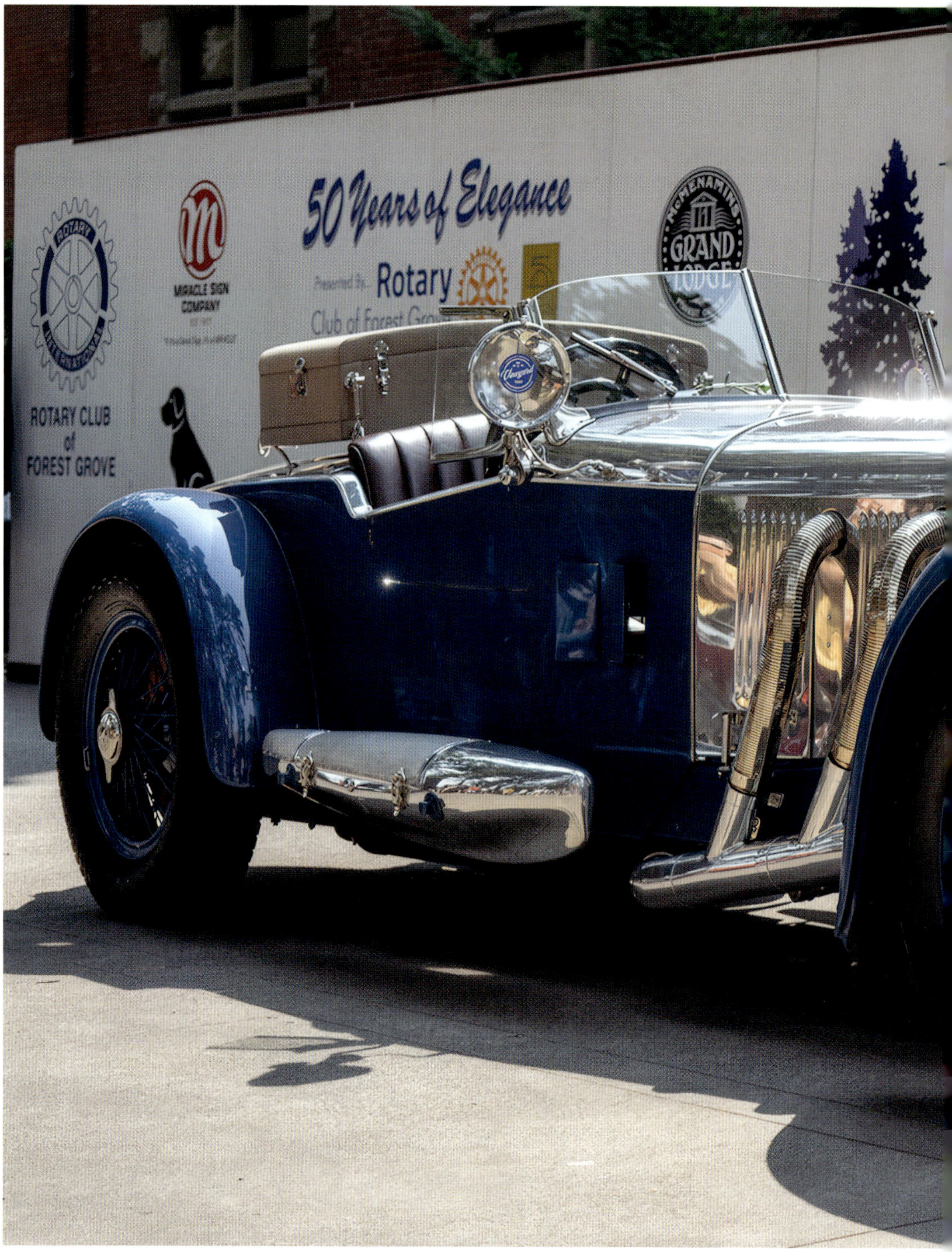


*The Forest Grove Concours d'Elegance has been presented on the campus of Pacific University every July for over fifty years. The Concours marks the highlight of summer for the Pacific Northwest classic car enthusiast community, often drawing cars that debuted at Pebble Beach the previous year. With a mix of professionally prepared and amateur-driven vehicles, Forest Grove offers something for everyone. The timing in mid-July offers the best chance of pleasant weather. The ivy-covered halls of the university provide the perfect backdrop for elegance under the shade of mature trees. Shown here is the 2024 Best in Show winner: 1929 Mercedes-Benz 680 S Barker Tourer, owned by Vintage Racing Motors from Redmond, Washington. (Courtesy of the Forest Grove Concours d'Elegance)*

The Forest Grove Rotary Club established an event to fund scholarships and support community service projects. "The first year's concours wasn't much of a success," recalls Geoff Johnston, son-in-law of the late Allen Stephens Sr., the event's founder, and whose daughter Alisa Stephens Johnston is married to Geoff. "But by the second and third years, the concept gained momentum, and from then on, there was no turning back."

"My wife, Alisa Stephens, has been involved in the show since the beginning," explains Johnston. "I believe the first job she had was to drive owners who had overindulged at the Saturday night cocktail party to their overnight lodging. She is younger than I, and I know how old you must be to drive in Oregon, so I am not sure just how she managed to get away with providing taxi service. She has remained involved in the show. She has chaired many committees and has been the general chair of the event several times. The show would not happen without her work."

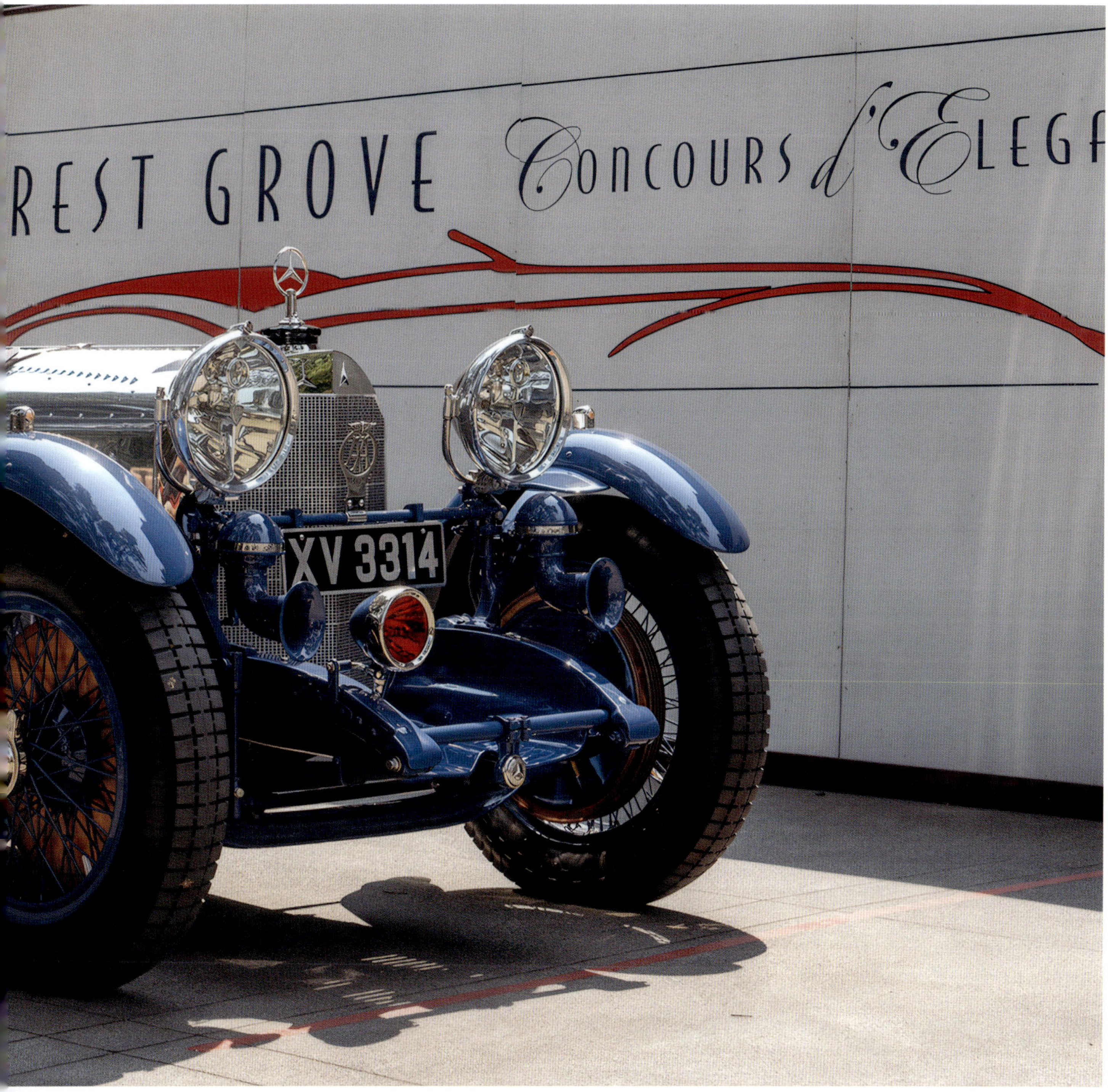

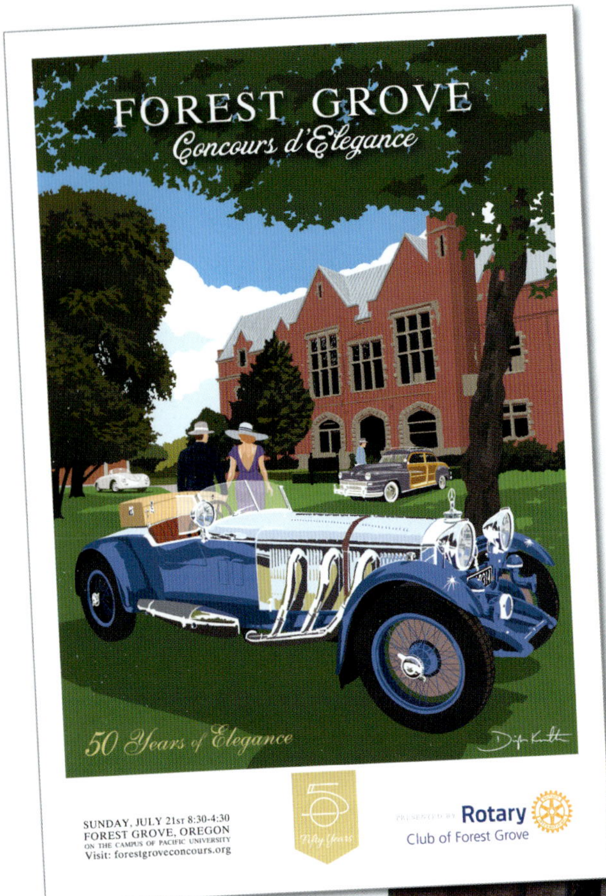


"In the early years, the winner was usually a CCCA Classic [referring to the Classic Car Club of America]," Johnston explains. "As the event grew, the number of entries swelled to over 300, though we've since reduced that number to around 250 cars, with about 90 in competition. The rest are primarily for display. This was done to raise the quality of cars in the concours. Modified cars and hot rods are now mostly on display and not part of the competition."

"The classes are always evolving," says Johnston. "In 2023, we celebrated the Corvette's $70^{th}$ anniversary, while in 2024, we introduced a Mustang class and the Art of Gandini class, honoring the late legendary designer. Jon Shirley's Lamborghini Miura P 400 won the Gandini class, while Best of Show went to Bruce McCaw's 1928 Mercedes-Benz 680 S Barker Tourer. McCaw even made a brief appearance."

Johnston notes that the judging at Forest Grove is based on a simplified version of the ICJAG (International Chief Judge Advisory Group) system, with 60 to 70 judges, typically three per class. Most entries come from

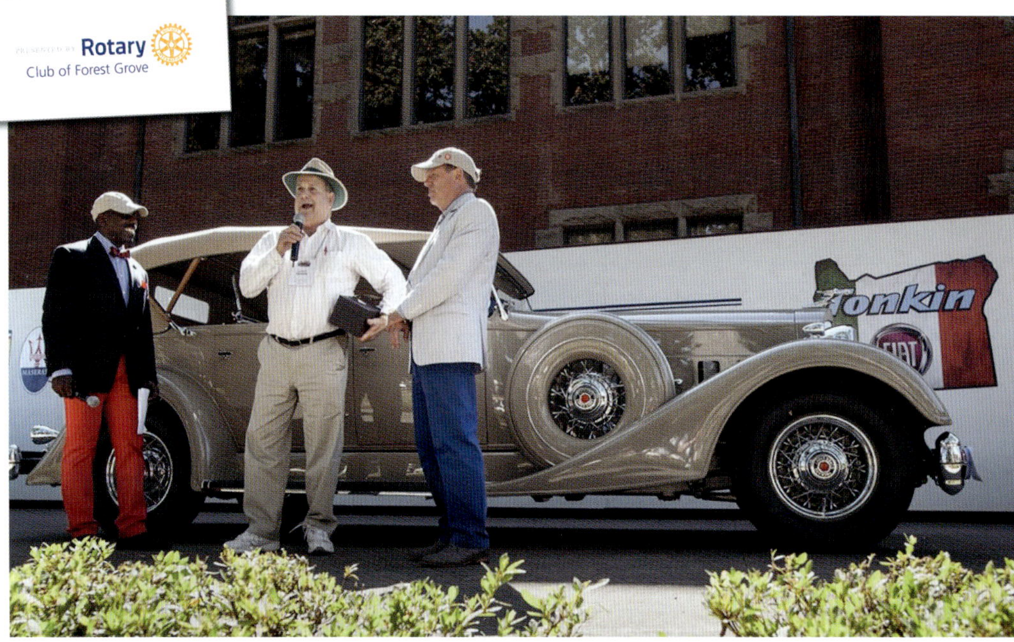

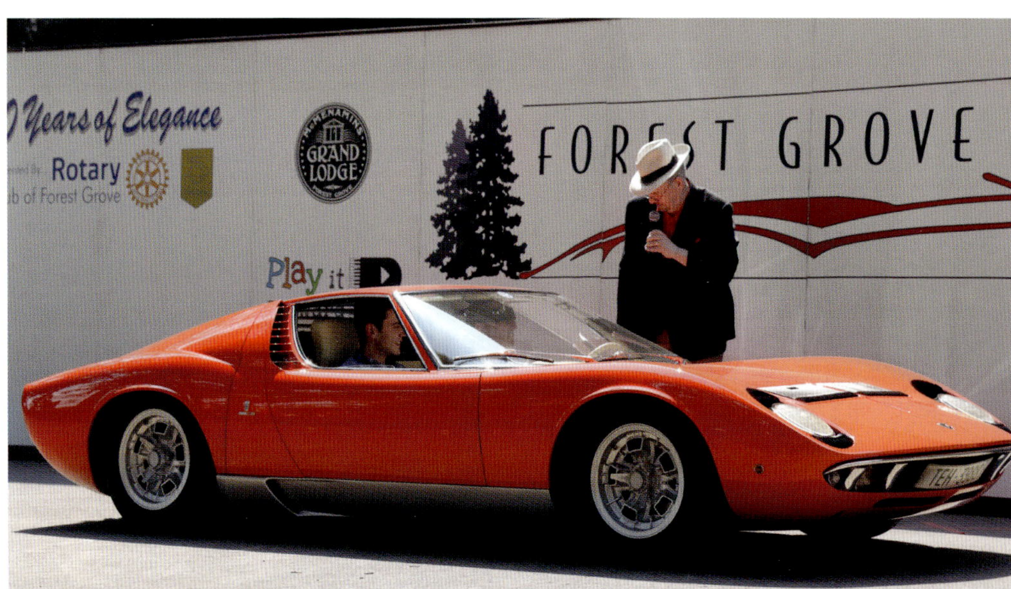


**ABOVE:** *2024 Forest Grove Concours d'Elegance $50^{th}$ anniversary poster.*

**RIGHT:** *Donald Osborne (left) and Keith Martin (right) present the Best in Show award to Larry Nannini for his 1934 Packard 1104 Super 8 in 2016. (Courtesy of the Forest Grove Concours d'Elegance)*

**BELOW:** *2024 'The Art of Gandini Design' winner: 1968 Lamborghini Miura P 400, owned by Jon Shirley of Anderson Island, Washington. (Courtesy of the Forest Grove Concours d'Elegance)*

the West Coast and Northwest, including cars from California, Washington, Colorado, Oregon, and occasionally Canada, though the majority hail from Portland and Oregon.

With an average of 5,000 paying attendees each year, at US$25 per ticket, along with sponsorships and partnerships, the event's proceeds are shared among various causes. The concours has contributed over US$1 million to the community, supporting more than 700 Washington County students with scholarships.

The event is powered by the dedication of volunteers, including Rotary members, their families, and non-Rotarians. Today, Allen Stephens Jr., son of the founder (and the brother of Alisa Stephens Johnston), leads the team that organizes the concours, which is held annually on the third Sunday in July at Pacific University's historic, tree-lined campus. With a legacy spanning more than 50 years, the Forest Grove Concours d'Elegance is one of the longest-running events of its kind in the United States.

The other prominent concours event from the 1970s was the Meadow Brook Concours d'Elegance, inaugurated in 1979. Held annually in August on the grounds of Meadow Brook Hall in Rochester Hills, Michigan, the concours grew into one of the world's largest collector car shows. The event celebrated automotive design, craftsmanship, and history over a week-long period. Additionally, it served as a fundraiser for the preservation of Meadow Brook Hall. However, on July 20, 2010, it was announced that the event would move to the Inn at St. John's in Plymouth, Michigan, marking the end of one of America's longest-running concours d'elegance competitions.

*One of the most successful events in the American Midwest, the Meadow Brook Concours d'Elegance moved to the Inn at St. John's in Plymouth, Michigan following the 2010 edition. This image of post-war sports cars was taken at the final event held at the original Meadow Brook location. (Cape Coventry Archives)*

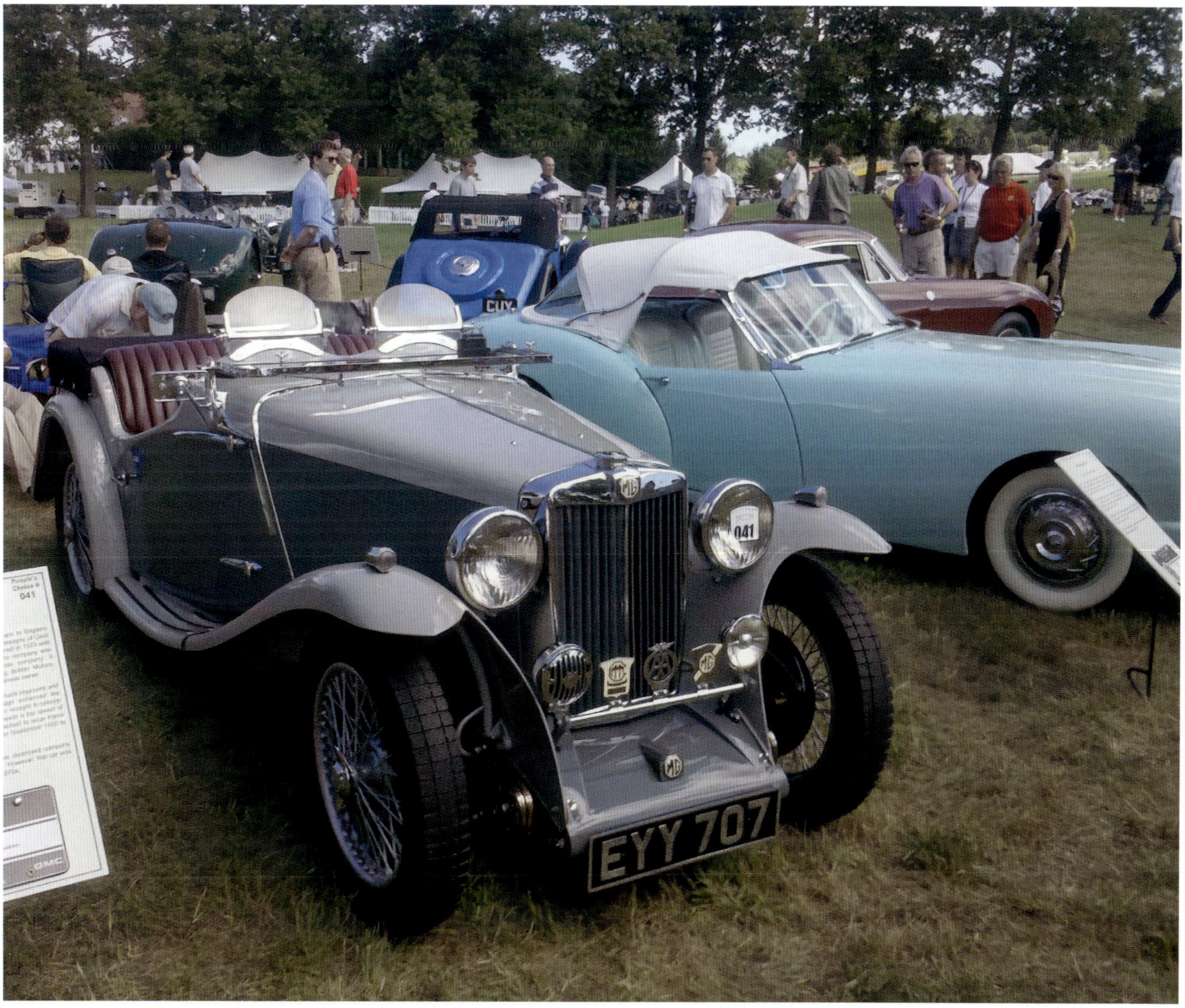

**THIS PAGE:** *With an emphasis from the start on broader marque representation, the Huntington Beach Concours provides a rare opportunity for more accessible enthusiast vehicles to shine in the famous surfing town's Central Park. (Cape Coventry Archives)*

## The other long-standing concours

Established in 1985 as part of Monterey Car Week, Concorso Italiano has been one of the key events leading up to the Pebble Beach Concours d'Elegance. Claiming to be the world's largest assembly of Italian historic vehicles, the event brings together over a thousand Italian cars alongside a celebration of Italian culture, including wine, food, jewelry, fashion, art, and automobiles. Organized by the non-profit Concorso Italiano Foundation, the event is more of a festive gathering than a traditional concours d'elegance competition, yet it remains an integral part of the Pebble Beach-associated events.

Another long-standing American concours event is the Huntington Beach Concours, established in 1986. Held on the scenic lawns of Huntington Beach Central Park in California, adjacent to the Huntington Beach Public Library and Cultural Resource Center, the event has always featured a diverse selection of American, British, European, and Asian vehicles, both vintage and contemporary, including automobiles, motorcycles, and more recently, vintage and racing bicycles.

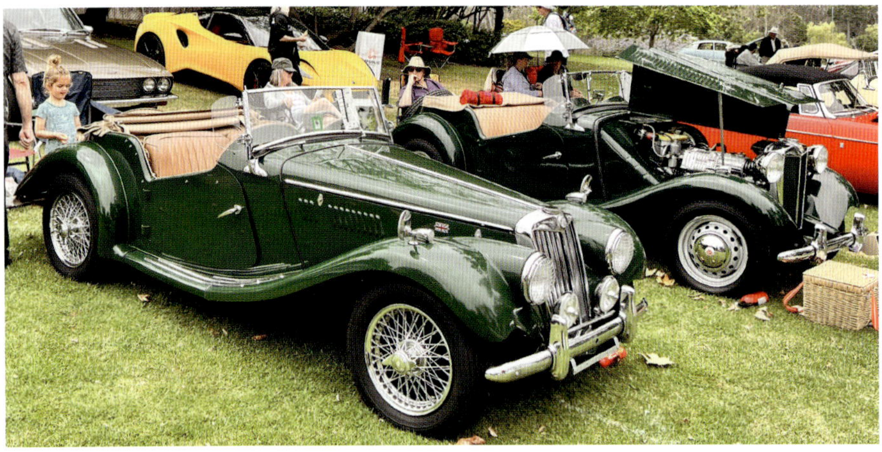

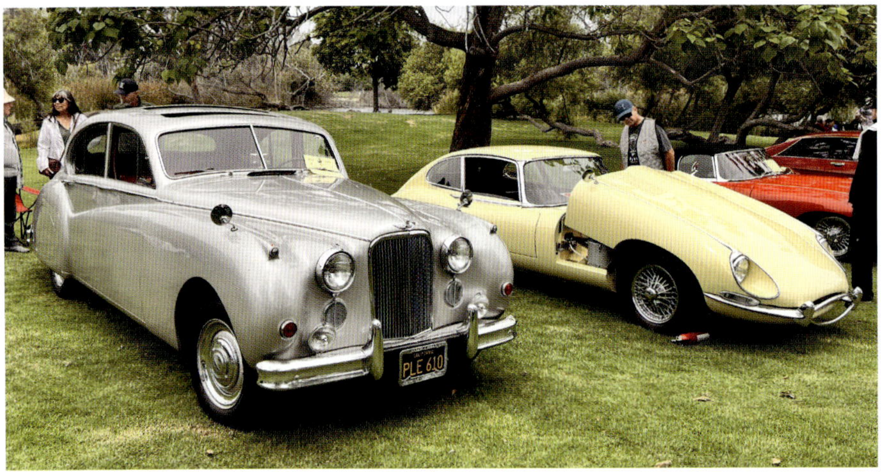



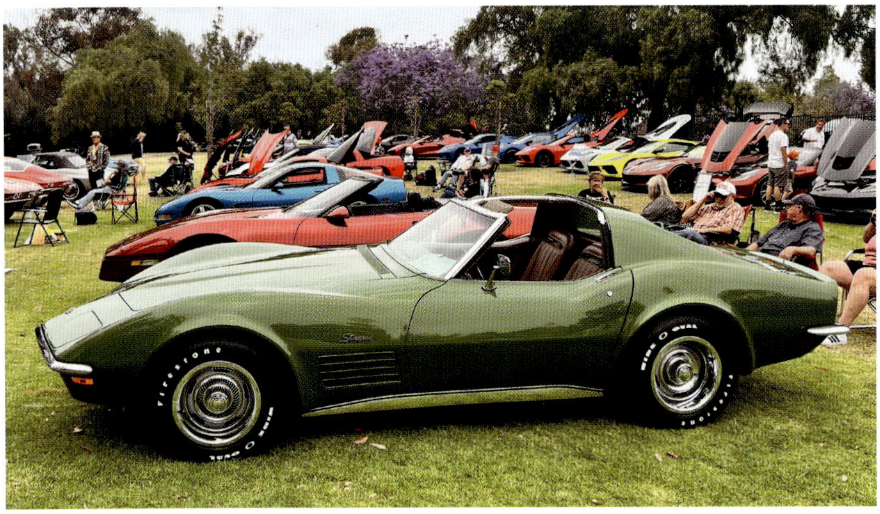

When the event began in 1986, it departed from the usual concours focusing on elite automotive marques. Led by founder Bart P. McGrath, the Steering Committee envisioned an event that celebrated all automotive makers, whether Ferrari or Ford, MG or General Motors. They recognized that enthusiasts, regardless of their resources, invested significant time, effort, and expense in their vehicles, and all entries deserved recognition.

At the inaugural event in 1986, there was no specific manufacturer marque celebrated, and the field included a wide range of vintage to contemporary vehicles from both American and foreign manufacturers. With over 30 Ferrari entries among a total field that exceeded 85 automobiles, the event attracted notable figures such as Jack Kelly, co-star of the *Maverick* TV series, and then Mayor of Huntington Beach, who assisted in the trophy presentations.

Over the years, the Huntington Beach Concours committee introduced changes to the traditional judging criteria by creating a "concours" class for entries driven less than 1,000 miles per year and a "concours street" class for regularly driven vehicles. This adjustment acknowledged that it was unfair to ask all entrants to compete at the same level and aimed to encourage broader participation by accommodating different usage patterns of the vehicles.

John Bowman, the General Manager of the Huntington Beach Concours, also pointed out that "This early June two-day event has been operating for forty years with the Rod &

Custom Show (150 vehicles) on Saturday and concours (275 vehicles) on Sunday and the venue has always been the scenic Huntington Beach Central Park grounds."

Bowman also added that as regards the judging system, "we use Speed Judge, an online system. Judges can submit scores of each vehicle using their mobile phones or use an identical manual judging card that is later entered. The system allows for immediate results for each car class and awards placement. This online system replaced an earlier, manually intensive system."

As concours d'elegance gained popularity in the United States, with events like Pebble Beach, Forest Grove, Meadow Brook, and Huntington Beach capturing magazine covers and headlines, the tradition had all but disappeared in its country of origin, France. However, one car enthusiast, Christian Philippsen, sought to change that.

"I had read about Pebble Beach and decided in the early 1980s to visit," Philippsen recalled. "What I saw amazed me. It was so impressive that I returned the following year, expecting to be disappointed, but once again, it was fabulous. I realized that the term Concours d'Elegance is French, yet the tradition had vanished from Europe. So, I thought, "Why not bring it back?"

This idea led to the creation of a new Concours d'Elegance in Paris. Philippsen teamed up with his friend Antoine Prunet, then editor-in-chief of the now-defunct magazine *Automobiles Classiques*. Together, they organized the first Bagatelle event in 1988, held at the Parc de Bagatelle in the Bois de Boulogne near Paris. The event was branded Automobiles Classiques à Bagatelle, later evolving into Automobiles Classiques et Louis Vuitton à Bagatelle and finally, the Louis Vuitton Classic à Bagatelle.

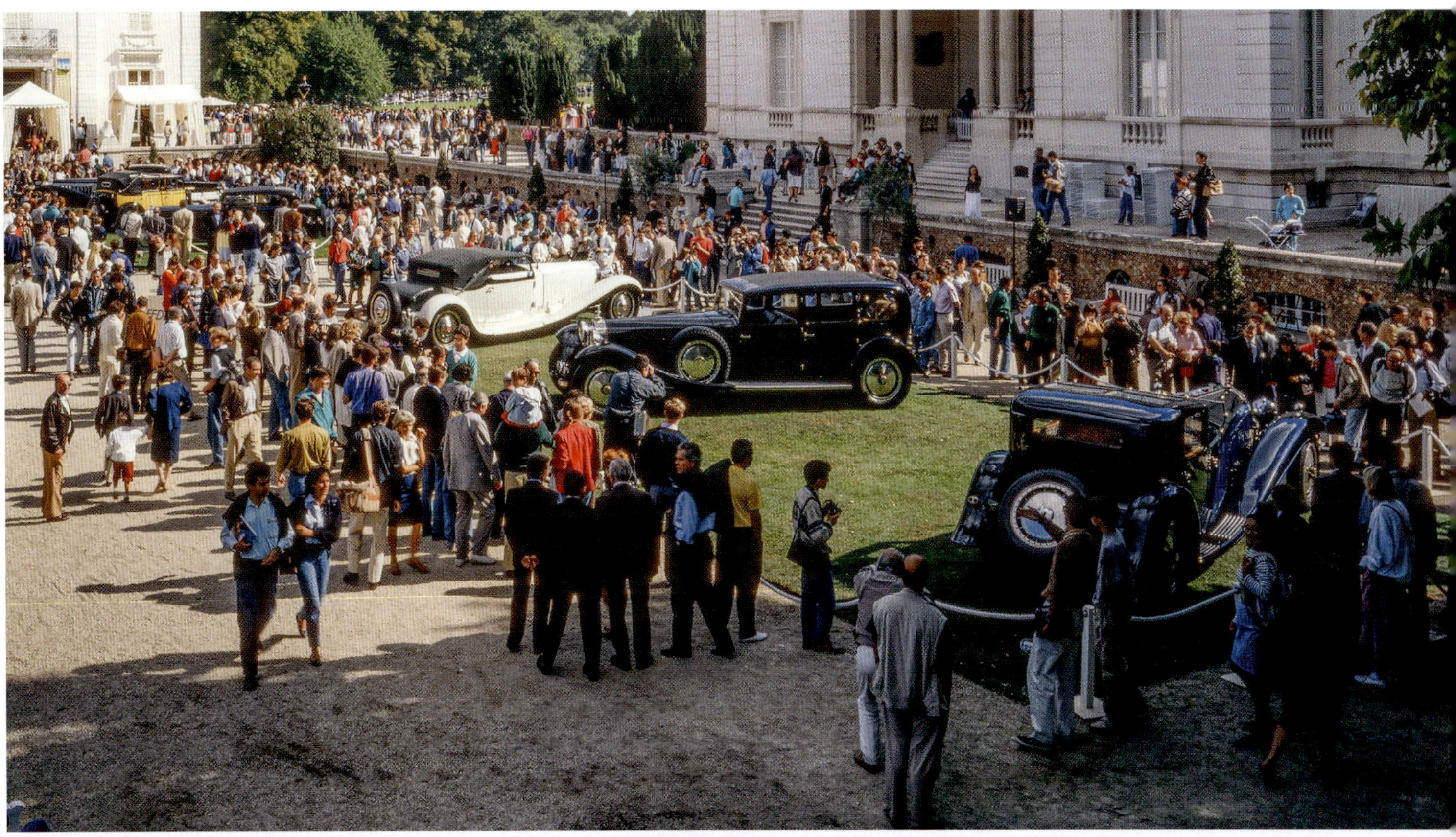

*With the Concours d'Elegance at Parc de Bagatelle from 1988 a new era of concours was relaunched in Europe. (Alberto Martinez)*

Philippsen explained how Louis Vuitton became involved: "We approached them to be a sponsor. The CEO, Henry Racamier, told us he would attend the event and, if he liked it, would support us from the second year onward. And that's exactly what happened. The event was a tremendous success."

He continued, "It was the first weekend of September, if I remember correctly. The weather was gorgeous, and we had about 18,000 visitors that first year. Press coverage was excellent, with *Le Figaro* dedicating a full page to the event and its themed display dedicated to the gold and silver cars of the maharajahs. Traffic around the Bois de Boulogne was chaotic. We started with around 60 cars

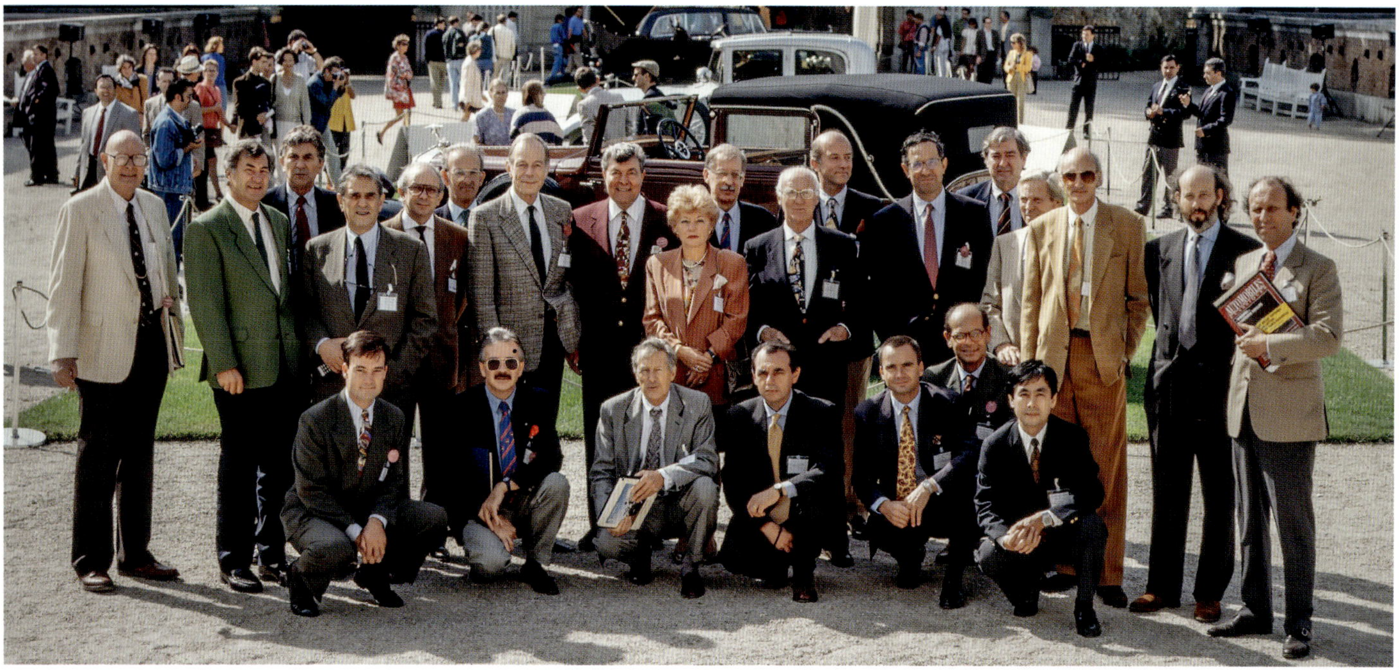

*In 1992 at Bagatelle, the jury was a who's who of the historic vehicle and auto industry world. Standing, from left to right: Lorin Tryon, Maurice Sauzay, Marc Nicolosi, Edgar Bensoussan, Dick Ruzzin, Anthony Hussey, Uwe Bahnsen, Dave Holls, Paul Lecoq, Paul Bracq, Carlo Felice Bianchi Anderloni, Christian Philippsen, Mauro Forghieri, Ed Gilbertson, Paul Frère, Yves Dubernard, Serge Bellu and Antoine Prunet. Front row, left to right: David Gooding, Patrick le Quément, André Binda, Fiorenzo Fantuzzi, Jean-Philippe Peugeot, Lorenzo Ramaciotti, Tetsuya Mizuno. (Alberto Martinez)*

and kept it at that level. From the beginning, we included classic cars across various classes, and we took great care to create consistent classes, thereby avoiding the mixing of styles, ages, and periods. We also had a class for new cars of special interest, such as the Ferrari F40, for we didn't just want to look back; we also wanted to look to the future."

Following the event's success, Philippsen reached out to Pebble Beach's organizers. "I wrote a letter to Jules Heumann, who was one of the co-chairmen of Pebble Beach, along with Lorin Tryon. I suggested exchanging winners between our events, hoping it would put us on the international map. To my surprise, I received a letter back, saying they thought it was a brilliant idea. For several years, we exchanged winners – Pebble Beach winners were invited to Bagatelle and vice versa. This brought us international recognition."

Philippsen also noted that they began exchanging judges, further enhancing Bagatelle's reputation. Louis Vuitton eventually took ownership of the event, rebranding it as the Louis Vuitton Classic. By the 1990s, Bagatelle had become known as the Pebble Beach of Europe, the most prestigious concours event on the continent. But then, it abruptly ended.

"In 2002, a new mayor was elected in Paris, and one of his first decisions was to prohibit cars from entering public gardens and parks," Philippsen explained. "This forced us to leave Bagatelle. We tried relocating to the Domaine National de Saint-Cloud, but we couldn't secure exclusivity for car events. Soon after, another event was held there, which confused the public. Louis Vuitton decided not to return to Saint-Cloud, and that marked the end of a classy concours in Paris."

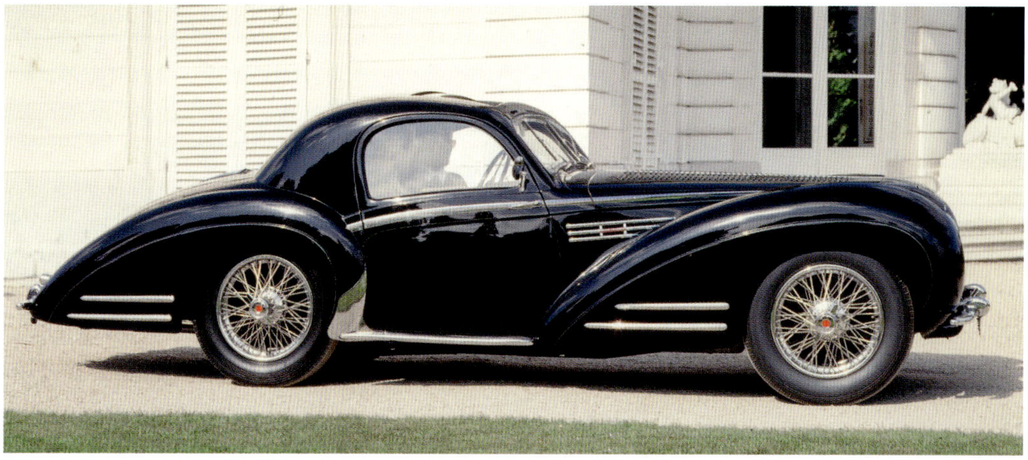

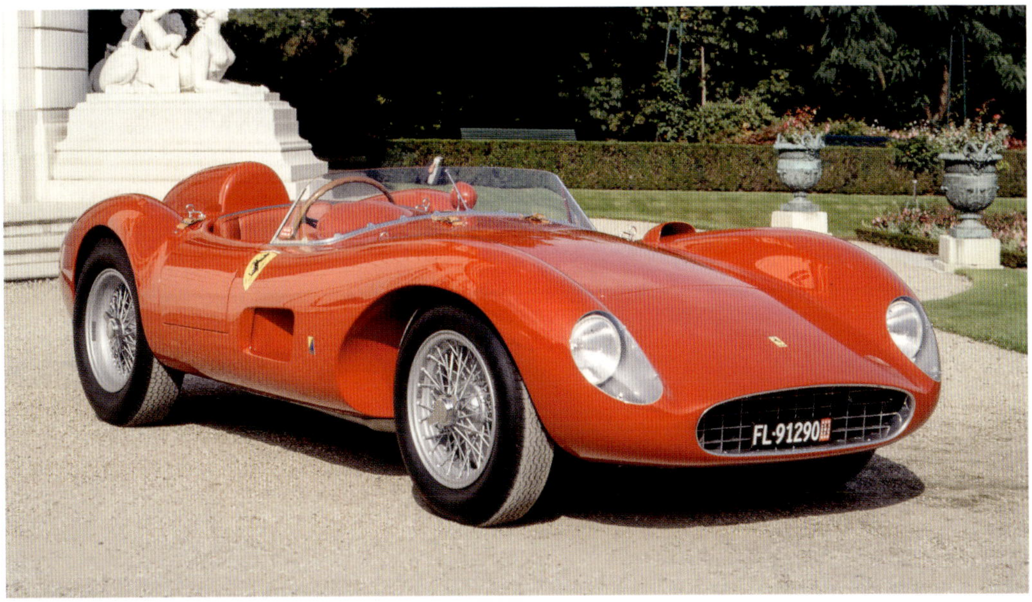


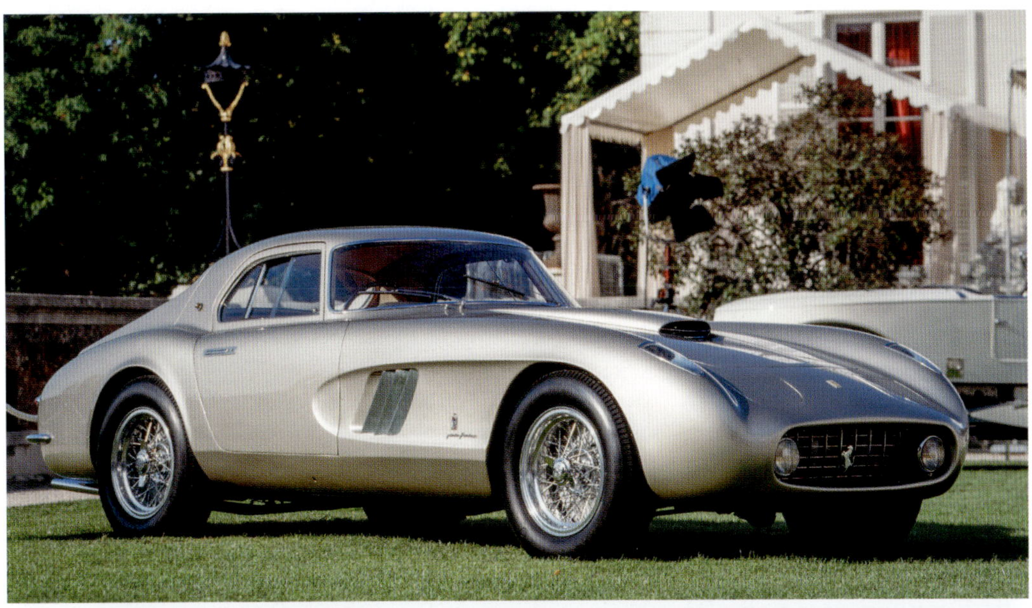

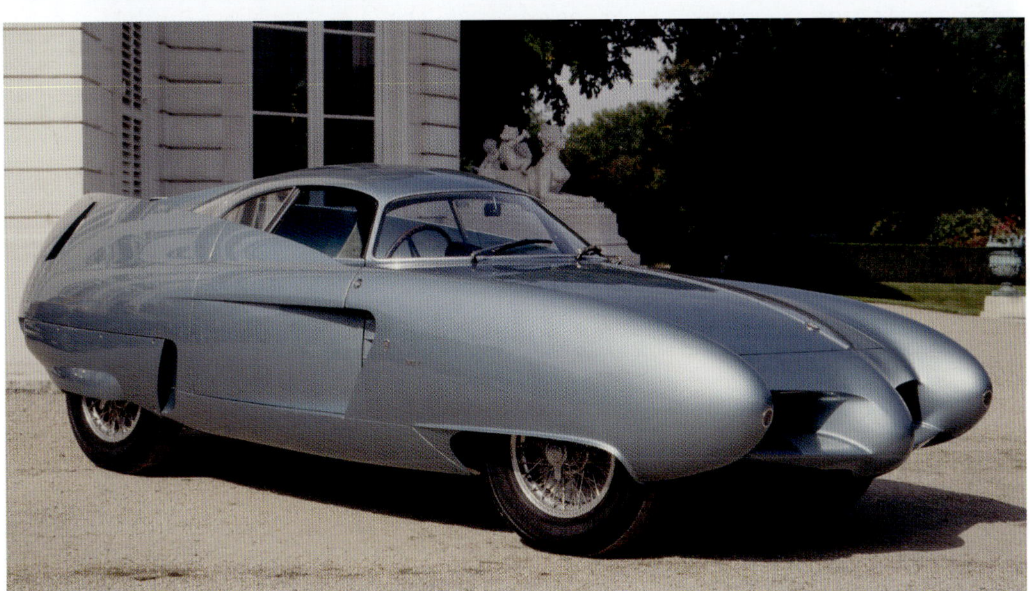

**LEFT TOP & MIDDLE:** *Bagatelle showcased both pre- and post-war automobiles, including several Ferraris. (Alberto Martinez)*

**BOTTOM:** *In 1992, Bagatelle also succeeded in bringing together the three legendary Alfa Romeo B.A.T. concept cars, following their acclaimed appearance at Pebble Beach. Shown here is B.A.T. 7. (Alberto Martinez)*

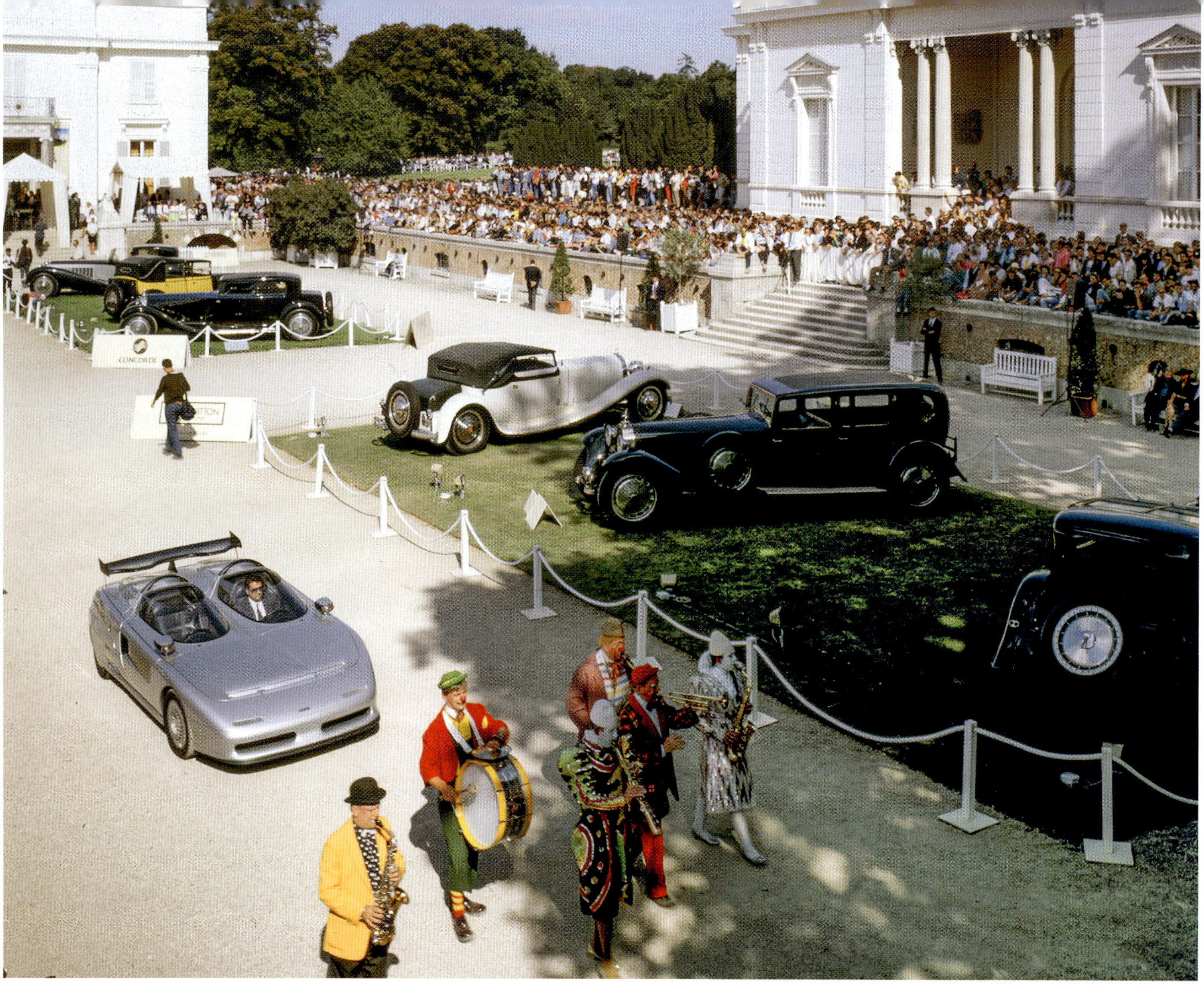

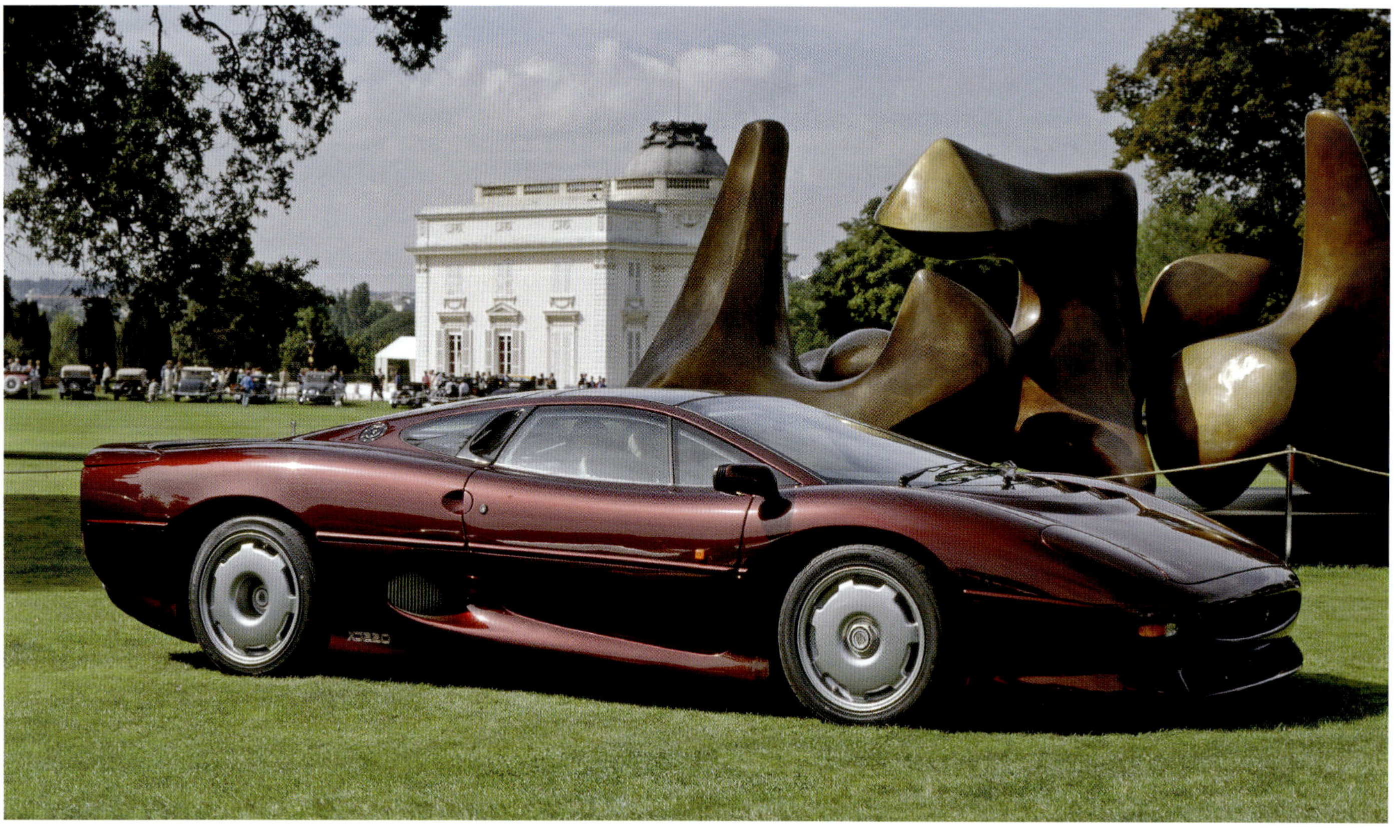

**TOP:** *Bagatelle also exhibited modern concept cars such as Giorgetto Giugiaro's Aztec concept from 1988.* (Alberto Martinez)

**BOTTOM:** *Even Jaguar's brand new XJ220 was featured at Bagatelle, juxtaposed with contemporary sculpture.* (Alberto Martinez)

## The reemergence of Villa d'Este

France's loss became Italy's gain, as another concours event with a rich history experienced a gracious revival. As recounted in Chapter 7, the Concorso d'Eleganza Villa d'Este rose to prominence in the 1930s, culminating in its last iteration in 1937 before World War II brought it to a halt. However, new management revived the event in 1947, two years after the war ended, reinstating the traditional Coppa d'Oro di Villa d'Este. The revival celebrated "a century of superb bodies," highlighting a transitional period in automotive design heavily influenced by American trends. During this era, styles fluctuated between past and future until the "Italian line" solidified as a definitive art form in coachbuilding.

The prestigious Coppa d'Oro Villa d'Este was awarded to a car designed by Stabilimenti Farina on a Lancia Astura chassis, recognized for its distinctive covered wheels. Pinin Farina's Cisitalia 202 Berlinetta also received a trophy, alongside a Fiat 1100 cabriolet, illustrating the evolving and often unpredictable tastes of the post-war automotive landscape. Additionally, an Alfa Romeo 6C 2500 Sport claimed the Grand Prix Referendum.

**THIS PAGE:** *No doubt the location of Villa d'Este next to Lake Como makes for a spectacular concours d'elegance. These images are from 2006. (Serge Cordey)*

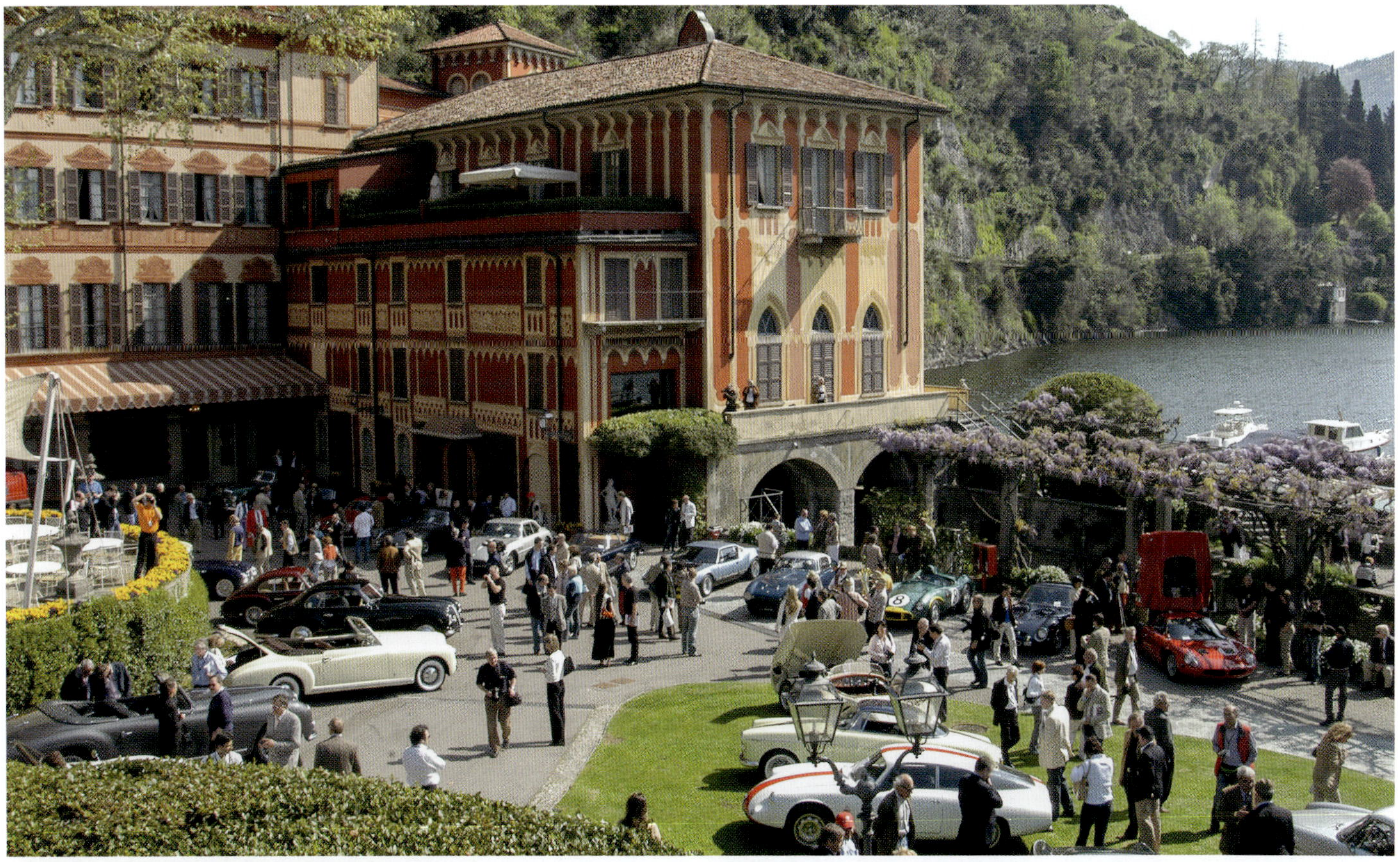

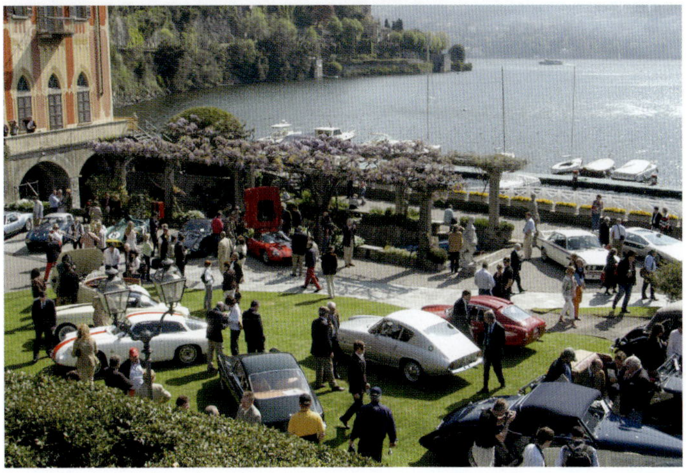

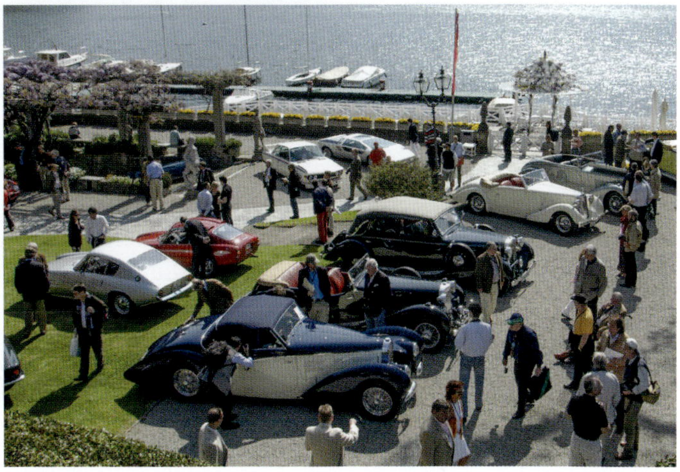

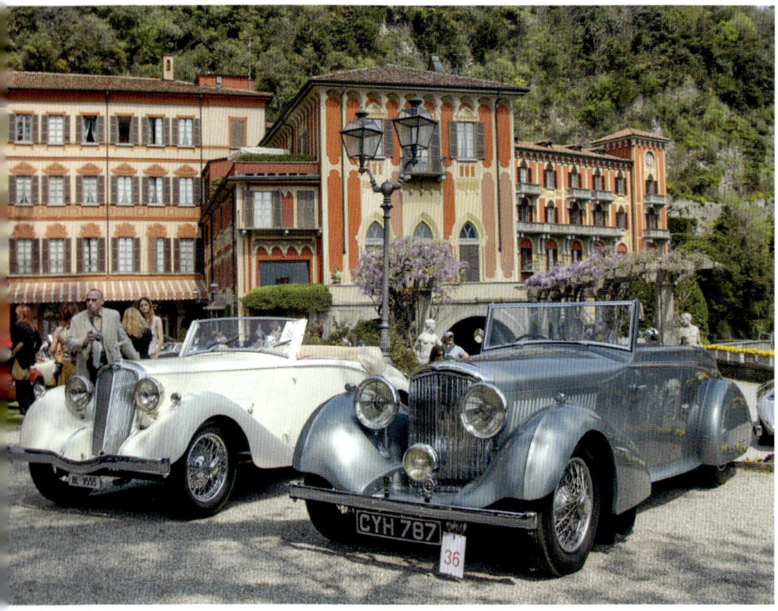


After a hiatus in 1948, the concours returned to Villa d'Este in 1949, where an Alfa Romeo 6C 2500 Sport again won the Coppa d'Oro Villa d'Este, and another Alfa Romeo 6C 2500 SS Coupé Touring took home the *Grand Prix Referendum*. However, war-ravaged Italy was in no position to sustain most concours events, and for all intents and purposes, the Concorso d'Eleganza Villa d'Este was consigned to history – at least for the time being.

In the 1980s, the renowned Italian author and historic vehicle expert Angelo Tito Anselmi, like Christian Philippsen, sought to revive the concept of the concours d'elegance in Europe. However, this revival, like others, focused on featuring historic vehicles rather than contemporary automobiles, which had been the original focus of all the European concours.

Collaborating with Villa d'Este, Anselmi aimed to resurrect the once-celebrated Concorso d'Eleganza Villa d'Este after a 37-year hiatus. Their initial effort in 1986 marked a significant moment, but the event was not held again in 1987. It wasn't until nine years later,

**THIS PAGE:** *It is not just the place, but the cars at Villa d'Este have been outstanding. (Serge Cordey)*

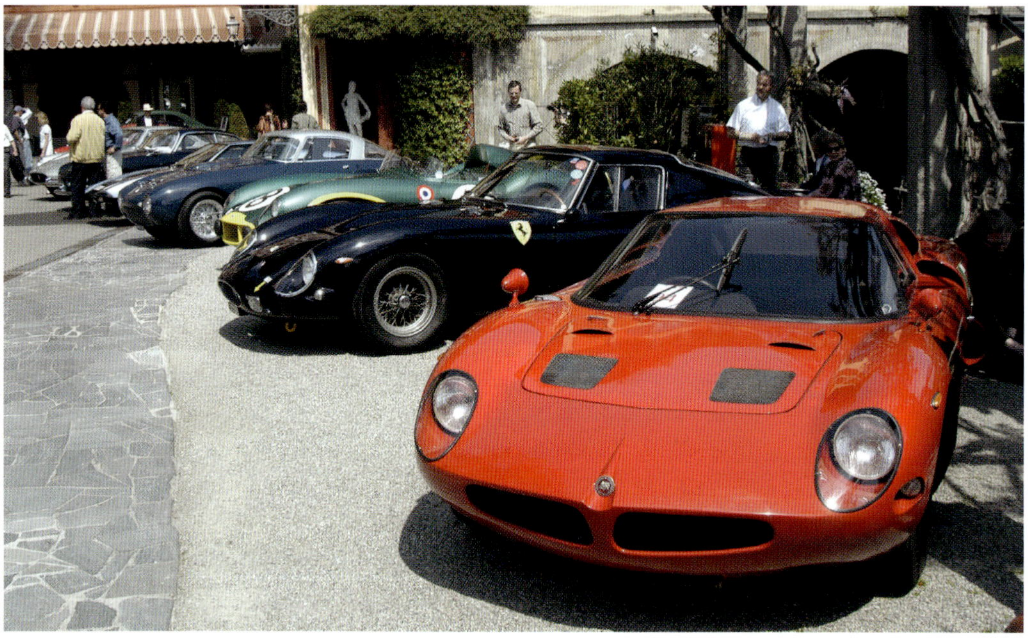

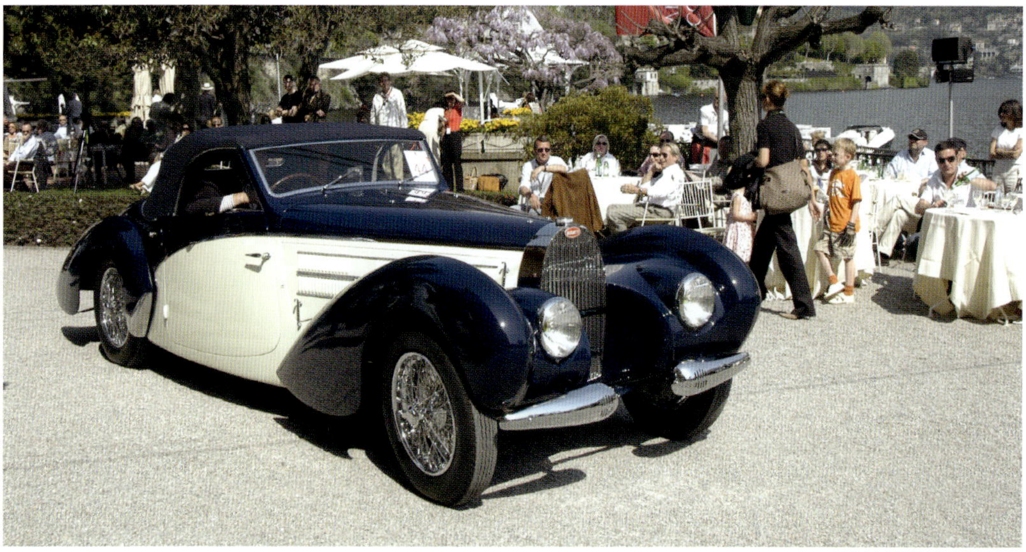

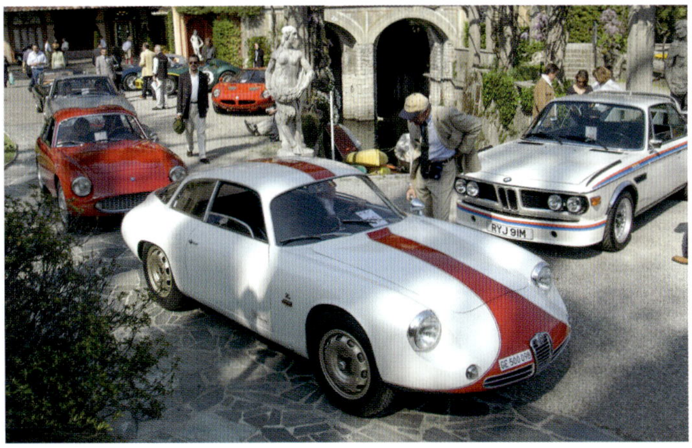

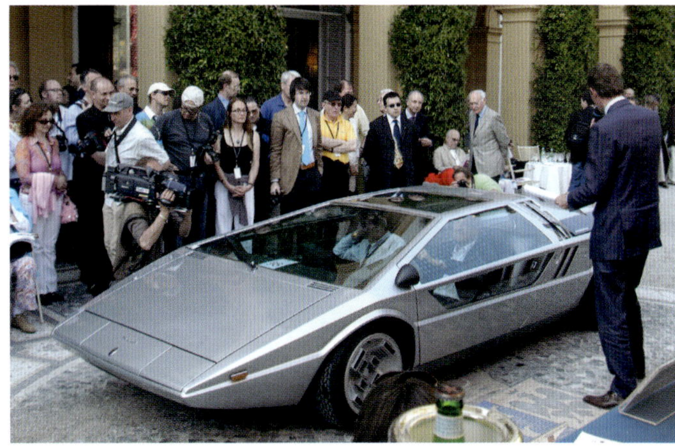

**ABOVE (BOTH):** *Seen at the 2016 event, cars like the one-off Maserati Boomerang and rarities such as Alfa Romeo Giulietta Zagatos have always been regulars at Ville d'Este. (Serge Cordey)*

in 1995, that the Concorso d'Eleganza Villa d'Este was revived, followed by events in 1996 and 1997.

However, this renewed momentum was short-lived, and the concours was discontinued once again in 1998. In 1999, BMW Group Mobile Tradition stepped in as a sponsor, supporting the event for the next two years. Thanks to the enthusiasm of historic vehicle owners, the Concorso d'Eleganza Villa d'Este saw a revival with events held in 2002, 2003, and 2004. By 2005, BMW took full control of organizing the event, transforming it into a prestigious Concours d'Elegance under their management, where it remains today.

Since 1999, when Christian Eich, then head of BMW Group Mobile Tradition, recognized the potential of the Concorso d'Eleganza Villa d'Este and provided patronage, a series of dedicated BMW enthusiasts have managed or provided leadership for the event. From 2021 onwards, Helmut Käs has been orchestrating the concours as the President of the event. Before this role, he held leading positions in sales at BMW, as head of MINI Munich, head of the BMW branch in Munich-Fröttmaning, and head of BMW Welt.

The evolving nature of the Concorso d'Eleganza Villa d'Este is highlighted by Helmut Käs, who remarked that the "return to the roots" of the event in 2002 marked a significant milestone: "At that time, an award for the latest design studies for Concept Cars & Prototypes was introduced to complement the competition for classic vehicles. Today, newer BMW Group cars are also launched during each edition of the event."

Another notable development has been the shift toward featuring more post-war classes. Käs noted that there is now "a good mix between pre- and post-war cars, reflecting the changing interests of younger generations. This shift also extends to the juries and committees involved in the event."

Helmut Käs emphasized the importance of preserving historic automotive heritage. He explained that in recent years, "the Concorso d'Eleganza Villa d'Este has focused more on the restoration status of the vehicles. Over-restoration is discouraged, and the event has even introduced a preservation class to recognize the best unrestored vehicles." This emphasis on preservation was highlighted by the Best of Show award at the 2024 event, which went to a beautifully preserved 1932 Alfa Romeo 8C 2300.

Käs also discussed efforts to make the event more inclusive: "The Concorso d'Eleganza Public Day at Villa Erba allows the public to experience automotive passion firsthand, and the 'Wheels & Weißwürscht' car and community event has become a regular feature. Despite these inclusive measures, the Concorso d'Eleganza Villa d'Este remains exclusive, with a tough selection process limiting participation to around 50 cars."

Finally, Käs mentioned that the "evaluation of vehicles has evolved to meet the needs of an international field of participants, emphasizing that the process must be fair and not purely subjective. While elegance remains a key criterion, the diversity of vehicles, especially post-war models used in competition, necessitates a broader focus."

He also acknowledged the variety of judging styles at concours events worldwide, such as the one used at Dinard Elegance, but affirmed that the Concorso d'Eleganza Villa d'Este will continue to maintain its unique "Italian style."

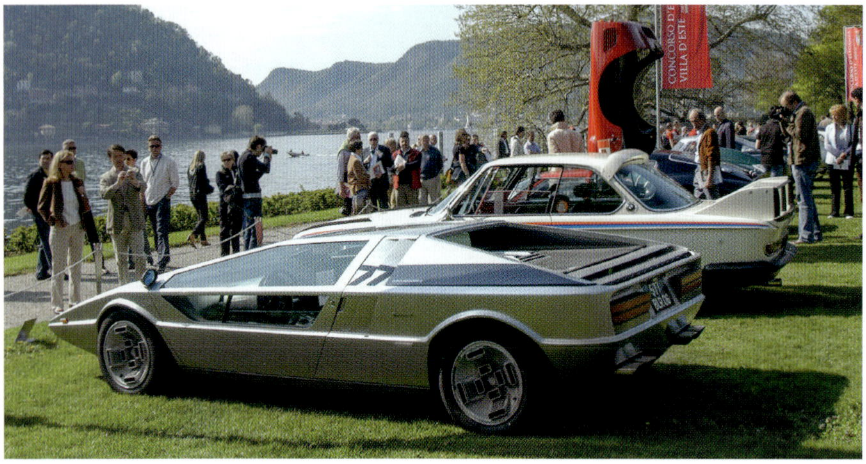

**LEFT & OPPOSITE TOP:** *All kinds of cars can be seen at Villa d'Este... from the Maserati Boomerang concept (front) designed by Giorgetto Giugiaro of Italdesign (left) to a De Tomaso Pantera (opposite top). (Serge Cordey)*

**BELOW:** *BMW's Helmut Käs with Nichole and Bill Lyons and the Best of Show winning Duesenberg SJ with J Gurney Nutting bodywork at the 2023 Concorso d'Eleganza Villa d'Este. (Courtesy BMW)*

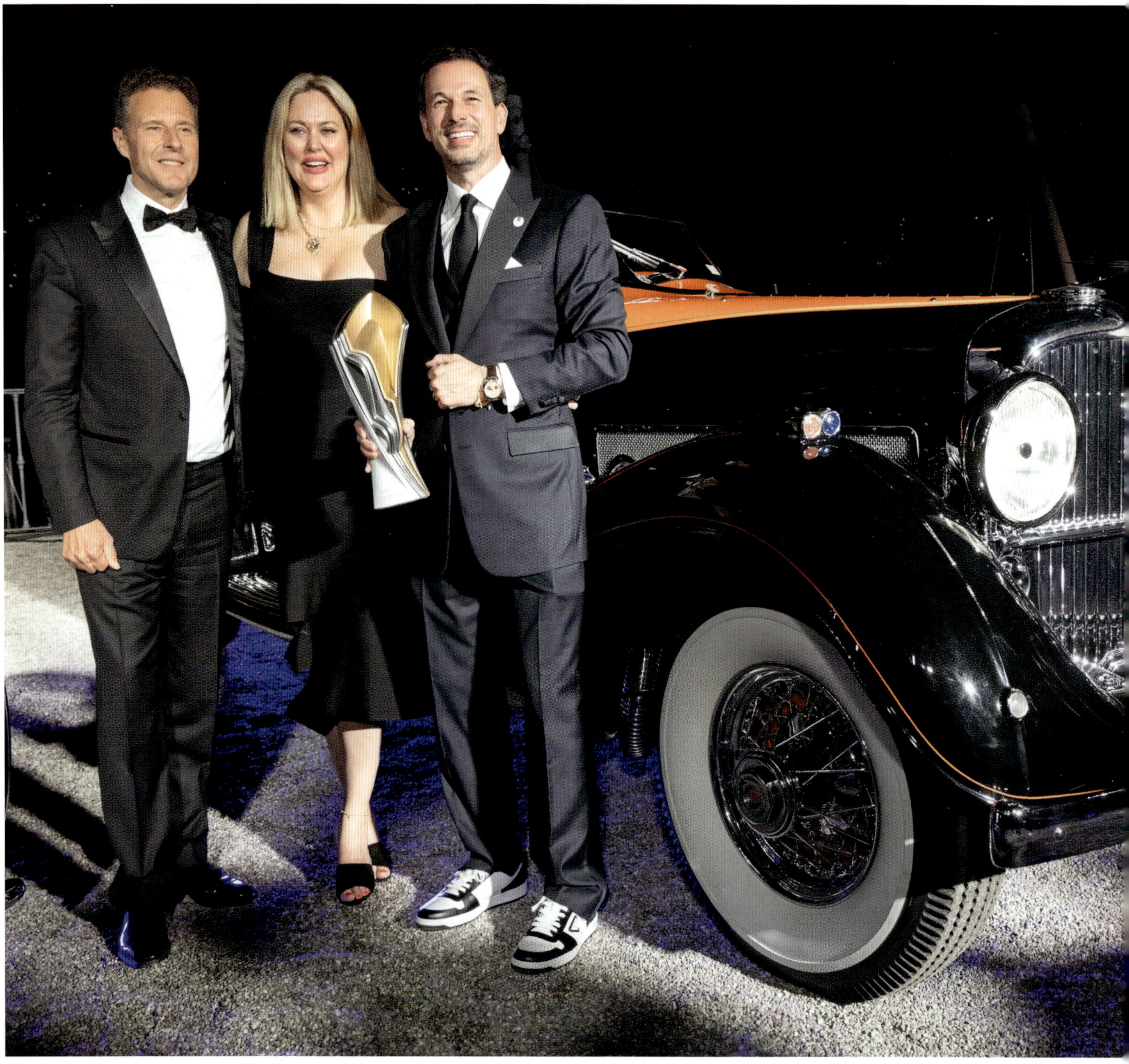

**MIDDLE:** *Trophies await the results. (Kimball Studios)*

**BOTTOM:** *Ferraris and a Ford Model A line up at Villa Erba in 2022. (Kimball Studios)*

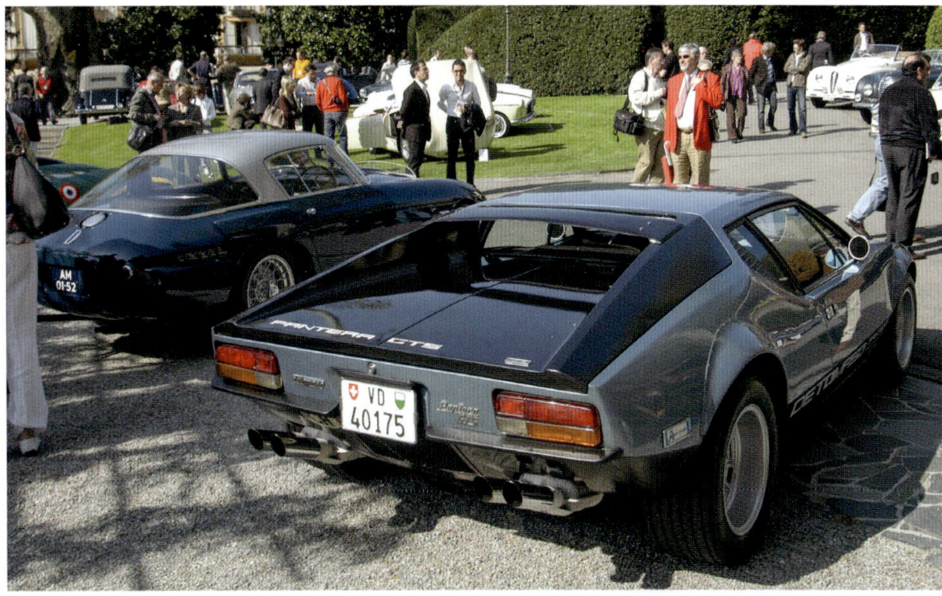


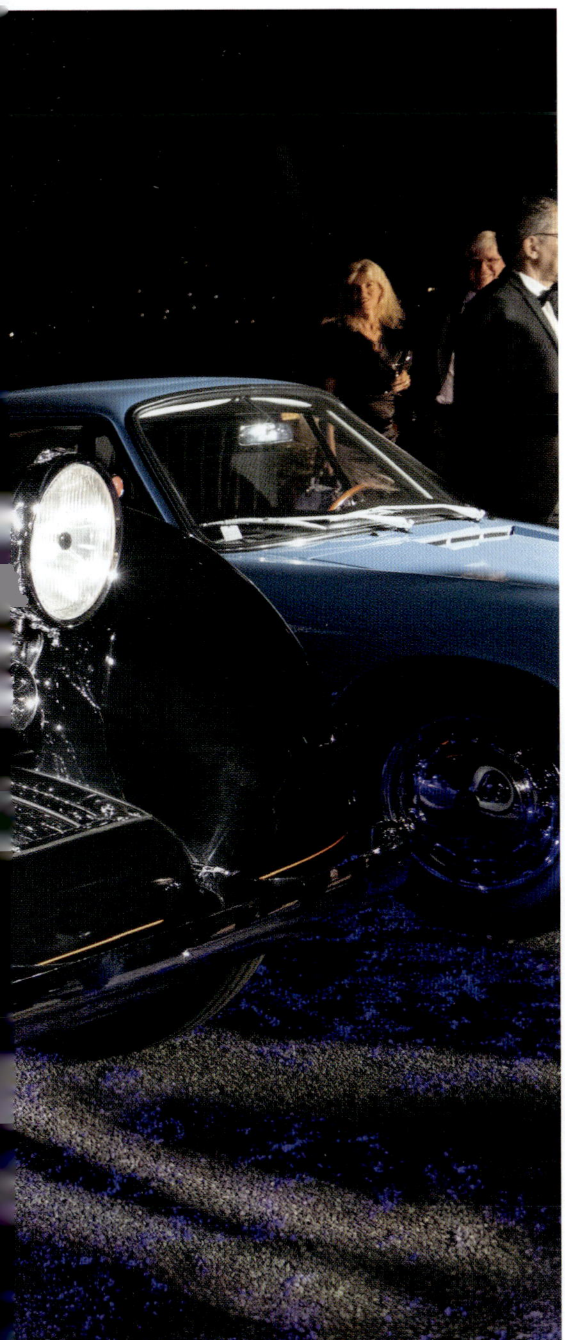

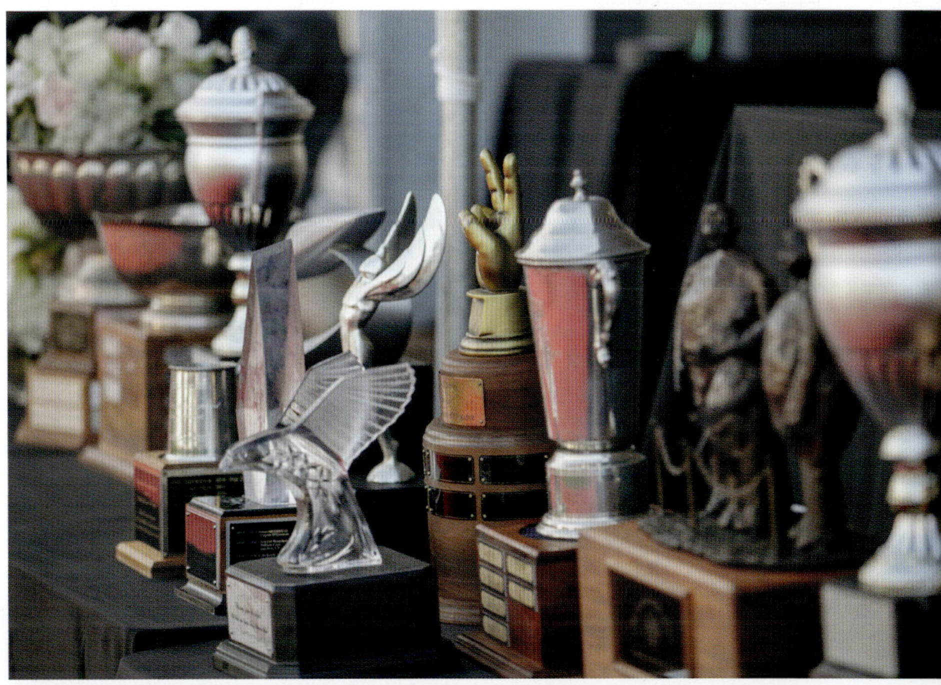

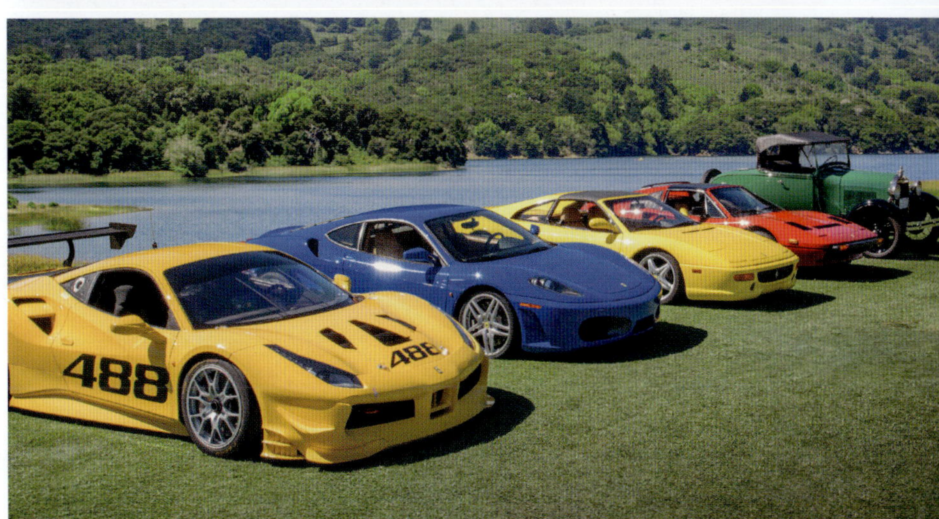

# *Concours in the Americas*

## A THRIVING SCENARIO

As Villa d'Este regained prominence alongside events like the Louis Vuitton Classic à Bagatelle, which revived the concept of concours d'elegance in Europe, new events began emerging across the United States throughout the 1980s and 1990s. While the Pebble Beach Concours d'Elegance set the standard for such events, others like Forest Grove, Meadow Brook, and Huntington Beach followed suit, creating a new wave of American-style concours. Among these, The Amelia Island Concours d'Elegance soon rose to prominence, rivaling Pebble Beach in both significance and prestige.

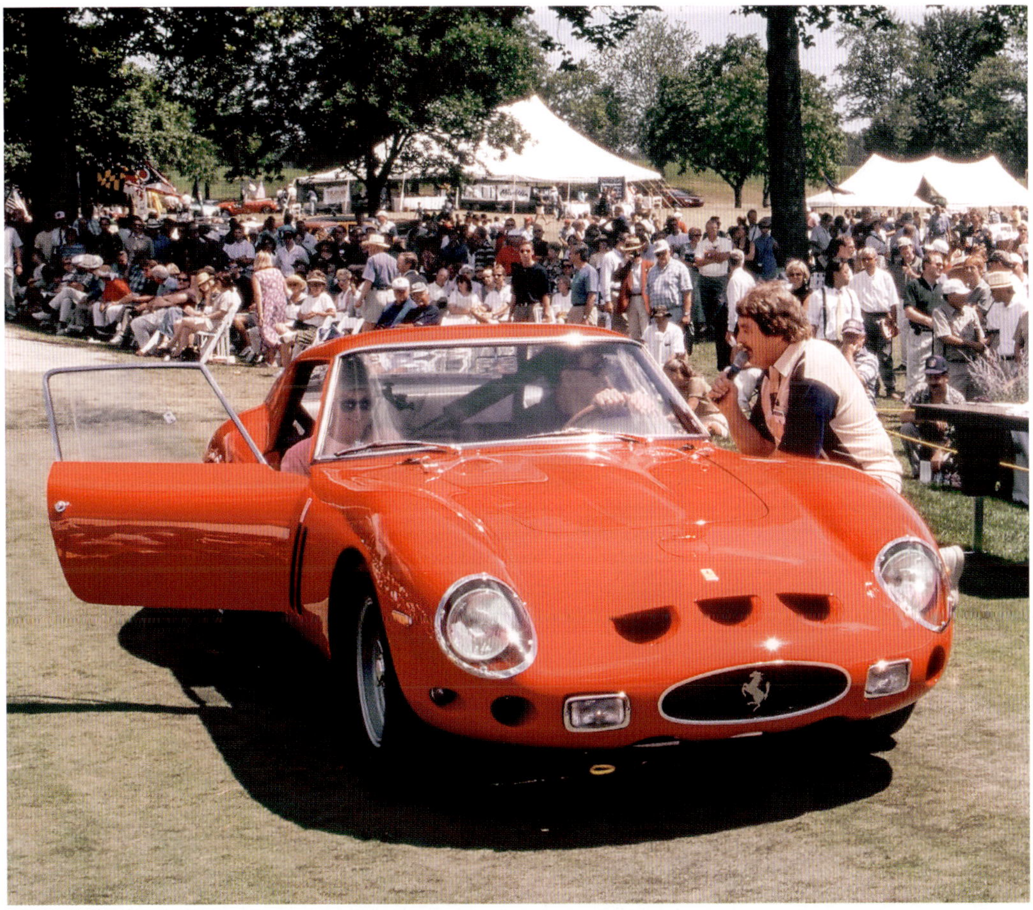

**LEFT:** *Arguably one of the most famous of the concours from the late 20th century, Meadow Brook Concours d'Elegance was one of the most prestigious and attracted cars like Ferrari 250 GTOs. (Xavier de Nombel)*

**BELOW:** *As well as very rare Alfa Romeo 8C 2900s. (Xavier de Nombel)*

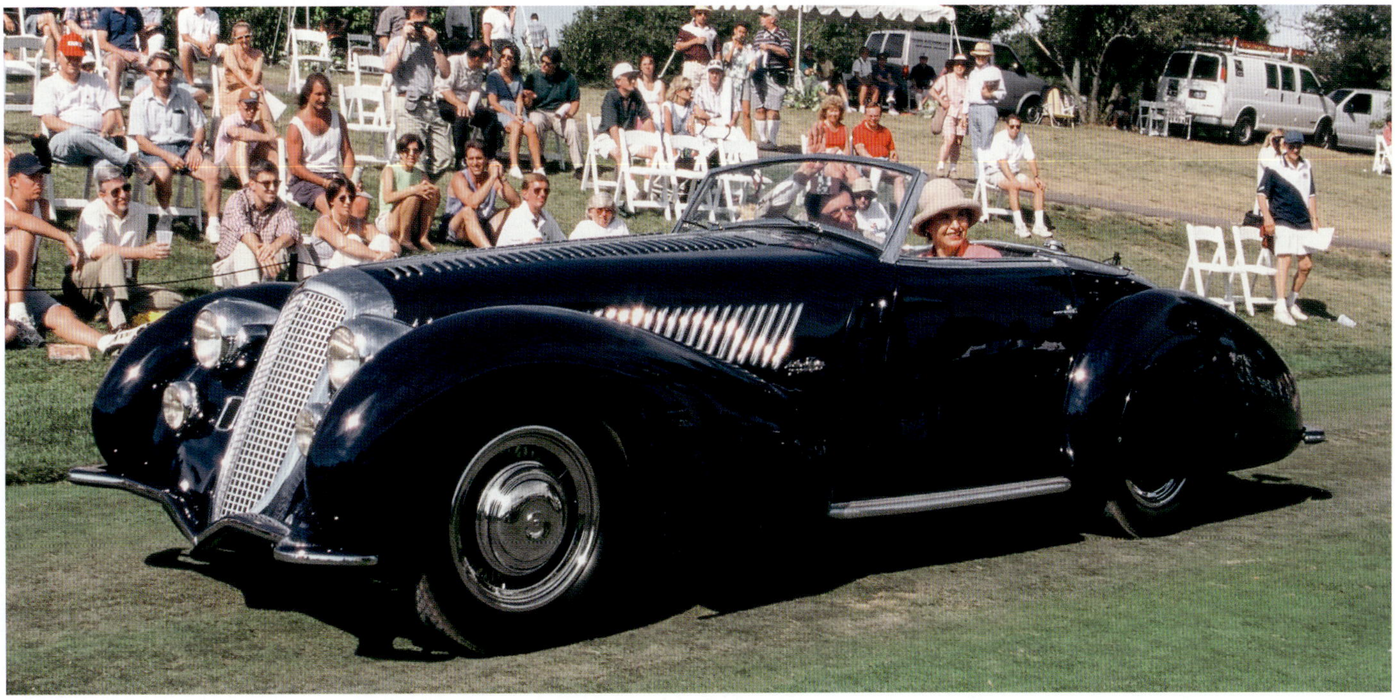

**RIGHT, BELOW RIGHT & BOTTOM:** *With the location being in the Detroit area, Meadow Brook was the place to check out the best from the Big Three. (Xavier de Nombel)*

**BELOW LEFT:** *Even the one-off Auburn Speedster de Cabina from 1929 was on show at the 1998 Meadow Brook Concours d'Elegance. (Xavier de Nombel)*

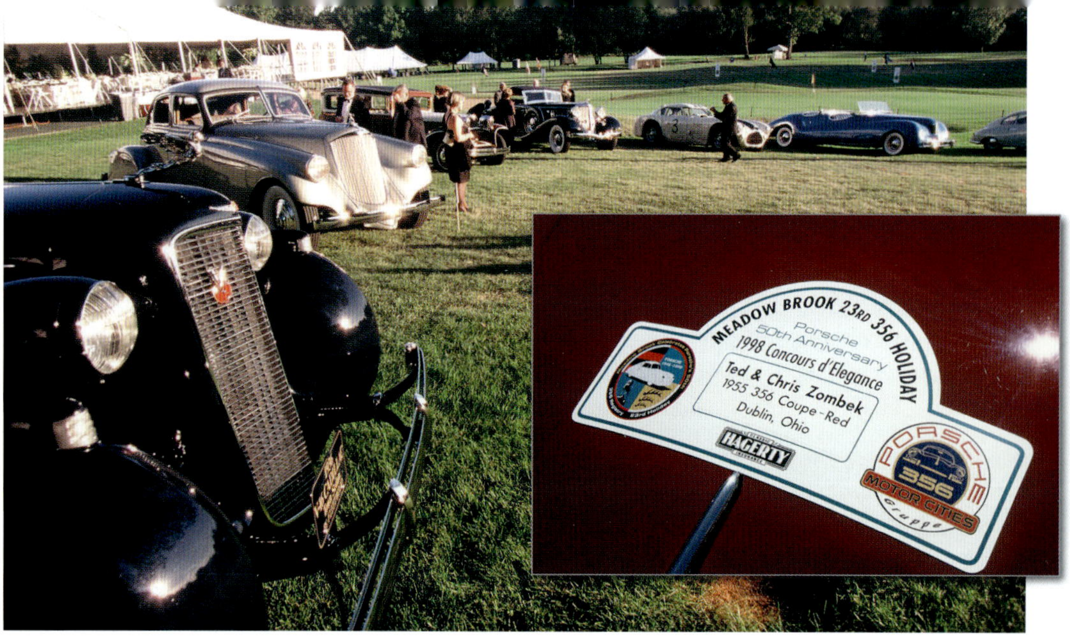


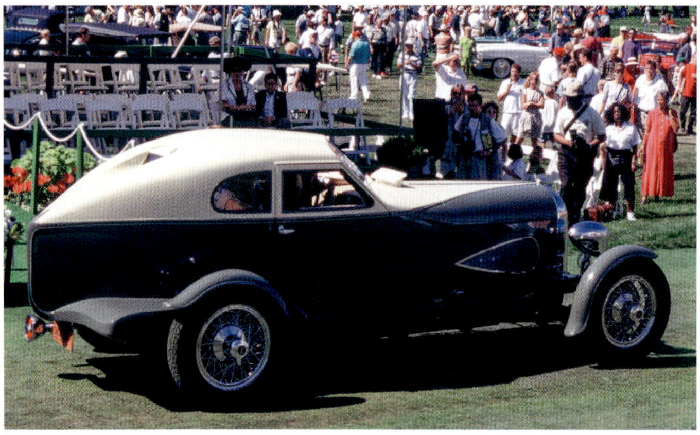

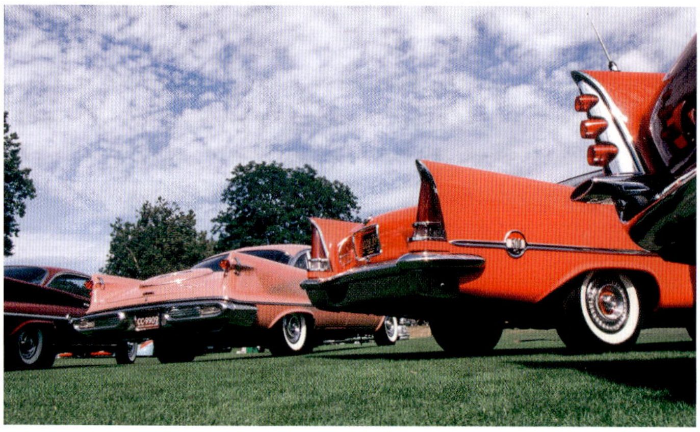

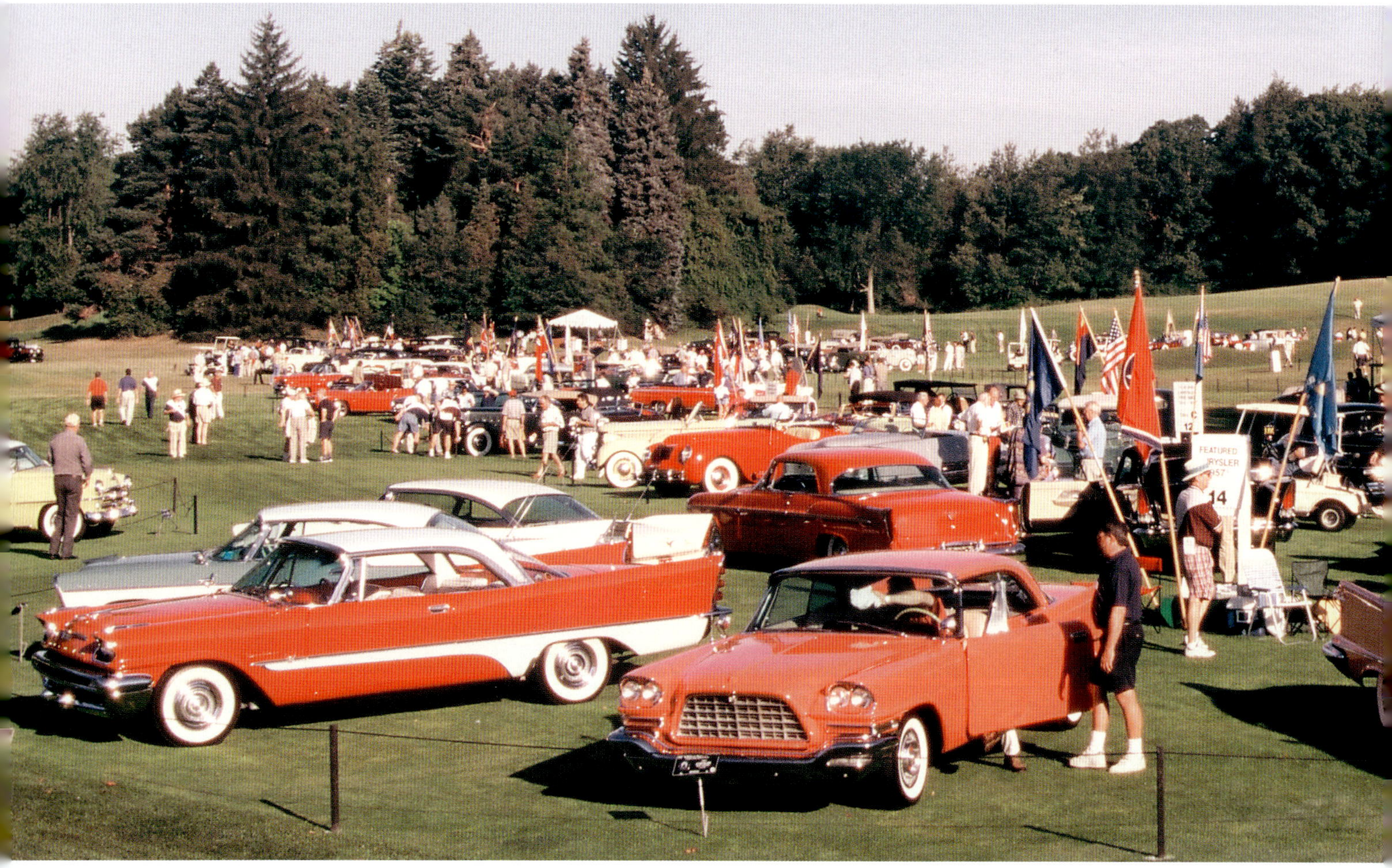

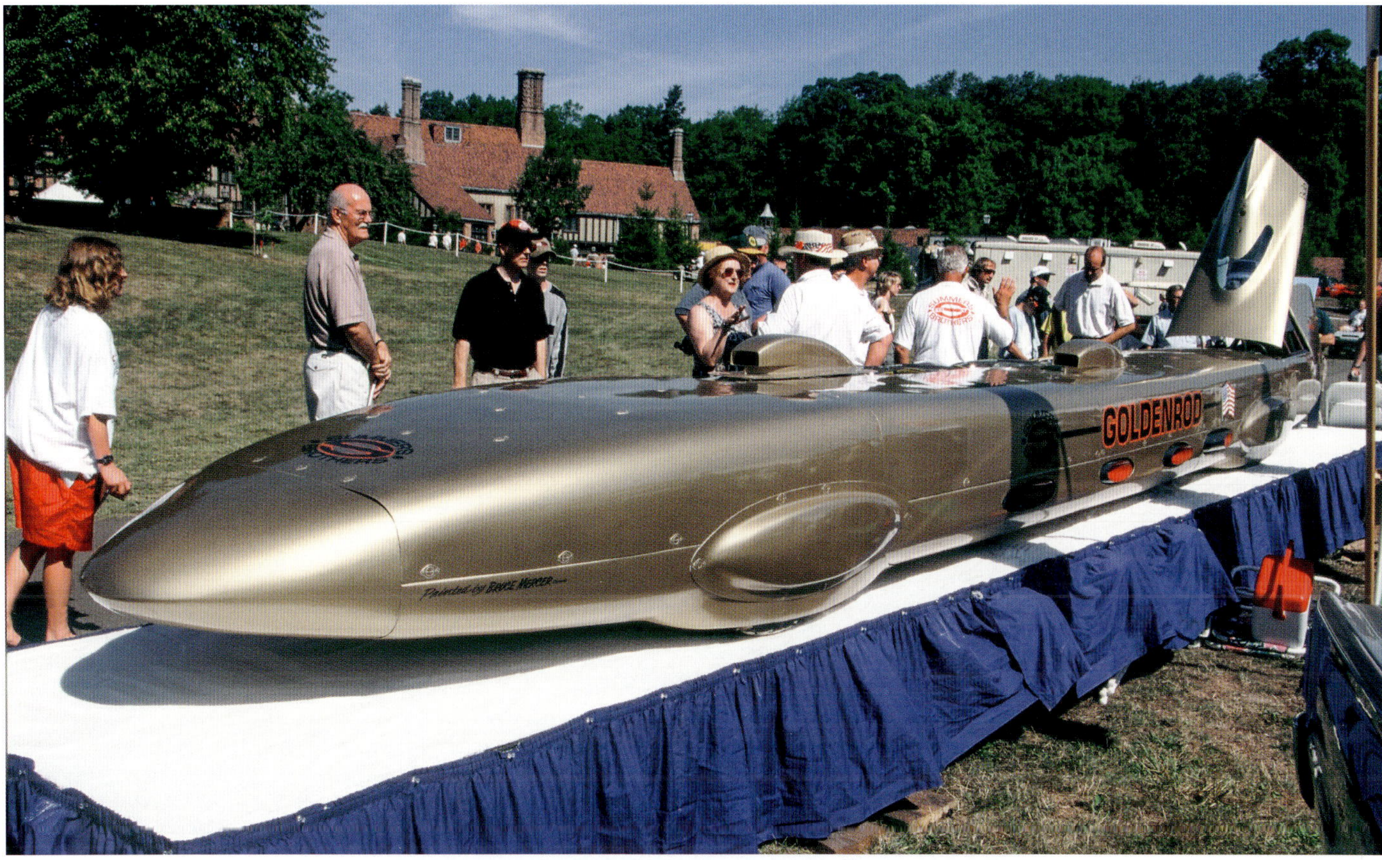


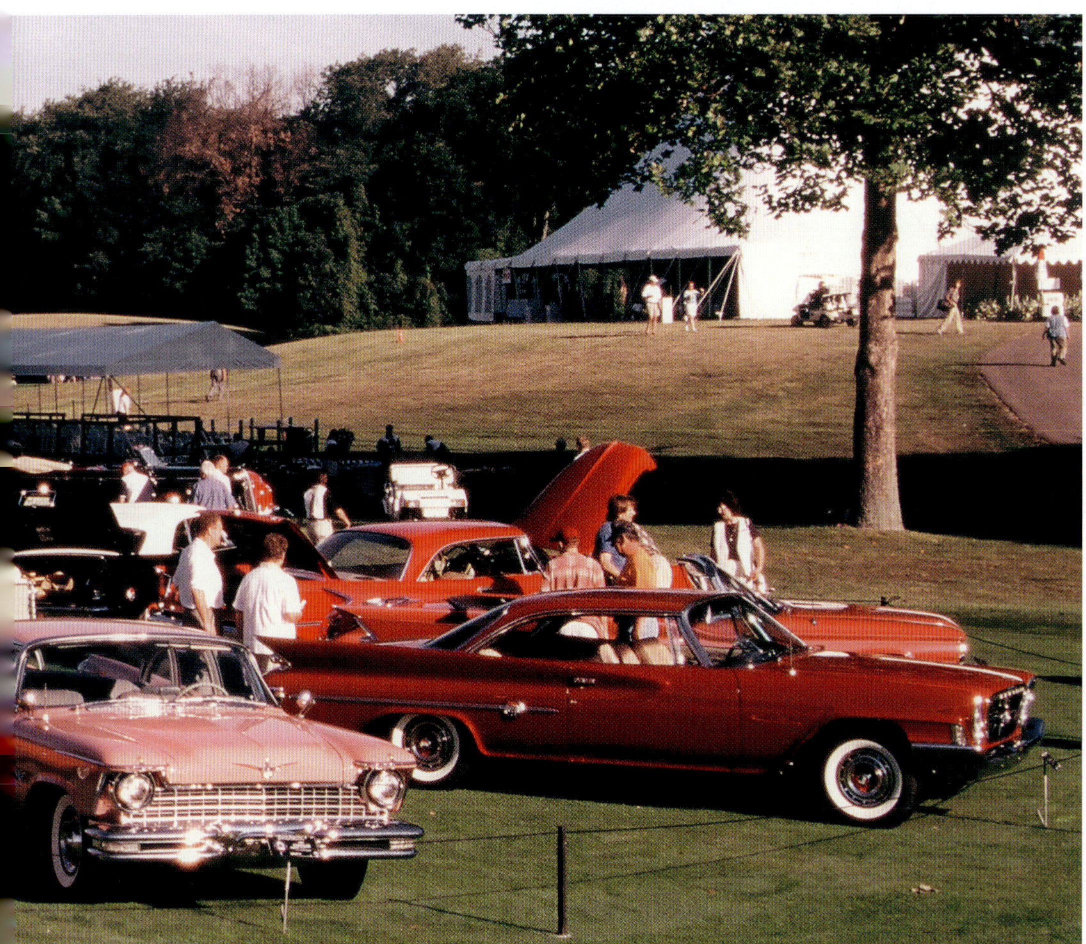

**ABOVE:** *Special displays in 1998 included cars like the famous Goldenrod wheel-driven land speed record car.* (Xavier de Nombel)

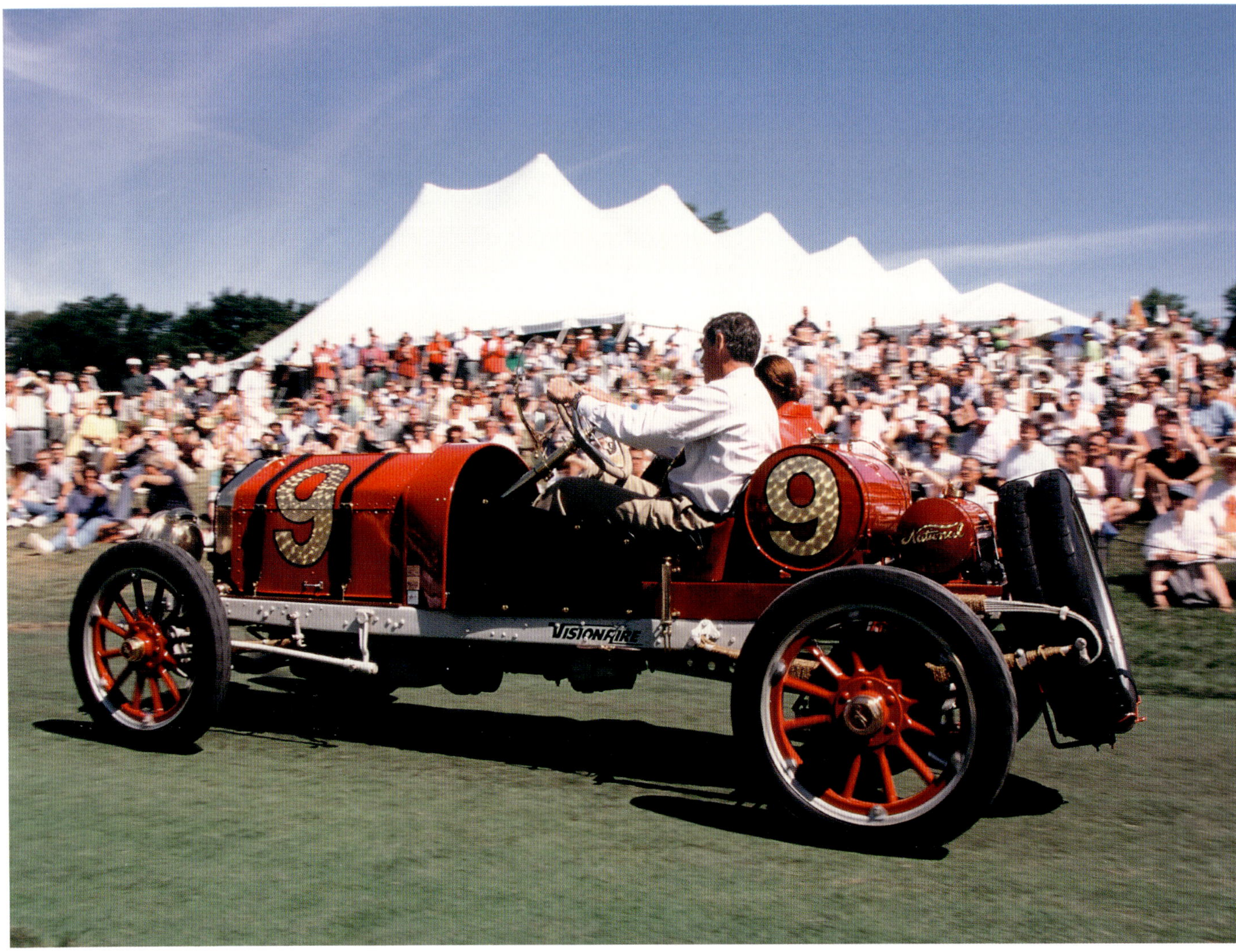

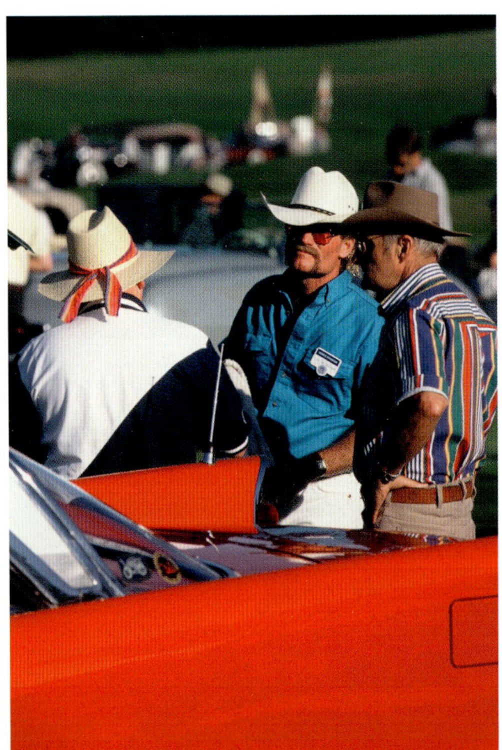

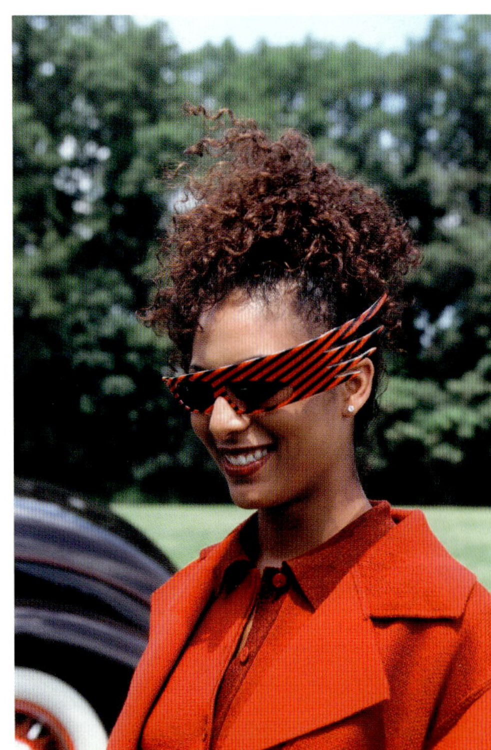

**THIS PAGE:** *Despite setting the tone for the American concours d'état style, Meadow Brook had to close its doors in 2010. (Xavier de Nombel)*

The inaugural Amelia Island Concours d'Elegance took place in 1996, spearheaded by automotive photographer, gentleman racer, and historic vehicle enthusiast Bill Warner. Held at The Ritz-Carlton in Amelia Island, Florida, the event featured Sir Stirling Moss as Honorary Chairman and Mercedes-Benz as a key sponsor. The first event attracted approximately 2,000 attendees and raised US$35,000 for Community Hospice Northeast Florida.

In 1997, Phil Hill, America's first Formula One champion, took over as Honorary Chairman. That year, the event expanded to spotlight specific marques, beginning with Ferrari, in celebration of the brand's 50th anniversary. The following years saw other legendary figures in motorsport take on the role of Honorary Chairman, including Hurley Haywood in 1998 to celebrate Porsche's 50th anniversary, Carroll Shelby in 1999 with a focus on Packard's centennial, and Brian Redman in 2000 to feature the Mercer marque.

*Amelia Island Concours d'Elegance has become one of the world's most important concours events. (N. Deremer)*

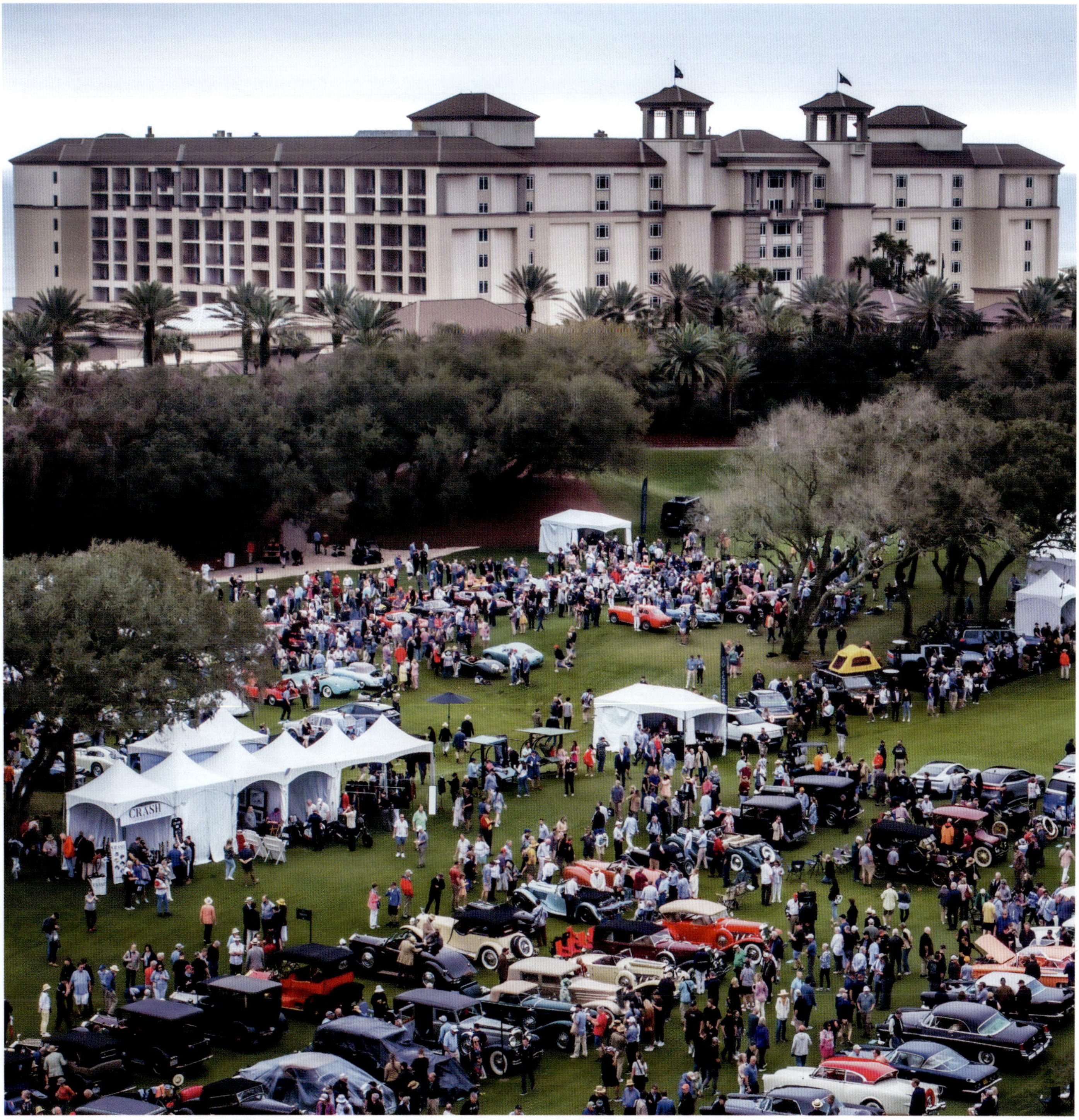

**BELOW & OPPOSITE TOP:** *Renamed as The Amelia when Hagerty took over in 2018, the event is held over Saturday and Sunday. Saturday is dedicated to Cars & Community, and features the sports car class. The Amelia also follows the concept of two Best of Show awards. One is for sports cars and the other for concours cars. For 2024, Best of Show for sports cars was a 1962 Ferrari 250 GTO (opposite top, left) and for concours cars was a 1947 Delahaye 135 MS Narval with Figoni et Falaschi bodywork (right). (N. Deremer)*

As the event gained traction, it continued to honor notable motorsport figures such as John Surtees (2001), Dan Gurney (2002), Jim Hall (2003), and Bobby Unser (2004). Iconic marques like Pierce-Arrow, Cadillac, and Corvette (in its 50th anniversary year, 2003) were also featured. By 2005, The Amelia Island Concours d'Elegance drew nearly 18,000 visitors and contributed US$1.4 million to Community Hospice since its inception.

What began as a unique celebration of automobiles in Florida has since become one of North America's premier automotive events. In 2021, the event was briefly moved from its usual March date to May but returned to its traditional late February / early March schedule the following year. That same year, Hagerty acquired the concours, as Bill Warner stepped down from his leadership role. Hagerty has continued Warner's tradition of honoring a motorsports figure each year, along with a featured marque or a special theme.

Today, The Amelia Island Concours d'Elegance includes a variety of events such as golf tournaments, car auctions, art and fashion shows, and charity dinners. The event culminates in judging and awards for different classes in the prestigious "Field of Dreams" exhibition. Top honors include the Best of Show Concours d'Elegance and Best of Show Concours de Sport, alongside corporate and class-specific awards, including the coveted People's Choice award.

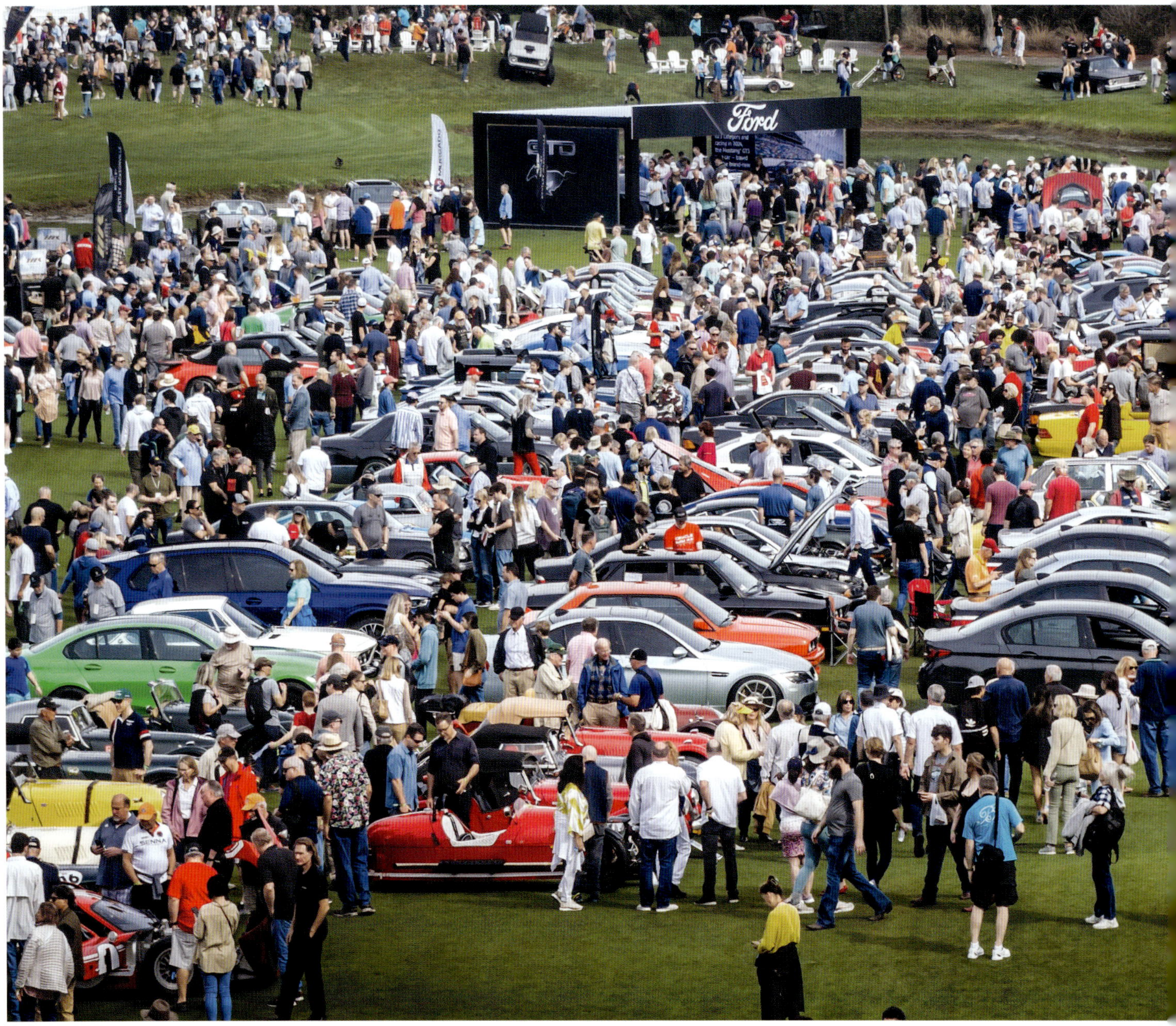

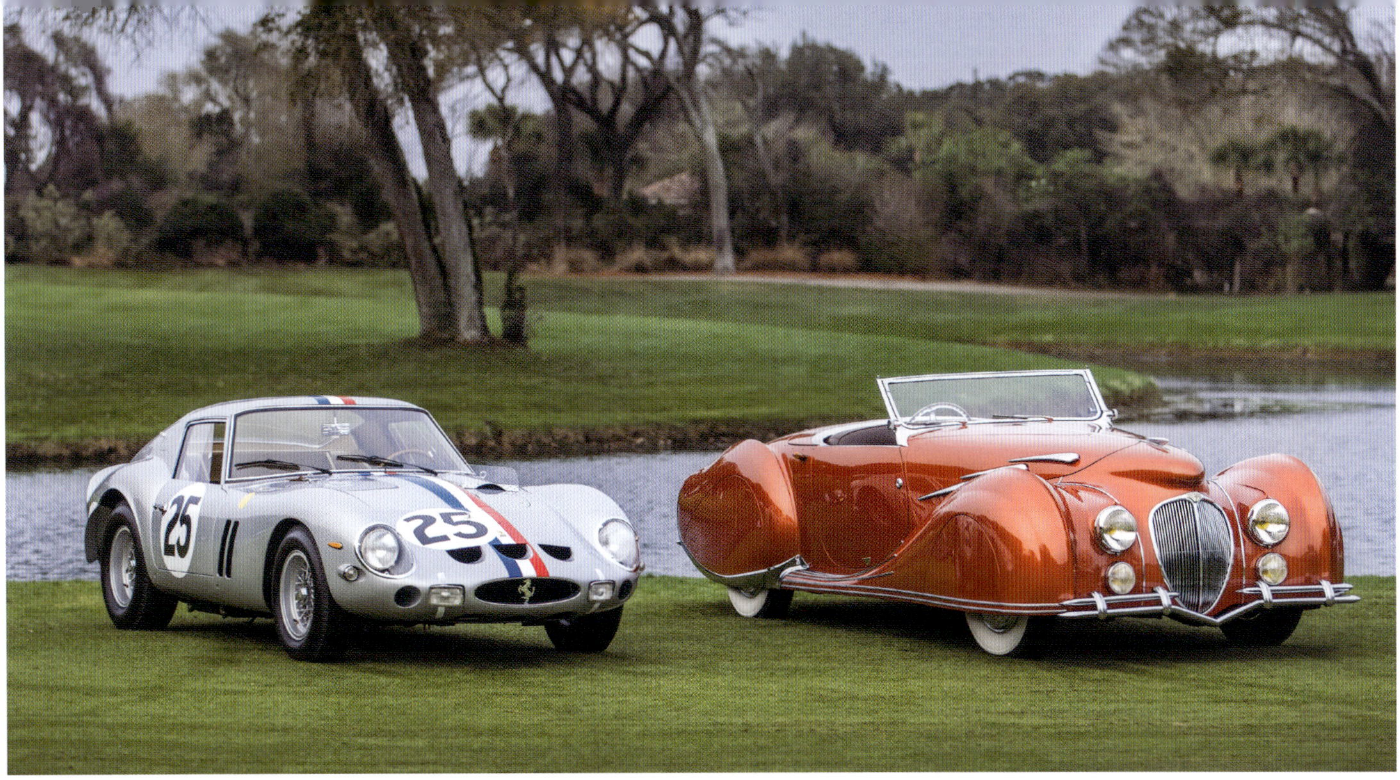

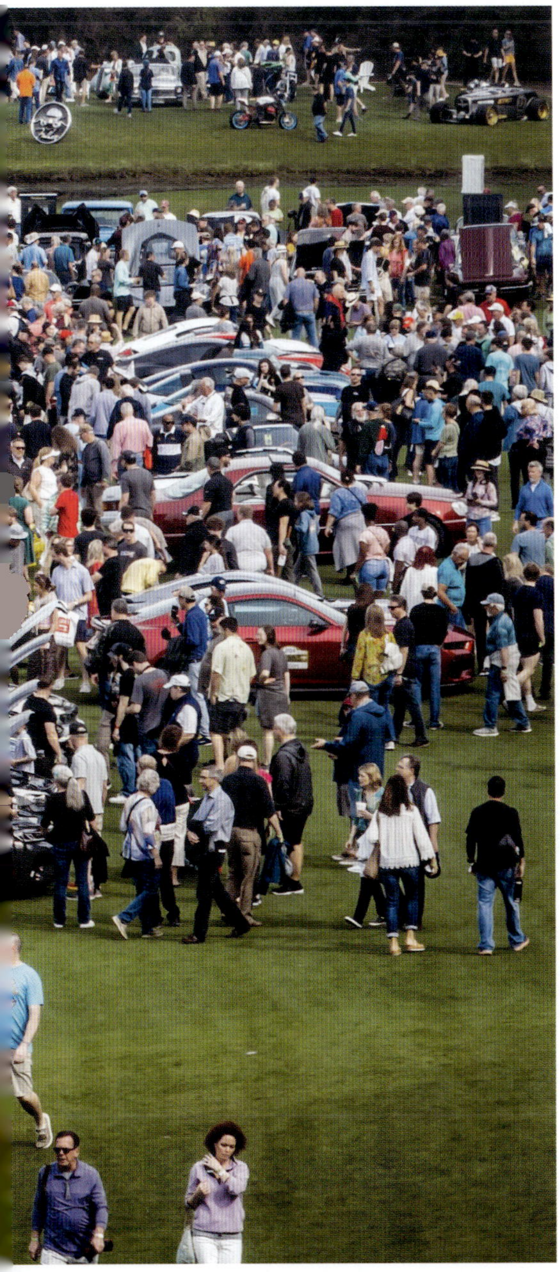

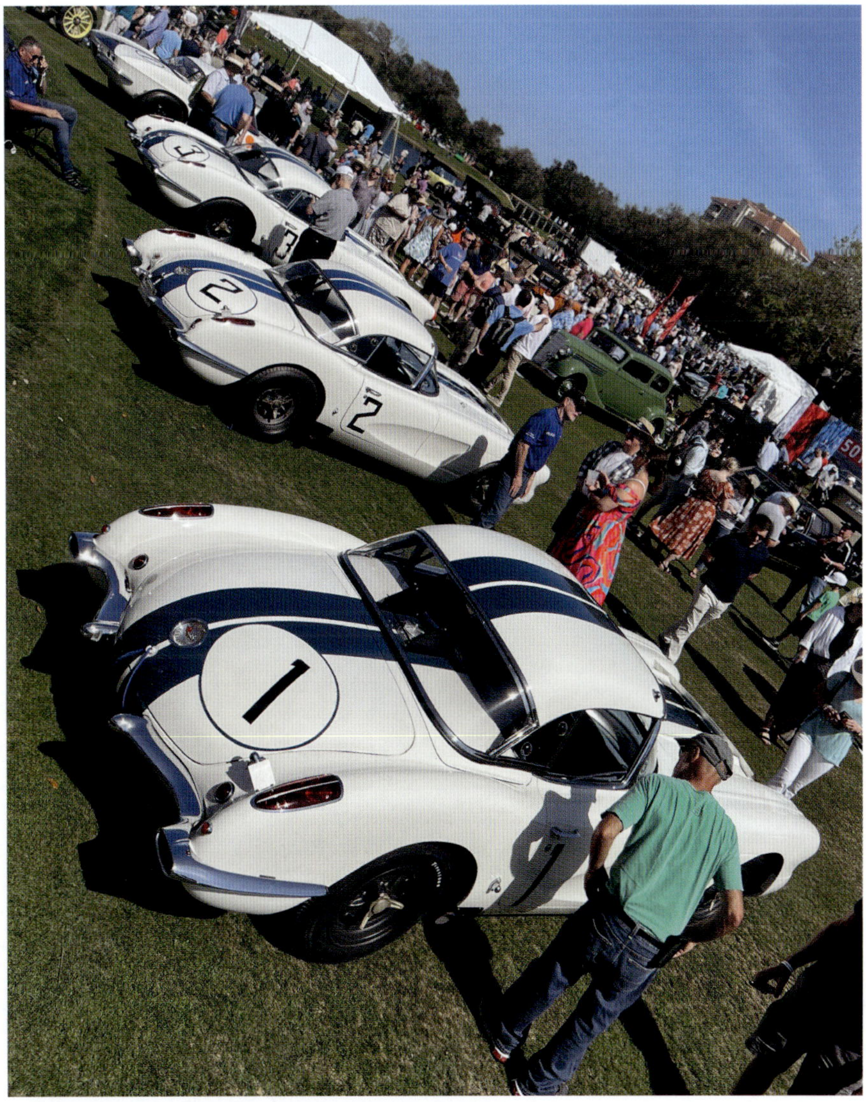

**ABOVE:** *Given event founder Bill Warner's long-held emphasis on competition and sporting machines, The Amelia honors Best of Show awards in both sport and concours categories. The Amelia show draws some of the world's biggest collectors to northern Florida, including this special collection honoring racing and prototype versions of Chevrolet's Corvette. (Cape Coventry Archives)*

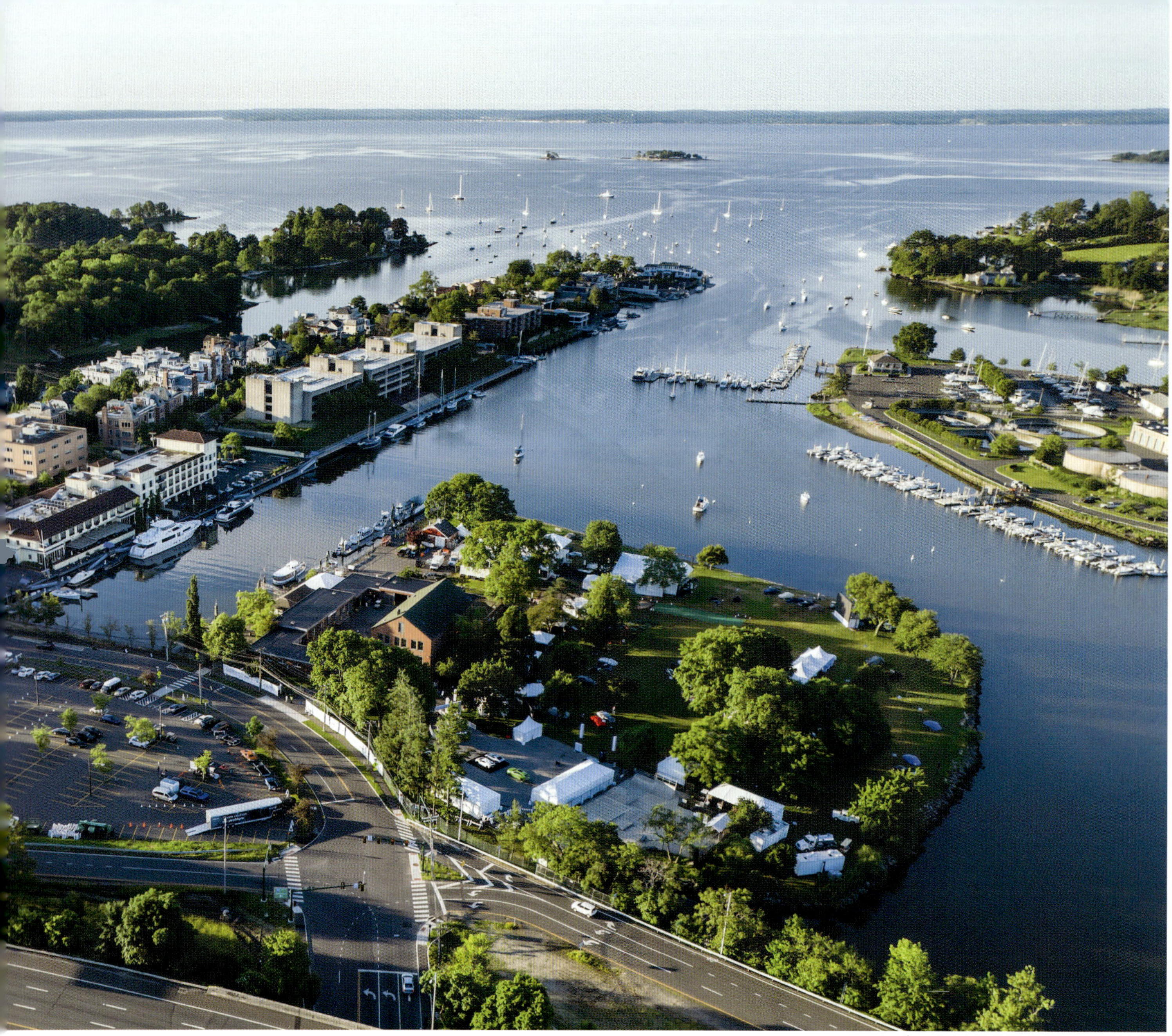

**ABOVE:** *Established in 1996, the Greenwich Concours d'Elegance is not particularly big as space restrictions limit the number of cars. (N. Deremer)*

**RIGHT:** *A Packard being judged at the 2023 Greenwich Concours. (X. Cesari)*

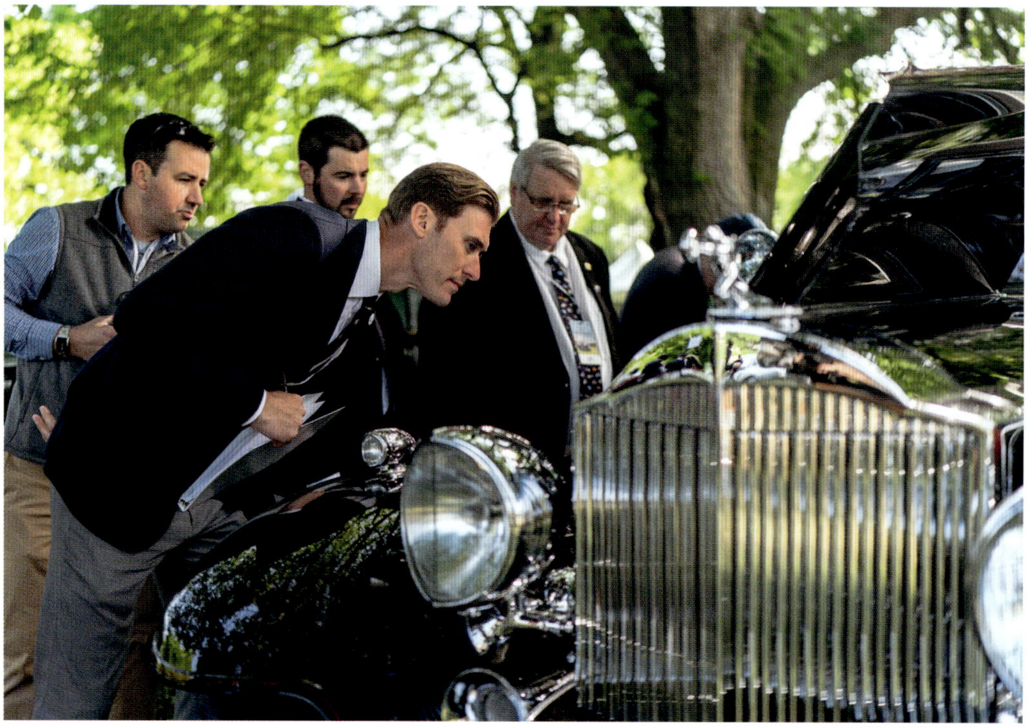

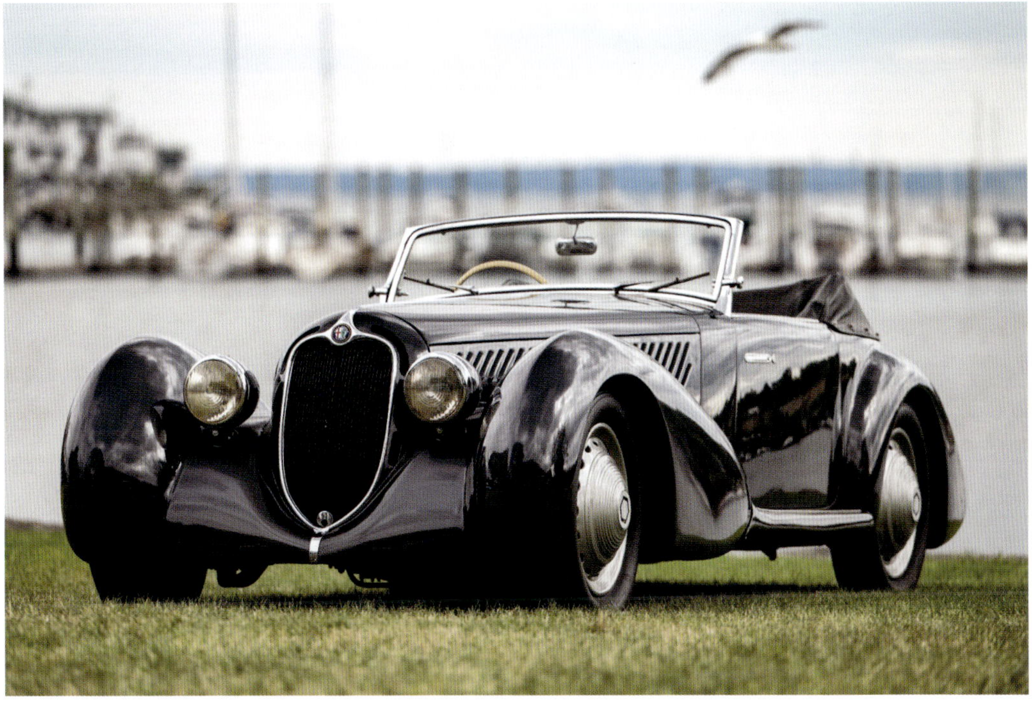

**LEFT:** *This 1937 Alfa Romeo 8C 2900 B was Best of Show at the 2023 Greenwich Concours. (J. Sweeney)*

**BELOW:** *Like most concours d'état, Greenwich has a Grand Tour to prove that most concours entrants are not just trailer queens. (J. Sweeney)*

The Greenwich Concours d'Elegance, founded in 1996, is another prestigious event of similar vintage to The Amelia Island Concours d'Elegance. Held annually around late May or early June, this concours was established by Bruce and Genia Wennerstrom, long-time residents of Greenwich, Connecticut (not to be confused with Greenwich Village, the bohemian neighborhood in New York City). Their passion for presenting world-class automobiles, combined with Greenwich's rich heritage and strong community spirit, helped build one of the top five concours events in North America.

Located just 30 miles from New York City and set against the picturesque backdrop of Greenwich Harbor, the event offers an ideal setting with its proximity to the renowned Delamar Hotel, upscale restaurants, and boutique shops along Greenwich Avenue. The concours' scenic venue and convenient location have contributed to its growing popularity over the years.

By 2010, the Greenwich Concours had expanded to showcase over 300 cars during its weekend event, which took place on Saturday, June 5, and Sunday, June 6, at Roger Sherman Baldwin Park in Greenwich. The event was divided into two themes: Saturday's Concours Americana, exclusively featuring American-made vehicles, and Sunday's Concours International, which highlighted imported marques.

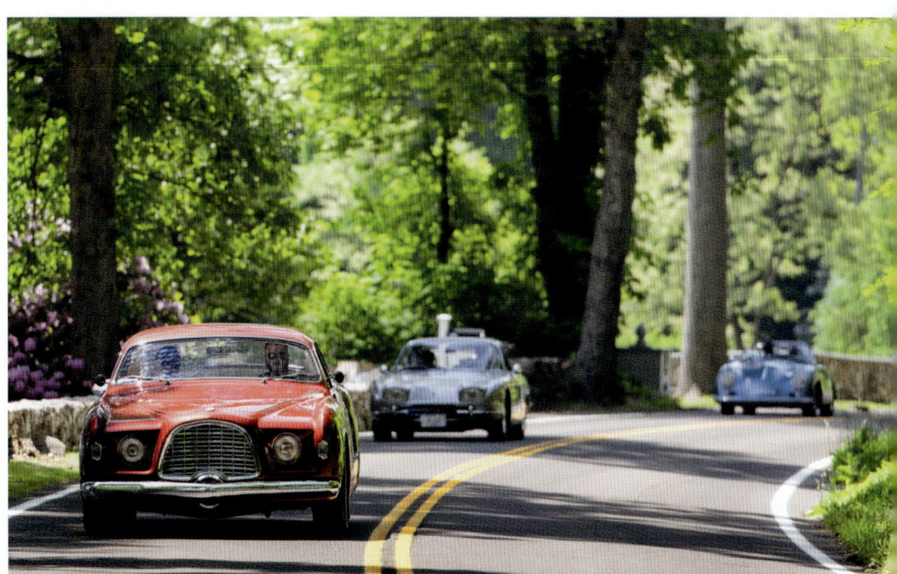

Over the two days, attendees were treated to a remarkable display of American and international cars, including pre-World War I brass-era vehicles, grand classics from the late 1920s and 1930s, and iconic marques like Rolls-Royce, Marmon, Pierce-Arrow, Lagonda, Cord, Auburn, Packard, Alfa Romeo, and Maserati. Post-World War II cars, known for their prominent fins and powerful engines, were also a highlight.

At the 2010 event, the Best of Show Americana award was presented to a 1938 Lincoln K-12 Convertible Sedan from the Marano Collection, while the Best of Show International award went to a 1933 Rolls-Royce Phantom II Continental owned by Frank E. Ricciardelli.

For more on the Greenwich Concours, see pages 298-307, and McKeel Hagerty's interview.

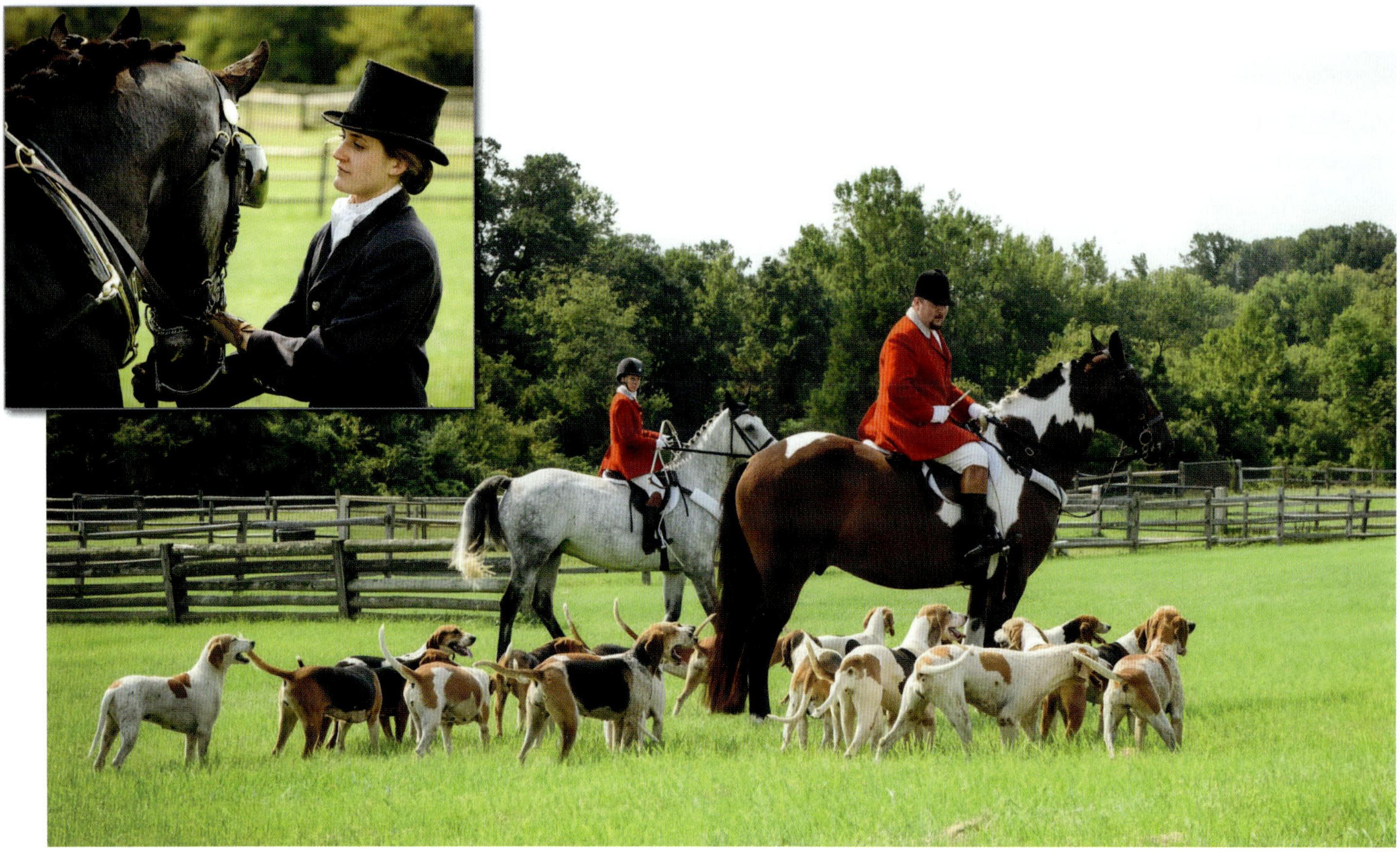

**ABOVE:** *Traditionally a fox hunting club, dating from 1883, Radnor Hunt decided to start a concours d'elegance event to promote the club, as early as 1997. Since then, it has grown to be one of the most important North American concours. (Bryan McCarthy)*

**OPPOSITE:** *Limited to around 100 cars, which suits the idyllic lawn next to the clubhouse, the event is also called the 100 Motor Cars of Radnor Hunt. (Bryan McCarthy)*

## The scoring system

Following the acquisition of The Amelia Island Concours d'Elegance, Greenwich has been taken over by Hagerty, and since 2019, the insurance company has been running the concours, with both having followed the International Chief Judge Advisory Group (ICJAG) system (until 2024) of deducting points from a 100 point maximum (or rather a 103-point scale), which is similar to the scoring system followed by the Classic Car Club of America (CCCA). Further details regarding the scoring system are on page 364.

Another concours that adheres to the ICJAG scoring system and is of a similar age is the Radnor Hunt Concours d'Elegance. Radnor Hunt, a fox hunting club established in 1883, was founded by a group of fox hunters who began hunting with hounds owned by Thomas Mather in Radnor Township during the late 19th century. The club's picturesque clubhouse, originally an expanded 1857 farmhouse, along with the stables and kennels, was renovated in 1931 under the direction of Arthur Meigs, a prominent Philadelphia architect and club member.

"We started the concours in 1997," says Thomas "Tobey" Ross, Co-Chairman and President of the Radnor Hunt Concours d'Elegance. "So, yes, it's one of the older events. Mike [Tillson] and I began organizing it around 1989, initially for a vintage sports car race in Philadelphia's city park, which was run from 1989 until 1995. When that became financially unfeasible, we transitioned towards organizing concours. By 1996, we weren't sure what to do next, and Mike, a long-time Radnor Hunt member, suggested we hold a small car show at the club. That's how it all began, and here we are, 27 years later, still hosting the same show."

Ross reflects on how the team has evolved: "The core group has been together since 1989, though we've seen changes in membership over the years. But we've stayed strong and have built one of the more established concours in the country. We're proud to be ranked among the top-tier concours events in the US."

In the 1990s, Ross explains, there weren't many concours: "You had Amelia Island and Pebble Beach, for sure. By 1997, maybe Meadow Brook Hall was around, but events like Cobble Beach in Canada came much later. We've persevered and grown, largely because we set solid guidelines from the start that worked well."

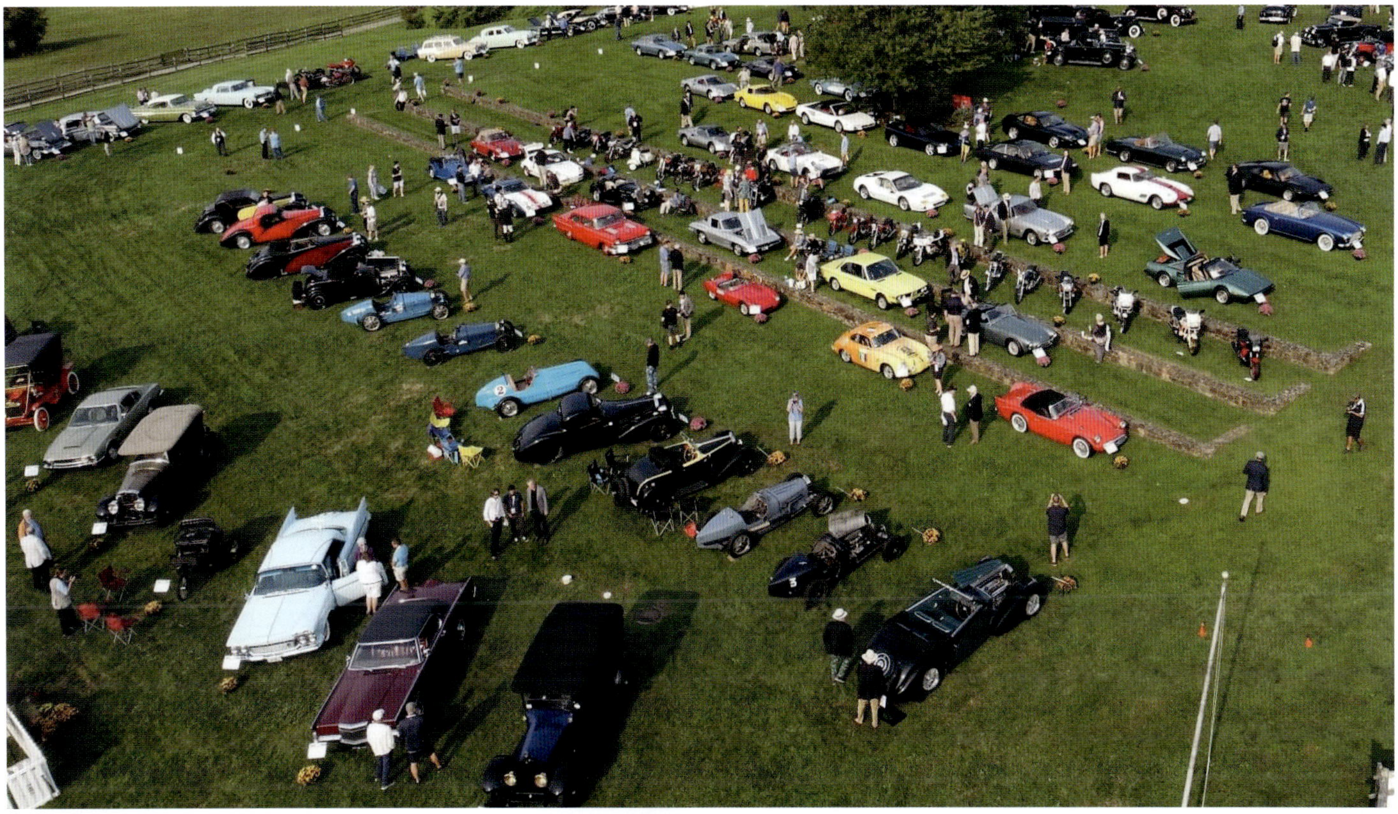

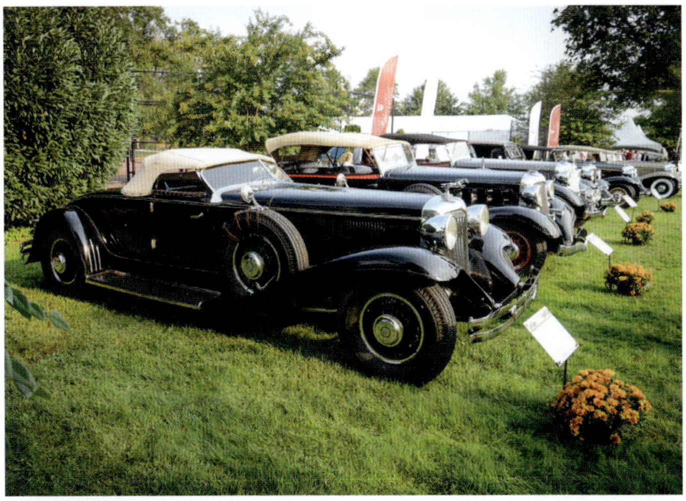

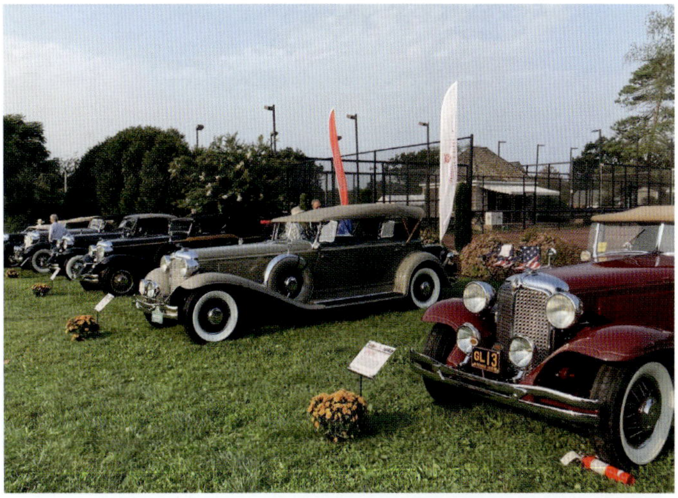

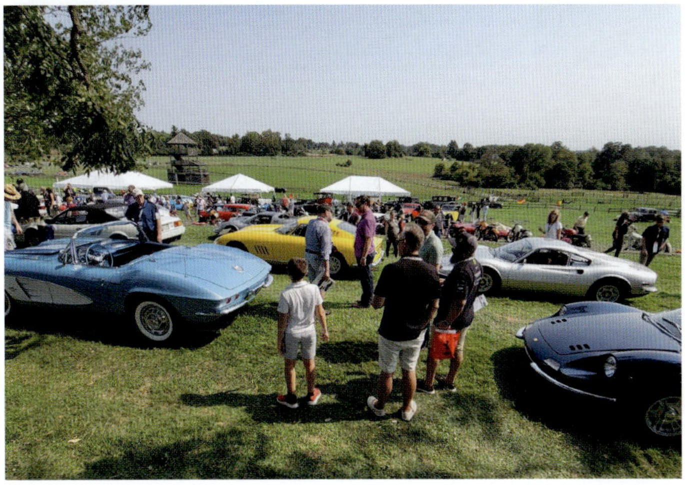

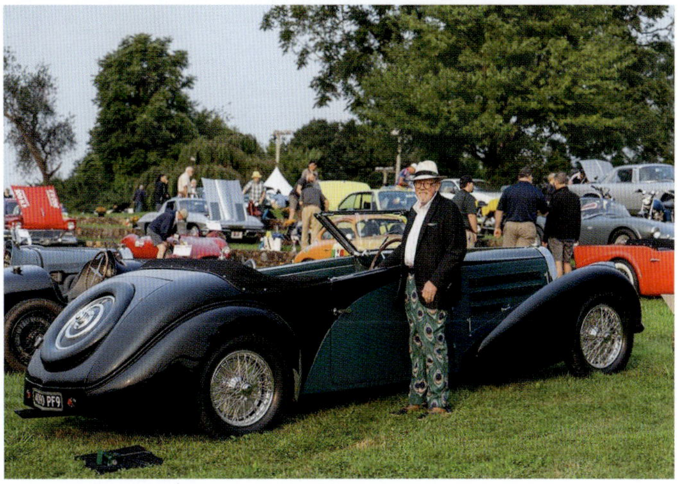

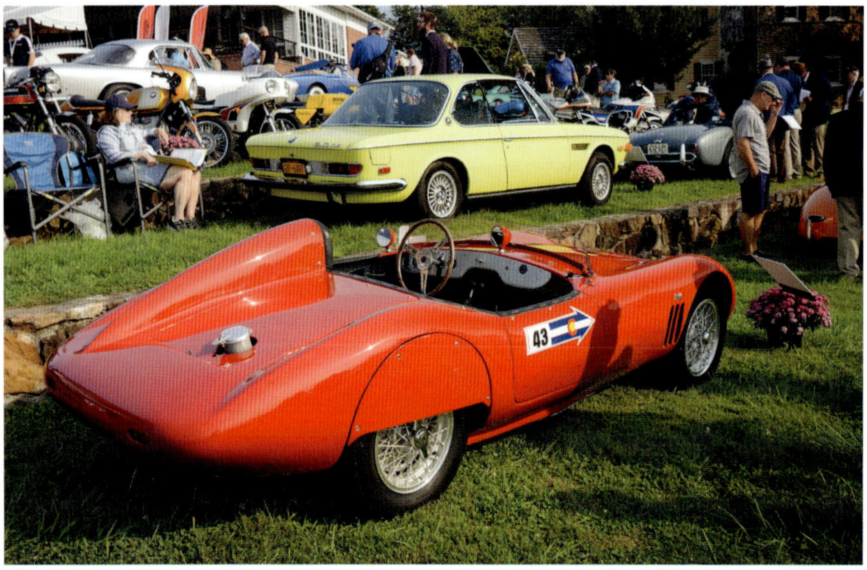

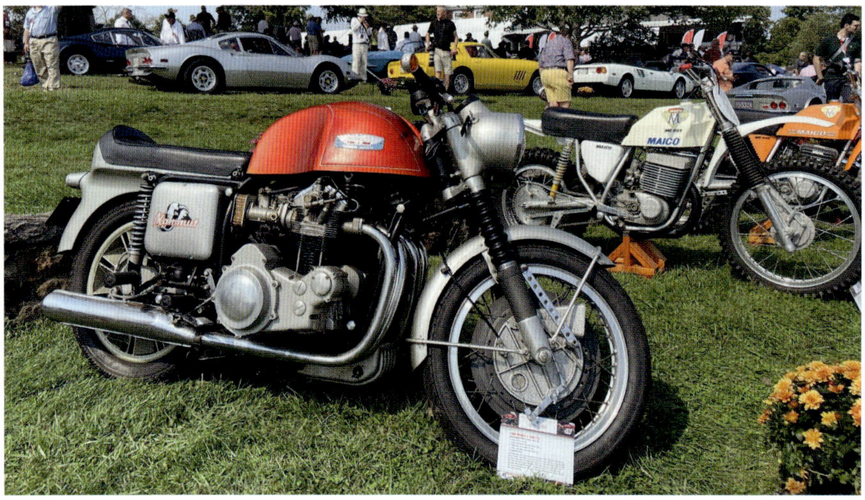

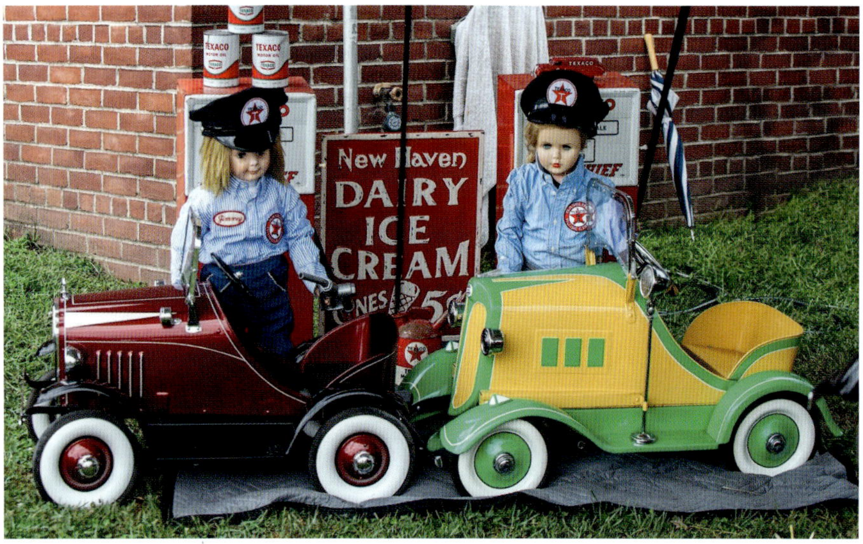

**TOP:** *Following the ICJAG scoring system, Radnor Hunt is a strict concours d'état type of event. (Bryan McCarthy)*

**MIDDLE:** *There are classes for a few motorcycles too. (Bryan McCarthy)*

**ABOVE:** *No, there are no classes for pedal cars... (Bryan McCarthy)*

Ross emphasizes what makes Radnor Hunt unique: "We're not just a golf course like many other concours venues. Being based at a fox hunting club gives us a distinctive atmosphere. Our show grew quickly, and we eventually limited the number of cars to around 100, which suits the idyllic lawn next to the clubhouse. We coined the term '100 Motor Cars of Radnor Hunt,' which adds a sense of exclusivity. If you don't submit your entry early, you might miss out on a spot."

Michael Tillson, founder and Chairman of Radnor Hunt Concours, added: "The classes change yearly, but we also have a separate MotorSport Park for newer cars that appeal to our younger enthusiasts and for great cars that are not concours prepared. That part of the event has grown to as many as 250 cars, and it's a show in itself. Some first-time attendees even think that's the concours until they realize the main event is on the adjacent field."

Tillson notes, "The invitational concours focuses on cars from the mid-1970s or older, unless there's a specific featured marque that justifies including newer models. In such cases, we might accept cars from the 1990s to help tell a more complete story."

Ross further explained the event structure: "Each year we have a core set of classes, with around eighteen judged categories and eight or nine special awards. Additionally, we feature two or three marques each year."

The Radnor Hunt Concours also features motorcycles. "We typically have around 100 cars plus 30 to 40 motorcycles, which are divided into several classes. Just like the cars, the motorcycles have a Best in Show award," Ross noted, comparing it to similar events like The Amelia.

Michael Tillson also points out that the "two events unique to the Radnor Hunt Concours are a parade of horse-drawn carriages, which makes the connection between carriage and automobile coachwork, as well as a parade of the Radnor Fox Hounds."

To judge the concours, there's a panel of about 50 jurors, with Tillson's partner, Susan Tatios, serving as the chief judge. Ross elaborated: "We've had international judges as well, and many of our jurors are former car designers and executives, much like other major concours events. We're very proud of our judging team."

Regarding the judging process, Ross added: "We follow ICJAG rules but take full advantage of the subjective categories, such as over-restoration or the car's provenance. We encourage judges to consider the overall impact a car has on the field, rather than getting bogged down in technical details like every headlight or a switch."

"Events are increasingly struggling to attract cars and engage the next generation of participants," says Nigel Matthews. "There is a clear need to modernize judging standards to better appeal to younger enthusiasts and newcomers. At the same time, we must protect our automotive heritage by allowing adaptations that make these cars suitable for modern traffic."

To address this, serial judges Chris Kramer and Nigel Matthews spent the past three years developing a new judging approach called the Modernized 100-Point System (M100), which has now been implemented at events worldwide. The M100 system aims to balance technical excellence with subjective qualities such as design, provenance, impact, and historical significance. Deductions are only applied for actual, verified flaws - not assumptions - allowing certain modifications to be accepted.

The M100 uses a similar 20 judged line-items categories as the age old CCCA (Classic Car Club of America) system. Each car starts with a perfect score of 100 points, with deductions applied only for deviations from original delivery condition. Up to 95 points are allocated to objective technical criteria, while judges have a five-point subjective allowance to recognize attributes like design, rarity, provenance, overall impact, or technical innovation. These subjective points function as bonus recognition.

"Although it uses a 100-point scale, M100 is fundamentally different from any other judging system and stands on its own merits," emphasizes Chris Kramer.

In 2025, at least nine major events worldwide plan to adopt the M100 system. Kramer and Matthews will serve as chief judges at seven of them. Among the events adopting M100 for 2025 are The Amelia, La Jolla Concours d'Elegance, Greenwich Concours d'Elegance, Radnor Hunt Concours, Lime Rock Concours, and Las Vegas Concours at the Wynn in the USA, Icons Mallorca in Spain, Salon Privé Concours d'Elegance in the UK, Lugano Elegance in Switzerland and 21 Gun Salute in India.

**BELOW:** *Set against the dramatic fountain and lakeside backdrop of Fountain Hills, Arizona, Concours in the Hills draws thousands of enthusiasts each year for a vibrant, family-friendly celebration of exotic, classic, and competition cars– all in support of Phoenix Children's Hospital. (Concours in the Hills)*

## The dozen odds

Nigel Matthews, chairman of ICJAG, states that "Around a dozen of the top concours events in North America now follow the ICJAG judging system at various Concours d'Elegance." These events include notable names like The Amelia, Greenwich, Radnor Hunt, La Jolla Concours d'Elegance, San Marino Motor Classic (both in California), Philadelphia Concours d'Elegance, Chattanooga Concours d'Elegance in Tennessee, and Cobble Beach Concours in Ontario, Canada.

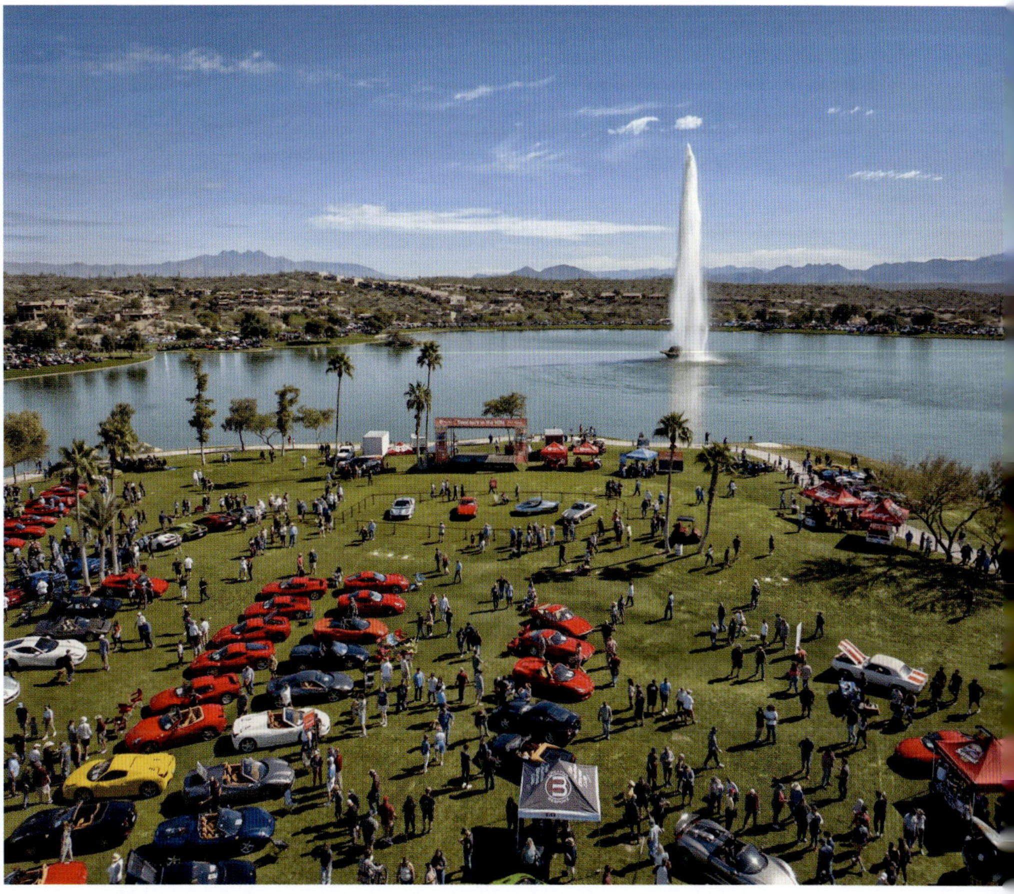

Since the late 2000s, the number of Concours d'Elegance has grown significantly, with more than a hundred events across North America for historic vehicle enthusiasts to attend throughout the year. The season kicks off in February with the Motorcar Cavalcade Concours d'Elegance (Miami), followed closely by the Cavallino Classic (Palm Beach) and the Concours in the Hills (Arizona). Then these two Florida events: Boca Raton Concours d'Elegance, and The Amelia.

**THIS PAGE:** *A selection of images from the 2007 Concorso Italiano, typically held at the Bayonet Black Horse Golf Course in Monterey, California. This event has grown into one of the most significant concours d'elegance gatherings during Monterey Car Week, leading up to the prestigious Pebble Beach Concours d'Elegance. Featured vehicles include De Tomaso Panteras, Alfa Romeo Giulietta Spiders, Maserati Meraks, and even classic motorcycles like MV Agustas. (J. Ellis)*

April features the La Jolla Concours d'Elegance in California, while May brings the Greenbrier and Greenwich Concours d'Elegance. June is packed with several shows including Huntington Beach (California), Cincinnati (Ohio), Rodeo Drive (California), Philadelphia, and the historic Hillsborough in California. July highlights Misselwood Concours d'Elegance in Beverly, Massachusetts, Keeneland in Kentucky, and Forest Grove in Oregon.

August is dominated by Pebble Beach and Monterey Car Week, which includes The Quail and Concorso Italiano. After Pebble Beach, three major concours in the US draw attention: Morgan Adams in Denver, Colorado, Geneva Concours in Illinois, and the San Marino Motor Classic in California. September and October are equally busy with events such as Radnor Hunt Concours d'Elegance in Malvern, Pennsylvania, The Bridge (Bridgehampton, New York), Ironstone in Murphys, California, St. Michaels at Chesapeake Bay in Maryland, The Boston Cup, and Hilton Head in South Carolina.

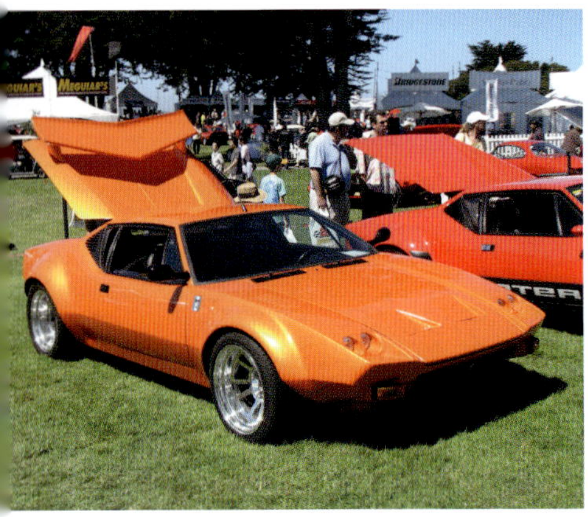

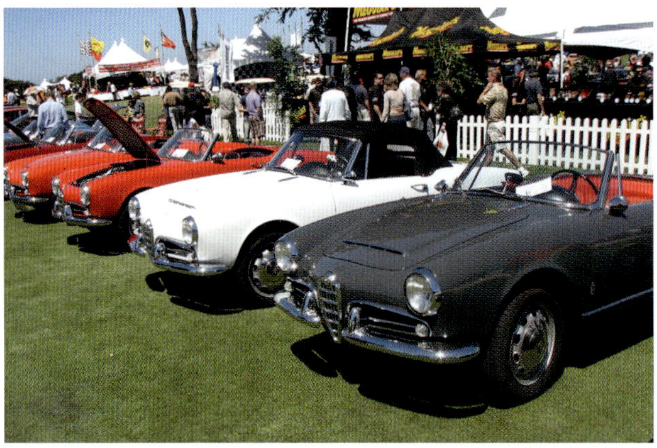

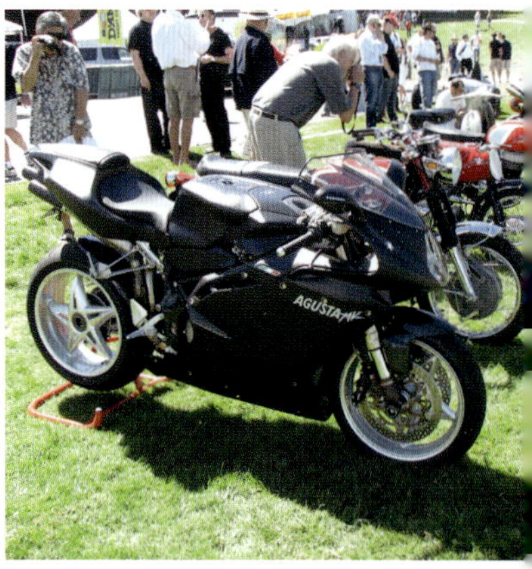

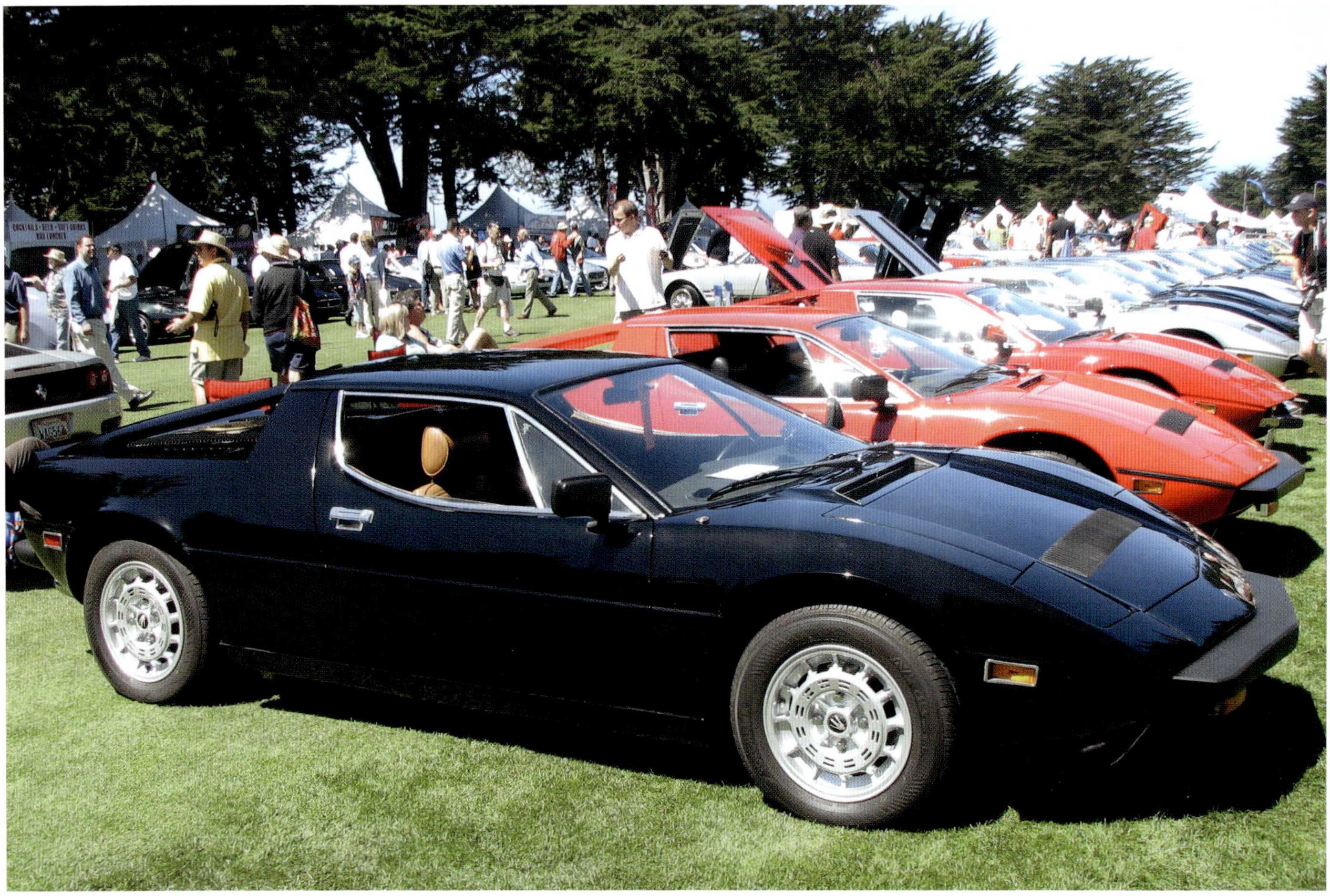

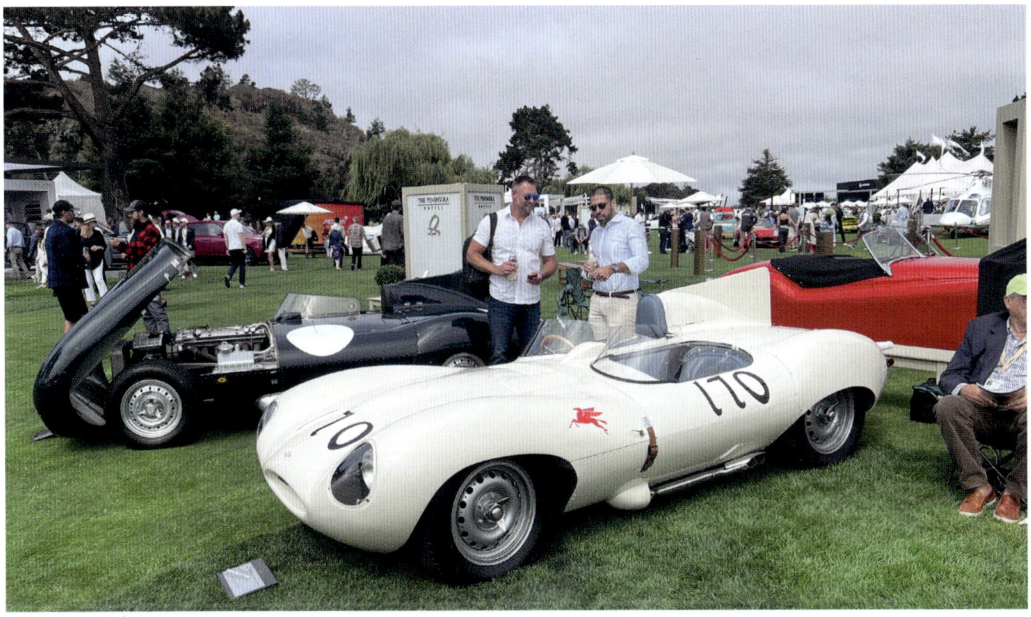

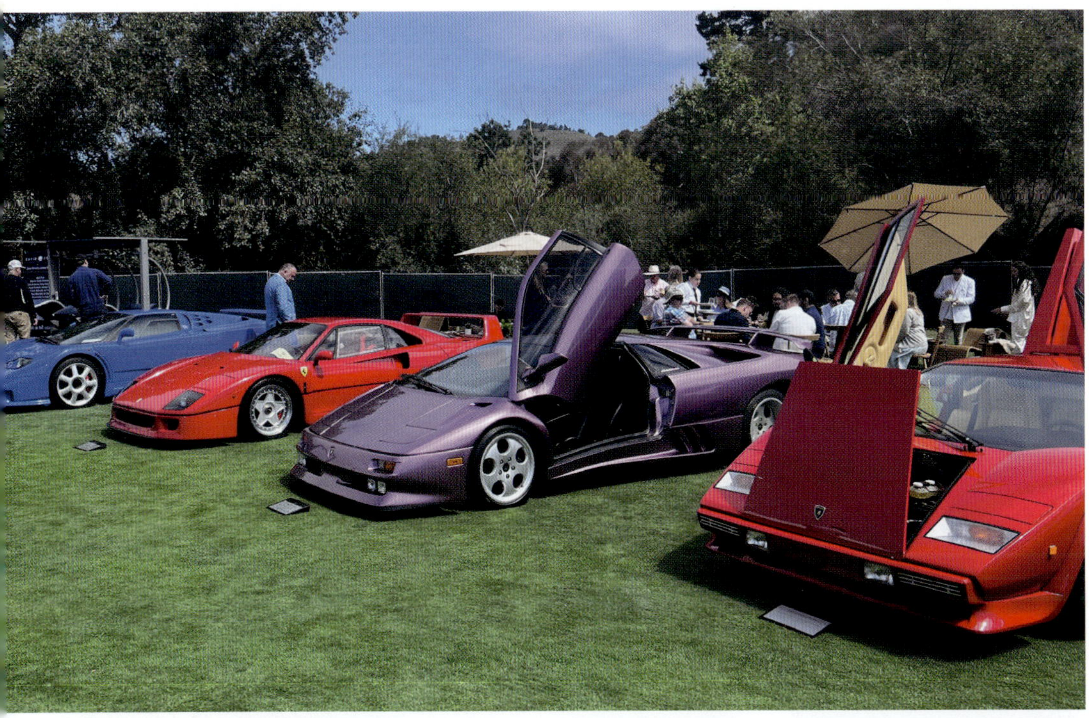

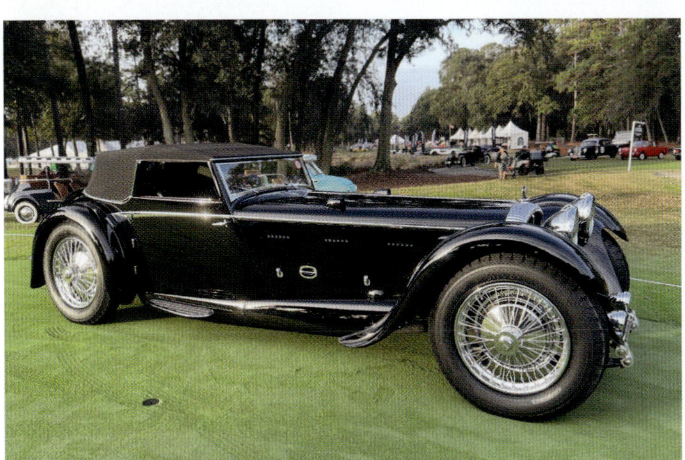

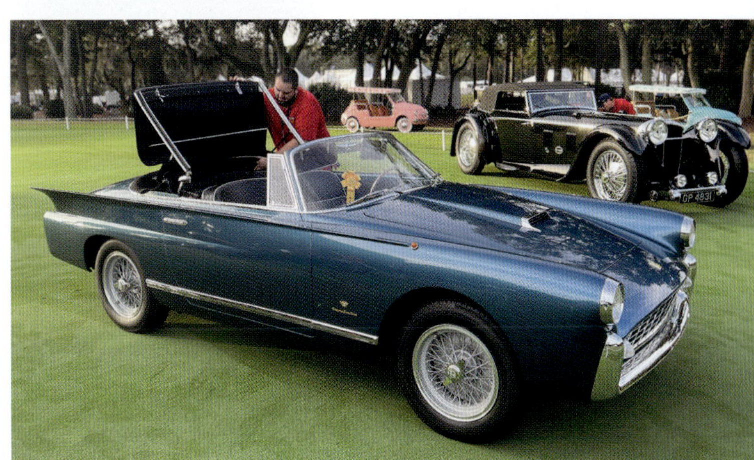

**LEFT:** *Known for hosting a number of eclectic classes and numerous debuts for new vehicles, The Quail: A Motorsports Gathering is one of the most popular and prestigious events that take place during the (now) 10-day automotive celebration known as Monterey Car Week. (Cape Coventry Archives)*

**MIDDLE:** *One of the first concours in the world to recognize the shift toward newer vehicles, The Quail has long dedicated a number of classes for modern supercars and sports cars. (Cape Coventry Archives)*

**BOTTOM LEFT:** *The Hilton Head Concours d'Elegance in South Carolina is able to leverage its position as one of the few major events in the American Southeast to attract entrants like Robert M. and Anne Brockinton Lee's 1931 Daimler Double Six 50 Corsica Drophead Coupe, a former Best of Show winner at the Pebble Beach Concours d'Elegance. (Cape Coventry Archives)*

**BOTTOM RIGHT:** *A 1956 Ferrari 250 GT Boano cabriolet pictured at the Hilton Head Concours d'Elegance, just one example of the truly special vehicles that this event has hosted in recent editions. (Cape Coventry Archives)*

**THIS PAGE:** *Set at one of American motorsport's most iconic venues, the Lime Rock Concours is renowned for its strong classes of competition vehicles—and unlike traditional lawn events, its entrants are displayed right on the racing surface of Lime Rock Park, an especially fitting stage for the many competition cars that gather each year. (Cape Coventry Archives)*

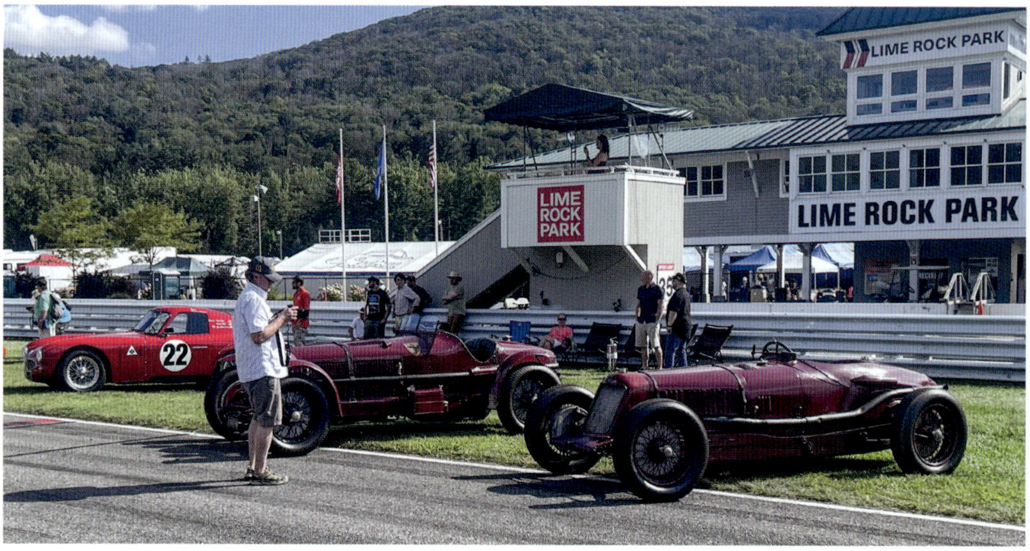


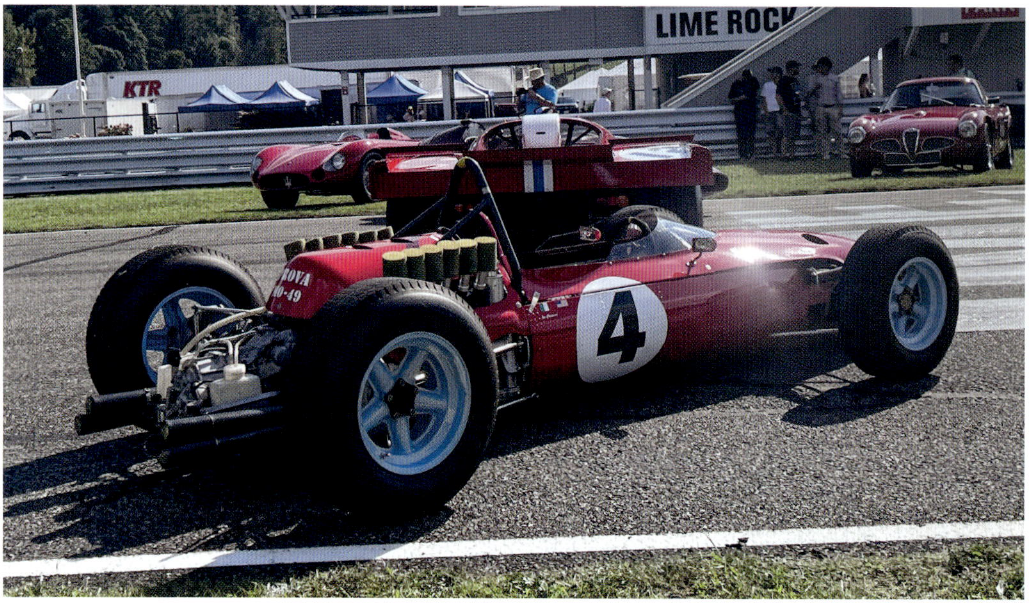


Additionally, noteworthy concours include the Concours d'Elegance at Copshaholm in South Bend, Indiana, the Lime Rock Concours in Connecticut, and the prestigious Audrain Newport Concours d'Elegance in Rhode Island.

Launched in 2019 from an idea conceived by Nicholas Schorsch, Donald Osborne, and Jay Leno, the Audrain Concours & Motor Week has quickly risen to become one of the most significant concours events in North America. Set in historic Newport, Rhode Island – a city with deep automotive roots tracing back to the Gilded Age – the event draws on a legacy that began with the first Vanderbilt Cup trophy race on September 6, 1900, where Willie K. Vanderbilt and his friends raced their newly imported automobiles at a local horse track.

The centerpiece of the concours is hosted on the grand lawn of The Breakers, the iconic summer "cottage" of the Vanderbilt family, making it a truly picturesque venue befitting the elegance of the automobiles on display.

What began with just 15 judged classes in 2019 has grown into a multi-day celebration featuring more than 20 classes, encompassing everything from pre-war American and European vehicles to Racing and Post-war Sports and Touring cars. Special featured classes are a hallmark of the event, with the 2025 edition spotlighting Cadillac, Chrysler, Ferrari, and the original 1965 Ford GT road cars.

Each year, over 180 outstanding cars and motorcycles are exhibited on the lawn of The Breakers during

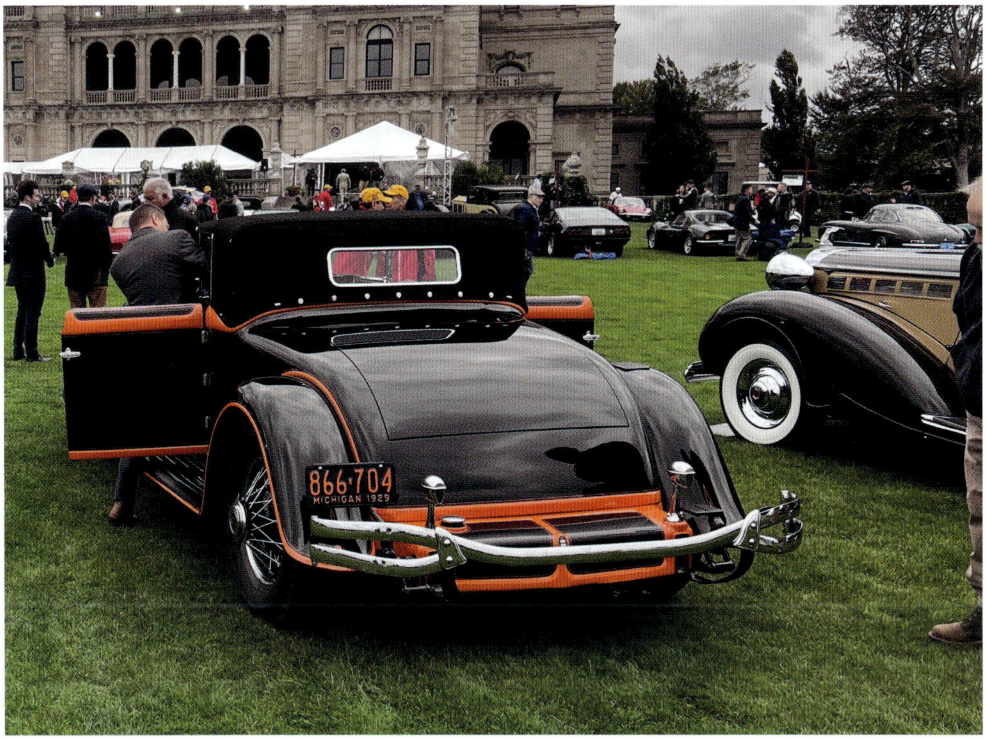


**LEFT:** *Held on the grounds of The Breakers, built for Cornelius Vanderbilt II and the most celebrated of Newport, Rhode Island's "summer cottages," the Audrain Concours d'Elegance offers one of the world's most bucolic settings for an automotive event. (Cape Coventry Archives)*

**BELOW:** *With a veritable plethora of the world's leading collectors located in the American Northeast, the Audrain Concours d'Elegance has been able to populate the show field with a number of automotive luminaries including the 1968 Alfa Romeo 33 Stradale pictured in the foreground with a pair of gems from ATS visible in the background. (Cape Coventry Archives)*

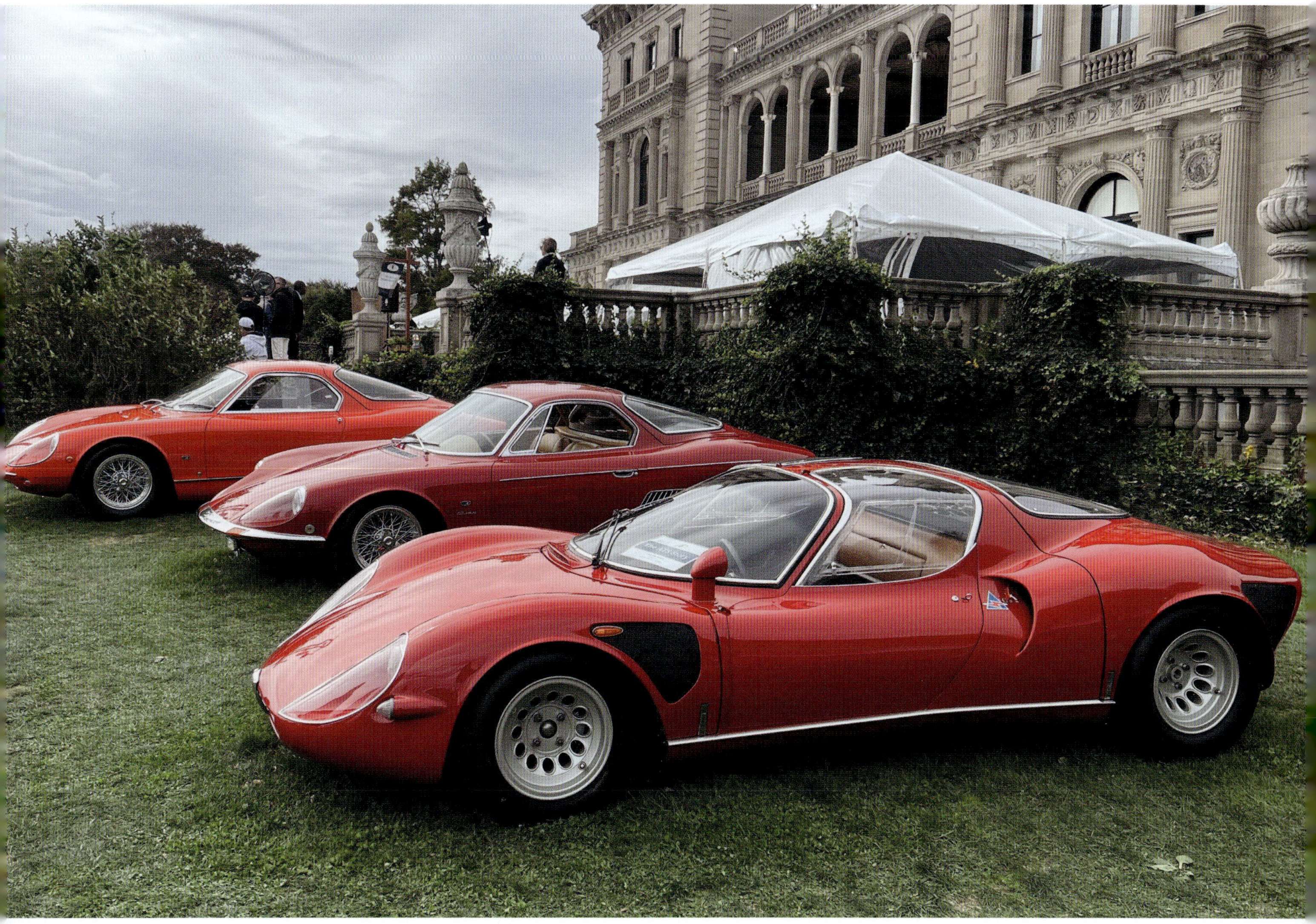

the culminating event of the Audrain Newport Concours & Motor Week. These vehicles compete across a range of categories, including the popular "Thirty Under 30" class, with one outstanding entry earning the coveted "Best in Show" title.

The "Thirty Under 30" class was the brainchild of Jay Leno, aimed at welcoming younger enthusiasts into the concours world. To qualify, owners must be under 30 years old and have invested no more than $30,000 into their vehicles. The concept has been so well received that it has expanded into as many as three distinct classes.

The Audrain Newport Concours d'Elegance uses the "French" judging style, which blends objective assessment with subjective impression. Reflecting Newport's identity – "History, Luxury & Sport" – the international panel of judges evaluates how well each vehicle embodies these values.

"The judges seek an emotional connection," explains Nic Waller, President of the Audrain Group. "While originality and condition are key, it's often a car's history and the owner's role in its preservation that truly resonate. Sometimes, these factors create a stronger bond than a flawless, untouched restoration."

Waller emphasizes the purpose behind the event: "Our goal is to share these vehicles with the broader public, to introduce new audiences to the passion and stories behind them."

Appointed Executive Director of the Concours in 2021, Nic Waller has led the vehicle selection process ever since. Each year, he collaborates with a small selection committee to review submissions. For the 2025 event, Audrain received more than 500 applications, from which 180 of the finest entries were chosen.

Now serving as President of the Audrain Group, Waller works closely with Donald Osborne, who continues to support the organization as Consulting Director. "We award First and Second in Class honors for each category, as well as an Honorable Mention," Waller notes. "In addition, we present ten special awards, including Most Elegant, Chairman's Award, and People's Choice."

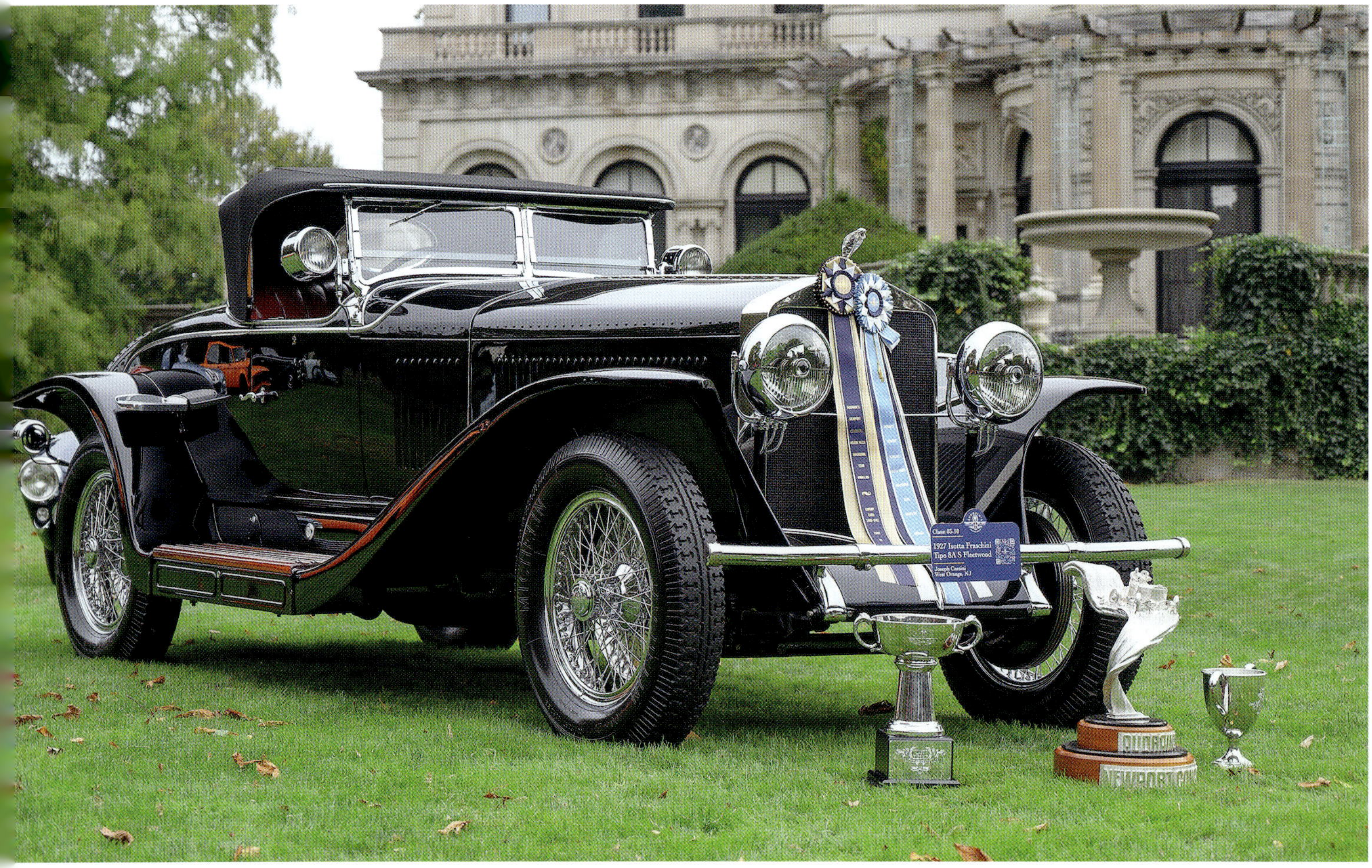

**OPPOSITE:** *The array of cars at the 2024 Audrain Concours ranges from an original Mini to a De Tomaso Mangusta by way of a Lincoln Continental, Maserati Khamsin and an Iso Grifo. (Courtesy Audrain Concours & Motor Week)*

**BELOW:** *An Isotta Fraschini Tipo 8A S from 1927, with rare Fleetwood coachwork at the 2019 Audrain Concours & Motor Week. (Courtesy Audrain Concours & Motor Week)*

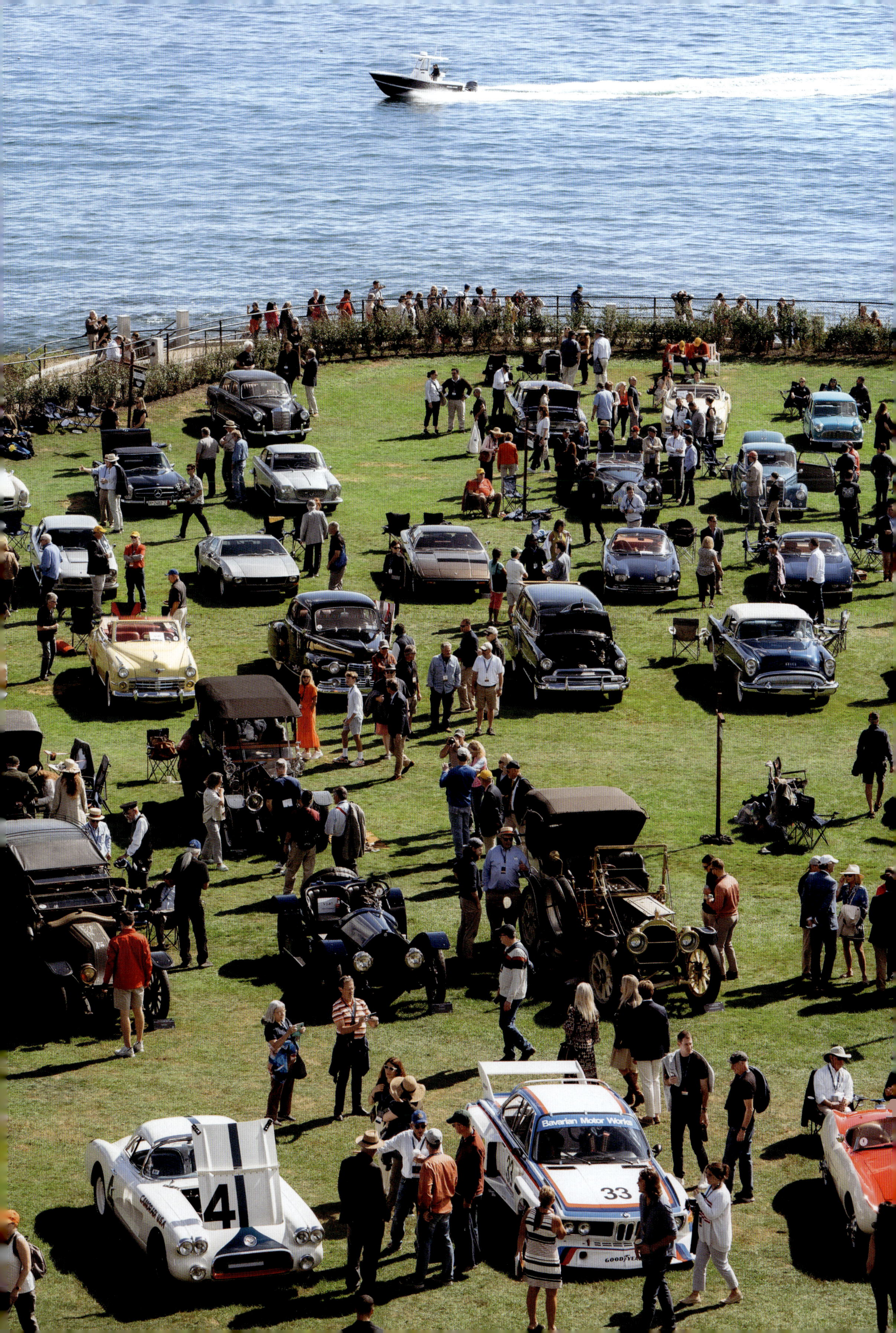
Bavarian Motor Works
33
41

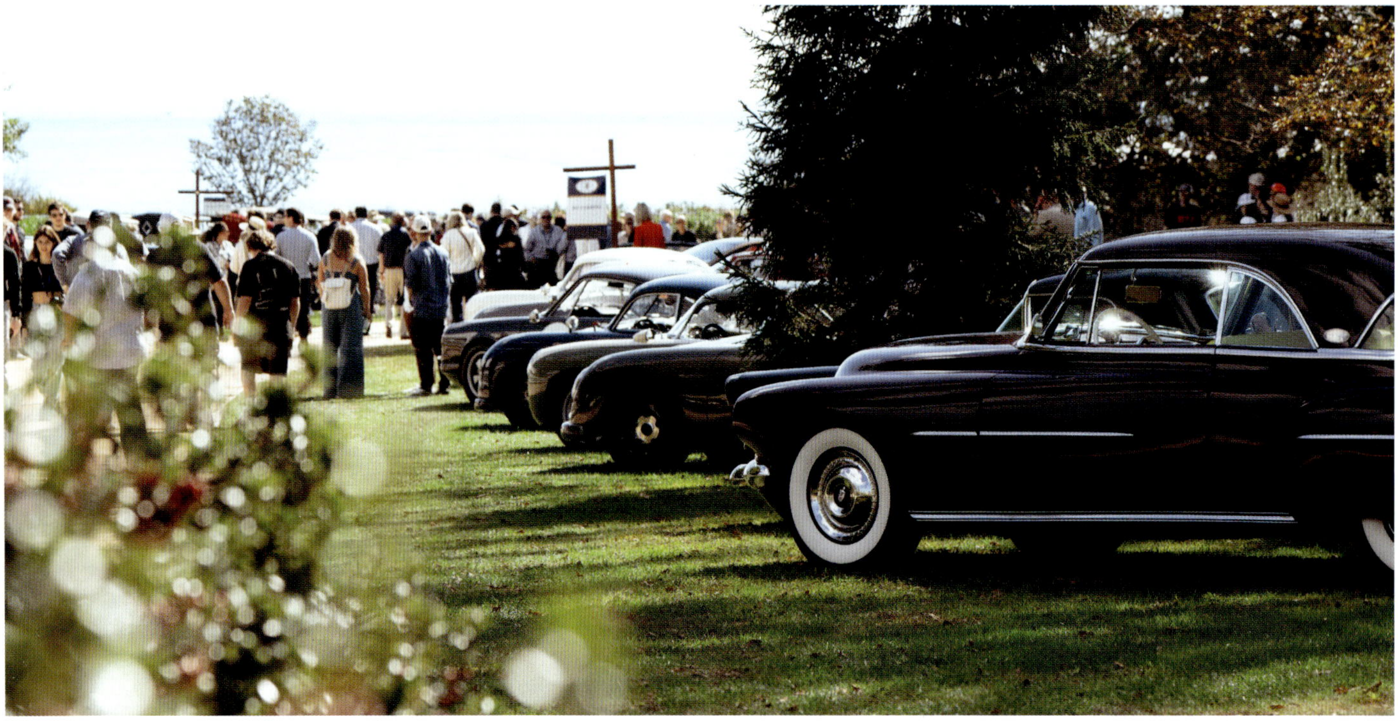

**ABOVE:** *An interesting line-up of cars at the 2024 Audrain Concours & Motor Week. (Courtesy Audrain Concours & Motor Week)*

**OPPOSITE TOP:** *For the 2023 Audrain Concours, among the rarities, the Ghia Gilda was a star attraction. (F. Dolnier)*

**BELOW:** *An eclectic lineup of cars at the Audrain Concours & Motor Week. (F. Dolnier)*

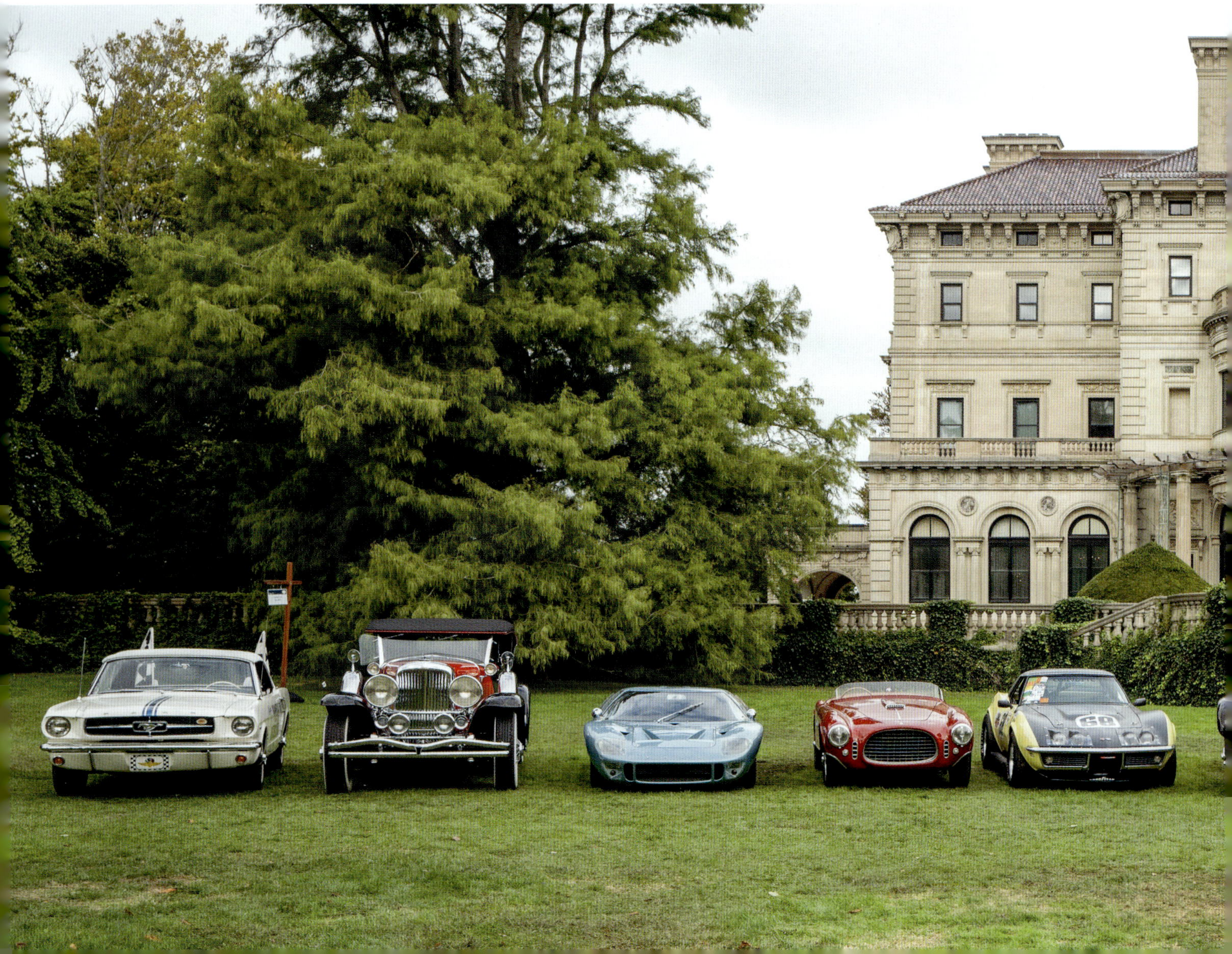

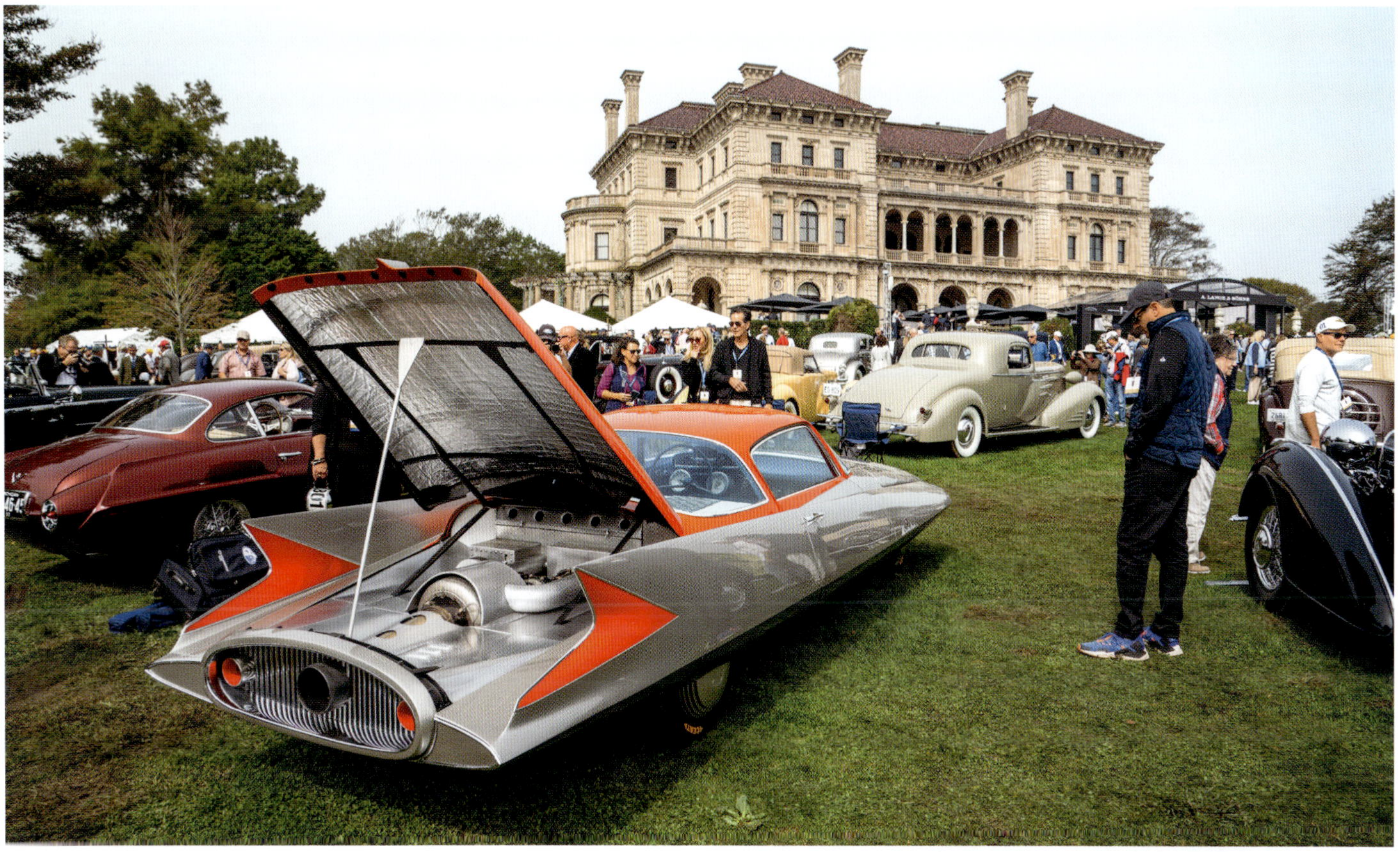

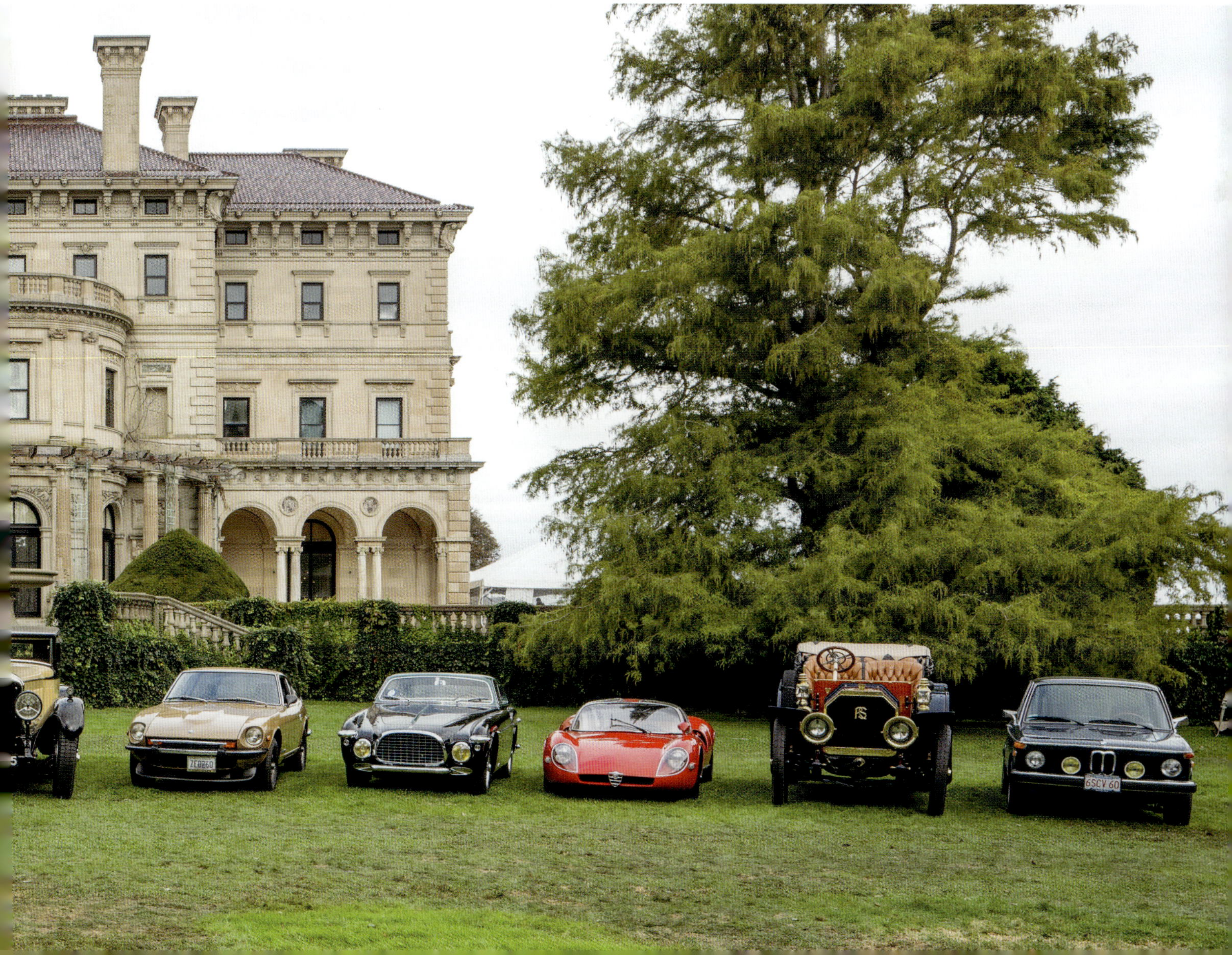
6SCV 60

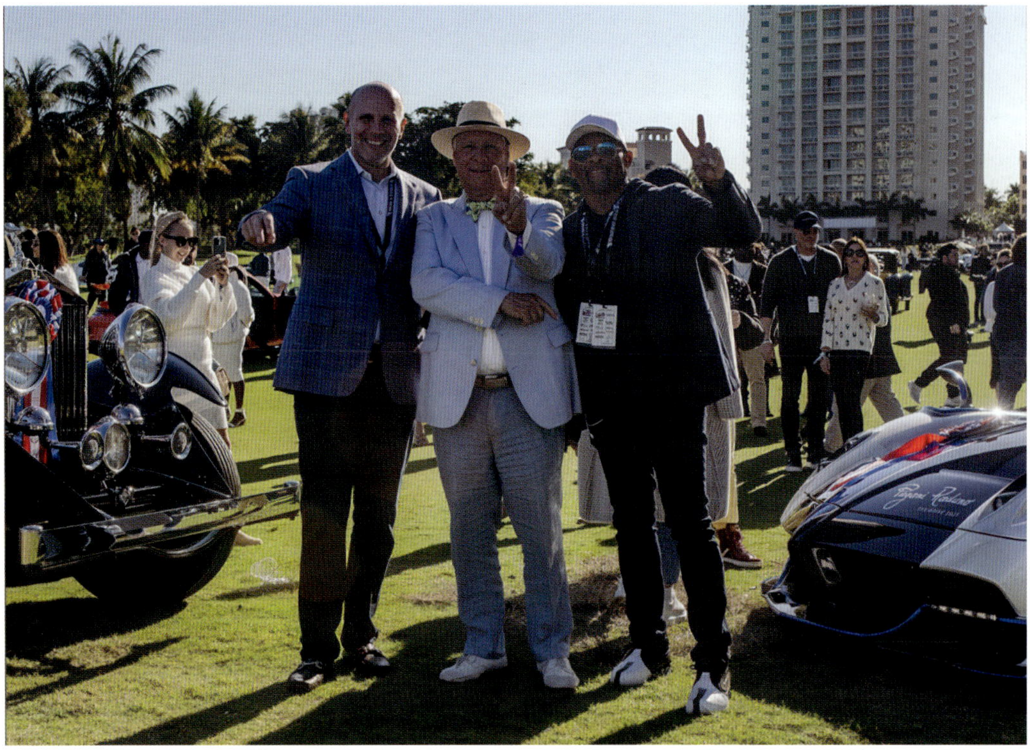

**RIGHT & MIDDLE (BOTH):** *Conceived and managed by Jason Wenig, Motorcar Cavalcade is backed by the Warren Henry Auto Group whose chairman Warren Zinn wanted to have a concours like Pebble Beach, but Wenig convinced him to do it differently. (Steve Cucinotta)*

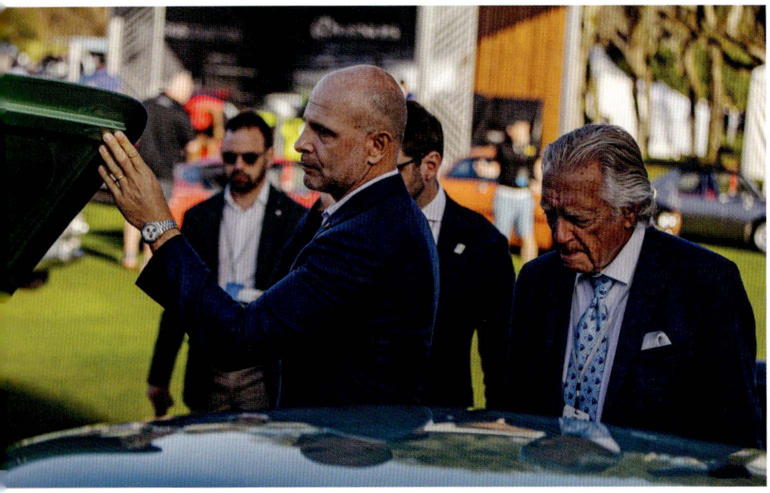

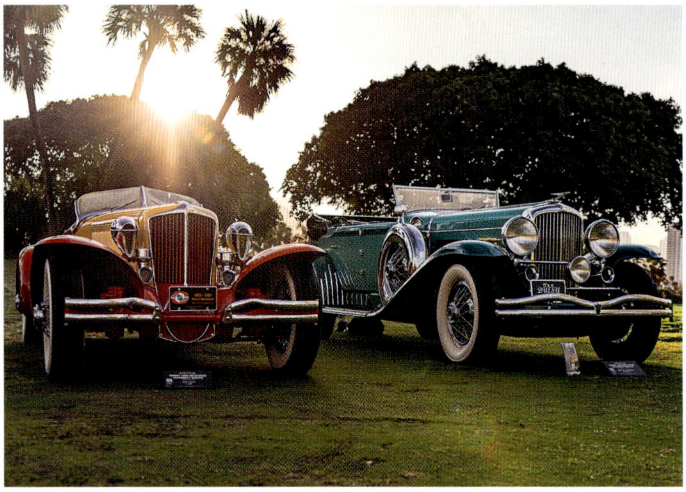

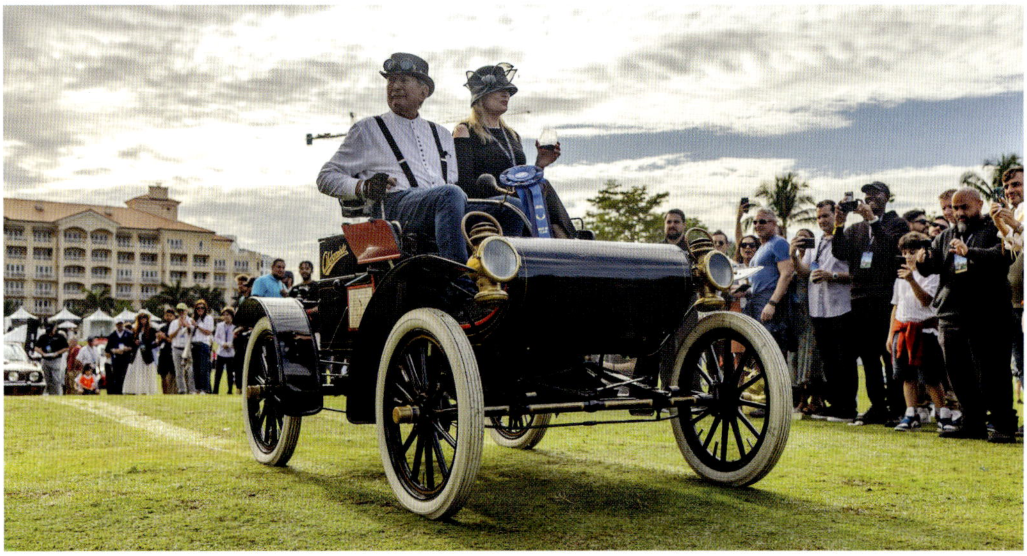

**RIGHT:** *An Oldsmobile Curved Dash at the 2024 Motorcar Cavalcade. (Steve Cucinotta)*

***With so many concours events throughout the year, how do they all maintain their uniqueness?***

According to Jason Wenig, who has been running the Motorcar Cavalcade Concours d'Elegance in Miami, Florida since 2022, it's all about creating a distinct identity. "The key is to be different, to have your own character," says Wenig. One of his challenges came from Warren Zinn, founder and chairman of the event and owner of Warren Henry Auto Group, who wanted to build something like Pebble Beach. However, Wenig pushed back, believing it was more important to carve out a unique niche.

"Warren, being a car dealership owner, wanted to incorporate modern cars, which posed a challenge for me as a classic car restorer," Wenig explains. "We deliberated, and the solution became showcasing both classic and modern cars in a way that tells the story of the automobile's evolution." The event now highlights innovations in car design from 1896 to 2025, focusing on different elements such as lighting, doors, and finishes, inviting attendees to compare and explore the advancements over time.

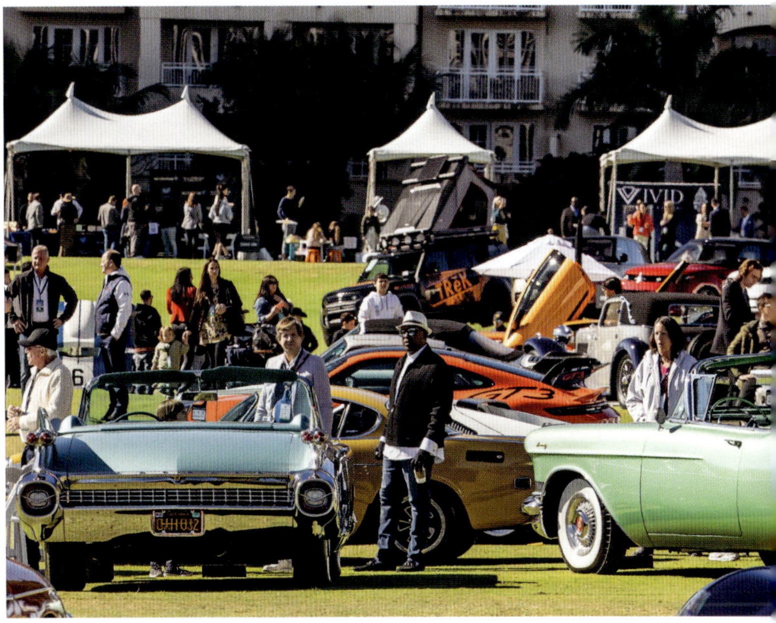

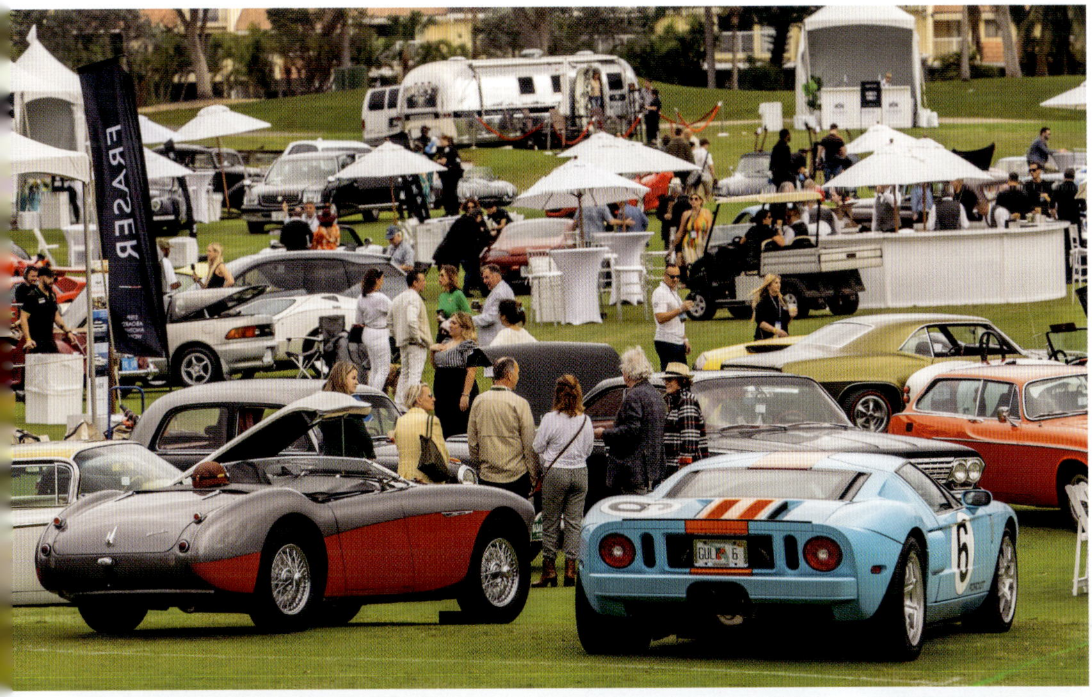

**THIS PAGE:** *Wenig seems to have found the right mix and with the French-style scoring system has injected a sense of casualness to the concours. (Steve Cucinotta and Charlton Ochoa)*

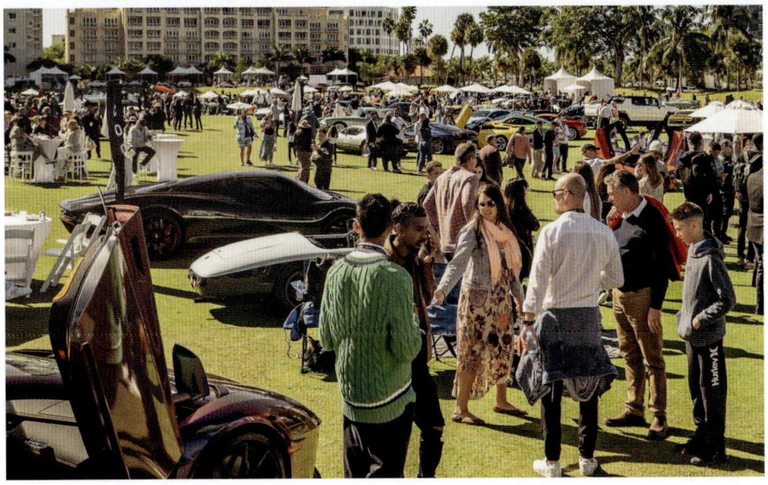

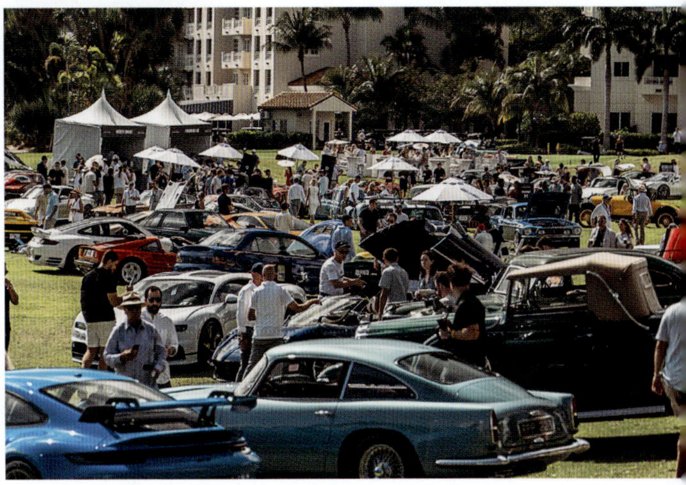

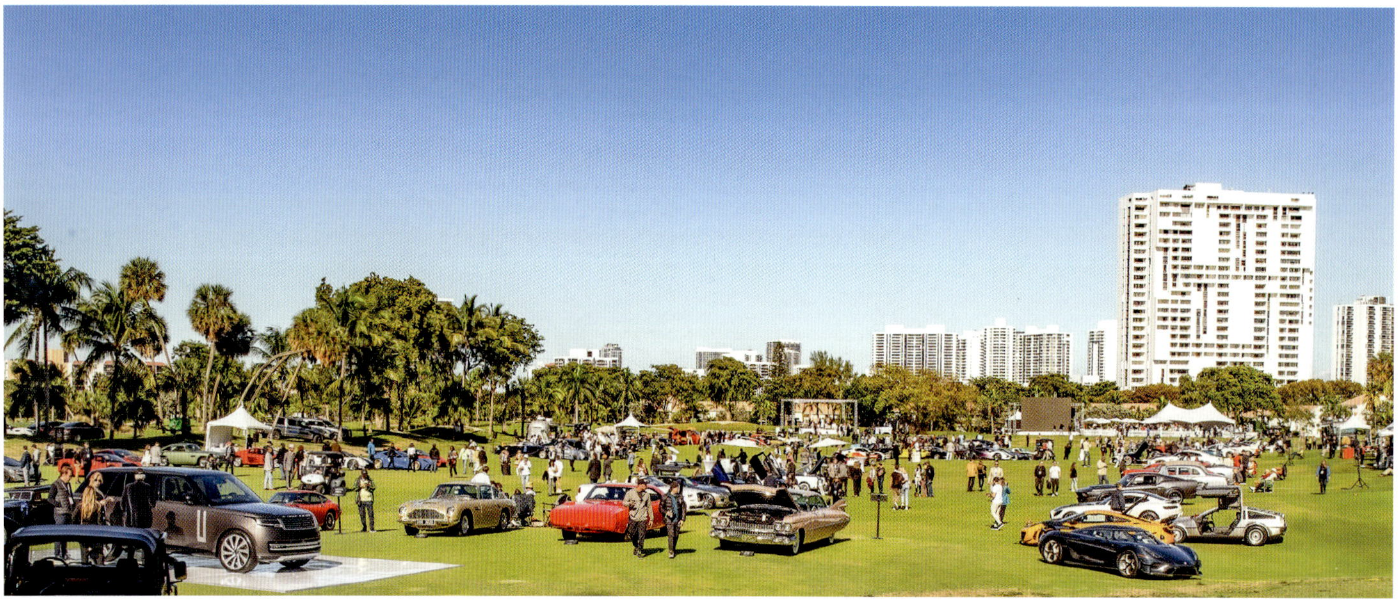

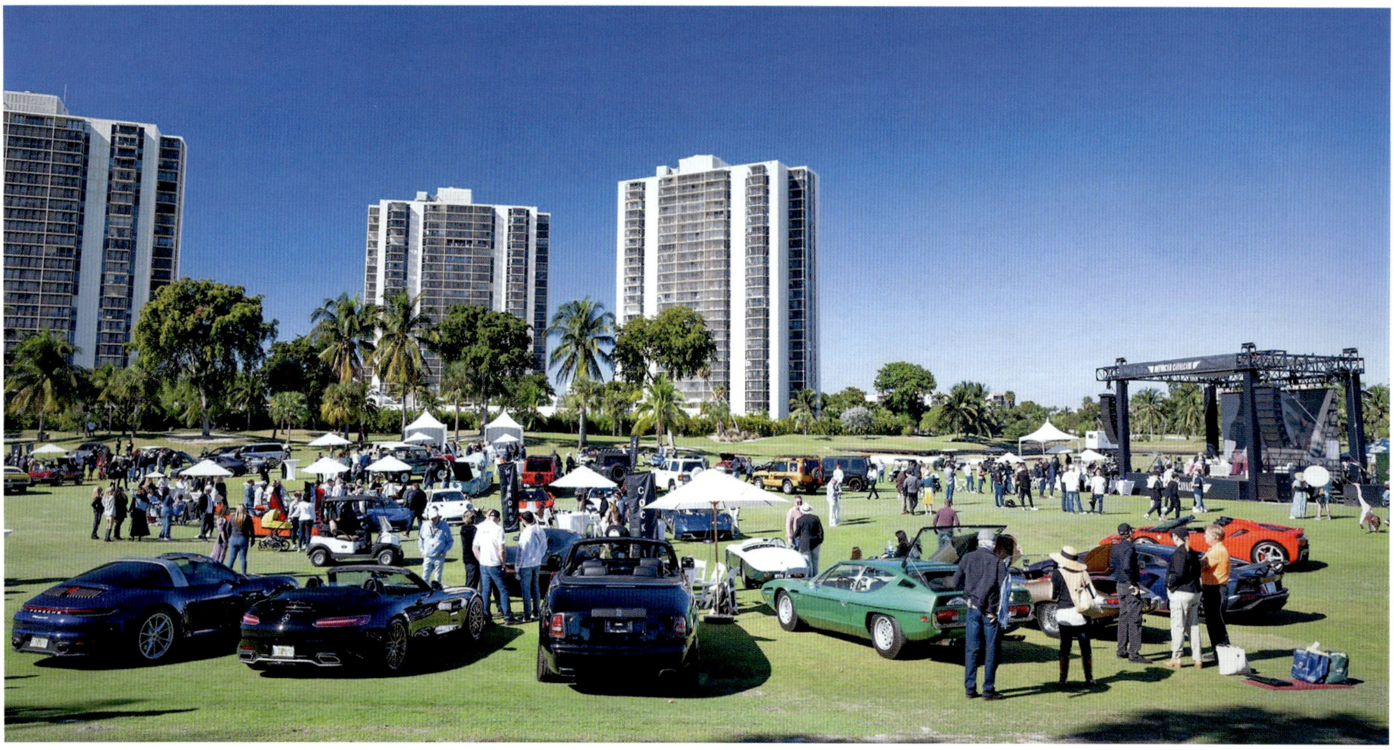

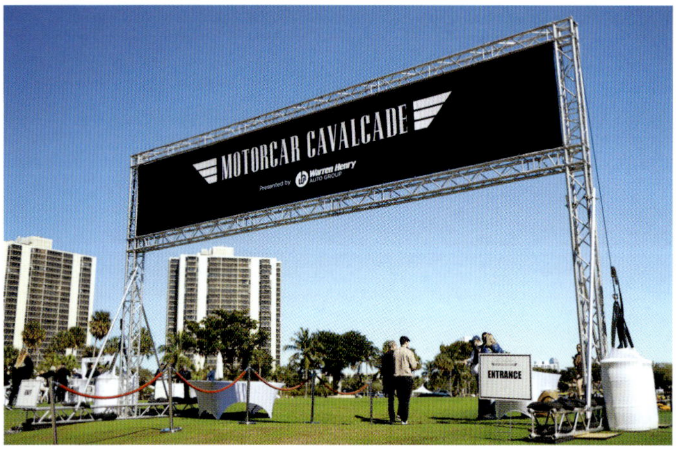
MOTORCAR CAVALCADE
ENTRANCE

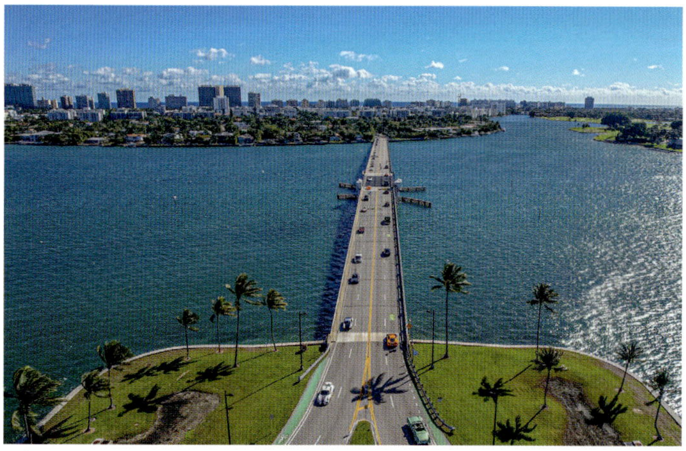

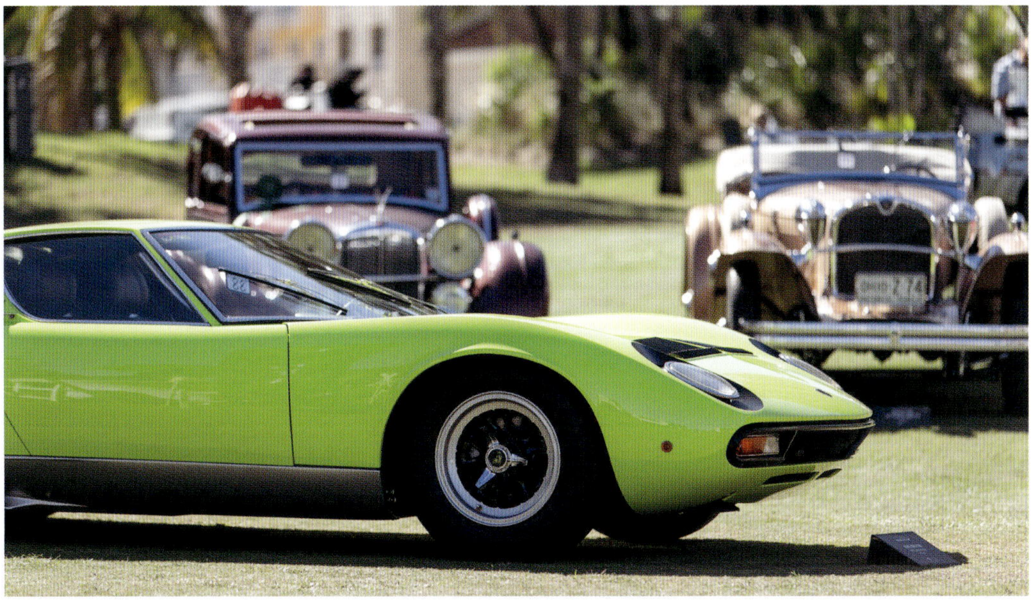

**THESE TWO PAGES:** *Miami's Motorcar Cavalcade has the right kind of car. And the right kind of glamour. (All photos ©Motorcar Cavalcade)*

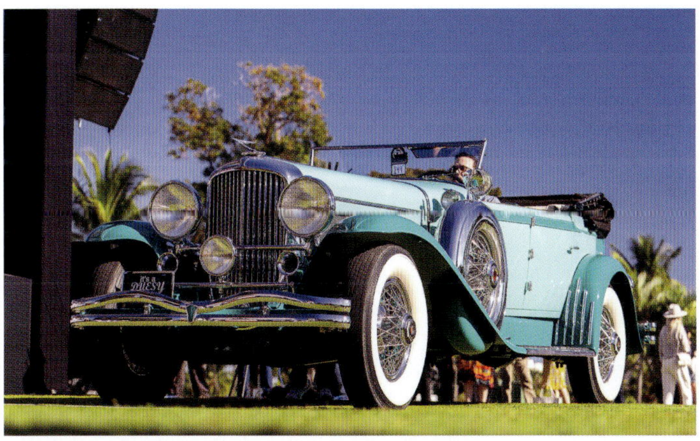

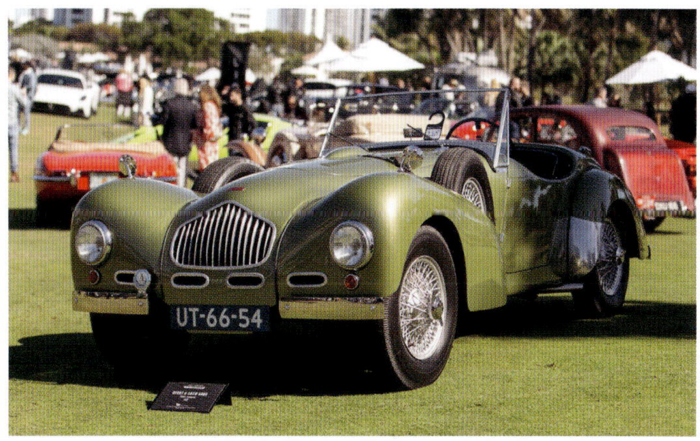


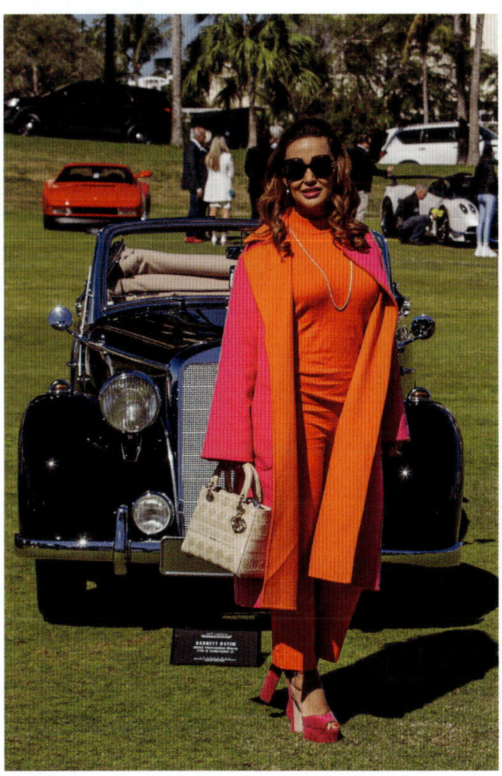

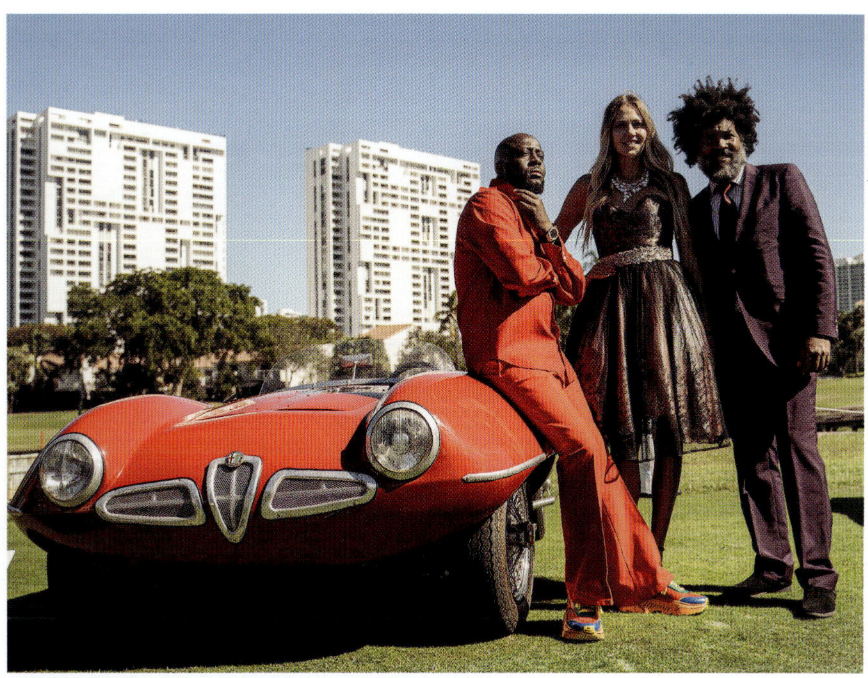

Wenig describes how they built the event around this concept: "We created classes and invited people to a party where everyone dresses up, blending elegance with fun." Participants can dress in period fashion or attire similar to the Kentucky Derby, turning the event into a celebration of both style and cars. Though the idea was initially met with skepticism by industry insiders, including directors from the AACA (Antique Automobile Club of America) and Pebble Beach – the event has since won them over. Many now serve as docents or advisory board members, helping to guide VIP celebrity judges through the show.

Motorcar Cavalcade takes a fresh approach by having docents, experienced concours judges, assist VIP judges, many of whom come from outside the car world, including designers, musicians, and athletes. Wenig notes that this broader perspective brings a sense of excitement and relatability to the event, appealing to people who might not be car experts but simply appreciate beautiful automobiles.

Wenig also acknowledges that many classic car enthusiasts have grown weary of the elitism and intensity surrounding traditional concours. "Some people are just tired of the lack of emotion and the competitive spirit that comes with these high-stakes events," he says. "They want to enjoy incredible cars without the pressure."

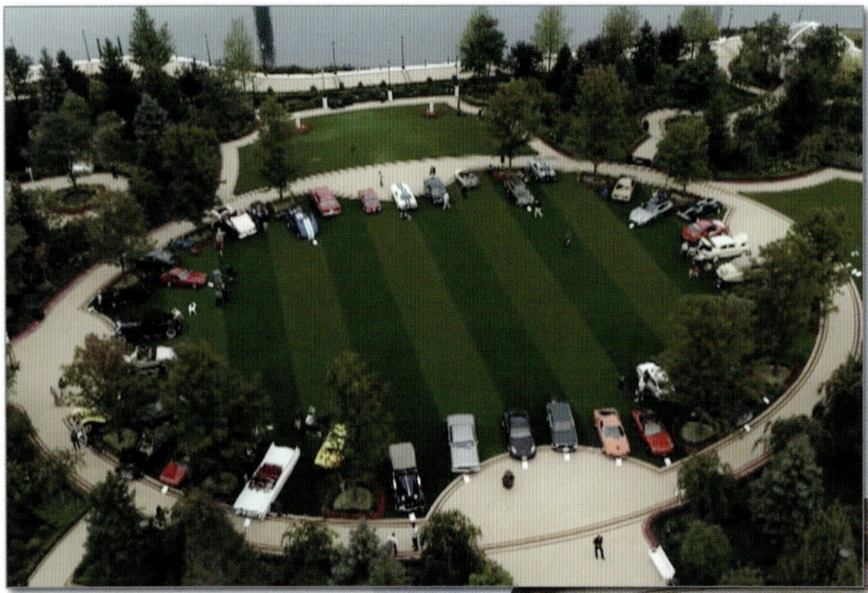

Despite this, Wenig emphasizes the importance of technical concours like Pebble Beach, which adhere to strict judging systems like ICJAG to maintain the highest restoration standards. However, he believes not every concours needs to follow this model. Motorcar Cavalcade uses the French judging system, with about 45 to 50 judges, including docents, evaluating the roughly 200 cars on display. This approach has been a key factor in the event's growing popularity.

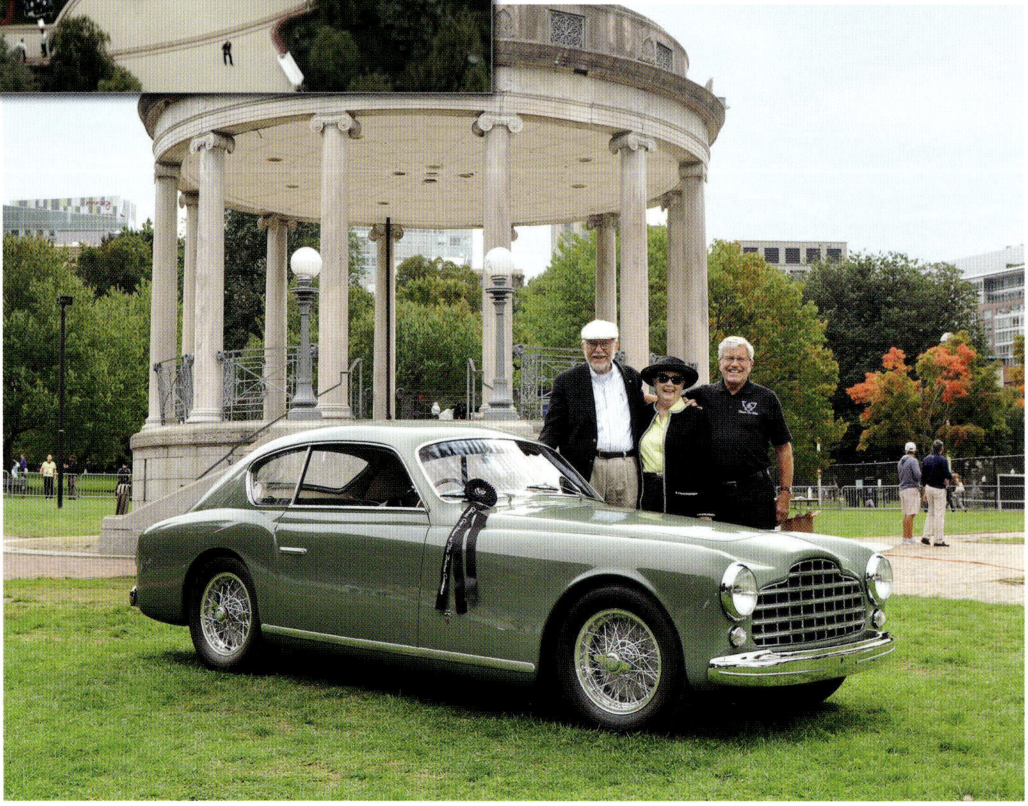

**ABOVE:** *The Boston Cup's unusual layout of arranging the cars in a circle with the exterior circle accessible for all and sundry, has made this concours one of the most popular with very high scores for eyeballs. (The Boston Cup)*

**RIGHT:** *The 2022 Best of Show at The Boston Cup was this distinctive Ghia-bodied 1950 Ferrari 195 Inter owned by Sally and Roger Demler. (The Boston Cup)*

## The French system

The "French judging system," as Jason Wenig describes it in the context of the Motorcar Cavalcade, is quite different from the system currently used in France, though we'll delve into that distinction later. Interestingly, an increasing number of concours in the US are adopting this American-style "French judging system." This approach, now followed by events such as the Misselwood Concours d'Elegance in Beverly, Massachusetts, Ironstone in Murphys, California, the St. Michaels Concours d'Elegance at Chesapeake Bay, Maryland, and several others, places much more emphasis on emotion and aesthetic appeal rather than on the technical perfection of every nut and bolt.

This shift in focus allows for a more subjective appreciation of beauty and design, creating a more relaxed and enjoyable atmosphere, which contrasts with the more stringent, detail-oriented judging criteria of the "now traditional" American-style concours d'état.

The Boston Cup, a popular concours event held in its namesake city since its debut on September 23, 2012, exemplifies a very different approach to judging and event organization. "Our system is peer judging," explains Richard "Rich" Doucette, Founding Chair of The Boston Cup. "With 100 owners and 100 ballots, we ask participants to select the best American, British, German, Italian, and other cars. We also have subcategories within these groups."

Doucette highlights the distinctiveness of the Boston Cup, noting its location in America's oldest public park, where they cannot charge the 30,000 spectators who attend for free. "Bill Warner advised me years ago not to rely on admission fees for revenue. If it rains, and only 50 cars show up, without paid admission, you'll be in financial trouble. He didn't make a profit for at least 10 years. Our goal is just to break even. Anyone thinking they'll get rich from organizing a concours is mistaken; car shows aren't about making money."

According to Doucette, the event is driven by passion. "It's a reunion – a gathering of like-minded hobbyists who see each other just once a year. I have volunteers who I don't even speak to all year, but they show up at seven o'clock on the day of the show, go straight to their stations, and have the time of their lives. Then, we shake hands, and I see them again the next year."

"There's a loosely connected fraternity of men and women who enjoy this hobby. As McKeel Hagerty says, we need to inspire the next generation, because these cars are like art on wheels. Without support, they'll end up rusting in junkyards, and we want to prevent that. All you need is a great venue, great cars, and great sponsors. It's a simple formula – a three-legged stool."

*Another recently established concours event is the Chattanooga Motorcar Festival, which prides itself in featuring serious forums on historic vehicle topics, thanks to the chairmanship of well-known author Ken Gross. (Chattanooga Motorcar Festival / Patrick Tremblay)*

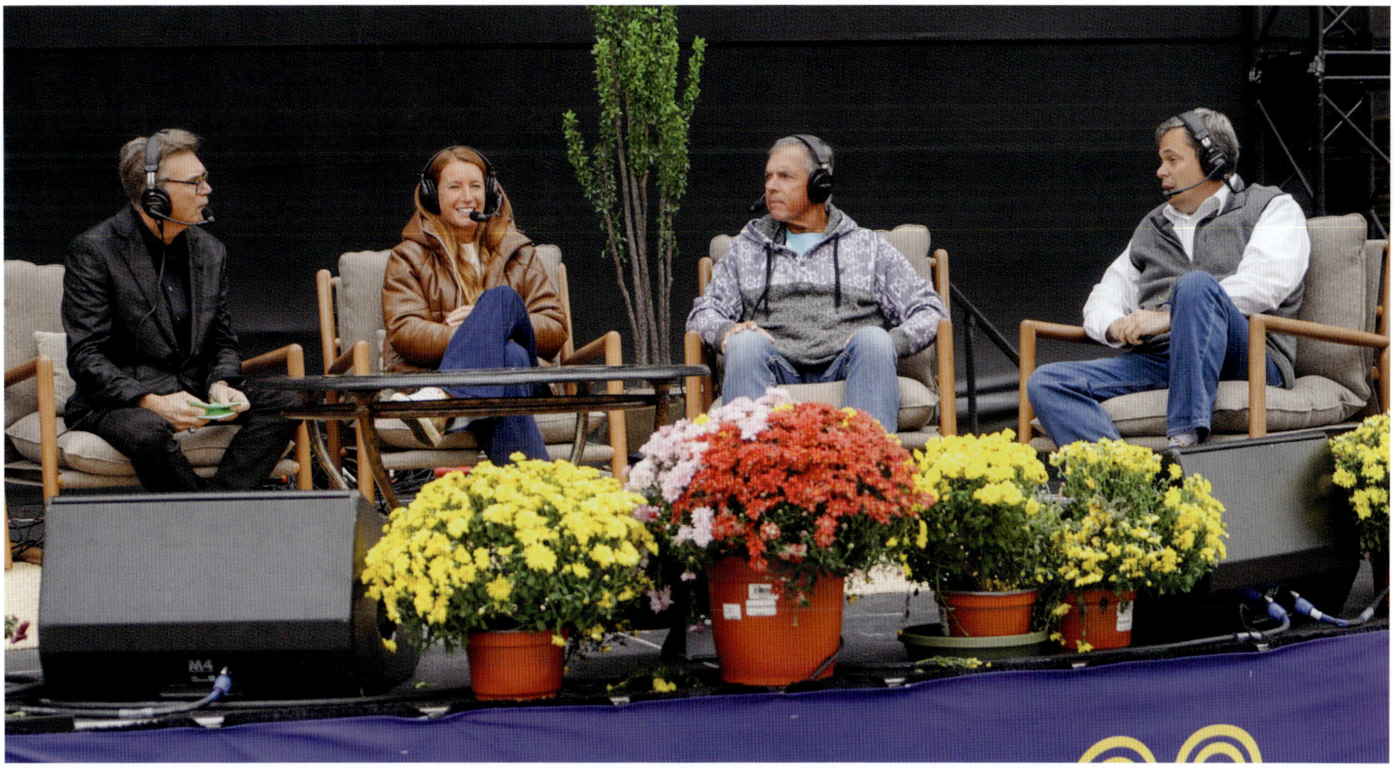

Doucette recalls the success of Ralph Lauren's 2005 exhibit at the Museum of Fine Arts in Boston, which was one of the museum's most successful non-art shows. Inspired by that event, Doucette and his car enthusiast friends decided to combine the art and car worlds in a prominent Boston location. After some time, they secured permission to host The Boston Cup at the Parkman Bandstand on Boston Common, a site that attracts thousands of visitors on a sunny summer weekend.

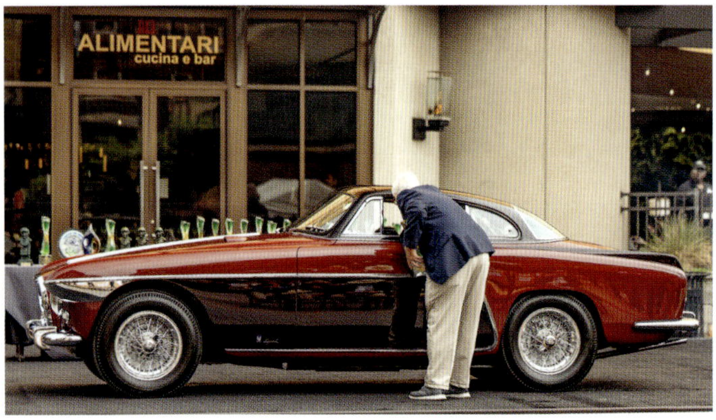


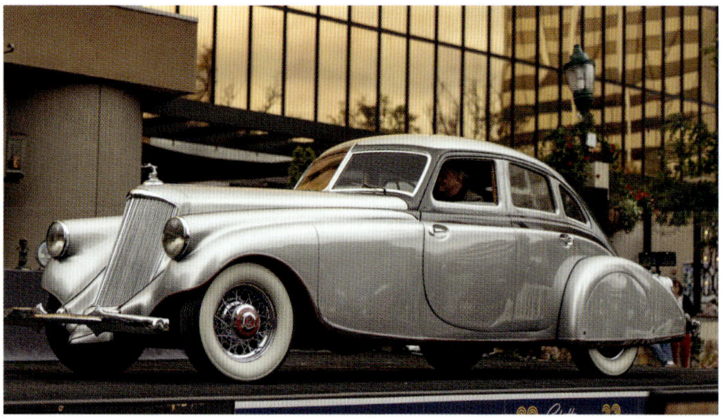

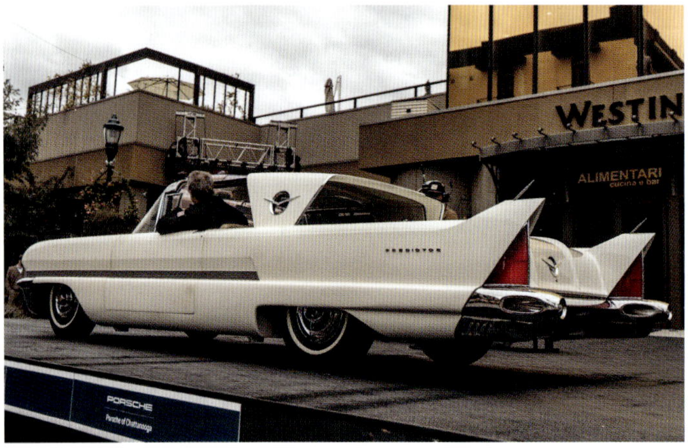


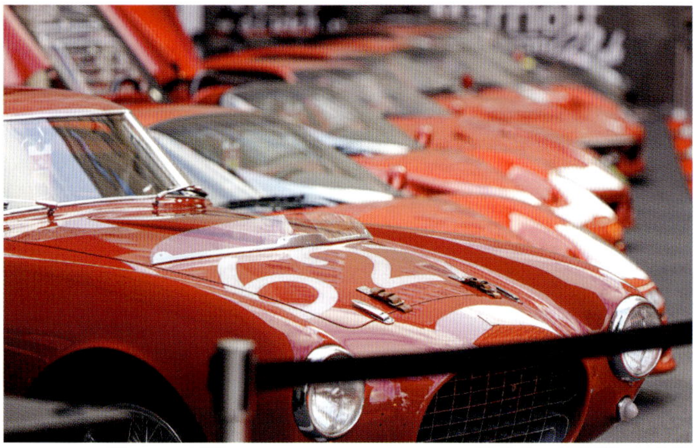

**TOP (BOTH):** *Thanks to the contacts of Ken Gross, Chattanooga has been attracting some outstanding entries such as this Ferrari 250 Europa bodied by Vignale (left) from 1953, as well as this 1933 Pierce-Arrow Silver Arrow (right). (Chattanooga Motorcar Festival)*

**MIDDLE LEFT:** *This 1957 Packard Predictor Hardtop Coupe Concept was a hit on the stage. (Daniel Morris)*

**MIDDLE RIGHT & RIGHT:** *Of course, no self-respecting concours is complete without a significant line up of Ferraris, like these at Chattanooga. (Chattanooga Motorcar Festival)*

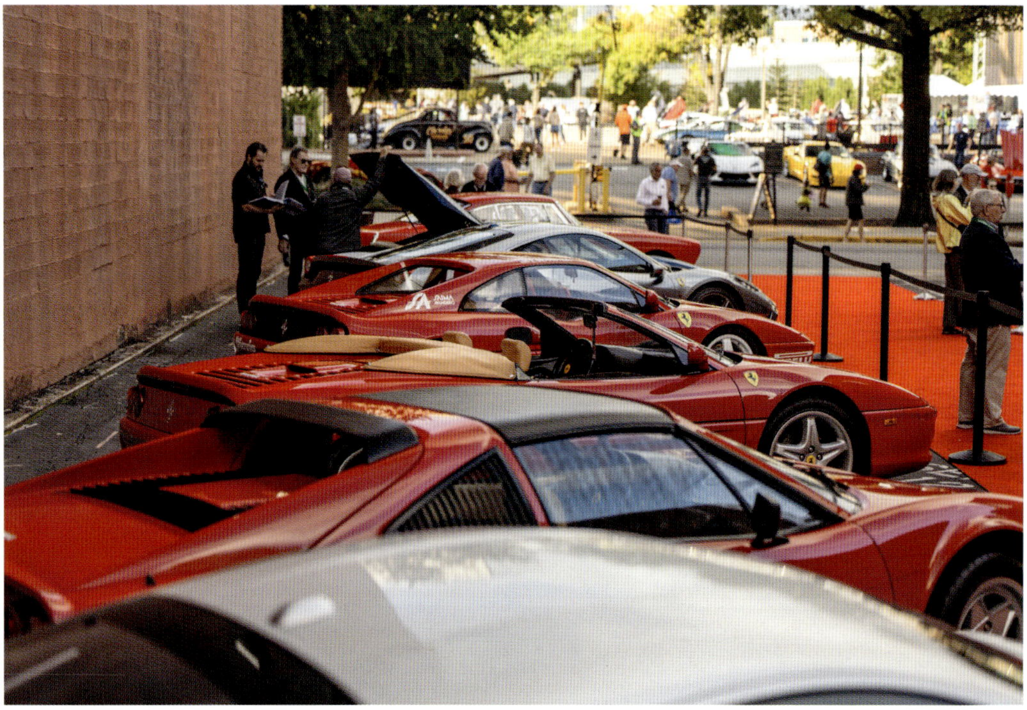

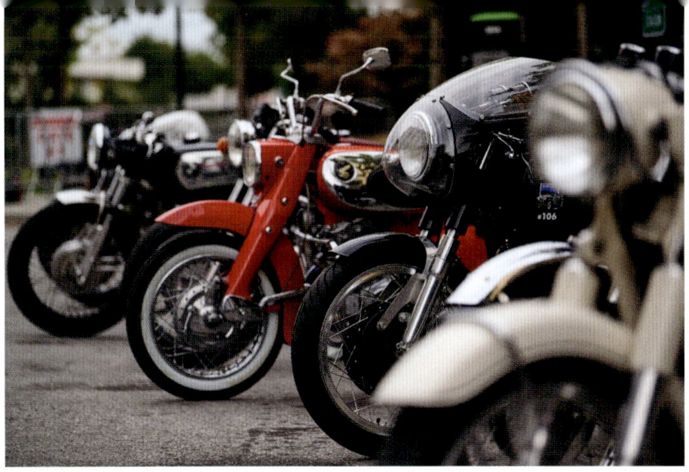

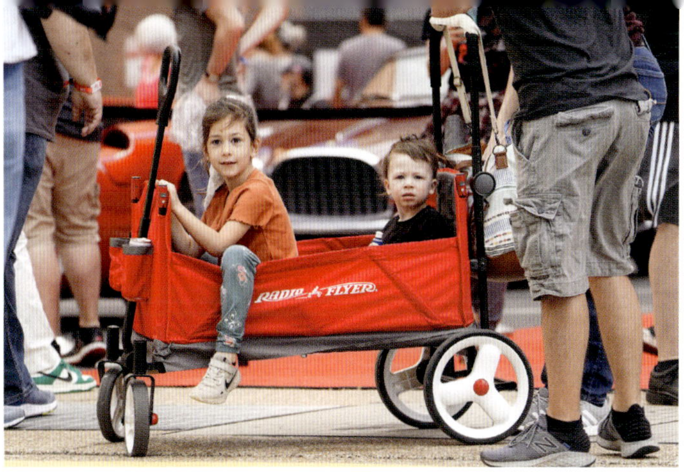


## For a wider public

To manage the event, police barriers are set up, allowing only car owners or those who pay $50 or $100 to enter. However, just three or four feet away, passersby can view the cars for free. "We get 30,000 people walking by and seeing the cars, while 1,000 are inside. We're on the Freedom Trail, right in the heart of Boston. The weather's great, the event is now well-known, and parking is free on Sundays. That's why we call ourselves America's largest free car show – because these 30,000 people wouldn't get to see these cars otherwise."

A similar mindset drives the relatively new Chattanooga Concours d'Elegance, held every October since 2019 (with the exception of 2020 due to COVID-19). The event draws 20,000 to 25,000 visitors each year. With an affordable ticket price of just $25, the Chattanooga Concours aims to be more inclusive than exclusive.

**THIS PAGE:** *At the other end, the Chattanooga Concours also features motorcycles. Despite the casual friendly atmosphere that attracts more than 25,000 visitors every year, judging follows the rather rigorous and technically demanding ICJAG system. (Chattanooga Motorcar Festival)*

"We've reduced the fees to a very accessible level," explained Ken Gross, a renowned author, long-time Pebble Beach judge, and the Concours Chairman for the Chattanooga event. "This allows families to attend and ensures all proceeds go to charity."

Gross went on to describe the origins of the event: "The first Chattanooga Concours took place in 2019. Byron DeFoor and his brother Ken, respected developers in Chattanooga, are the driving forces behind it. They own several businesses, including nursing homes, insurance companies, and hotels like Westin and Indigo. They also developed a section of Chattanooga called the West Village, transforming what was once a neglected area into a vibrant neighborhood, complete with restaurants and a stunning hotel, the Westin Chattanooga."

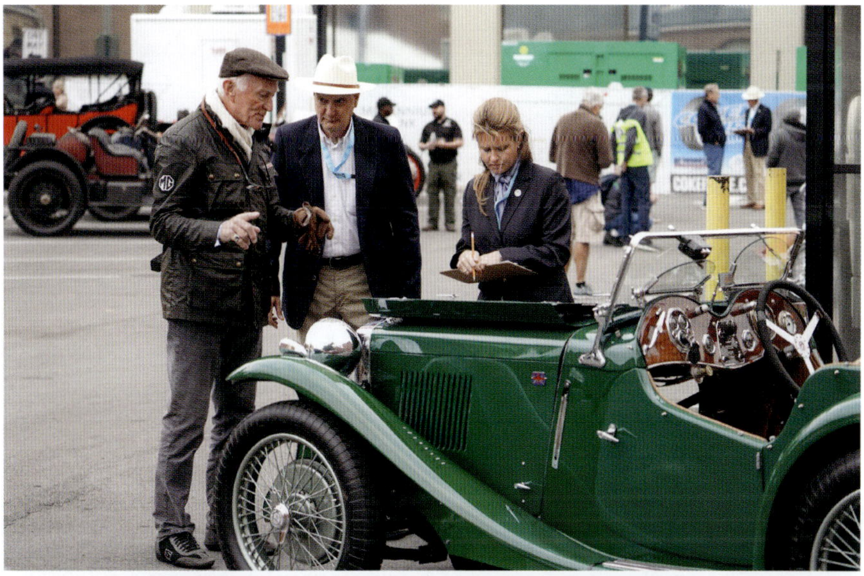

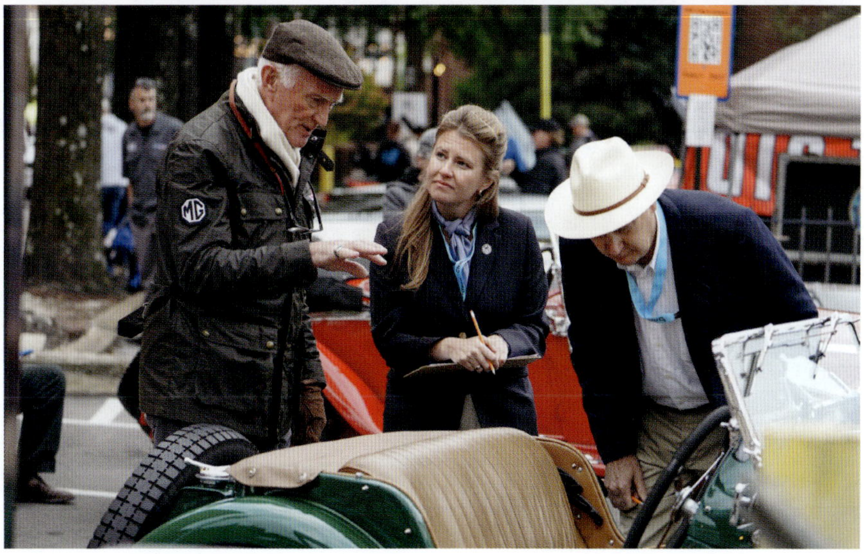

"Byron's interest in vintage racing played a key role in the event's creation," Gross continued. "He purchased a Lola T70 and, within a year, won a racing championship. During that time, he became friends with rock band AC/DC's Brian Johnson, another racing enthusiast. The two bonded over their shared experiences with loved ones affected by Alzheimer's, stroke, and Parkinson's disease. Byron used a concours d'elegance as a way to raise funds for research into these illnesses, benefiting the Chattanooga community."

Gross also highlighted the event's connection to healthcare advancements: "There's a higher incidence of stroke in the middle Tennessee and mid-Atlantic regions. Researchers there are doing world-class work, developing procedures that allow real time, remote support for stroke patients admitted to any hospital. Byron believed this was a cause worth supporting, so all proceeds from the Chattanooga Concours go toward this research."

Gross emphasized the event's inclusive atmosphere: "Byron didn't want the event to feel stuffy or exclusive, so it's called the Chattanooga Motorcar Festival. In recent years, they even built a track in the city, hosting races for historic cars, though this didn't happen in 2024. The city is able to block off areas downtown to charge admission."

**THIS PAGE:** *Sponsorships and advertising help to support research on Alzheimer's and Parkinson's disease. (Chattanooga Motorcar Festival)*

"Byron also enjoys mingling with celebrities," Gross noted. "We've had people like Wayne Carini, Luigi Chinetti Jr., Donald Osborne, Linda Vaughn, and race legends like Lyn St. James and Brian Redman. Hot rod designer Chip Foose has also attended. David Hobbs will attend again in 2024, along with Patrick Long and his Luftgekühlt Porsche display, as well as Bill Warner and racer Jochen Mass. We hold seminars and forums on a large central stage, much like Pebble Beach, where attendees can watch and engage. It's been a great success."

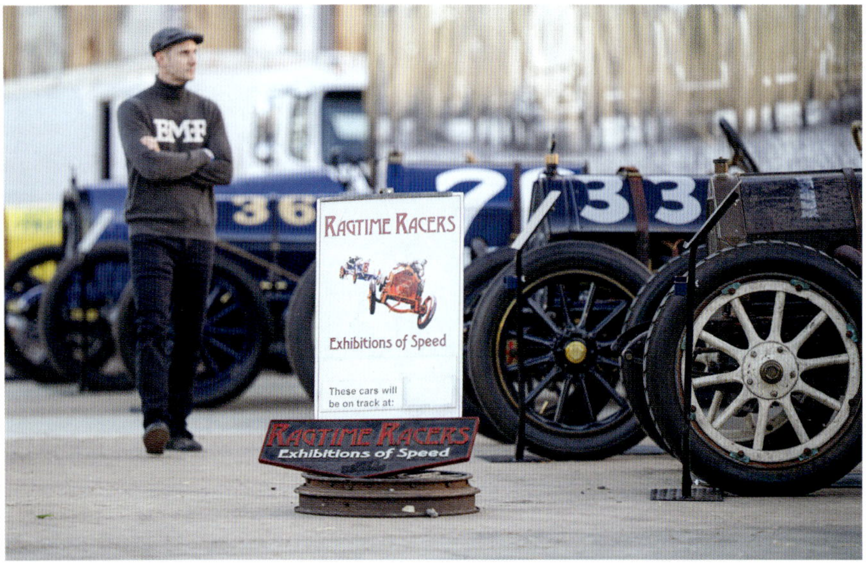


Although Gross is modest about his own contributions, he has been instrumental in attracting some of the most prestigious cars to the event, including Bill Marriott's Talbot-Lago, a Ferrari 250 Testa Rossa, the 1965 24 Hours of Le Mans-winning Ferrari 250 LM, and concept cars from the GM Heritage Museum. These rare vehicles have helped make the Chattanooga Concours one of the most talked-about events in recent years.

Despite the relaxed setting, Gross has maintained a high level of seriousness in judging, using the ICJAG scoring system, with Michael Tillson of Radnor Hunt serving as the chief judge.

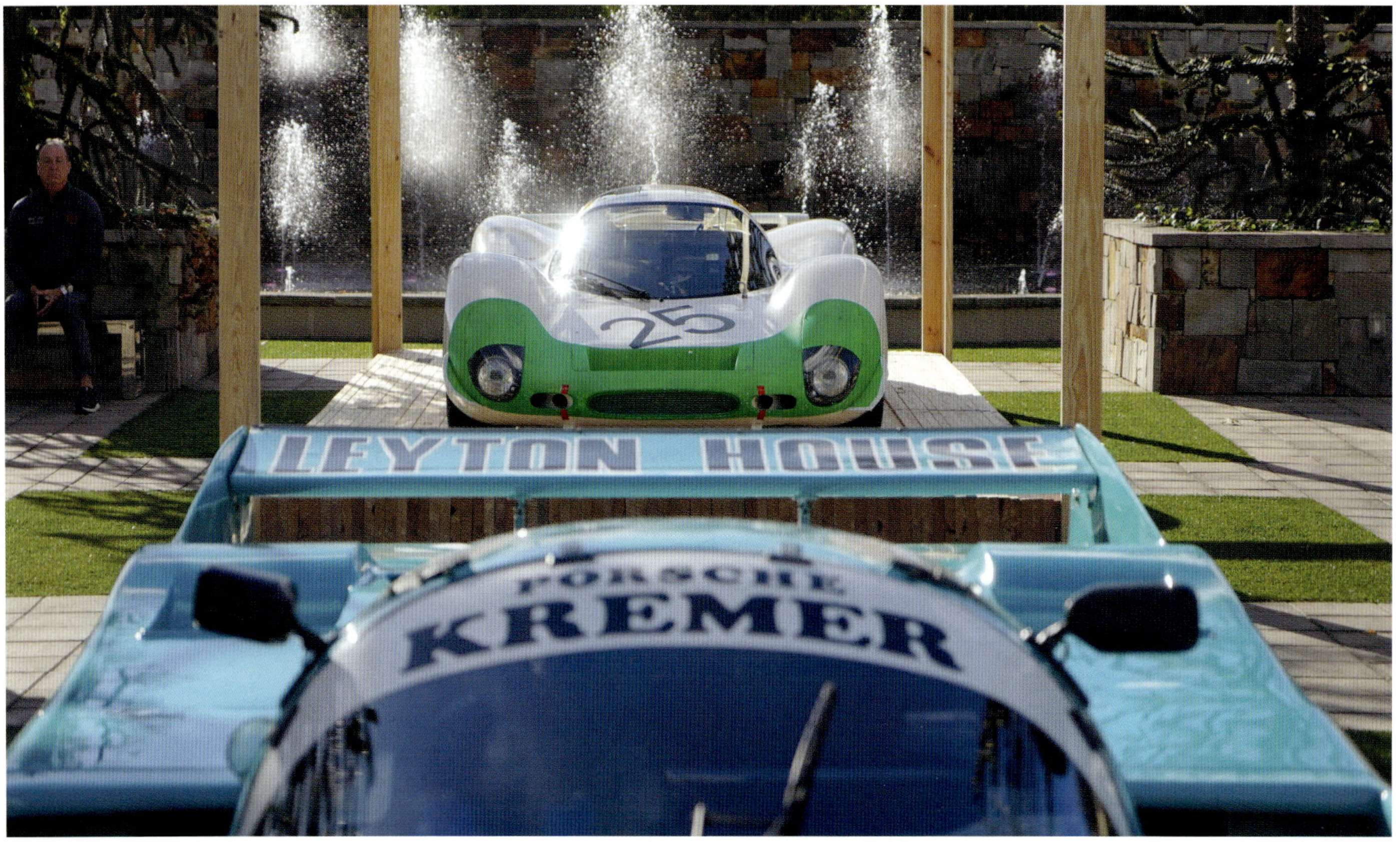

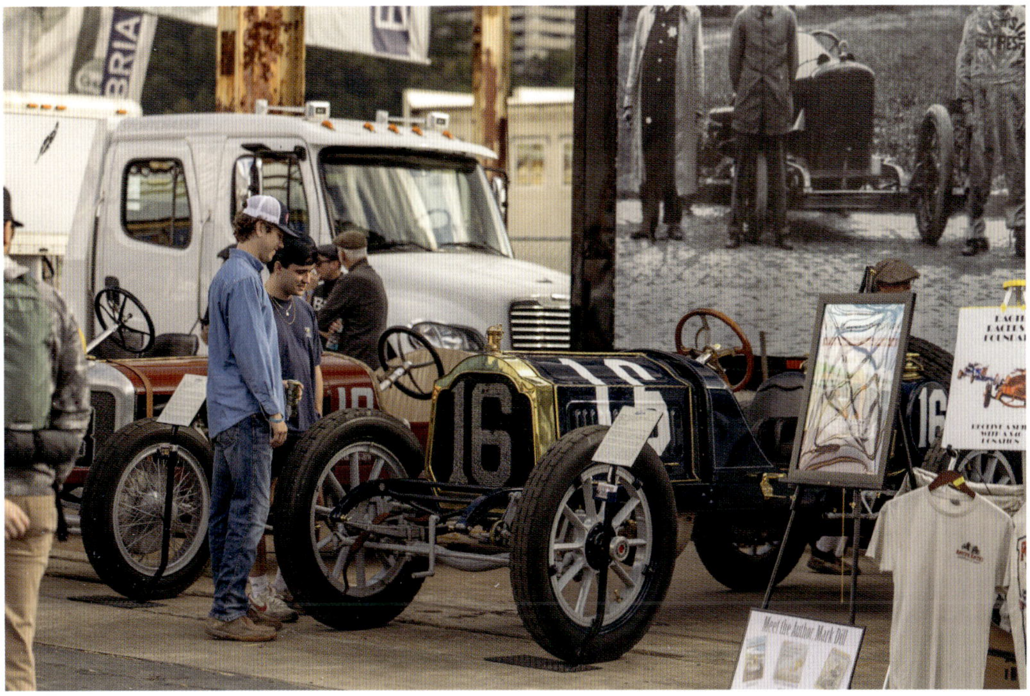


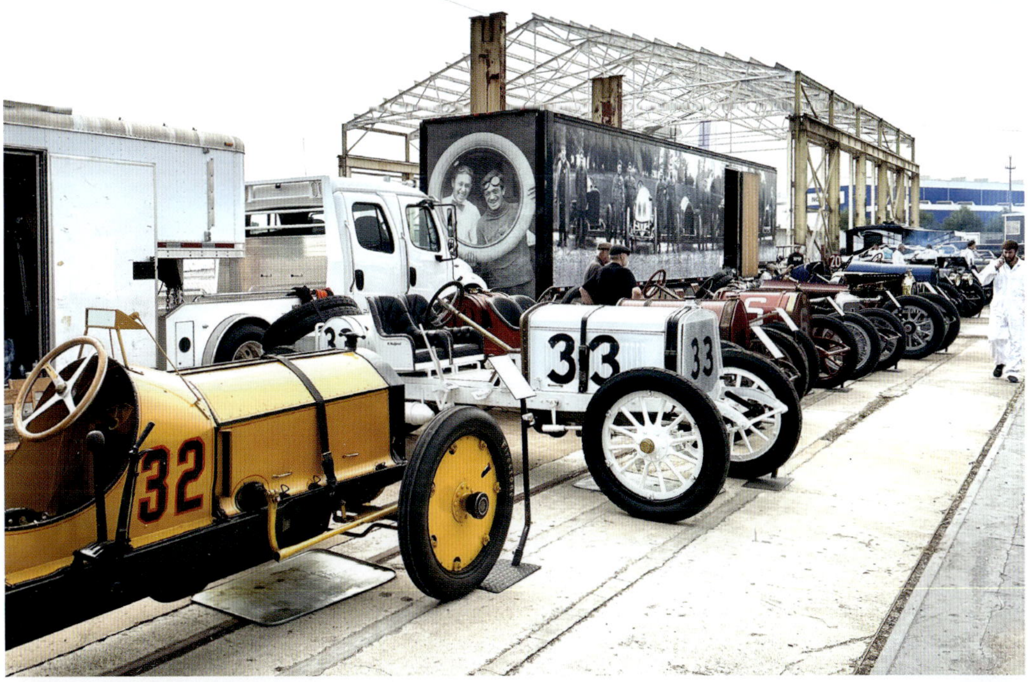


**LEFT, MIDDLE AND BOTTOM LEFT:** *The Chattanooga Concours takes full advantage of the cityscape for a busy tight setting for all the displays. (Chattanooga Motorcar Festival)*

**BOTTOM RIGHT:** *It comes as no surprise that one of the primary goals of the concours d'elegance event is to promote the Westin Hotel, owned by the event's founders, Byron and Ken DeFoor. (Chattanooga Motorcar Festival)*

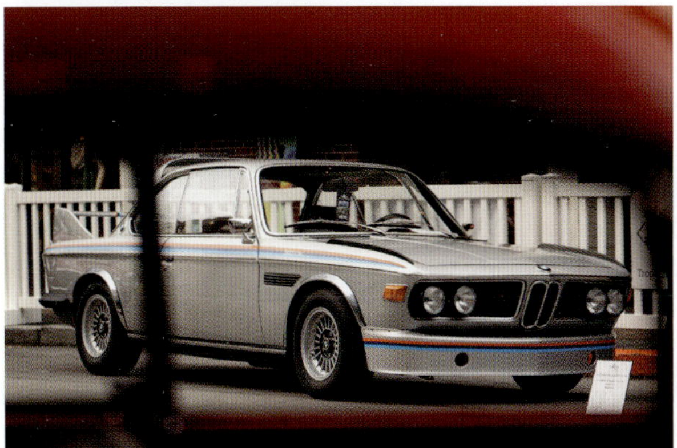

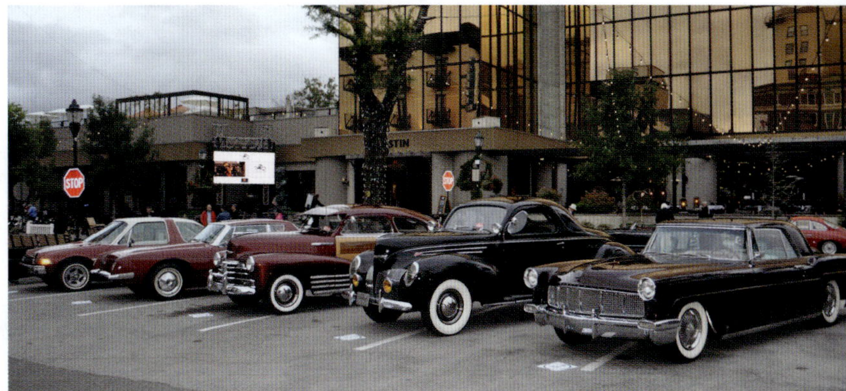

**RIGHT:** *McKeel Hagerty steers an Aston Martin as well as he has steered Hagerty.* (M. Poehlman)

**BELOW:** *At his core McKeel Hagerty is a Porsche fanatic.* (K. Keenan)

**OPPOSITE:** *For Hagerty, The Amelia is the flagship event among all that they have acquired over the last few years.* (N. Deremer)

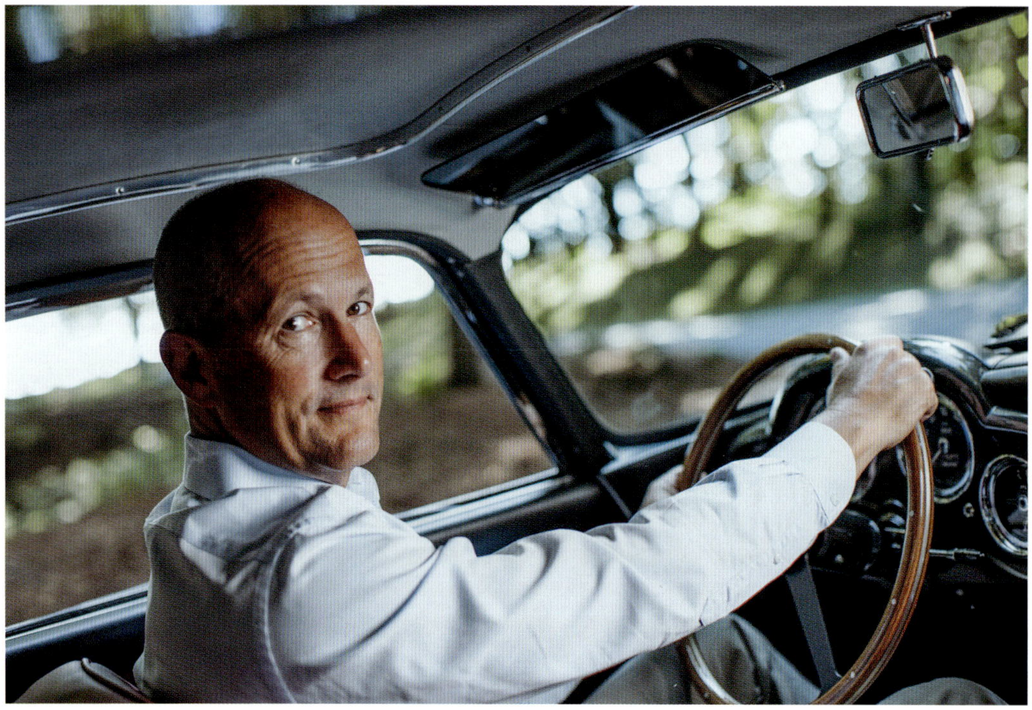

## Interviewing McKeel Hagerty

***GS: How and why did Hagerty and you get involved in concours d'elegance?***

**MH:** We have been attending concours events for over 30 years and have sponsored many of them during that time. Personally, I have been involved as a judge at concours events worldwide for nearly three decades. One of my proudest moments was being part of the inaugural group of preservation class judges for FIVA at the Pebble Beach Concours. Through our creation of the Historic Vehicle Association (HVA) in the US (with the help of Horst Brüning and Mark Gessler), which served as the FIVA ANF in North America (Autorité Nationale de la FIVA), we began promoting the concept of preservation class judging at concours events.

We have been corporately involved in these events for a long time as part of a marketing strategy, but I have also been engaged as a judge and organizer. My agenda has been to explore this idea of preservation class cars as part of US concours, which often didn't have preservation class cars. They were all about restored cars. It has been a happy evolution for me to watch more concours recognize unrestored vehicles. Historic vehicles that have not been restored are an important part of automotive history and they are worthy of preservation.

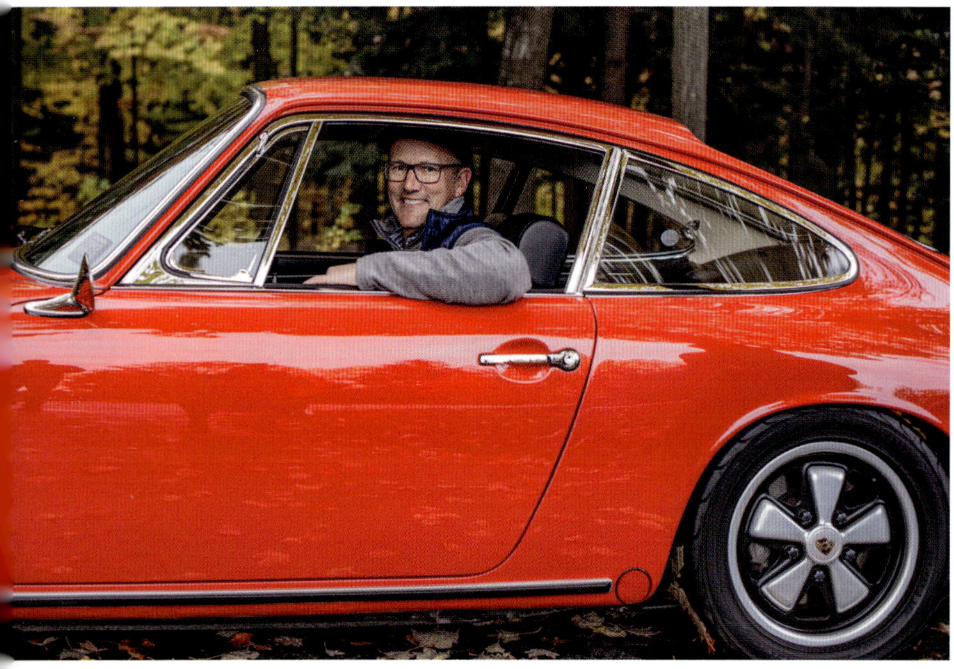

Now, in terms of concerns about the future of concours, there are real questions about the next generation of car collectors. What will they think of these events? How do we make these events relevant to them? Some people take the view that the next generation doesn't care about concours events or vintage cars. So, the presumption of some people is that they're just going to go away and that these concours will fail.

Having been a long-time sponsor, I've developed close relationships with the founders and organizers of many concours events worldwide.

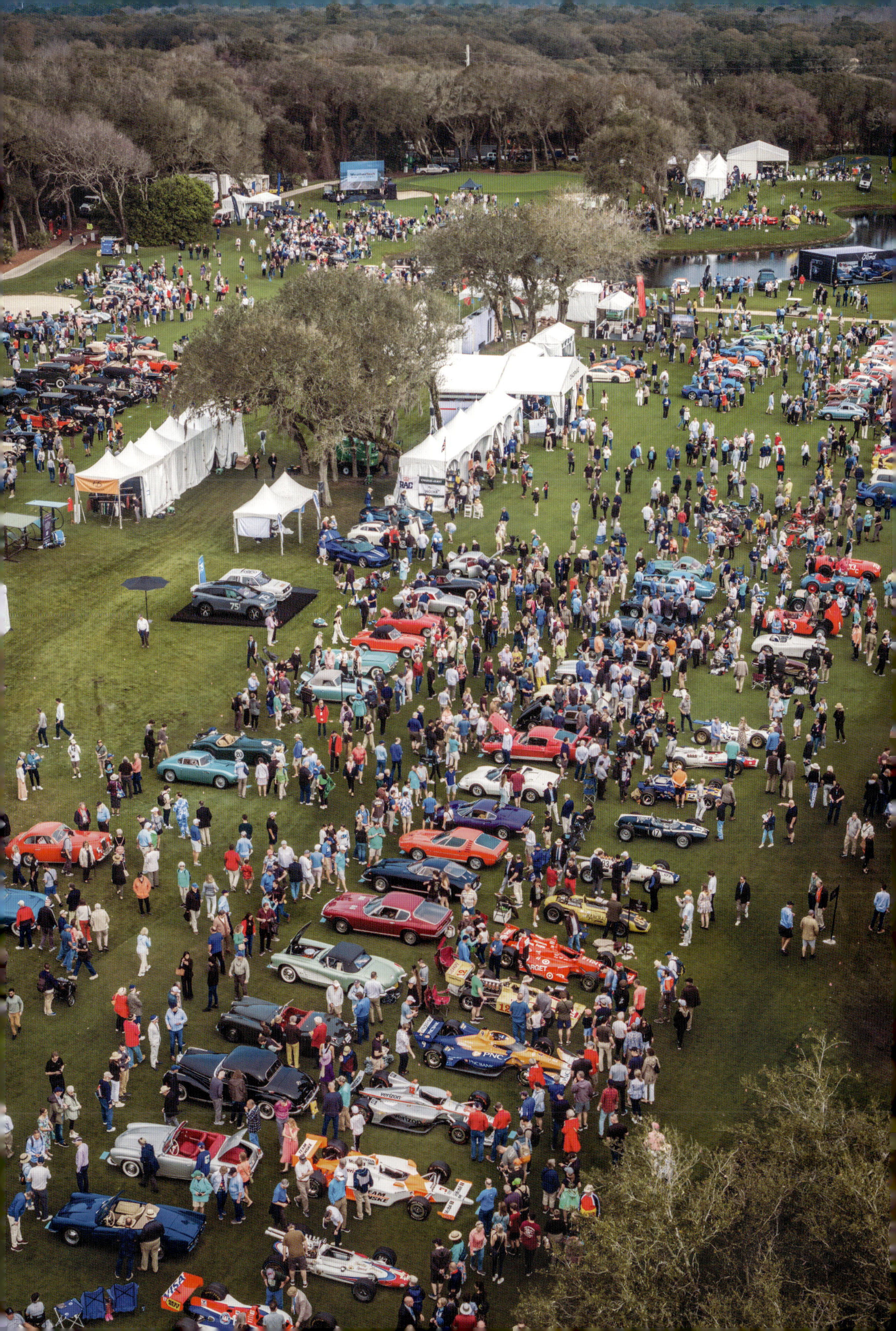

I don't see myself just as a sponsor but as a peer – professionals working together in the automotive space. I can tell you that it is challenging and expensive to run a concours. No one is getting rich running these events, and I have had many conversations with organizers about the struggles they face.

There is often no clear succession plan for what happens when the founder or long-time operator of a concours steps down. Over time, some of these organizers have approached us, asking if we would consider taking over their events, given our experience running and participating in so many. That's how we became more involved.

Greenwich and The Amelia are, of course, our big concours. The Detroit Concours, formerly known as the Concours of America, and before that Meadow Brook, is currently on pause. It was a big event around the Detroit area because of its proximity to the Big Three automakers there, but it struggled for a long time. And, despite the incredible renaissance of downtown Detroit, we have honestly struggled with it, too. So, this year, we produced one of our RADwood events there to build some momentum.

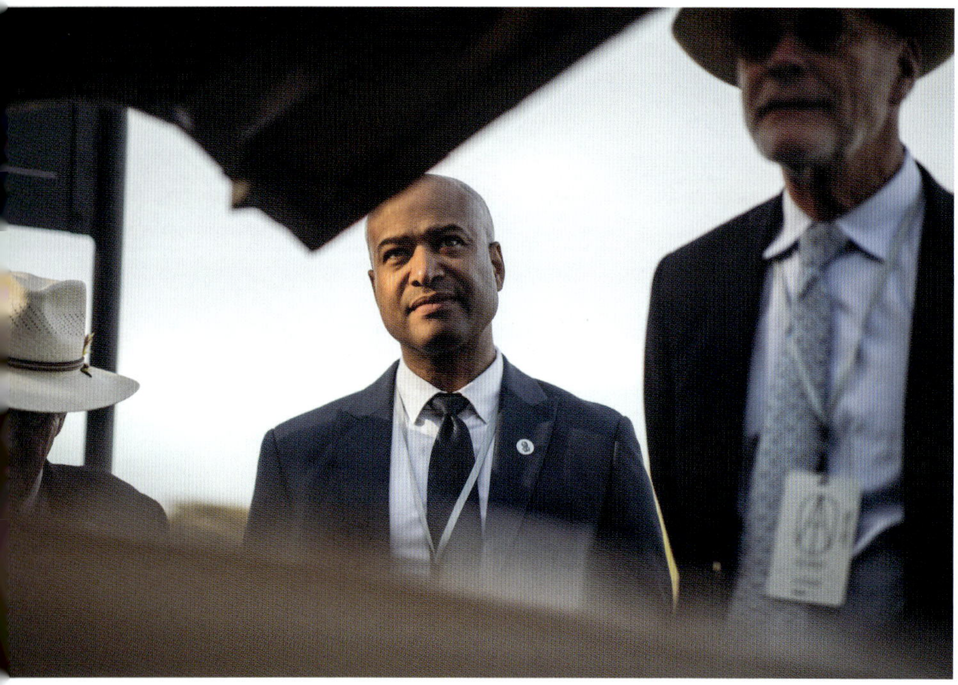

Ultimately, we got involved because we were asked to help steward these events. My business hypothesis was that by managing multiple concours, we could share the operational costs and challenges across several events, making them more sustainable. The key, however, is to maintain the unique character of each event – what makes Villa d'Este different from Pebble Beach, or Amelia distinct from Hampton Court or Salon Privé. Our challenge has been to create economies of scale while preserving the individual charm of each concours, which is what makes them special. That's the background behind why we took on this role.

**ABOVE:** *Ralph Gilles (middle), the design head of Stellantis checking out historic vehicles with barn find hunter Tom Cotter (right). Ken Gross is on the far left, partially out of sight, at the 2023 Amelia event. (M. Tierney)*

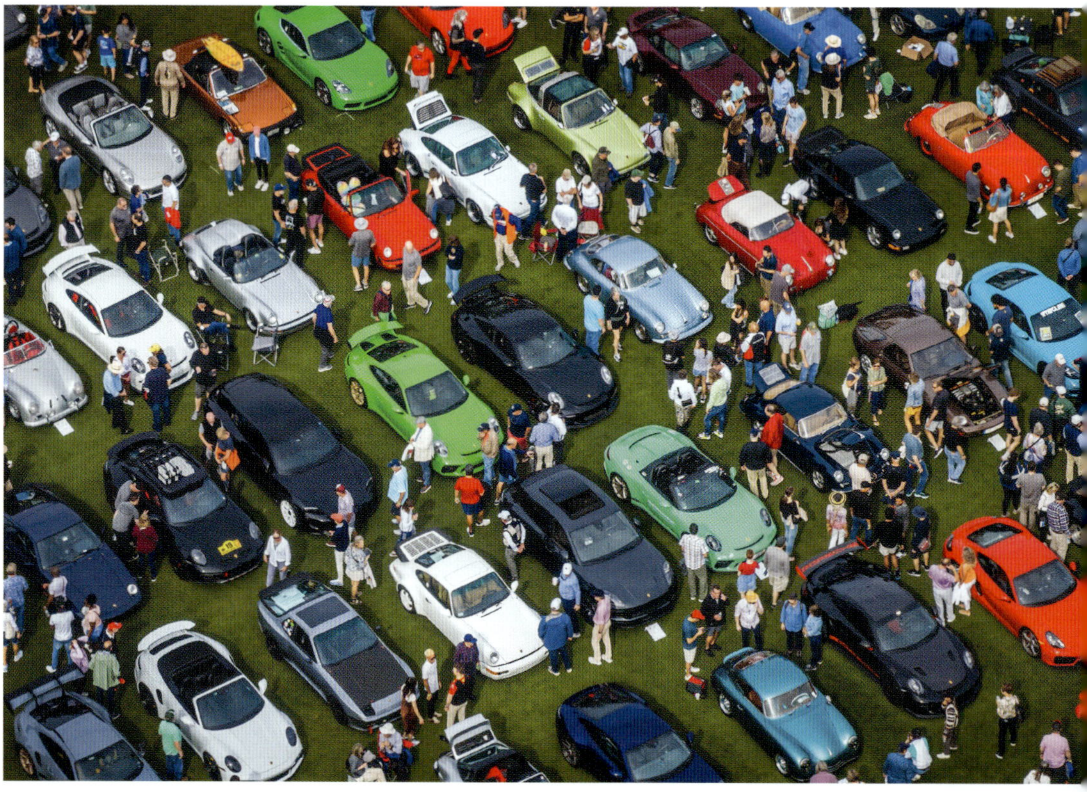

**RIGHT:** *The Amelia's Saturday Cars & Community event has consistently been a great success. Additionally, if the weather on Concours Sunday proves unfavorable, there is always the option of rescheduling the concours to Saturday. (N. Deremer)*

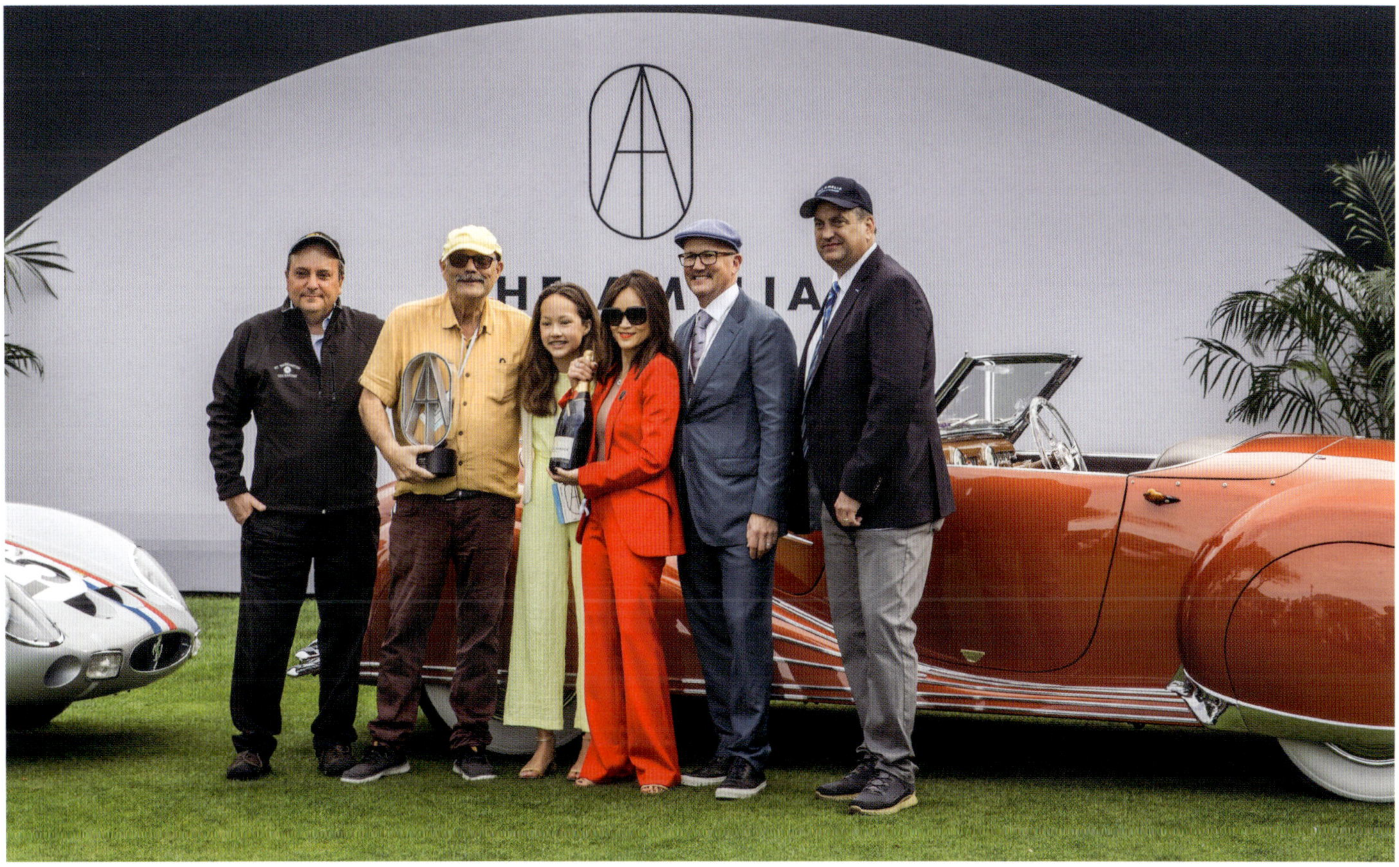


*McKeel Hagerty with The Amelia's 2024 Best of Show winners, David MacNeil and Dana Mecum, who presented a 1962 Ferrari 250 GTO and a 1947 Delahaye 135 MS Narval cabriolet, respectively. Pictured left to right: David MacNeil, Dana Mecum, Ava Hagerty, Soon Hagerty, McKeel Hagerty, and Matthews Orendac (Vice Chairman of The Amelia). (N. Deremer)*

**GS: How do you differentiate between The Amelia and Greenwich Concours? And how do they stand out from others?**

**MH:** Events have a kind of DNA to them. The Amelia, for instance, was founded over 30 years ago by Bill Warner, who is beloved in the industry. Bill had a deep passion for concours events and motorsport, so Amelia uniquely reflects that dual love. To this day, The Amelia is known as the racer's concours, featuring both a Concours de Sport and a traditional Concours d'Elegance. Keeping that emphasis on racers is important, which is why we have continued to have an Honoree from the racing world each year, such as Rick Hendrick and Jeff Gordon recently, and many of the truly great racers from the past. We bring some of their cars together. We take photos with them. We have panels with them where they talk to other racers, team owners or people connected to their careers. That is a big part of the DNA of The Amelia. I will never forget bumping into Stirling Moss in the hallway of that hotel in 2015.

There is also an interesting story that shaped the event's format. Years ago, a terrible rainstorm hit the usual concours Sunday, forcing the cancellation of the event. Bill realized then that having the option to move the show to Saturday could be crucial. So, he started a "community day" on Saturday, much like Villa d'Este's public day, which gave him flexibility. On rainy weekends, Amelia could shift the main event to Saturday, and he did it, I think, at least once or twice.

Greenwich has a different feel than Amelia. It takes place in a small, picturesque park on the harbor in Greenwich, Connecticut, right in the heart of one of the wealthiest areas in the country. The park is so small that the event can only accommodate 100 to 150 cars, which makes it more intimate than other concours. The founder, Bruce Wennerstrom, handled this by splitting it into a two-day event. We have kept that format but now dedicate day one to a Concours de Sport, featuring sports cars and modern race cars and day two to the more traditional Concours d'Elegance. There is judging on both days. The goal for Greenwich is to be recognized as THE Concours d'Elegance for the larger New York area.

One thing we are proud of is our youth judging program, something I am very passionate about. We started this nearly 20 years ago at Amelia with Bill Warner's support. The idea was to get kids involved in concours events, which traditionally have not been very family friendly. We now help run youth judging at around 75 events across the US and Canada each year. We pre-select a few cars and

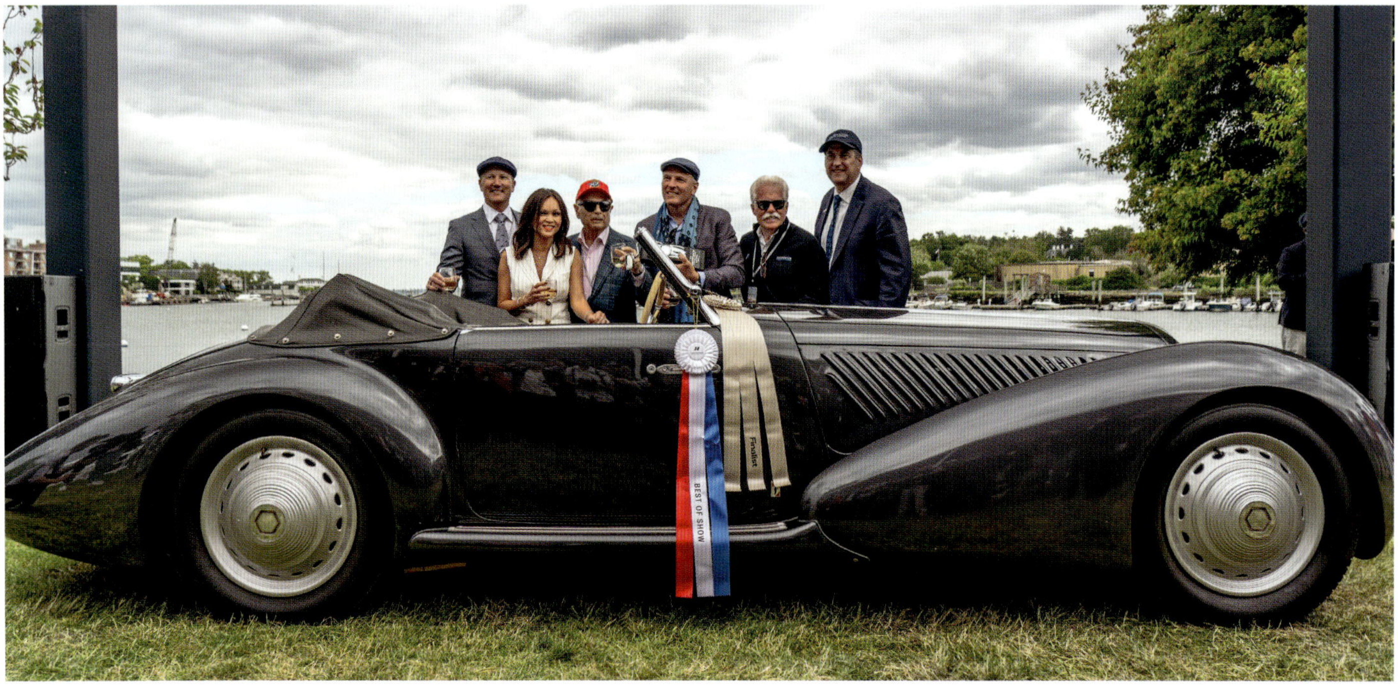

**ABOVE:** *The other major concours that Hagerty acquired is the Greenwich Concours, which remains popular. The 2023 Best of Show at Greenwich was an Alfa Romeo 8C 2900. McKeel Hagerty, Soon Hagerty, Alfa Romeo owner Lawrence Aurania, the Aurania collection Manager Michael Di Pleco, Wayne Carini, Matthews Orendac (Vice Chairman of Greenwich Concours). (X. Cesari)*

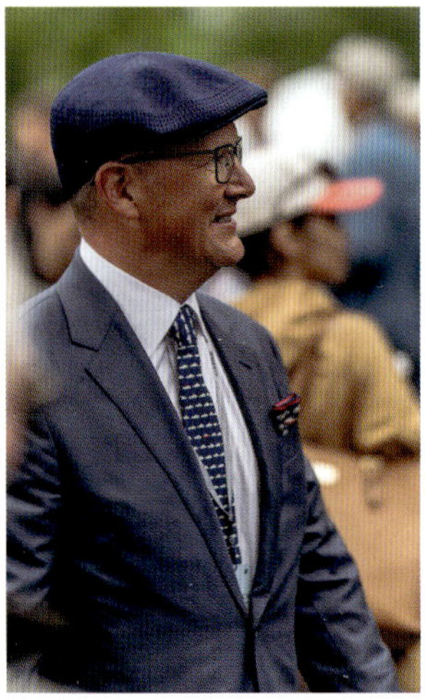

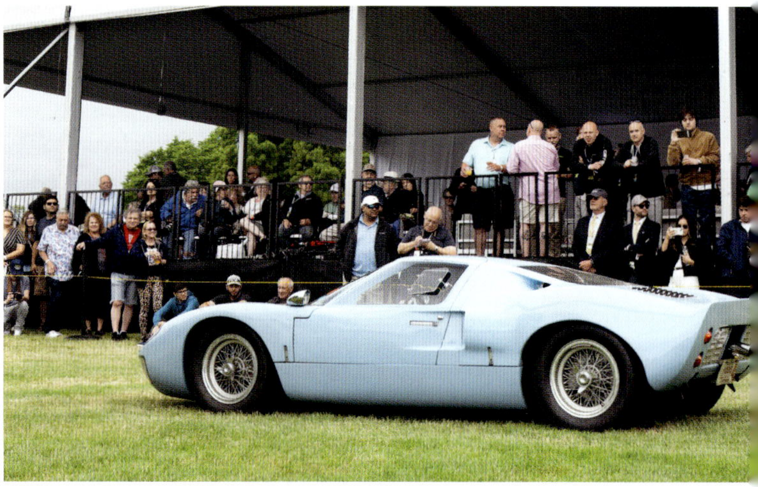

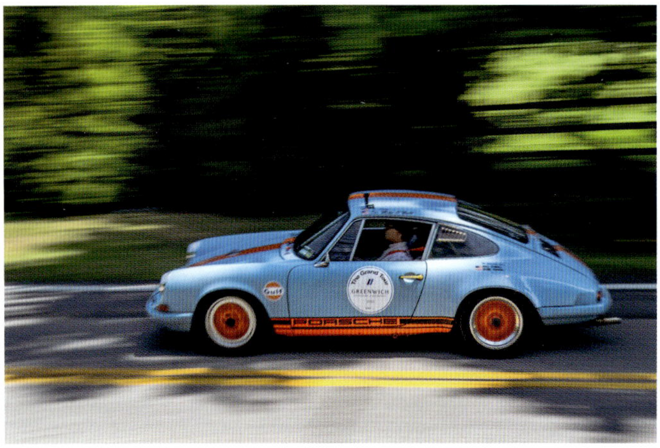

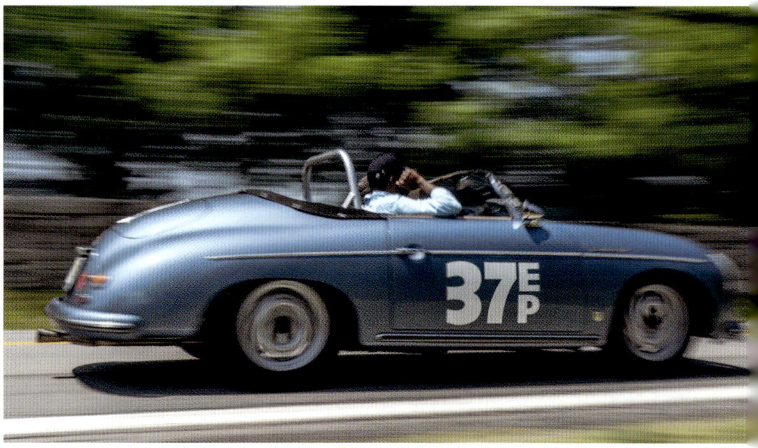



**MIDDLE LEFT:** *McKeel Hagerty embraced the spirit of the event at the 2023 Greenwich Concours de Sport. (X. Cesari)*

**MIDDLE RIGHT:** *With limited space, the Greenwich Concours focuses on quality. At the 2023 Concours de Sport, a Ford GT40 emerged as the winner. (J. Sweeney)*

**ABOVE (BOTH):** *Greenwich also organizes a Grand Tour the day before the concours. (X. Cesari)*

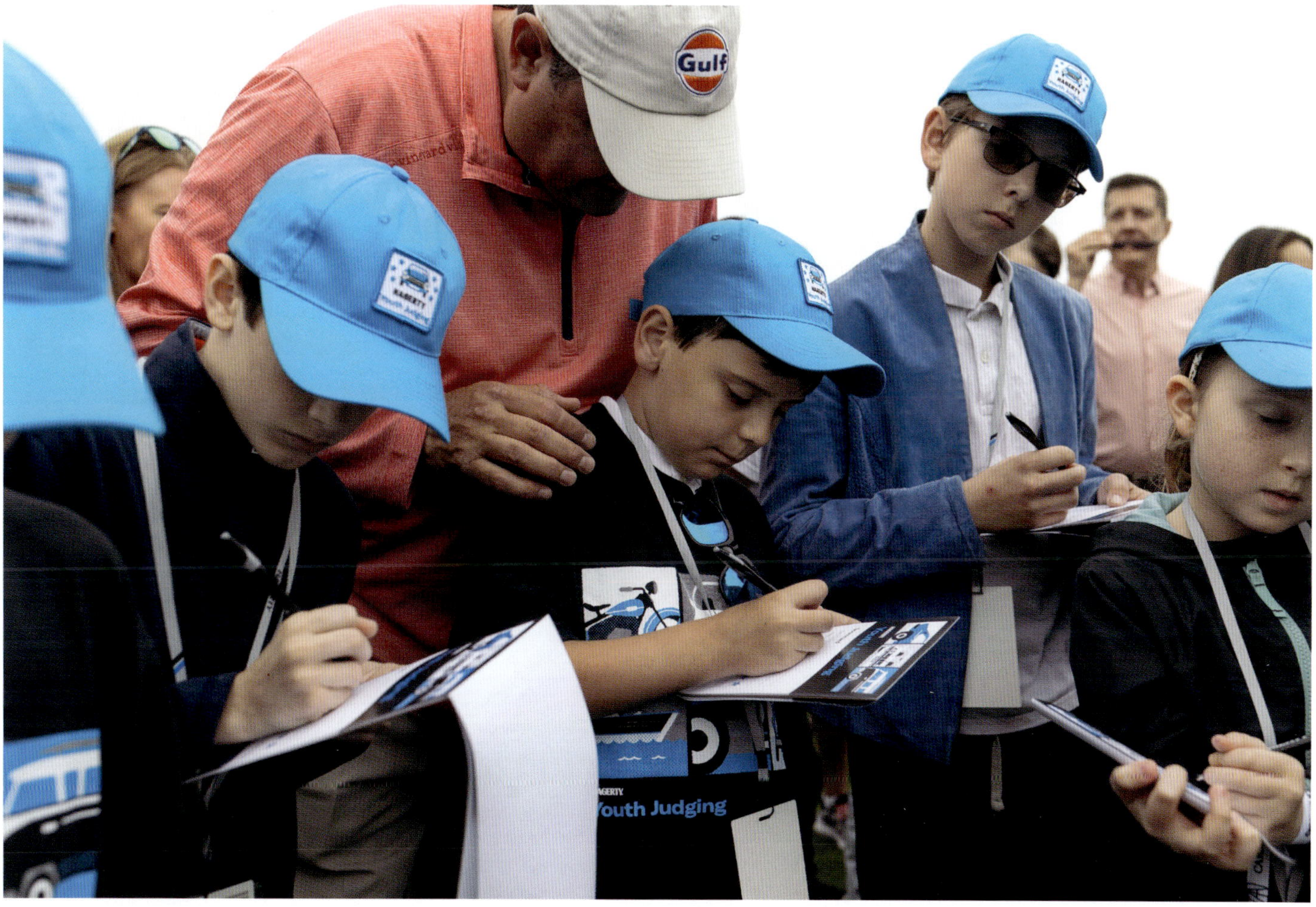


*One of the most intriguing initiatives at The Amelia has been the youth judging system, a program that Hagerty is also supporting at several other concours events. (K. Keenan)*

involve 20 to 30 kids, often with owners who are more approachable. It is a way of teaching the next generation about these cars and making the event more engaging for families.

Another improvement we made when sponsoring events was to remove the velvet ropes around the cars. Some events, including Greenwich, had velvet ropes around the vehicles, and that made the event feel less accessible. If we can't teach people, especially kids, how to respect cars up close, then that's on us. No car is too valuable to be kept behind ropes – except something like the Mercedes-Benz 300 SLR that Stirling Moss drove!

**GS: The Amelia and Greenwich focus on showcasing exceptional cars, but you also support events like the Festival of the Unexceptional, which is on the other end of the spectrum. Why are you involved in such events?**

**MH:** The Festival of the Unexceptional and similar events like the Concours d'Lemons represent the opposite end of the spectrum from high-end concours. The idea behind Concours d'Lemons was created by Alan Galbraith, who envisioned an anti-concours out in Monterey, up against Pebble Beach. The idea was for people to bring the worst examples of anything they could find and hold the event in a dusty parking lot rather than a leafy golf course. It was all about having fun with humorously named classes like "Needlessly Complicated Italians" and "Swedish Meatballs." We got involved with it to show that while we sponsor prestigious events, we also appreciate the lighter side of car enthusiasm. It's a way to have fun and not take ourselves too seriously.

In the UK, we found that the term "lemon" didn't really resonate, so we developed the idea of the Festival of the Unexceptional, which celebrates practical, everyday cars – what many call "practical classics." These are cars that people still drive, not because they are valuable, but because they love them. Many of these cars have become rare over time, even more so than some exotic vehicles, simply because they were everyday cars that have now disappeared. The event is held at Grimsthorpe Castle, which is a stunning venue, making it feel like a proper concours – but for very ordinary cars.

**TOP:** *When was the last time you spotted a first-generation Ford Fiesta? The Festival of the Unexceptional has been a remarkable success in the United Kingdom.* (M. Kimberley)

**MIDDLE:** *Or an Austin Metro…* (M. Kimberley)

**BOTTOM:** *Set against the stunning backdrop of Grimsthorpe Castle, the Festival of the Unexceptional has been a tremendous success for Hagerty.* (M. Kimberley)

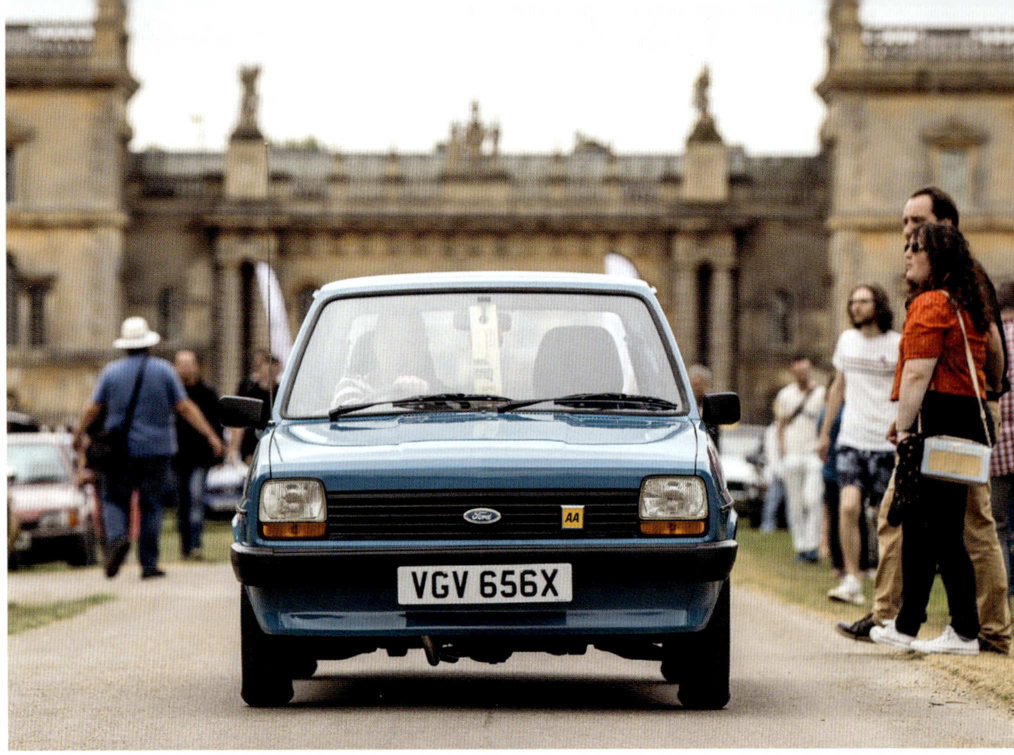


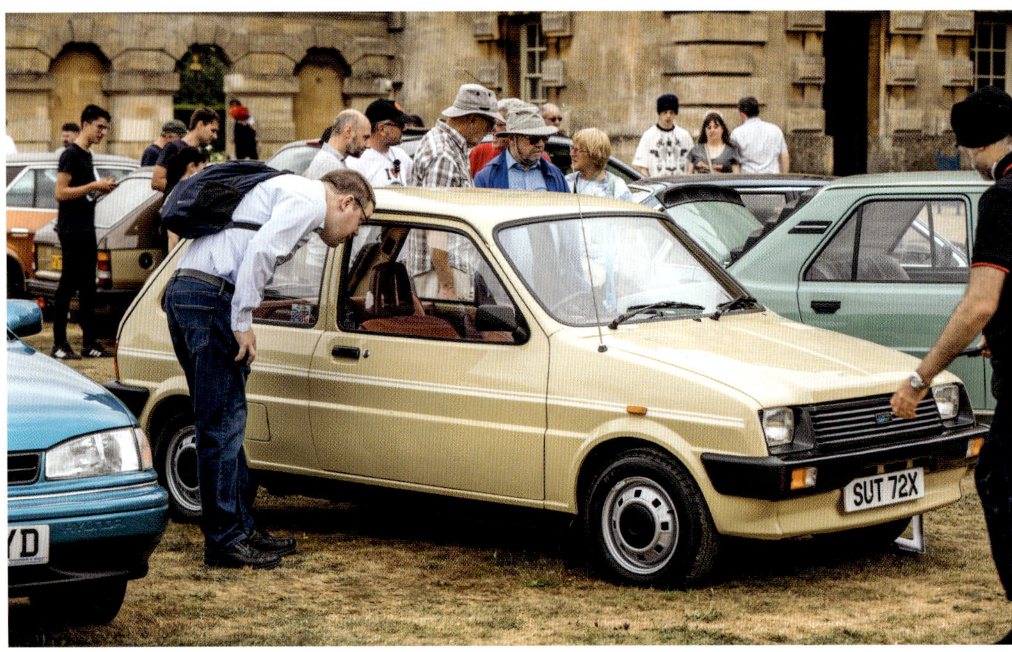


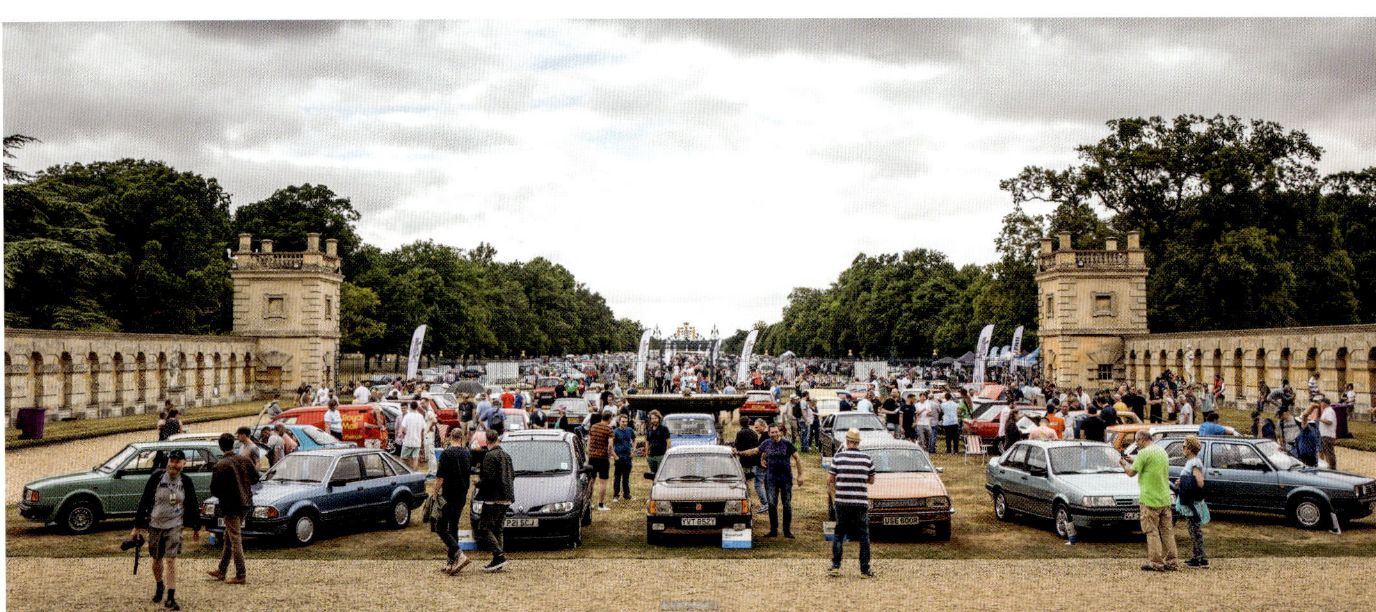

And what's amazing is that people show up with well-preserved, older cars that might have been inexpensive when they were new and still are not worth much, but they hold a lot of sentimental value. That's what we're celebrating. I love the spirit of this event.

***GS: These unexceptional cars are mostly ordinary vehicles that have become rarer than some exotics. Would you consider taking this concept to other parts of the world?***

**MH:** Yes, I think we would like to do that. It makes so much sense. There are probably more Lamborghini Miuras than there are Chrysler K cars left. It couples with another venture of ours, RADwood, which focuses on cars from the 1980s and 1990s. These cars are not old enough to be considered true classics yet, but they are certainly not modern either. RADwood events bring together people who love these cars, and there is a lot of nostalgia and humor involved. People come dressed in '80s fashion with '80s music playing in the background. It is light-hearted but not cynical. We are running six or seven of these RADwood events now. While they are not judged in a formal concours style, they still encourage people to bring the best of the best.

**BELOW (BOTH):** *Hagerty also introduced a RADwood on Saturday at The Amelia which has been a resounding success. (N. Deremer)*

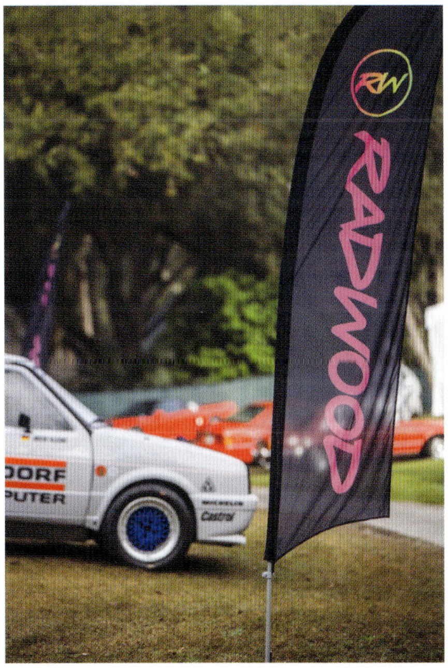


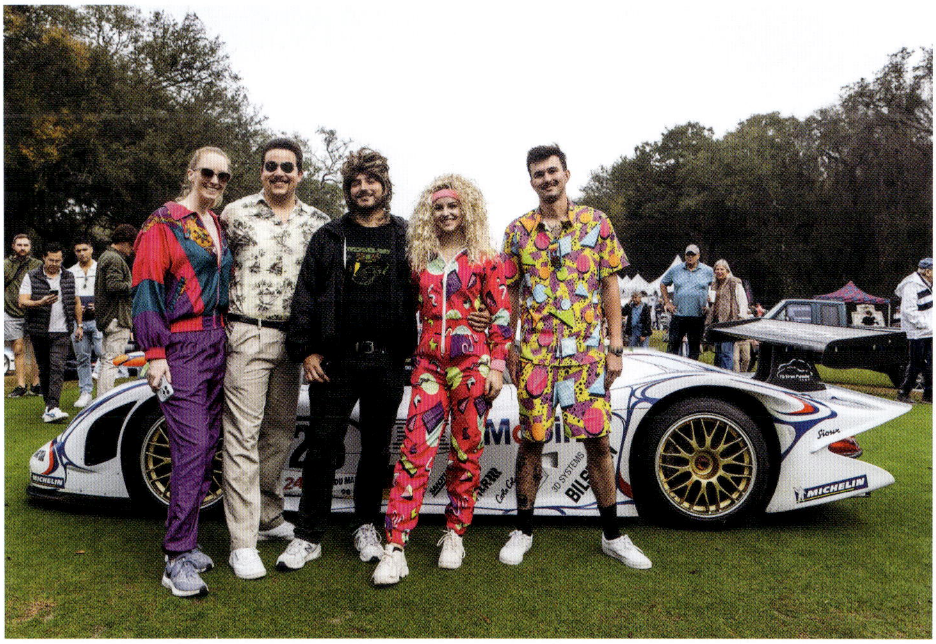


***GS: To remain popular in the modern age, how can concours events adapt? And which ones set the standard?***

**MH:** The two events that set the highest possible standard still are Pebble Beach and Villa d'Este. They get the best of the best. I like to think that The Amelia is just one tier down from that. But those are the best, and everybody copies them for what they are. I think the challenge all of them are facing is, how do we keep running the same playbook every year? Or do they start introducing other elements into them to make them a little bit more approachable? You see everybody wrestling with it. But it is unevenly decided. So, for example, do you play classical music at the event, or do you play other kinds of music? Do you start introducing some classes of cars that are a little newer and appeal to more people, or do you just stick to the older cars?

The Amelia is bigger every year. But I am worried that we need to make it even friendlier for the next generation. For example, many of the best events are not always couples-friendly. There are a lot of events that are great places for a group of guys to go to experience cars and go to the auction at night and do things. But it really was not much of a couples event, and I am intrigued by some of the events that address that. The Quail, for example, has really gotten that right. It is the most couples-friendly event in Monterey.

There is also a significant financial challenge in that sponsorship support is challenging for these events.

With so many challenges in the larger industry, it is difficult to get automotive OEM sponsorship at a high level. The car companies just do not support these events now. They will come to certain events,

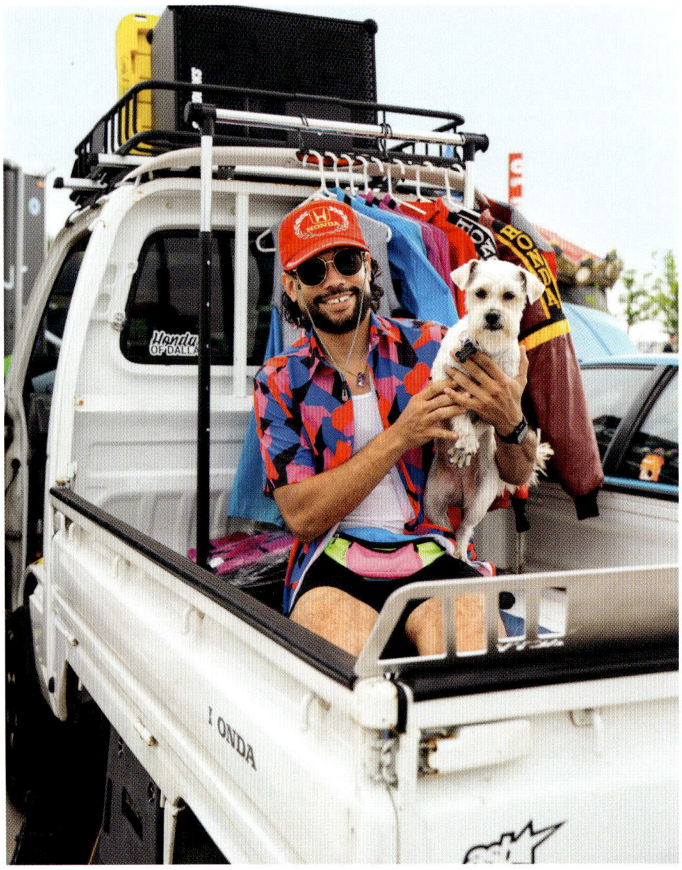


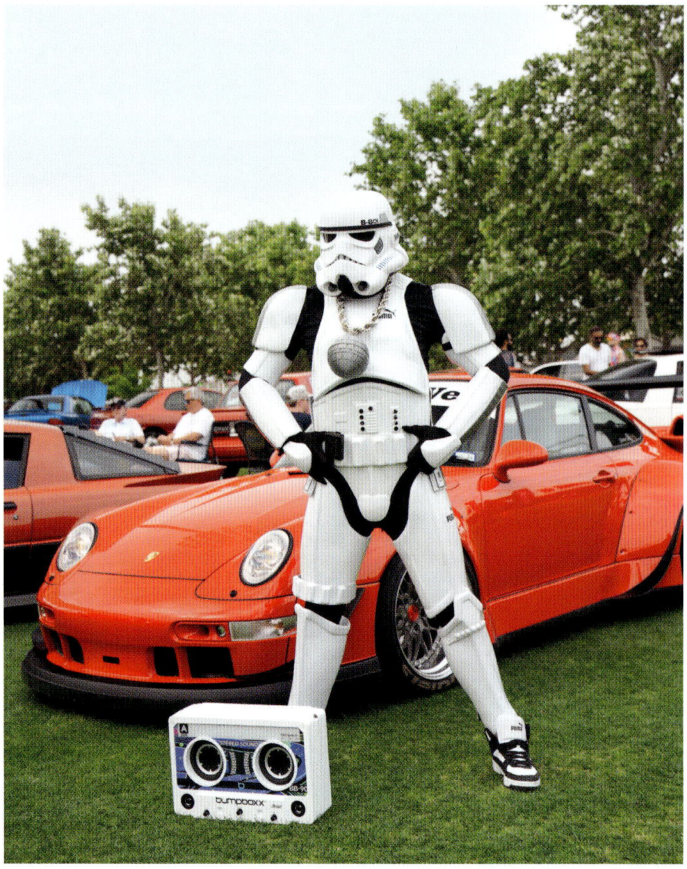


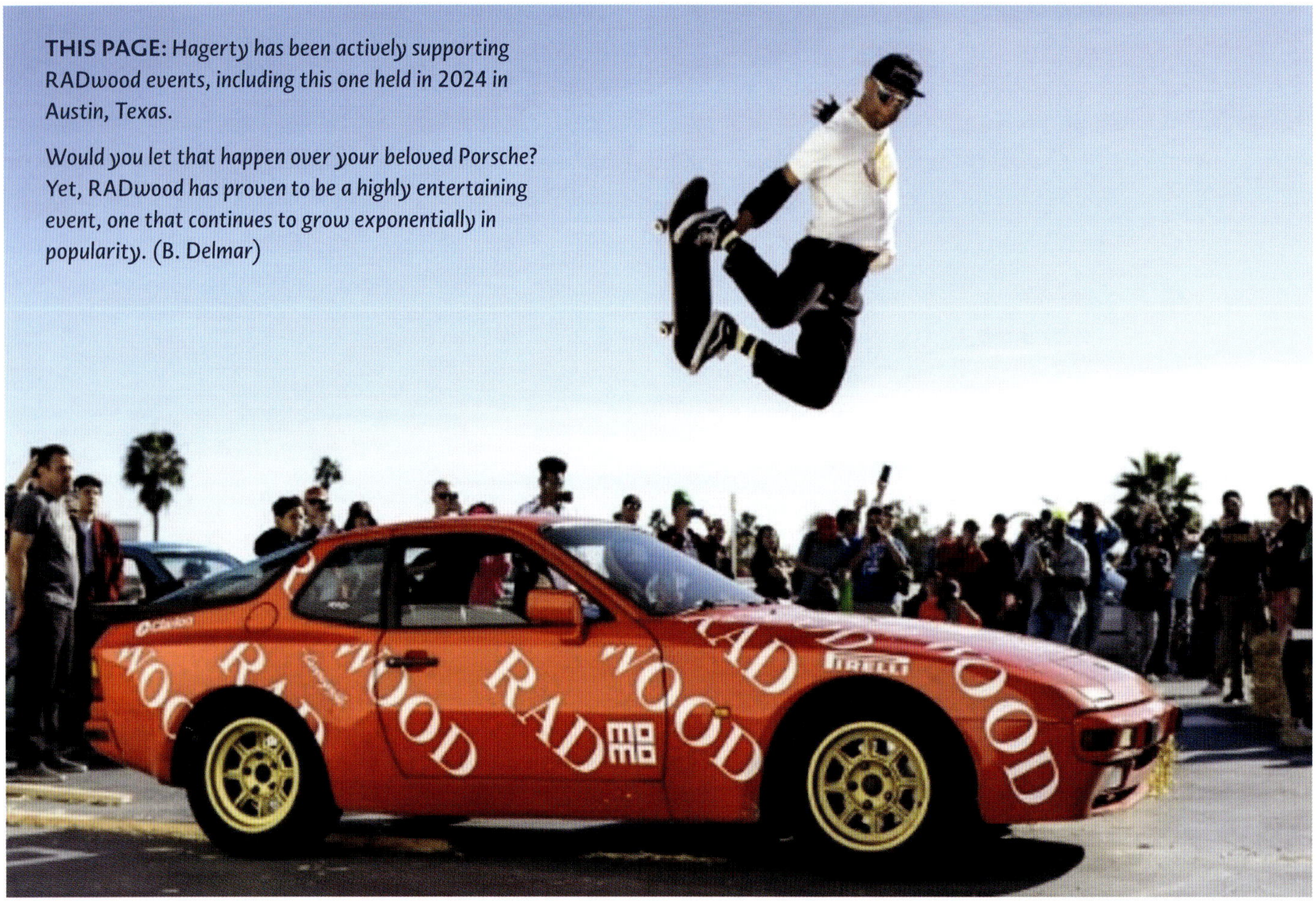


**THIS PAGE:** *Hagerty has been actively supporting RADwood events, including this one held in 2024 in Austin, Texas.*

*Would you let that happen over your beloved Porsche? Yet, RADwood has proven to be a highly entertaining event, one that continues to grow exponentially in popularity.* (B. Delmar)

say Pebble Beach, where it is easier for them to entertain dealers and talk to customers. So, there is a sponsorship platform for these things, but they are difficult financially.

There's also a lot of fatigue for people who have car collections. A lot of the older collectors just do not send their cars to events as often as they used to because of the expense. It is common for us, when we invite a car, for somebody to say, 'Sure, I'll send the car, but you're paying for it," but it just won't work if cars must be paid for. Big collections are going to have to come to grips with the fact that if they are ever going to be seen, they are going to have to spend a significant amount of time and money each year shipping them around. I think it will mean that it will be increasingly challenging for certain events to survive.

**GS: This year (2024) at Villa d'Este, Pebble Beach, and eventually Chantilly, we saw preservation class cars winning Best of Show. Does this validate what you and FIVA have advocated for, and do we see a significant shift?**

**MH:** I would like to think our work has been effective. We have more judges now who can assess preservation. We have concours organizations that recognize that preservation cars are a little different and they either need to be in their own class or must be recognized a little bit differently. At The Amelia, we have two Best of Shows: one for concours d'elegance and another for concours de sport, which focuses on racing cars. I think more concours will add either a second or a third category for best-preserved car or best historic car. That was my dream from the beginning. It really wasn't to have preserved cars compete with restored cars. It was to recognize that they are different. The minute you restore a car, you re-manufacture it to some standard, sometimes better than when it was born. And I love them. They are beautiful. Who doesn't love a beautifully restored car? But a preserved car is a different beast.

Take for example the Bugatti Type 59 that won Best of Show at Pebble Beach. I know that it is not necessarily a great-looking car to some. It is not an elegant car. It may not even have been the best-preserved car. But, as the process goes, it captured the hearts of enough Chief Class Judges to win the big prize.

A perfectly restored car is about how it's presented, and a perfectly preserved car is about its history, its story, and about how people cared for it. For the best-restored car, the trophy goes to the owner, and for the best-preserved car, it goes to the car's history. It goes to everybody who had something to do with that car. I don't think it sells a lot of tickets to events, though.

**GS: As a judge, entrant and concours operator, what are your views on the Modernized 100-point judging system (M100)?**

**MH:** I like it. It's critical for our world to evolve and keep pace with the times. ICJAG is a good system for scoring completely restored cars of similar quality and character, but less suited for preservation cars where history and originality mean more. The modernized system strikes a good balance between creativity and technical excellence. The idea is that we should be looking at a vehicle as a whole. That means it's not only an engineered machine but also has a certain design, provenance and impact. For instance, it may have technological significance or a super interesting story. As judges, we need to take these things into account. M100's strength is it makes room for subjective factors. The result is a better, more interesting competition, in my view.

**GS: One last question: there is growing pressure to reduce the number of cars on the road, particularly older ones, especially in cities. Do you think concours d'elegance events might face push-back due to concerns over emissions and their elitist image?**

MH: There are two main pressures, which vary depending on the political climate. One is the environmental concern, and the other relates to income inequality. Motorsport events, for example, are experimenting with their environmental impact by exploring options like synthetic fuels. For our touring events, such as the California Mille, we have started using carbon offsets. We also want to experiment with the idea of offsetting participant trips.

Personally, I am glad we can take these steps. However, the income inequality issue in certain countries worries me. I had a negative experience during a car tour in Belgium, where people threw rocks at the cars because they represented the social abuses of wealthy people. We must be mindful to spend a little more to make these events environmentally friendly, as affordable as possible, and to demystify why these cars still exist. It needs to be about a balance of passion, culture, and only then, about the value of these cars.

At The Amelia, for instance, we do not serve bottles of water. Instead, we have water bottle stations. I am also planning to experiment with synthetic fuels in some of my cars at events to see how feasible it is. It costs about US$50 per gallon, but since we use fewer than 30 gallons a year for concours cars, it is manageable.

# Concours in the Rest of the World

## GOING INTERCONTINENTAL

This level of meticulous restoration-focused judging, down to the nuts and bolts, is also evident, with several European concours adopting the concours d'état scoring system. The concept has clearly gained traction in Europe, where the number of concours events rivals, if not exceeds those held in North America, which hosts over a hundred annually.

The United Kingdom was among the first European countries to embrace this post-war American passion for concours d'elegance, with the Salon Privé launching in 2006 at the Hurlingham Club in London and eventually ending up at Blenheim Palace. Until 2024, Salon Privé used the ICJAG scoring system, with founder and Chairman Andrew Bagley emphasizing that it is "one of only four ICJAG Plus-designated events globally, and we are immensely grateful to the 30 international experts who join us each year to judge." The chief judge for Salon Privé has been the erstwhile chairman of ICJAG, Nigel Matthews.

As Salon Privé grew in popularity and evolved into a high-profile event for also unveiling new luxury car models, a rival event, the distinctly British Concours of Elegance (a deliberate departure from the French term Concours d'Elegance used by others) launched its first event at the majestic Windsor Castle in 2012. It then moved to St. James's Palace in 2013, followed by Hampton Court Palace in 2014 and the

**BELOW (BOTH):** *One of the first to capitalize on the renewed interest in concours d'elegance was Salon Privé, in the United Kingdom. This is from September 2012. (Hugues Vanhoolandt)*

**BOTTOM:** *In no time Salon Privé had direct competition from the Windsor Concours of Elegance, which for several years was timed to be on the same date as Salon Privé, usually in September. This is from 2012. (Hugues Vanhoolandt)*

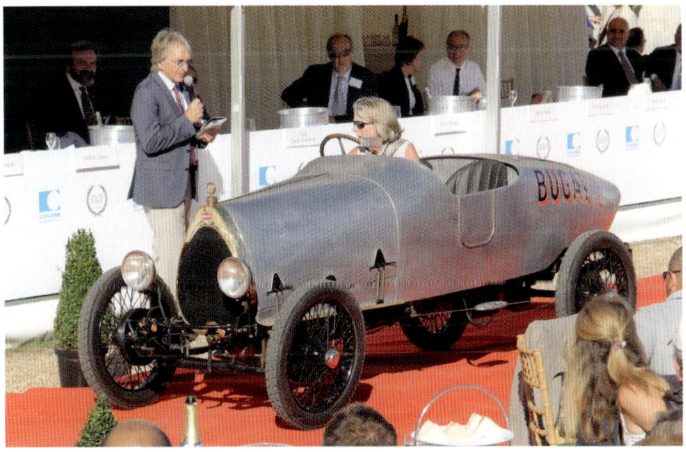

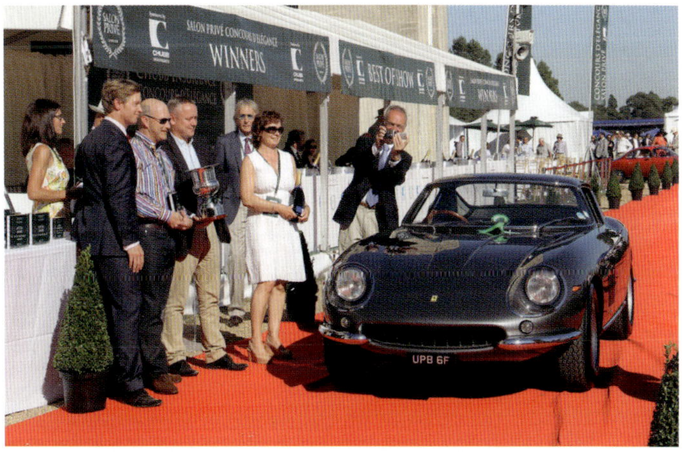


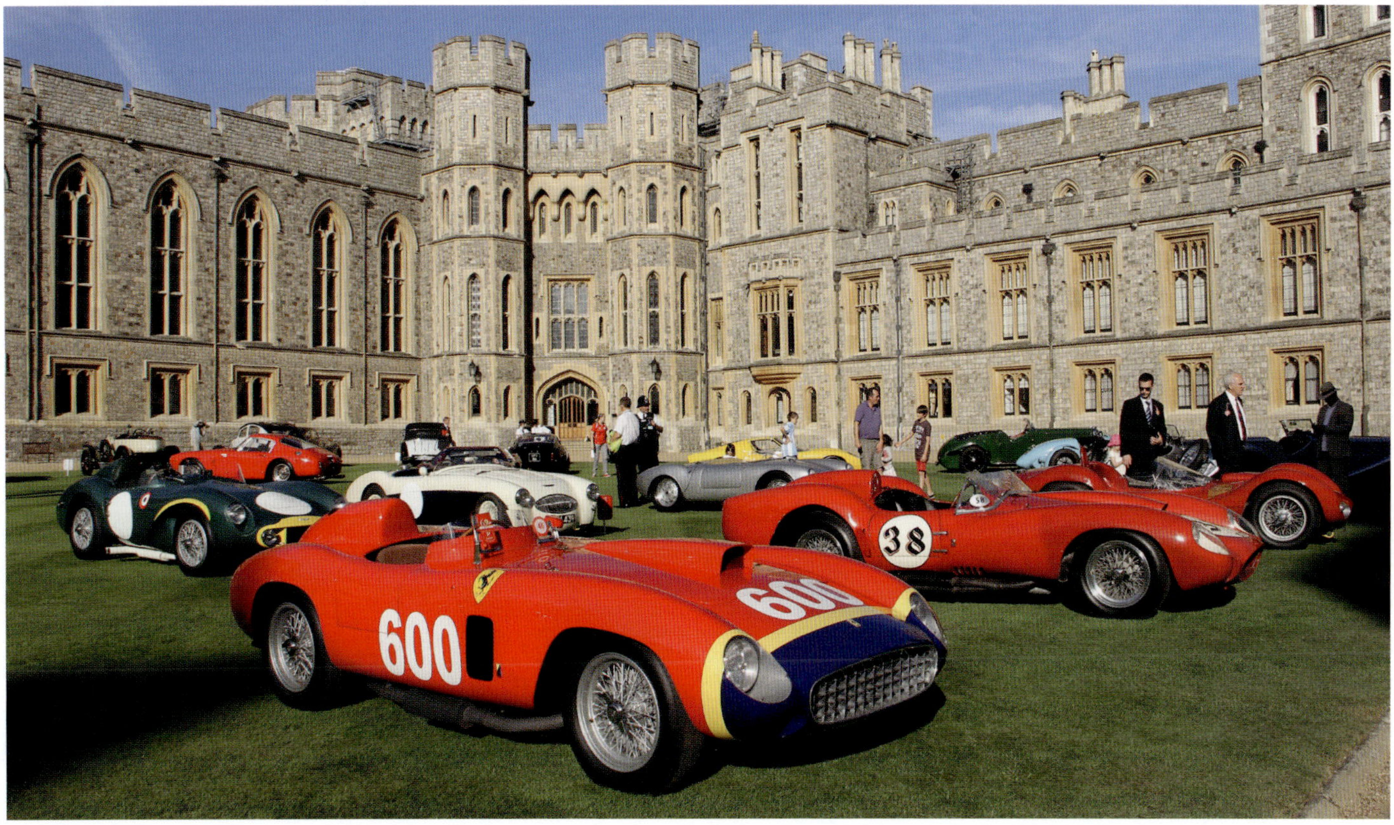

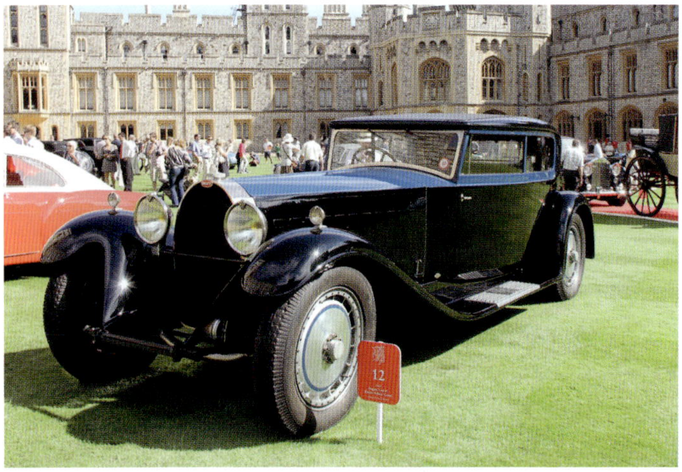

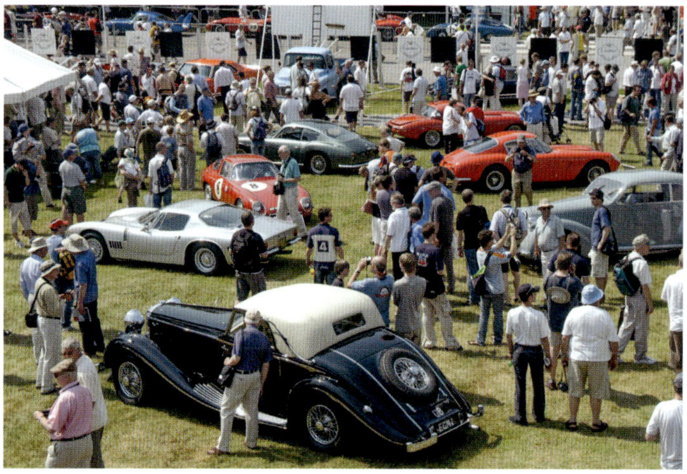

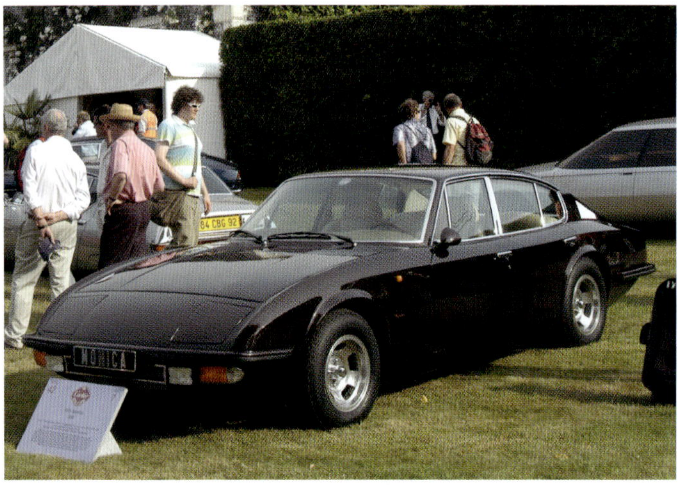
MONICA

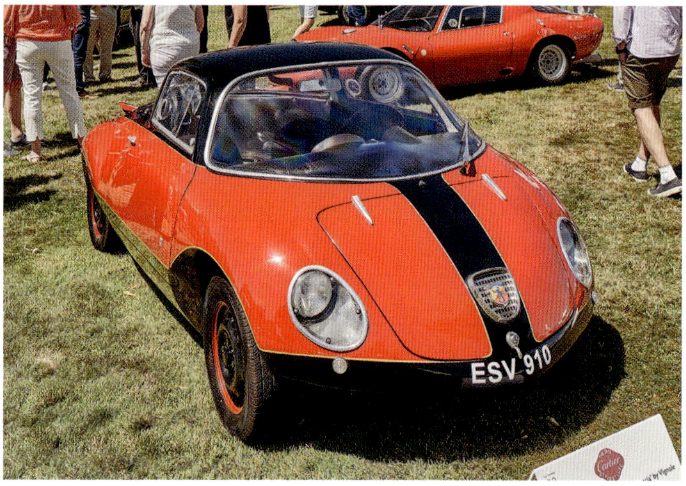
ESV 910

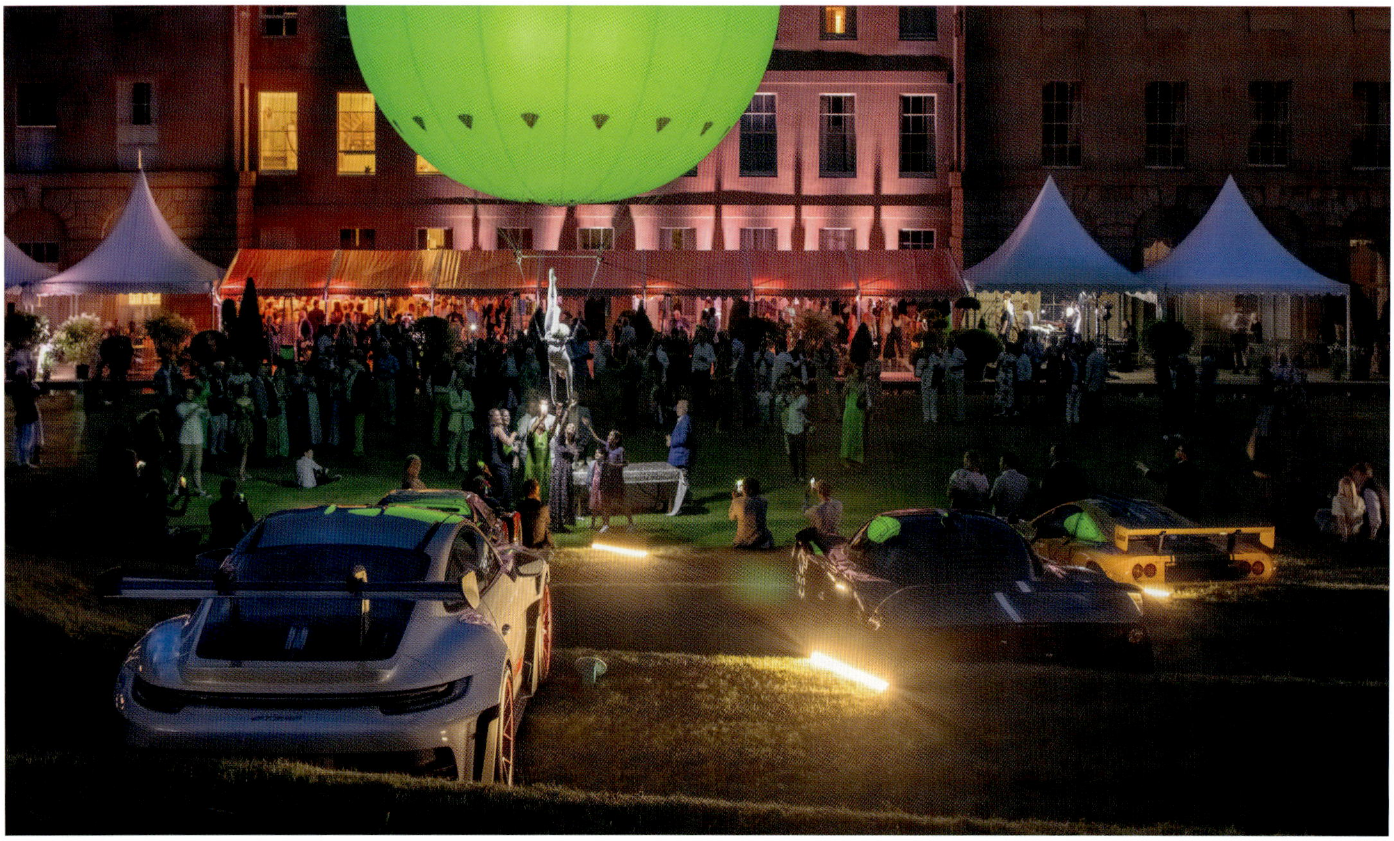

Palace of Holyroodhouse in Edinburgh (the official Scottish residence of His Majesty The King) in 2015. In 2016, the event returned to Windsor Castle before settling at Hampton Court Palace from 2017 onwards, where it has remained ever since.

Competing with these two events, there are several other concours d'elegance in the UK, such as the Concours on Savile Row, the London Concours in the heart of the Square Mile, and the Cartier Style et Luxe, which takes place at Goodwood, during the Festival of Speed, in July. Another notable event making headlines is the Heveningham Concours d'Elegance. Launched in 2016, it has quickly become one of the country's premier annual gatherings for historic vehicles and airplanes.

"Founded by Jon and Lois Hunt, the owners of Heveningham Hall in Suffolk, the event was designed as a refreshing alternative to the traditionally formal concours often seen in the UK," says Anthony Peacock, Managing Director of Mediaworld, which helps organize the event. "The goal was to attract a younger audience, a vision shared by the couple's sons, Harry and Max, who are deeply involved in organizing it. Typically, around 50 historic vehicles are invited to participate, along with about twenty aircraft, many from the World War II era or earlier."

"The planes land on a grass strip and fly in to compete," Peacock adds. "The emphasis is on design, with jury members often including renowned designers like Peter Stevens. The scoring system is less rigid compared to other events, valuing the history and design of the cars as much as their condition. Another highlight is Horsepower Hill, an eighth mile sprint up the estate's main drive, which attracts both classic cars and modern supercars. There's also the Heveningham Tour, a 50-mile drive through the Suffolk countryside for concours participants, with a lunch stop along the way. The event coincides with the annual Heveningham Hall Country Fair, creating a weekend of automotive excitement for both younger audiences and serious car enthusiasts."

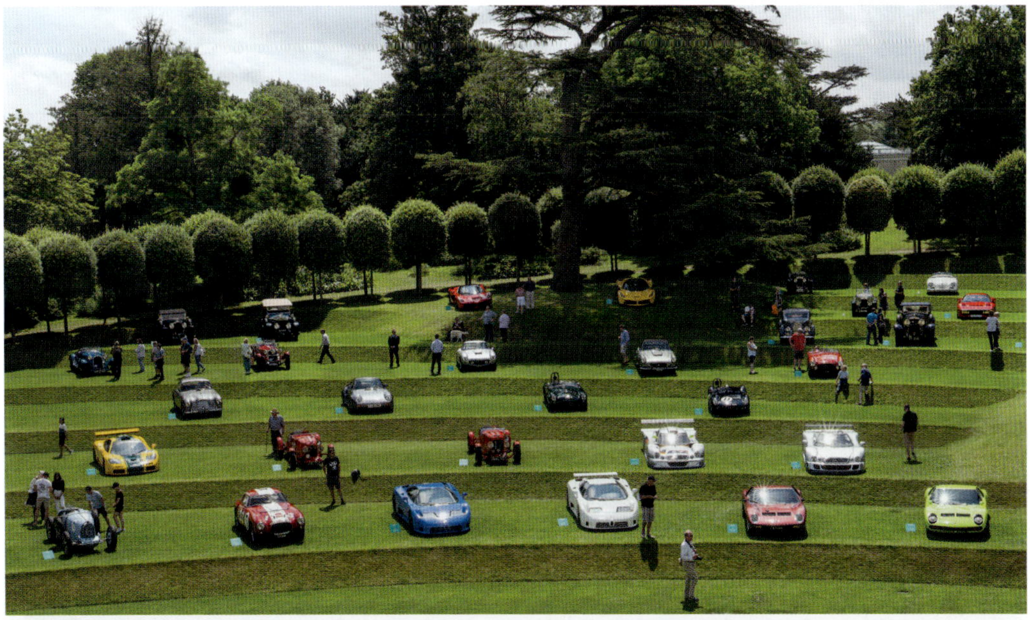

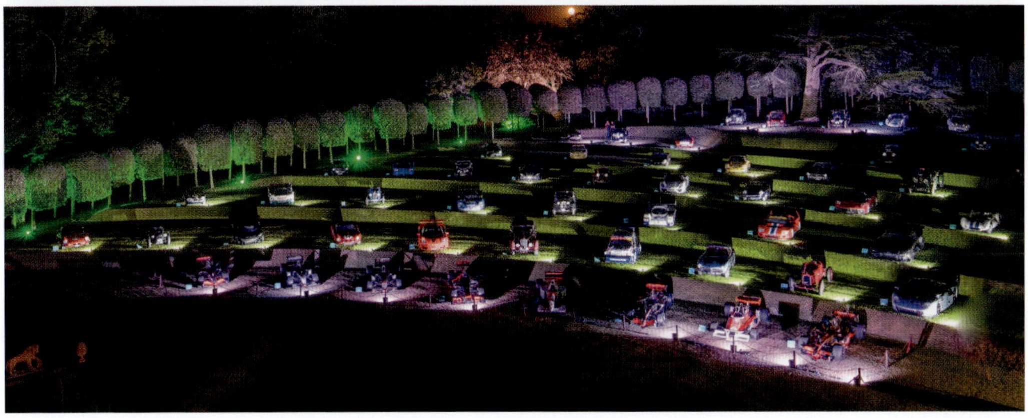

**OPPOSITE TOP LEFT:** *The Windsor Concours d'Elegance also managed to get some rare and spectacular cars such as this Bugatti Royale Kellner coupé chassis number 41141, in an unknown private collection at the time (2012) and presented by Lükas Hüni. (Hugues Vanhoolandt)*

**OPPOSITE TOP RIGHT:** *It is to the credit of Cartier that they have been holding a concours d'elegance at the Goodwood Festival of Speed for the best part of two decades. This image is from 2005. (Serge Cordey)*

**OPPOSITE MIDDLE LEFT & RIGHT:** *Despite including rarities such as this French Monica sports sedan (middle left) and an Abarth (right), the Cartier Concours d'Elegance has not received so much attention, as it has been part of a much larger event. (left: Serge Cordey; right: Makarand Baokar)*

**OPPOSITE MIDDLE RIGHT:** *Until as recently as 2018, the Cartier Concours d'Elegance at Goodwood was still considered a side event, despite featuring some remarkable automobiles. (Makarand Baokar)*

**OPPOSITE BOTTOM:** *Established in 2016 by Jon and Lois Hunt, the owners of Heveningham Hall, the concours bearing the name of the manor has since become one of the most popular in the United Kingdom. (C. Maynard)*

**THIS PAGE:** *With the cars displayed on the stepped lawns, the Heveningham Hall Concours manages to have a very distinctive look about it. This was the line-up for 2017. With appropriate lighting the display looks even more spectacular at night. (C. Maynard)*

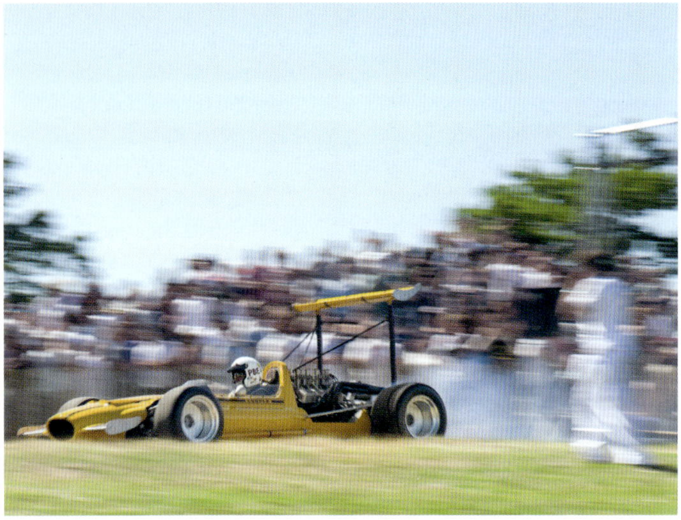

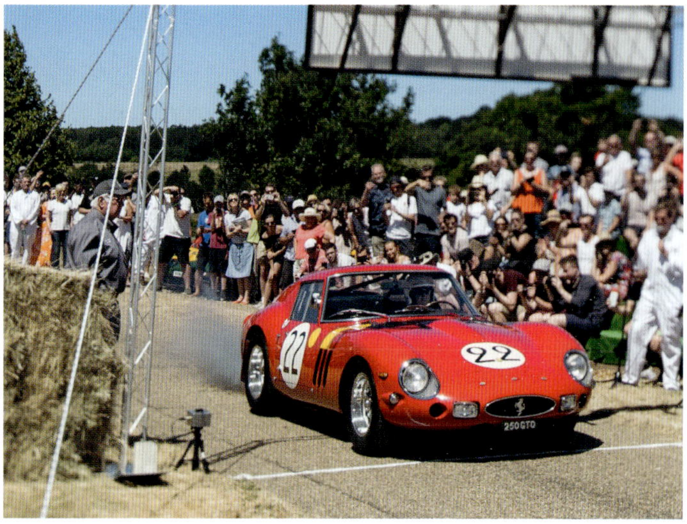


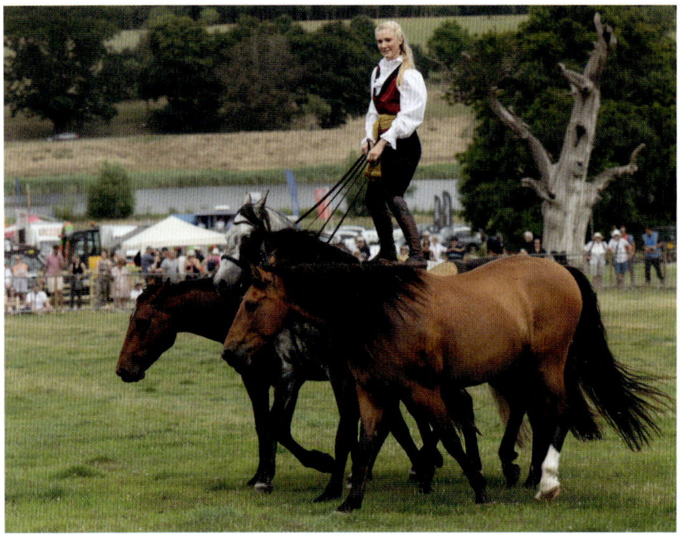

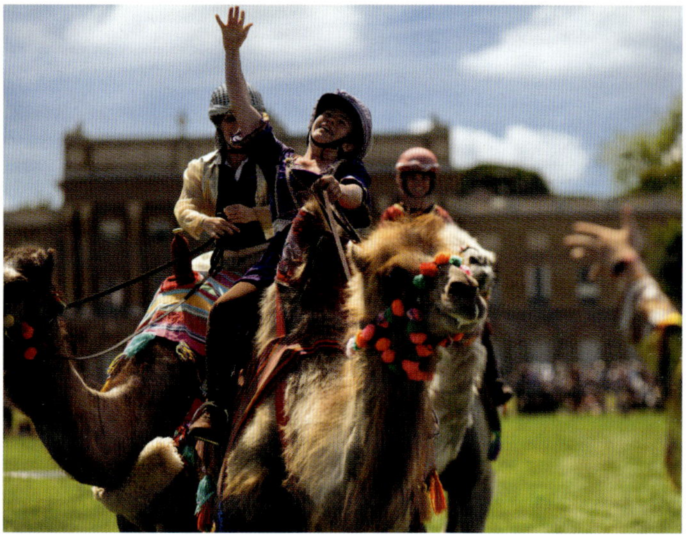

**THIS PAGE:** *In addition to its main automotive events, Heveningham Hall offers a variety of other exciting attractions. (C. Maynard)*

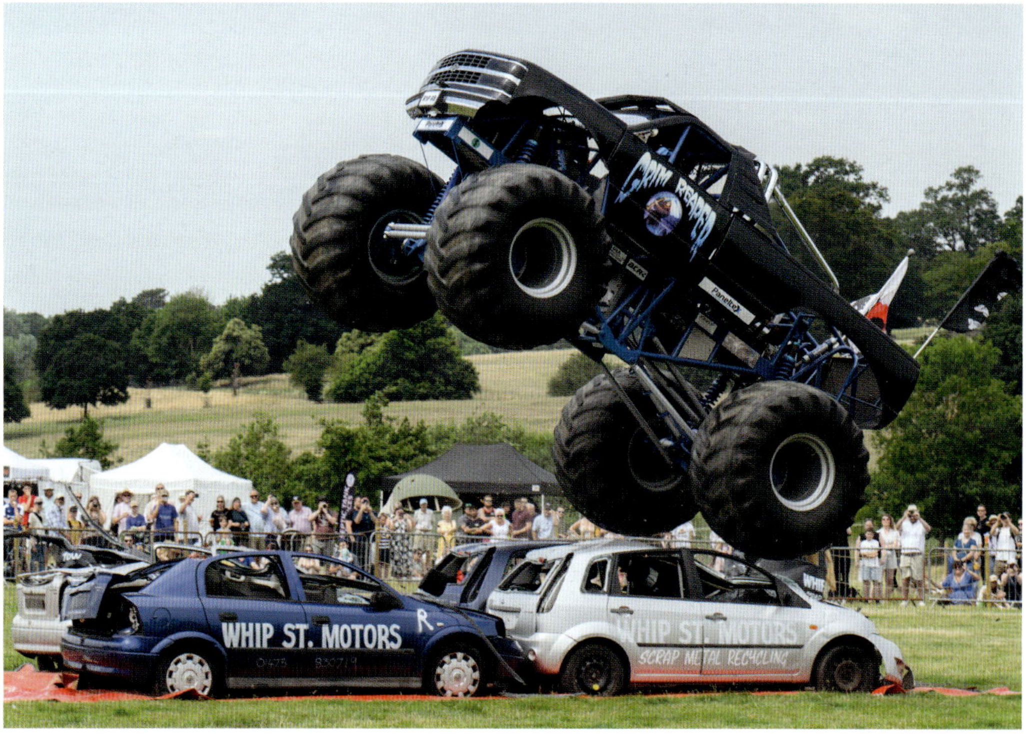

Inspired by flagship American concours d'elegance events like Pebble Beach, The Amelia, and Greenwich Concours – and following in the footsteps of Salon Privé and a revitalized Villa d'Este – the surge of concours d'elegance events has spread across Continental Europe. Italy, for instance, now hosts the Cavallino Classic in Modena, dedicated exclusively to Ferraris, as well as the Concorso d'Eleganza Varignana in Varignana, Poltu Quatu in Sardinia, and Lugano Elegance, among many others.

Belgium offers the Antwerp Concours at Belvedere-Wijnegem, while the Netherlands celebrates automotive history at the Concours d'Elegance Paleis Soestdijk. Germany has also embraced this trend, with the Concours d'Elegance Tegernsee and the Int'l Concours d'Elegance Gala Schwetzingen, held at the grand Schwetzingen Palace in the Baden-Württemberg region.

The European nation that has fully embraced the concept of concours events is Switzerland, where several major gatherings take place throughout the year. These include the Zurich Classic Car Award (ZCCA) in August, the Ascona Classic Car Award (ACCA) in September, and the spectacular International Concours of Elegance (ICE) in the snow-covered town of Saint-Moritz, typically held in February.

**BELOW:** *As Heveningham Hall has a grass airstrip, about twenty historic aircraft also turn up every year during the concours. This was in 2019. (C. Maynard)*

**BOTTOM:** *Antwerp also has an annual concours d'elegance. This image is from the 2018 concours. (Hugues Vanhoolandt)*

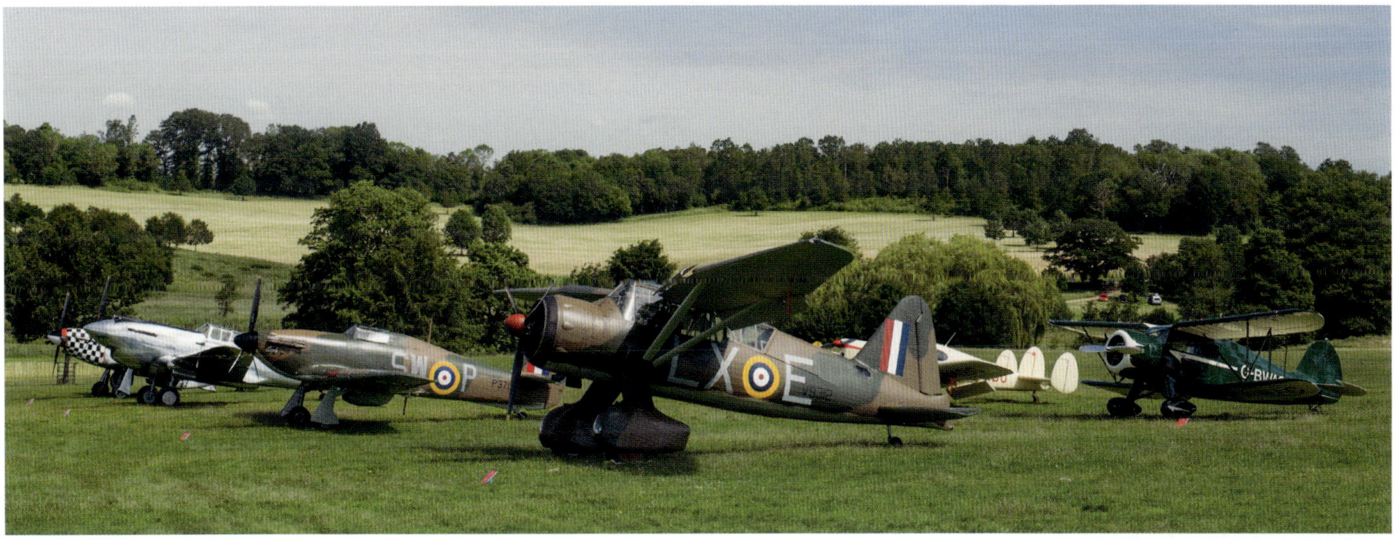

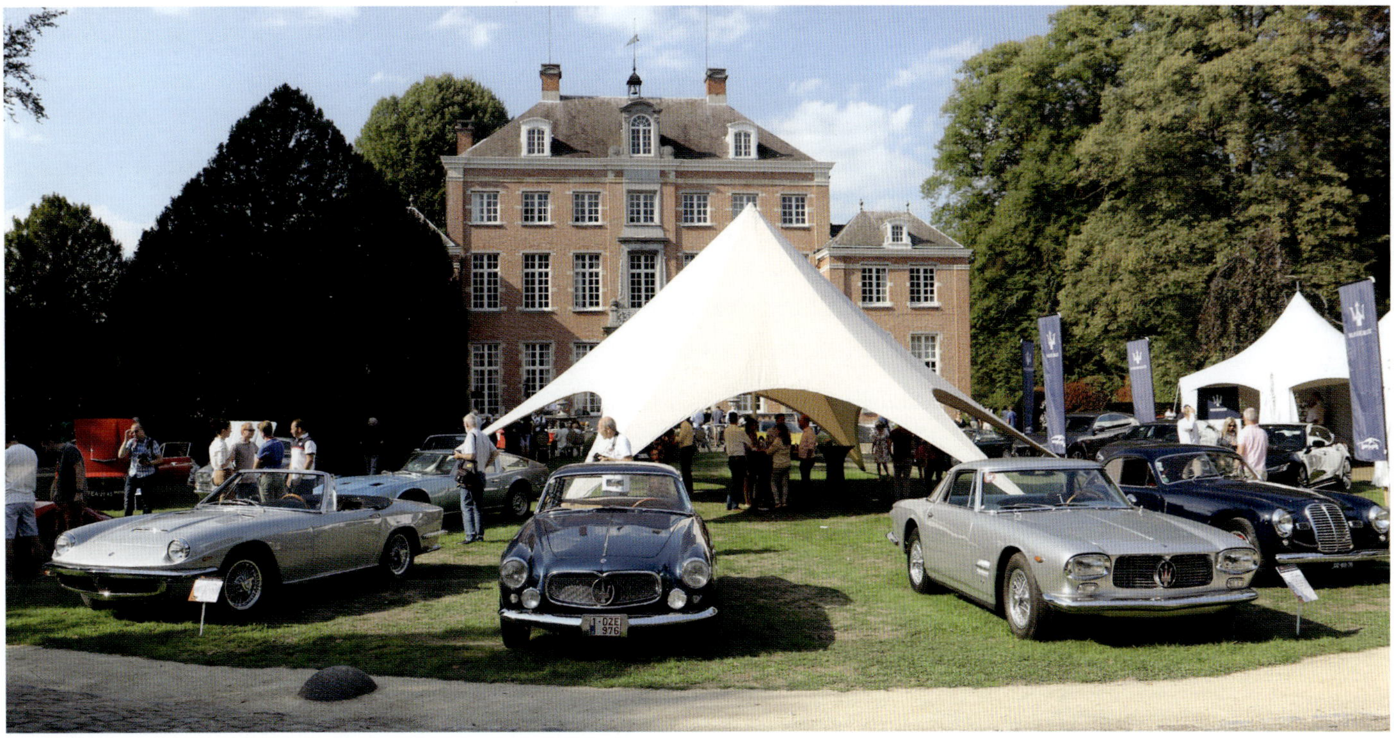

However, the premier event among Swiss concours is the Concours d'Elegance Suisse, which is usually held in June at the Château de Coppet, near Geneva. "As you may know, Geneva has a long history of hosting Concours d'Elegance events, dating back to 1927," explains Mathias Doutreleau, founder of the recently revived concours. "The event continued until around 1956, after which it disappeared. So, we brought it back in 2016, and except for 2020, we've had a successful run every year since."

"In our inaugural year of 2016, we featured 47 cars, with a Lamborghini Miura SV winning Best of Show," Doutreleau recalls. "This year, in 2024, we showcased about 90 cars and welcomed nearly 3,700 spectators. Our concours is focused on celebrating style, design, and elegance, and we use a unique judging system: 50 percent of the score is based on the car's condition, or 'état,' and the remaining 50 percent on its elegance, style, and design innovations. Our jury consists of around a dozen members."

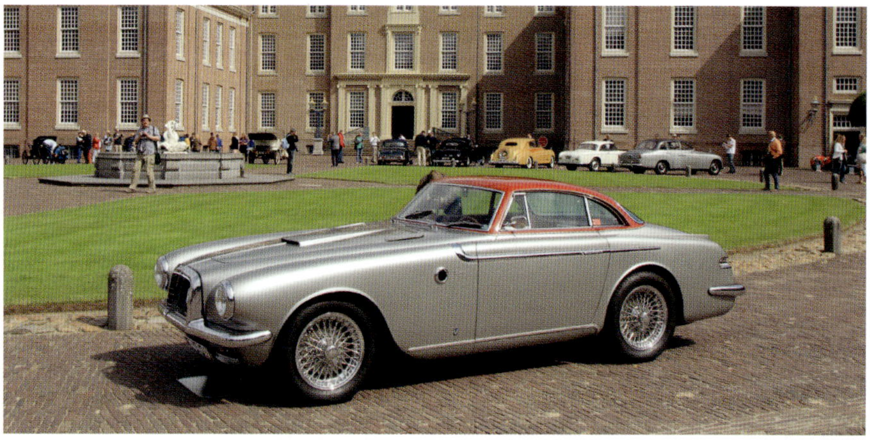

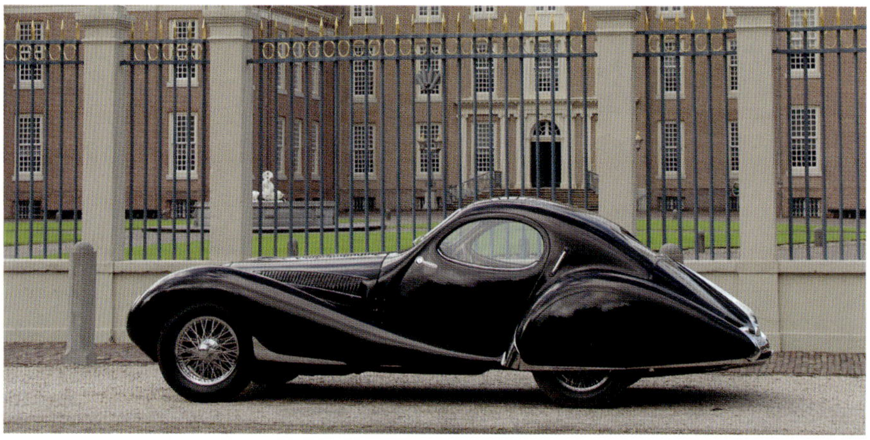

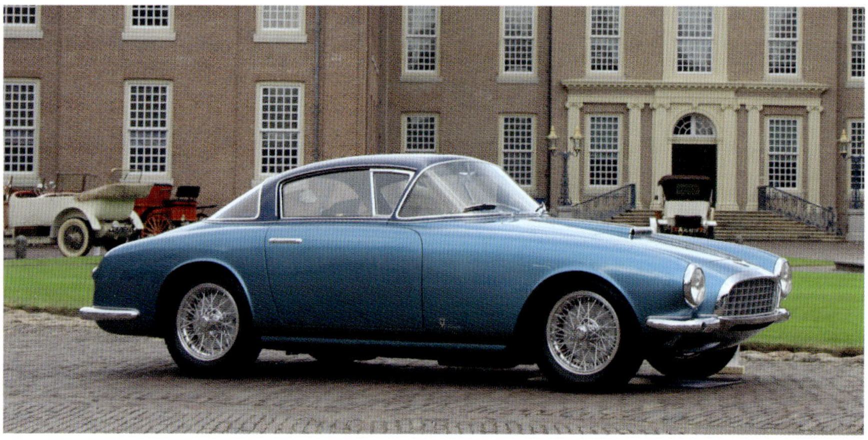

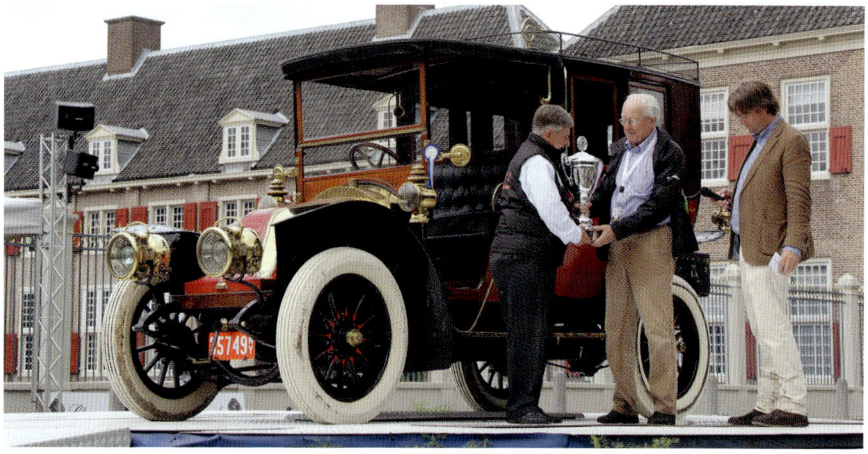

**THIS PAGE:** *One of the most elegant settings of all the European concours d'elegance is at the Paleis Het Loo in The Netherlands. These images are from the concours held on June 21, 2014. (Hugues Vanhoolandt)*

**OPPOSITE TOP:** *Undoubtedly, the most remarkable and unique of all European concours is the International Concours of Elegance (ICE) held in Saint-Moritz, standing apart from all the rest. Here, the breathtaking Lancia Stratos Zero roars across the ice, creating a striking image of a futuristic space mobile in motion. (Davide de Martis)*

**OPPOSITE BOTTOM RIGHT:** *When was the last time you saw a Ferrari 250 GTO slip sliding its way across ice? (Pietro Martelletti)*

**OPPOSITE BOTTOM LEFT:** *Or for that matter, a racing Ferrari Daytona in action on snow? ICE at Saint-Moritz is indeed spectacular. (Nicola Fornaciari)*

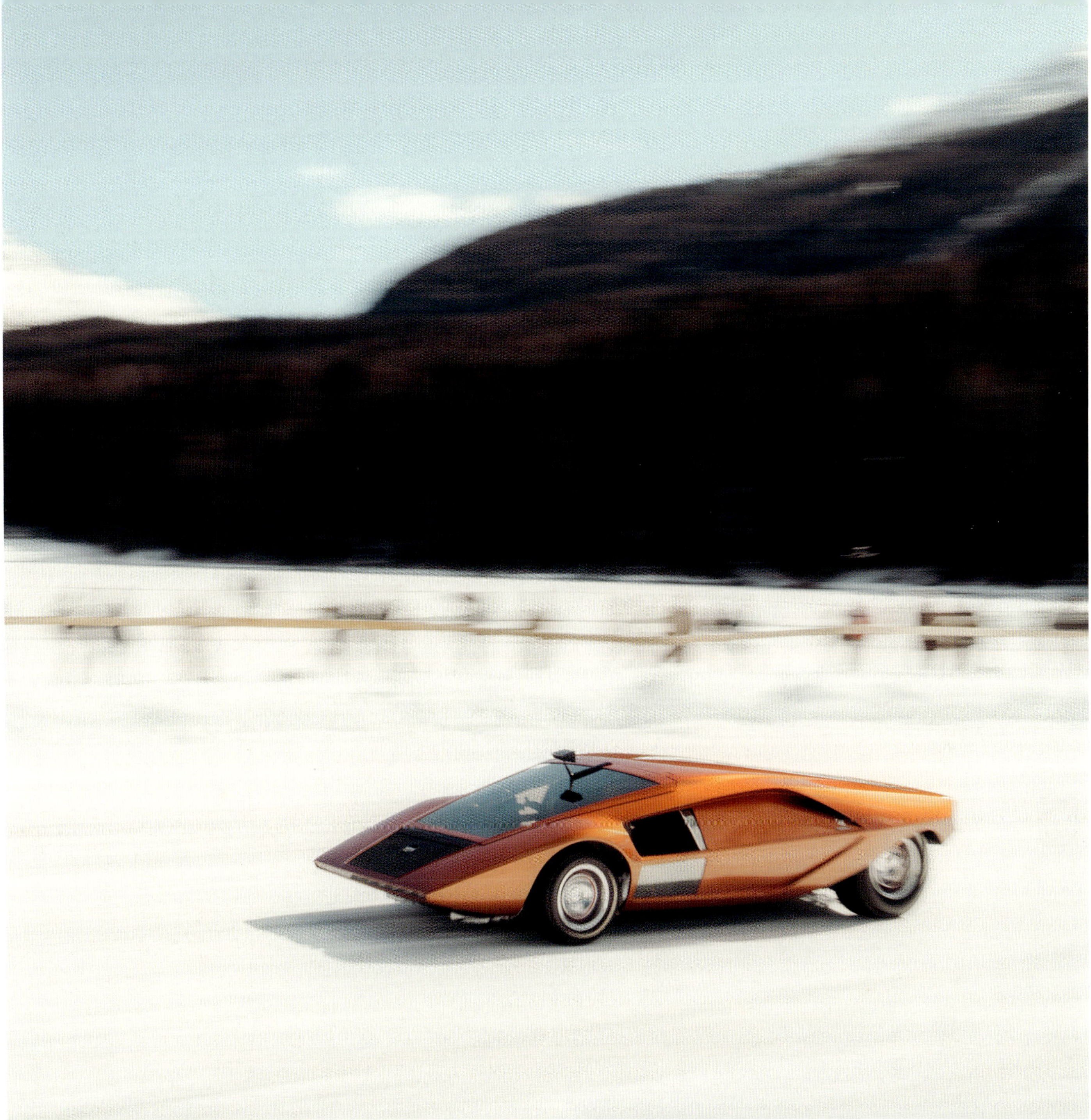
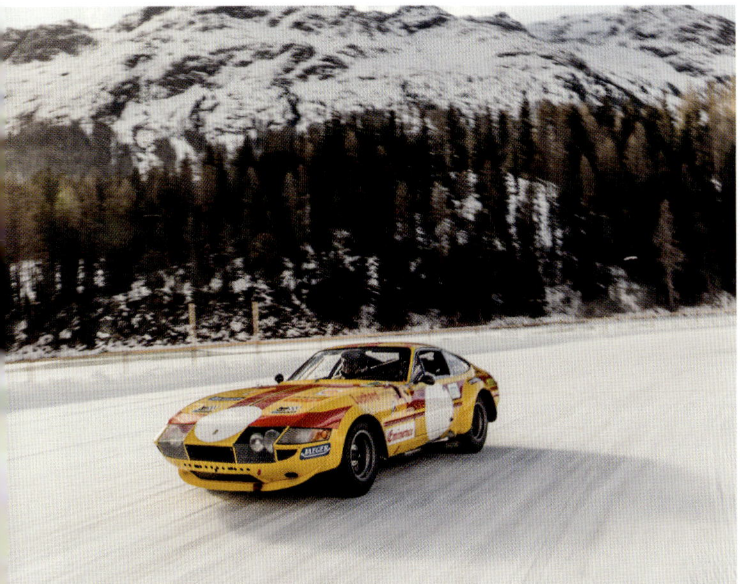
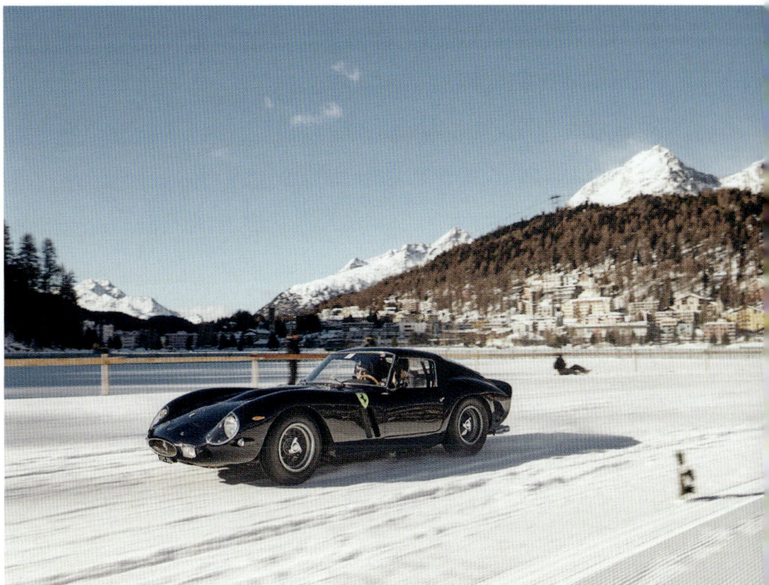

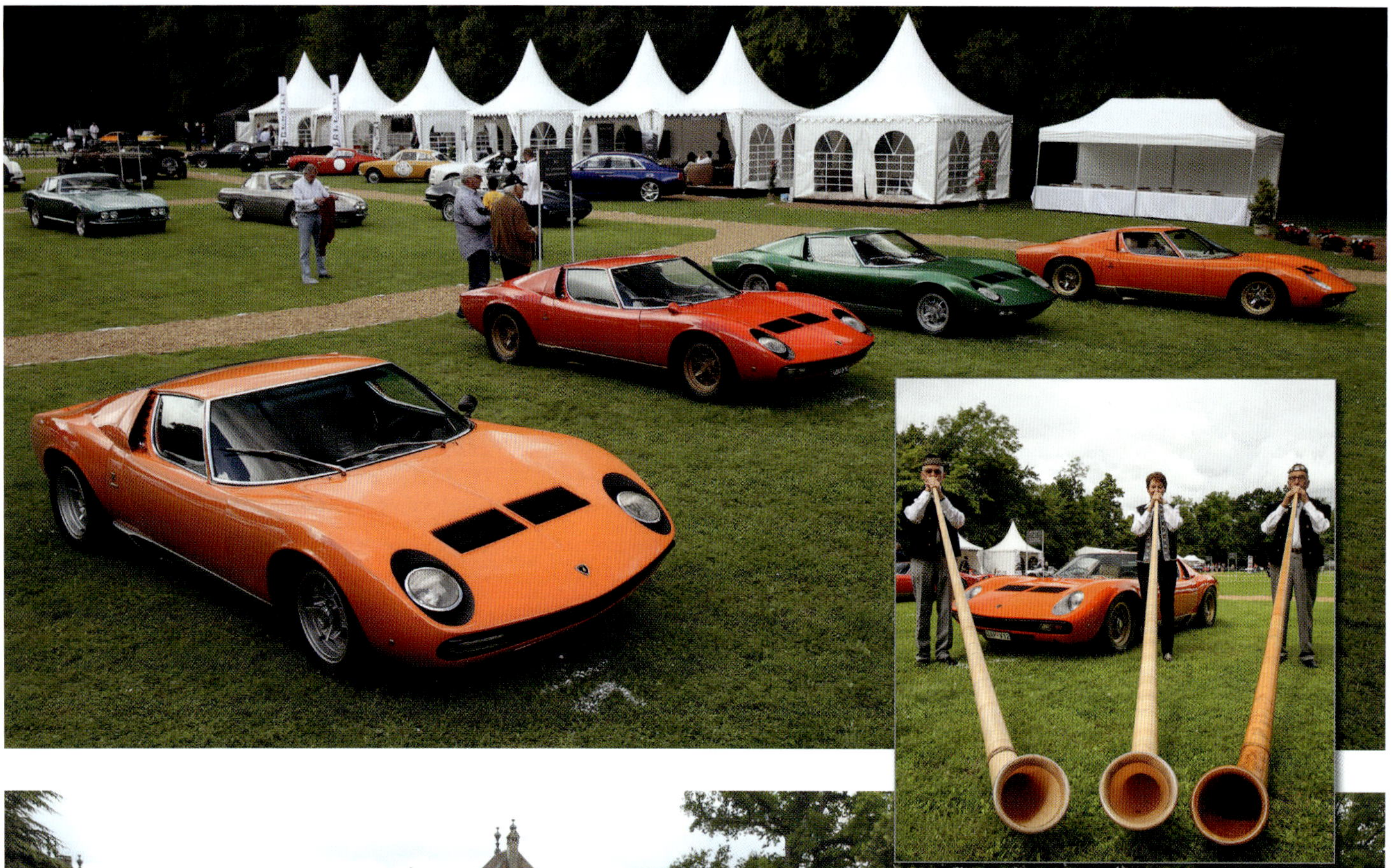

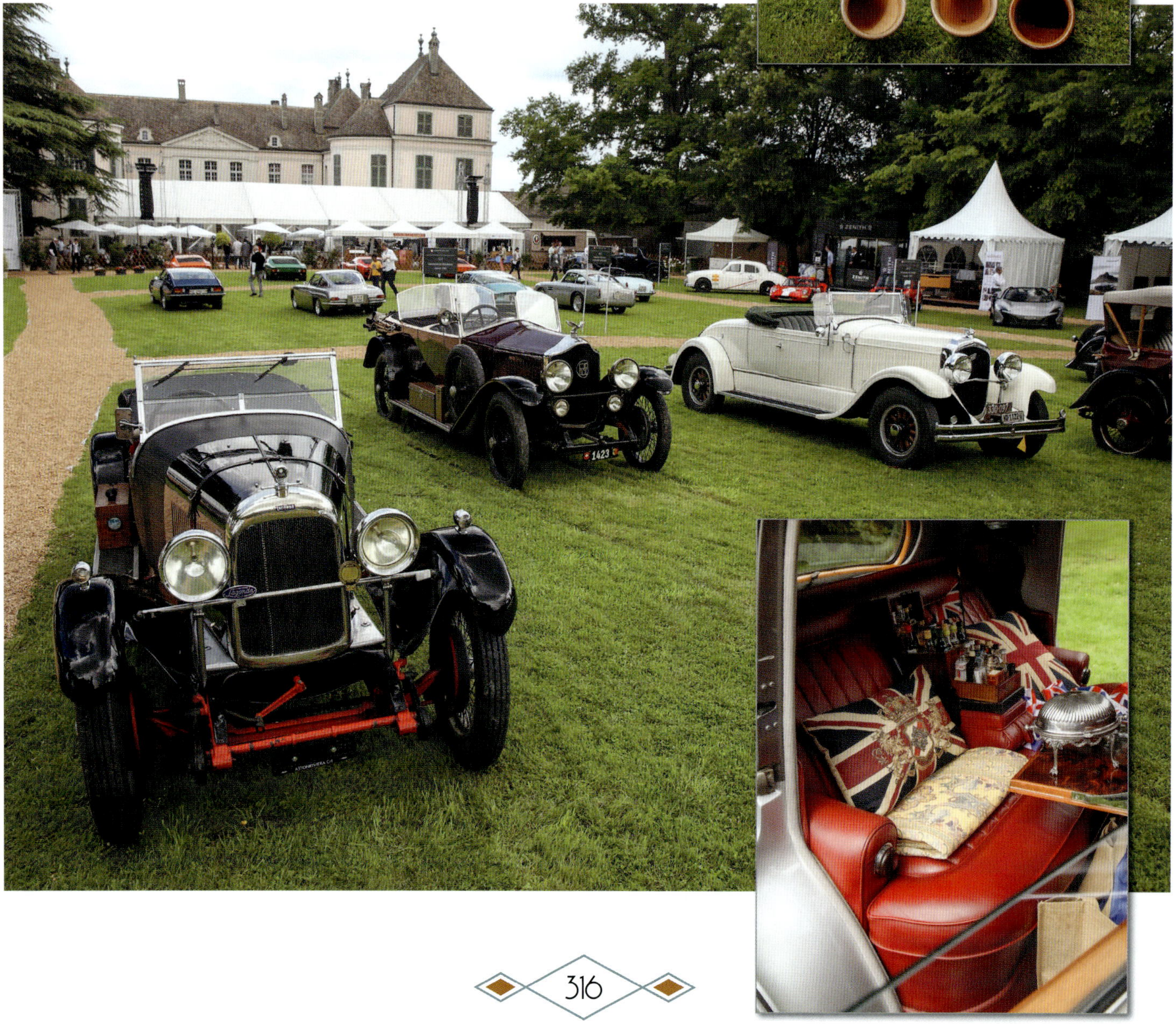

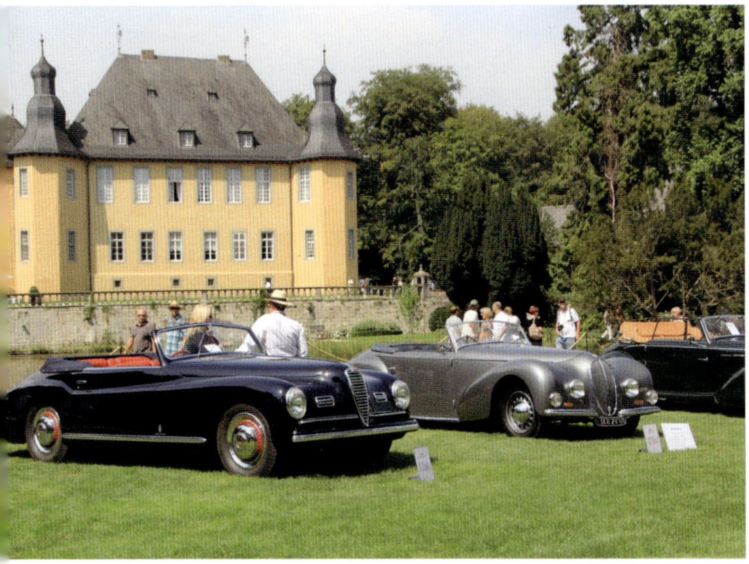

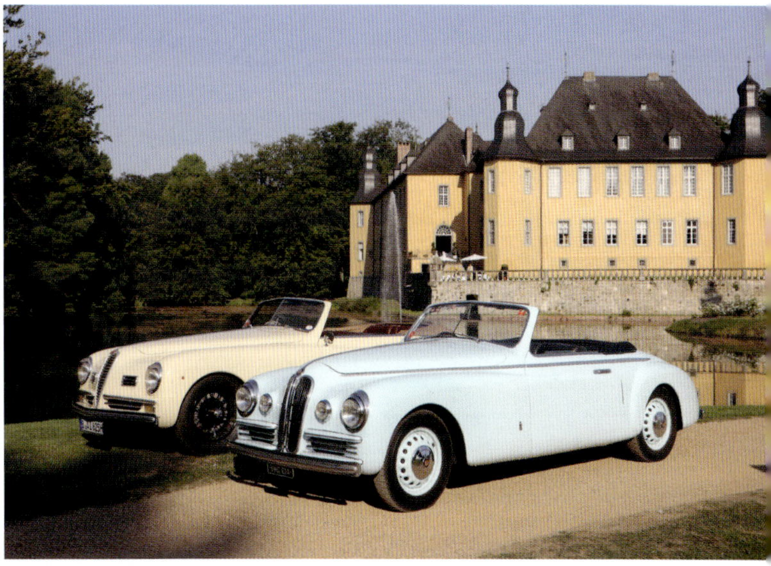

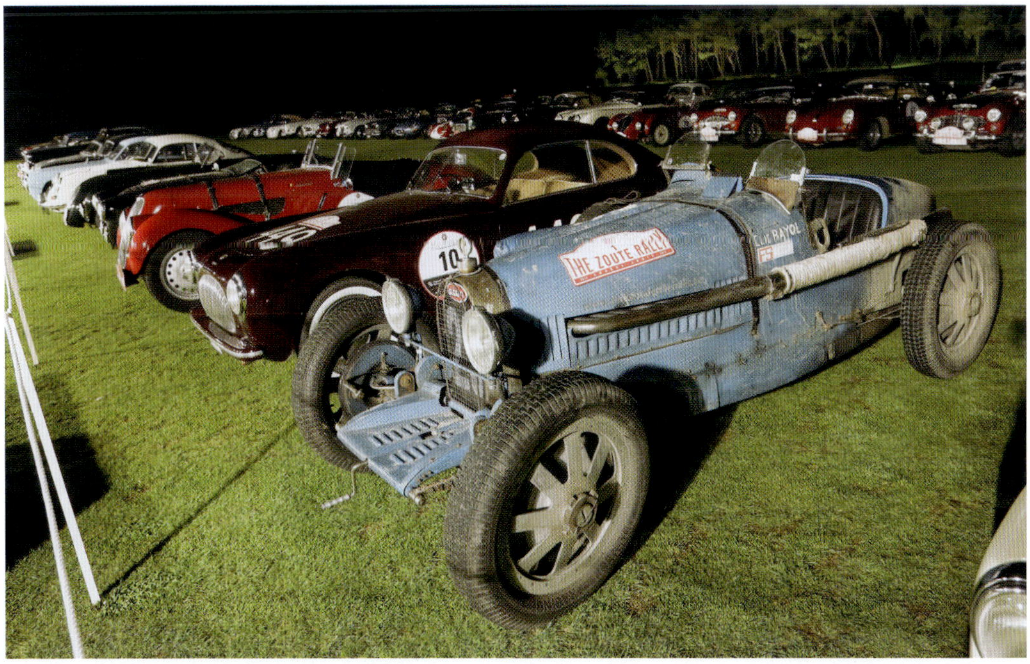


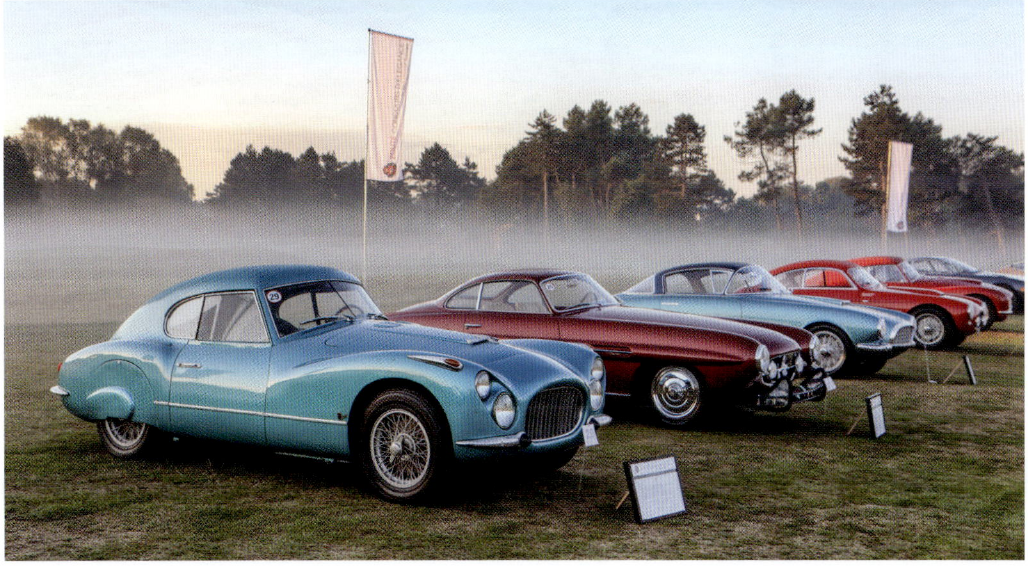

**ABOVE (BOTH):** *The Masterpiece & Style Concours d'Elegance at Schloss Dyck near Jüchen is another niche European concours. These images are from the August 2014 concours. (Hugues Vanhoolandt)*

**MIDDLE:** *The Knokke Zoute Concours d'Elegance is one of five different events taking place at the same time in that one location. These images are from the 2016 concours. (Xavier de Nombel)*

**BOTTOM:** *The Knokke Zoute Concours d'Elegance has its own special charm. Images from October 2016. (Hugues Vanhoolandt)*

**OPPOSITE (ALL):** *Mathias Doutreleau brought back the Concours d'Elegance Suisse to Geneva in 2016. As many as 47 cars were featured in the 2016 Concours d'Elegance Suisse with a Lamborghini Miura SV winning Best of Show. Blending an exquisite selection of pre-war rarities and post-war classics, the Concours d'Elegance Suisse has secured its own unique place in the European concours calendar. (Xavier de Nombel)*

**INSET:** *The concours at Sinaia is an interesting mix of judging the condition of a car à la concours d'état, as well as taking into consideration the design and the history of the vehicle. (Dragos Savu)*

**OPPOSITE:** *The Valletta Concours, set on the historic and picturesque island of Malta, is undoubtedly one of the most enchanting concours events. (Courtesy John Saliba)*

**BELOW:** *Even if it is not one of the bigger concours, the Valletta Concours in Malta attracts several senior judges such as Sandra Button, who has been the chief judge at the event. Mohammed Luqman Ali Khan (in hat) and Thomas Aigner (left) explaining the finer points of the Rolls-Royce Phantom I experimental car, chassis number 17EX. (Kaustabh Khare)*

## Romania again

As mentioned previously, the concept of concours d'elegance had spread to several Eastern European countries before World War II, with one of the most notable events being held in Sinaia, Romania (pages 188-191). In 2011, Retromobil Club Romania, the FIVA-recognized federation of all historic vehicle clubs in the country, revived the Concurs de Confort și Eleganță. Since then, the event has evolved into a unique and distinctive concours, setting itself apart from others in providing importance to the equipage.

"Our scoring system places great importance on the history and story behind each car," explains Gabriela Magureanu, President of Retromobil Club Romania. "We also focus on the quality of restoration, with a strong emphasis on authenticity. We are firmly opposed to over-restoration. Conservation and preservation are key principles of the Sinaia concours, and our jury, composed of a select group of about a dozen experts, carefully evaluates each vehicle. The cars are displayed according to their class, with each group parked in a different location, encouraging enthusiasts to explore the castle grounds, enjoy the fresh mountain air, and fully appreciate the spectacular setting of Sinaia's royal castle."

The two-day event begins with judging on the first day, followed by an awards ceremony on the second day, where the cars parade in front of spectators and jury members. The occasion also provides a delightful spectacle as owners and their companions step out of the cars, showing off attire carefully chosen to match the vehicle's color, period, or theme, adding a delightful, theatrical element to the event.

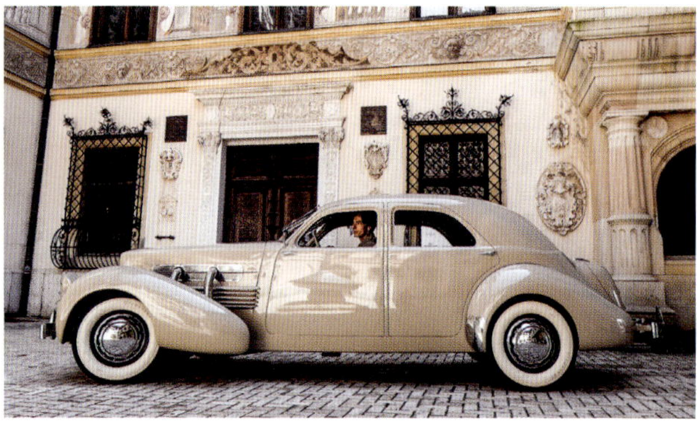

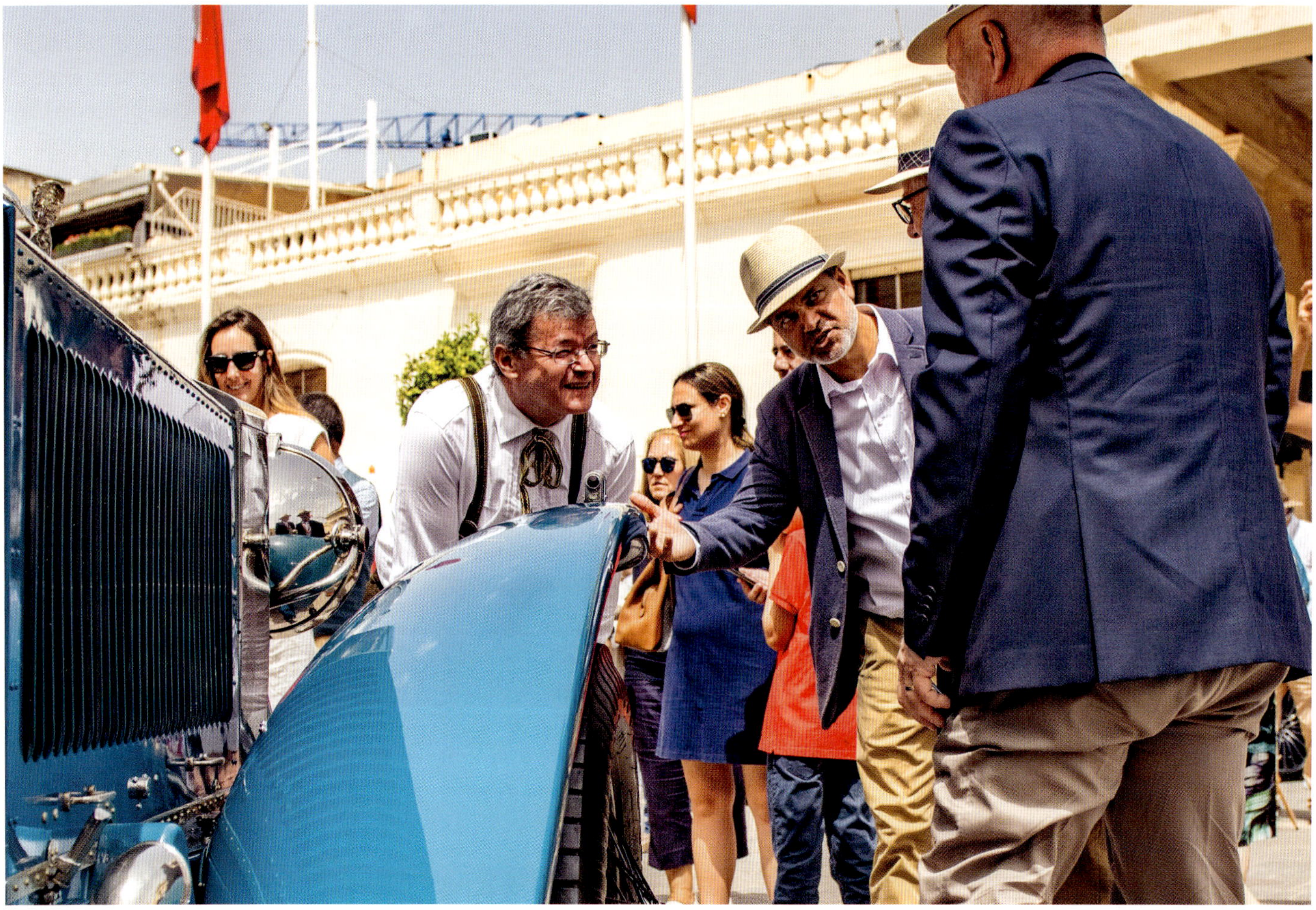

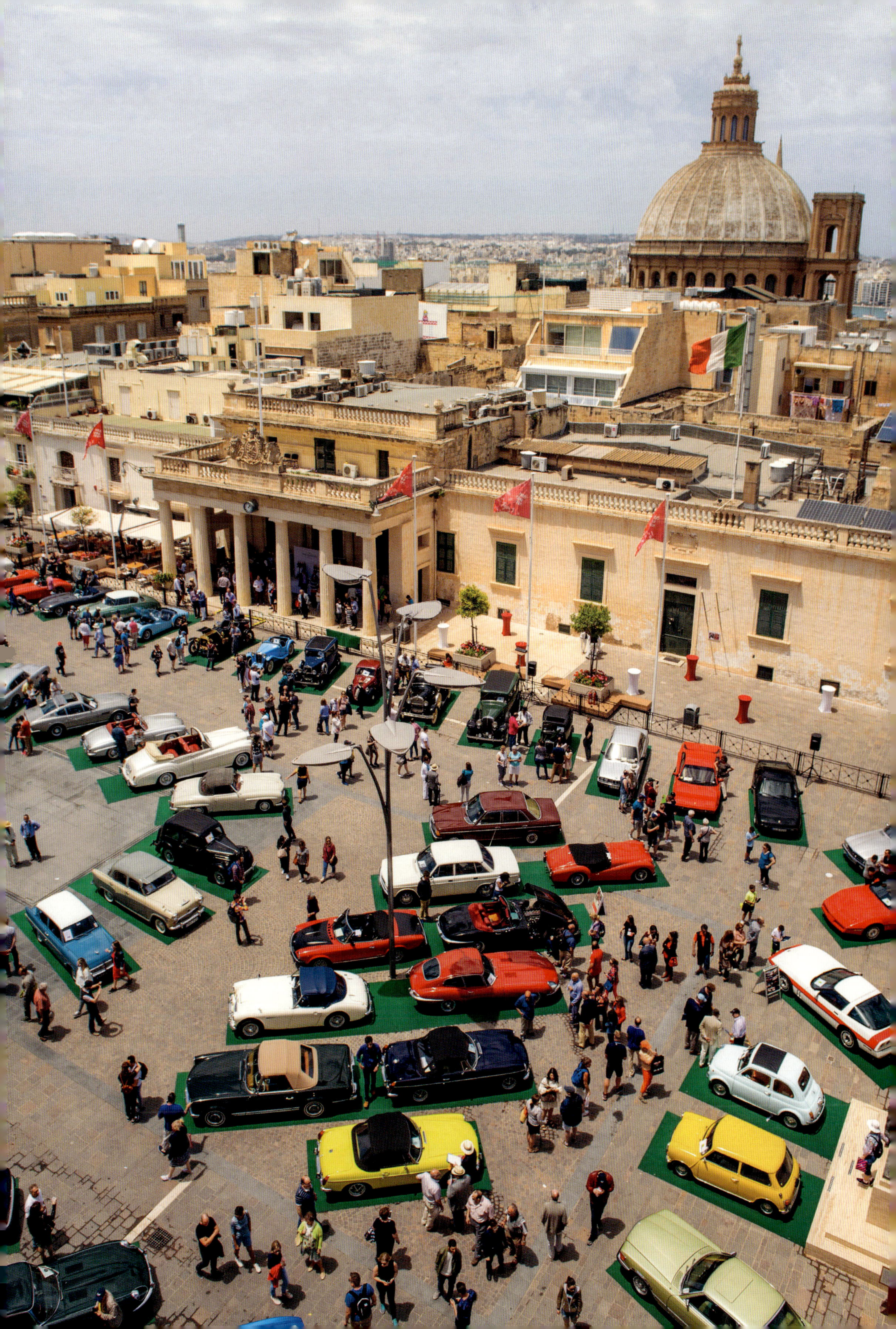

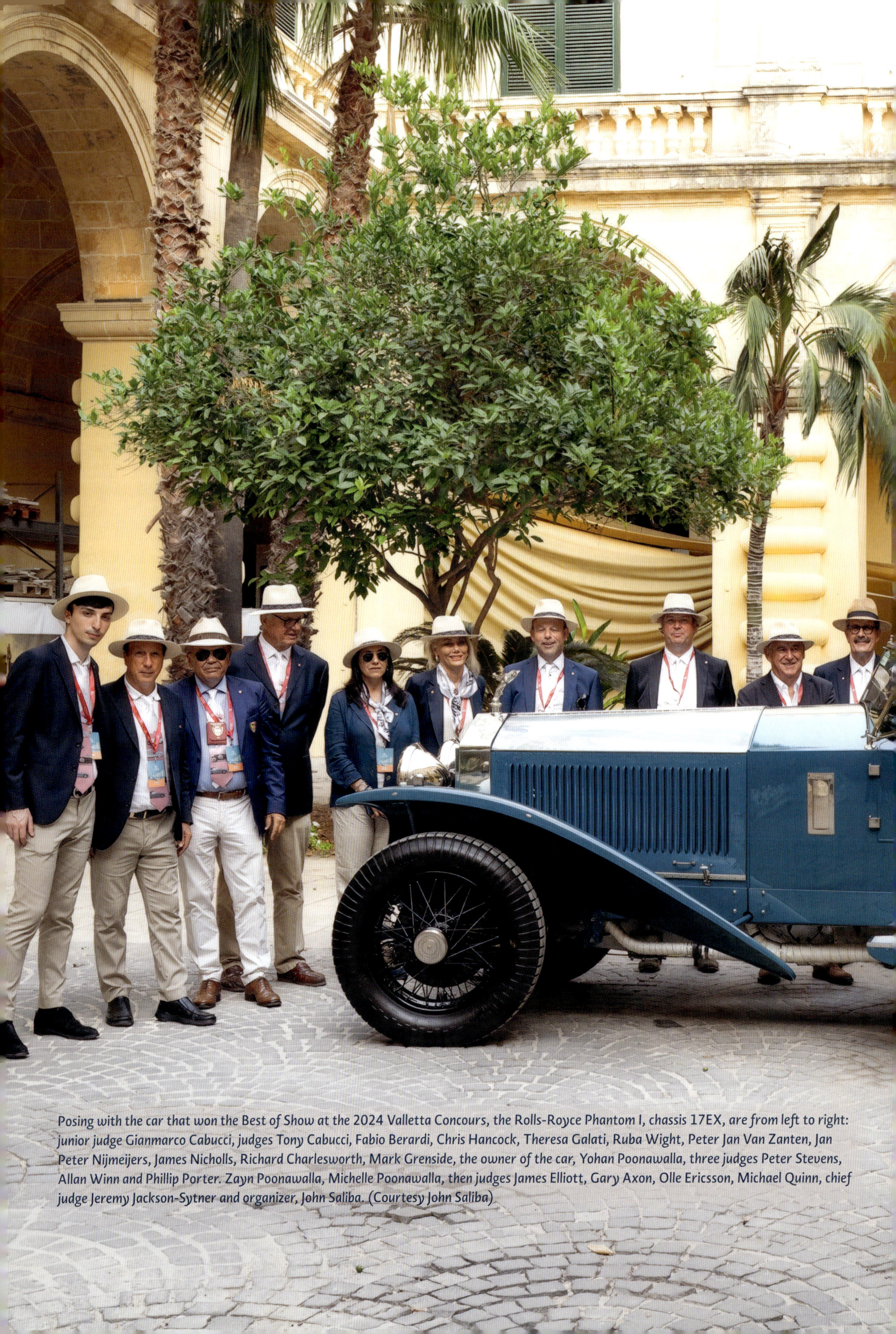

*Posing with the car that won the Best of Show at the 2024 Valletta Concours, the Rolls-Royce Phantom I, chassis 17EX, are from left to right: junior judge Gianmarco Cabucci, judges Tony Cabucci, Fabio Berardi, Chris Hancock, Theresa Galati, Ruba Wight, Peter Jan Van Zanten, Jan Peter Nijmeijers, James Nicholls, Richard Charlesworth, Mark Grenside, the owner of the car, Yohan Poonawalla, three judges Peter Stevens, Allan Winn and Phillip Porter. Zayn Poonawalla, Michelle Poonawalla, then judges James Elliott, Gary Axon, Olle Ericsson, Michael Quinn, chief judge Jeremy Jackson-Sytner and organizer, John Saliba. (Courtesy John Saliba)*

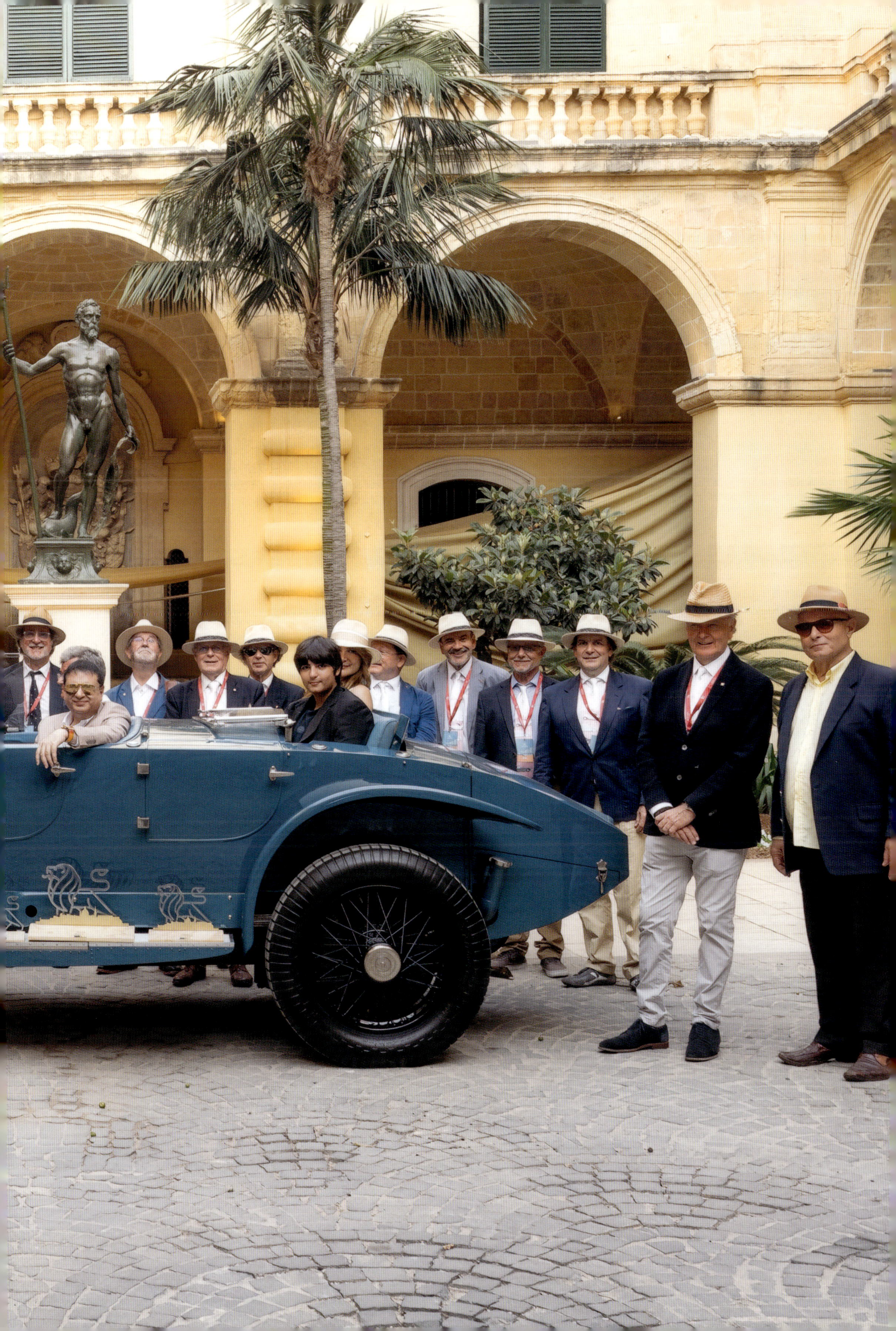

**THIS PAGE:** *Judges and participants discussion at the ICONS Mallorca at Palma de Mallorca in 2024. From left to right: Thomas Aigner and Mohammed Luqman Ali Khan, judges Jürgen Lewandowski and Lorenzo Ramaciotti, and Yohan Poonawalla, the owner of Rolls-Royce Phantom I, chassis 17EX, of which there is a glimpse of the front right fender. (Kaustabh Khare)*

The Balatonfüred Concours d'Elegance in Hungary exudes a similar mood and entertainment as its central theme. "What makes this concours truly special is its location," says Zoltan Papp, editor-in-chief of *Veterán Autó és Motor*, part of the same publishing house that owns *AutóMagazin*, which took over organizing the event in 2023. "The shores of Lake Balaton, particularly the scenic Tagore Promenade, provide an ideal setting. The event coincides with the opening of the sailing season, so visitors also get to see around a hundred classic boats and enjoy a grand ceremony. One unique aspect of our concours is that entry is free for visitors – in 2023, we attracted around 22,000 people!"

Papp further explains the judging process: "We have two juries, each with seven members, including photographers, editors, restorers, graphic artists, and classic car authors from our publishing house. The final jury's decisions are made through discussion, not a point system – just weighing the pros and cons. They know how many awards there are and choose the winners for each category and other honors. The 'Best of Show' doesn't have to be a category winner; it's simply the best of the best (in 2023, it was a Lincoln Continental, and in 2024, a Facel Vega II). Cars can win multiple awards, and the public also gets to vote for their favorite."

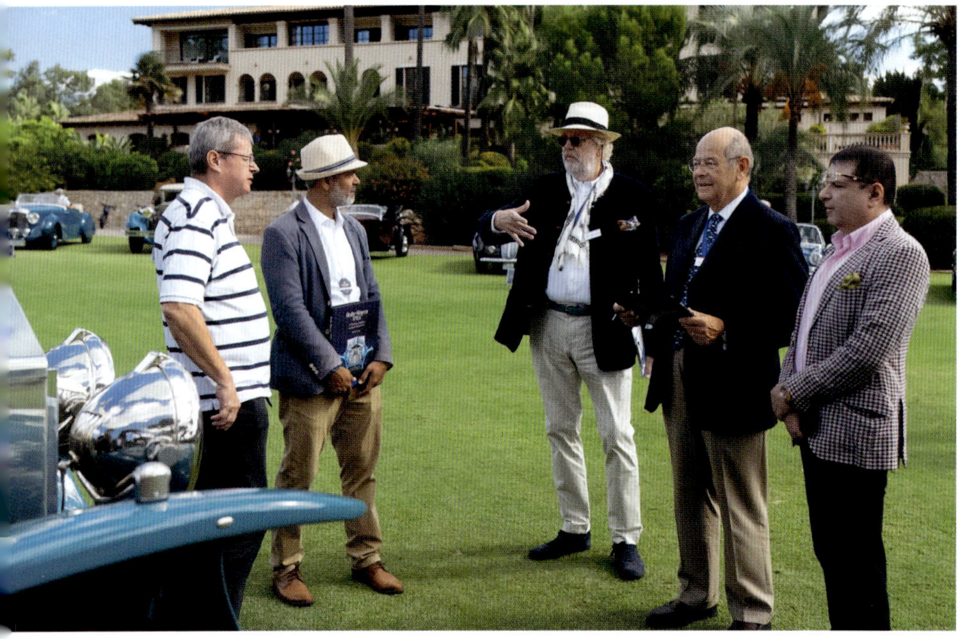

In addition to classic cars, the Balatonfüred Concours also features motorcycles and restomods. "I'll be reaching out to several European manufacturers to join us," Papp adds. "There's also a 150 km (90 miles) classic car tour on the Friday before the concours, taking participants along the most beautiful roads of the Balaton-Felvidék region."

Another distinctive concours d'elegance is the Valletta Concours d'Elegance, typically held in early June in Malta's capital, Valletta. Although one of the newer concours events, with its inaugural show in 2017 featuring Audrain's Nic Waller as chief judge, the Valletta Concours stands out by sharing the judging scoresheets with participants after the event. "I like to provide the judging results and the point sheet to all participants," says founder and organizer John Saliba.

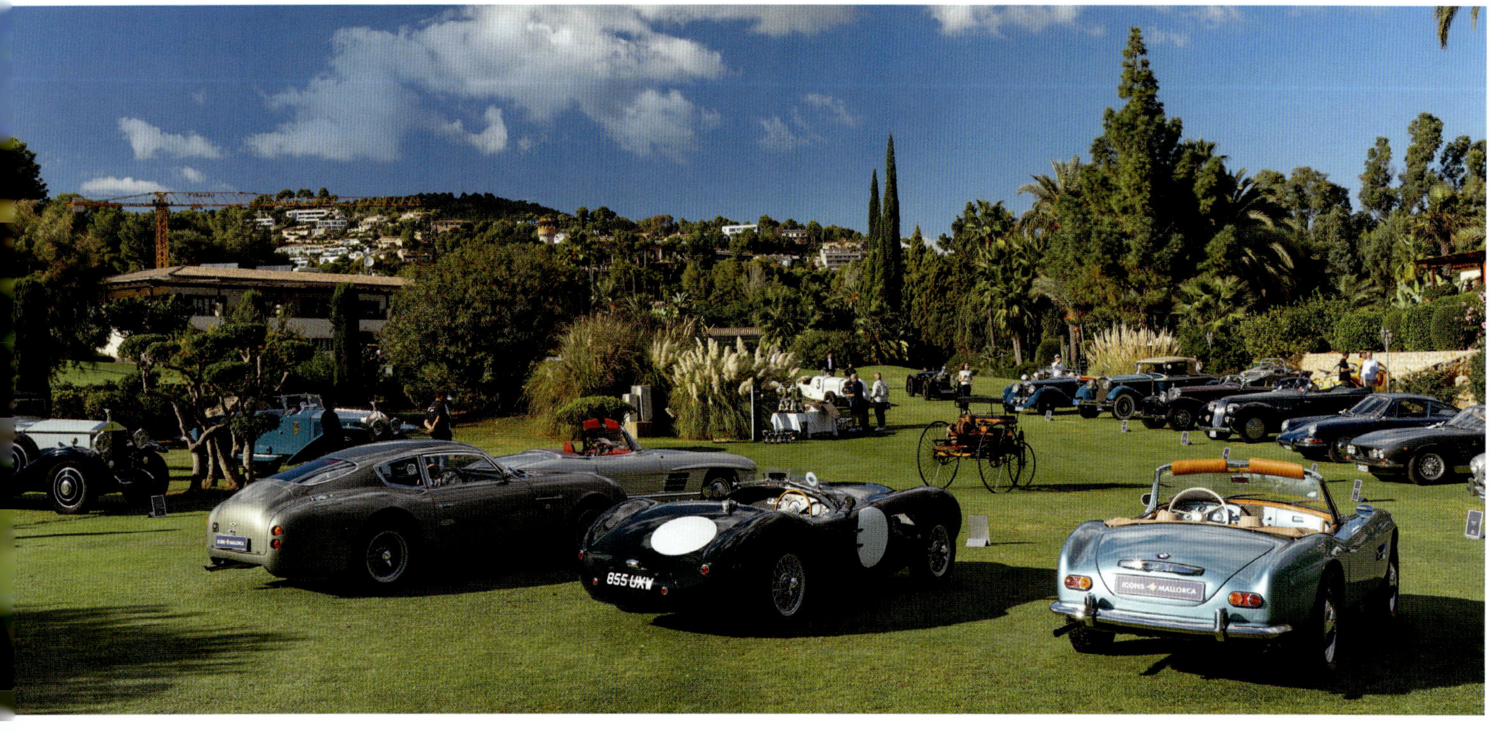

Saliba believes in transparency: "Participants should know where their car fell short," he explains. The event uses a 100-point scoring system, with notable judges like Nic Waller in 2017 and Sandra Button in 2018. Since 2019, Jeremy Jackson-Sytner has been the chief judge. Set against the backdrop of Valletta's 16th century architecture, the main square becomes a beautiful showcase for around 65 cars, 15 to 20 motorcycles, and even a class for pedal cars. The event also features a parallel exhibition of approximately 30 cars from a different marque club each year – 2022 highlighted the Ferrari club, 2023 featured the Maserati club, in 2024 it was the Porsche club, with the MG club in 2025. The Valletta Concours draws up to 15,000 spectators annually.

## Asia and beyond

Concours d'elegance events have gained popularity worldwide. Pakistan hosted its first concours on February 24, 2024, in Karachi. China's inaugural event took place from October 18-20, 2013, in Shanghai, with the Bund Classic Concours d'Elegance held at the iconic Waitanyuan No. 1 villa in the Huangpu district. Although two more editions followed in 2014 and 2015, the event did not continue.

In 2015, ABV Media Group, the publisher of *Autobild China,* organized a remarkable Concours d'Elegance on YanQi Island, near Beijing, paired with a stunning event known as the Great Wall Rally. "The first concours was a tremendous success, with the Best of Show going to a beautiful 1965 Lamborghini 350GT," recalls ABV Media Group's Marketing Director, Sherry Xu. The event continued on YanQi Island in 2016, 2017, and 2018, with Best of Show honors awarded to a 1964 Porsche 356SC, a 1957 Alfa Romeo Giulietta SS Prototipo, and a 1972 Audi 100 Coupé S, respectively. Unfortunately, the event has not been held since, though there are hopes to revive it in the future.

The Middle East too has had a few concours events in the recent past, in Kuwait, Qatar, and the UAE. Mohammed Luqman Ali Khan, a UAE-based motoring historian and consultant with expertise in concours events and collector cars, shared: "The Middle East has hosted several noteworthy concours events in the past, including in Kuwait, Qatar, and the UAE. The Kuwait Concours d'Elegance showcased iconic cars in 2011 and 2012, and in 2016 and 2017 the Gulf Concours was held at the renowned Burj Al Arab Hotel, the world's only seven-star hotel at the time. I curated the UAE Concours d'Elegance, which took place at the Bvlgari Hotel and Resorts in Dubai in 2019 but before that, there was the long-running and highly successful Emirates Classic Car Festival,

**THIS PAGE:** *The Great Wall Rally was organized by Wang Yin (left image) and Sherry Xu (right, right image) of AutoBild China. (Courtesy ABV Media)*

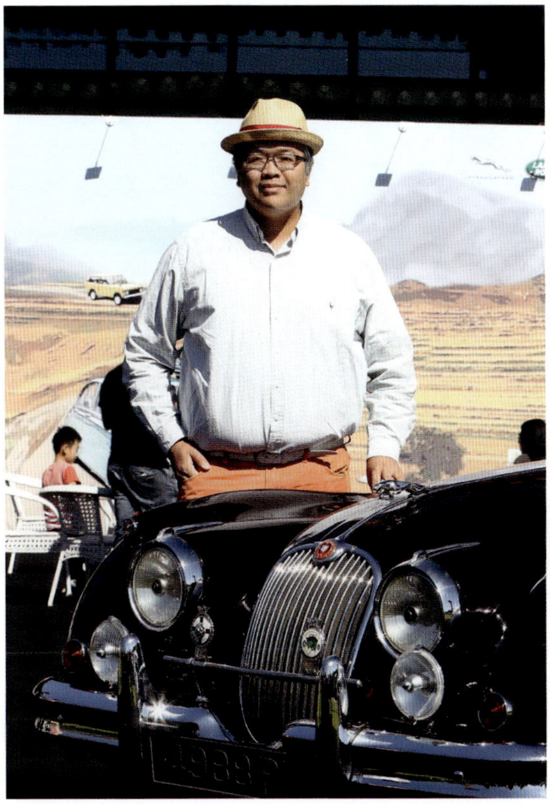

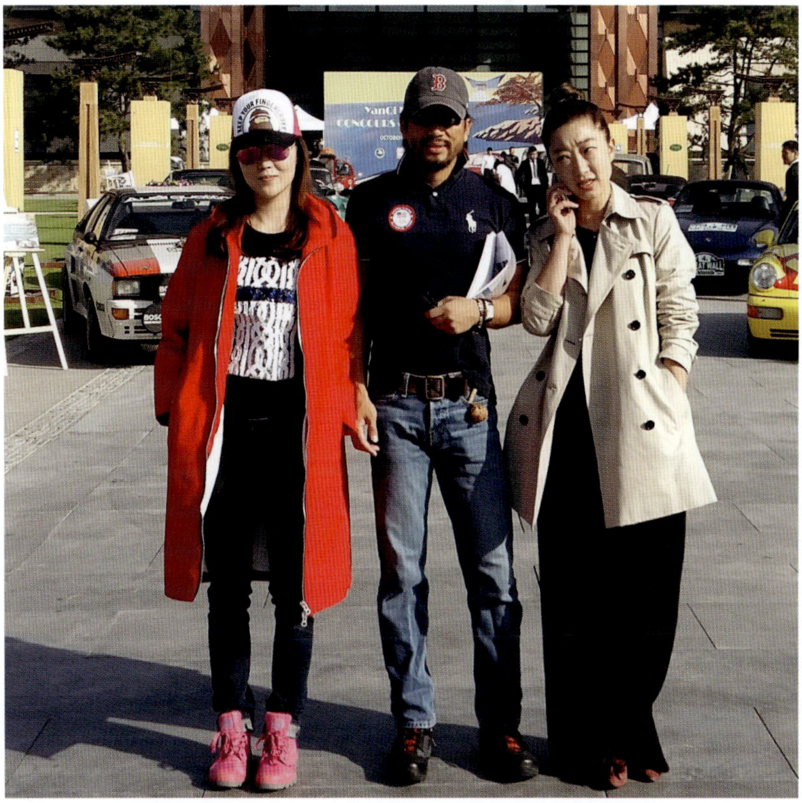

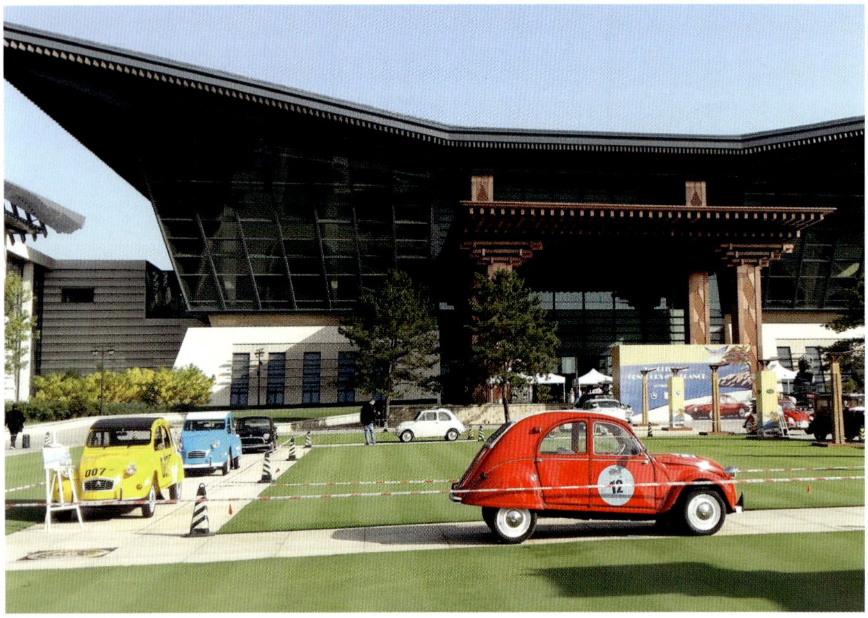

**ABOVE & BELOW:** *YanQi Island Concours d'Elegance was one of the more exotic of concours that combined a rally – The Great Wall Rally – and a concours to make for a great event. (Gautam Sen)*

**RIGHT:** *The Great Wall Rally was an added attraction. (Courtesy ABV Media)*

**OPPOSITE BOTTOM (BOTH):** *With cars like the Audi 100 Coupé, Auto Union 1000 SP, and Jaguar Mk IIs on display, the YanQi Concours d'Elegance stands out as a truly distinctive and engaging event. (Gautam Sen)*

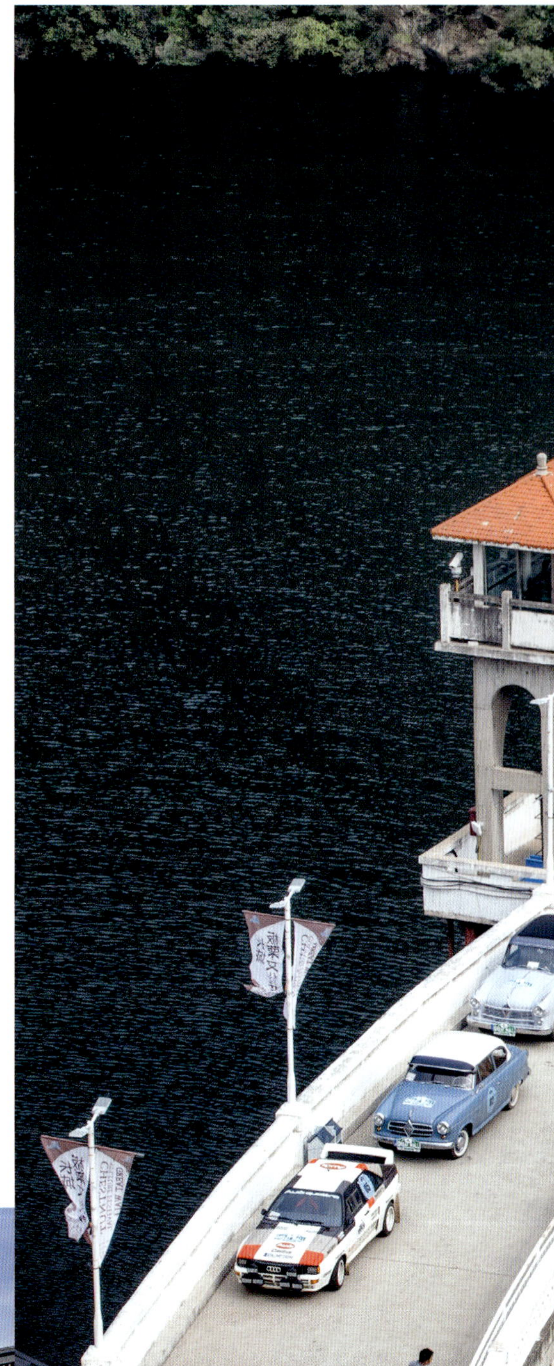

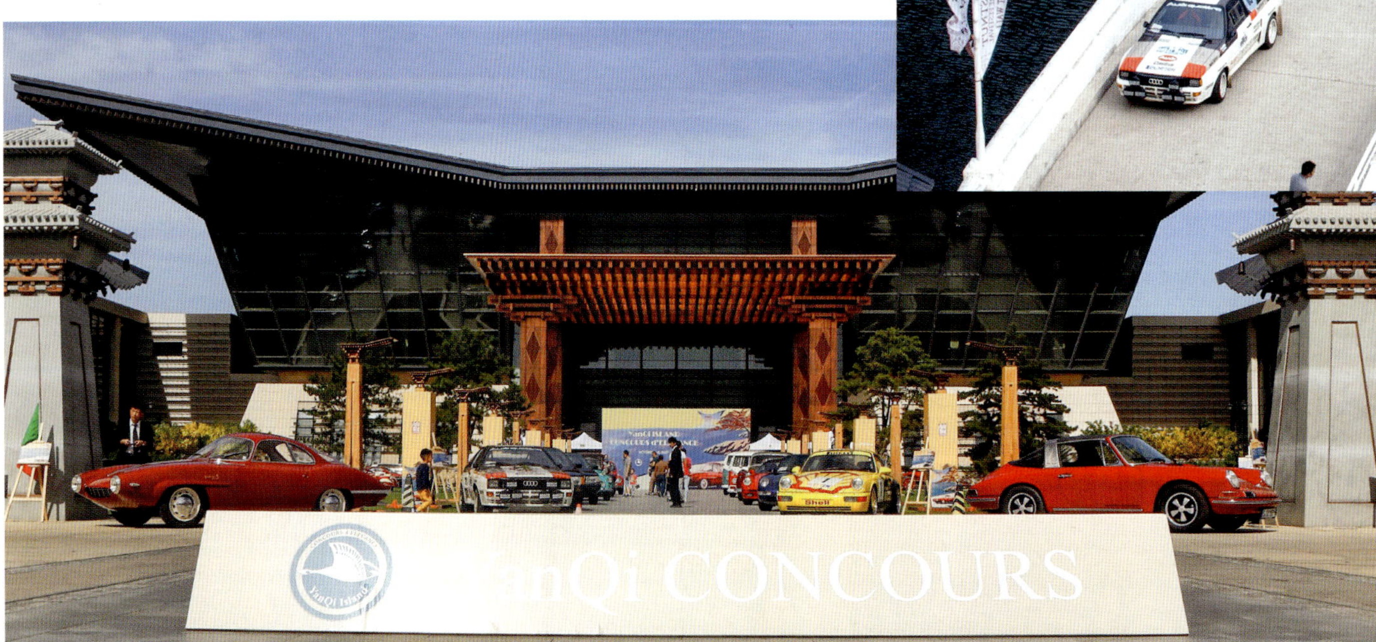

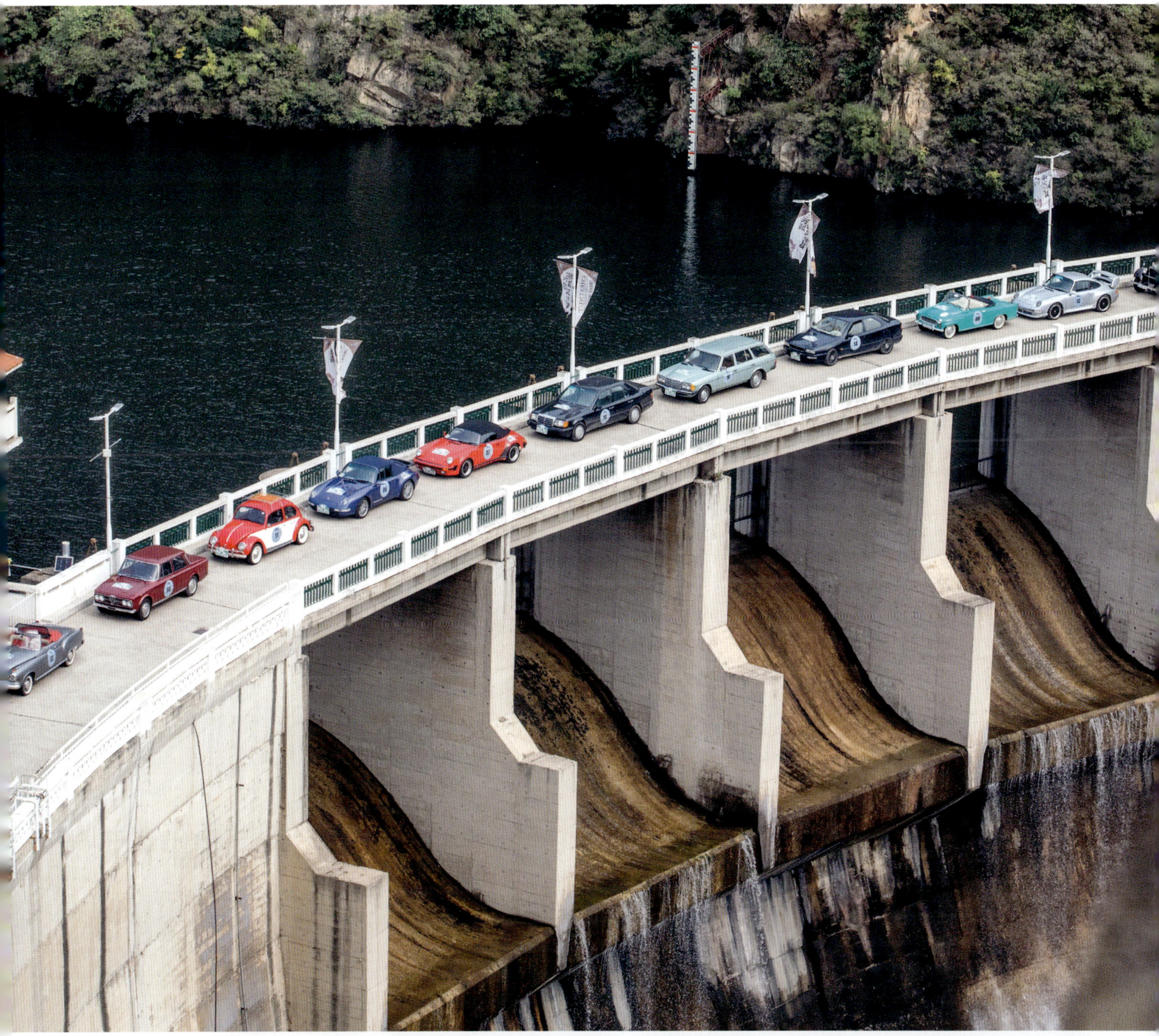

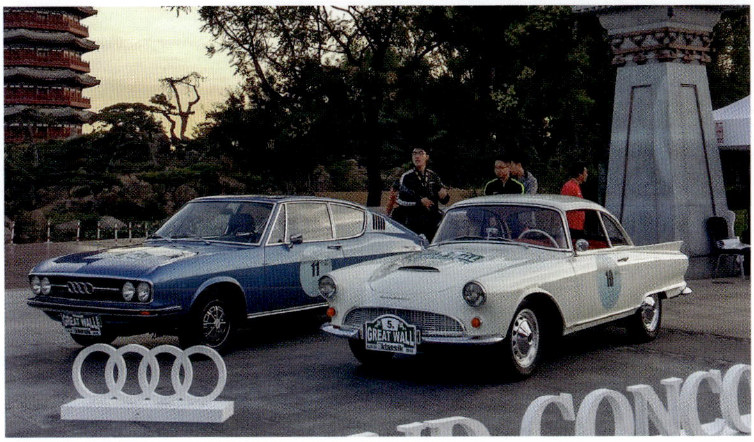
GREAT WALL

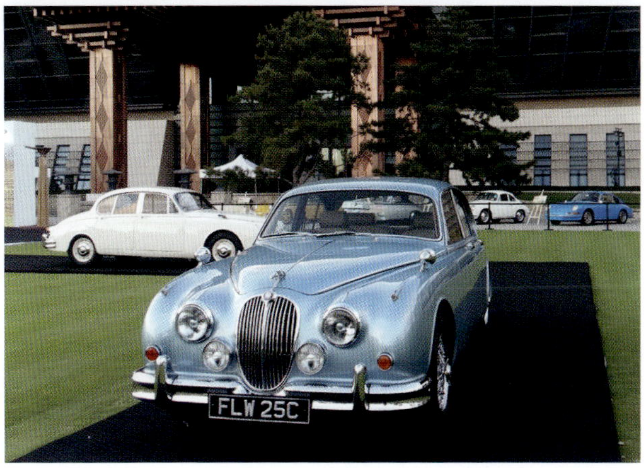
FLW 25C

**RIGHT:** *The YanQi concours attracted several top collectors and enthusiasts from Europe such as Corrado Lopresto, who brought rarities like this one-off prototype Alfa Romeo Giulietta Super Sprint. (Gautam Sen)*

**BELOW:** *The first important concours in India was the Cartier Travel and Style Concours d'Elegance in 2008. The second occurrence in 2011 in Delhi had the winner of the first concours, a Delahaye 135M with Figoni et Falaschi body, on display. (Gautam Sen)*

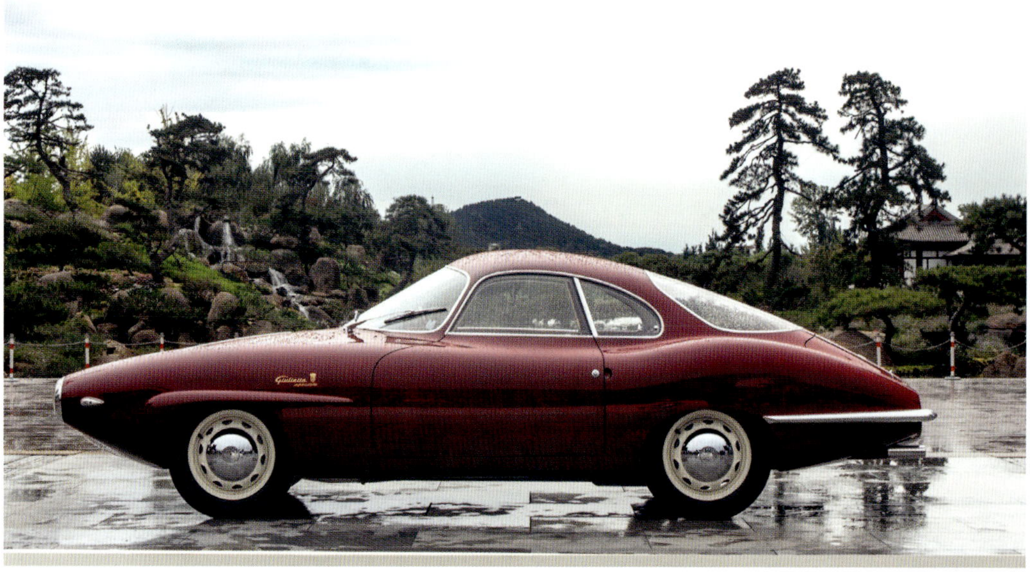

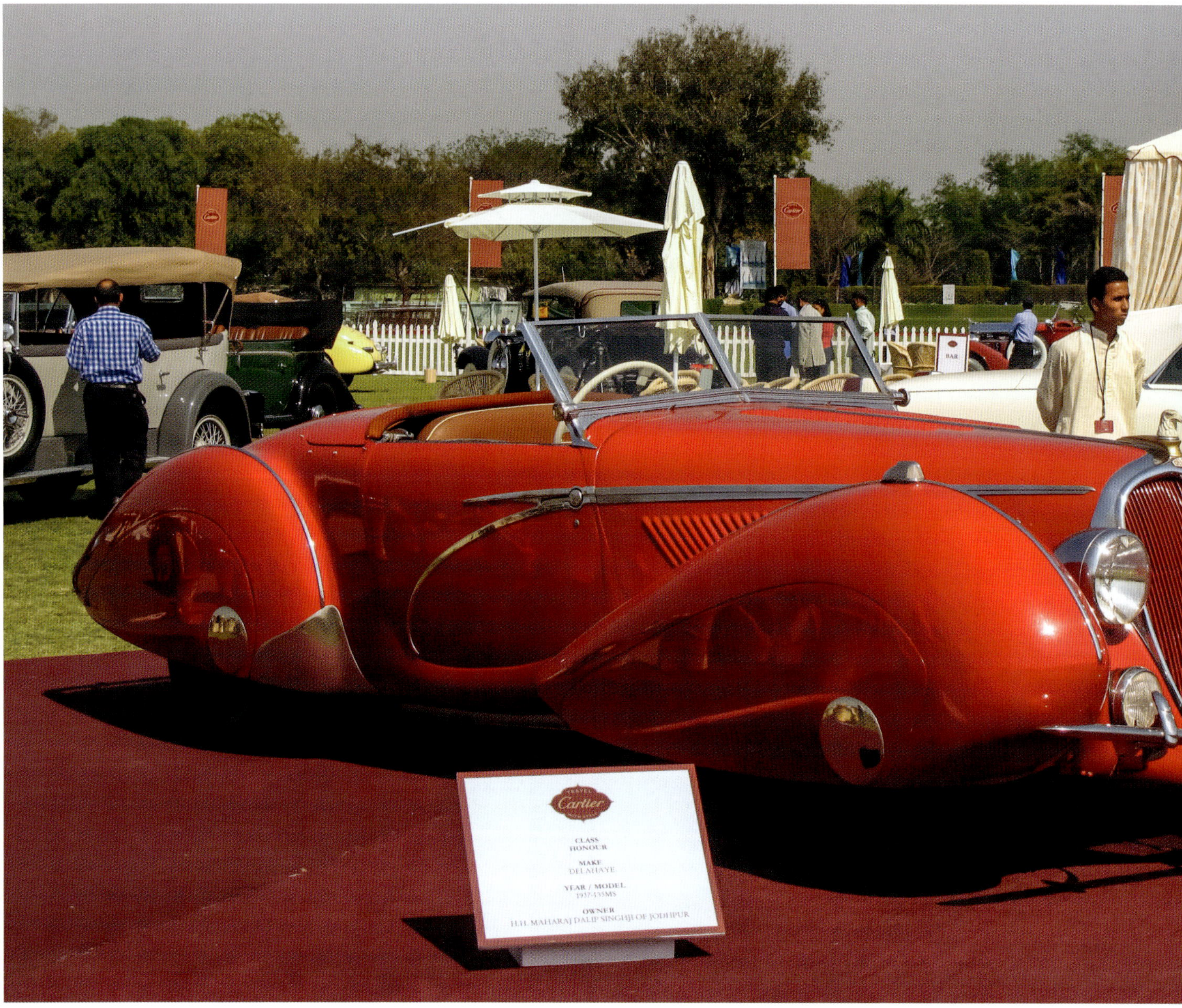

led by current FIA President Mohamed Ben Sulayem. Since 2016, Katara Cultural Village in Doha has been home to the Qatar Concours d'Elegance. The continued growth of the Seven Concours in Saudi Arabia's capital Riyadh highlights the opportunities and vast potential for such events in this region."

"Despite being home to many valuable and historically significant collector cars, most of which are kept hidden in private collections, the region still lacks a truly world-class concours event," remarked Mohammed Luqman.

But it is India, with its rich heritage of historic vehicles, which embraced the concours d'elegance trend earlier than most of the Asian countries. French luxury brand Cartier sponsored India's first major event, launching the Cartier Travel and Style Concours d'Elegance in Mumbai in 2008. After 2011, the event was held biennially, with the last show taking place in 2019 before Cartier withdrew from India. Another successful Indian concours is the 21 Gun Salute. The most recent addition is the Oberoi Concours d'Elegance, whose inaugural event was held from February 16-18, 2024, at the luxurious Oberoi Udaivilas in Udaipur.

"It all stems from my love for automobiles, particularly classic models," explained Arjun Oberoi of The Oberoi Group. "I owned an Alfa Romeo 2000 GTV at university, which I brought back to India after completing my studies in the UK. My uncle, Biki Oberoi, had a collection of vintage and classic cars, and The Oberoi Group had a fleet of custom-designed Hindustan Ambassadors for our guests, created by Manvendra Singh."

He continued, "I also owned a Royal provenance Indian Chief motorcycle. After attending the Cartier Concours, I decided to enter the motorcycle at the 2017 Hyderabad event and had renowned restorer Gurmukh Singh Salh bring it back to life. It won Best of Show. That experience sparked my passion. David McKirdy, who oversees Sir Michael Kadoorie's collection, encouraged me to enter The Quail event in the US, where the Indian Chief also won its class. At the 2019 Cartier Concours in Jaipur, my Alfa Romeo 2000 GTV won as well. These three wins deepened my interest in concours events."

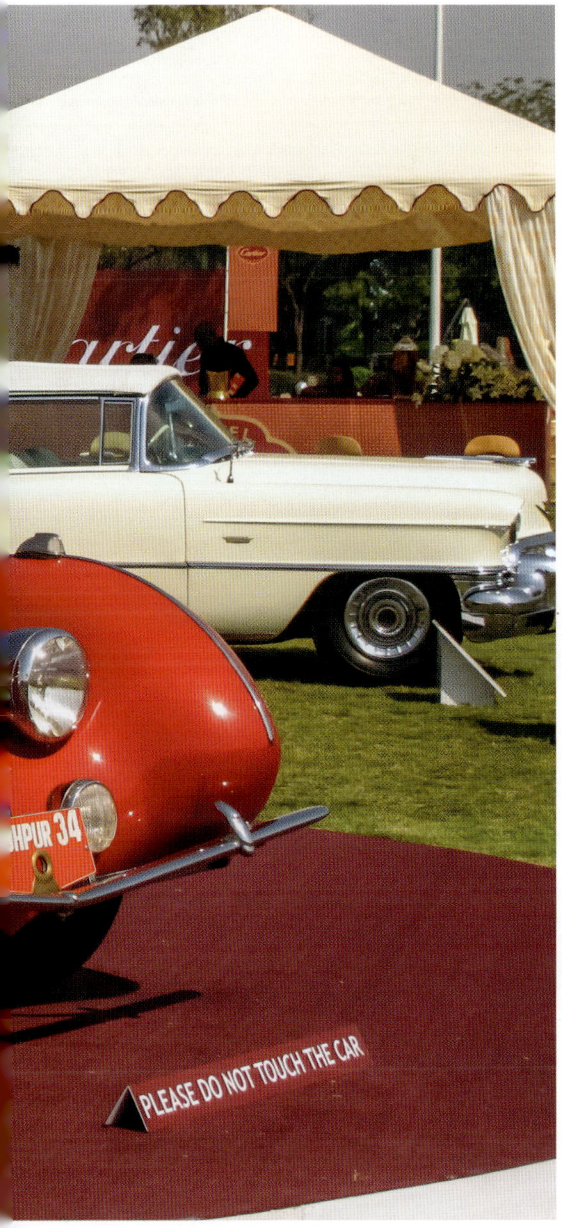


Oberoi then discussed the evolution of their concours venture: "When the Cartier Concours came to an end, we saw an opportunity to create a high level event similar to Cartier's. After discussions with Manvendra, we realized the potential. Manvendra had a network of judges, Sandra Button among others, and knew where the cars were. Oberoi delivered the hospitality and iconic venue, which is our strength. BMW came on board as a sponsor, and everything came together."

Arjun Oberoi also highlighted that 2024 marked the 90th anniversary of The Oberoi Group's founding: "This was another reason we organized the concours and a celebration of this important milestone. We also had the opportunity to display some of the family's cars, including a Bentley 3½ Liter once owned by the Maharaja of Indore, Yashwant Rao Holkar II, along with a 1967 Cadillac and a Lincoln Continental."

Reflecting on the inaugural event in February 2024, Oberoi said, "We believe the concours was well received and a success. We're planning our second edition in 2026 at the same venue – Oberoi Udaivilas in Udaipur. The setting is stunning, Udaipur is easily accessible, and the format worked well. We may consider using other Oberoi properties in the future, like our Agra hotel near the Taj Mahal. Also, as the event grows, it plans to attract a wider international audience and participation and will add a tour of Rajasthan for those who would wish to explore India and enjoy their cars beyond the concours."

Regarding the financial aspect, Arjun Oberoi added, "We didn't turn a profit in the first year, but with increased sponsor interest, we hope to break even by the second event."

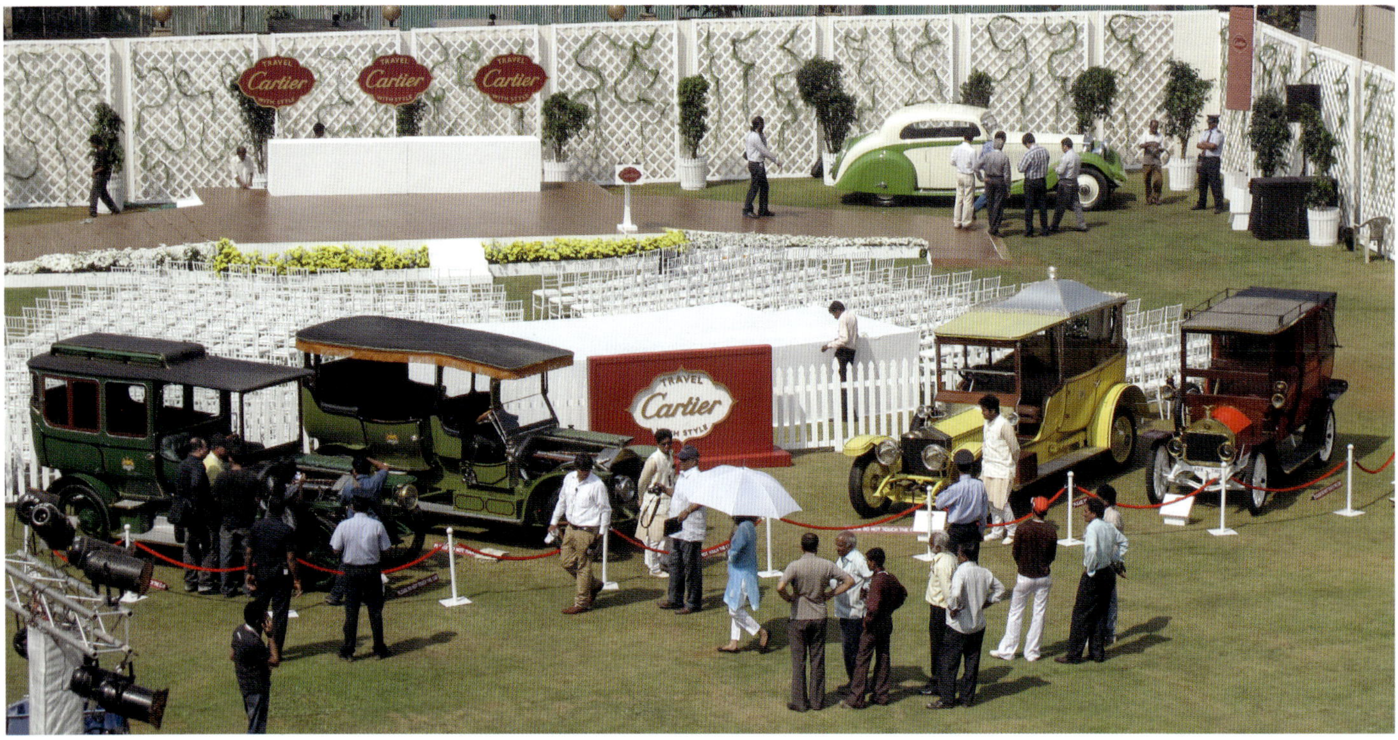
Cartier
Cartier
Cartier
Cartier

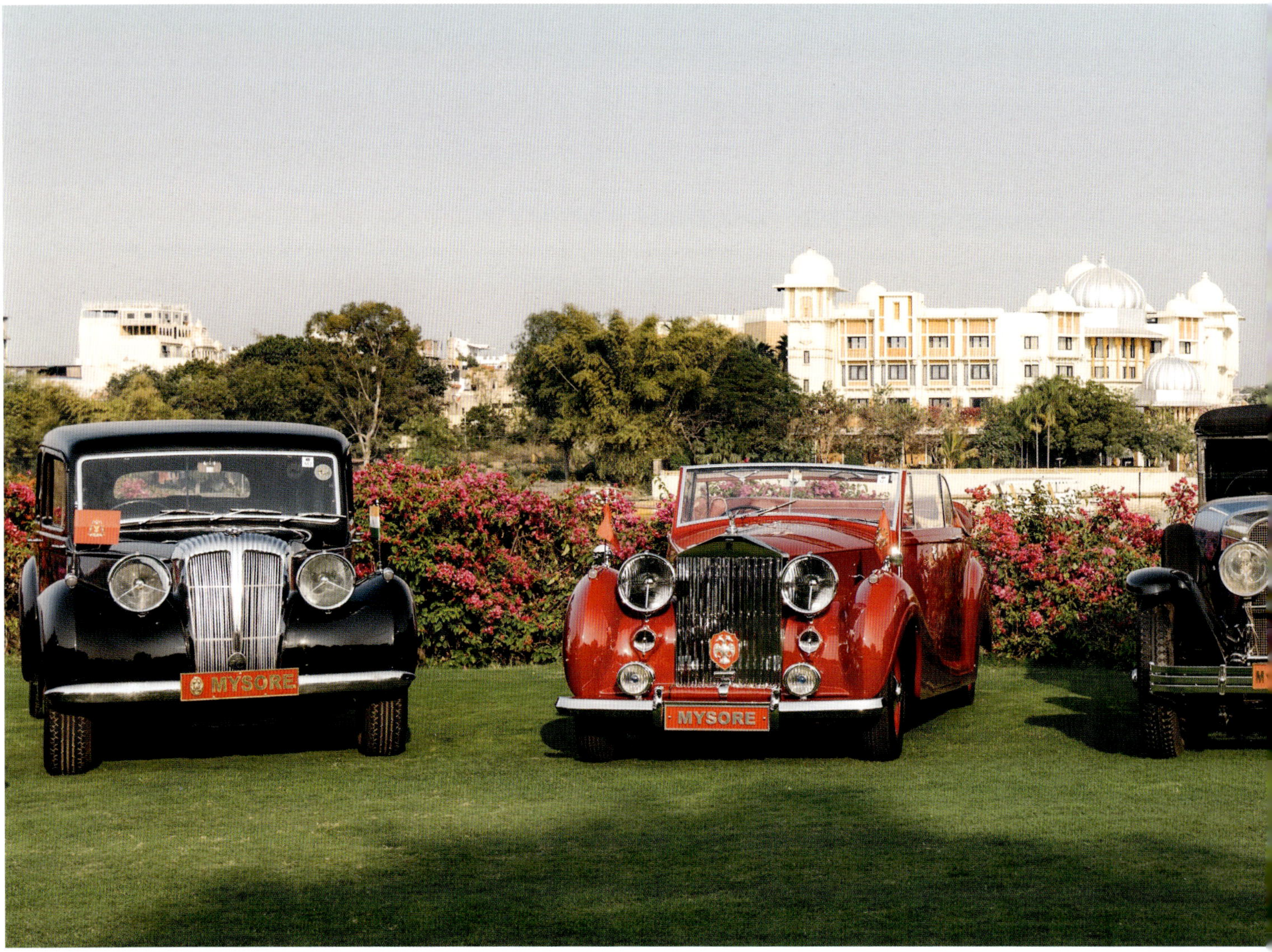
MYSORE
MYSORE

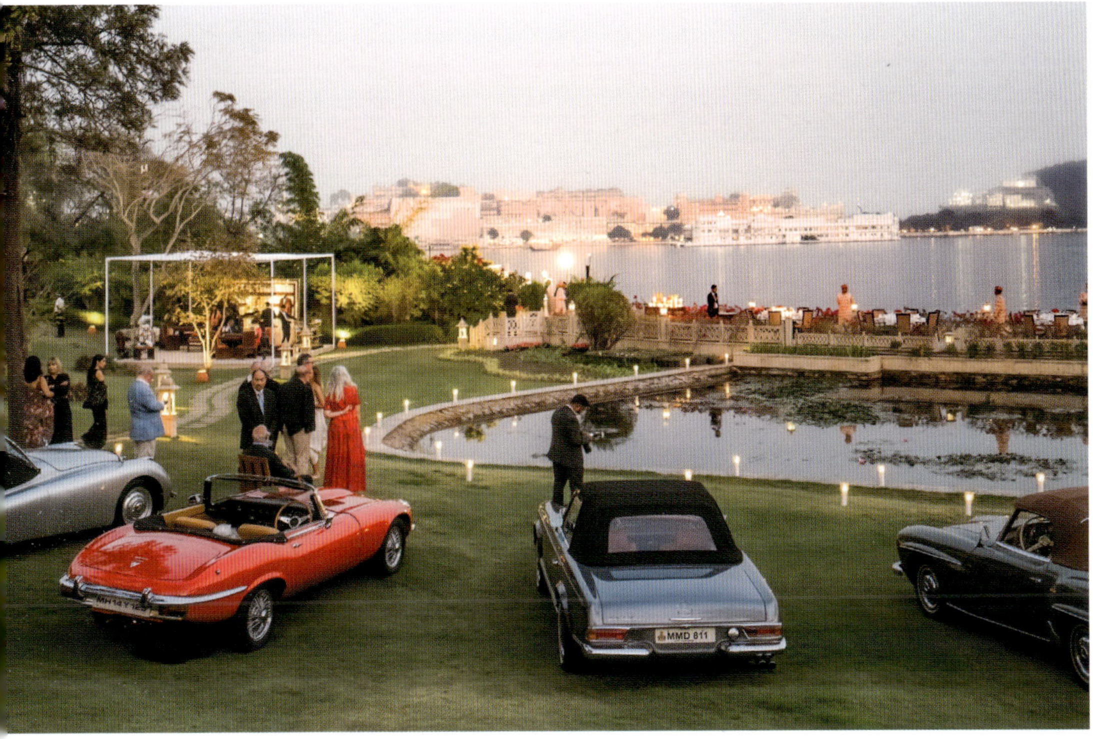


**OPPOSITE TOP:** *The third Cartier was in 2013, when the concours was back in Mumbai. (Gautam Sen)*

**THIS PAGE:** *The new event that caught the attention of enthusiasts was the Oberoi Concours d'Elegance held in Udaipur in February 2024. The Oberoi Concours managed to gather some of India's finest historic vehicles such as this Bentley Mk VI (second from right in bottom image) from 1949, which used to belong to the Maharaja of Mysore, and is now a part of the Yohan Poonawalla collection, as well as this unique Hispano-Suiza H6C (middle), another ex-Maharaja of Mysore car. (Sanket Patel)*

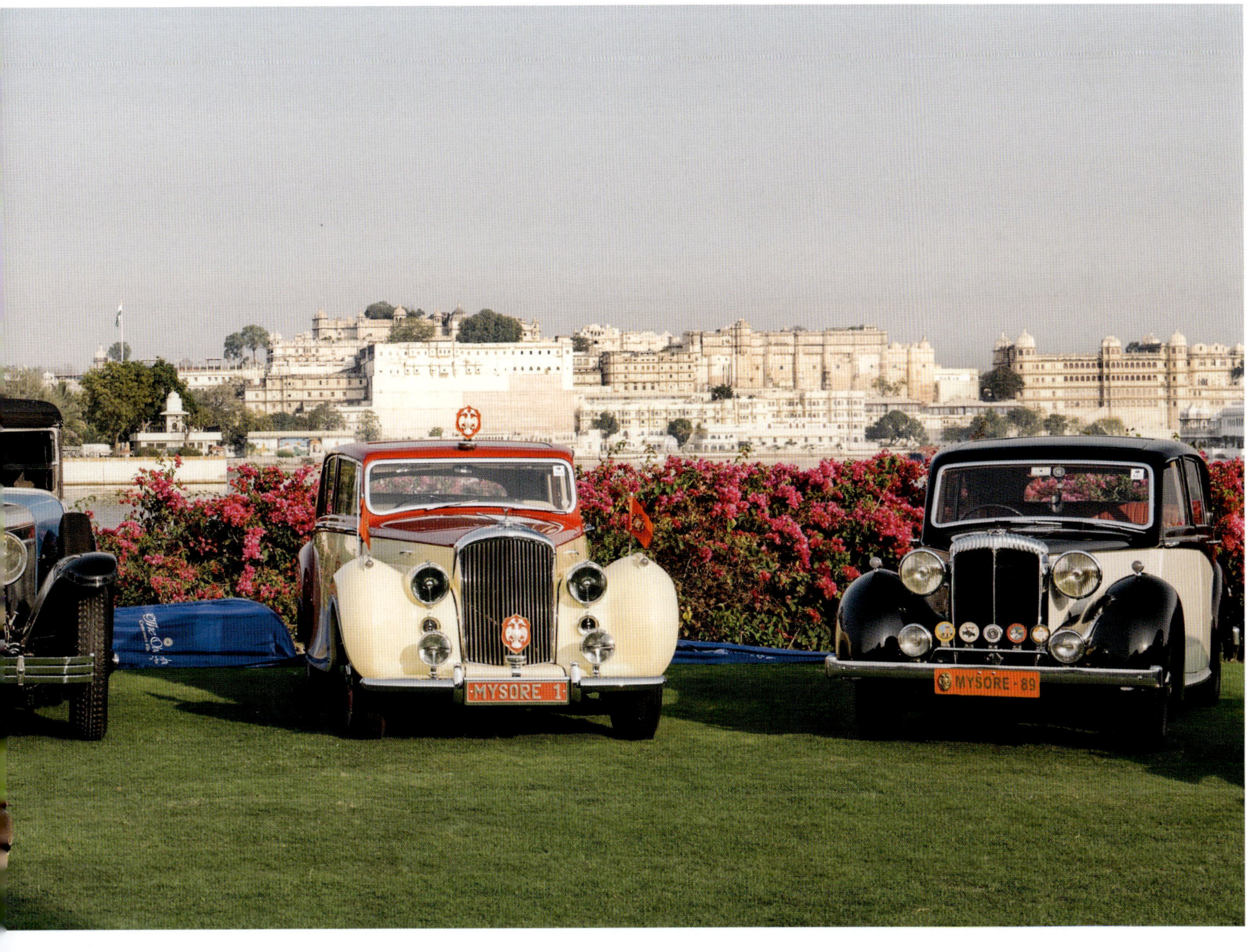

**RIGHT:** *Conceived by Arjun Oberoi (right) and executed by a team headed by Manvendra Singh and his son Siddhraj, the Oberoi Concours d'Elegance has rapidly built up a fine reputation. Prince Michael of Kent (left) gives a speech at the Concours. (Sanket Patel)*

**BELOW:** *India's only Mercedes-Benz 300 SL Roadster was also displayed. (Sanket Patel)*

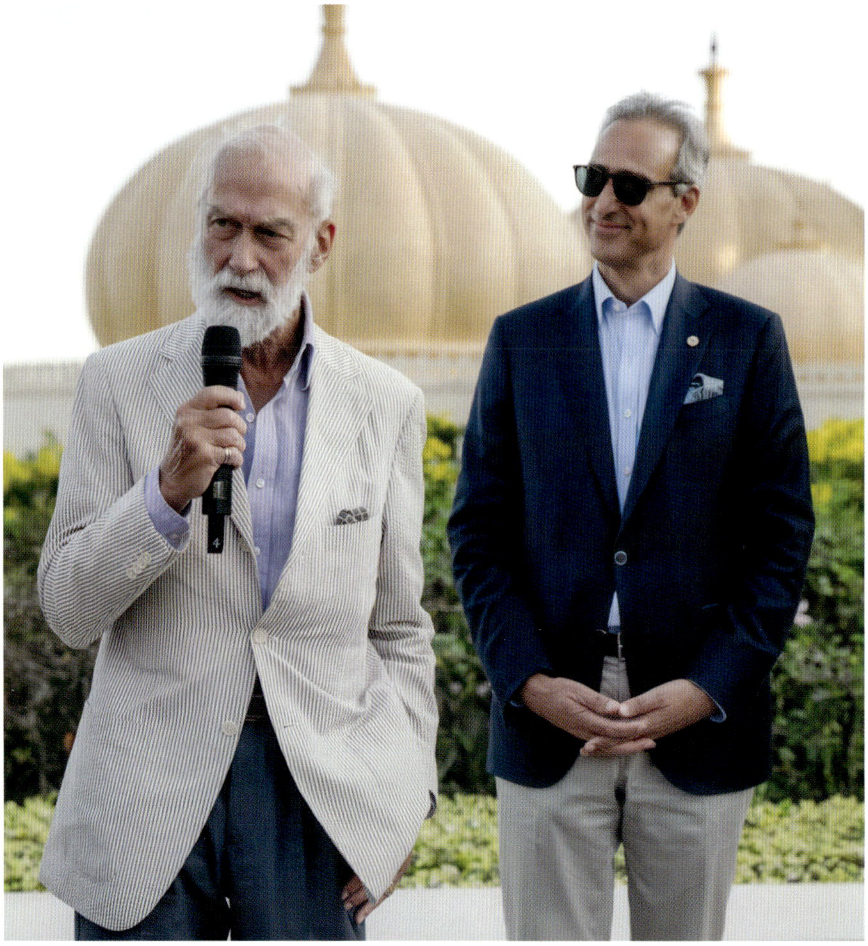

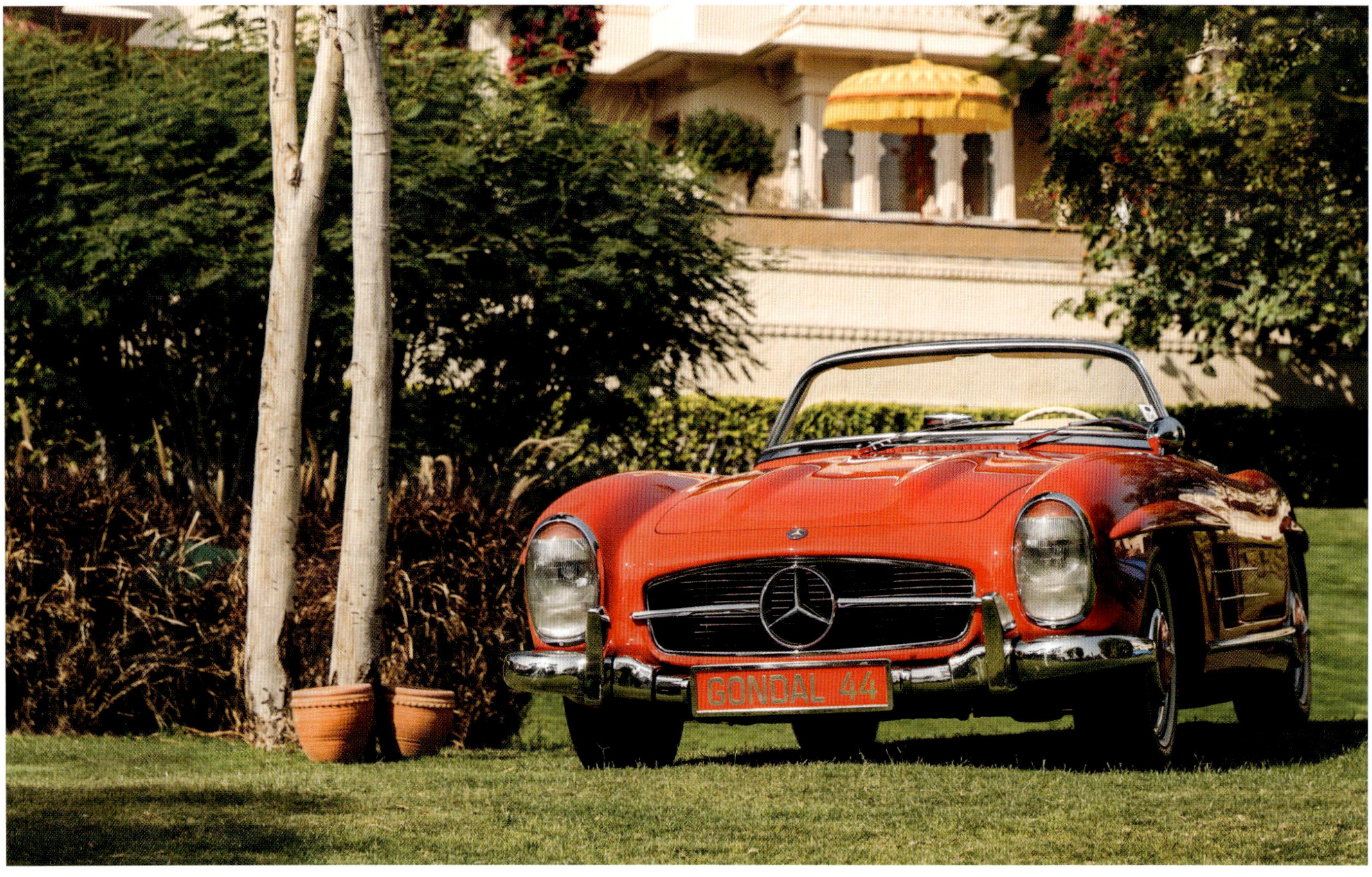

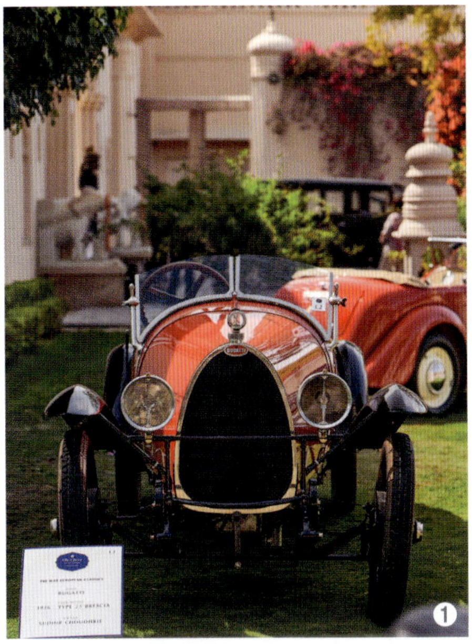

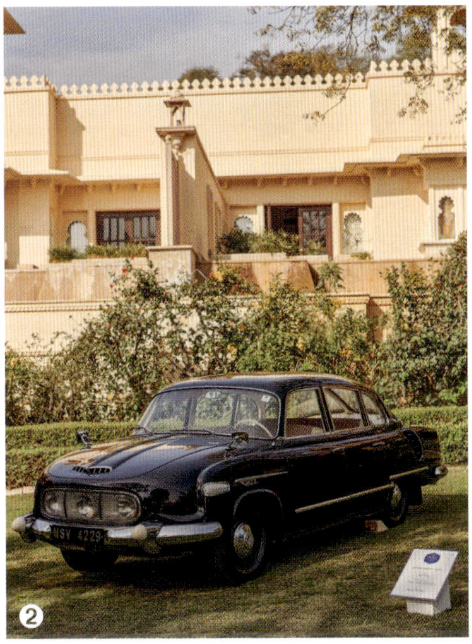

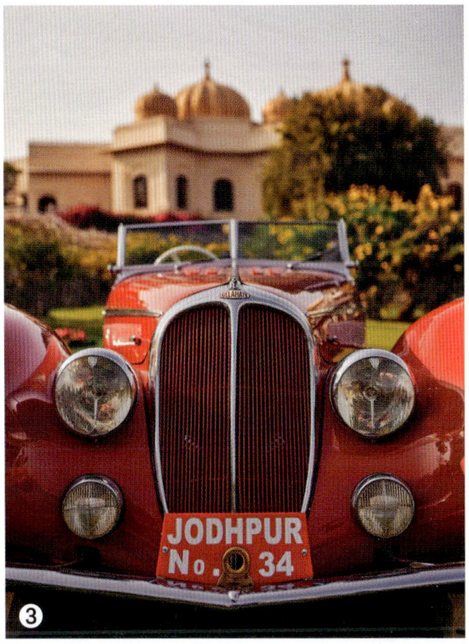


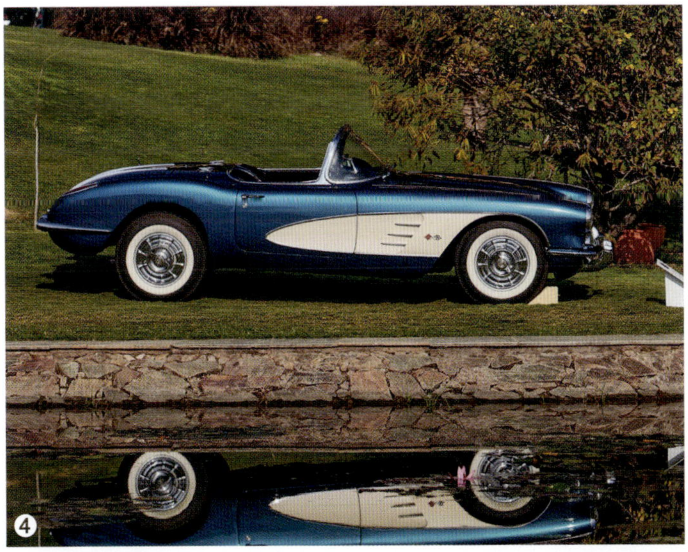

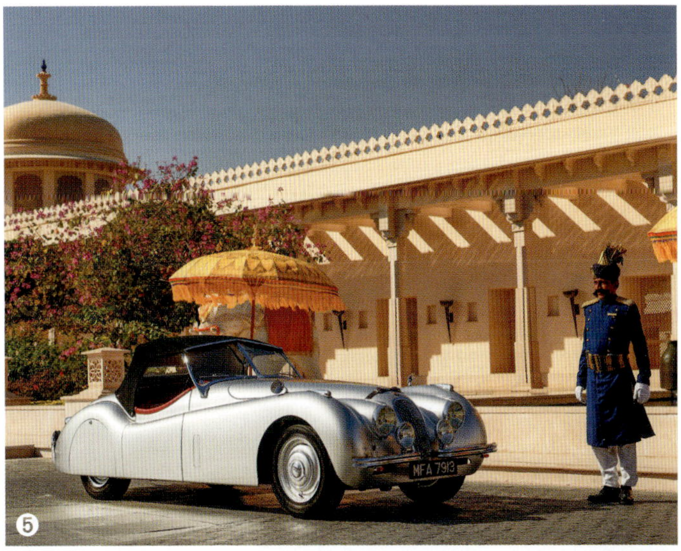

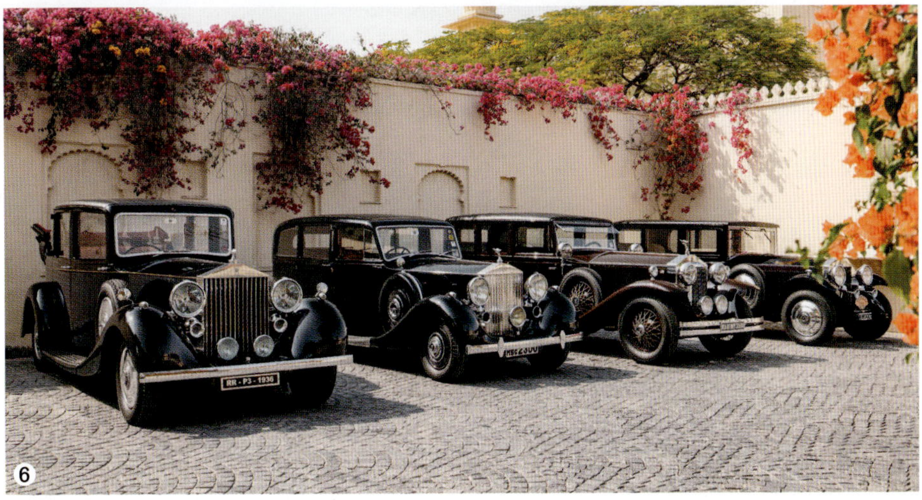

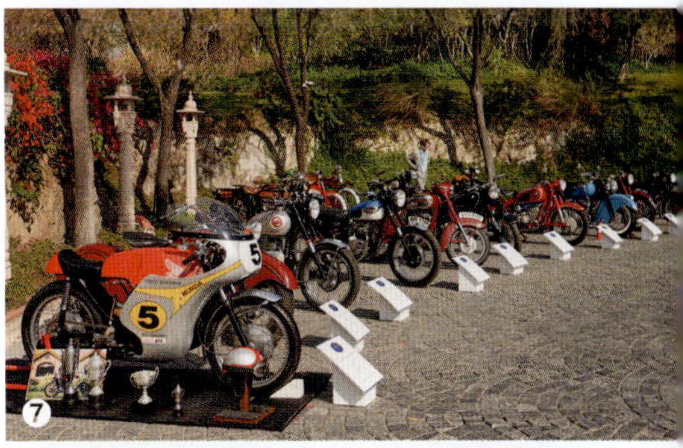

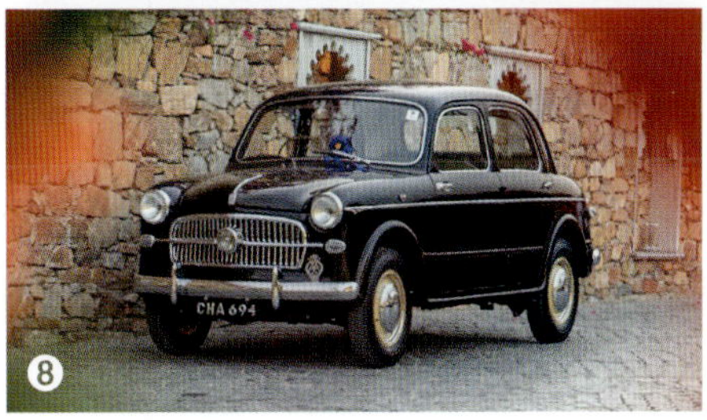

*❶The Oberoi Concours also managed to get one of the handful of Bugattis in India. ❷A recently restored Tatra 603 ❸1939 Delahaye 135M Figoni & Falaschi. ❹An early Corvette. ❺Jaguar XK120. ❻A fleet of Rolls-Royces. ❼There was even a class for motorcycles. ❽And a class for cars which had been assembled or made in India. (Makarand Baokar)*

Outside of Europe, New Zealand may boast one of the oldest concours d'elegance events in the world. The Intermarque Concours d'Elegance, an integral part of the Ellerslie Car Show, traces its origins to December 1, 1972, when the first concours was held at the Sunken Gardens in Cornwall Park, Auckland, organized by the MG Car Club. From the start, it was a collaborative, inter-club event, with MG winning the inaugural Inter-Club Challenge Shield. Other early winners included Alvis, Studebaker, Jowett, Riley, and Citroën car clubs. MG and Jaguar dominated the competition until 1981, when Porsche claimed the Shield. That same year, the event was relocated to Ellerslie, a suburb of Auckland. By 2024, over 80 car clubs, major vehicle dealerships, sponsors, and the restoration industry participated in the event.

Australia has a long history of hosting concours d'elegance events, such as the one-off Ten Ten Ten and the now-defunct Motor Classica Melbourne, though many older ones have faded. The most notable in recent years is the Sydney Harbour Concours d'Elegance. "The inaugural Sydney Harbour Concours took place in March 2019 at Lindesay, a National Trust property dating back to 1834," explains James Nicholls, the Anglo-Italian founder and curator of the event. Highlights of the debut included the first Bentley to race at Le Mans in 1923 and the 1933 Mille Miglia-winning MG K3, driven by legends like George Eyston, Count Johnny Lurani, and Tazio Nuvolari.

"Due to space constraints, with only 35 cars at Lindesay, we moved the event to the larger Swifts estate, known for its beautiful Victorian lawns and gardens," Nicholls recounts. "In 2022, we relocated to the UNESCO World Heritage site of Hyde Park Barracks, a remarkable testament to Australia's colonial legacy, and returned there in 2023."

In 2024, the event moved once again, this time to Cockatoo Island, another UNESCO World Heritage site located a 20-minute boat ride from Sydney Harbour. "Transporting cars, equipment, and

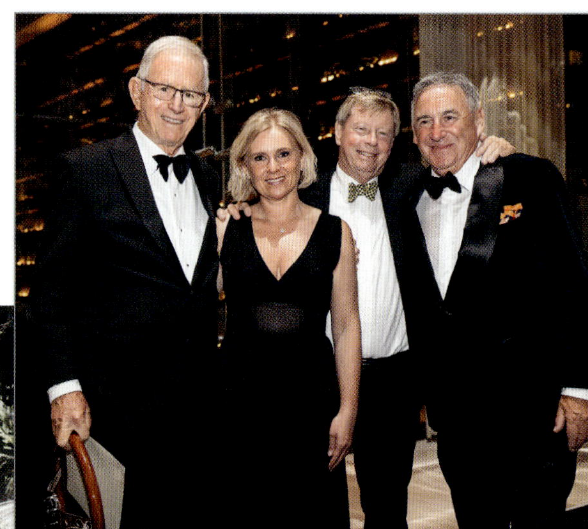

**INSET:** *Founded in 2019 and managed by James Nicholls (pictured on right), the Sydney Harbour Concours is now recognized as one of the top events East of Suez. With James is Peter Harburg (left), the owner of the 2024 Best of Show winning Ferrari 250 LM (see photograph opposite), chief judge Alessandra Giorgetti, and Steven Gregg (second from right), the Chairman of Ampol. (Sydney Harbour Concours)*

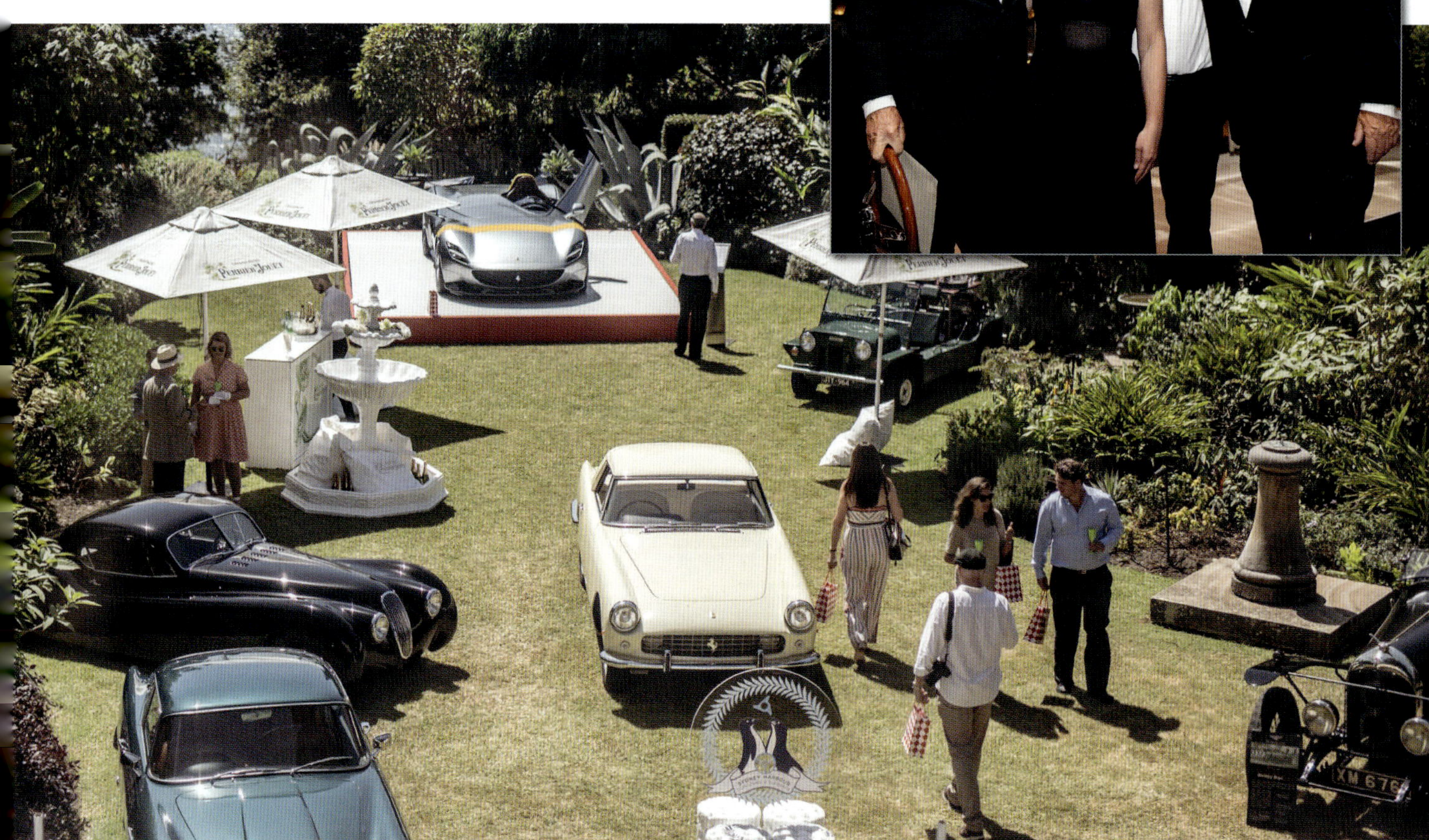

**BELOW:** *With the 2024 Sydney Harbour Concours d'Elegance shifting to Cockatoo Island, organizer James Nicholls has found a wonderful location for his concours. (Sydney Harbour Concours)*

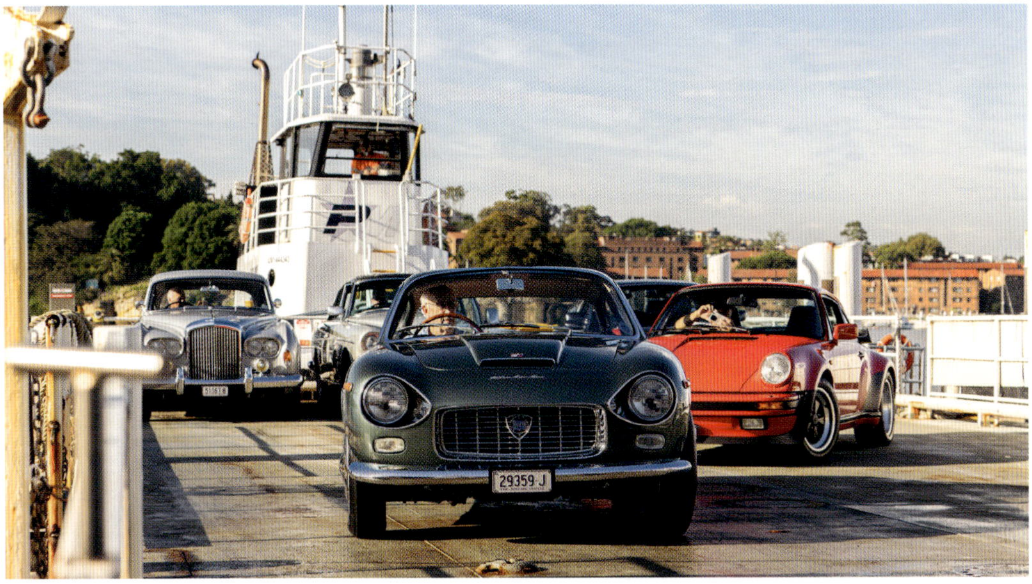

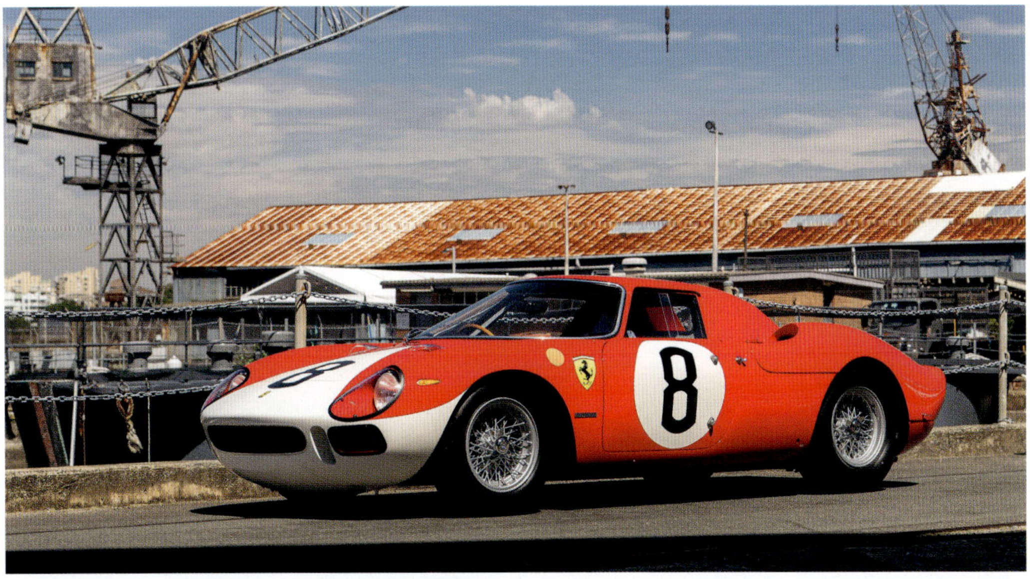

**TOP:** *Cockatoo Island has its costs, as transporting all the cars and people on barges to the island was quite an operation. (Sydney Harbour Concours)*

**MIDDLE:** *Despite being far away from the usual concours circuit, James Nicholls has managed to get some excellent entries over the years, including Peter Harburg's 1964 Ferrari 250 LM, the 2024 Best of Show winner. (Sydney Harbour Concours)*

**BELOW:** *The storing of the cars and the preliminary judging is at a Sydney Harbour warehouse. (Sydney Harbour Concours)*

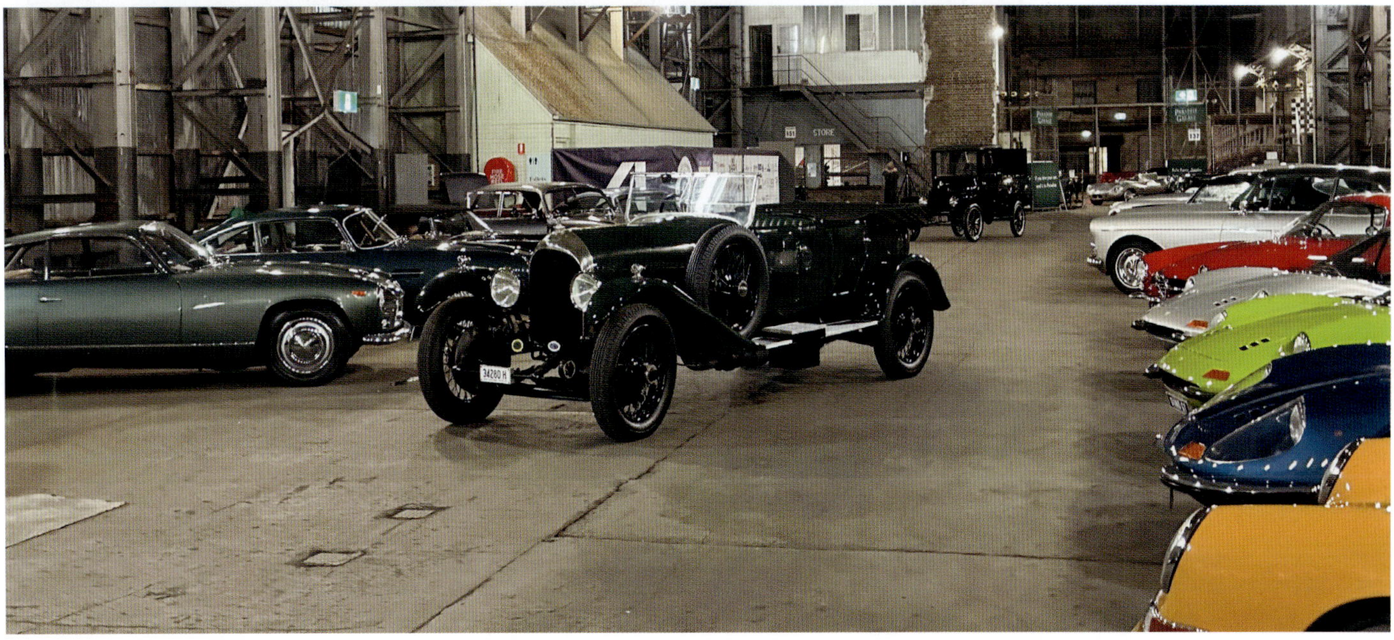

guests was a massive undertaking, requiring barges for the cars and VIP boats for participants, but the effort paid off, and everyone loved the experience. We plan to return to Cockatoo Island in 2025," Nicholls added.

Nicholls also highlighted that the event's chief judge, Alessandra Giorgetti (Vice President to the Registro Internazionale Touring Superleggera), oversaw a more subjective scoring system, avoiding the restrictions of the ICJAG system.

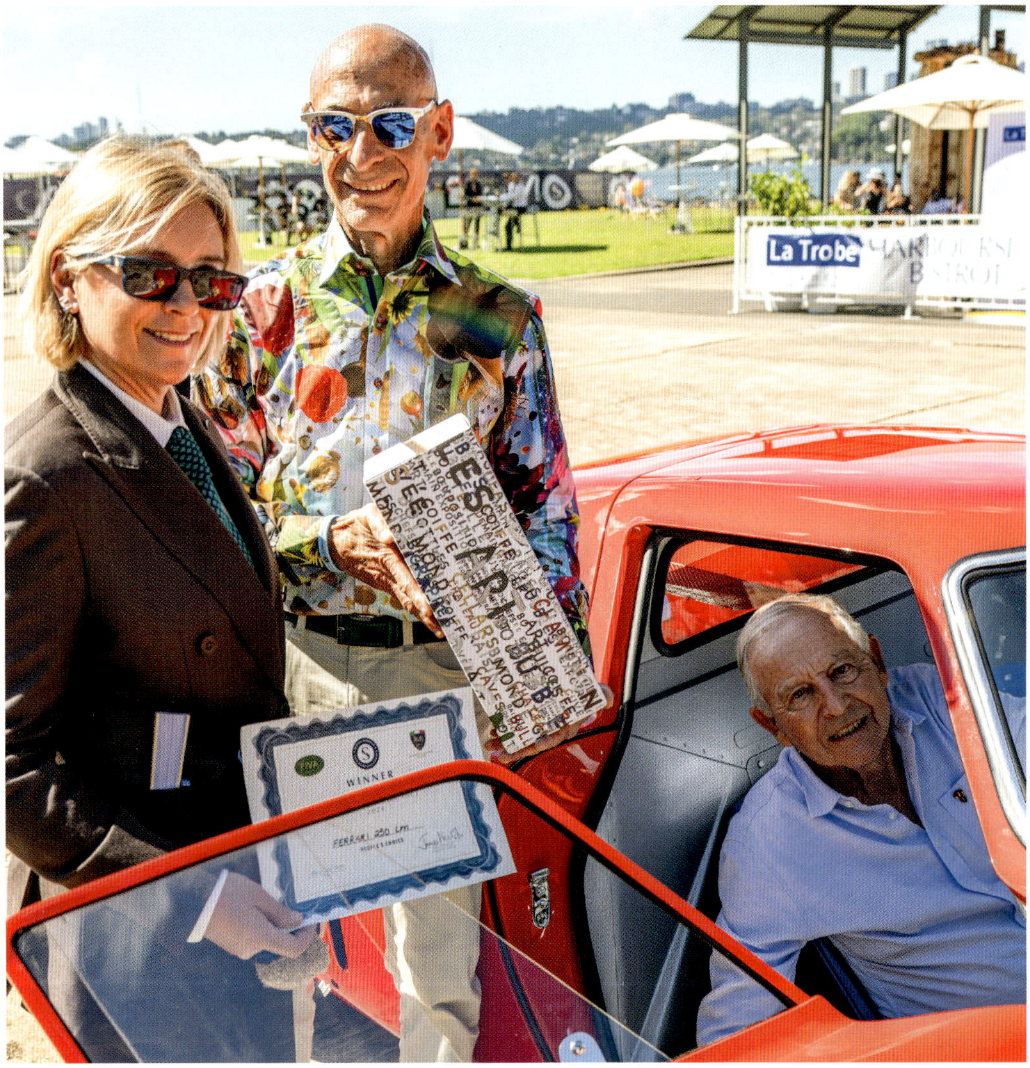


**TOP:** *Chief Judge Alessandra Giorgetti from Italy with fellow judge from the USA, Stuart Field. Driving the 250 LM is Spencer Martin, well known Australian racing driver who drove a 250 LM among other cars in period. (Sydney Harbour Concours)*

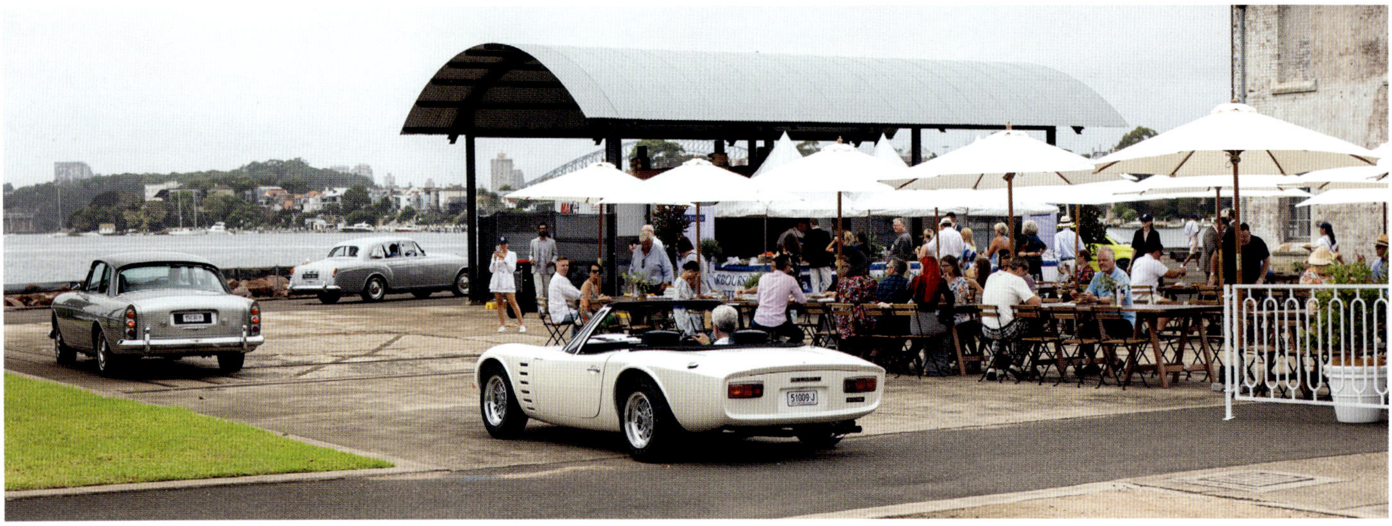

**BELOW:** *Among the rarities that James Nicholls has managed to get to his concours, this is one of a limited number of the Australian Bolwell Mk VIII Nagari roadsters ever made. (Sydney Harbour Concours)*

## Concours d'Elegance reinvented

What about the country that first pioneered the concept? After the Louis Vuitton Classic à Bagatelle ended in 2002, France experienced a lengthy pause until the Chantilly Arts & Elegance made its debut on September 7, 2014. However, when the French re-entered the scene, they did so with impressive flair, showcasing elegant style and sophistication.

The event was masterminded by the renowned Patrick Peter and his team at Peter Auto. Chantilly Arts & Elegance broke away from traditional concours d'elegance norms by introducing the term concours d'état to judge older vehicles. As Patrick Peter explained, "Globally, the historic vehicle community uses the term concours d'elegance for what the French correctly refer to as concours d'état – essentially, a competition of condition for older vehicles."

Peter further elaborated, "The original concours d'elegance originated in France and focused entirely on elegance. It was about the vehicle's design, its aesthetics, and the woman presenting it – her outfit, hat, gloves, even the pedigree dog. It was all about style, finesse, and distinction, with a jury composed of both automotive experts and fashion designers."

"In contrast, what the Anglo-Saxons call a concours d'elegance involves a panel of automotive experts meticulously assessing the authenticity of a historic vehicle, whether restored or preserved," Peter added. "These are usually static events, often not open to the wider public."

He acknowledged the importance of these concours in maintaining originality and preventing improper modifications, comparing it to preserving the historical integrity of an 18th century manor house. Experts and historians play a crucial role in keeping cars and motorcycles authentic for future generations.

*The setting of Château de Chantilly beats just about every other concours when it comes to sheer class and elegance. (Anne Castagnos-Sen)*

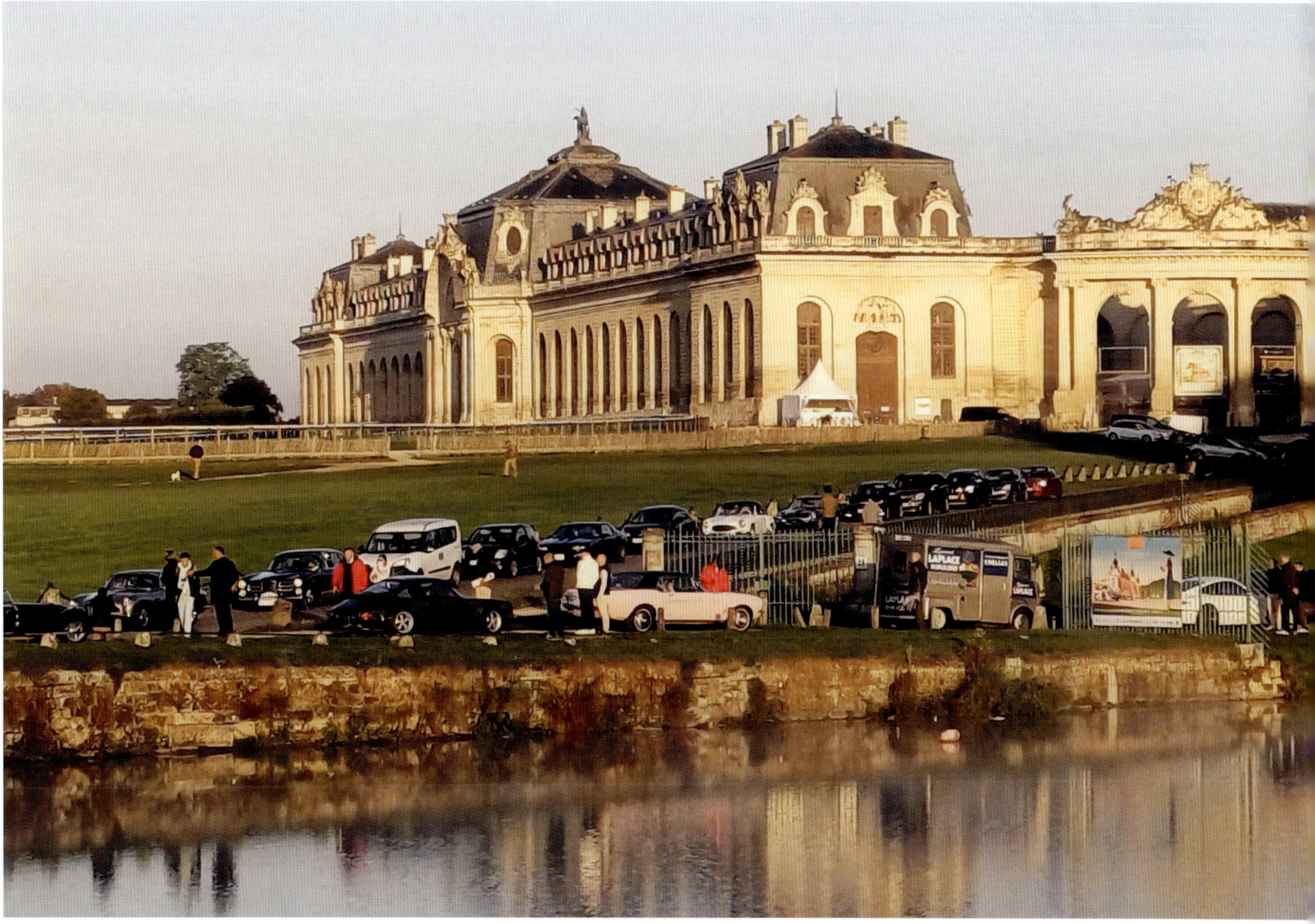

The elegance, style and flair continues inside the château for the masked ball. (Anne Castagnos-Sen)

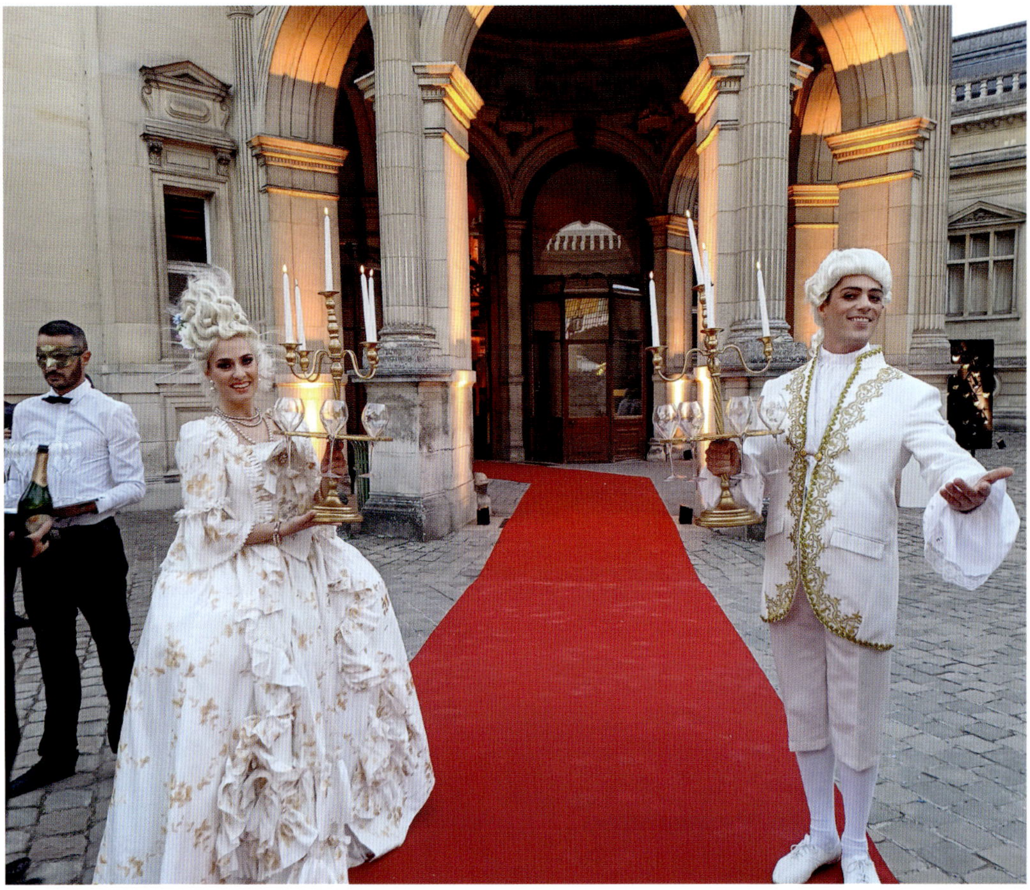

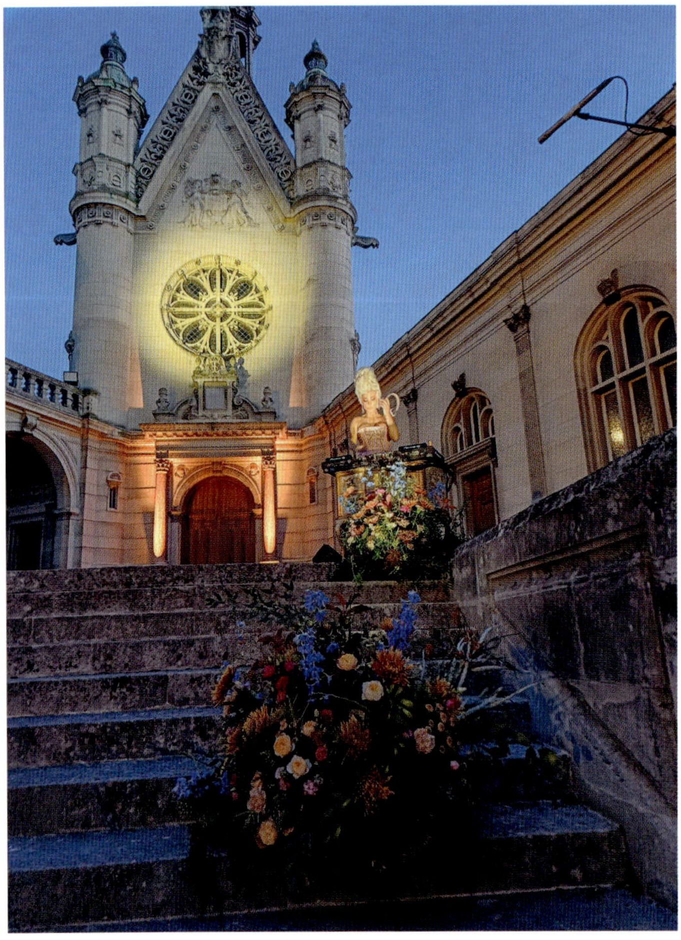

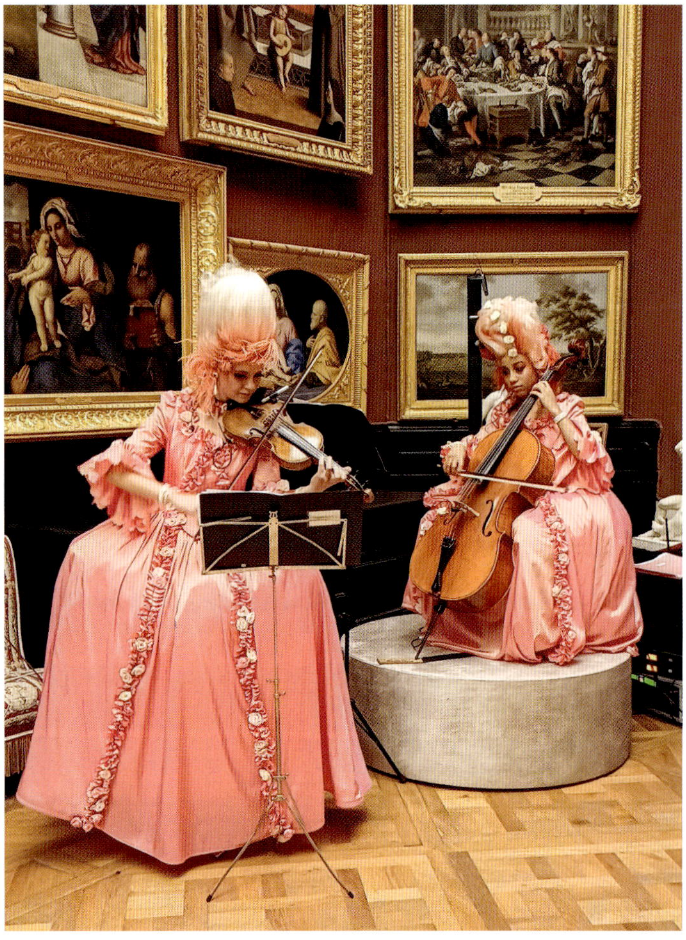

However, Patrick Peter's Chantilly Arts & Elegance shifted the focus. Authenticity accounted for only 20 out of 100 points, with equal weight given to design, history, condition of the vehicle's exterior and interior, and its functionality.

But Peter's central point was this, "For a Frenchman like me, it's a bit odd that the term 'elegance' is used to evaluate the size, type, origin, or condition of engines, metal parts, and nuts and bolts. While these elements are integral to a vehicle, I would hardly call them 'elegant'."

Thus, the Chantilly Arts & Elegance became the first to acknowledge the term concours d'état. They took it a step further by establishing two separate Best of Show categories, one for pre-war vehicles and another for post-war, recognizing that comparing them was akin to comparing apples to oranges.

Chantilly also introduced a class dedicated to brand-new concept cars, one-offs, and prototypes, aptly named Concours d'Elegance. In this category, a different judging system was used to evaluate new cars based on their design, elegance, and presentation, with models adorned in outfits from renowned haute couture designers.

Since 2014, the biennial Chantilly Arts & Elegance has awarded three Best of Show titles – one for the Concours d'Elegance for new cars, and two for the Concours d'État in the pre-war and post-war categories.

**UPPER & MIDDLE (ALL):** *More importantly Chantilly has always managed to line up some extraordinary cars, such as these Bugatti 101s in 2014, or these Ballots and the cars designed by design ace Marcello Gandini, in 2019. (Xavier de Nombel and Makarand Baokar)*

**BOTTOM (ALL):** *The optional tour the day before and the mandatory drive by at the fountain ensures that there are no trailer queens at the Chantilly Arts & Elegance. (Gautam Sen)*

**OVERLEAF:** *The 20-odd clubs which are invited to exhibit at the Château de Chantilly grounds also put up a great show featuring some of the finest cars from among the clubs. This is from 2014. (Xavier de Nombel)*

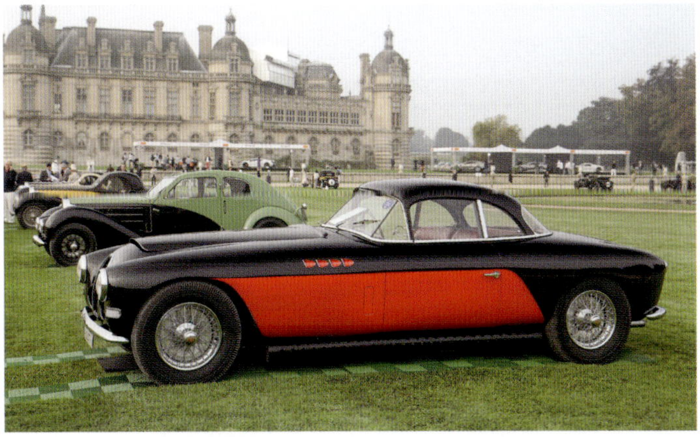

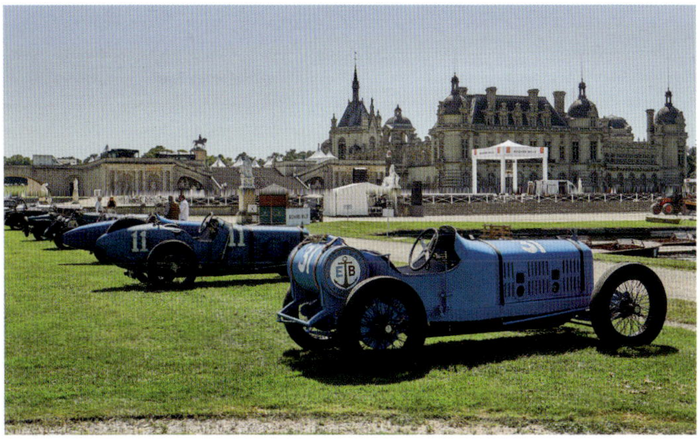

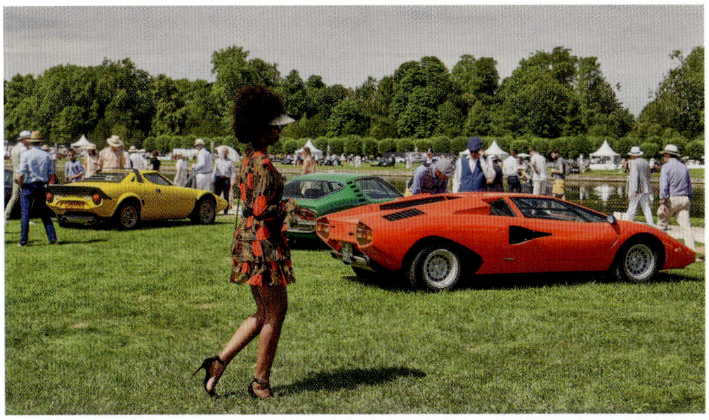

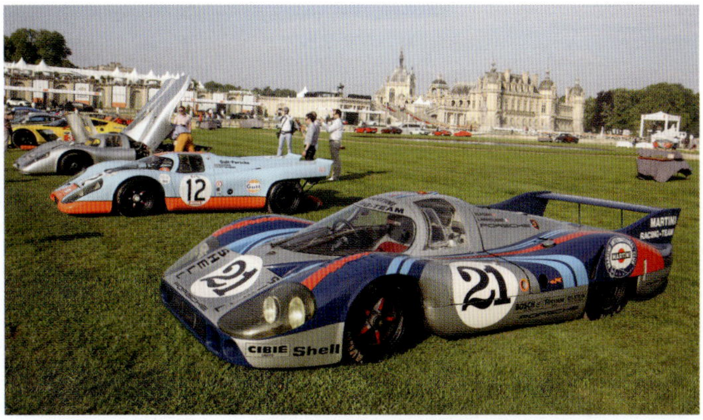


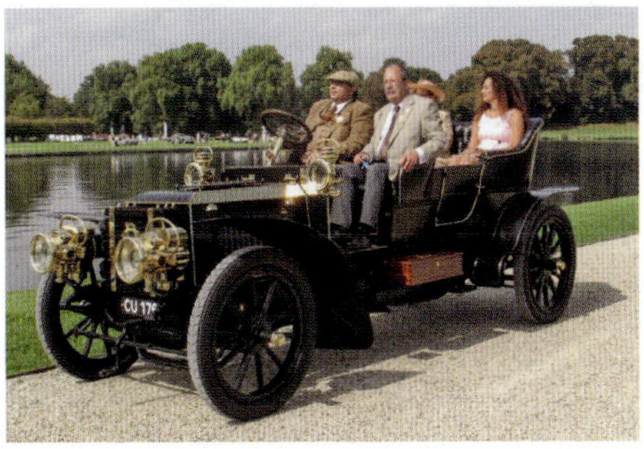

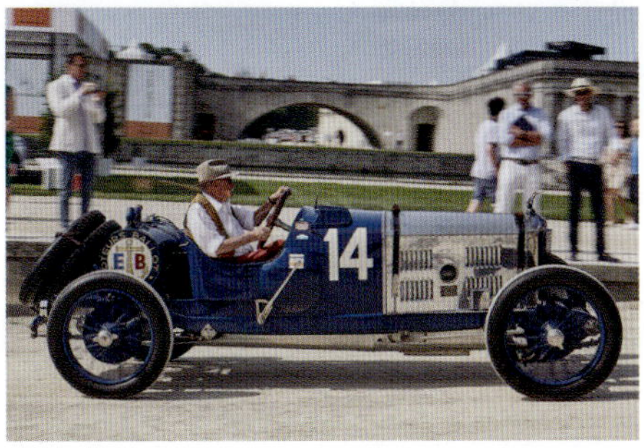


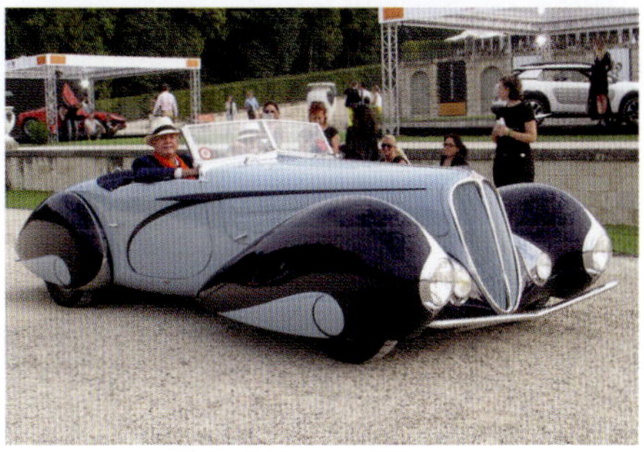

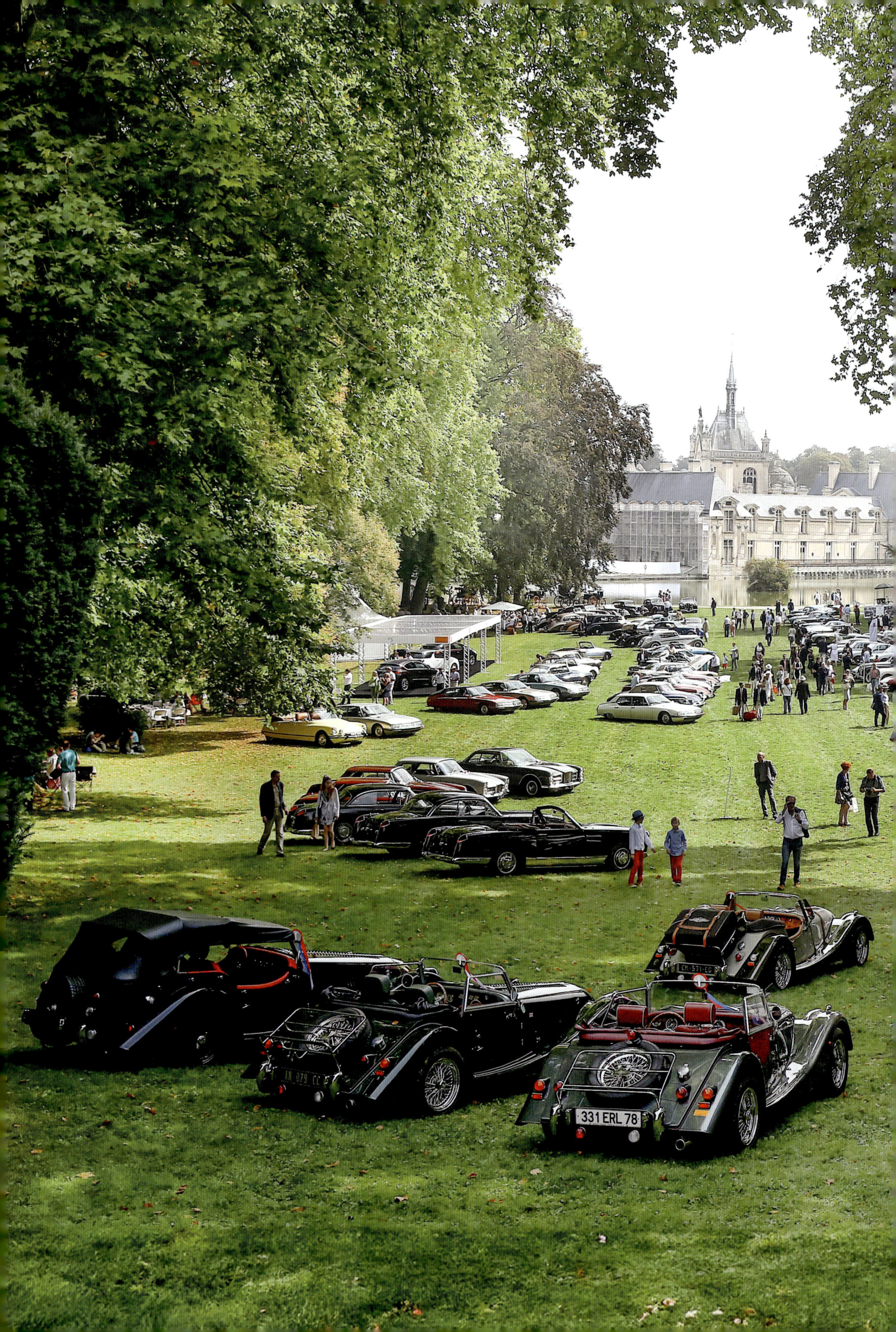
331 ERL 78

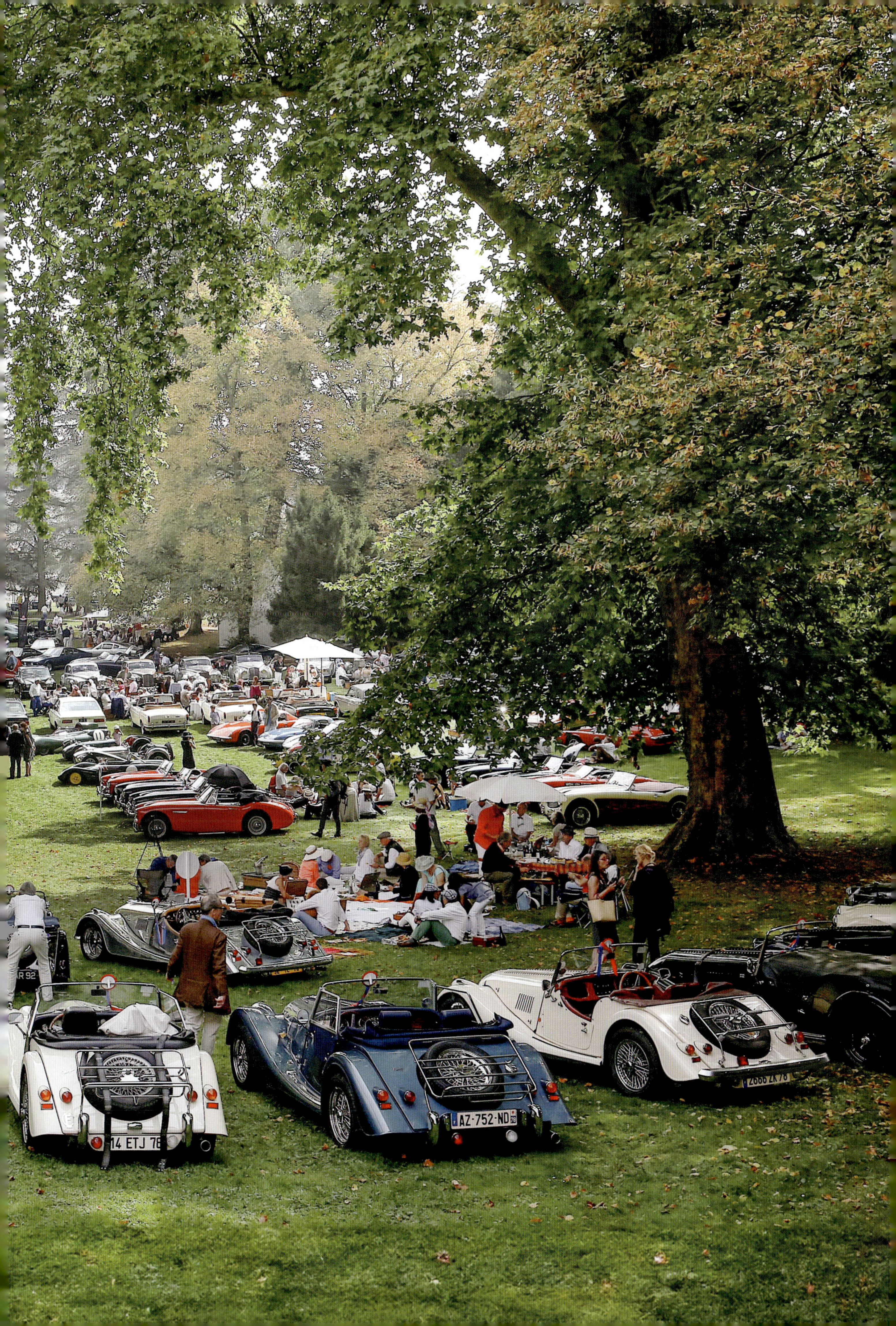
14 ETJ 78
AZ-752-ND

**RIGHT (BOTH):** *Chantilly has also always boasted a distinguished and international jury including personalities such as Nick Mason (in 2014) and Yasmin Le Bon (2019). (Xavier de Nombel & Makarand Baokar)*

**MIDDLE (ALL):** *The entertainment and the side show at Chantilly are most attractive for members of the family who are less interested in the historic vehicles. (Gautam Sen)*

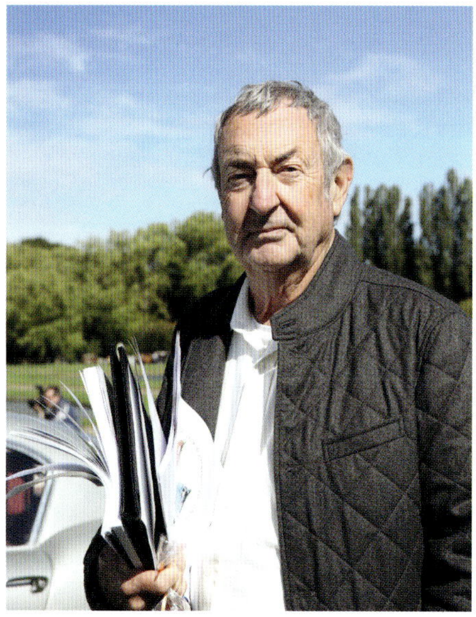

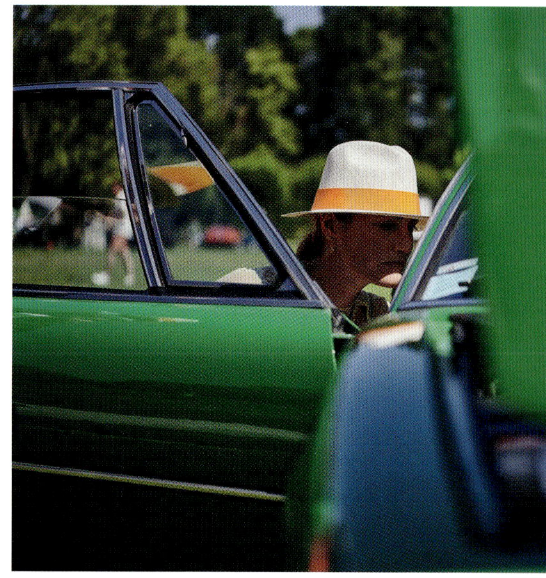

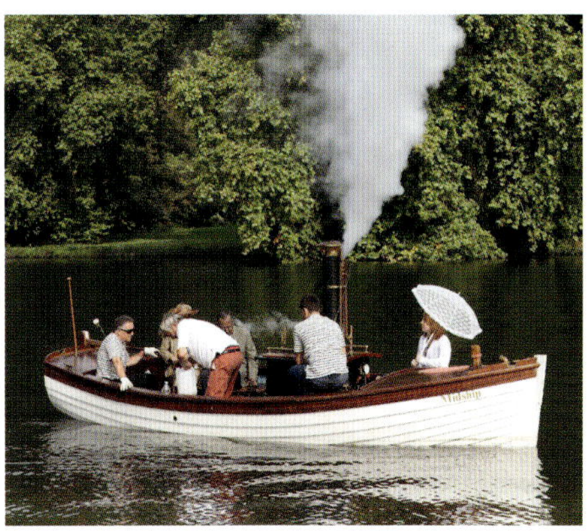

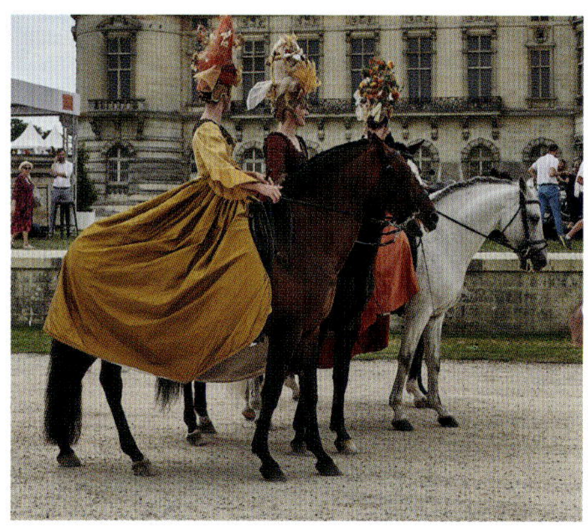

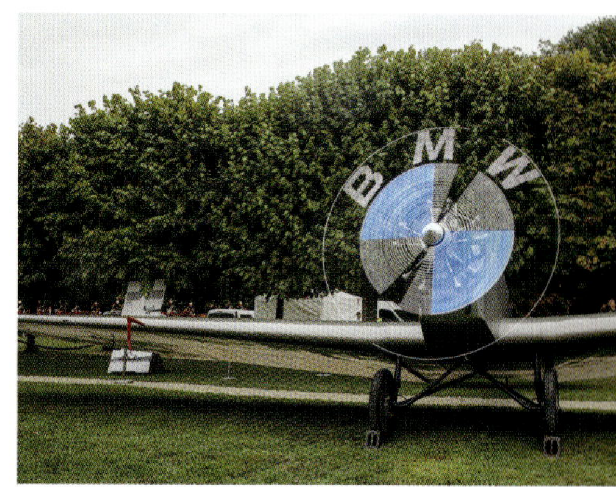


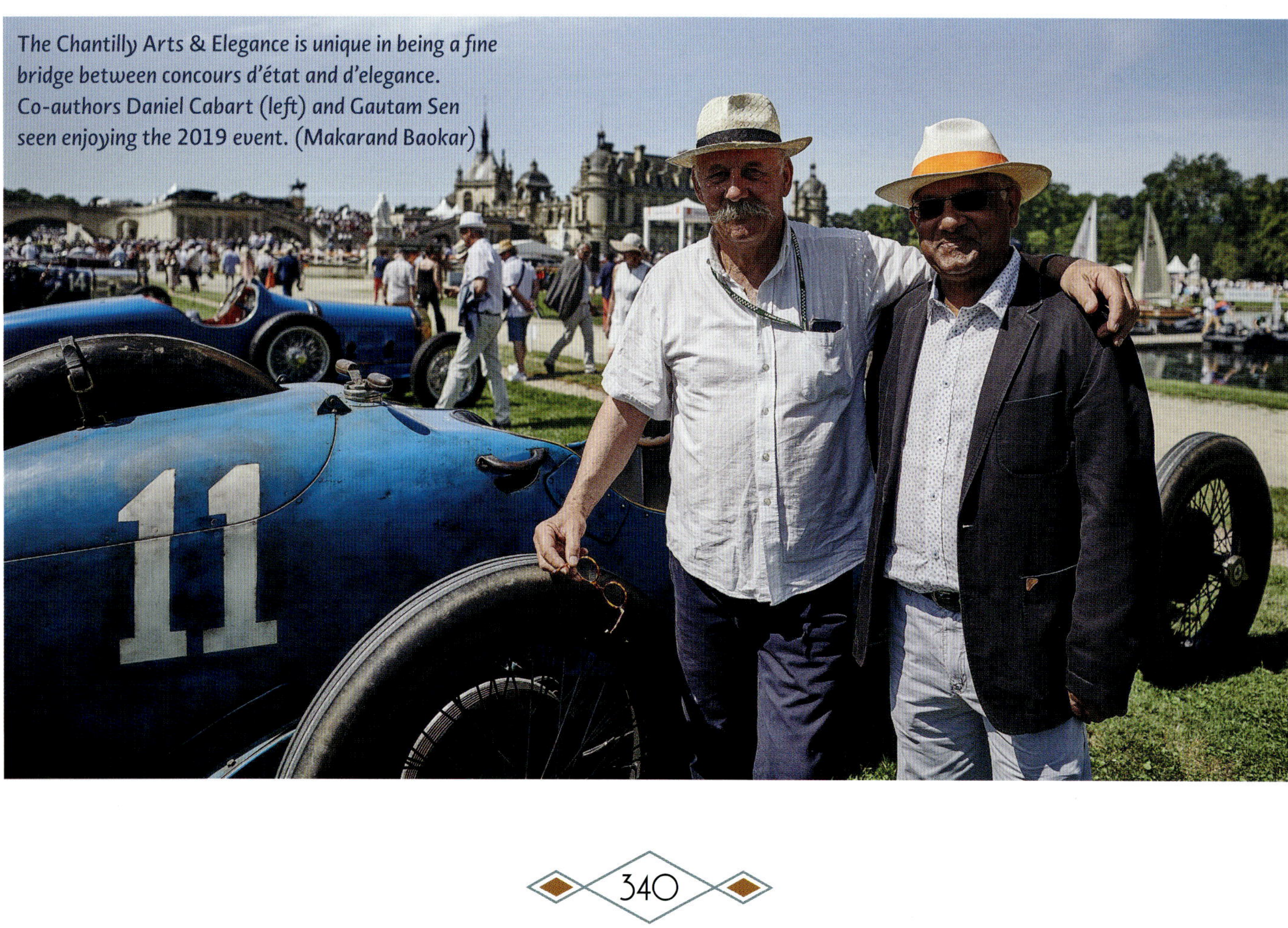

*The Chantilly Arts & Elegance is unique in being a fine bridge between concours d'état and d'elegance. Co-authors Daniel Cabart (left) and Gautam Sen seen enjoying the 2019 event. (Makarand Baokar)*

With Sylviane and Patrick Peter deciding to retire, Peter Auto has been sold to Amaury Sport Organization, the same group that organizes the renowned Tour de France, and the Automobile Club de l'Ouest, which runs the 24 Hours of Le Mans and other events at the legendary circuit. The 2024 Chantilly Arts & Elegance Richard Mille marked the transition to the new management, led by Marc Ouayoun. Ouayoun envisions expanding the event significantly, stating, "We need to make Chantilly Arts & Elegance a household name, one that any enthusiast of historic vehicles must attend and enjoy."

"This vision has already started to take shape, with this year's (2024) event attracting around 28,000 visitors over the weekend," pointed out Marc Ouayoun. "Attendees had the chance to view over a thousand cars, including a hundred competing in the three concours and many more showcased in the tour and club displays."

The French Federation of Historic Vehicles (Fédération Française des Véhicules d'Epoque, or FFVE) has embraced the idea of making events more inclusive and popular among enthusiasts and the public by officially recognizing around thirty events as concours d'elegance à la française. This follows the success of renowned events like Chantilly Arts & Elegance and Dinard Elegance.

Dinard Elegance, in particular, played a key role in reviving the concept of the French-style concours, now widely referred to as concours d'elegance à la française. As mentioned previously, Dinard was the site of one of the earliest French-style Concours d'Elegance, as defined by André de Fouquières. The modern revival of this concours updated the traditional pre-war presentations, setting a new trend for displaying historic vehicles.

Rather than a static display where judges move from car to car, Dinard Elegance introduced a dynamic presentation in which the vehicles paraded on a ramp. The audience sat on one side, while the jury sat on the other, evaluating not only the car but also the presentation of the ensemble. The participants – be they a gentleman and a lady, two ladies, or two men, often accompanied by a dog or children – are dressed in period-appropriate costumes that complement the colors of the vehicle.

On September 4, 2021, Dinard Elegance held its second modern concours to mark the 100th anniversary – to the date – of its first Concours d'Elegance in 1921. The event was a remarkable spectacle, with the cars parading in the evening under floodlights, creating a magical atmosphere. The 35 cars were presented in front of an audience and a jury that included Dinard's mayor, a magazine editor, the granddaughter of the sculptor François Bazin, representatives from FFVE and FIVA, and other notable figures.

*In a French-style Concours d'Elegance such as Dinard Elegance, the audience and judges remain seated while the cars and their teams parade on a ramp to showcase the vehicles. (Patrick Chevalier)*

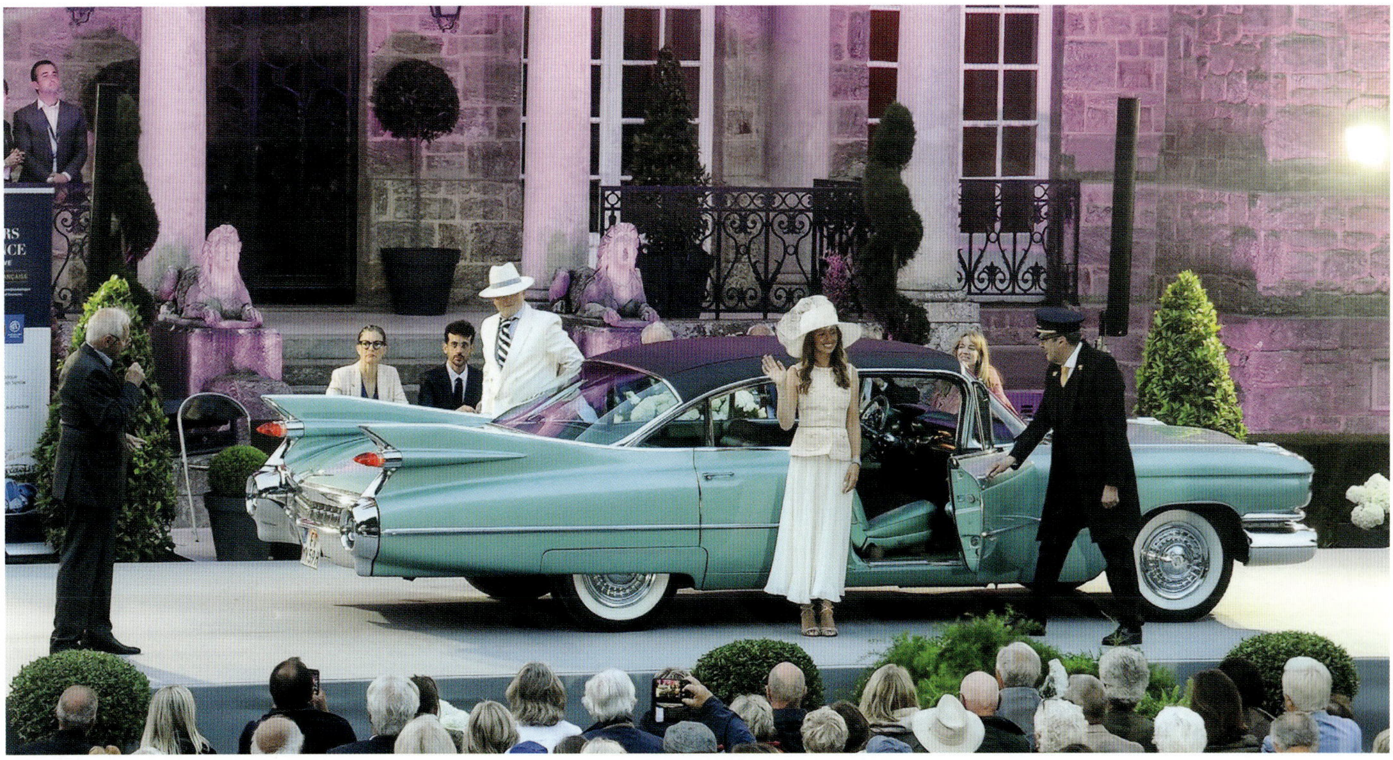

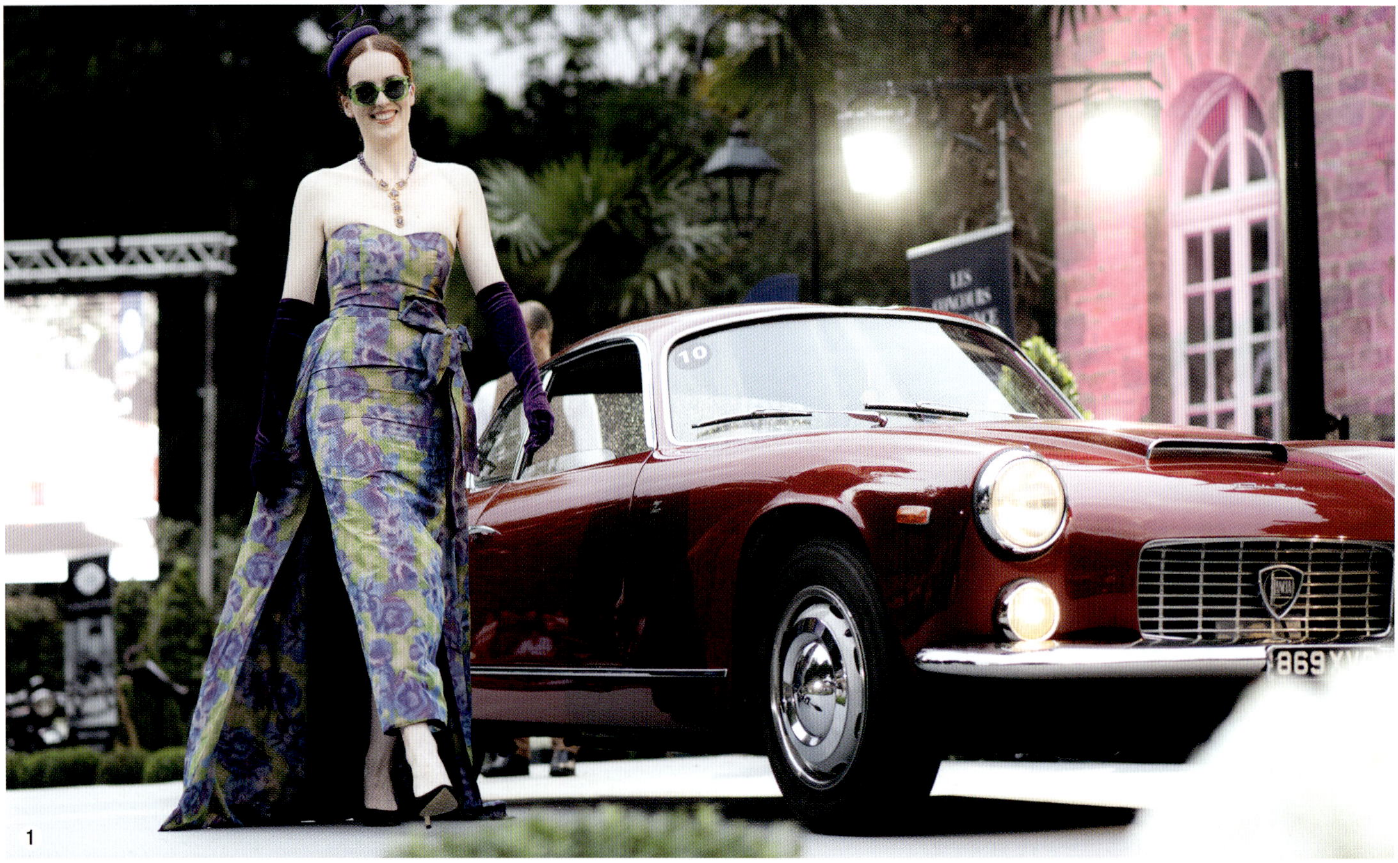

1

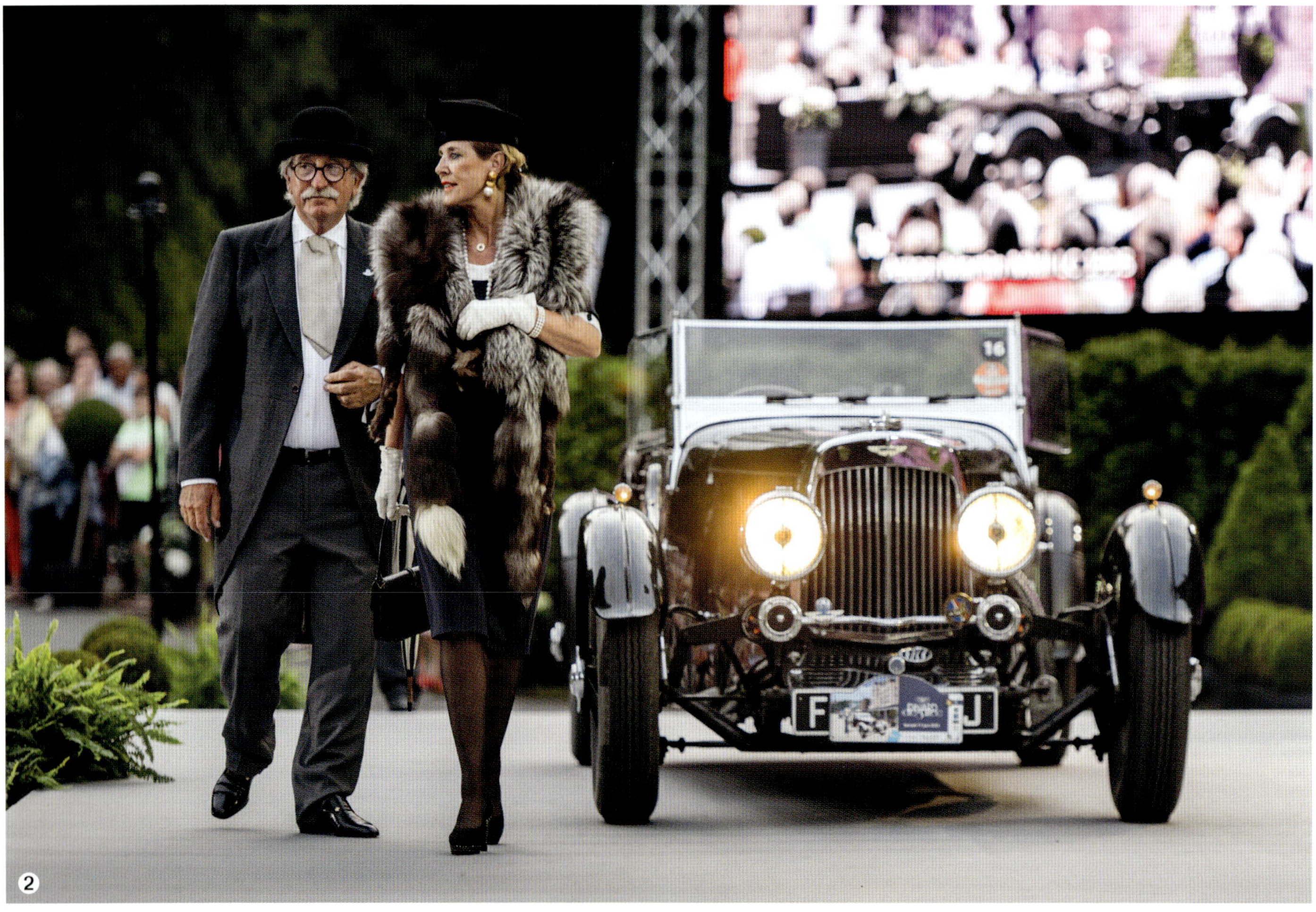

2

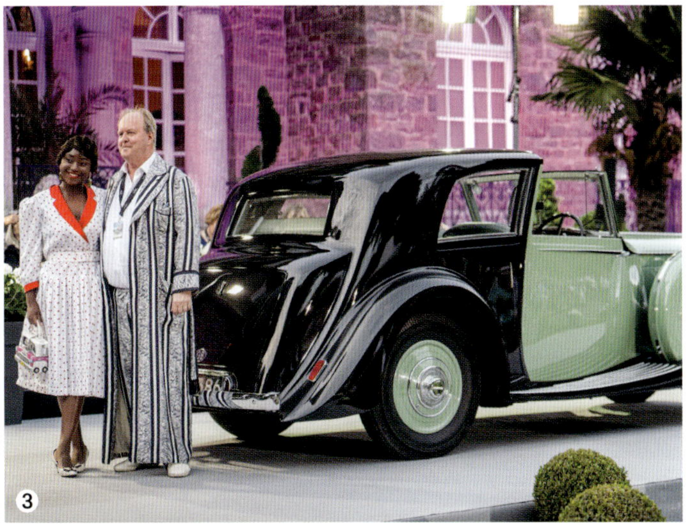

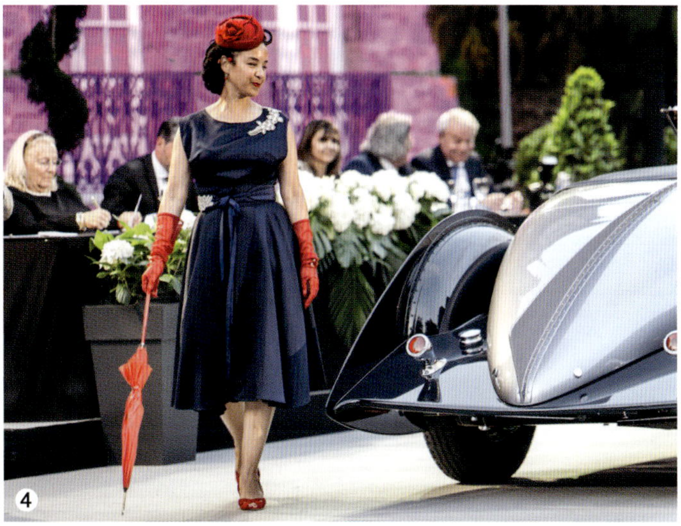

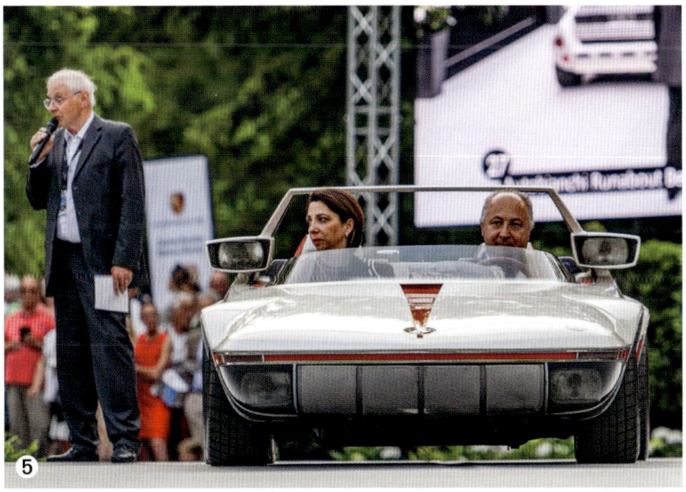

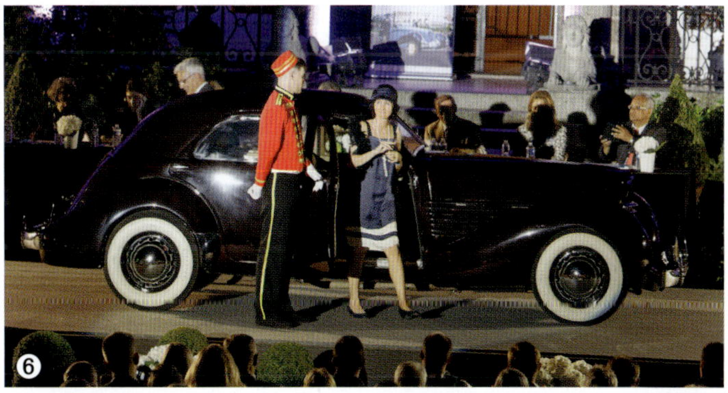

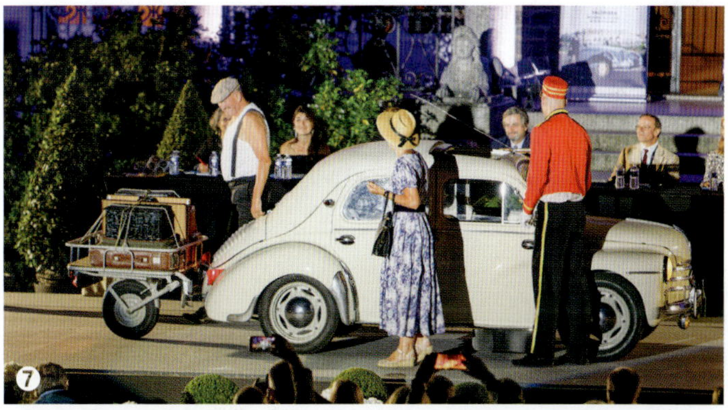

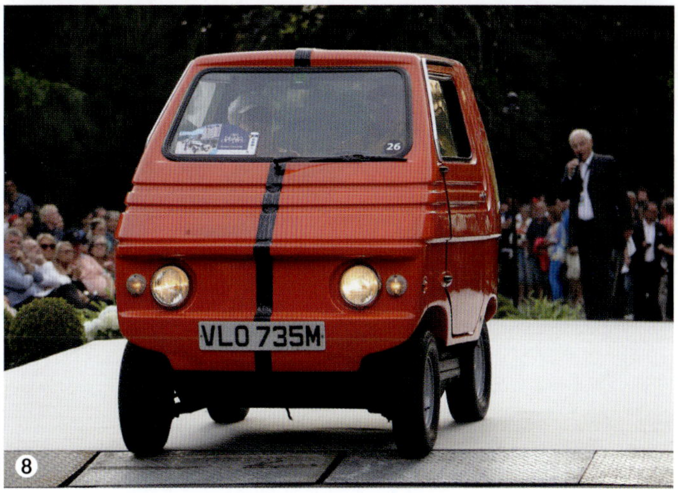


❶ ❷ One-third of the score is based on how the team presents the car, including their attire and mannerisms, which should align with the car and its era. This is exemplified in this bottom image featuring a class-winning 1935 Aston Martin Mk II, presented by Mr. and Mrs. Jean-Luc Despierre. (top: Ambre Clément, bottom: Jérôme Cadoret)

❸ Occasionally, this can be done with a touch of originality. ❹ Or with a sense of sheer class. ❺ The master of ceremonies plays a crucial role in both engaging and entertaining the audience and jury while providing insightful explanations. (Louis Monnier)

❻ The jury, seated behind, evaluates and scores the car, its design, and the presentation. In 2021, the Grand Prix d'excellence was awarded to this Cord 810, owned by Xavier Jenvrin and Marie-Line Delaunay. That year, the concours celebrated the 100th anniversary of the first concours d'elegance held in Dinard. (Alain Barreau)

❼ Since the entire presentation is crucial, even modest or everyday cars have a fair chance of making an impression and earning recognition. (Alain Barreau)

❽ The selection of vehicles includes a diverse mix, such as this rare and beautifully preserved Zagato Zele EV from 1972, belonging to the President of the Federation of British Historic Vehicle Clubs, David Whale. (Jérôme Cadoret)

Also shown at the Dinard Concours d'Elegance, a *one-off* Alfa Romeo 6C 1750 GS Spider *from* 1931, *owned by the one-off specialist* collector Corrado Lopresto. (Jérôme Cadoret)

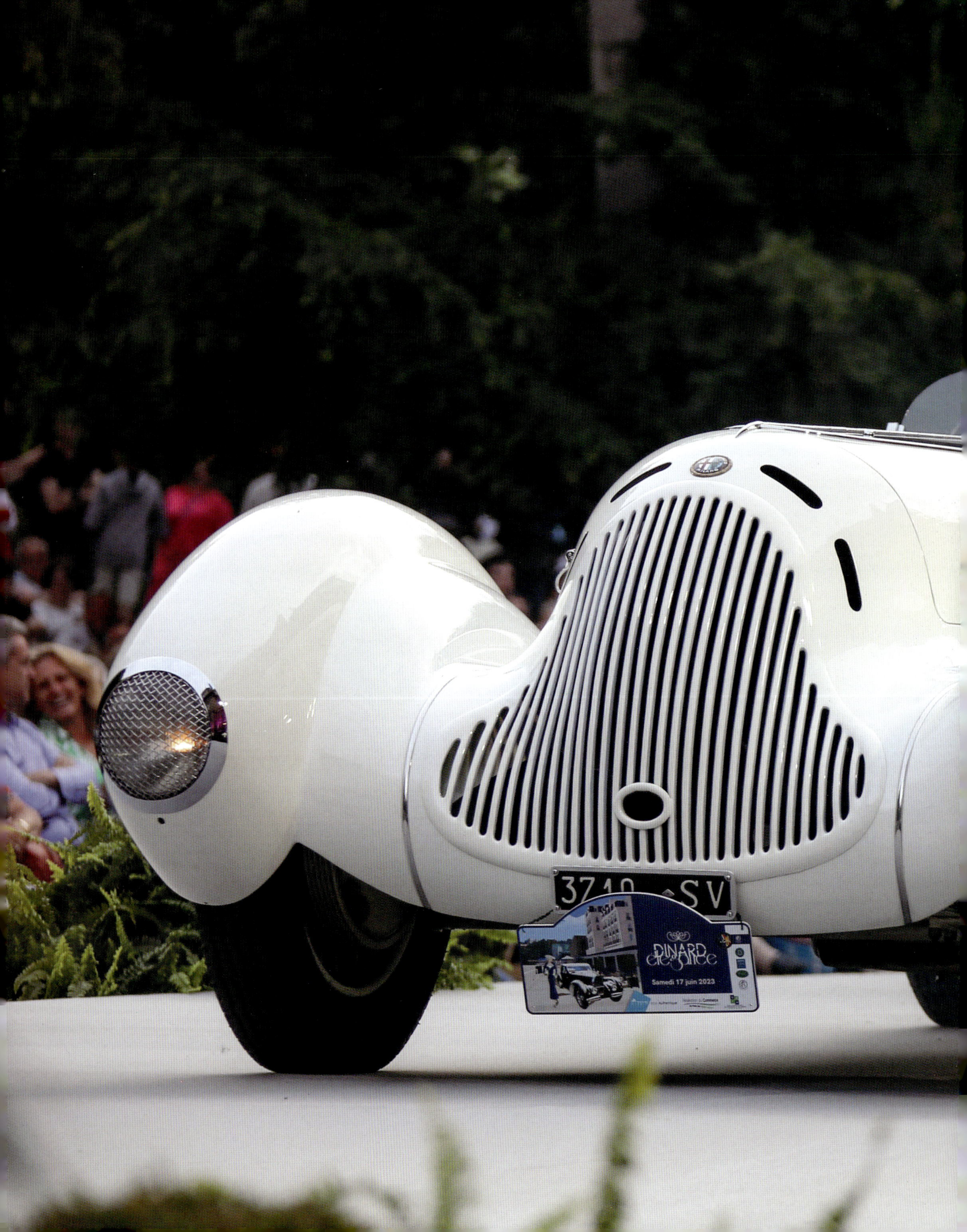

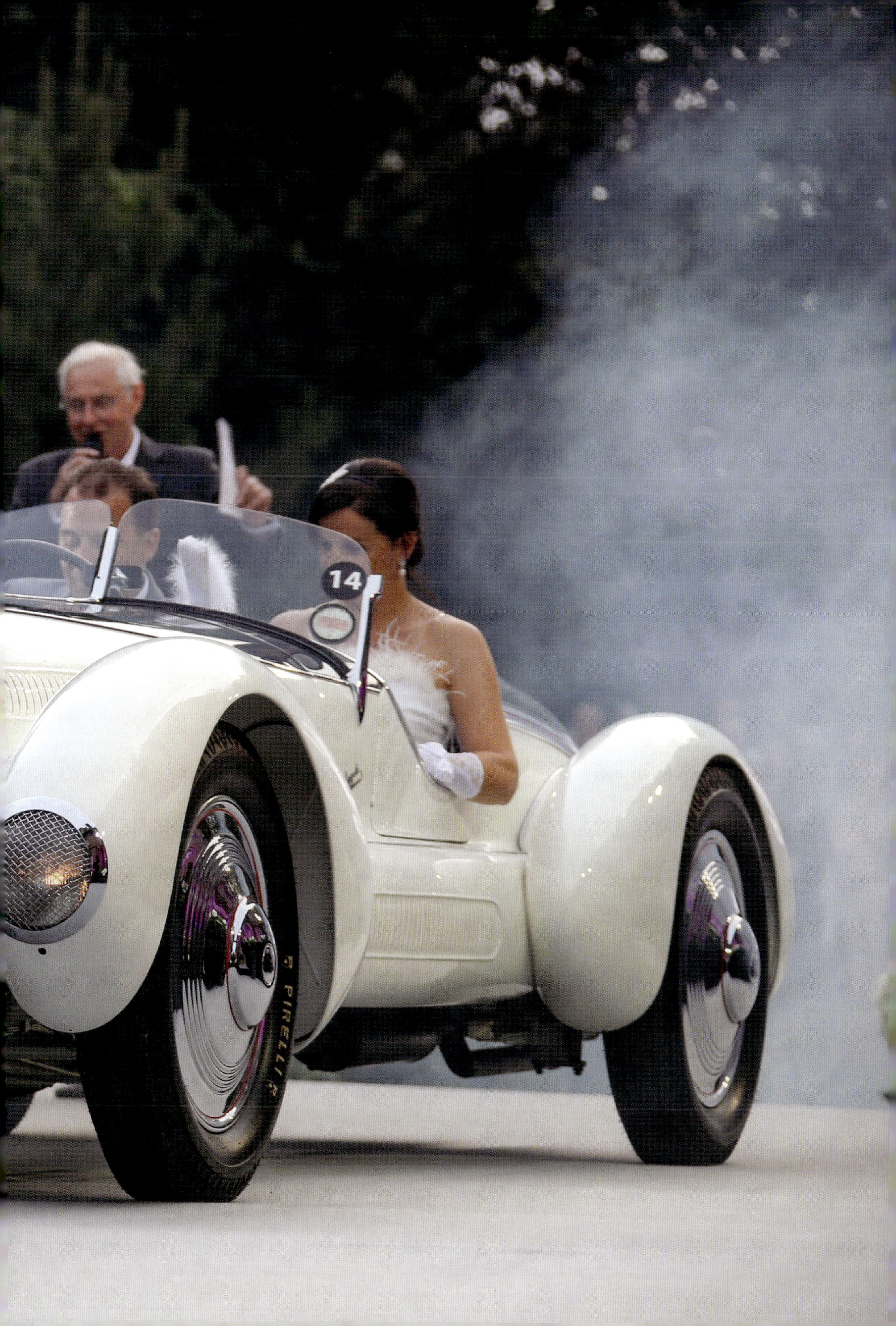
14
PIRELLI

Under the colorful spotlights, the cars and their presenters created a stunning visual display. Points were awarded based on three criteria; a third for the car's condition, a third for the elegance of its design, and the final third for how well the car was presented. This included the attire of the driver and partner, their coordination with the vehicle, and the story they chose to enact. The presentation was enhanced by period-appropriate music from a live band, while the master of ceremonies shared anecdotes and the histories of the cars.

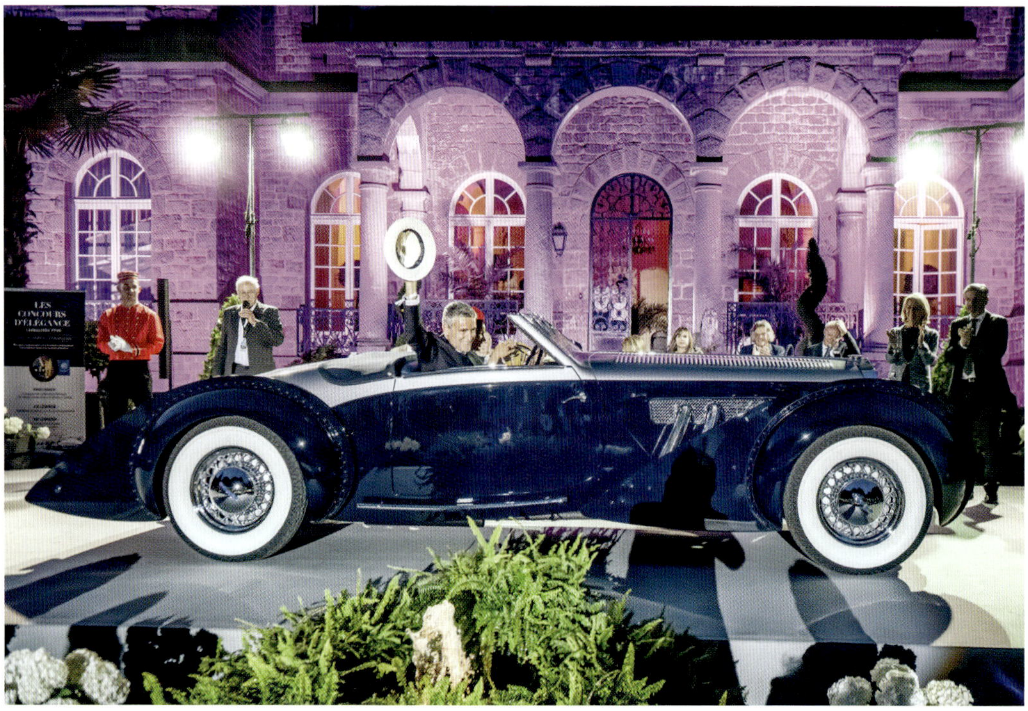

**RIGHT:** *Ultimately, the most exotic and beautifully presented car often claims the Best of Show title, as exemplified by this stunning Delage D8-120 De Villars roadster. (Louis Monnier)*

**BELOW:** *The Tour d'Elegance in Dinard features not only the cars participating in the Concours d'Elegance, but also others joining for the drive. It's not unusual for thousands of spectators to line the streets of Dinard and neighboring towns to catch a glimpse as the cars pass by. (Ville de Dinard)*

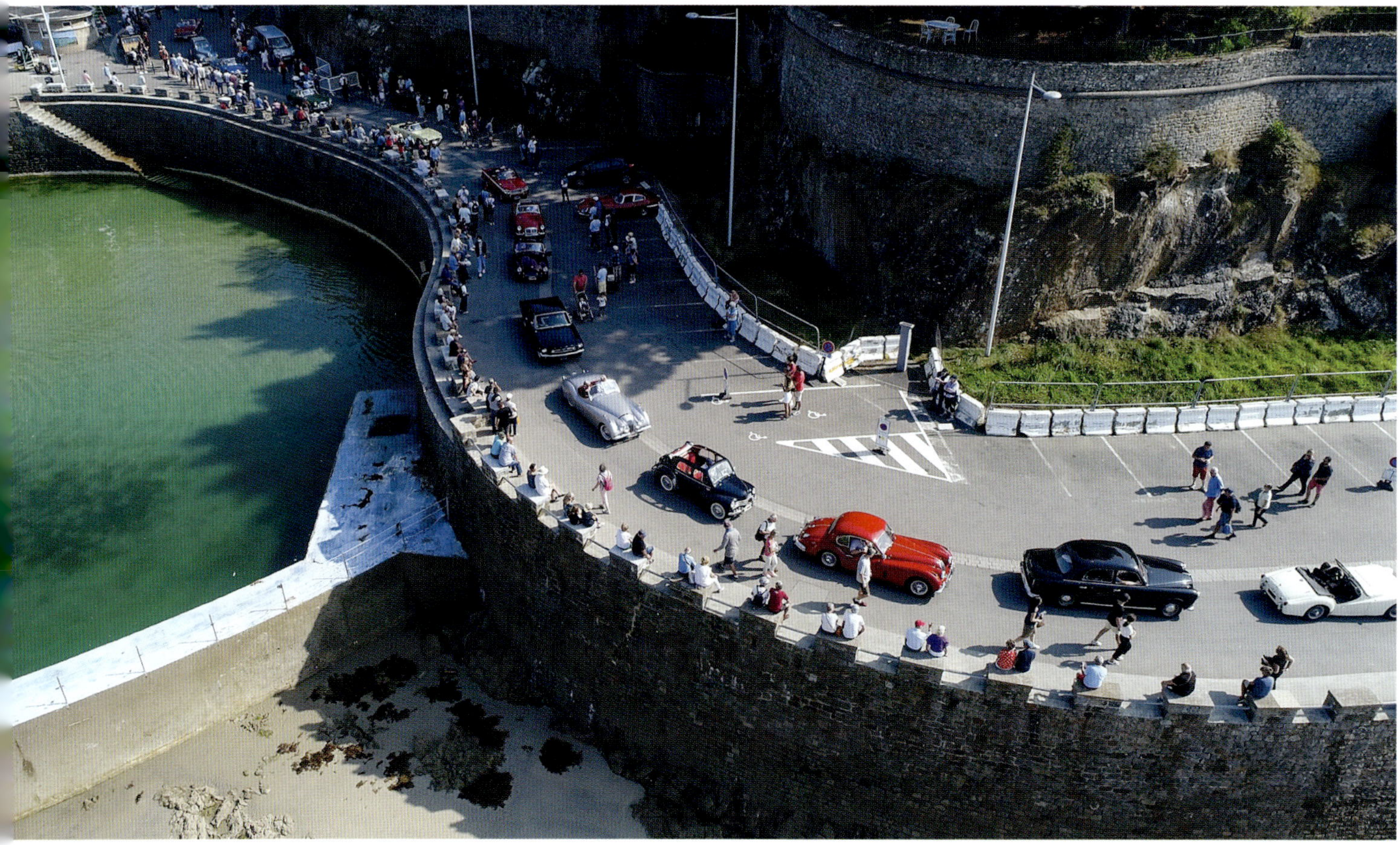

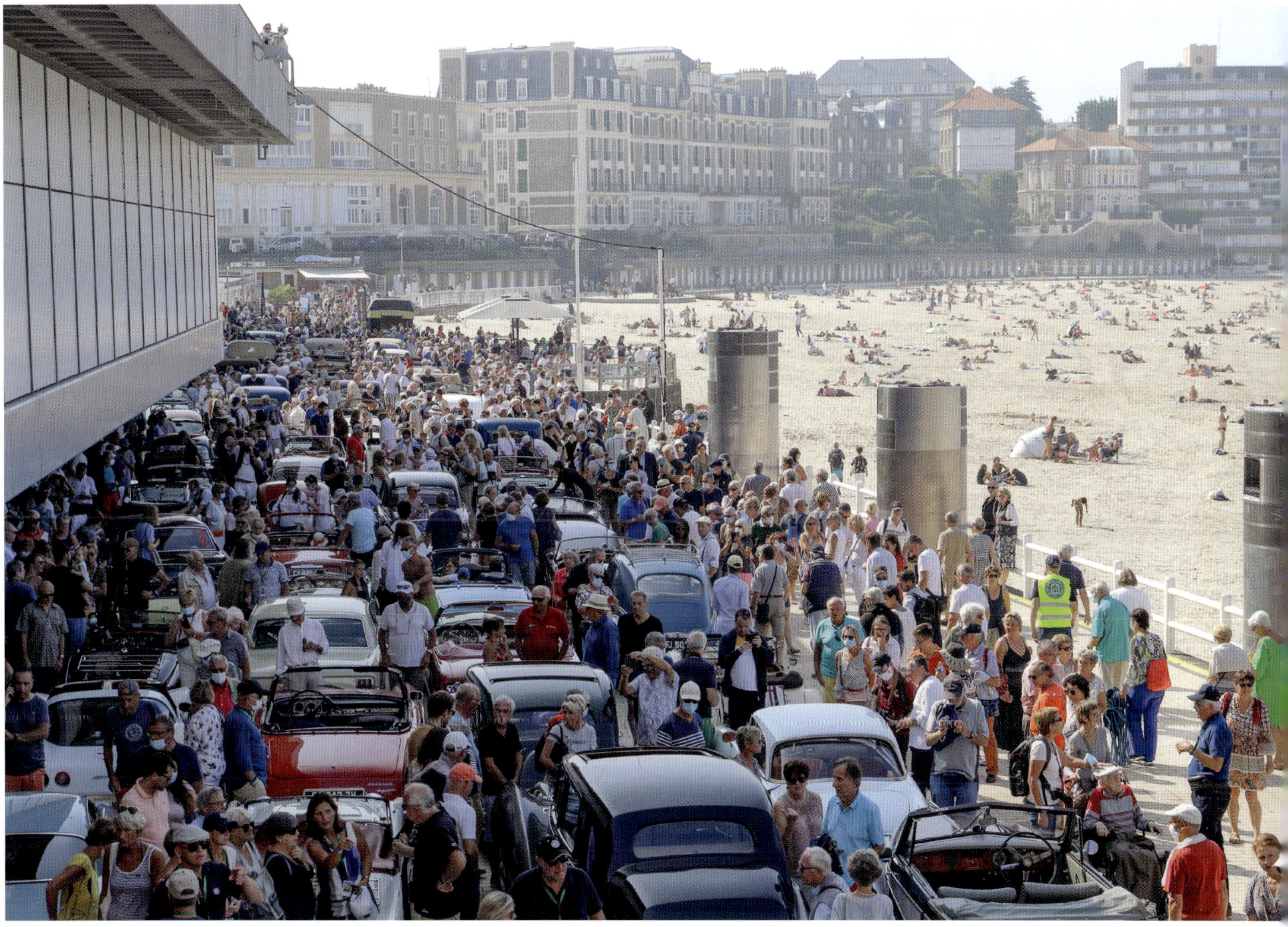

*A large crowd gathers to closely admire the cars at the Plage de l'Ecluse esplanade in Dinard. (Gautam Sen)*

Dinard Elegance's success set the standard for future concours d'elegance à la française. Resorts such as La Baule, Vittel, Vichy, and Nice revived their pre-war concours in this new French format, while towns like Tonneins, south of Bordeaux, adapted to the modern style. In 2024, the FFVE recognized 30 concours across France, awarding its official label to events that meet specific guidelines.

According to FFVE President Jean-Louis Blanc, the concours d'elegance à la française, which rose to prominence in the 1930s, are distinct from international events by focusing not only on car restoration but also on cultural elegance: "To earn the FFVE label, organizers must adhere to a charter that includes scoring the car and the crew's attire. Additionally, vehicles must be at least 30 years old, and the jury must meet specific composition rules. Winners receive trophies, with a Prix d'excellence awarded overall."

Blanc further explained that "organizers must apply for the FFVE label, meet various criteria, and promote automobile elegance globally. The event venue must be elegant and accessible, and photos of the winning vehicles must be submitted afterwards."

Many of these concours have been highly successful. Dinard Elegance regularly attracts more than 20,000 visitors, Tonneins draws about 10,000, and La Baule in 2024 saw over 30,000 attendees. "Overall, hundreds of thousands of people may have likely attended in concours d'elegance à la française, highlighting the popularity of this format," explains Jean-Louis Blanc. "It's a way to bring the historic vehicles to the people, instead of the other way around."

Despite some criticism from American concours experts who have dismissed the French-style as "theatre," there is no doubt that this format is more engaging, educational, and enjoyable for a broader audience.

# *Differentiating between Concours d'État and d'Elegance*

## CONDITION VERSUS ELEGANCE

## The Original Concept

The term 'concours d'elegance' is today used globally, yet in French there is a subtly different meaning to the phrase, thereby creating two rather distinct concepts. This warrants clarification.

In its French origin, concours d'elegance was an event inspired by aristocrat André de Fouquières in the early 1920s. It was a prestigious automobile showcase where the highlight was a jury of experts evaluating the harmonious collaboration of a brand-new design produced by a car manufacturer, a coachbuilder, and a couturier – the so-called 'three Cs'.

Women dressed by renowned haute couture houses, would present rare automobiles designed on special chassis by master coachbuilders. Often, these women were accompanied by dogs, adding to the scene of opulence. These events were officially known, starting in 1926, as the Concours d'Elegance Féminine en Automobile.

Before World War II, these competitions were a core part of the social 'season', attended by the aristocracy and the elites from various sectors, including industry, finance, arts, and entertainment. The events were held in vacation hot spots like the Tennis Club of La Baule, the Grand Hôtel de Biarritz, Boulevard Cornuché in Deauville, Plage de l'Ecluse in Dinard, outside the Westminster at Paris Plage, or at the Vichy casino, among other elegant locations in France's many spa towns and seaside resorts.

Today, these French-style concours still take place, especially in their home country. However, they are often recreations of a bygone era, featuring historic vehicles accompanied by period-appropriate fashion and accessories. Primarily organized by clubs and volunteer groups and supported by the French Federation of Historic Vehicles, FFVE, these events have become more inclusive and popular. Participants are no longer exclusively women, and the vehicles on view range from popular historic runabouts to sports and luxury models. The focus has shifted from targeting an elite social class to appealing to a broader audience.

One of the few remaining events where the original spirit of the concours d'elegance endures is the Chantilly Arts & Elegance Richard Mille. This event celebrates contemporary excellence, featuring recent concept cars from top coachbuilders and manufacturers, alongside models dressed in haute couture by modern-day designers.

Bournemouth Concours d'Elegance from July 1, 1928. (Motorsport Images)

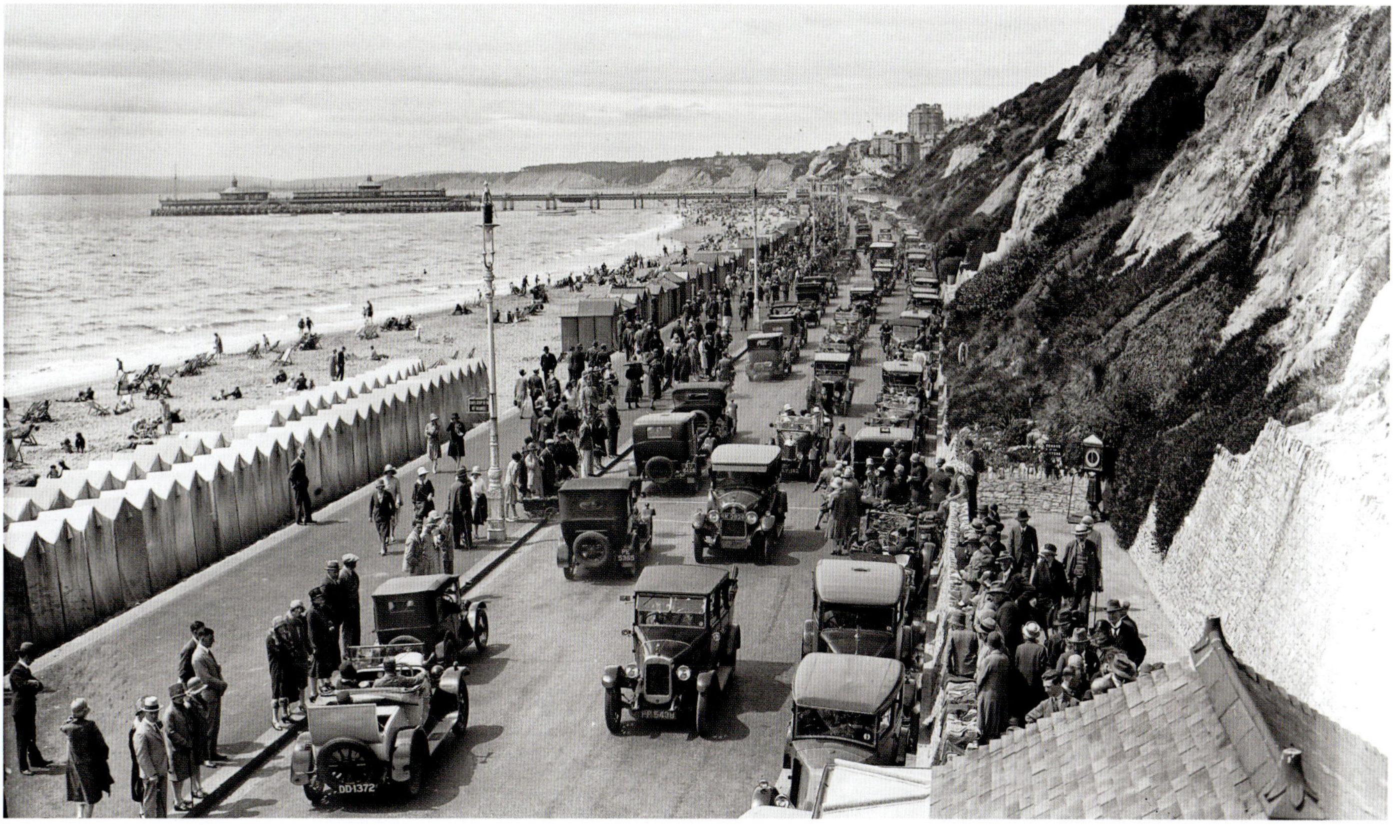

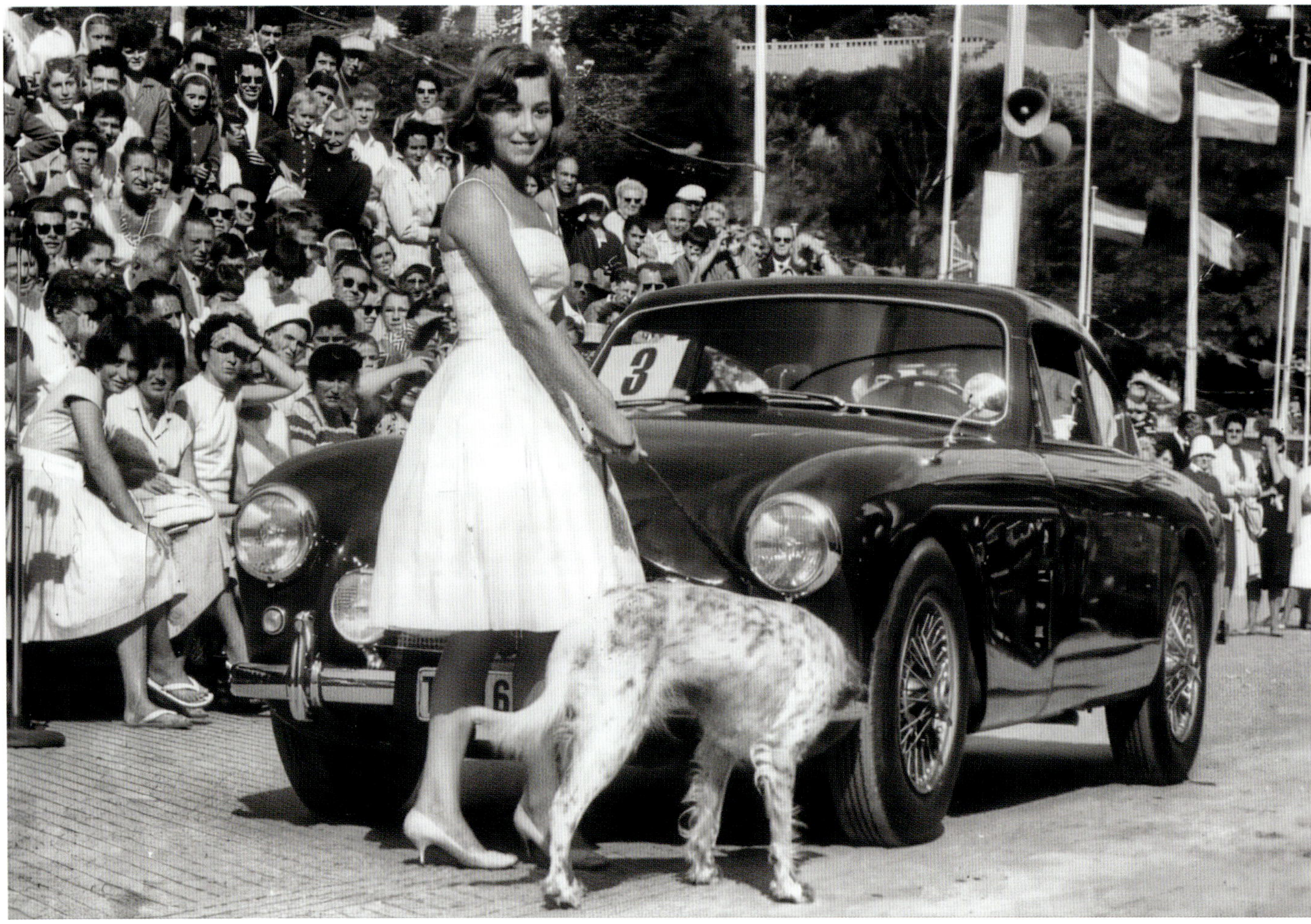
3

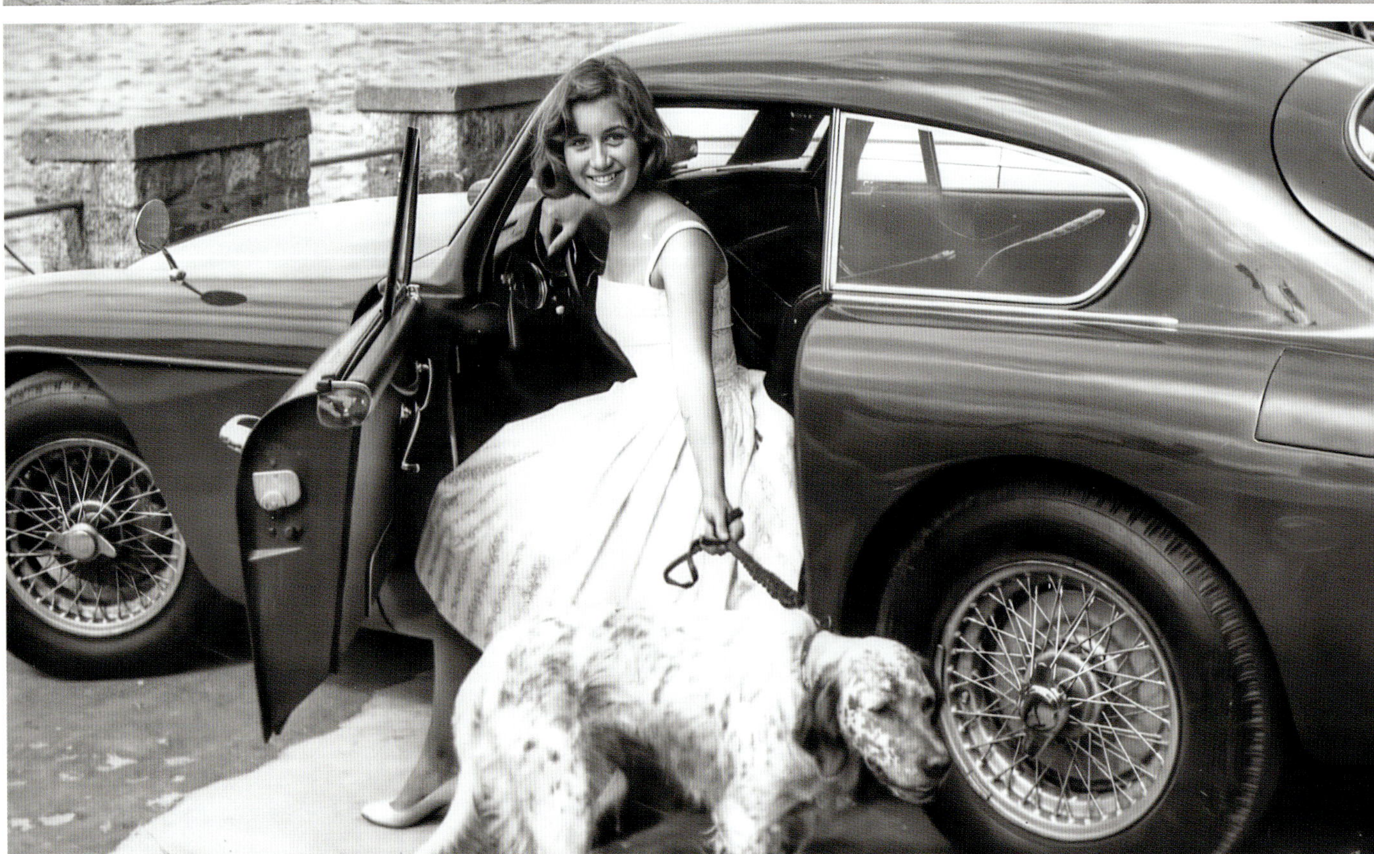

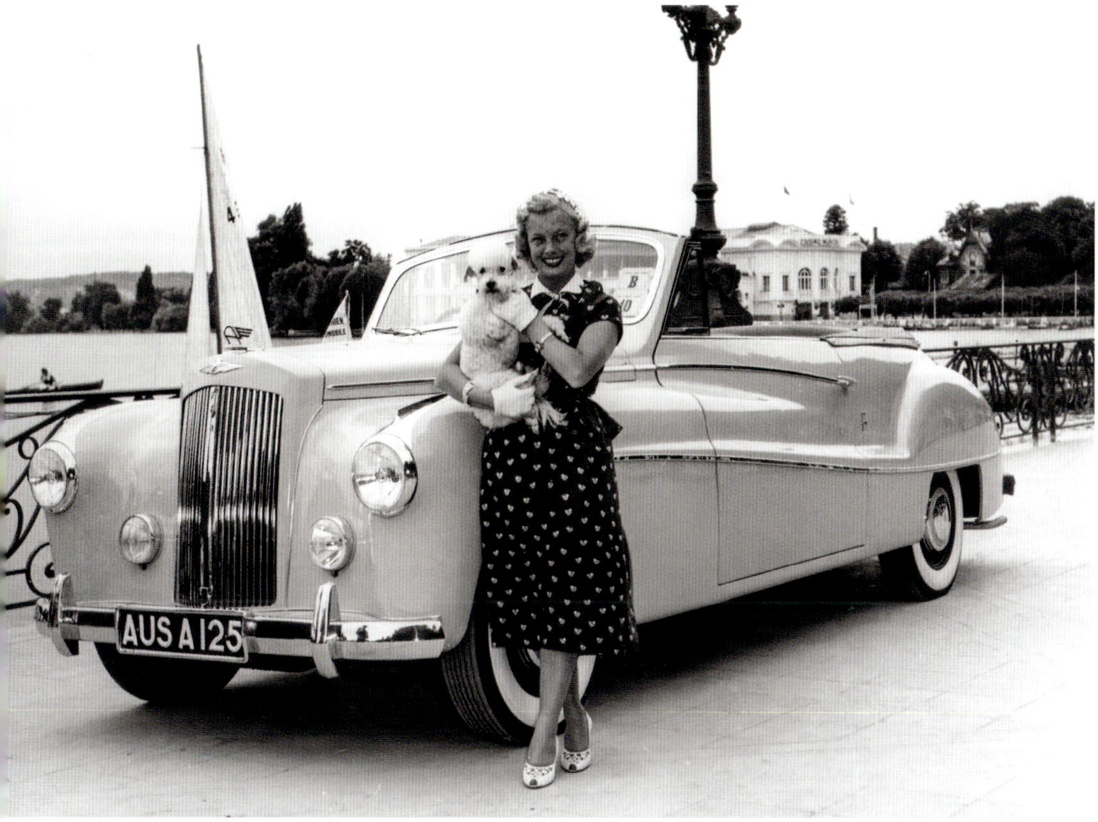


**OPPOSITE:** *Even after the war, French concours d'élégance continued to feature young women and charming dogs, in perfect harmony with a 1957 Aston Martin DB2/4 Mk III, as seen here at Dinard.* (Dinard Elegance)

**LEFT:** Austin A125 Sheerline cabriolet at Enghien-les-Bains. (Gilles Blanchet Archives)

**BELOW:** Actress Maud Loty, her dog Mitsou, and her Fiat at the 1928 Automobile Club of Nice Concours d'Elegance. (Gilles Blanchet Archives)

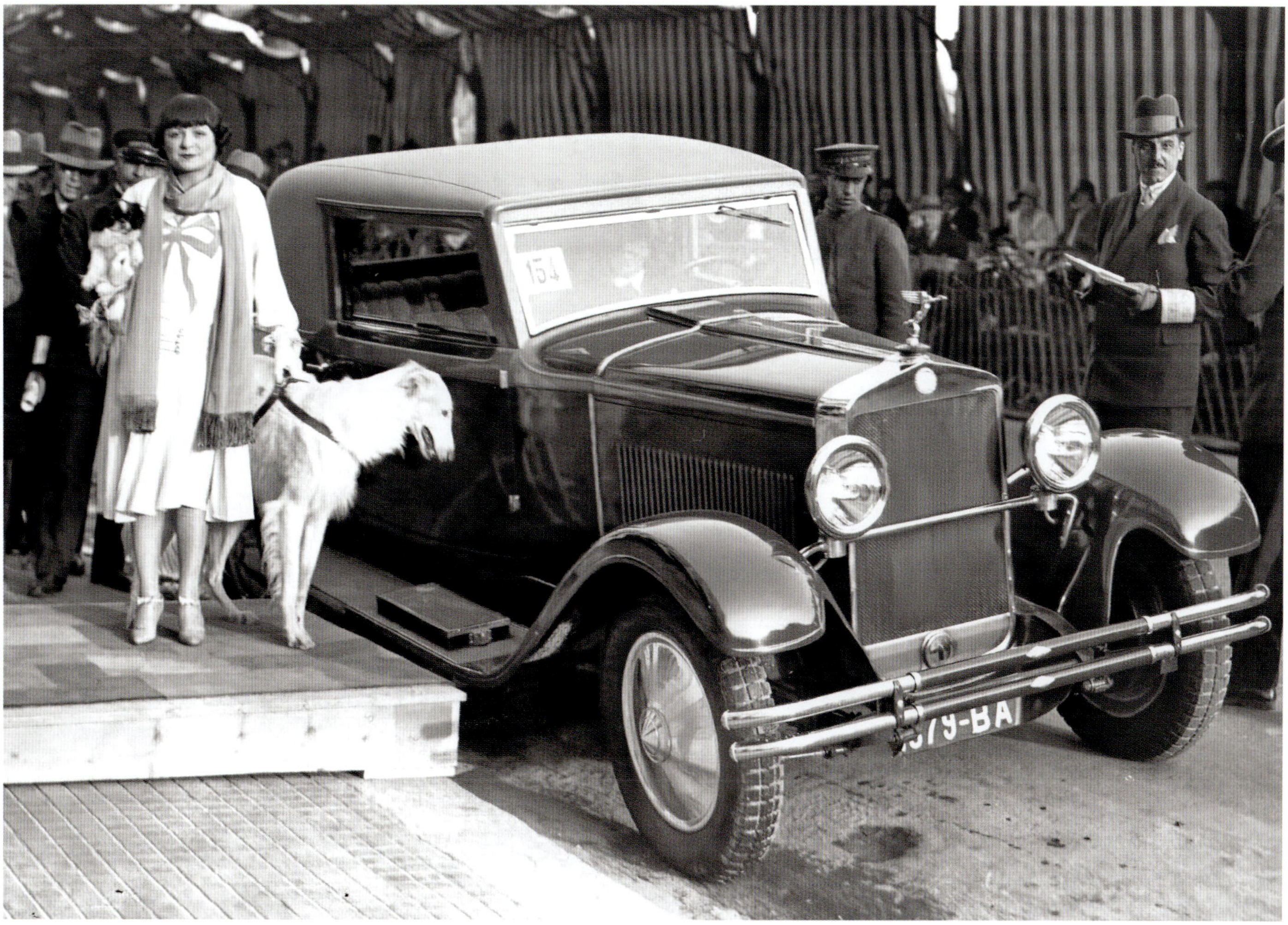

## Landmark Moments: The development of Concours d'Elegance events in France during the 1920s and 1930s.

| Publication, Date | Place | Name |
|---|---|---|
| *L'Ouest-Éclair*, July 1924 | Dinard | Concours d'Elegance |
| *L'Intransigeant*, June 1925 | Paris | Concours d'Elegance Féminine en Automobile |
| *Biarritz Illustré et Côte basque Pyrénées*, September 1925 | Biarritz | Concours d'Elegance Féminine en Automobile |
| *L'Auto-Vélo*, 1928 | Le Touquet | Concours d'Elegance et de Confort |
| *Fémina*, June 1929 | Paris | Concours d'Elegance Féminine en Automobile |
| *L'Auto-Vélo*, June 1932 | Paris | Le Concours d'Elegance de L'Auto |
| *La Carrosserie*, June 1933 | Paris | Le Concours d'Elegance des Champs-Élysées |
| *La Vie Automobile*, June 1937 | Paris | Concours d'Elegance en Automobile |
| *L'Art Vivant*, January 1937 | General article | Concours d'Elegance Féminine en Automobile |

| Extracts from articles / rules | Observations |
|---|---|
| *"The Jury's evaluation will emphasize the elegance of design, the comfort, the innovation in the coachbuilder's arrangements, the layout and uniqueness of the accessories, as well as the overall harmony and refinement of the ensemble."* | The Dinard competition, first held in 1921, focused primarily on the automobile, although the concept of "elegance of the entire movement" added a dynamic aspect to its presentation. Only a few women participated individually, and that included Madame Citroën. |
| *"The jury's evaluation emphasizes both the elegance of the car, including the originality of its details and innovative fittings, and the elegance of the drivers, as the competition is exclusively open to cars driven by women."* | Feminine elegance was officially recognized by the jury, with women taking the wheel – an innovation described in the article as an "additional attraction." This marked the emergence of a new concept: the "Feminine Concours d'Elegance events in automobiles." |
| *"The competitors paraded past the jury twice in their cars. The outfits were greatly admired, and the presented cars were deemed very charming."* | The Concours d'Elegance included a special prize recognizing the elegance of the women seated in the car. |
| *"The ranking will be determined based on the following criteria: Elegance of the design (minimum 20 points) Comfort (maximum 20 points) Ingenuity and originality of the fittings and accessories (maximum 10 points) For convertible cars, the speed and ease of the transformation maneuver will be assessed, with one point deducted for each minute taken from the 20 points awarded to each competitor."* | It remained a competition focused on elegance and the comfort of the automobile, evaluated using a points system. This concept continued to coexist for years alongside the Concours d'Elegance for women and the automobile. |
| *"As in every year, the jury was tasked with judging the design of the cars and bodies, the elegance of the driver or passenger, and the attire of the mechanic, if applicable."* | Although the *Fémina* and *L'Intransigeant* Concours d'Elegance competition had been held since 1924, and the triptych of automobiles, bodywork, and women's fashion was now well-established, it was still not yet widespread in France. |
| *"On a bright morning, set against the stunning backdrop of the Bois de Boulogne, the eleventh Concours d'Elegance et de Confort organized by* L'Auto *took place yesterday. This was complemented by the Concours des Carrosseries de série presented by manufacturers, on one hand, and the Coupe d'Elegance de L'Auto, introduced this year, on the other hand, as well as the fourth Trophée de l'Elegance Féminine."* | The Automobile d'Elegance et de Confort concours had been held since 1921, complemented by a competition recognizing coachbuilders and another for manufacturers. In 1932, an innovation was introduced with a concours focused on the elegance of automobile design. The Trophée de l'Elegance Féminine was created in 1929, with two separate juries coexisting: one for the technical aspects and another for fashion. |
| *"Three concours were on the program: the coachbuilders' concours, the constructors' concours, and the women's automobile elegance concours. As for the women's elegance concours, the usual factors were considered: the elegance of the women's attire, the car, and the overall harmony, while still giving attention to the grace and beauty of the driver, which accounts for the popularity of this type of competition."* | The Champs-Élysées Concours, organized by the *Chambre Syndicale de la Voiturette*, followed a nearly identical format. It separated the categories of coachbuilders, manufacturers, and feminine elegance, offering three distinct concours. However, the "factory" bodies appeared to be excluded from consideration. |
| *"One could say that to succeed there, you need four things at once: a beautiful car, a beautiful woman, a beautiful dress, and a beautiful dog."* | The tone in car magazines was sometimes ironic or even sexist, in contrast to the tone found in fashion magazines. |
| *"In women's automobile elegance competitions, it is the overall presentation that catches the eye, with several elements coming into play: the engine, the bodywork, and the driver. Couture plays a key role, as the woman must wear an outfit that harmonizes with the overall presentation, and even the mechanic must contribute by sporting an impeccable outfit."* | The women's Concours d'Elegance competition had gained widespread popularity and had become the standard. |

## The Evolution

After World War II, a new concept of the concours emerged in the United States. This idea took shape in 1950 with a gathering of automobile enthusiasts on the Monterey Peninsula in California, eventually evolving into what is now the Pebble Beach Concours d'Elegance (written in English without an accent). Unlike the original French-style event, this version, referred to in French as a 'concours d'état' or a 'competition of condition,' focuses on evaluating the preservation or quality of restoration of the vehicles rather than their elegance.

Today, this model dominates concours events worldwide. For example, *Concours Year* 2023, an annual compilation of the 50 most prestigious concours globally, edited by David Lillywhite and his team at Hothouse Media, lists 49 concours d'état, with only one French-style concours d'elegance, Dinard Elegance, as an exception.

In France, despite the efforts of the French Federation of Historic Vehicles, FFVE, and noted expert Robert-Louis Brézout-Fernandez, as well as the discontinuation of the Parc de Bagatelle competition in 2003, few concours d'état events are held in the country today. When they do occur, they often run alongside a French-style concours d'elegance à la française, as seen in Le Touquet, Angoulême, or Deauville.

It's worth mentioning that both forms of concours – concours d'état and concours d'elegance – still exist for motorcycles in several countries, including Eastern Europe, Italy (Villa Erba), and France (Nice and Angoulême).

In concours d'état events, alongside historic vehicles, newer models are not excluded. These could include 'youngtimers' (newer historic vehicles), prototypes, concept cars, and sometimes even popular vehicles, such as those featured in the UK's Festival of the Unexceptional. In contrast, the concours d'elegance à la française remains focused on a retrospective approach, featuring vehicles with an emphasis on elegance and historical significance.

*At the Hagerty-backed Festival of the Unexceptional it is not uncommon to see mostly youngtimers on display. (M. Kimberley)*

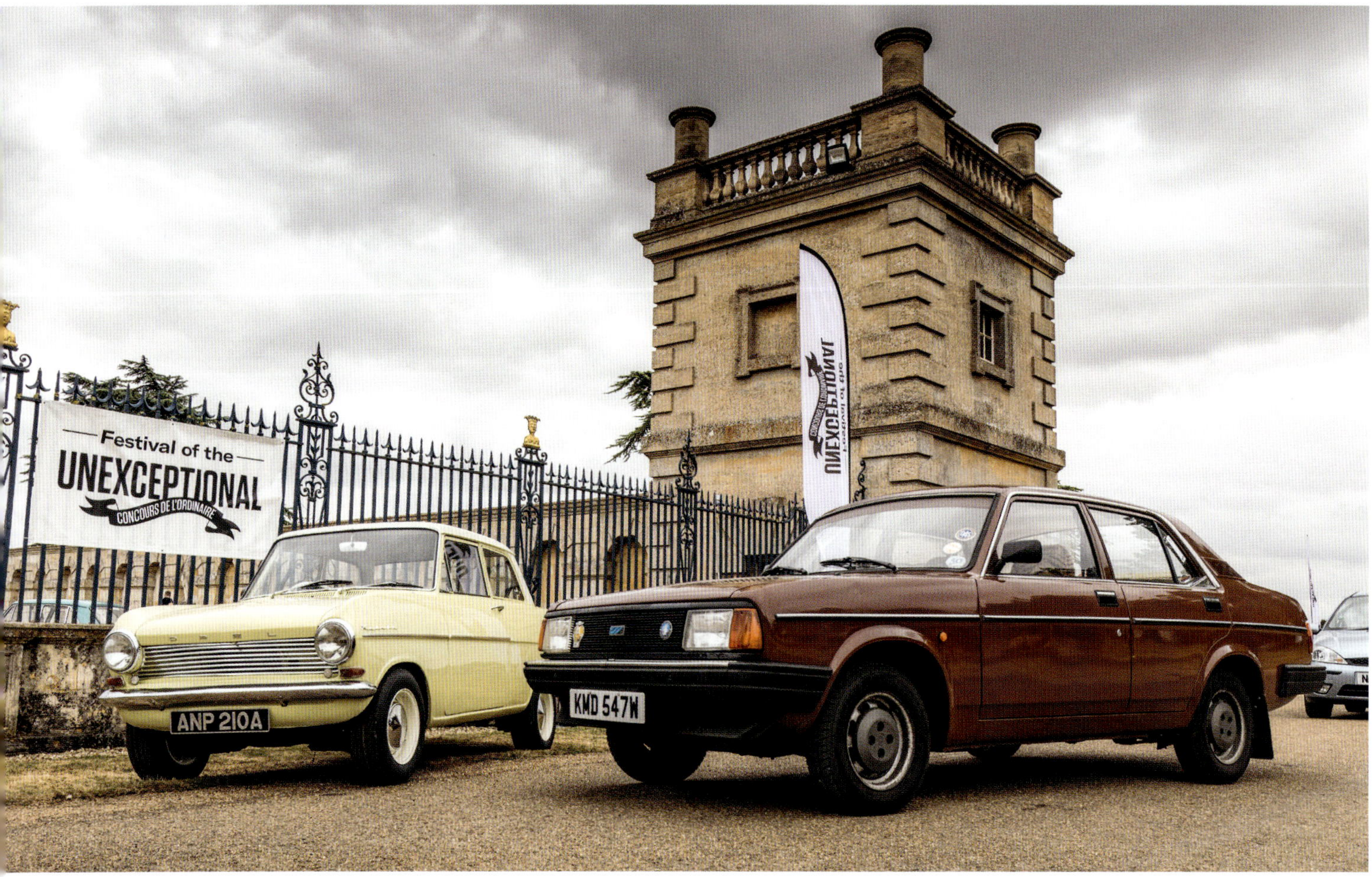

## Differences and Similarities

The primary difference between a concours d'état and a concours d'elegance à la française lies in the focus: the concours d'état evaluates only the vehicle, while the concours d'elegance à la française judges the vehicle and its crew (the driver and passenger, often with attention to their attire).

In a concours d'état, the event is "static." Once the vehicles arrive, they are parked in designated spots, remaining there for most of the competition while the jury moves around to assess them. To streamline judging, vehicles are grouped into categories (or "classes") – such as sports cars, sedans, convertibles, post-war coupés, or specific marques – based on their characteristics.

However, the concours d'état has seen some excesses emerge. One such excess is over-restoration, where historic vehicles are presented in a condition that surpasses their original state when they left the factory. Sometimes there is too much chrome, oversized tires, lowered suspensions, and flashy paintwork – additions that compromise the authenticity of the vehicle.

To address this, the International Federation of Historic Vehicles, FIVA urged the organizers of Pebble Beach to introduce a special class for unrestored (or preserved) vehicles, which was done in 2000, emphasizing the value of maintaining a vehicle's original state, provided it is not in ruinous condition. This trend gained prominence in 2024, when three renowned concours d'état – Pebble Beach, Villa d'Este, and Chantilly – awarded their 'Best of Show' honors to preserved vehicles, sparking intense debates.

**THIS PAGE:** *Concours d'état events tend to be rather "static" such as the 4th European Concours d'Elegance at Düsseldorf, held at the Galopprennbahn Grafenberg horse race track in 2004. Despite the presence of several truly remarkable cars on display, they remain stationary as crowds gather to admire them. (Serge Cordey)*

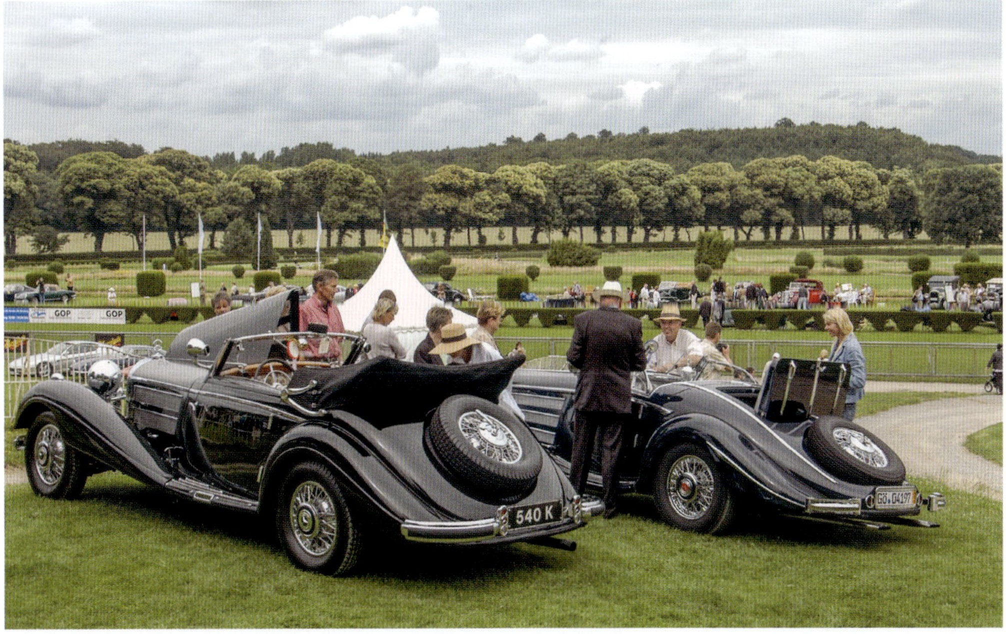

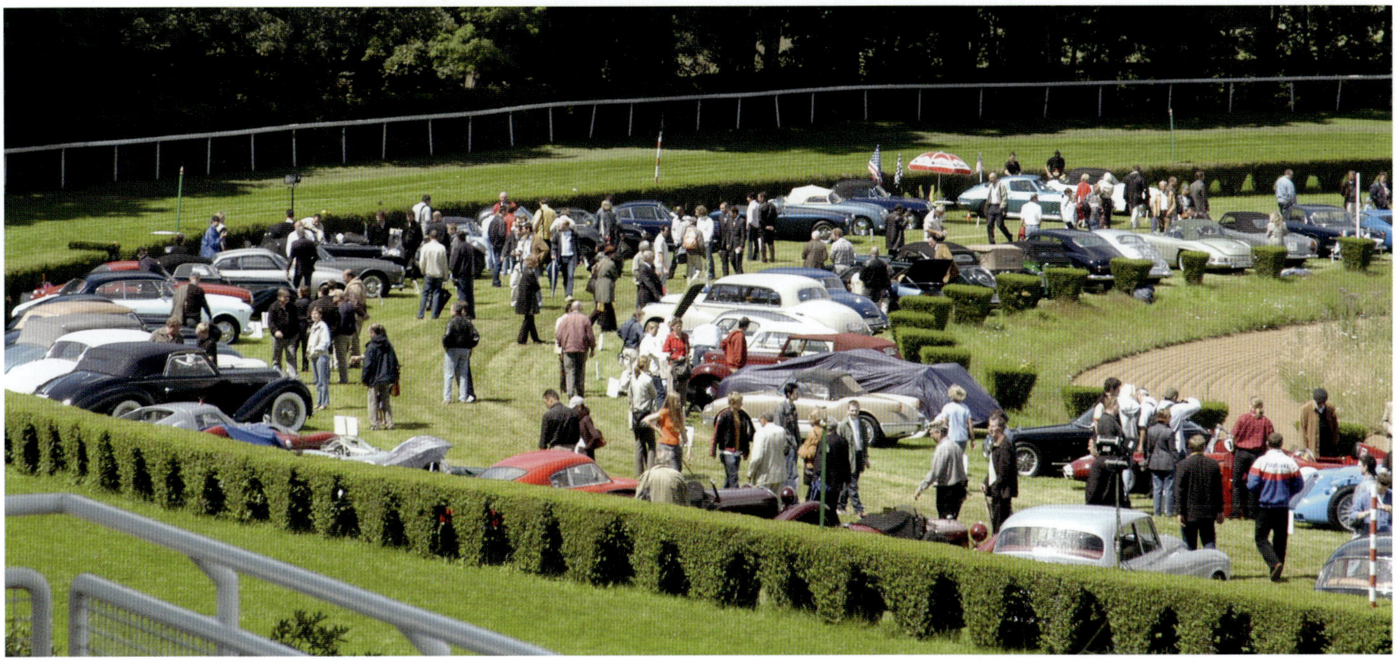

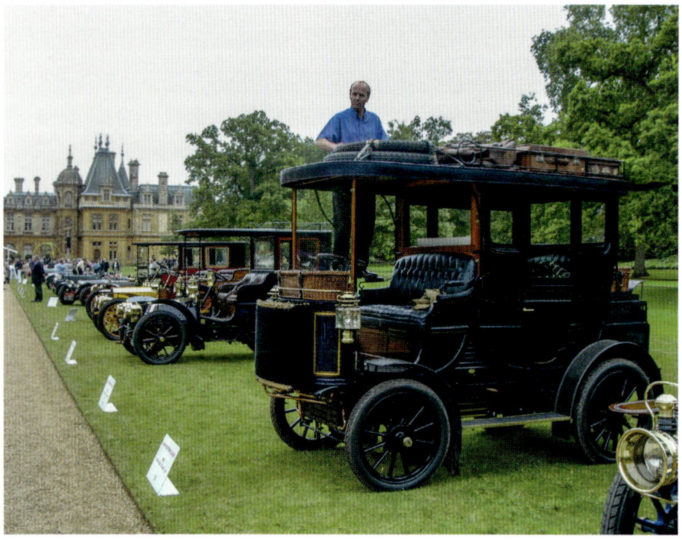

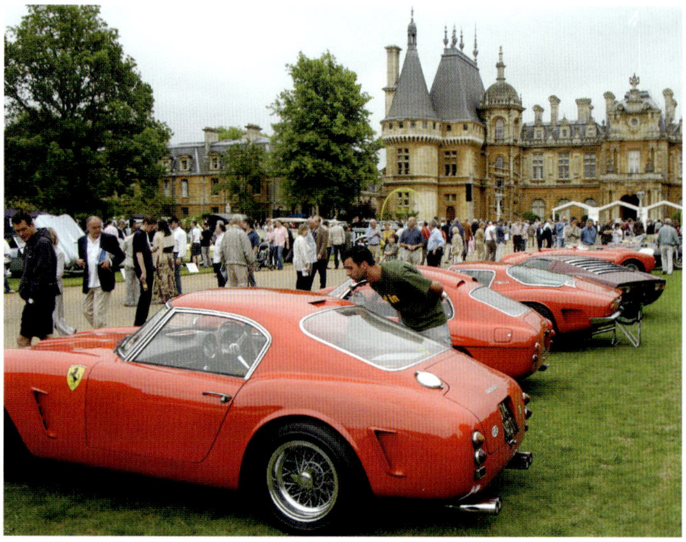

**ABOVE (BOTH):** *Even when Louis Vuitton hosted a Concours d'Elegance at Waddesdon Manor in June 2004, it adhered to the now-classic static format. (Serge Cordey)*

**RIGHT & OPPOSITE TOP:** *Among modern concours, Chantilly is the one that strives to capture the ambiance of the pre-war events. Like a few others, Chantilly seeks to blend tradition and modernity by presenting the latest Bugatti model alongside over a hundred historic vehicles competing across various classes. (Gautam Sen)*

**OPPOSITE RIGHT:** *Fritz Burkard's Bugatti Type 59 made history as the first preservation class car to win Best of Show at Pebble Beach. The same car was featured at the 2022 Concorso d'Eleganza Villa d'Este, where it received the FIVA Preservation Award. (Gautam Sen)*

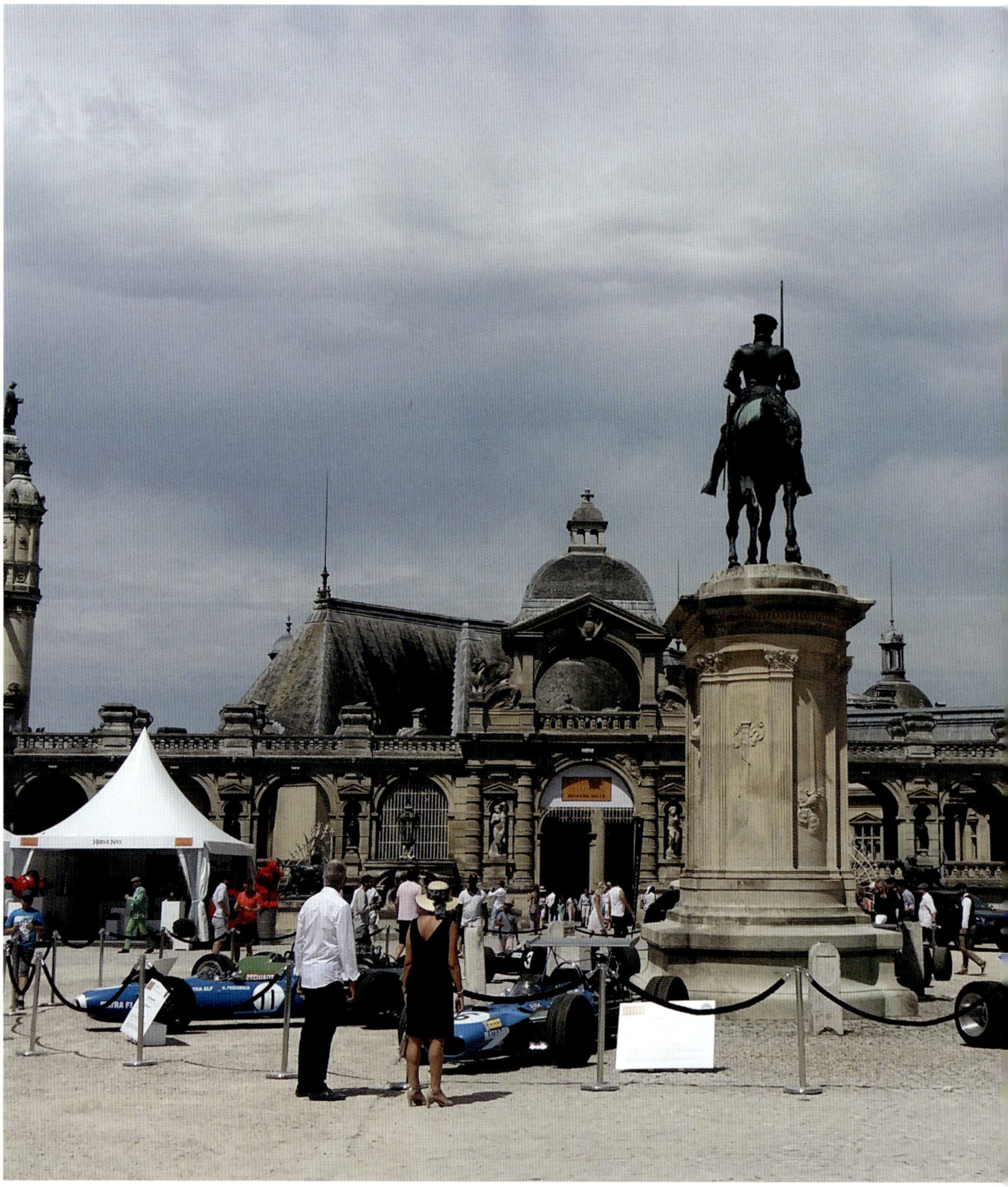

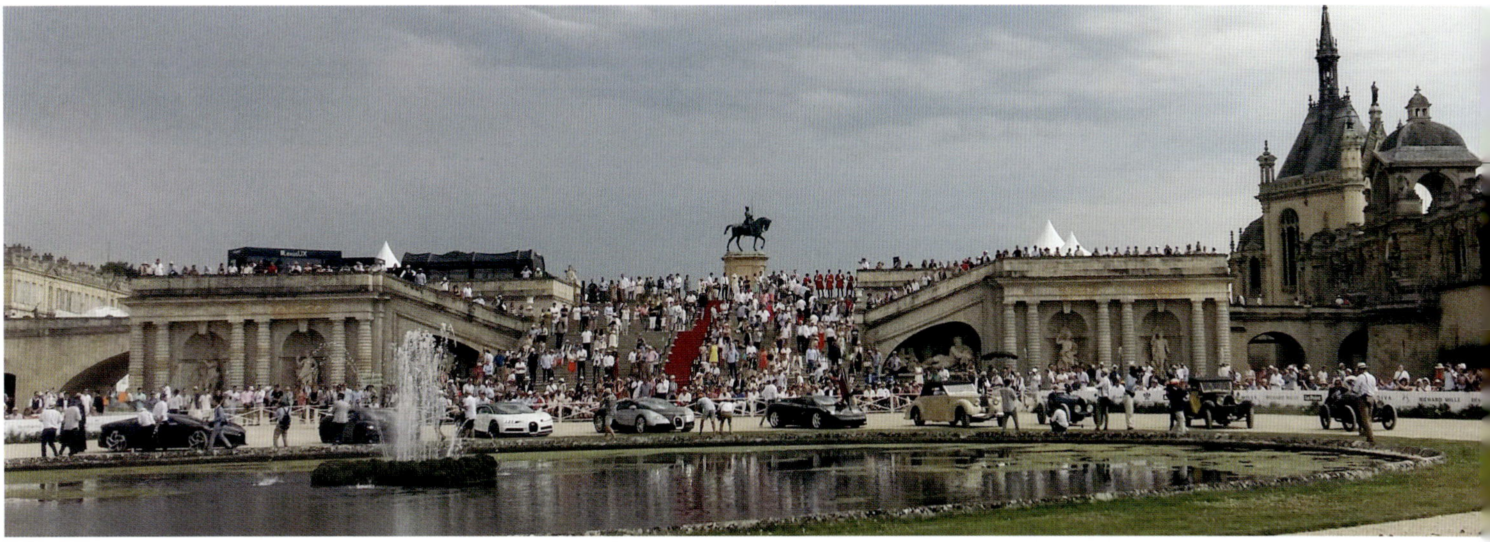

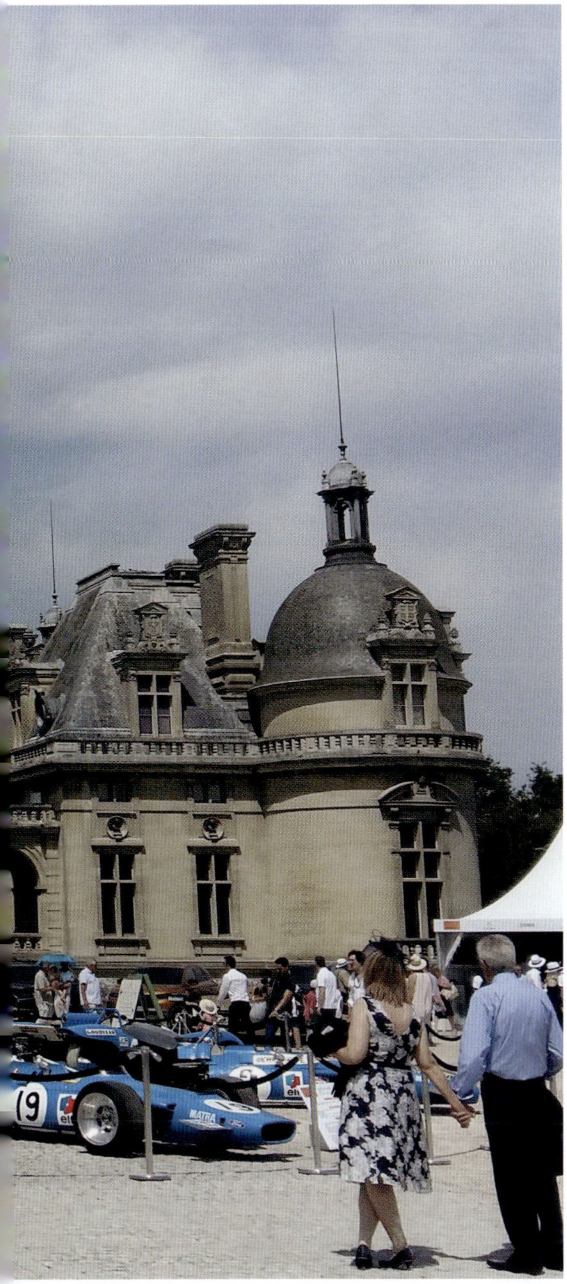

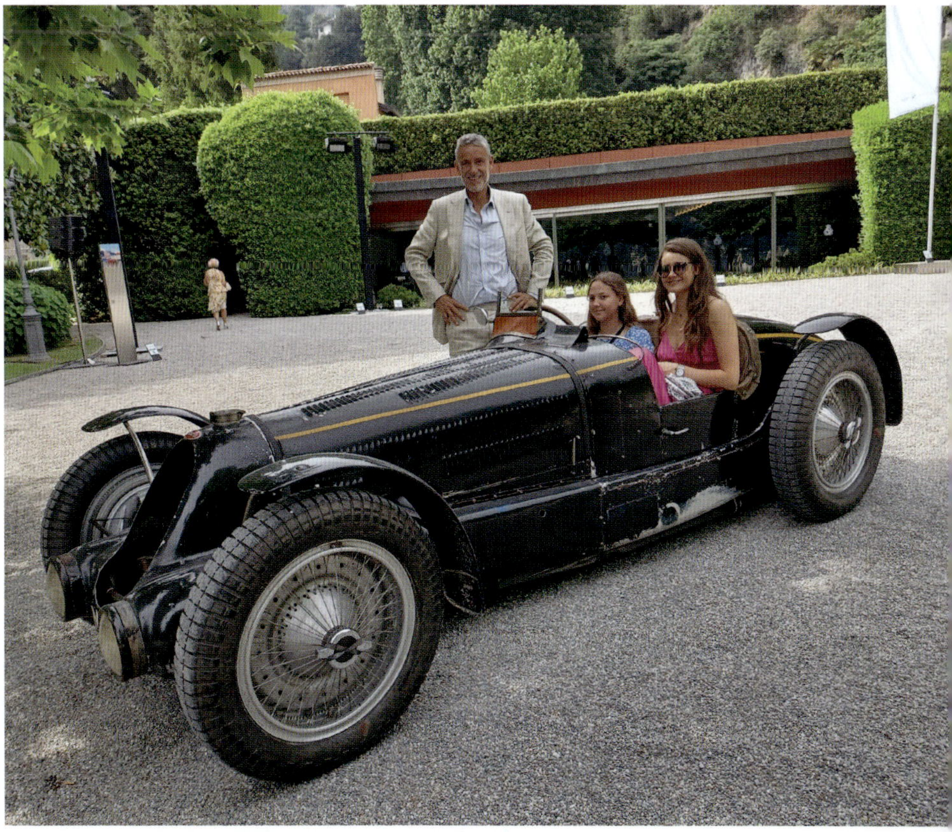

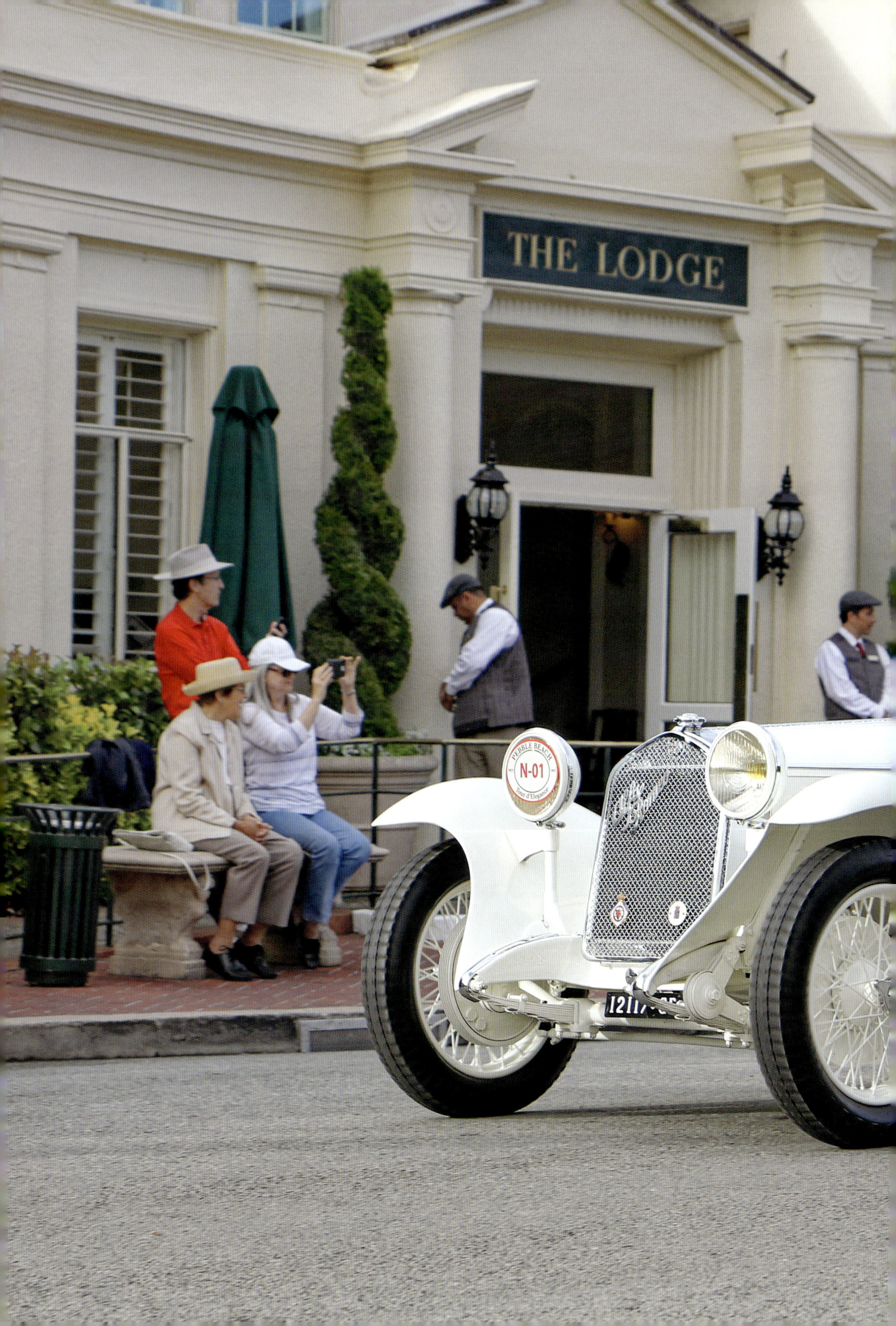
THE LODGE
PEBBLE BEACH
N-01
Tour d'Elegance

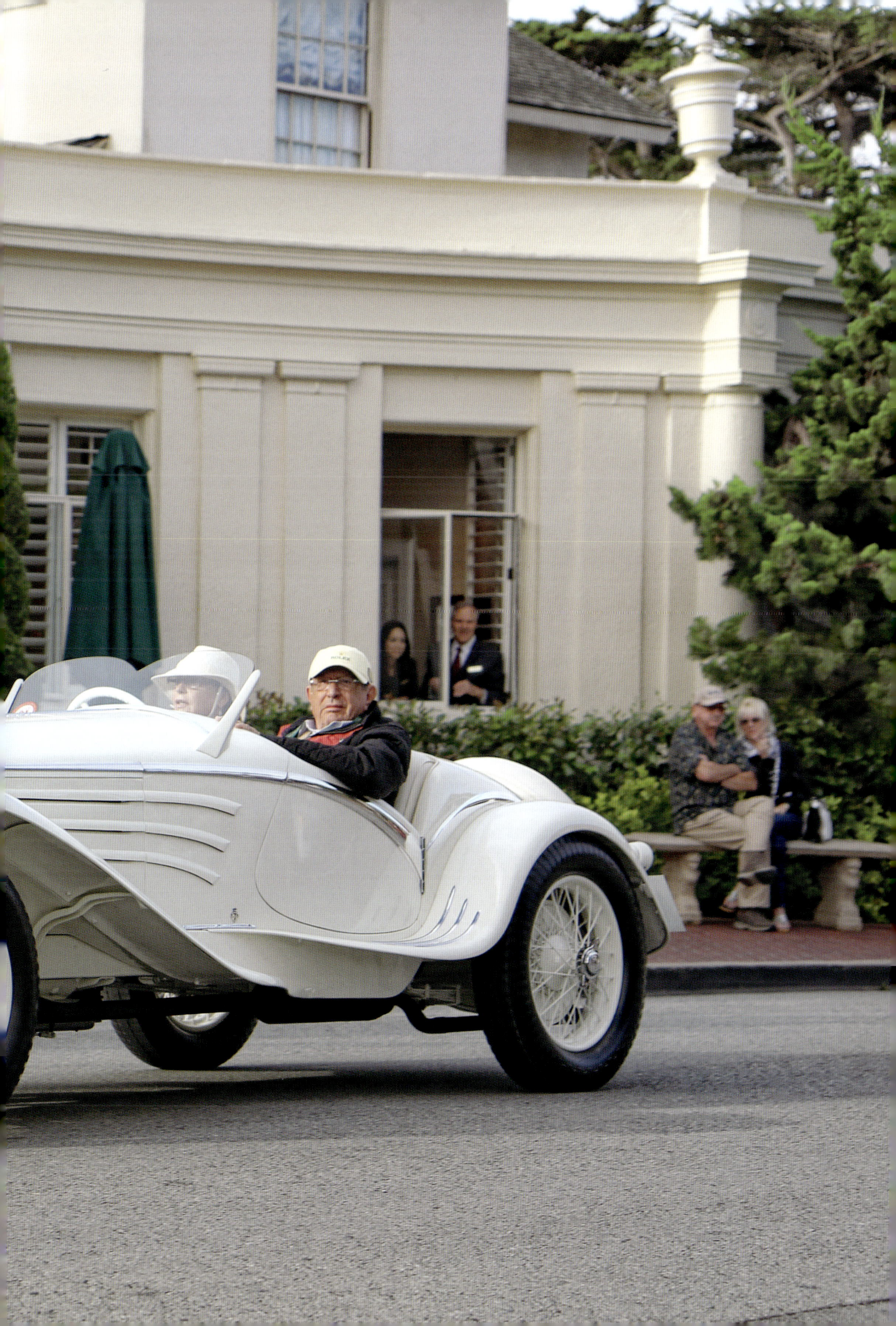

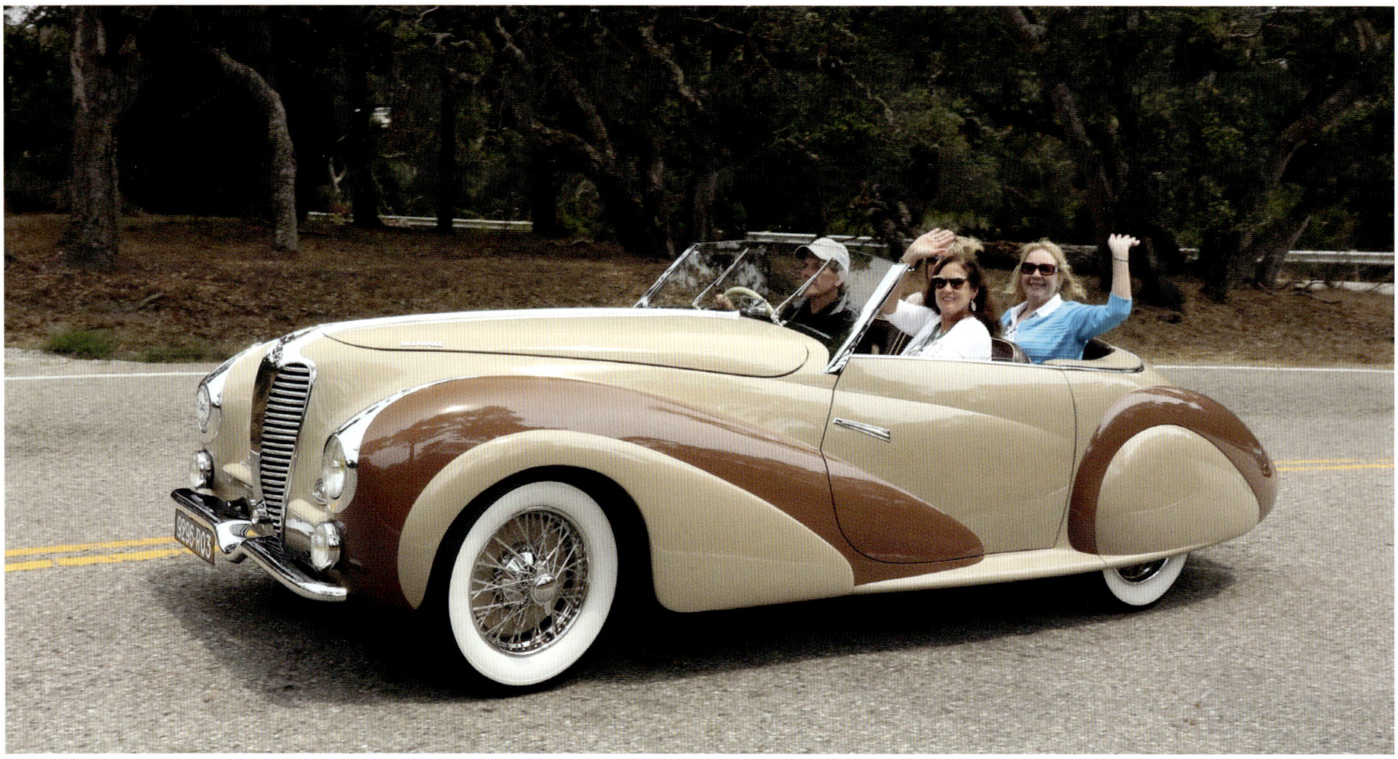

**ABOVE & PREVIOUS TWO PAGES:** *To address criticism of "trailer queens," events like Greenwich and Pebble Beach include a Tour for participants. (Hugues Vanhoolandt)*

Another issue is the phenomenon of 'trailer queens' – pristine historic cars that arrive in climate-controlled trailers and are meticulously positioned on the competition grounds (sometimes even with imported grass!). These cars, polished beyond factory condition, are often perceived as overly staged. To counter this, some concours d'état organizers encourage participating cars to join a 'Tour d'Elegance', a driving route of at least 50 kilometers, with additional points awarded for completing the tour, ensuring the cars are not merely static displays.

In contrast, the concours d'elegance à la française is dynamic, as the vehicle drives up to a jury and audience, who remain seated. Once the car stops, a gentleman (or groom) opens the door for 'madame,' who steps out, revealing an outfit that complements the car's design. The couple circle the car, greeting the judges, audience, and photographers, before getting back into the vehicle and driving off.

Leonardo Fioravanti, a prominent figure at Pininfarina and the designer of several Ferraris, voiced his preference for seeing cars in motion. As a member of the jury at the Concorso d'Eleganza Villa d'Este, he remarked, *"We design them to move, to live – that's how they reveal their beauty, their style, their heart..."* This captures the essence of why motion is crucial in a concours d'elegance, offering a full display of the vehicle's elegance and purpose.

Another significant difference between the concours d'elegance à la française and the concours d'état lies in their judging methods. The French-style concours employs a straightforward approach, without complicated regulations. The criteria recommended by the French Federation of Historic Vehicles, FFVE, consist of three broad categories, with a slight emphasis on the vehicle over the crew:

- The make and model of the automobile (rarity, historic significance, etc.)
- Presentation of the car (condition, color, history, etc.)
- Presentation and elegance of the crew

This simplicity is suited to a chic yet relaxed atmosphere, where results are typically available almost in real time as a courtesy to the audience. However, this method invites criticism for being superficial, subjective, and largely based on emotional judgment, raising the risk of what is often termed 'romantic amateurism.' Additionally, the jury's composition may be questionable, as judges are typically experts in only one aspect of the criteria. The panel usually includes individuals with diverse sensibilities, such as automotive specialists, fashion experts, and socio-economic figures (including elected officials, sponsors, and celebrities).

Judges can coordinate before the competition begins, during the mandatory review of the participating vehicles, throughout the presentations, and finally during the concluding debate prior to announcing the results. While the final decisions remain subjective, it's hard to predict when evaluating something as nuanced as elegance.

In stark contrast, the concours d'état operates in a different realm. Here, subjectivity is minimized, as judges are predominantly automotive professionals – top restorers, historians with extensive knowledge, renowned designers, senior executive from major organizations, respected dealers, racing drivers, and presidents of automotive federations, both past and present, as well as specialized journalists.

With such esteemed individuals on the judging panel, the evaluation criteria cannot be vague; they are notably rigorous and detailed, necessitating a thorough, nuts and bolts examination of each vehicle. Some judges go to the extent of using mirrors and electric lamps to inspect the undercarriage of cars, ensuring no detail is overlooked. This meticulous approach reflects the high standards and expertise required in a concours d'état, underscoring its focus on authenticity and restoration quality.

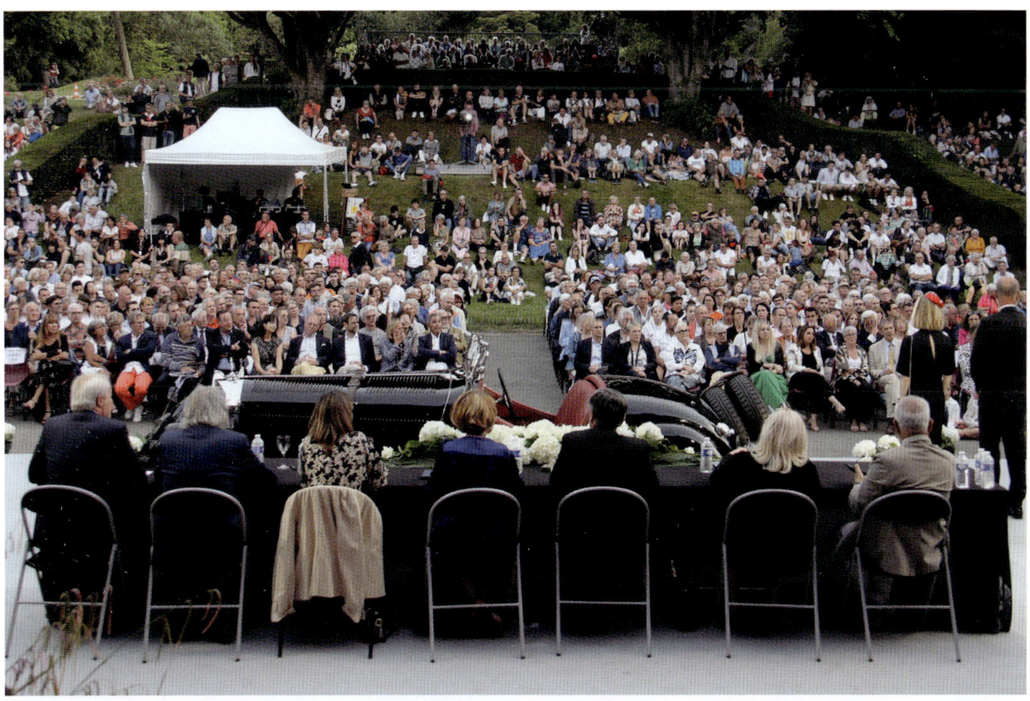

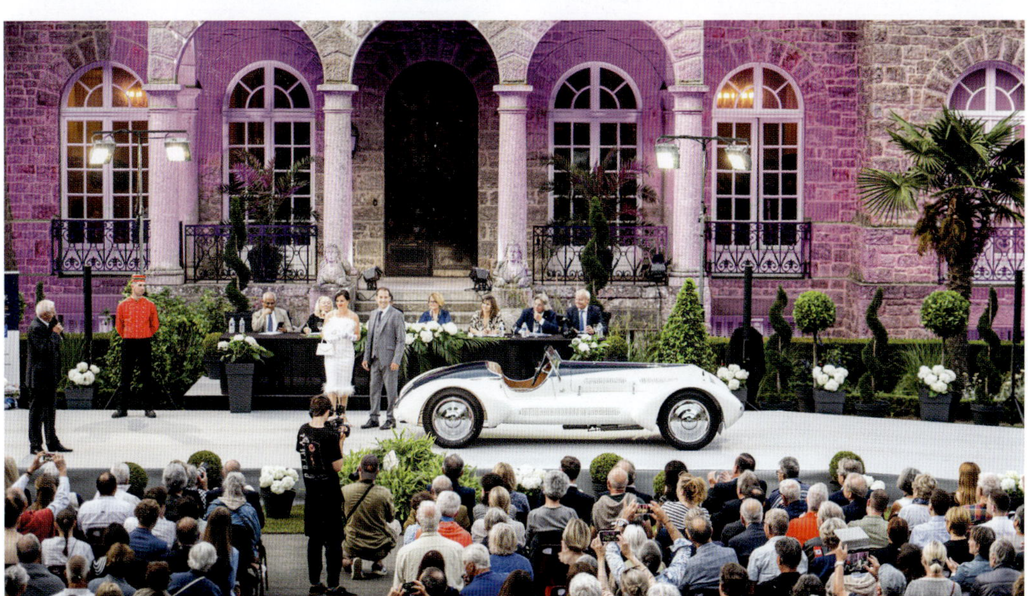

**UPPER:** *In the French concours system, the jury is seated in a row along one side of a ramp, as seen at the Dinard Elegance event in 2023. (Jérôme Cadoret)*

**LOWER & OVERLEAF TOP:** *The audience is positioned on the opposite side of the ramp, watching (and voting) as the cars parade by with their respective owners and teams. (Louis Monnier)*

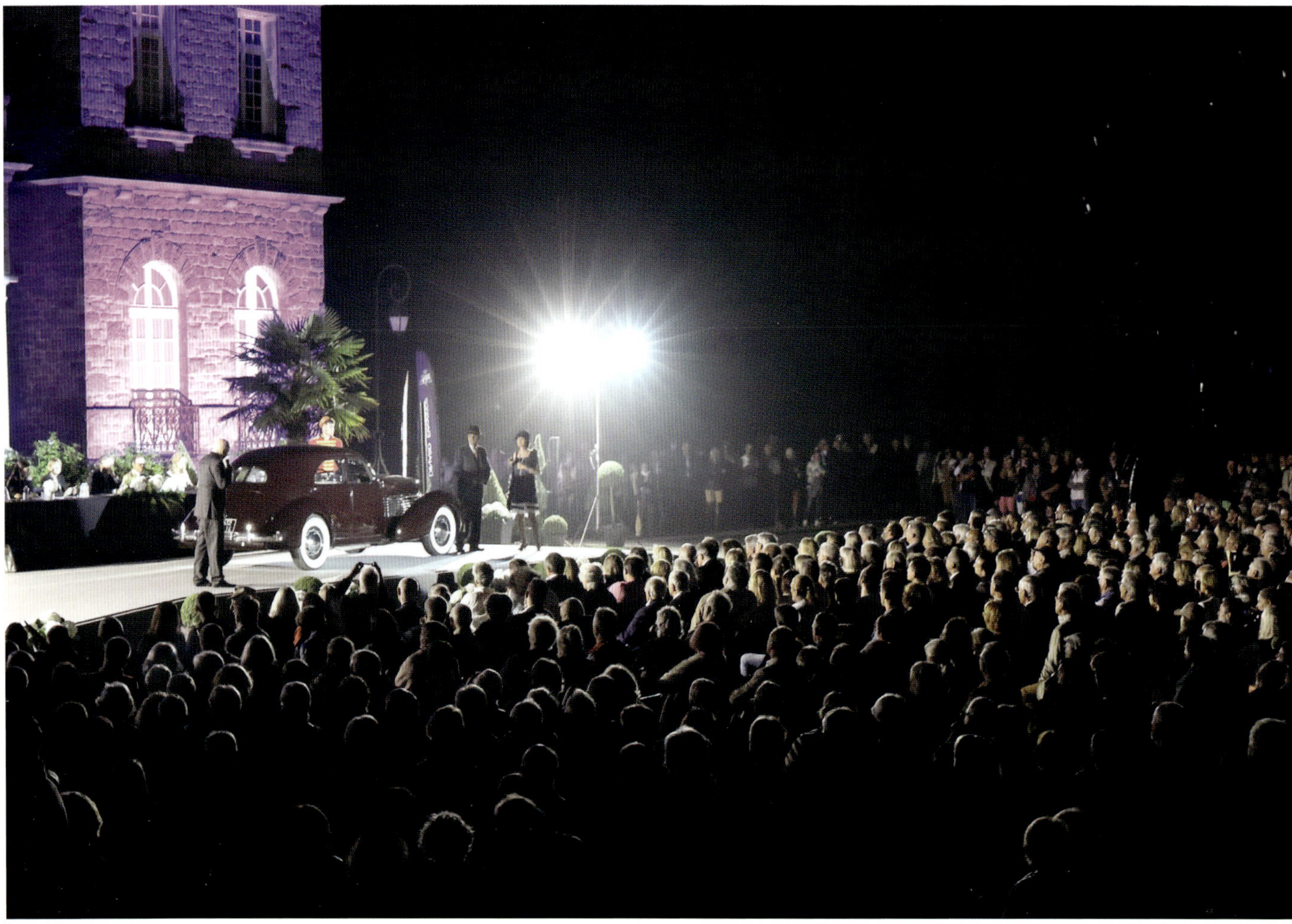

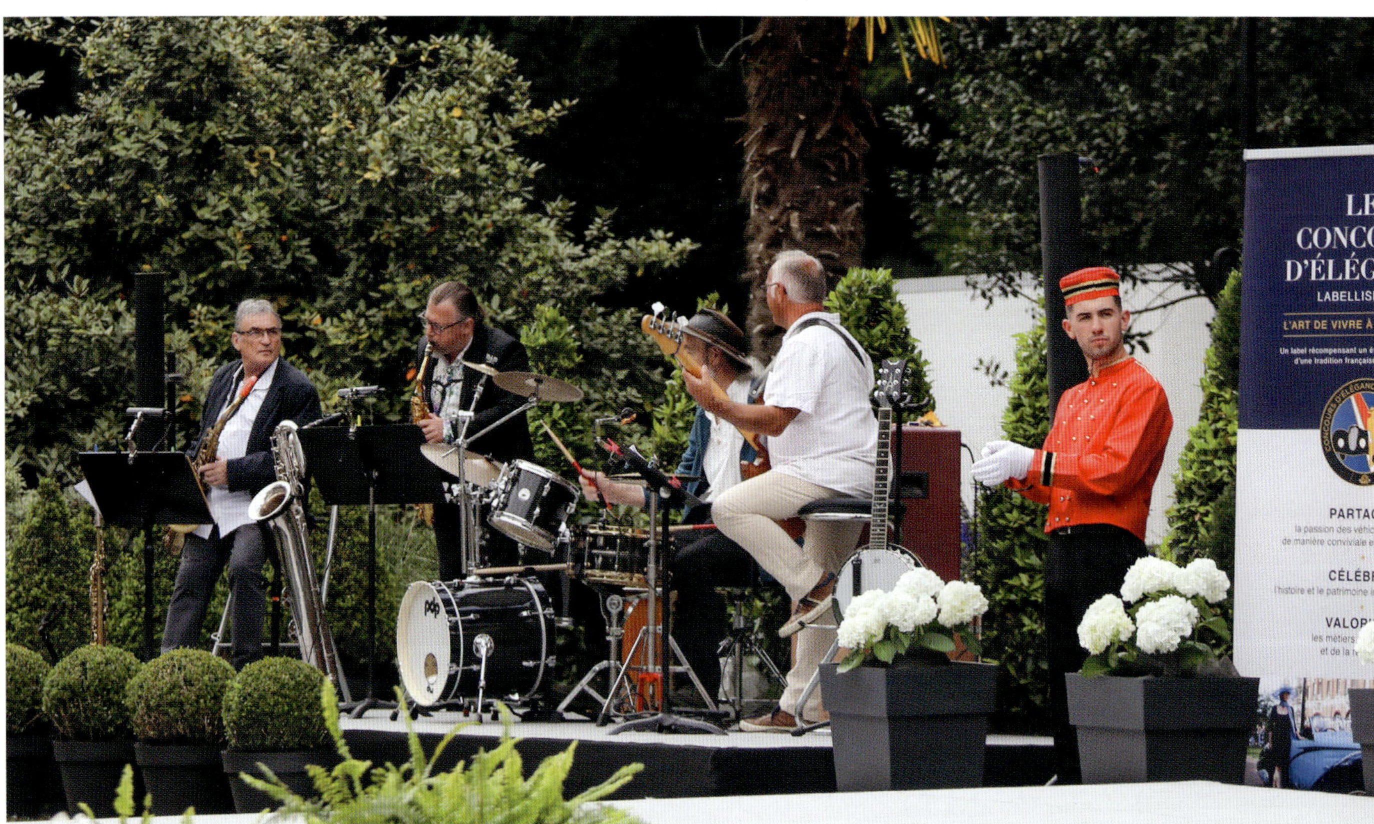
LES
LABELLISÉS

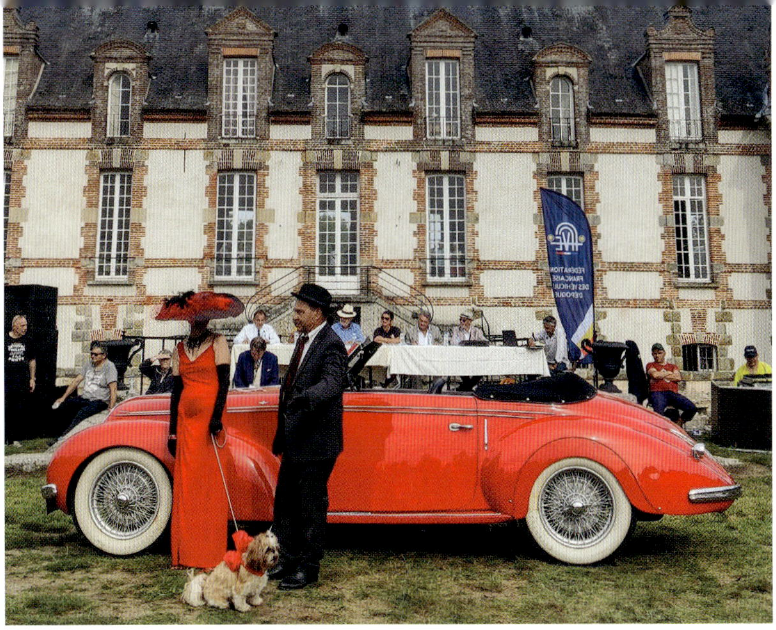

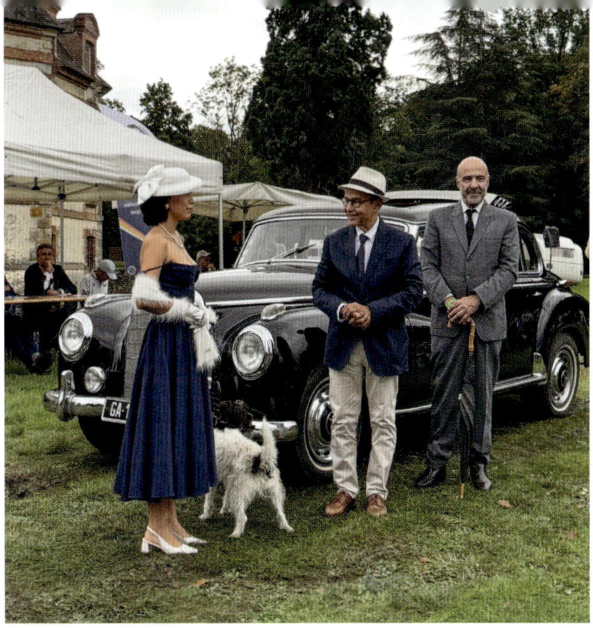

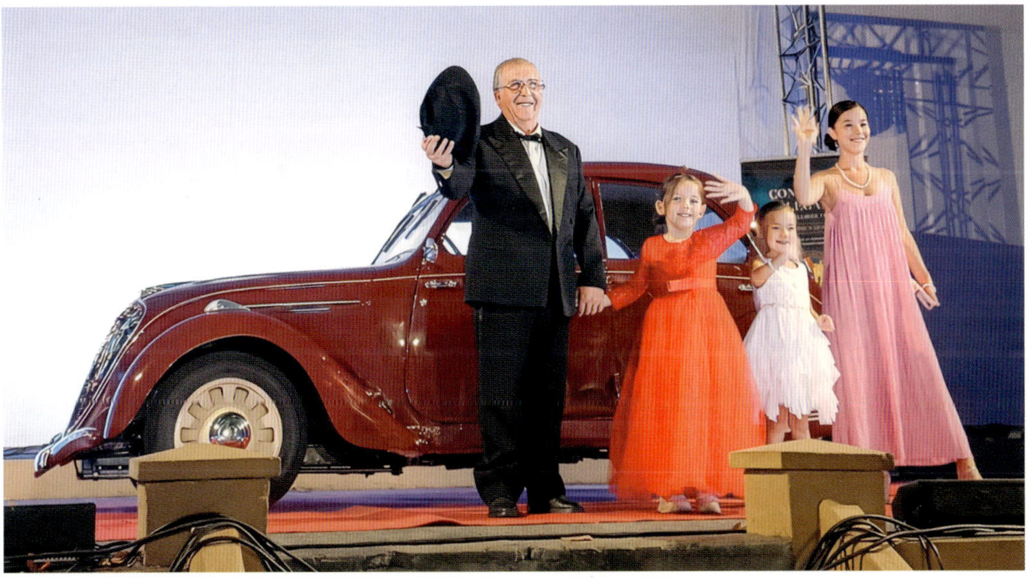

**OPPOSITE TOP:** *The car that won the Grand Prix d'Excellence at the 2021 Dinard Elegance, an immaculate Cord 810, presented beautifully by Marie-Line Delaunay and Xavier Jenvrin. (Guillaume Nedelec)*

**OPPOSITE BOTTOM:** *Meanwhile, a live band performs tunes in keeping with the era, period, and theme presented by the team, as seen at Dinard Elegance 2023. (Jérôme Cadoret)*

**TOP (BOTH):** *An equipage presentation can be observed at the Art Automobile event in Gambais, near Paris. (Ameya Mishra)*

**MIDDLE:** *The Concours d'Elegance in Tonneins, a town situated south of Bordeaux, also follows a similar presentation style to Dinard and Gambais. (Gautam Sen)*

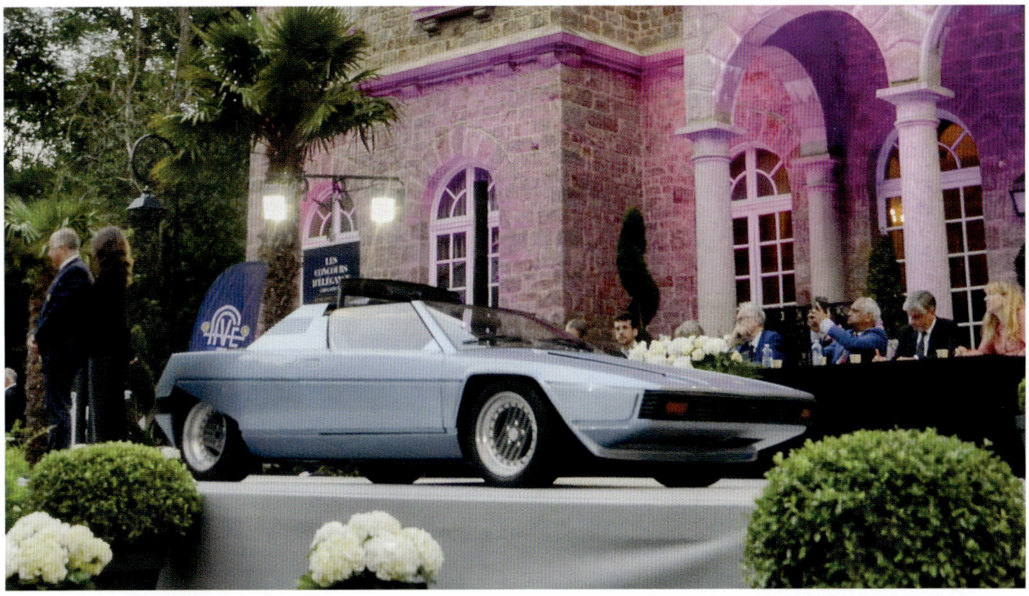

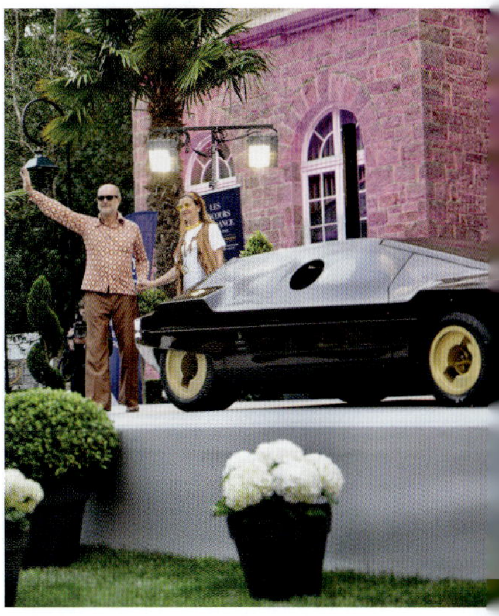

**ABOVE:** *The Marcello Gandini-designed Ferrari Rainbow from the ASI Collection was a big hit with the spectators at Dinard 2025, receiving the Prix Public. The car was presented by the young enthusiast Kalina Benhamou and Giuseppe Dell'Aversano, from ASI. (Dominique Torchut)*

**RIGHT:** *Another Marcello Gandini-designed concept car, the Lancia Sibilo, owned by famous Italian collector Corrado Lopresto, won its class at Dinard 2025. The car was presented by Chatuna and Emmanuel Bacquet. (Dominique Torchut)*

## Judging Systems

For a concours – a competition in English – a scoring system has always been essential. In the pre-war era, when Concours d'Elegance featured brand-new cars straight from coachbuilders or manufacturers, the judging criteria focused on the car's design and elegance, the attire of the model accompanying it, and the overall presentation of the ensemble with the vehicle.

Today, however, concours events primarily feature vehicles that are decades old, judged long ago for their original design and elegance. Over time, new scoring systems have been developed, each tailored to different goals and objectives. The reality is that most cars in modern concours have spent many years in use before undergoing restoration – sometimes to conditions even better than when they were new. Several well-preserved, unrestored cars have gained increased recognition and importance in recent years. Additionally, some events include categories for modern cars, concepts, one-offs, and limited-edition or series production models.

Given this diversity, can a single scoring system apply to all types of vehicles? Should there be uniform standards, or should there be variations in judging criteria and event formats to reflect the differences between vehicles?

Below, we outline some of the scoring systems currently in use:

### ICJAG (International Chief Judge Advisory Group) and M100

Until recently, the ICJAG (International Chief Judge Advisory Group) judging system was widely adopted by several of the most prominent American-style Concours d'Elegance, as well as a select number in Europe and India. Known for its rigorous, detail-oriented approach, the ICJAG system emphasized restoration quality and authenticity above all else. Judges were required to possess a high degree of expertise and specialization to effectively apply the system's stringent standards.

However, one key limitation of the ICJAG method was its limited consideration for preservation-class vehicles and its lack of emphasis on the historical and aesthetic significance of an automobile. In response to growing calls for a more holistic judging philosophy, a small but meaningful shift occurred when 10 percent of the total score was allocated to account for a vehicle's emotional and visual appeal.

Using a 100-point scale, ICJAG relied primarily on point deductions for flaws or inaccuracies found during meticulous inspections. By 2024, eleven concours events across the United States, Europe, and India had officially adopted the ICJAG methodology.

In 2025, former ICJAG Chairman Nigel Matthews, in collaboration with Chris Kramer, introduced a revised system: the M100 (Modified Professional Judging 100). This updated format retains the foundational rigor of ICJAG but introduces a dedicated five-point category evaluating a car's elegance, provenance, and field presence – acknowledging the broader emotional and historical dimensions of concours-level vehicles.

Several concours events are expected to transition to this updated M100 system beginning in 2025:

- The Amelia (Amelia Island, FL, USA)
- La Jolla Concours d'Elegance (La Jolla, CA, USA)
- Lugano Elegance (Lugano, Canton Ticino, Switzerland)
- Greenwich Concours d'Elegance (Greenwich, CT, USA)
- Salon Privé (Woodstock, Oxfordshire, UK)
- Festival of the Unexceptional (Lincolnshire, UK)
- ICONS Mallorca (Mallorca, Spain)
- 21 Gun Salute Concours d'Elegance (New Delhi, India)

### FIVA (Fédération Internationale des Véhicules Anciens)

FIVA, the global federation that represents historic vehicle clubs and associations, recommends a scoring system developed by Mark Gessler, former president of the now-defunct American Historic Vehicle Association (HVA). This system is similar in nature to the ICJAG system, with a focus on the vehicle's condition, making it suitable for a concours d'état format.

Using a 100-point system, FIVA's criteria assess various aspects of the vehicle, including the condition of the coachwork, paint, tires, bumpers, chrome trims, seats, upholstery, dashboard, trunk, engine, chassis, fuel and exhaust systems, and cooling systems. Additionally, points are awarded for the vehicle's provenance and age, although no points are given for presentation.

Some of the concours events that have adopted the FIVA system for evaluating historic vehicles include:

- Ascona Classic Car Award / ACCA (Ascona, Switzerland)
- Concours d'Elegance Tegernsee (Tegernsee, Germany)
- Int'l Concours d'Elegance Gala Schwetzingen (Germany)
- Sydney Harbour Concours d'Elegance (Sydney, Australia)
- Zurich Classic Car Award / ZCCA (Zurich, Switzerland)

### IAC / PFA (International Advisory Council for the Preservation of Ferrari Automobiles)

The IAC / PFA is an independent organization based in the United States, composed of twenty Ferrari enthusiasts, researchers, and authors, who believe in focusing on the correct maintenance of the originality and authenticity of Ferraris. The judging criteria have been designed to ensure that these cars remain as close as possible to their original condition, emphasizing preservation over restoration, even if the scoring system gives an impression of focusing on the nuts and bolts of each vehicle. The judging system is followed by events such as the Ferrari Club of America Annual Experience, Regional and Chapter FCA judged events, and all the Cavallino Classics, such as:

- Cavallino Classic Middle East (Abu Dhabi)
- Cavallino Classic Palm Beach (Florida, USA)
- Cavallino Classic Modena (Italy)

### Peer Review

In the Peer Review system, the judging is carried out by the entrants themselves. Participants in each class evaluate and select the vehicle they believe to be the best within their category. Once the class winners are determined, the participants then vote for the overall Best of Show from among these class winners. This approach fosters a more communal and participant-driven judging process.

Several concours events employ the Peer Review system, allowing participants to play an active role in selecting the winners:

- Concours in the Hills (Fountain Hills, AZ, USA)
- Cartier Style et Luxe (Goodwood Festival of Speed, UK)
- Las Vegas Concours at Wynn (Las Vegas, NV, USA)

### People's Choice

The People's Choice system, widely used in the UK, allows the general public to vote for their favorite cars. Like the pre-war tradition, spectators are invited to cast votes for the vehicles that appeal to them the most. There is no jury of experts to decide the winners. One downside of this system is that cars lacking in authenticity may still win their class or even Best of Show, as public opinion is based on personal preference rather than technical accuracy. In some concours, particularly in France, as well as Villa d'Este (the Coppa d'Oro), the people's choice is awarded as an additional prize alongside more formal judging categories.

Two of the most well-known events using this system of letting the spectators decide are:

- Concours on Savile Row (London, England)
- Concours of Elegance (Hampton Court, England)

### French-style

The French-style judging system is a more relaxed and freeform approach, where the car's condition is evaluated alongside its presentation, design, and overall elegance. Unlike more rigid and formal systems, this method allows for a broader interpretation of what makes a car stand out. However, it has faced criticism for its lack of strict criteria, which sometimes results in less serious or less authentic cars winning class awards or even Best of Show.

Several concours events adopt the French-style system, with the following three being notable examples:

- Misselwood Concours d'Elegance (Beverly, MA, USA)
- St. Michaels Concours d'Elegance at Chesapeake Bay (Kent Narrows, MD, USA)
- Audrain Newport Concours d'Elegance (Newport, RI, USA)

### FFVE (French Federation of Historic Vehicles)

The FFVE has developed an intriguing judging system that adapts the concept of Concours d'Elegance to a 21st century context. This system emphasizes the presentation of older historic vehicles alongside an accompanying equipage (crew or occupants who accompany the vehicle). To reflect this modern approach, the judging scoresheet allocates twenty points each for three main categories of Chassis (including originality, authenticity, and quality of restoration), Carrosserie or the coachbuilder (the beauty of the design), and Couture (the presentation of the crew).

This scoring system is well-suited for smaller concours d'elegance à la française. As of 2025, the FFVE has officially endorsed events throughout France that utilize this framework:

- Les Légendes de Nice (Nice)
- Concours d'Elegance du Rallye d'Aumale (Chantilly)
- Rétro Cœur / AutAu Cassis (Cassis)
- Concours d'Elegance des 40 ans du TRAC Albi (Albi)
- Concours d'Elegance Pau – Creasud (Pau)
- Dinard Elegance (Dinard)
- Vittel Elegance Automobile (Vittel)
- Concours d'Elegance de la traversée de l'Hurepoix (Nozay)
- Elegance Automobile à Bourg-en-Bresse (Bourg-en-Bresse)
- Elegance à Sury-aux-Bois (Sury-aux-Bois)
- Puy Mary Historic (Vic-sur-Cère)
- Concours d'Elegance d'automobiles de collection de la ville de Tonneins (Tonneins)
- Concours d'Elegance de Saint-Cado (Belz)
- Saint-Jean-Cap-Ferrat Prestige Motors (Saint-Jean-Cap-Ferrat)
- Le Touquet-Paris-Plage Historique By JPM (Le Touquet)
- Concours d'Elegance du Circuit des Remparts d'Angoulême (Angoulême)
- Automobile et Patrimoine au Château de Bouthéon (Andrézieux-Bouthéon)
- Elegance sur le Port (Saint-Pierre de la Réunion)
- Sainte-Maxime Rêves Auto (Sainte-Maxime)
- Concours d'Elegance Cap Rétro (Le Cap d'Agde)
- Agora du Véhicule de Collection (Tours)
- Concours d'Elegance à la Française en Automobile-Art (Gambais)

Concours d'Elegance for motorcycles:

- Concours Moto FFVE (Lyon)
- Les Légendes de Nice (Nice)
- Concours d'Elegance du départ de La France en motos anciennes (Flixecourt)
- Concours d'Elegance du Circuit des Remparts d'Angoulême (Angoulême)
- Concours d'Elegance Moto Pau Rétro Show (Pau)

In addition to the aforementioned systems, it is worth highlighting the unique approaches used by Retromobil Club Romania for the Concurs de Confort și Eleganță and the Chantilly Arts & Elegance Richard Mille.

For the Concurs de Confort și Eleganță, the scoring system is a blend of traditional concours d'état criteria, focusing on the quality of restorations, combined with a smaller emphasis on the presentation of the owner and their partner. This mix creates a balanced approach, valuing the vehicle's condition and the elegance of its presentation.

At Chantilly Arts & Elegance Richard Mille, the scoring system is more nuanced, with equal weight given to several attributes, such as authenticity, the vehicle's specific history, design and elegance of its shape, exterior and interior condition and working condition of the vehicle.

Additional points are awarded for participating in the drive and for having a FIVA ID card, ensuring a comprehensive evaluation of the vehicle's authenticity and performance. Notably, Chantilly's concours focuses on concours d'état for historic vehicles, with a separate scoring system for brand-new vehicles, models, and fashion presentations in the Concours d'Elegance category.

While the judging style of French-style concours is straightforward and flexible – sometimes too much so – the push for precision and objectivity in concours d'état can go too far, becoming overly detailed and rigid. When any value is pushed to an extreme, it risks becoming a counter-value. Recognizing this potential pitfall, some judges of concours d'état express concern over the lack of emotional engagement in their evaluations. They lament the absence of the inevitable 'love at first sight' or that elusive 'je ne sais quoi' that sets apart an entry of real distinction. To address this, there have been suggestions to incorporate an emotional component into the scoring system, potentially accounting for up to 10 percent of the total score. This debate continues, highlighting the contradiction between objectivity and emotional resonance in judging.

The primary reason for these differing approaches stems from the distinct objectives and challenges faced by each type of concours. The French-style concours, often organized by clubs or volunteer groups, aims to promote historic vehicles by providing the general public with an "educational relaxation" experience. Participants can rediscover the rich world of automobiles encompassing automotive manufacturing, coachbuilding, haute couture, millinery, leather goods, and glove making, all brought together in a stylish setting.

For participants in the concours d'elegance à la française, the focus is on enjoyment and camaraderie, fostering a relaxed atmosphere among enthusiasts. The stakes are low, with the primary rewards being a trophy or kind words from a knowledgeable presenter regarding their vehicle and crew. This light-hearted approach strives to maintain a chic and tasteful ambiance.

Whereas the concours d'état in several cases could be seen as a "business" with substantial economic implications. These events are typically organized by professional event specialists, major automotive or automotive-related marques, foundations, or high-profile patrons. Of course there are exceptions too, such as the Hillsborough Concours d'Elegance, which is organized by volunteers, but these are few and far between. The focus though for most concours d'état is not only on the celebration of automotive excellence but also on the economic and branding opportunities associated with presenting high-value vehicles. This professional orientation, no doubt, shapes the rigorous standards and detailed evaluation methods employed in concours d'état events.

The outcomes of concours d'état have a profound impact on the history and economic value of a vehicle. Securing a class win, particularly the prestigious Best of Show title at renowned events like

**THIS PAGE:** *At concours events like Chantilly Arts & Elegance Richard Mille, in addition to the tour, there is a brief parade of all the cars during the prize distribution ceremony. (Gautam Sen)*

**OPPOSITE:** *Most French concours, such as Dinard, are free and open to everyone, making them highly effective in promoting the legacy of historic vehicles to the general public. (Gautam Sen)*

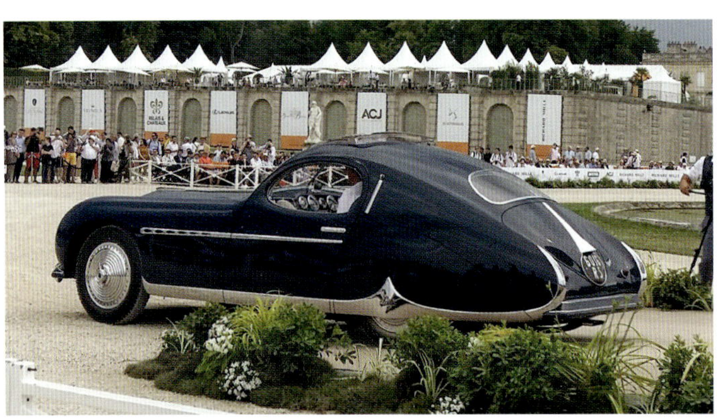


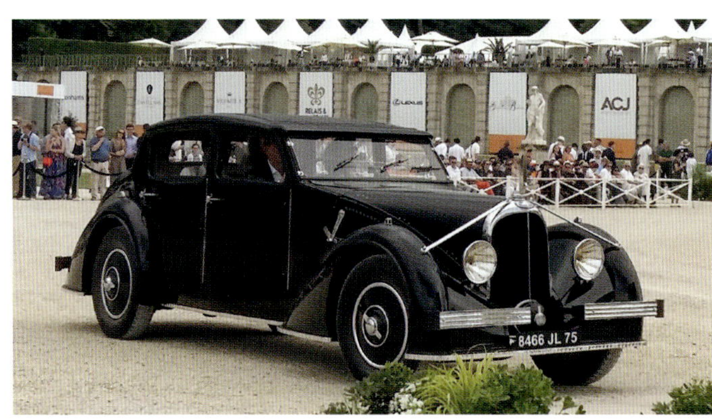


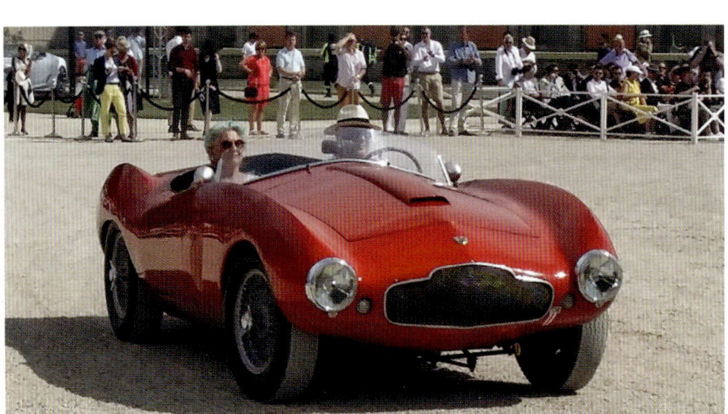

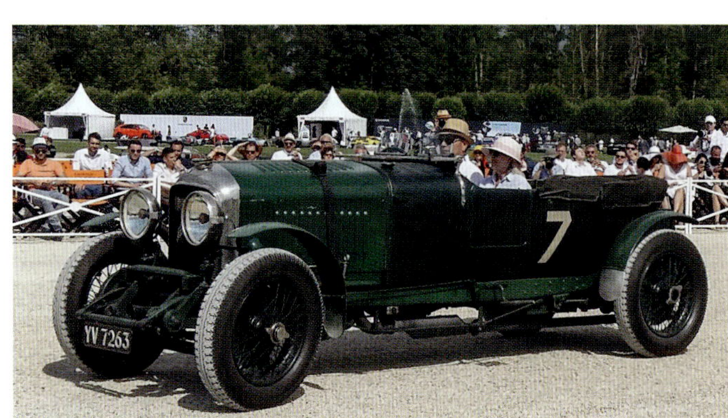

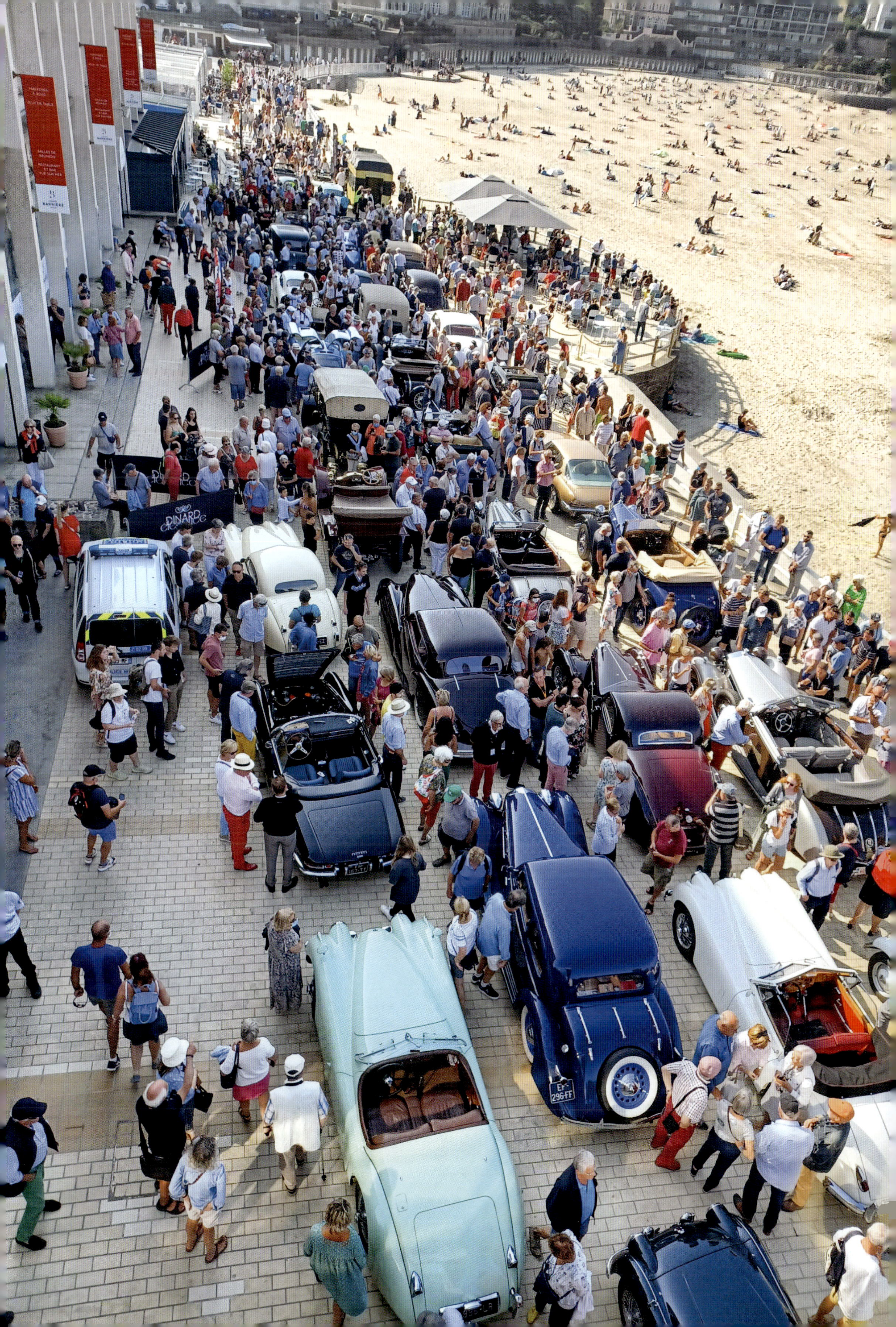
MACHINES À SOUS
JEUX DE TABLE
SALLES DE REUNION
RESTAURANT ET BAR VUE SUR MER
BARRIERE
DINARD

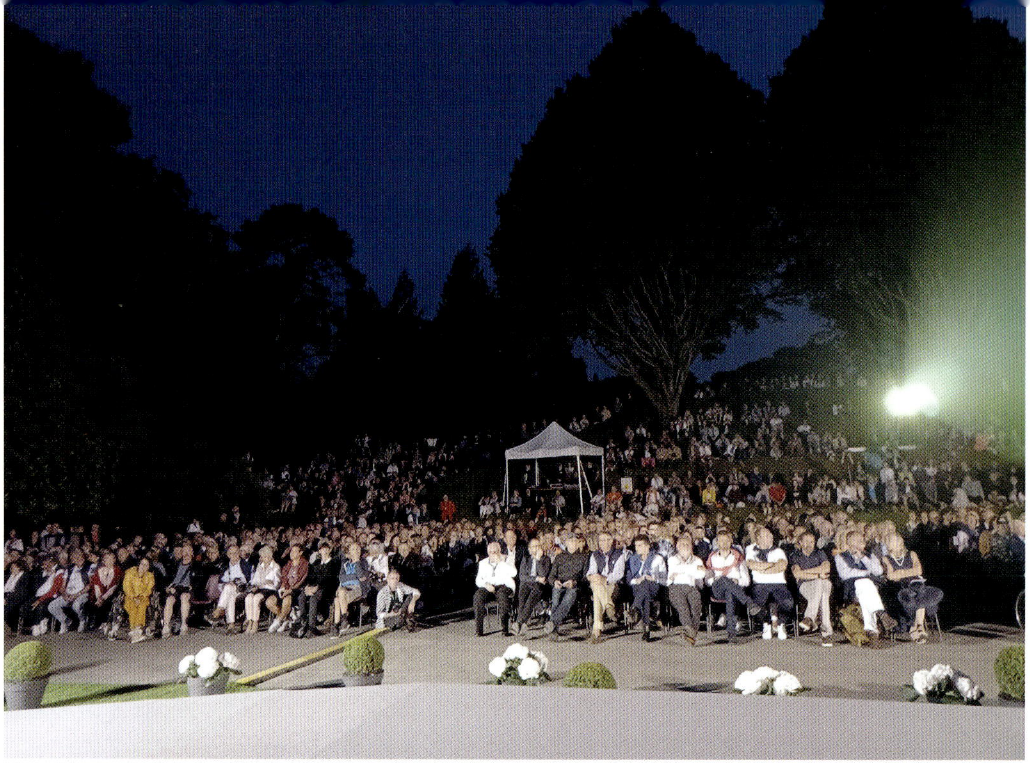

**RIGHT:** *Even when the audience at the car presentation is by invitation, attendance is typically free for them as well. (Gautam Sen)*

**MIDDLE (BOTH):** *It's not uncommon to see several ordinary cars, but what truly sets them apart is the way they are presented. (Gautam Sen)*

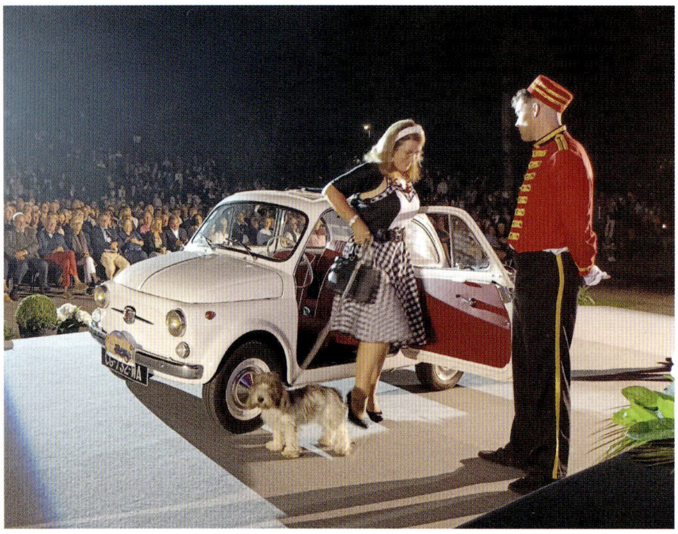

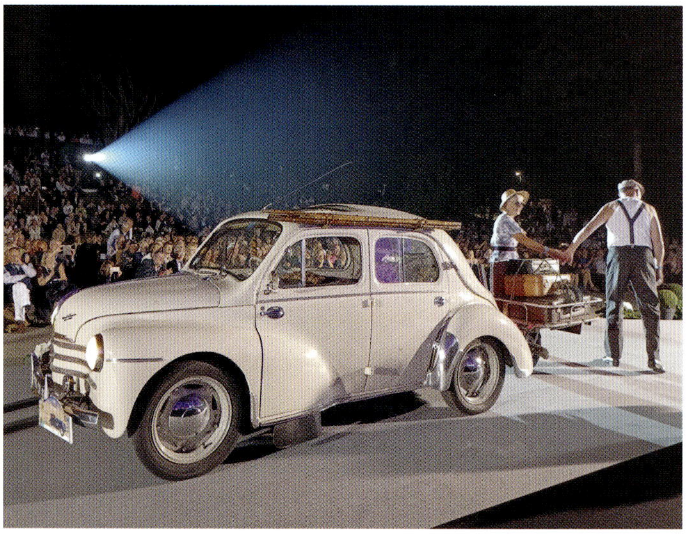

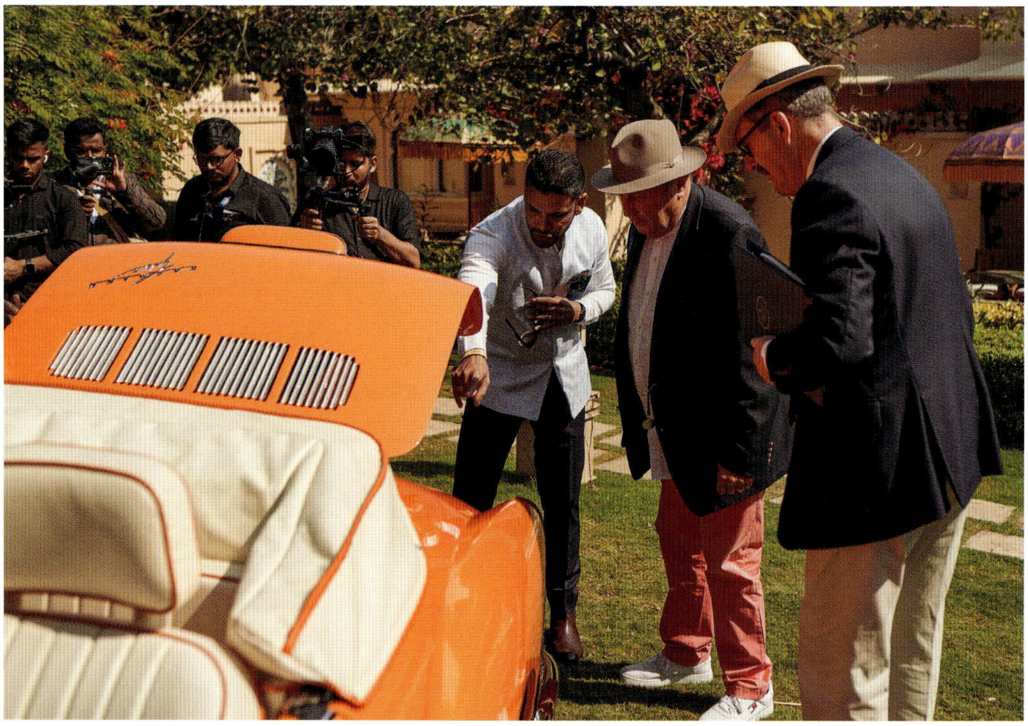

**RIGHT & OPPOSITE TOP (BOTH):** *In stark contrast, at a concours d'état like The Oberoi Concours, the judges visit the static cars, meet the owners, and thoroughly assess the condition of each vehicle. (Makarand Baokar)*

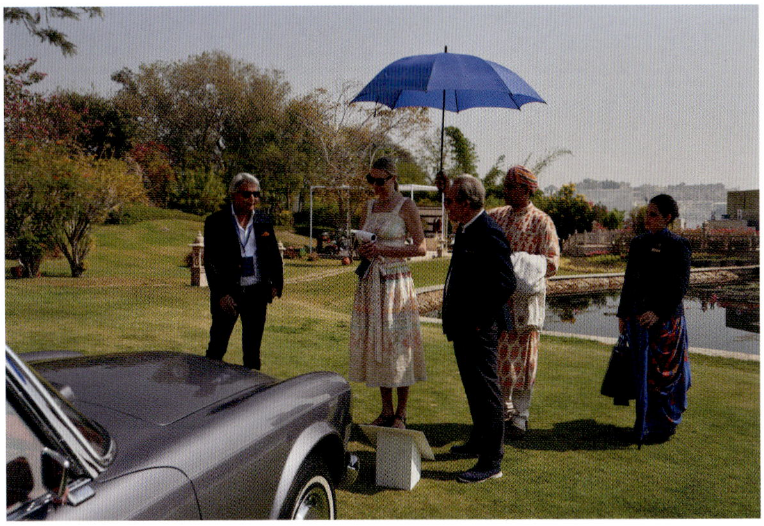

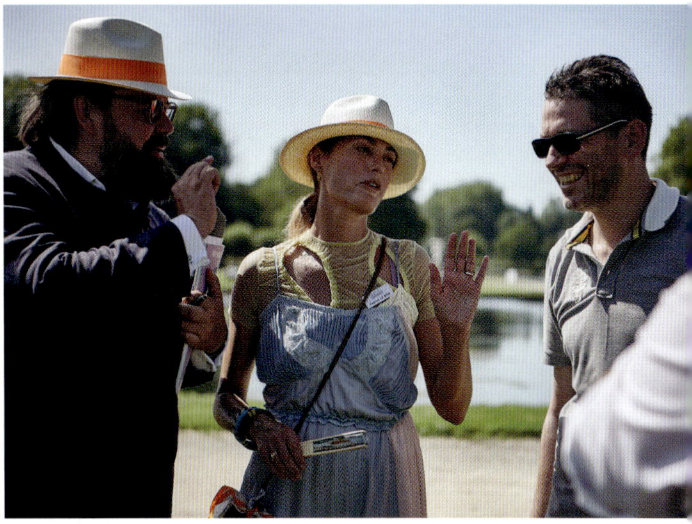

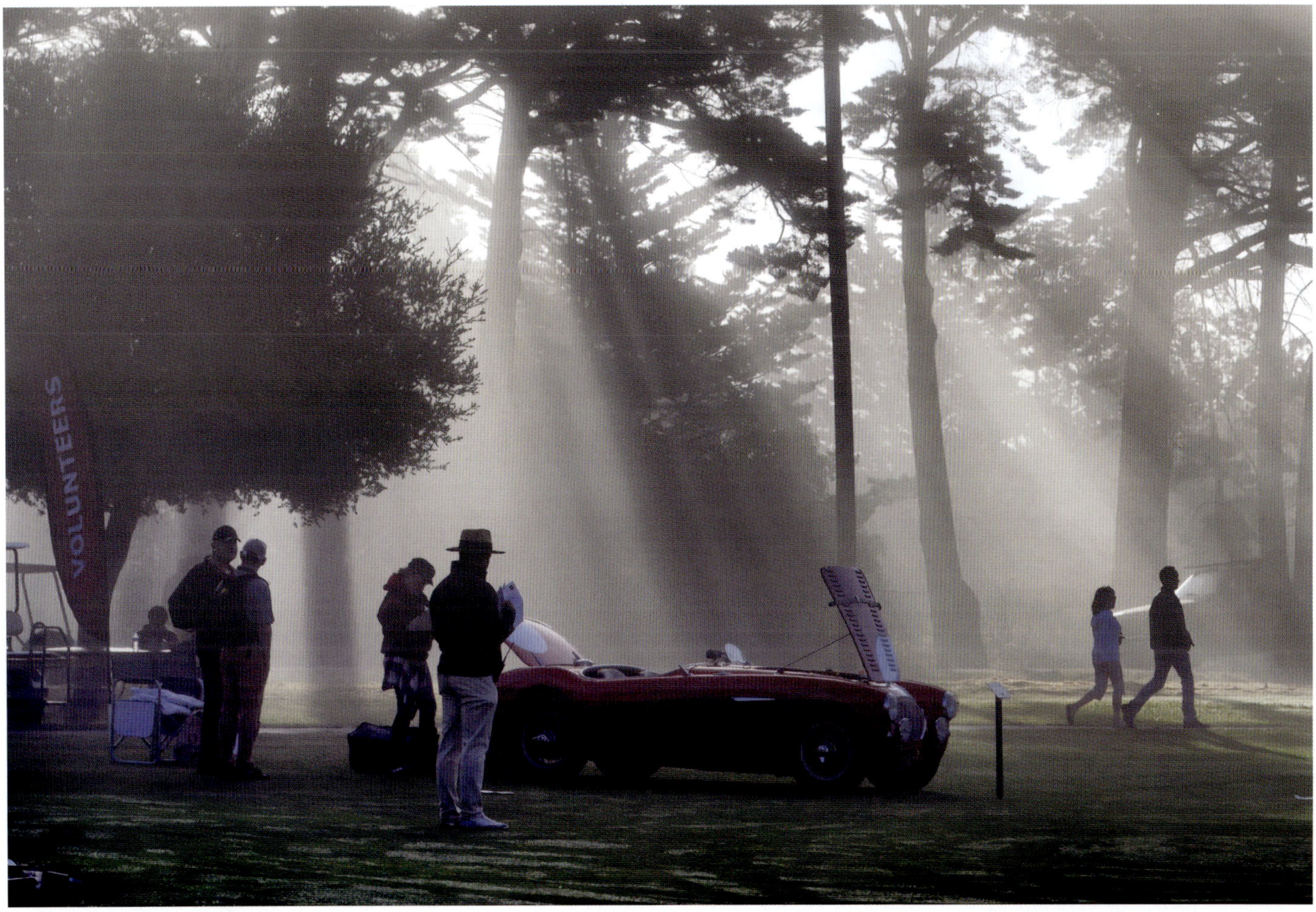


Not all concours d'état are professionally managed; many, like the Hillsborough Concours, are organized by volunteers. This particular scene is from the 2022 event. (Kimball Studios)

Pebble Beach, significantly enhances a vehicle's pedigree and market value. The stakes associated with winning or losing are far from neutral – both economically and psychologically – making the selection of competent judges an absolute necessity in an increasingly litigious environment.

The economic implications of concours d'état are substantial. Owners may invest considerable sums to present a vehicle in '100-point condition', often described as being 'in concours condition'. Before even registering, owners end up spending significant amounts on restorations by top professionals. Additionally, there may be high registration fees, accompanied by significant costs for travel and accommodation expenses for themselves, their families, and their teams of mechanics.

At the same time, the organizers also bear substantial expenses, including renting high-quality venues and accommodating many international judges in luxury hotels, as well as hosting VIPs and partners accustomed to premium experiences. Concours d'état events typically cater not only to participants but also to invited guests, which may include clients, partners, and influential figures from cultural, economic, or political spheres. Several concours d'état are by invitation only, and for the ones where a paying public is allowed, the entry fees tend to be substantial. Of course, there are exceptions too, such as The Boston Cup, where access to the "outer ring" is free.

In contrast, French-style concours d'elegance are usually free for the public, with modest participation costs for entrants. However, these events can still offer quality reception conditions, thanks to sponsorship from private or public partners.

Despite their differences, all concours share two essential elements: the expertise of a knowledgeable presenter and the quality of the venue. Whether the audience is hand-picked or larger in number, spectators must learn about the vehicles through engaging commentary from a presenter capable of narrating the unique histories, anecdotes, and notable figures behind each car, beyond just technical specifications like carburetors and valves. Each vehicle has its own story, and the public is eager to discover what that story is.

Concours events serve as a powerful platform for promoting historic vehicles and facilitating interactions between owners and the general public. They allow car enthusiasts to step outside their usual circles, fostering a connection to the broader community and encouraging each participant to champion the heritage they represent.

*A daring attempt to create another world-class concours in the United States, the Audrain Concours d'Elegance draws a lineup of competitors to Newport, Rhode Island, rivaling those at the nation's top automotive events. (Cape Coventry Archives)*

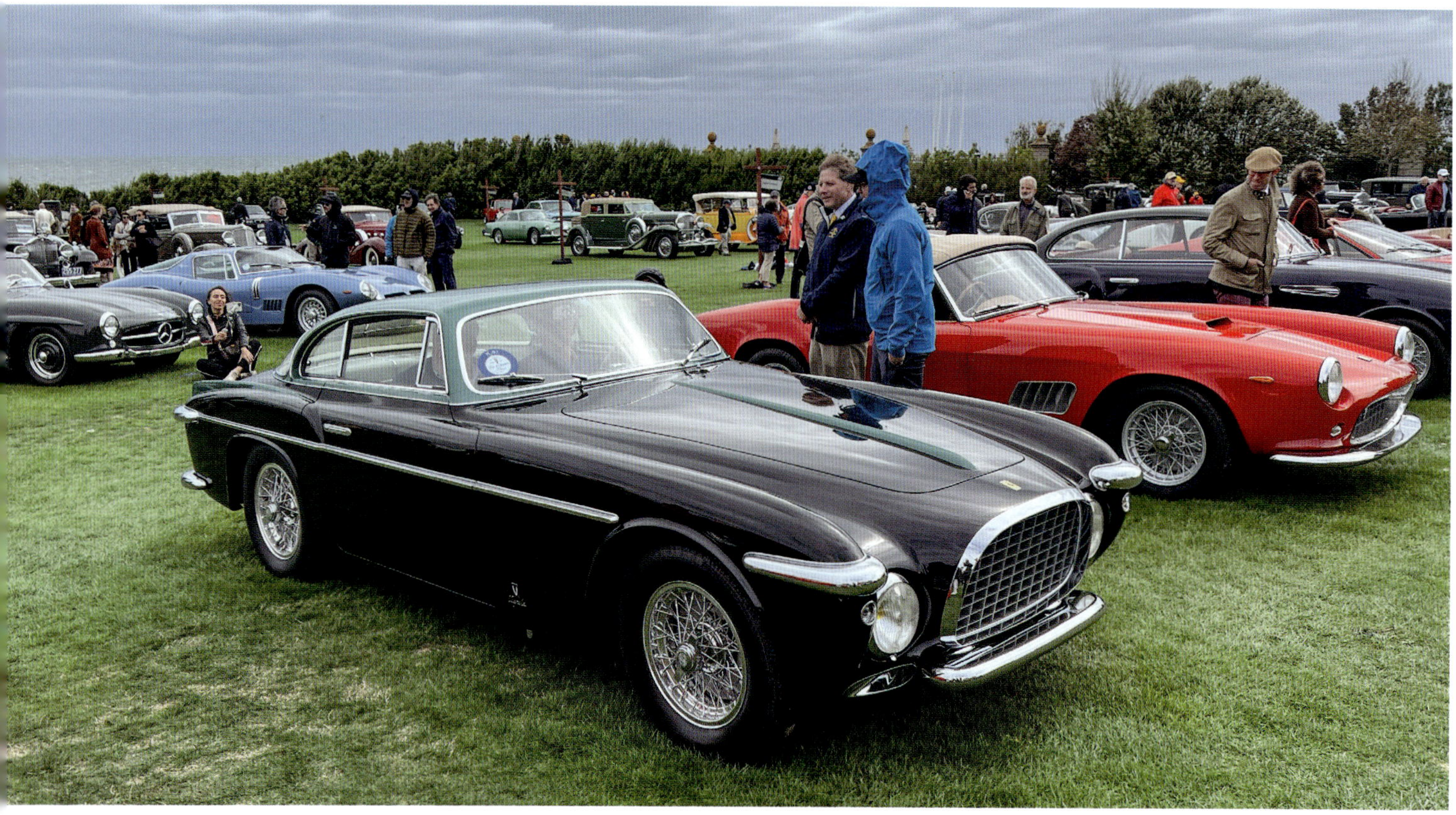

The venues selected for these events must be 'elegant', elevating the vehicles and their crews while inspiring spectators to aspire toward sophistication and good taste – qualities that need not be prohibitively expensive. Ideal locations include castle parks, golf or tennis clubhouses, spa casinos, historic city squares, and tree-lined boulevards or parks, all of which provide a fitting backdrop for the celebration of automotive history and elegance.

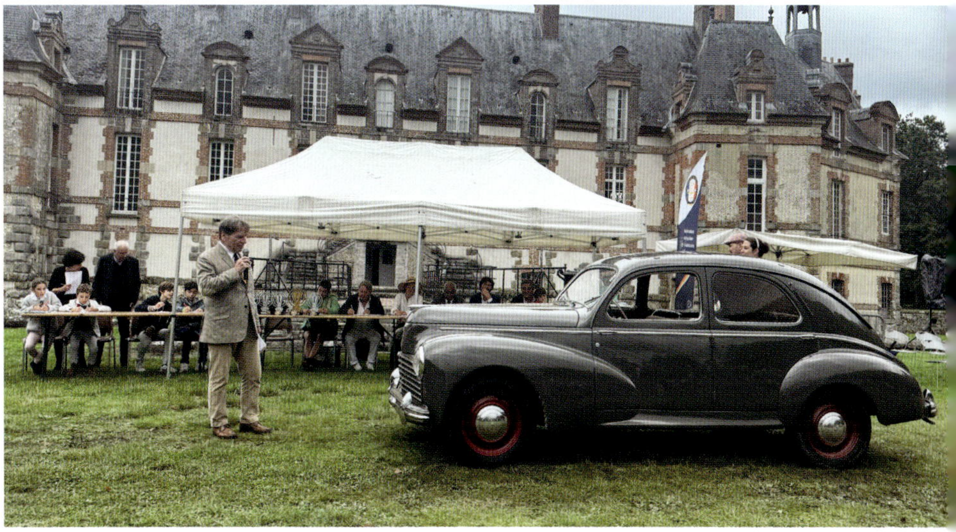

*At the Art Automobile Gambais, a Peugeot 203 parades past the junior jury, seen at left. (Ameya Mishra)*

Two significant sociological trends are emerging in the world of concours: the movement toward gender parity and the introduction of younger members into judging panels for the future.

While the proportion of female judges in concours d'état remains low, it is more common in French-style concours, which typically employ less technical criteria, such as the quality of crew presentation and attire. Some events have even successfully implemented entirely female juries, like the one at Angoulême in 2023.

In the United States, a notable initiative led by McKeel Hagerty, the Youth Judging Program was established in 2007. This program has trained nearly 20,000 children to evaluate historic vehicles. Equipped with a judging kit, a name lanyard, and simplified observation sheets, these young judges take their roles seriously and express great enthusiasm, which is shared by their parents.

Similarly, French-style concours are experimenting with junior juries, as seen in Gambais and Montlhéry in 2024, with plans for further expansion of these initiatives. Engaging young people in this way can only be beneficial for the future of the automotive community.

### So, which is better... concours d'état or concours d'elegance à la française?

The fact of the matter is that these two concepts are quite distinctive, and neither is superior to the other; they simply serve different purposes. As philosopher Alain (otherwise known as Émile-Auguste Chartier) famously stated, "Fairness consists in treating differently what is not equal." This perspective underscores the unique contributions and values of each type of concours, enriching the automotive landscape and broadening participation across diverse demographics.

## A summary:

| | Concours d'elegance | Concours d'état |
|---|---|---|
| ***Res judicata*** | The vehicle and crew | The vehicle |
| ***The jury*** | Mix of automobile, fashion, sponsors, elected officials and specialists | A large group of automotive experts |
| ***The criteria*** | Three criteria, each rated out of 20: historic significance, elegance of design, rarity / state of preservation-authenticity / presentation by the crew | Multi-criteria grid for judging authenticity or quality of restoration. Theoretical total of 100 points |
| ***The public*** | Wide public | Often by invitation (VIPs, clients, sponsors, etc.) |
| ***Cost (to the public)*** | Free or a very nominal price | Paid, mostly expensive or "à la carte" (low entry fee, with "options" added) |
| ***Cost (for entrants)*** | Free or modest | Expensive, reflecting high-level services |
| ***Organization*** | Most often voluntary | Commercial entity |

# Concours d'Elegance~ A Reflection of its Era

## THE FUTURE – WHAT CAN WE ANTICIPATE?

As seen in this book, the concours d'elegance events have always celebrated the automobile, embracing both technological progress and artistic beauty. However, these events have never been isolated from broader societal trends. Early concours d'elegance et confort events responded to society's fascination with this innovative object, the automobile. Initially, static displays evolved into dynamic presentations, enjoyed by an aristocracy that saw the concours as a continuation of refined and elegant living.

By the mid-1920s, the advent of women's concours d'elegance in automobiles signaled a shift, reflecting a new social role for women. Later, the post-war concours revealed that neither the creative flair nor the extravagance of master coachbuilders could hold back an industry focused on productivity and mass industrial production.

In 1950, the concours d'elegance experienced a revival on the shores of Pebble Beach, embracing a new purpose centered on the appreciation and preservation of heritage. This Californian model has since inspired numerous events in the United States, Europe and Asia. Understanding the concours d'elegance phenomenon requires looking beyond the traditional triad of manufacturers, coachbuilders, and fashion designers. The history of the concours d'elegance is deeply cultural but also inherently political, economic, and social.

## Concours d'Elegance: A New Golden Era?

Today's environment seems highly favorable for the growth of concours d'elegance, with many indicators suggesting a fresh momentum. The passion for historic vehicles remains strong, and revenues from this market are projected to rise by almost ten percent through 2026 (Statista 2024). The number of competitions continues to expand, with the Hothouse Media published *Concours Year* featuring some 50 major events worldwide in 2024.

Organizations such as FIVA and national federations are actively promoting concours events. As we know from the previous chapter, the French Federation of Historic Vehicles, FFVE, has introduced its own label and officially selects thirty concours every year to encourage concours d'elegance à la française. Many enthusiasts today view modern automobiles as merely functional, a commodity, and instead tend to gravitate towards historic vehicles such as cars and two-wheelers that embody a richer sense of history and evoke the thrill of the past, a yearning for authenticity.

For several years, motor shows have struggled to maintain their audiences in the West, with the traditional allure of these events waning, and manufacturers now recognize the need to explore alternative ways to connect. The Geneva Motor Show, a historic event since 1905, may have closed its doors forever – a significant moment in automotive history. Meanwhile, shows focused on historic vehicles are drawing a growing audience. Many car marques have recognized the importance of being present at these events, which provide a valuable platform to showcase their heritage – a particularly pressing issue in the face of industry-wide move to alternative energy vehicles.

**BELOW (BOTH):** *Once upon a time not so long ago one of the most important motor shows was that of Geneva. (Gautam Sen)*

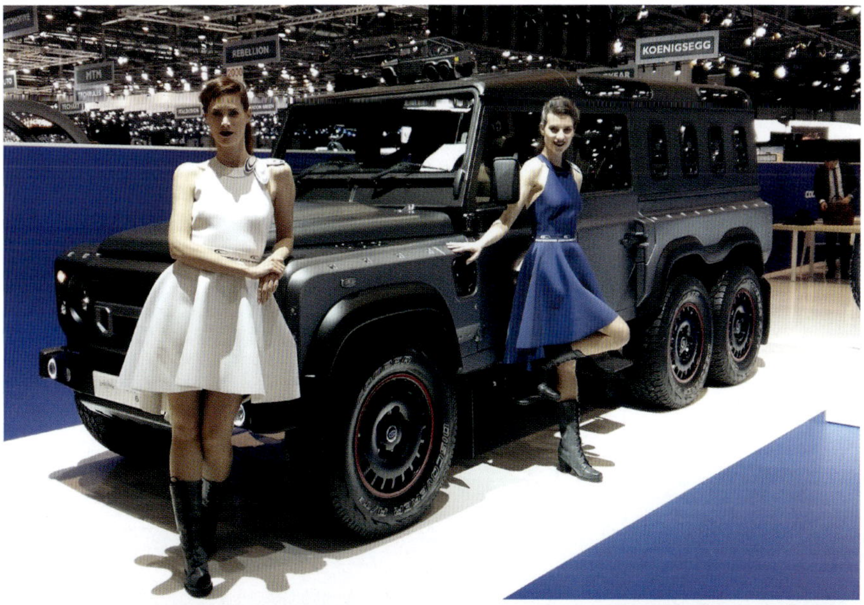


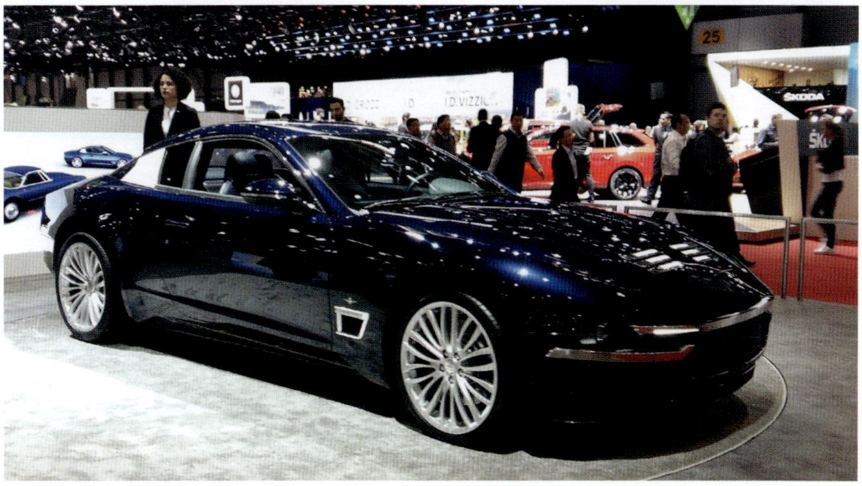

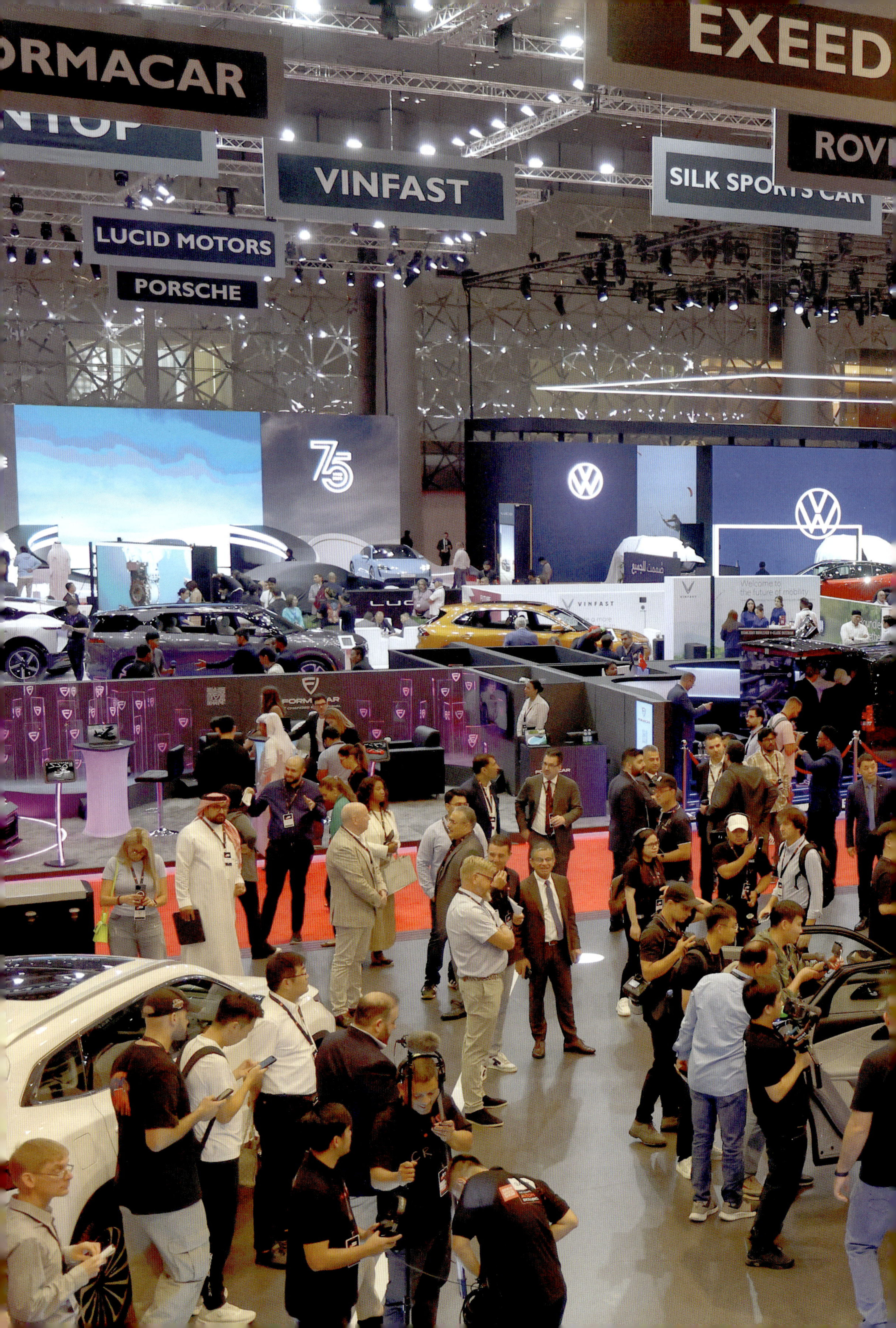
EXEED
VINFAST
SILK SPORTS CAR
LUCID MOTORS
PORSCHE

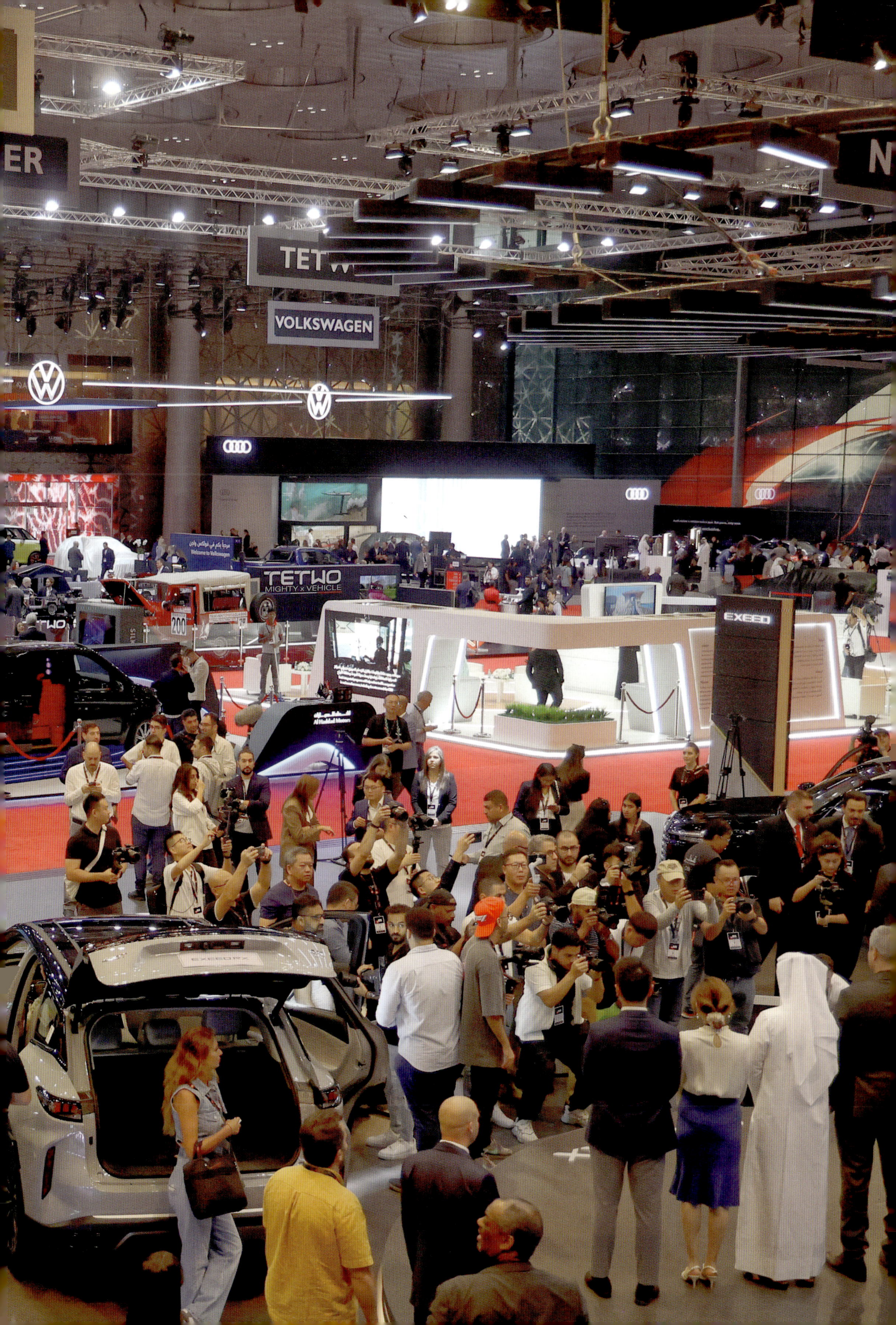
TETWO
VOLKSWAGEN
TETWO
MIGHTY x VEHICLE
EXEED

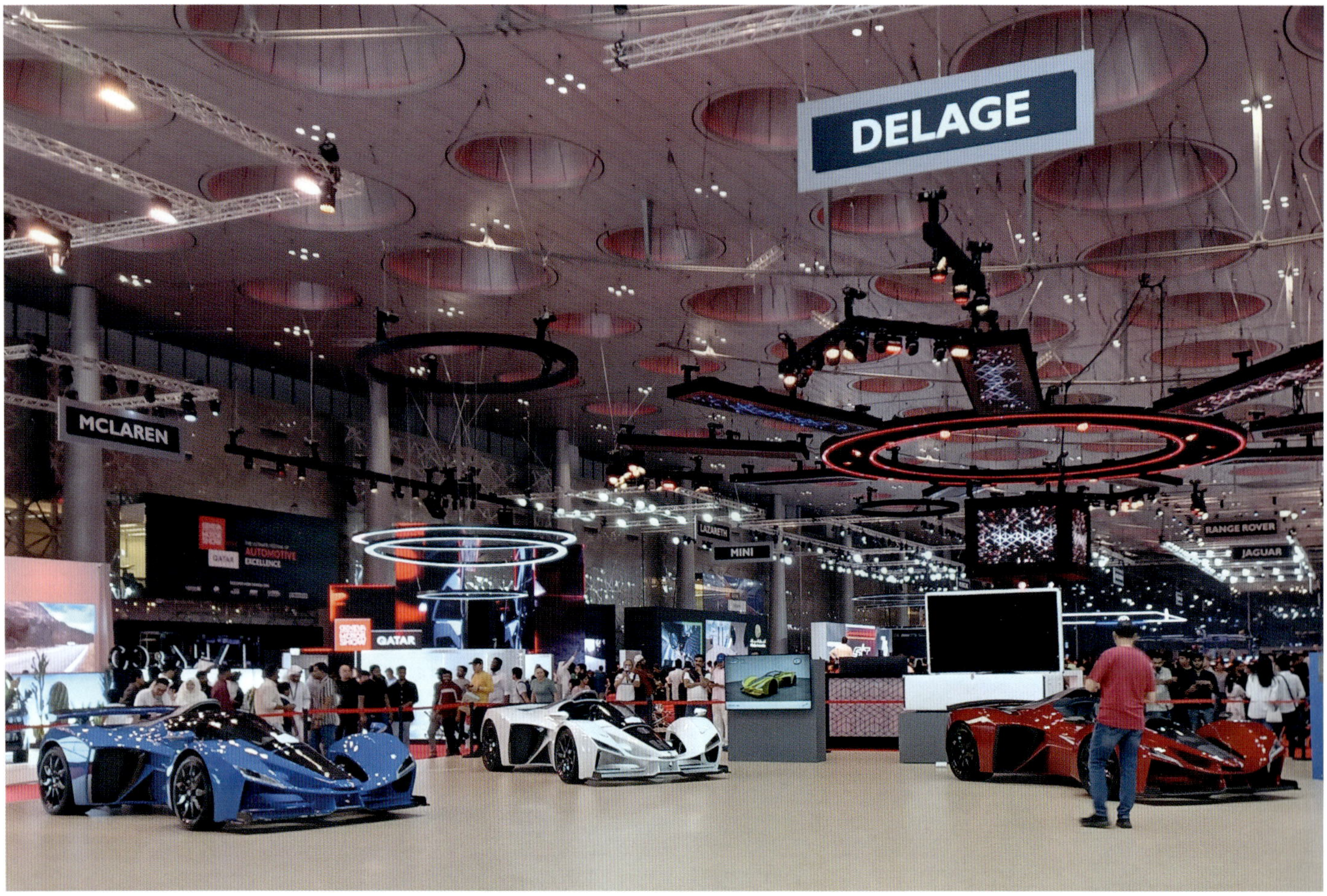


**ABOVE & PREVIOUS TWO PAGES:** *As the Geneva International Motor Show fades into history, the Geneva Motor Show Qatar carries the torch forward in another part of the world. (Gautam Sen)*

In a world where high-performance electric vehicles are easy to develop for most manufacturers, mechanical prowess may no longer serve as a unique differentiator. Brands will instead rely on design, legacy, and distinctive histories to set themselves apart. The success of events like Salon Rétromobile in Paris demonstrates this trend, with marques such as Volkswagen, Renault or MG highlighting their legacy as they celebrate the anniversary of a model or use the salon as an occasion to reintroduce a model, or for a brand to draw on its history to establish a modern presence, and so on.

Could manufacturers once again become the driving force behind concours d'elegance events, as they were in the 1920s? In recent years, events like the Concorso d'Eleganza Villa d'Este, hosted by the BMW Group Classic, have embodied a blend of tradition and innovation, offering a fascinating lineup of iconic models, automotive treasures, and supercars. Similarly, The Quail's Concours d'Elegance event highlights luxury and sports vehicles, drawing manufacturers into the spotlight.

At the 2024 Chantilly Arts & Elegance Richard Mille, attendees saw a selection of concept cars and new models, including the extraordinary Bugatti Tourbillon, proving that the blend of iconic classics and forward-looking designs is highly effective. Although the concours audience remains niche, digital platforms and social media amplify these events, generating significant visibility and engagement for all involved.

## The Concours d'Elegance of Tomorrow: Balancing Heritage and Innovation

While the tribute to automotive heritage remains a core cultural foundation, there is room to elevate how this heritage is presented. Beyond traditional static displays, the reimagining of the concours d'elegance in the spirit of the 1930s, enhanced by modern projection and lighting technology, transforms these events into live spectacles that resonate with a broad audience. Highlighting the individual stories of each vehicle and crew is essential, as well as weaving a narrative around the manufacturer, the model, its past owners, and the fashion trends of its era.

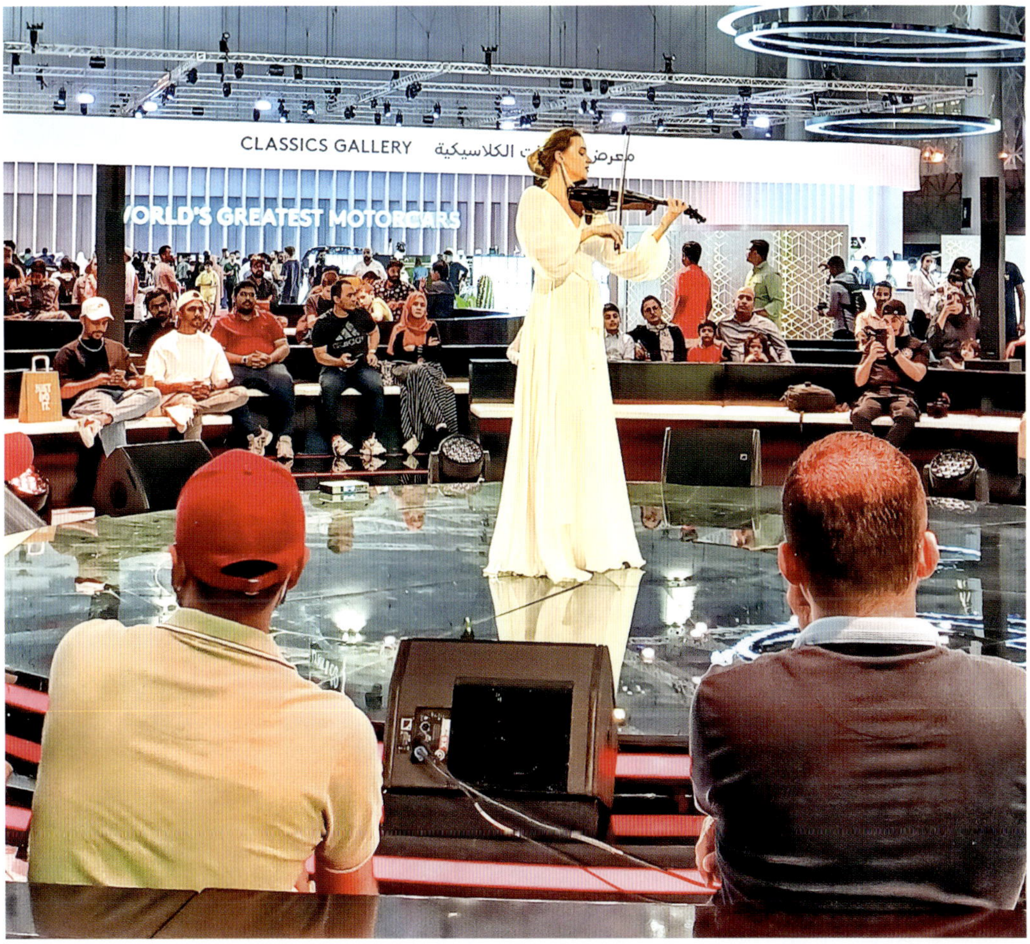


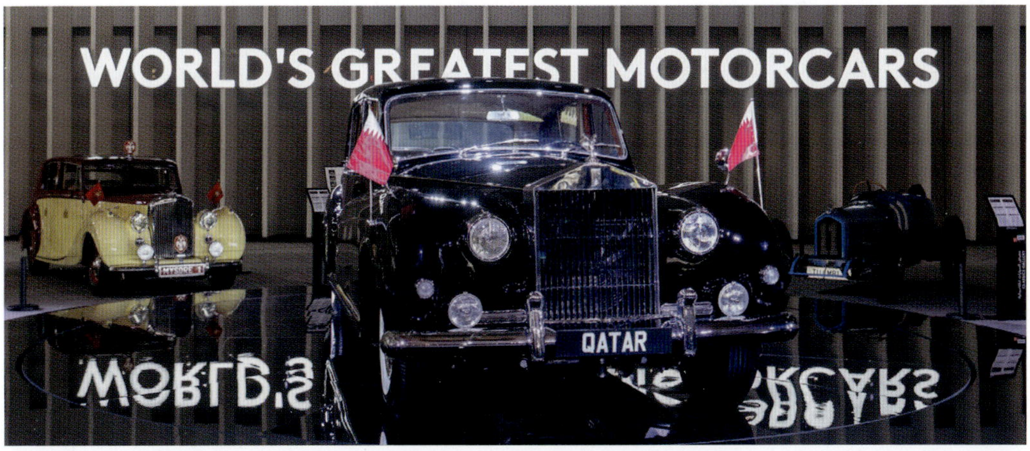


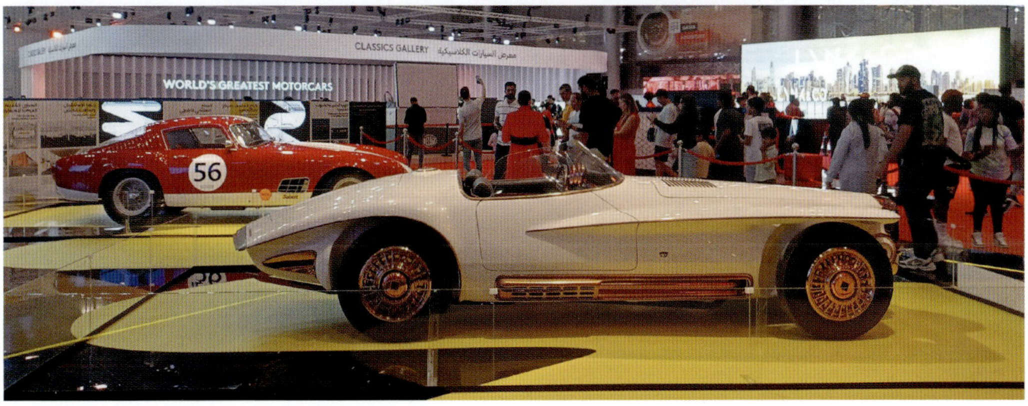

**LEFT:** *The brand-new motor show in Qatar has taken on a different form, incorporating entertainment like Western classical violin recitals alongside the automotive displays. (Gautam Sen)*

**MIDDLE & BOTTOM:** *In addition, a significant section is devoted to the display of historic vehicles, with plans to host a concours d'elegance in the future. (Gautam Sen)*

**RIGHT:** *Further east, in the world's largest car market – China – motor shows continue to thrive, particularly in second-tier cities like Chongqing. (Gautam Sen)*

**MIDDLE:** *As the country develops a growing appreciation for historic vehicles, as seen in the lavish Sanhe Classic Car Museum in Chengdu, the interest in vintage cars continues to expand. (Gautam Sen)*

**BOTTOM:** *Of course, it may take some time before China fully appreciates its automotive history and legacy. (Gautam Sen)*

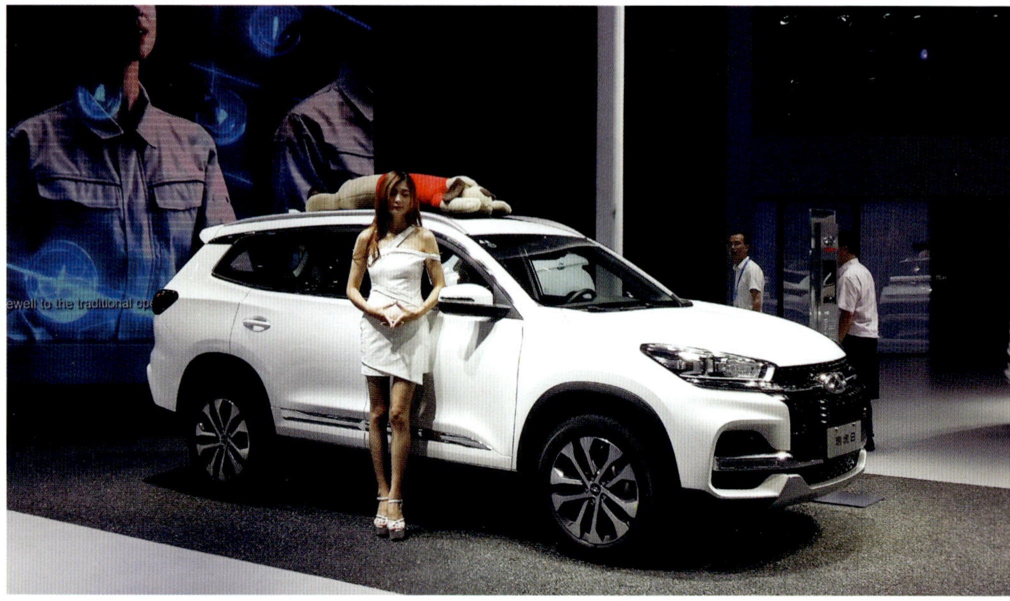

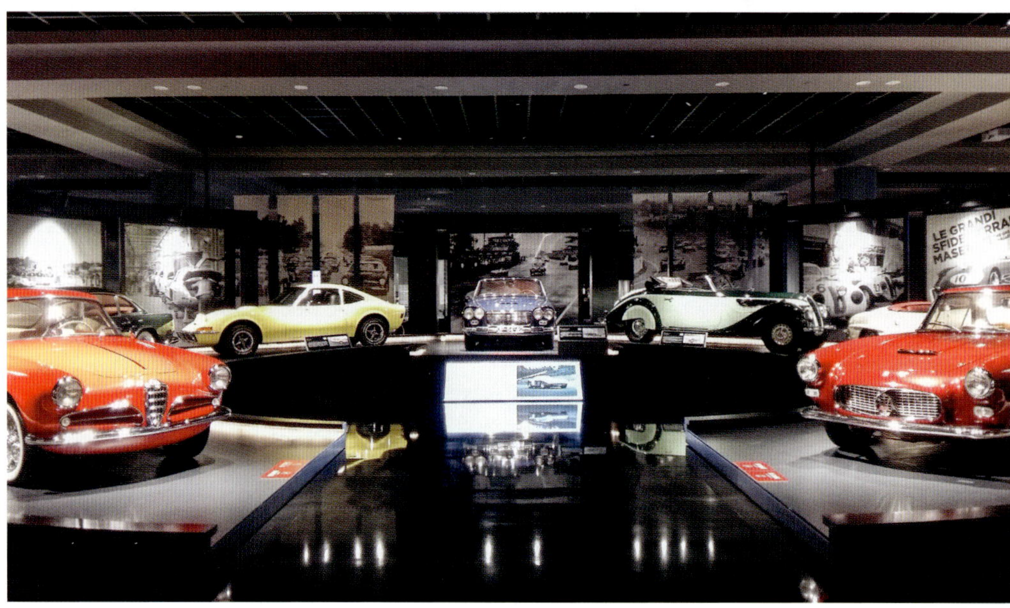

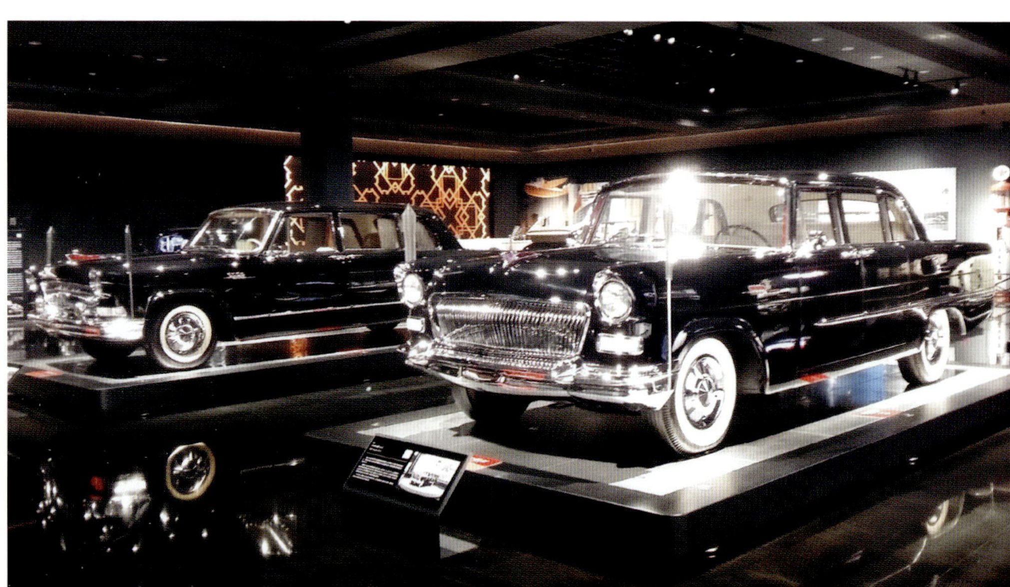

This storytelling approach adds historical and cultural layers that enrich the passion for automobiles. Complementary events such as panel discussions and film screenings, organized as part of these competitions, consistently attract enthusiasts and the general public alike. Elevating the automobile's story and sharing knowledge serve an educational purpose, reinforcing the concours as an esteemed "temple" of automotive history and culture.

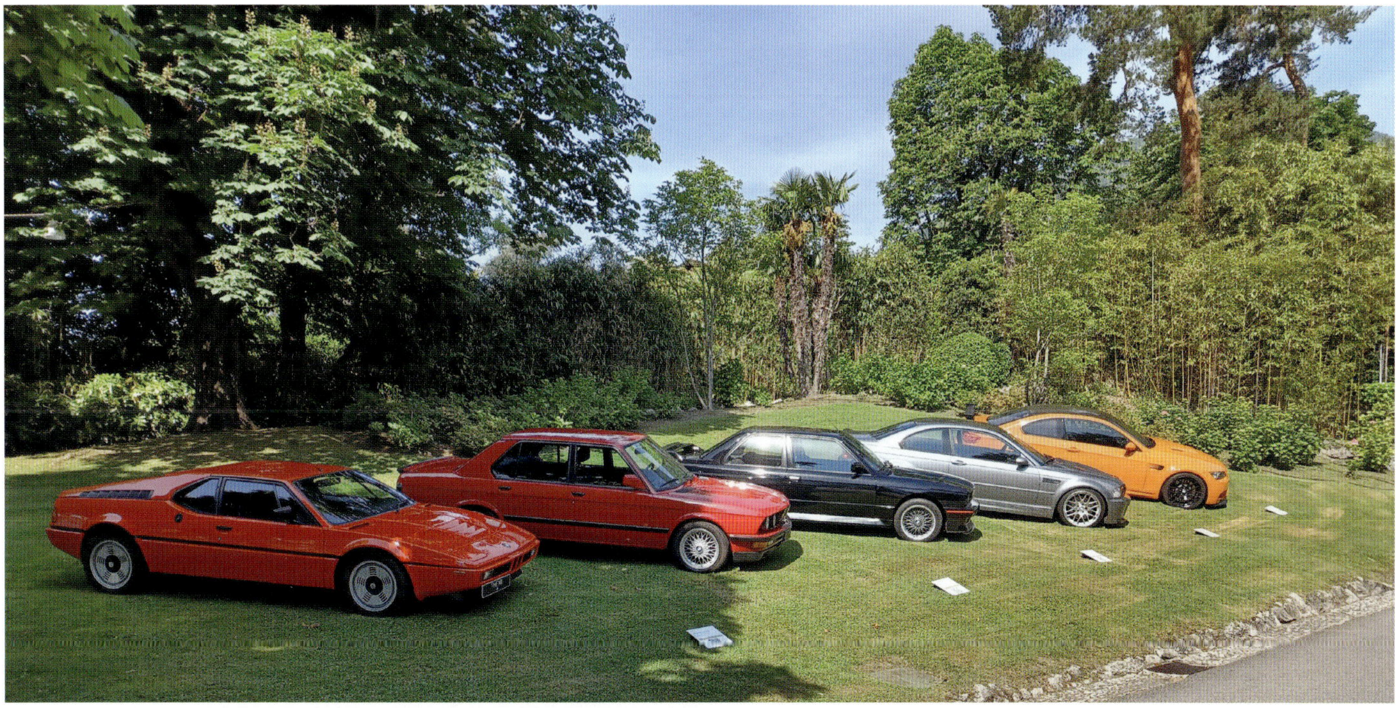

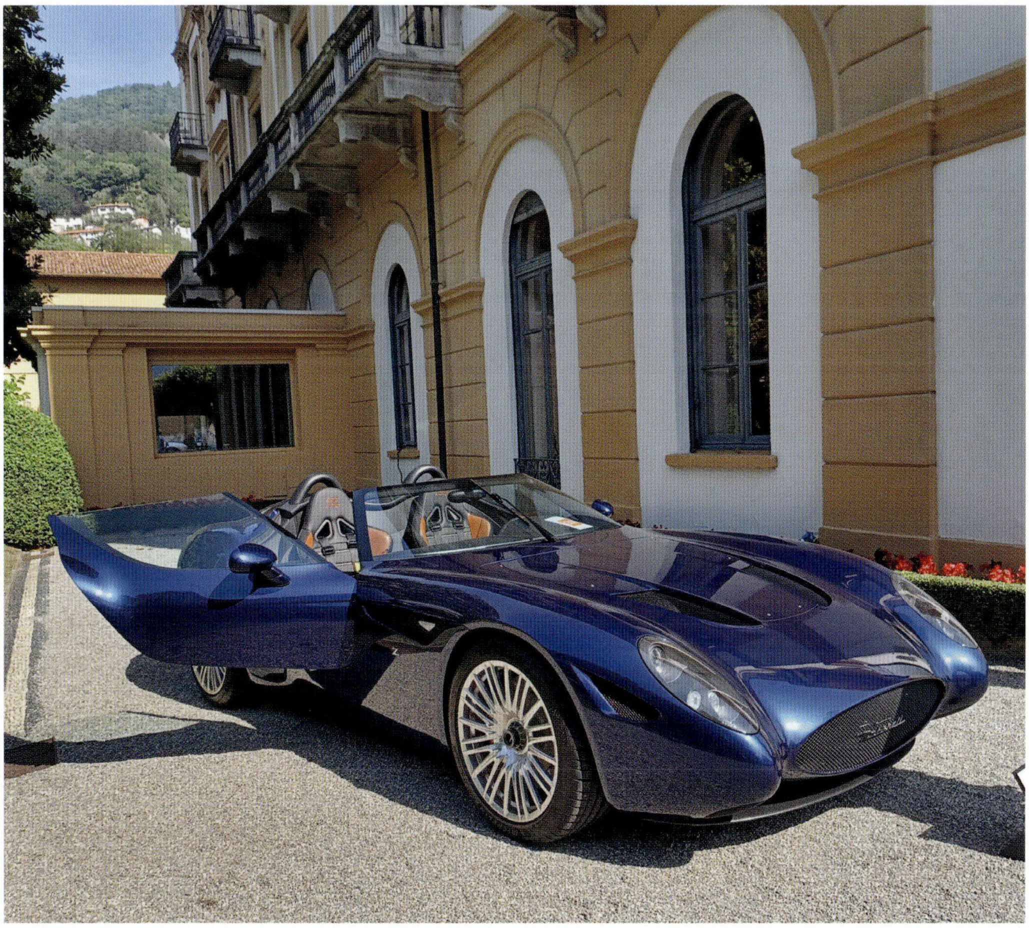

**ABOVE:** *In the meantime, the now traditional concours d'états have evolved to feature modern cars, such as BMWs, at events like the Concorso d'Eleganza Villa d'Este. (Gautam Sen)*

**LEFT:** *Zagato featured the Maserati Mostro Barchetta at the 2022 Concorso d'Eleganza Villa d'Este. (Gautam Sen)*

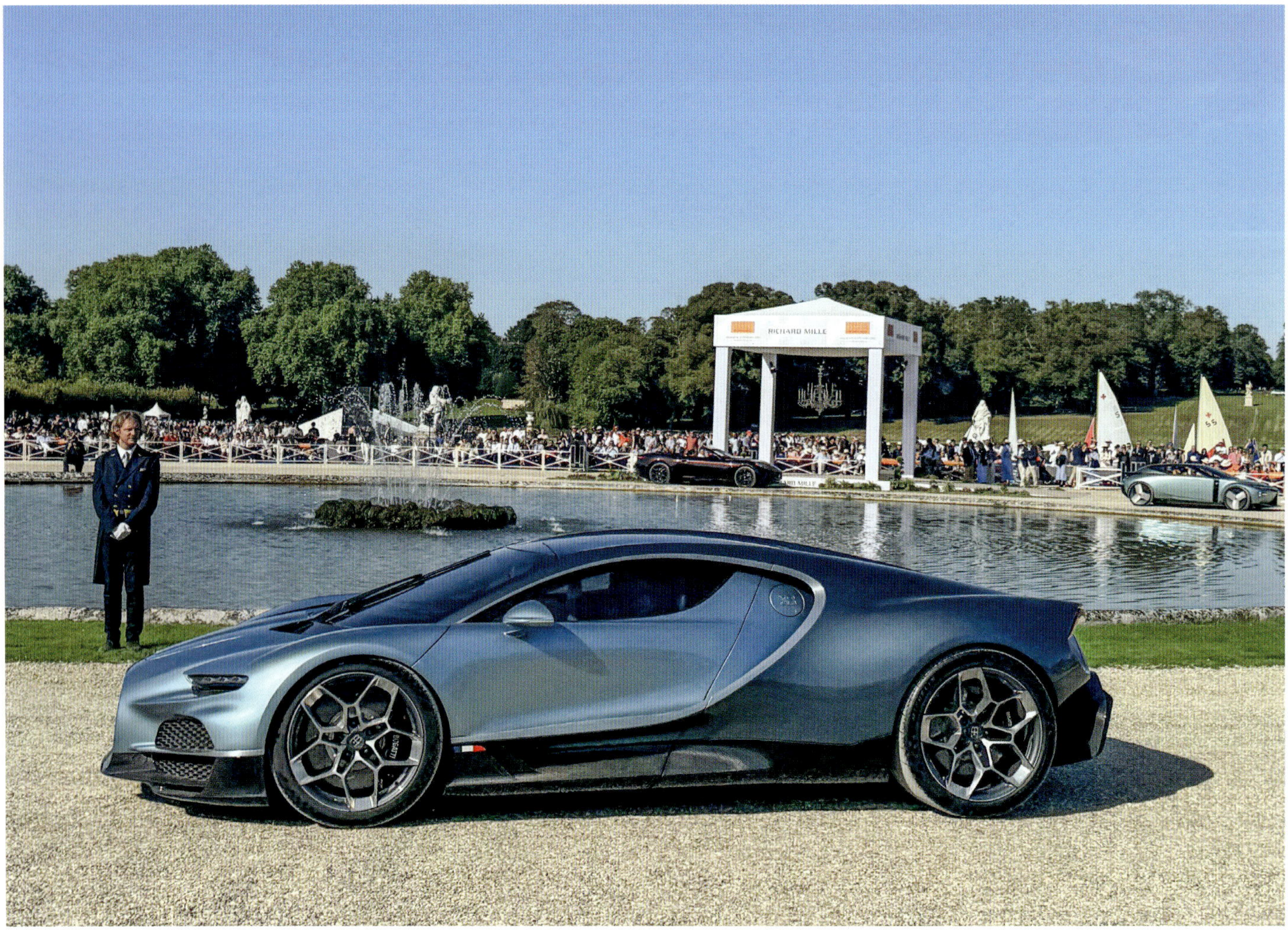


**ABOVE:** *Luxury car makers prefer launching their latest models at prestigious Concours d'Elegance. Bugatti brought the Tourbillon to the 2024 Chantilly Arts & Elegance Richard Mille. (Ameya Mishra)*

**RIGHT:** *Concours events need to modernize and become more interactive, such as incorporating voting for the People's Choice, as seen at the 2023 Dinard Elegance. (Louis Monnier)*

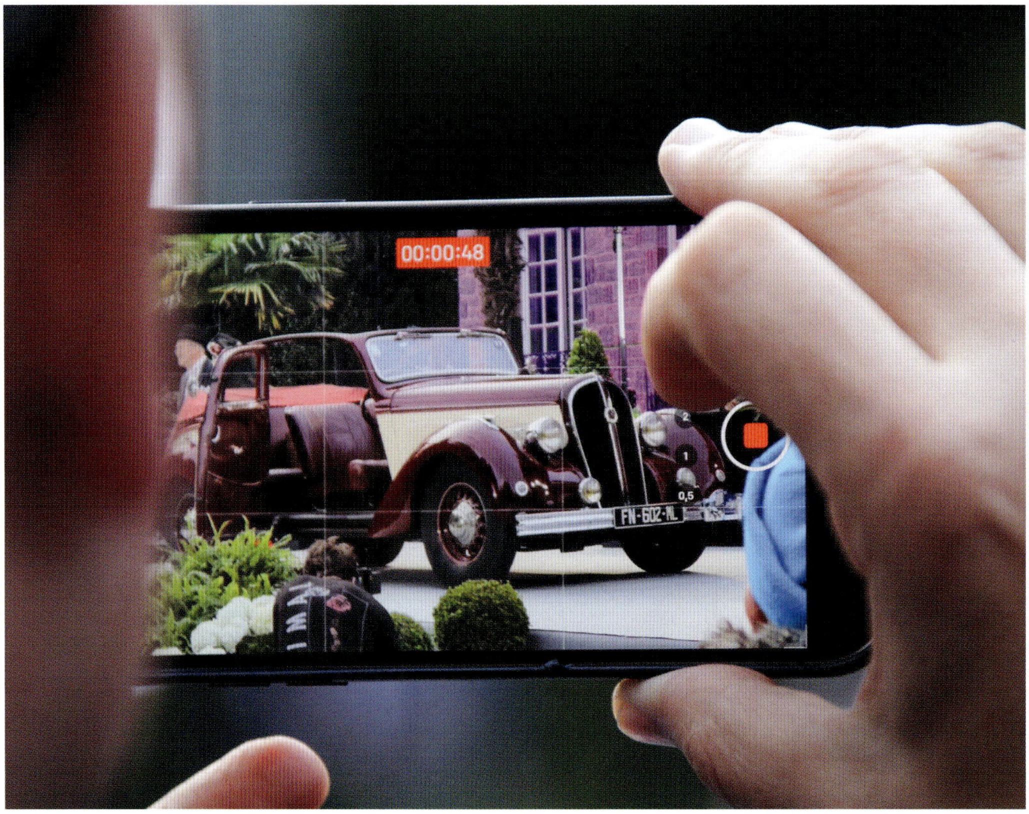

## Expanding the Audience

Preserving the essence of concours d'elegance events does not mean limiting their reach. In fact, expanding their audience is crucial not only for growth but also for long-term sustainability. The collector car community is diverse, often separated into distinct cultural groups, and the concours d'elegance stand at the intersection of this diversity, presenting an opportunity to bridge these worlds.

Organizers can choose to enhance their unique positioning or broaden their scope with new formats. Engaging younger audiences by featuring more recent vehicles is both a challenge and a necessity, considering the age demographics of many collectors. The Villa La Massa Excellence concours in Florence, which highlights iconic supercars from the 1990s onward, exemplifies this approach. It's not hard to imagine a future concours d'elegance dedicated to just "youngtimers", cars that millennials now see as iconic.

The range of potential categories is vast: racing cars, beach vehicles, early electric models, concept vehicles, and even customized utility vehicles, harking back to the 1920s. Historic vehicles of all types, regardless of value, have their audiences. Dividing automotive history into "elite" versus "popular" is unnecessary, as it has always thrived on this balance. Inspiration can be drawn from music festivals or art exhibitions, which often blend genres, styles, and generations. Generational inclusivity is essential; for example, creating opportunities for young people to engage could involve inviting owners to display their own cars, integrating age diversity among crews, or establishing youth juries.

Additionally, women remain underrepresented at automotive events. By celebrating women's contributions as drivers, engineers, and designers, and by highlighting them more fully within concours d'elegance events, their long-standing role in automotive history is also honored. Greater gender inclusivity, such as increasing female jurors and encouraging women to drive - as was common in the 1930s - could attract more women attendees and foster a sense of belonging. Collaborating with schools of art, design, and engineering can help bring young men and women into the heart of these events, fostering fresh perspectives and energy in the concours community.

## Revitalizing Tradition through Immersive Experiences

Exploring the world of immersive experiences reveals a powerful way to engage visitors by involving them on a sensory level. While this approach may seem futuristic, integrating new digital technologies at events offers a promising direction. By blending experience with emotion, it may be possible to create pre-event sequences that engage audiences even before they attend. Tools like virtual reality (VR) and augmented reality (AR) could allow visitors to experience historic vehicles in a new way.

For example, they may be able to explore the detailed process of starting a vintage car - from engaging the ignition to adjusting the timing. Imagine driving a 1930s Bugatti or Hispano-Suiza with authentic sounds and sensations or learning to double-clutch on a "crash" box or experiencing the unique vibrations of a post-war British car. Virtual films, interactive information, guided tours, and public voting enhance involvement, personalizing each visitor's experience and adding genuine educational value.

This immersive approach also extends to physical, hands-on experiences. For newcomers, sitting inside a historic vehicle and discovering its unique interior, instruments, and craftsmanship could be a memorable, multi-sensory encounter. Sharing the history of a bygone brand awakens curiosity and fosters a warm, communal atmosphere. Additionally, hosting live restoration workshops and spotlighting artisanal techniques can play a vital role in preserving and passing down automotive craftsmanship.

Traditionally, concours d'elegance events have focused on the vehicles and the narrated presentations, but creating more interactive opportunities enriches the visitor experience. While honoring the primary mission of celebrating automotive heritage, expanding these practices would attract a broader, younger audience. To ensure a vibrant future, the movement must embrace trends, engage proactively with the public, address environmental concerns, and communicate these values widely on social media. The next challenge is to find harmony between modern innovation and heritage, respecting the core values of the concours d'elegance while evolving with the times.

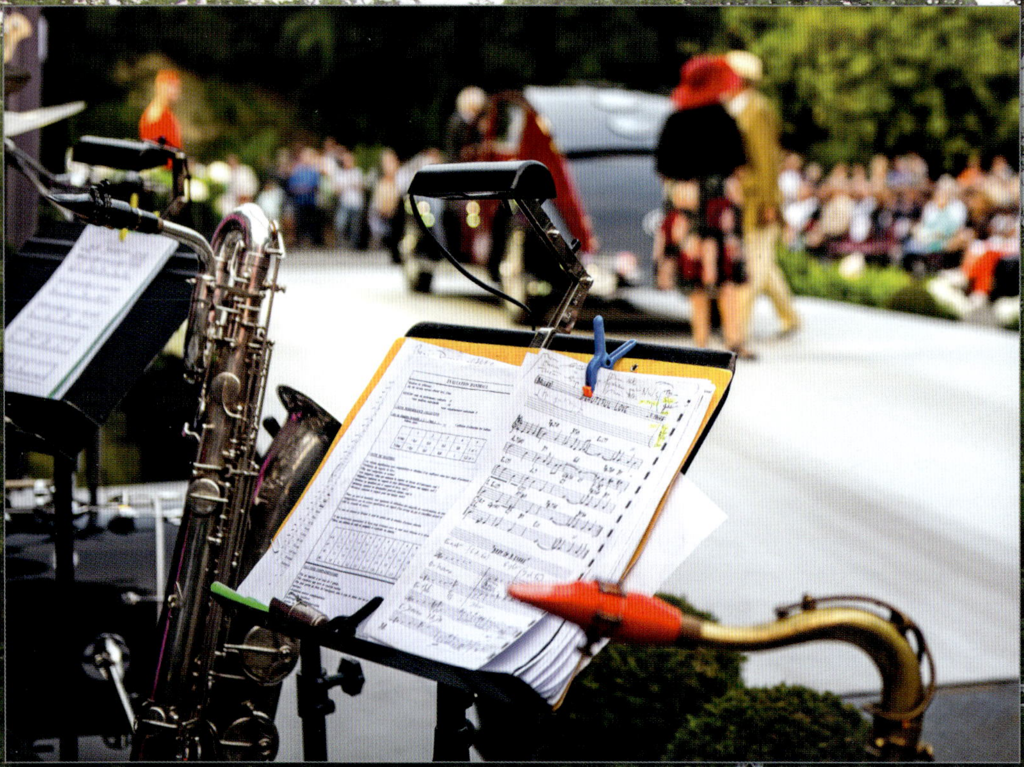

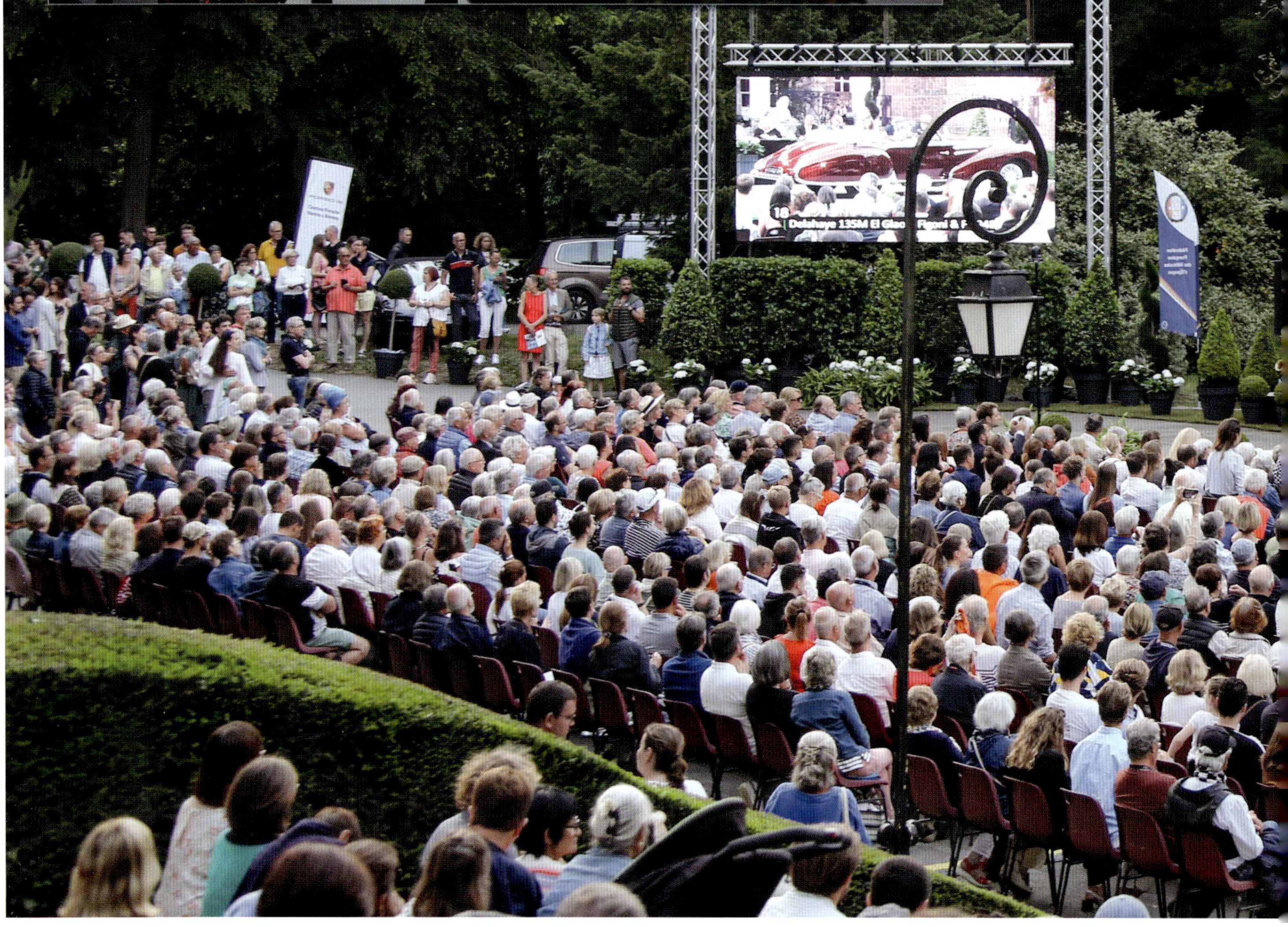

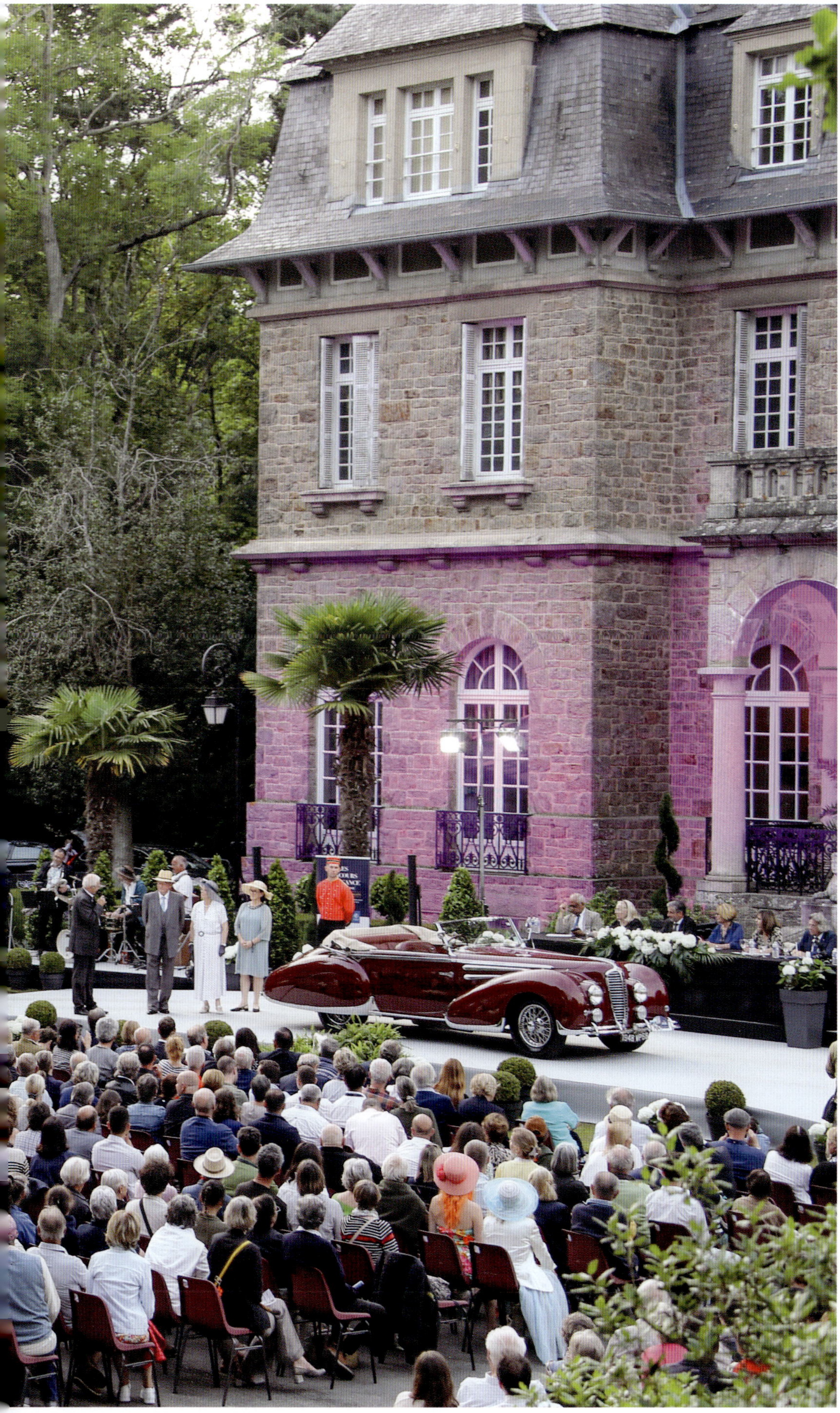

*No doubt, future occurrences of concours d'elegance and d'état will make even better use of videos, sound, and light shows to enhance the experience.* (Louis Monnier)

**INSET:** *Additionally, live music plays a significant role in setting the perfect atmosphere for these events.* (Louis Monnier)

# APPENDIX: SCORESHEETS

## ICJAG

The scoring system traditionally used in American-style concours d'elegance – more accurately, concours d'état – was most prominently represented by the ICJAG (International Chief Judge Advisory Group) system, conceived by veteran Pebble Beach judge Ed Gilbertson. Until 2024, nearly a dozen of the premier concours events followed the ICJAG methodology, making it arguably the most widely accepted standard.

Under the ICJAG system, up to 20 specific items are evaluated, encompassing details such as bodywork, fender and panel fit, and functional elements like glass, side curtains, top boots, and window operation. These are grouped under five broader categories: coachwork, interior, engine, chassis, and operations. Each item is scored out of a maximum of five points, with deductions made for imperfections. The vehicle with the fewest deductions – thus the highest remaining score – is declared the class winner. See page 386 for an example of the scoresheet.

## M100

In 2025, several leading concours d'état events began transitioning to the newly introduced Modified Professional Judging 100 (M100) scoring system. While still based on a system of deducting points for flaws, M100 introduces a more flexible and less dogmatic approach to judging. The system is designed to give judges more latitude in recognizing the subtleties and nuances of high-level restorations and original-condition vehicles.

Under M100, 95 points are allocated across five broad categories: exterior, interior, drivetrain, chassis, functionality / state of restoration. However, judges are given greater discretion in how they apply these points within the categories, allowing them to weigh significance and execution with more judgment-based nuance.

An additional 5 points are reserved for elegance, provenance, and field presence – intangible qualities that elevate a car beyond its technical merit and acknowledge its overall impact and historical importance. An example of this scoresheet is shown opposite bottom.

## Chantilly Arts & Elegance scoring system

Chantilly Arts & Elegance Richard Mille has, since its inception, adopted a distinctive judging system that sets it apart from the more technically rigid frameworks used elsewhere. Rooted in the ethos of a true concours d'elegance – which prizes both artistic expression and historical integrity – the event's main section is a concours d'état, where the judging criteria are carefully balanced to reflect both aesthetic and historical considerations.

The Chantilly scoring system assigns equal weight to the five key criteria of authenticity (how faithful the car remains to its original specification), historical significance and provenance, including its ownership, competition, and exhibition history, design merit, which takes into account the creativity, influence, and visual appeal of the coachwork, the exterior and interior condition – judged not just on perfection, but also on period-correct preservation or restoration quality – as well as the operational condition, ensuring the vehicle is not only presentable but also roadworthy.

Additional considerations that can influence the final evaluation include whether the car participated in the Tour d'Elegance – demonstrating functionality – and whether it holds a FIVA Identity Card, serving as an official validation of historical accuracy and originality. See page 387 for an example of the scoresheet.

## Retromobil Club Romania

Retromobil Club Romania, the FIVA-recognized national federation for historic vehicles in Romania, has adopted a simplified version of the Chantilly Arts & Elegance judging system for use in its national concours d'état events. This format remains firmly rooted in the principle of balanced evaluation, granting equal importance to the five foundational pillars of concours judging, that of authenticity, history and provenance, design, exterior and interior condition, with each of these five categories scored equally. While this format is streamlined for broader national application, it retains the spirit and integrity of the Chantilly Arts & Elegance system, emphasizing elegance, historical value, and preservation over pure restoration perfection:

## FFVE

The French Federation of Historic Vehicles (FFVE) has championed the revival and promotion of the concours d'elegance à la française, emphasizing style, heritage, and presentation over technical scrutiny. In this tradition, the judging criteria are elegantly simple yet meaningful, reflecting the cultural origins of the concours d'elegance as a celebration of art, fashion, and motoring sophistication.

The scoring system is composed of a total of 60 points, divided evenly across three main categories with 20 points for the make, model, and rarity of the automobile, as well as its historical significance and originality, another 20 points for the overall condition, appropriateness of color, and the documented history or provenance of the vehicle, and a last 20 points for presentation, which includes not only the elegance of the vehicle itself but also the attire, bearing, and harmony of the crew accompanying the car.

This format encourages a holistic appreciation of the automobile as both cultural artifact and style statement. Vehicles are not merely judged for their mechanical or restoration perfection, but also for their evocative presence, narrative, and ability to capture the spirit of an era.

**EXCELLENCE COMPETITION 2025**

CATEGORY
MAKE/MODEL
YEAR

| | | | |
|---|---|---|---|
| **Authenticity** | *Traceability* | | /5 |
| | *Original configuration* | | /10 |
| **History** | *Rarity of the model* | | /10 |
| | *Interest of the model* | | /3 |
| | *Prize list* | | /2 |
| **Design and Elegance** | *Innovation* | | /5 |
| | *Elegance* | | /15 |
| | *Presentation by owner* | | /5 |
| **Exterior** | | | /20 |
| **Interior** | | | /20 |
| **Working order** | | | /5 |
| **Total** | | | 100 |

Comments:

**LEFT:** *Using the scoring system of the Chantilly Arts & Elegance as the starting point, the Retromobil Club Romania has developed a simpler system that addresses the issues of authenticity, history, design, and the quality of restoration / preservation.*

**BELOW:** *The M100 scoresheet is structured like the below.*

2025 Car Judging Form - M100

Class:

Chief Class Judge and team:

Vehicle / year and model:

starting number

Each section shall be evaluated for authenticity and originality.
Mark category with a dash ( - ) if there is no deduction.
Deductions in 1/2 point increments to 2 points MAX per line.
If required more, contact Chief Judge prior deducting !
Please always fill in comments in blank area.
Consult suggested deduction list if unsure about deductions !
The deductions will be taken from the starting score of 95 points.
Then the up to 5 awarded points for subjective factors such as elegance, design, provenance etc. are added to arrive at the final Score !

| position | Judging area | Judges Notes | SCORE |
|---|---|---|---|
| | **Body Exterior** | | |
| 1 | Body, Fenders and Panel Fit | | |
| 2 | Bumpers, Emblems, Lights, Rubber, Trim | | |
| 3 | Door(s), Hood, Trunk Fit and Operation, Proper Soft Top | | |
| 4 | Paint Finish, Period Color, Striping | | |
| | **Body Interior** | | |
| 5 | Covers, (Seat, Floor) Headliner | | |
| 6 | Dash, Instruments, Manuals, Pedals, Shifter, Steering Wheel | | |
| 7 | Glass, Side Curtains, Window Operation | | |
| 8 | Rubber, Weather Protection | | |
| | **Chassis** | | |
| 9 | Frame, Suspension, Brakes, Steering, Drivetrain | | |
| 10 | Muffler(s), Manifold, Pipe(s), Exhaust Pipe(s),Tip(s) | | |
| 11 | Wheels, Rims, Tires, Spare Tire | | |
| | **Drivetrain** | | |
| 12 | Condition of Engine Bay / Correct Type | | |
| 13 | Cooling System, Radiator, Hoses, Pump, Fan | | |
| 14 | Electrical System Battery, Battery, Relays, Overall Condition | | |
| 15 | Fuel System /Carburetor(s),Injector(s), Pump(s), Linkage, Plating | | |
| 16 | Oil, Fuel and Water Lines, Belt(s) | | |
| | **Functionality / Status of Restoration** | | |
| 17 | Engine Run, Leaks | | |
| 18 | Exterior Lights, Turn Signals, Brake and Backup Lights | | |
| 19 | Instruments, Interior Lights, Horn, Radio | | |
| 20 | Over Restoration and/or Applied Patina | | |
| | | **Points deducted total for sections 1 to 20 Add up deducted points of all 20 sections.** | |
| | | Basic Score ( originality and authenticity) 95 points minus deductions | |
| | | Awarded BONUS points on a scale from 1 to 5 for Elegance, Provenance, Field Presence etc. ! | |
| | | **Final Score - Basic Score plus BONUS points - max 100 !** | |

©Chris Kramer concoursjudging.com 2025 - M100

***Forest Grove Concours d'Elegance 2024***

**Entry Number: F4 - 2**

**Vehicle / Year Model: 2011 BMW M3 Convertible**

**Chief Class Judge and Team:______________________**

***Evaluate each item for elegance, authenticity and condition.***

**Mark category with a dash ( - ) if there is no deduction. Evaluate each item for elegance, authenticity and condition. Points shall be deducted from 1/2 to 2 points - maximum is 5. Please always fill in remarks when you deduct on back of this scoresheet.**

Criteria

| COACHWORK | Points |
|---|---|
| 1. Body, Fenders and Adjacent Panel Fit | |
| 2. Door(s), Hood, Trunk Fit and Operation, Proper Soft Top | |
| 3. Paint Finish, Period Color, Striping | |
| 4. Bumpers, Lights, Trim, Emblems, Plating, Rubber | |
| **INTERIOR** | |
| 5. Glass, Side Curtains, Top Boot, Window Operation | |
| 6. Upholstery, Trim, Floor Covering, Headliner, Soft Top Well | |
| 7. Dash, Instruments, Steering Wheel, Pedals, Shifter, Manual(s) | |
| 8. Weather Stripping, Welting, Rubber, Running Board Material | |
| **ENGINE** | |
| 9. Engine Type, Overall Condition of Engine Bay | |
| 10. Oil, Fuel and Water Lines, Belt(s) | |
| 11. Carburetor(s), Injector(s), Pumps(s), Linkage, Plating | |
| 12. Battery, Relays, Overall Condition of Electrical System | |
| **CHASSIS** | |
| 13. Radiator, Hoses, Pump, Fan | |
| 14. Wheels, Rims, Tires, Spare Tire, Tools, Jack | |
| 15. Frame, Suspension, Brakes, Steering, Drivetrain | |
| 16. Manifold(s), Muffler(s), Pipe(s), Hanger(s), Exhaust Tip(s) | |
| **OPERATIONS** | |
| 17. Smoothness of engine run, leaks | |
| 18. Instruments, Interior Lights, Horn, Radio, Wipers NOT Checked | |
| 19. Exterior Lights, Turn Signals, Brake and Backup Lights | |
| 20. General over Restoration and/or Applied Patina - | |
| **POINTS DEDUCTED TOTAL ITEMS 1-20** | |
| Total of Points from Lines 1 through 20 = | |
| **100 minus total deducted points above equals TOTAL SCORE:** | |

*This example of the ICJAG scoresheet comes from the 2024 Forest Grove Concours d'Elegance.*

**FICHE DE NOTATION**
***SCORE CARD***

CHANTILLY ARTS & ELEGANCE RICHARD MILLE

29 & 30 Juin/June 2019

## LES PLUS BELLES VOITURES DU MONDE
## *THE MOST BEAUTIFUL CARS IN THE WORLD*

CLASSE/CLASS..........................................................................................

NOM DU JUGE/*NAME OF THE JUDGE*..........................................................................................

Présenté par / *Owner* : ..........................................................................................

### VOITURE / *CAR*

Marque/*Make*.................................... Modèle/*Model*.................................... Année/*Year*....................

N° Chassis : .................................... Carrossier / *Bodywork* : ....................................

Moteur / *Engine* & C.C. : ..........................................................................................

| | | | |
|---|---|---|---|
| **Authenticité** *Authenticaty* | Traçabilité / *Traceability* | _____ /10 | Note_____ /20 |
| | Configuration d'origine / *Original configuration* | _____ /10 | |
| **Historique** *History* | Rareté du modèle / *Rarity of the model* | _____ /5 | Note_____ /20 |
| | Palmarès / *Prize list* | _____ /5 | |
| | Pedigree | _____ /5 | |
| | Intérêt du modèle / *Interest of the model* | _____ /5 | |
| **Design et Elegance du Dessin** *Design and Elegance* | Innovation | _____ /5 | Note_____ /20 |
| | Intemporalité / *Timelessness* | _____ /5 | |
| | Elégance du dessin / *Elegance of the drawing* | _____ /10 | |
| **Etat Extérieur et Intérieur** *Exterior and Interior Condition* | Etat Extérieur / *Exterior Condition* | _____ /10 | Note_____ /20 |
| | - Point bonus si d'origine/*Bonus points if origin* | _____ /1 | |
| | Etat Intérieur / *Interior Condition* | _____ /10 | Bonus_____ /2 |
| | - Point bonus si d'origine/*Bonus points if origin* | _____ /1 | |
| **Etat de Fonctionnement** *Working Order* | Moteur / *Engine* | _____ /10 | Note_____ /20 |
| | Equipement Electrique / *Electrical equipment* | _____ /5 | |
| | Equipement Carrosserie / *Bodywork equipment* | _____ /5 | |
| **Participation au Rallye** *Participation in the rally* | | | Bonus_____ /1 |
| **Carte FIVA / *FIVA Card*** | | | Bonus_____ /1 |
| | | **TOTAL** | _______ **/100** |

**COMMENTAIRES / *COMMENTS*:**

Chantilly Scoring System.

# ACKNOWLEDGMENTS


*Richard Adatto*
*Gaël Archimbaud*
*Makarand Baokar*
*Denis Berthauld*
*Jean-Louis Blanc*
*Gilles Blanchet*
*Christian Boyer*
*Bernard Brulé*
*David Burgess-Wise*
*Sandra Button*
*Marc Caveng*
*Bogdan Coconoiu*
*Jodi Ellis*
*Francesco Fiordelisi*
*Rob Fisher*
*Laurent Friry*
*McKeel Hagerty*
*Kandace Hawkinson*
*Ken Gross*
*Andy Heller*
*Helmut Käs*
*Christian Kramer*
*Gabriela Magureanu*
*Alberto Martinez*
*Nigel Matthews*
*Cameron Maynard*
*Hubert Moisy*
*Glyn Morris*
*Jean Morris*
*James Nicholls*
*Xavier de Nombel*
*Arjun Oberoi*
*Adolfo Orsi*
*Marc Ouayoun*
*Sanket Patel*
*Anthony Peacock*
*Patrick Peter*
*Christian Philippsen*
*Francis Piquera*
*Thomas Ross*
*Jean-François Ruchaud*
*John Saliba*
*Olivier De Serres*
*Marc Thiesbuerger*
*Michael Tillson*
*Hugues Vanhoolandt*
*Jason Wenig*
*Xiaorui Xu*


# BIBLIOGRAPHY

## Books


*70 Years of Pebble Beach Concours d'Elegance* (Hothouse Media, 2020)

*L'Atlas des voitures de collection* (Editions Atlas, 2002)

Adatto, Richard. *From Passion to Perfection* (Editions SPE Barthélémy, 2002)

Anselmi, Angelo Tito. *Villa d'Este: The Italian Concours* (Le Edizioni dell'Opificio, 2004)

Auroy, Carole; Meynard, Cécile; Pinto-Mathieu, Elisabeth. *La vie élégante* (Presses universitaires de Rennes, 2017)

Balzac, Honoré (de). *Traité de la vie élégante* (Librairie nouvelle, 1854 [1830])

Barthes, Roland. *Mythologies* (Paris, Editions du Seuil, 1957)

Beau, Jules, Photography albums

Bellu, Serge. *A French Touch of Class* (Editions Nicolas Chaudun, 2011)

Bellu, Serge, présenté par. *Les 100 Belles du Siècle* (Solar, 1999)

Bellu, Serge. *Reines de Vitesse* (Solar, 2003)

Buckley, Martin. *The Car: A Book of Classic Motors* (Southwater, 1997)

Cabart, Daniel and Rouxel, Claude and Burgess-Wise, David. *Delage France's Finest Car* (Dalton Watson Fine Books, 2007)

Casiraghi, Michele P. *Best in Show* (Skira, 2014)

Cheetham, Craig. Classic Cars: *Legendary Automobiles Design and Development* (Metro Books, 2008)

Clarkson, Jeremy. *Hot 100: Cars That Make You Go Phwoar!* (Virgin Books, 2004)

Cohignac, Denis. *L'entre-deux guerres, sport automobile et féminisme* (Revue histoire et patrimoine, 2023)

Colin, Marc-Antoine. *Delahaye 135* (E.T.A.I., 2003)

Cortinovis, Genevieve. *Framing Parisian Chic at the Concours d'Elegance 1936-1939* (Thesis, Cortinovis, 2010)

Desanti, Dominique. *La Femme au temps des Années Folles* (Stock, Laurence Pernoud, 1984)

Design Museum. *Fifty Cars That Changed the World* (Conran Octopus)

Devlin, Robert T. with Hawkinson, Kandace. *Pebble Beach Concours d'Elegance The Art of the Poster* (Dalton Watson Fine Books, 2003)

*Fifty Years of the American Automobile* (Beekman House, 1989)

Flonneau Mathieu. *Les cultures du volant. Essai sur les mondes de l'automobilisme, XXe-XXIe siècles* (Autrement, collection Mémoires / Culture, 2008)

# BIBLIOGRAPHY, CONT.


Fouquières, André (de). *Cinquante ans de panache* (Editions Pierre Horay Flore, 1951)

Fouquières, André (de), *De l'art, de l'élégance, de la charité* (Editions Fontemoing, 1910)

Henri-Labourdette, Jean. *Un siècle de carrosserie française* (Édita Lazarus, Suisse, 1972)

Huet, Christian. *Atlantic Bugatti Type 57S chassis n°57473* (Editions Christian Huet, 2011)

Garçon, Anne Françoise, Beaussaron Gaëlle. *L'automobile, Son monde et ses réseaux* (Presses universitaires de Rennes, 1998)

Garnier, Guillaume. *Paris Couture, Années Trente* (Éditions Palais Galliera, Guillaume Galliera, 1987)

Lesueur, Patrick. *Concours d'élégance automobile – Le rêve automobile* (E.T.A.I., 2011)

Néret, Gilles, Poulain, Hervé. *l'art, la femme, et l'automobile* (EPA, 1989)

Pascal, Dominique. *Les Concours d'élégance* (MDM, 1998)

Pochna, Marie-France. *Nina Ricci* (Editions du Regard, 1989)

Poiret, Paul. *En habillant l'époque* (Paris, Bernard Grasset, 1986 [1930])

Renucci, France. *Autos-femmes: Union imposée, rupture impensable* (Les cahiers de médiologie, Editions Gallimard, 2001)

Reynolds, John. *André Citroën: The Man and the Motor Cars* (Sutton Publishing, 1996)

Reynolds, John. *Citroën: Daring to be Different* (Haynes, 2004)

Rim, Carlo. *Le grenier d'Arlequin: journal* 1916-1940 (Denoël, 1981)

Rouff, Marcel. *Une industrie motrice: la haute couture parisienne et son évolution* (Annales Economies, sociétés, civilisations, 1946. pp. 116-133)

Sparrow, Andrea and David. *Alfa Romeo Colour Family Album* (Veloce, 1999)

Stewart, Mary Lynn. *Dressing Modern Frenchwomen: Marketing Haute Couture* 1919-1939 (Johns Hopkins University Press, 2008)

*The Concours Year 2023* (Hothouse Media, 2024)

*The Concours Year 2024* (Hothouse Media, 2025)

Tumminelli, Paolo. *Car Design Asia: Myths, Brands, People* (teNeues, 2014)

Wherry, Joseph H. *Concours d'Elegance* (Howell-North, 1969)


## Magazines, newspapers


*Adam*
*Allgemeine Automobil-Zeitung*
*Art Goût Beauté*
*Autobild China*
*Auto Italia*
*Auto Italiana*
*Auto Moto Ciclo*
*Automobile Club Féminine Bulletin*
*Automobilia*
*Biarritz Illustré*
*Bulletin de l'Automobile Club de Nice*
*Bulletin de l'Automobile Club du Rhône*
*Cannes, ses deux saisons.*
*Dimanche-Auto – Le Miroir de la Route*
*Dinard mon pays, Revue mondaine illustrée de la Côte d'Émeraude*
*Excelsior*
*Fémina*
*L'Art Nouveau*
*L'Art Vivant*
*L'Auto*
*L'Auto-Carrosserie*
*L'Auto-Vélo*
*L'écho des modes*
*L'Éclaireur du DImanche Illustré*
*L'Équipe*
*L'Equipement Automobile*
*L'Intransigeant*
*L'Officiel de la Couture et de la Mode, Paris*
*L'Ouest-Éclair (édition de Caen; édition de Rennes)*
*La Carrosserie*
*La Locomotion Automobile*
*La Mode Illustrée*
*La Revue des Sports et du Monde (Ford / Mathis)*
*La Saison de Cannes et des Alpes Françaises*
*La Saison de la Côte d'Azur et des Alpes*
*La Vie au Grand Air*
*La Vie Automobile*
*Le Figaro*
*Le Journal*
*Le Journal des Etrangers*
*Le Petit Journal*
*Le Soir Illustré*
*Les Modes*
*Littoral Mondain*
*Marseille Auto (1938/07)*
*Motor Sport*
*Omnia, Revue pratique de la locomotion*
*Renaissance de l'art français et des industries du luxe*
*Revue de la Carrosserie*
*The Autocar*
*The Automotor Journal*
*The Menton and Monte Carlo News*
*The Motor*
*The Riviera Herald*
*Vogue*

# MEET THE AUTHORS

## Daniel Cabart

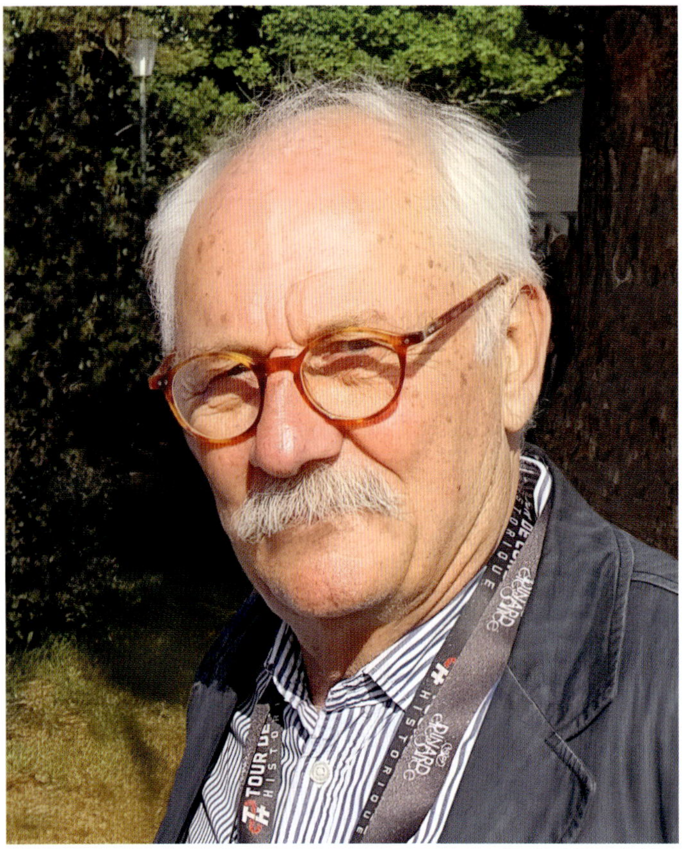

Daniel Cabart Archives

Daniel Cabart is a renowned expert on the Delage brand, with an impressive archive built over more than 40 years. His collection, gathered from the pioneers of Delage's history, remains an essential reference to this day.

Cabart has co-authored several acclaimed books on the brand, including *Delage, la belle voiture française*, *Delage Champion du Monde* and *Delage, Records et Grand Prix*. These works have earned prestigious awards, including the Prix Bellecour (France), the Royal Automobile Club Award (UK), and the SAH Cugnot Award (USA).

In addition to his work on Delage, Cabart co-authored a key reference book on the Ballot brand in 2019 and regularly contributes to the documentation and illustration of numerous books on the history of automobiles and coachbuilding.

Delage, having been crowned "World Champion of Manufacturers" in 1927, showcased its commitment to automotive excellence by adopting the title of "Specialist in Concours d'Elegance." Given this, it's no surprise that Cabart has become deeply involved in the history and evolution of these prestigious events, which have always been intertwined with the broader automotive legacy.

## Denis Cohignac

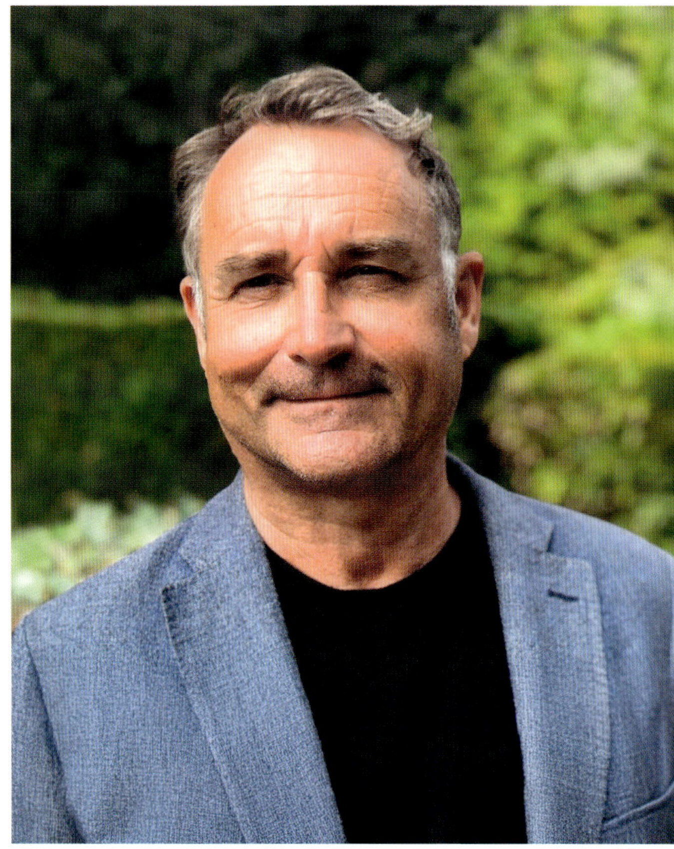

Dinard Elegance

Driven by a lifelong passion for automobiles, Denis Cohignac began his career in sales management at the Renault Group, eventually rising to lead the Boulogne-Billancourt subsidiary.

An enthusiast of high-level challenges, he has competed in prestigious international endurance races, including 1,000-kilometer events and 24-hour competitions. His doctoral thesis, focused on the corporate values of the automotive industry, reflects his ability to combine rigorous analysis with his deep love of cars.

As the founder of Formatis International, Cohignac has spent more than two decades providing training and strategic consulting to major manufacturers such as Audi, Porsche, Renault, and Harley-Davidson, supporting their transformation initiatives.

In 2018, as President of Dinard Elegance, he partnered with the City of Dinard and a team of dedicated enthusiasts to revive the historic Dinard Elegance Concours – an event dating back to 1921 and celebrating the enduring beauty and sophistication of the automobile.

## Patrick Rollet

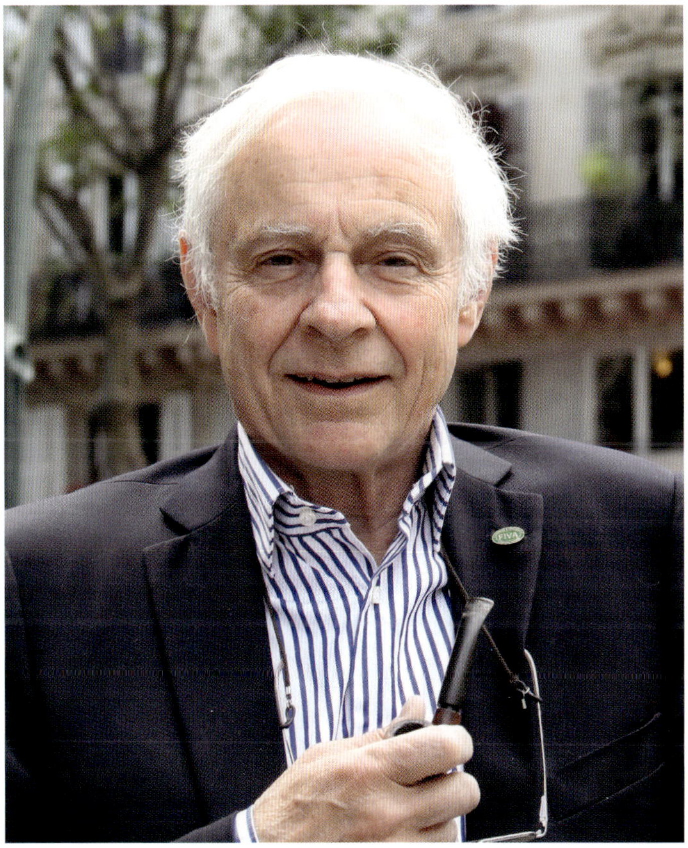

Mark Dixon, Octane

With over 30 years of experience in the world of historic vehicles, Patrick Rollet has made significant contributions to the field. He co-founded the international Fougères Rally and later became deeply involved in advocacy for classic cars. His leadership roles include serving as Director (2005) and Vice President (2006-2009) of the French national federation, FFVE, and as Vice President (2006-2013) and President (2013-2019) of FIVA, the International Federation of Vintage Cars.

Rollet is also an active committee member for several prestigious concours d'elegance in France, including Dinard, Gambais, and La Baule, and has served as a judge at concours events across Europe, the Middle East, and Asia. In 2004, he published *Automobiles & Patrimoine* and has contributed articles to various historic car magazines.

Honored for his extensive work in the automotive heritage field, Patrick Rollet holds the title of Chevalier de l'Ordre National du Mérite and Chevalier des Arts et des Lettres. He is also the Honorary President of both FIVA and the Club de L'Auto.

## Gautam Sen

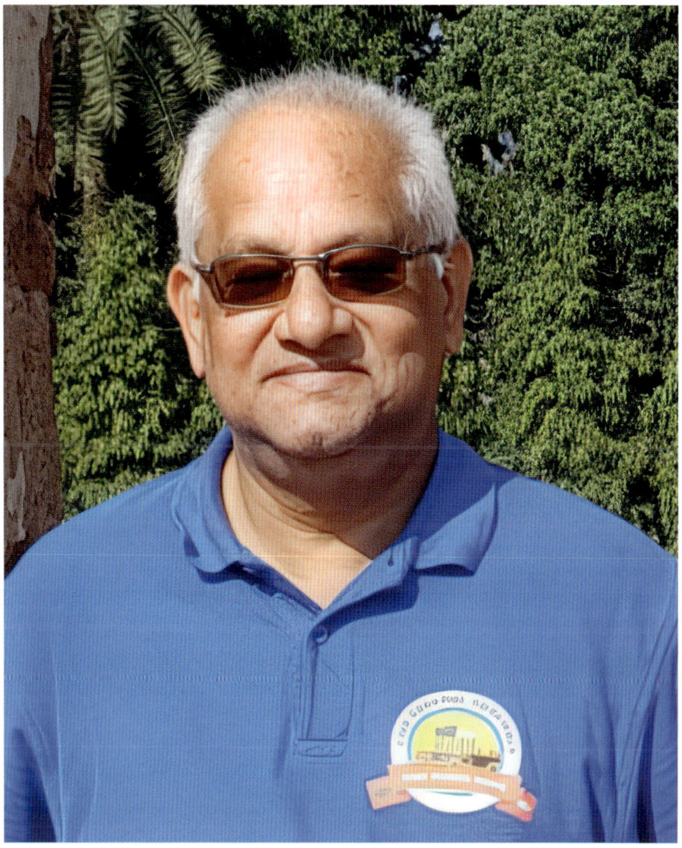

Anne Castagnos

As a frequent concours judge, author and design consultant, as well as a co-conspirator with prominent designers such as Marcello Gandini, Tom Tjaarda, and Gérard Godfroy on several vehicle projects, Gautam Sen has been the founder or editor of several Indian automotive magazines, establishing the country's first car magazine in 1986. In addition to designing several motorcycles, Sen played a key role in creating India's first sports car, the San Storm. His experience spans extensive work with the automotive industry, where he has advised major companies such as Hero Motors, Hindustan Motors, Mahindra & Mahindra, Maruti Suzuki, Tata Motors, and TVS.

A Vice President with FIVA during 2015-2024, Sen has also authored several award-winning books: *Marcello Gandini: Maestro of Design, Ballot, The Bertone Collection* and *Tom Tjaarda: Master of Proportions,* as well as *Lamborghini: At the Cutting Edge of Design, Bugatti: The Italian Decade, Alfa Romeo SZ Coda Tronca: The Art of Conservation, When Wedge Had the Edge,* and *Lancia Stratos Zero: The Eternal Futurist* (all published by Dalton Watson Fine Books) and others.

# INDEX

## VEHICLES

## COACHBUILDERS

# EVENTS

## COMPANIES AND ORGANIZATIONS



## PEOPLE

## PLACES

## MEDIA



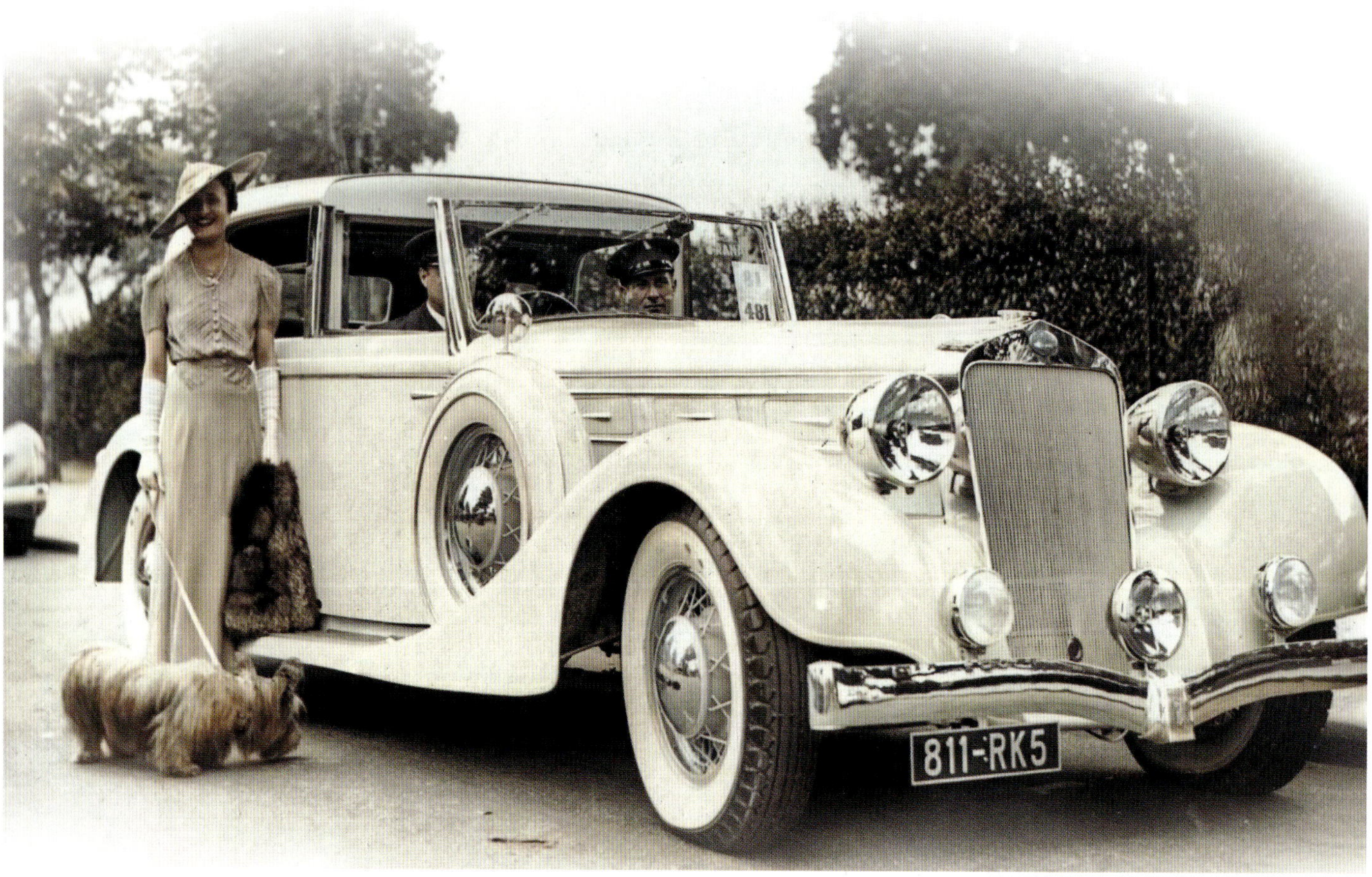



| | |
|---|---|
| **DESIGN** | Jodi Ellis Graphics |
| **PRINTER** | Interpress Ltd. |
| **PAGE SIZE** | 219 mm x 304 mm |
| **TEXT PAPER** | 150gsm Magno Gloss |
| **END PAPER** | 170gsm Woodfree Offset |
| **DUST JACKET** | 170gsm Glossy Artpaper |
| **CASING** | Foil stamping on front and spine, on Azul Mar Geltex, over 3 mm board |
| **CHAPTER HEADS** | 25pt Broadacre Regular |
| **SUBHEADS** | 14pt Novel Display XBold |
| **MAIN BODY TEXT** | 9pt Novel Display Regular |
| **CAPTIONS** | 9pt Novel Display SemiBd It |